HETEROCYCLIC CHEMISTRY

Fourth edition

J.A. Joule
& K. Mills

b

Blackwell
Science

© 1972, 1978 J.A. Joule and G.F. Smith; 1995 J.A. Joule, K. Mills and G.F. Smith; 2000 J.A. Joule and K. Mills

Blackwell Science Ltd
Editorial Offices:
Osney Mead, Oxford OX2 0EL
25 John Street, London WC1N 2BL
23 Ainslie Place, Edinburgh EH3 6AJ
350 Main Street, Malden
  MA 02148 5018, USA
54 University Street, Carlton
  Victoria 3053, Australia
10, rue Casimir Delavigne
  75006 Paris, France

Other Editorial Offices:

Blackwell Wissenschafts-Verlag GmbH
Kurfürstendamm 57
10707 Berlin, Germany

Blackwell Science KK
MG Kodenmacho Building
7-10 Kodenmacho Nihombashi
Chuo-ku, Tokyo 104, Japan

The right of the Authors to be identified as the Authors of this Work has been asserted in accordance with the Copyright, Designs and Patents Act 1988.

First edition published by Chapman & Hall 1972
Second edition published by Chapman & Hall 1978
Third edition published by Chapman & Hall 1995
Fourth edition published 2000

Typeset by Gray Publishing, Tunbridge Wells, Kent
Printed and bound in Great Britain at the University Press, Cambridge

The Blackwell Science logo is a trade mark of Blackwell Science Ltd, registered at the United Kingdom Trade Marks Registry

DISTRIBUTORS

Marston Book Services Ltd
PO Box 269
Abingdon
Oxon OX14 4YN
(*Orders:* Tel: 01235 465500
           Fax: 01235 465555)
USA
Blackwell Science, Inc.
Commerce Place
350 Main Street
Malden, MA 02148-5018
(*Orders:* Tel: 800 759 6102
                    781 388 8250
           Fax: 781 388 8255)
Canada
Login Brothers Book Company
324 Saulteaux Crescent
Winnipeg, Manitoba R3J 3T2
(*Orders:* Tel: 204 837-2987
           Fax: 204 837-3116)
Australia
Blackwell Science Pty Ltd
54 University Street
Carlton, Victoria 3053
(*Orders:* Tel: 03 9347 0300
           Fax: 03 9347 5001)

A catalogue record for this title is available from the British Library

ISBN 0-632-05453-0

Library of Congress Cataloging-in-Publication Data:

Joule, J. A. (John Arthur)
    Heterocyclic chemistry. – 4th ed. J.A. Joule, K. Mills.
        p.   cm.
    Includes bibliographical references and index.
    ISBN 0-632-05453-0
    1. Heterocyclic chemistry. I. Mills, K. (Keith)
    II. Title
QD400.J59   2000
547'.69 – dc21
                      99-38976
                        CIP

For further information on Blackwell Science, visit our website:
www.blackwell-science.com

HETEROCYCLIC CHEMISTRY

Contents

Preface to Fourth Edition

Since the preparation of the Third Edition, George Smith, the principal author of the First and Second Editions, has died. Many generations of students in the Chemistry Department, The University of Manchester, benefitted from his inspirational teaching and his enthusiasm for Organic Chemistry, not the least the two authors of this Fourth Edition. We dedicate this book in its goal of instruction in the field of Heterocyclic Chemistry, to his memory.

Heterocyclic compounds are of particular interest in medicinal chemistry, and this has catalysed the discovery and development of much new heterocyclic chemistry and methods. The preparation of a Fourth Edition has allowed us to review thoroughly the material included in the earlier Editions, to make amendments in the light of new knowledge, and to include much recent work. Within the restrictions of scope which space dictated, we believe that all of the most significant Heterocyclic Chemistry of the 20th Century has been covered, or referenced.

Introduction to the Fourth Edition

We have maintained the principal aim of the earlier Editions – to teach the fundamentals of heterocyclic reactivity and synthesis in a way which is understandable by undergraduate students. However in addition, and in recognition of the level at which much heterocyclic chemistry is now normally taught, we have included more advanced and current material which make the book appropriate both for postgraduate level courses, and as a reference text for those involved in heterocyclic chemistry in industry.

1. The main body of factual material is to be found in chapters entitled 'Reactions and synthesis of ...' a particular heterocyclic system. Didactically vital material is to be found partly in a general discussion of heterocyclic reactivity (chapter 2), and partly in six short summary chapters (such as 'Typical reactivity of pyridines, quinolines, and isoquinolines'; chapter 4), which aim to capture the essence of that typical reactivity in very concise resumés. These last are therefore suitable either as an introduction to the chemistry of that heterocyclic system but they are insufficient in themselves and should lead the reader to the fuller discussions in the 'Reactions and synthesis' chapters. They will also serve the undergraduate student as a revision-summary of the typical chemistry of that system.

2. More than 3500 references have been given throughout the text: the references to original work have been chosen as good leading references and are, therefore, not necessarily the first mention of that particular topic or method; some are included as benchmark papers and others for their historical interest. These are most relevant to post-graduate teaching and to research workers, however we believe that the inclusion of references does not interfere with the readability of the text for the undergraduate student. Many review references are also included: for these we give the title of the article; titles are also given for the books to which we refer.

3. Exercises are given at the ends of most of the substantive chapters. These are divided into straightforward, revision exercises such as will be relevant to an undergraduate course in heterocyclic chemistry. More advanced exercises, with solutions given in an Appendix at the end of the book, are designed to help the reader to develop understanding and apply the principles of heterocyclic reactivity. References have not been given for the Exercises, though all are real examples culled from the literature.

4. We avoid the used of 'R' and 'Ar' for substituents in schemes, and instead give actual examples. We believe this makes the chemistry easier to assimilate, especially for the undergraduate reader. It also avoids implying a generality which may not be justified.

5. Structures and numbering for heterocyclic systems are given at the beginnings of chapters. Where the commonly-used name differs from that used in *Chemical Abstracts*, the name given in square brackets is the official *Chemical Abstracts* name, thus: indole [1*H*-indole]. We believe that the systematic naming of heterocyclic substances is of importance, not least for use in computer data

bases, but it serves little purpose in teaching or for the understanding of the subject and, accordingly, we have devoted little space to nomenclature. The reader is referred to an exposition on this topic[1] and also to the Ring Index of *Chemical Abstracts* in combination with the Chemical Substances Index, whence both standardised name and numbering can be obtained for all known systems.

6. There are several general reference works concerned with heterocyclic chemistry, which have been gathered together as a set at the end of this chapter, and to which the reader's attention is drawn. N.B. In order to save space, these vital sources are not referred to in particular chapters, however all the topics covered in this book are covered in them, and recourse to this information should form the early basis of any literature search.

7. The literature of heterocyclic chemistry is so vast that the series of five listings – 'The literature of heterocyclic chemistry', Parts I–V[2] – is of enormous value at the start of a literature search. These five listings appear in *Advances in Heterocyclic Chemistry*,[3] itself a prime source for key reviews on heterocyclic topics; the journal, *Heterocycles,* also carries many useful reviews specifically in the heterocyclic area. *Progress in Heterocyclic Chemistry*[4] published by the International Society of Heterocyclic Chemistry[5] also carries reviews, and monitors developments in heterocyclic chemistry over a calendar year. Essential at the beginning of a literature search is a consultation with the appropriate chapter(s) of '*Comprehensive Heterocyclic Chemistry*', original[6a] and its update,[6b] or, for a useful introduction and overview, to the 'handbook' which is Volume 9 of the original.[7]

8. There have been many instances of interconversion of heterocyclic systems, with changes of hetero atom components and/or ring size; these are gathered together in one useful volume.[8] Three comprehensive reviews of degenerate heterocyclic ring transformations comprise Vol 74 in *Advances in Heterocylic Chemistry*.[9] A book,[10] which has provided considerable inspiration for more than two decades, highlights the utility of heterocycles, often as 'carriers' of functionality, in a general synthetic context.

9. There are three comprehensive compilations of heterocyclic facts: the early series[11] edited by Elderfield discusses pioneering work. The still-continuing and still-growing series of monographs[12] dealing with particular heterocyclic systems, edited originally by Arnold Weissberger, and latterly by Edward C. Taylor, is a vital source of information and reviews for all those working with heterocyclic compounds. Finally, the heterocyclic volumes of '*Rodd's Chemistry of Carbon Compounds*'[13] contain a wealth of well-sifted information and data.

How to use this textbook

As indicated above, by comparison with earlier editions, this Fourth Edition of '*Heterocyclic chemistry*' contains more material, including more which is appropriate to study at a higher level, than that generally taught in a first degree course. Nevertheless we believe that undergraduates will find the book of value and offer the following *modus operandi* as a means for undergraduate use of this text.

The undergraduate student should first read chapter 1, which will provide a structural basis for the chemistry which follows. We suggest that the material dealt with in chapter 2 be left for study at later stages, and that the undergraduate student proceed next to those chapters (4, 7, 10, 12, 16, and 20) which explain heterocyclic principles in the simplest terms and which should be easily assimilated by students who have a good grounding in elementary reaction chemistry, especially aromatic chemistry.

The student could then proceed to the main chapters, dealing with 'Reactions and synthesis of ... ' in which will be found full discussions of the chemistry of particular systems – pyridines, quinolines, etc. These utilise many cross references which seek to capitalise on that important didactical strategy – comparison and analogy with reactivity already learnt and understood.

Chapter 2 is an advanced essay on heterocyclic chemistry. Sections can be sampled as required – 'Electrophilic substitution' could be read at the point at which the student was studying electrophilic substitutions of, say, thiophene – or chapter 2 can be read as a whole. We have devoted considerable space in chapter 2 to discussions of radical substitution, metallation, and palladium-catalysed reactions. These topics have grown enormously in importance since the last edition of this book, are of great relevance to heterocyclic chemistry, and are relatively poorly explained in general textbooks.

References

1. 'The nomenclature of heterocycles', McNaught, A. D., *Adv. Heterocycl. Chem.*, **1976**, *20*, 175.
2. Katritzky, A. R. and Weeds, S. M., *Adv. Heterocycl. Chem.*, **1966**, *7*, 225; Katritzky, A. R. and Jones, P. M., *ibid.*, **1979**, *25*, 303; Belen'kii, L. I., *ibid.*, **1988**, *44*, 269; Belen'kii, L. I. and Kruchkovskaya, N. D., *ibid.*, **1992**, *55*, 31; *idem, ibid.*, **1998**, *71*, 291.
3. *Adv. Heterocycl. Chem.*, **1963–1999**, *1–74*.
4. *Progr. Heterocycl. Chem.*, **1989–1999**, *1–11*.
5. http://euch6f.chem.emory.edu/ishc.html and the related Royal Society of Chemistry site: http://www.rsc.org/lap/rsccom/dab/perk003.htm
6. (a) 'Comprehensive heterocyclic chemistry. The structure, reactions, synthesis, and uses of heterocyclic compounds', Eds. Katritzky, A. R. and Rees, C. W., Vols 1–8, Pergamon Press, Oxford, **1984**; (b) 'Comprehensive heterocyclic chemistry II. A review of the literature 1982–1995', Eds. Katritzky, A. R., Rees, C. W., and Scriven, E. F. V., Vols 1–11, Pergamon Press, **1996**.
7. 'Handbook of heterocyclic chemistry', Katritzky, A. R., Pergamon Press, Oxford, **1985**.
8. 'Ring transformations of heterocycles', Vols. 1 and 2, van der Plas, H. C., Academic Press, New York, **1973**.
9. 'Degenerate ring transformations of heterocyclic compounds', van der Plas, H. C., *Adv. Heterocycl. Chem.*, **1999**, *74*.
10. 'Heterocycles in organic synthesis', Meyers, A. I., Wiley-Interscience, New York, **1974**; for a more recent article in the same vein see: 'Five-membered heteroaromatic rings as intermediates in organic synthesis', Lipshutz, B. H., *Chem. Rev.*, **1986**, *86*, 795.
11. 'Heterocyclic compounds', Ed. Elderfield, R. C., Vols. 1-9, Wiley, **1950–1967**.
12. 'The chemistry of heterocyclic compounds', Series Eds. Weissberger, A. and Taylor, E. C., Vols. 1-56, Wiley-Interscience, **1950-1999**.
13. 'Rodd's chemistry of carbon compounds', Eds., Coffey, S. then Ansell, M. F., Vols IVa-IVl, and Supplements, **1973–1994**, Elsevier, Amsterdam.

Definitions of abbreviations

Adoc	=	adamantanyloxycarbonyl $[C_{10}H_{15}OC=O]$
Aliquat®	=	tricaprylmethylammonium chloride $[MeN(C_8H_{17})_3Cl]$
p-An	=	*para*-anisyl $[4-MeO-C_6H_4]$
aq.	=	aqueous
9-BBN triflate	=	9-borabicyclo[3.3.1]nonyl trifluoromethanesulfonate
BINAP	=	2,2'-bis(diphenylphosphino)-1,1'-binaphthyl
BINOL	=	1,1'-bi(2-naphthol) $[C_{20}H_{14}O_2]$
Bn	=	benzyl $[PhCH_2]$
Boc	=	*t*-butoxycarbonyl $[Me_3COC=O]$
Bt	=	benzotriazol-1-yl $[C_6H_4N_3]$
i-Bu	=	*iso*-butyl $[Me_2CHCH_2]$
n-Bu	=	*normal*-butyl $[Me(CH_2)_3]$
s-Bu	=	*secondary*-butyl $[MeCH_2C(Me)H]$
t-Bu	=	*tertiary*-butyl $[Me_3C]$
BSA	=	*N,O*-bis(trimethylsilyl)acetamide $[MeC(OSiMe_3)=NSiMe_3]$
c.	=	concentrated
c	=	cyclo as in c-C_5H_9 = cyclopentyl
CAN	=	cerium(IV) ammonium nitrate
CDI	=	1,1'-carbonyldiimidazole $[(C_3H_3N_2)_2C=O]$
Chloramine T	=	*N*-chloro-4-methylbenzenesulfonamide sodium salt $[TsN(Cl)Na]$
ClMgTMP	=	(2,2,6,6-tetramethylpiperidino)magnesium chloride $[ClMgN(CMe_2(CH_2)_3CMe_2)]$
cod	=	cycloocta-1,5-diene $[C_8H_{12}]$
cp	=	cyclopentadienyl anion $[c$-$C_5H_5^-]$
m-CPBA	=	*meta*-chloroperbenzoic acid $[3-ClC_6H_4CO_3H]$
CuTC	=	thiophen-2-ylcarboxylic acid copper(I) salt $[C_5H_3CuO_2S]$
DABCO	=	1,4-diazabicyclo[2.2.2]octane $[C_6H_{12}N_2]$
dba	=	dibenzylideneacetone $[PhCH=CHCOCH=CHPh]$
DBU	=	1,8-diazabicyclo[5.4.0]undec-7-ene $[C_9H_{16}N_2]$
DCC	=	N,N'-dicyclohexylcarbodiimide $[c$-$C_6H_{11}N=C=N$-c-$C_6H_{11}]$
DDQ	=	2,3-dichloro-5,6-dicyano-1,4-benzoquinone $[C_8Cl_2N_2O_2]$
de	=	diastereomeric excess
DEAD	=	diethyl azodicarboxylate $[EtO_2CN=NCO_2Et]$
DIBALH	=	diisobutylaluminium hydride $[(Me_2CHCH_2)_2AlH]$
DMA	=	*N,N*-dimethylacetamide $[MeCONMe_2]$
DMAP	=	4-dimethylaminopyridine $[C_7H_{10}N_2]$
DME	=	1,2-dimethoxyethane $[MeO(CH_2)_2OMe]$
DMF	=	dimethylformamide $[Me_2NCH=O]$
DMFDMA	=	dimethylformamide dimethyl acetal $[Me_2NCH(OMe)_2]$

DMSO	=	dimethylsulfoxide [$Me_2S=O$]
DPPA	=	diphenylphosphoryl azide [$(PhO)_2P(O)N_3$]
dppb	=	1,4-bis(diphenylphosphino)butane [$Ph_2P(CH_2)_4PPh_2$]
dppf	=	1,1′-bis(diphenylphosphino)ferrocene [$C_{34}H_{28}FeP_2$]
dppp	=	1,3-bis(diphenylphosphino)propane [$Ph_2(CH_2)_3PPh_2$]
ee	=	enantiomeric excess
ESR	=	electron spin resonance
Et	=	ethyl [CH_3CH_2]
f.	=	fuming
FVP	=	flash vacuum pyrolysis
Het	=	general designation for a heterocyclic nucleus
HMDS	=	1,1,1,3,3,3-hexamethyldisilazane [$Me_3SiNHSiMe_3$]
hplc	=	high pressure liquid chromatography
HOMO	=	highest occupied molecular orbital
hν	=	ultraviolet or visible irradiation
LDA	=	lithium diisopropylamide [$LiNi\text{-}Pr_2$]
LiTMP	=	lithium 2,2,6,6-tetramethylpiperidide [$LiN(CMe_2(CH_2)_3CMe_2)$]
liq.	=	liquid
LR	=	Lawesson's reagent [$C_{14}H_{14}O_2P_2S_4$]
LUMO	=	lowest unoccupied molecular orbital
Me	=	methyl [CH_3]
MOM	=	methoxymethyl [CH_3OCH_2O]
MTBD	=	1,3,4,6,7,8-hexahydro-1-methyl-2H-pyrimido[1,2-a]pyridine [$C_8H_{15}N_3$]
MSH	=	O-(mesitylenesulfonyl)hydroxylamine [$H_2NOSO_2C_6H_2\text{-}2,4,6\text{-}Me_3$]
NBS	=	N-bromosuccinimide [$C_4H_4BrNO_2$]
NIS	=	N-iodosuccinimide [$C_4H_4INO_2$]
NMP	=	N-methylpyrrolidone [C_5H_9NO]
OXONE®	=	potassium peroxymonosulfate [$2KHSO_5.KHSO_4.K_2SO_4$]
Ph	=	phenyl [C_6H_5]
PhH	=	benzene
PhMe	=	toluene
Phosphorus oxychloride	=	$POCl_3$
Phosphoryl chloride	=	$POCl_3$
phth	=	phthaloyl [COC_6H_4CO]
PMB	=	p-methoxybenzyl [$4\text{-}MeOC_6H_4CH_2$]
PMP	=	1,2,2,6,6-pentamethylpiperidine [$C_{10}H_{21}N$]
ⓅⓅ	=	pyrophosphate [$OP(=O)(OH)OP(=O)OH$]
PPA	=	polyphosphoric acid
i-Pr	=	*iso*-propyl [Me_2CH]
n-Pr	=	*normal*-propyl [$CH_3CH_2CH_2$]
proton sponge	=	1,8-bis(dimethylamino)naphthalene [$C_{14}H_{18}N_2$]
py	=	pyridyl, as in 2-py = pyridin-2-yl, etc. [C_5H_4N]
rp	=	room (atmospheric) pressure
rt	=	room temperature
salcomine	=	N,N'-bis(salicylidene)ethylenediaminocobalt(II)

SDS	=	sodium dodecylsulfate [$C_{12}H_{25}SO_3Na$]
SEM	=	trimethylsilylethoxymethyl [$Me_3Si(CH_2)_2OCH_2$]
SET	=	single electron transfer
SOMO	=	singly occupied molecular orbital
TASF	=	tris(dimethylamino)sulfur (trimethylsilyl)difluoride [$(Me_2N)_3S(Me_3SiF_2)$]
TBAF	=	tetra-*normal*-butylammonium fluoride [$n\text{-}Bu_4N^+F^-$)
TBAS	=	tetra-*normal*-butylammonium hydrogen sulfate [$n\text{-}Bu_4N^+\ HSO_4^-$)
TBDMS	=	*tertiary*-butyldimethylsilyl [$Me_3C(Me)_2Si$]
TfO⁻	=	triflate [$CF_3SO_3^-$]
tfp	=	trifuran-2-ylphosphine [$P(C_4H_3O)_3$]
THF	=	tetrahydrofuran [2,3,4,5-tetrahydrofuran]
THP	=	tetrahydropyran-2-yl [C_5H_9O]
TIPS	=	tri-*iso*-propylsilyl [$i\text{-}Pr_3Si$]
TMEDA	=	N,N,N′,N′-tetramethylethylenediamine [$Me_2N(CH_2)_2NMe_2$]
TMP	=	2,2,6,6-tetramethylpiperidine [$C_9H_{19}N$]
TMS	=	trimethylsilyl [Me_3Si]
TMSOTf	=	trimethylsilyl triflate [$Me_3SiO_3SCF_3$)
o-Tol	=	*ortho*-tolyl [$2\text{-}Me\text{-}C_6H_4$]
p-Tol	=	*para*-tolyl [$4\text{-}Me\text{-}C_6H_4$]
TolH	=	toluene
TOSMIC	=	tosylmethyl isocyanide [$4\text{-}MeC_6H_4SO_2CH_2NC$]
triflate	=	trifluoromethanesulfonate [$CF_3SO_3^-$]
Ts	=	tosyl [$4\text{-}MeC_6H_4SO_2$]
(dR)	=	β-d-2-deoxyribofuranosyl
(R)	=	β-d-ribofuranosyl
(S)	=	a sugar, usually a derivative of ribose or deoxyribose, attached to heterocyclic nitrogen, in which the substituents have not altered during the reaction shown.
)))))	=	sonication

1 Structures and spectroscopic properties of aromatic heterocycles

This chapter describes the structures of aromatic heterocycles and gives a brief summary of some physical properties.[1] The treatment we use is the valence-bond description, which we believe is sufficient for the understanding of all heterocyclic reactivity, perhaps save some very subtle effects, and is certainly sufficient for a general text-book on the subject. The more fundamental, molecular-orbital description of aromatic systems, is still not so relevant to the day-to-day interpretation of heterocyclic reactivity, though it is necessary in some cases to utilise frontier orbital considerations,[2] however such situations do not fall within the scope of this book.

1.1 Carbocyclic aromatic systems

1.1.1 Structures of benzene and naphthalene

The concept of aromaticity as represented by benzene is a familiar and relatively simple one. The difference between benzene on the one hand and alkenes on the other is well known: the latter react by addition with electrophiles, such as bromine, whereas benzene reacts only under much more forcing conditions and then nearly always by substitution. The difference is due to the cyclic arrangement of six π-electrons in benzene: this forms a conjugated molecular orbital system which is thermodynamically much more stable than a corresponding non-cyclically conjugated system. The additional stabilisation results in a diminished tendency to react by addition and a greater tendency to react by substitution for, in the latter manner, survival of the original cyclic conjugated system of electrons is ensured in the product. A general rule proposed by Hückel in 1931 states that aromaticity is observed in cyclically conjugated systems of $4n + 2$ electrons, that is with 2, 6, 10, 14, etc., π-electrons; by far the majority of monocyclic aromatic, and heteroaromatic systems are those with 6 π electrons.

In this book we use the pictorial valence-bond resonance description of structure and reactivity. Even though this treatment is not rigorous it is still the standard means for the understanding and learning of organic chemistry, which can at a more advanced level give way naturally to the much more complex, and mathematical, quantum mechanical approach. We begin by recalling the structure of benzene in these terms.

In benzene, the geometry of the ring, with angles of 120°, precisely fits the geometry of a planar trigonally hybridised carbon atom, and allows the assembly of a σ-skeleton of six sp^2 hybridised carbon atoms in a strainless planar ring. Each carbon then has one extra electron which occupies an atomic p orbital orthogonal to the plane of the ring. The p orbitals interact to generate π-molecular orbitals associated with the aromatic system.

Benzene is described as a resonance hybrid of the two extreme forms which correspond, in terms of orbital interactions, to the two possible spin-coupled pairings of adjacent p electrons – structures **1** and **2**. These are known as canonical structures, have no existence in their own right, but serve to illustrate two extremes which contribute to the 'real' structure of benzene.

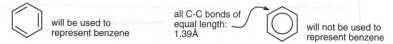

Sometimes, benzenoid compounds are represented using a circle inside a hexagon; although this emphasises their delocalised nature and the close similarity of the ring bond lengths (all exactly identical only in benzene itself), it is not helpful in interpreting reactions, and we do not use this method here.

will be used to represent benzene

all C-C bonds of equal length: 1.39Å

will not be used to represent benzene

Treating naphthalene comparably reveals three canonical structures, **3**, **4**, and **5**. Note the standard use of a double-headed arrow to interelate resonance contributors. This must never be confused with the use of opposing straight 'fish-hook' arrows which are used to designate an equilibrium between two species: resonance contributors have no separate existence; they are not in equilibrium one with the other.

This valence bond treatment predicts quite well the non-equivalence of the bond lengths in naphthalene: in two of the three contributing structures, C-1–C-2 is double and in one it is single, whereas C-2–C-3 is single in two and double in one. Statistically, then, the former may be looked on as 0.67 of a double bond and the latter as 0.33 of a double bond: the measured bond lengths confirm that there indeed is this degree of bond fixation, with values closely consistent with statistical prediction.

3   **4**   **5**   1.37Å
                                              1.41Å

1.1.2 Aromatic resonance energy[3]

The difference between the ground-state energy of benzene and that of hypothetical, non-aromatic, 1,3,5-cyclohexatriene corresponds to the degree of stabilisation conferred to benzene by the special cyclical interaction of the six π-electrons. This difference is known as aromatic resonance energy. Of course, quantification depends on the assumptions made in estimating the energy of the 'non-aromatic' structure, and for this reason and others, a variety of values have been calculated for the various heteroaromatic systems; perhaps their absolute values are less important than their relative values. What one can say with certainty is that the resonance energy of bicyclic aromatics, like naphthalene, is considerably less than twice that of the corresponding monocyclic system, implying a smaller loss of stabilisation energy on conversion to a reaction intermediate which still retains a complete benzene ring, for example during electrophilic substitution (section 2.2.2). The resonance energy of pyridine is of the same order as that of benzene, that of thiophene is lower, with pyrrole and lastly furan of lower stabilisation energy still. Actual values for the stabilisations of these systems vary according to assumptions made, but are in the same relative order (kJ mol^{-1}): benzene (150), pyridine (117), thiophene (122), pyrrole, (90), and furan (68).

1.2   Structure of six-membered heteroaromatic systems

1.2.1   Structure of pyridine

The structure of pyridine is completely analogous to that of benzene, being related by replacement of CH by N. The key differences are: (i) the departure from perfectly regular hexagonal geometry caused by the presence of the hetero atom, in particular the shorter carbon–nitrogen bonds, (ii) the replacement of a hydrogen in the plane of the ring with an unshaired electron pair, likewise in the plane of the ring, located in an sp^2 hybrid orbital, and not at all involved in the aromatic π-electron sextet; it is this nitrogen lone pair which is responsible for the basic properties of pyridines, and (iii) a strong permanent dipole, traceable to the greater electronegativity of the nitrogen compared with carbon.

It is important to realise that the electronegative nitrogen causes inductive polarisation, mainly in the σ-bond framework, and additionally, stabilises those polarised canonical structures in which nitrogen is negatively charged – **8, 9**, and **10** – which, together with contributors **6** and **7**, which are strictly analogous to the Kekulé contributors to benzene, represent pyridine. The polarised contributors imply a permanent polarisation of the π-electron system too (these equate, in the more rigorous molecular orbital treatment, to a consideration of the relative magnitudes of orbital coefficients in the HOMO and LUMO).

Because inductive and mesomeric effects work in the same sense in pyridine, there results a permanent dipole towards the nitrogen atom. It also means that there are fractional positive charges on the carbons of the ring, located mainly on the α- and γ-positions. It is because of this general electron-deficiency at carbon that pyridine and similar heterocycles are referred to as 'electron-poor', or sometimes 'π-deficient'. A comparison with the dipole moment of piperidine, which is due wholly to the induced polarisation of the σ-skeleton, gives an idea of the additional polarisation associated with distortion of the π-electron system.

1.2.2   Structure of diazines

The structures of the diazines (six-membered sytems with two nitrogen atoms in the ring) are analogous, but now there are two nitrogen atoms and a corresponding two lone pairs; as an illustration, the main canonical contributors (**11–18**) to pyrimidine are shown below.

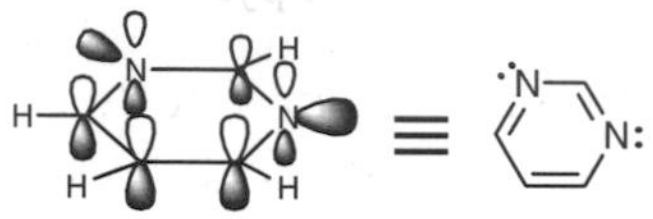

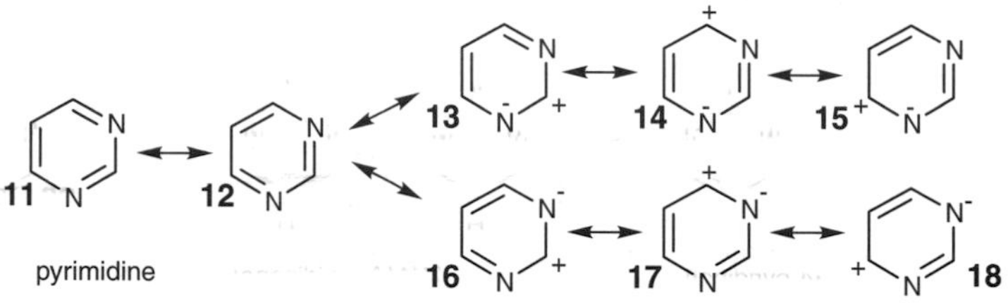

pyrimidine

1.2.3   Structures of pyridinium and related cations

Electrophilic addition to the pyridine nitrogen generates pyridinium ions, the simplest being 1H-pyridinium formed by addition of a proton. 1H-Pyridinium is actually isoelectronic with benzene, the only difference being the nuclear charge of nitrogen, which makes the system, as a whole, positively charged. Thus pyridinium cations are still aromatic, the diagram making clear that the system of six p orbitals required to generate the aromatic molecular orbitals is still present, though the formal positive charge on the nitrogen atom severely distorts the π-system, making the α- and γ-carbons in these cations carry fractional positive charges which are higher than in pyridine, with a consquence for their reactivity towards nucleophiles. Electron density at the pyridinium β-carbons is also reduced relative to these carbons in pyridines.

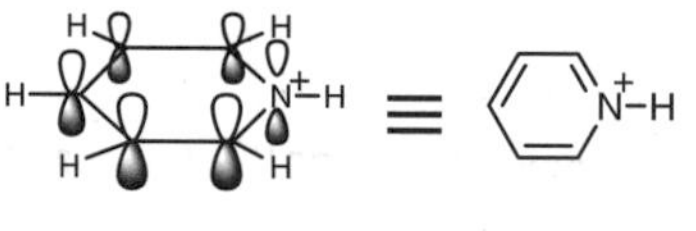

1H-pyridinium cation 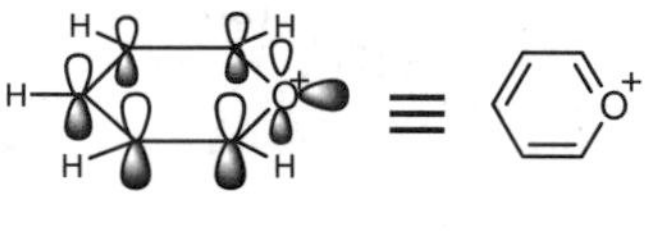   pyrylium cation

In the pyrylium cation, the positively charged oxygen also has an unshared electron pair, in an sp^2 orbital in the plane of the ring, exactly as in pyridine. Once again, a set of resonance contributors, **19–23**, makes clear that this ion is strongly positively charged at the 2-, 4- and 6-positions, in fact, because the more electronegative oxygen tolerates positive charge much less well than nitrogen, the pyrylium cation is certainly a less stabilised system than a pyridinium cation.

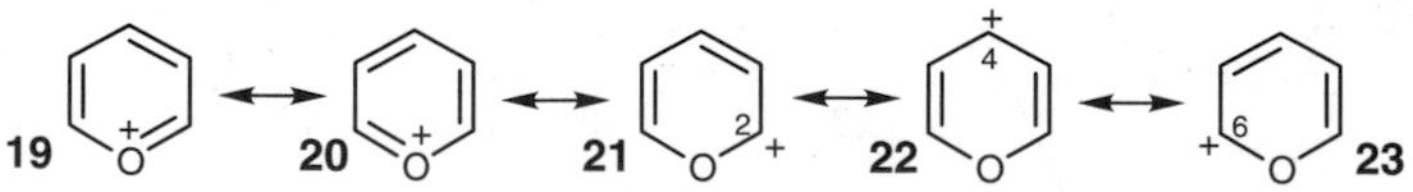

1.2.4   Structures of pyridones and pyrones

Pyridines with an oxygen at either the 2- or 4-position exist predominantly as carbonyl tautomers, which are therefore known as pyridones[4] (see also section 1.5).

In the analogous oxygen systems, no alternative tautomer is possible; the systems are known as pyrones. The extent to which such molecules are aromatic has been a subject for considerable speculation and experimentation, and estimates have varied considerably. The degree of aromaticity depends on the contribution which dipolar structures, **25** and **27**, with a 'complete' pyridinium (pyrylium) ring make to the overall structure. Pyrones are less aromatic than pyridones, as can be seen from their tendency to undergo addition reactions (section 8.2.2.4), and as would be expected from a consideration of the 'aromatic' contributors, **25** and **27**, which have a positively charged ring hetero atom, oxygen being less easily able to accommodate this requirement.

4-hydroxypyridine
[4-pyridinol]

4-pyridone
[4(1*H*)-pyridinone]

2-pyrone
[2*H*-pyran-2-one]

1.3 Structure of five-membered heteroaromatic systems[5]

1.3.1 Structure of pyrrole

Before discussing pyrrole it is necessary to recall the structure of the cyclopentadienyl anion, which is a 6-π-electron aromatic system produced by the removal of a proton from cyclopentadiene. This system serves to illustrate nicely the difference between aromatic stabilisation and reactivity, for it is a very reactive, fully negatively charged entity, and yet is 'resonance stabilised' – everything is relative. Cyclopentadiene, with a pK_a of about 14, is much more acidic than a simple diene, just because the resulting anion is resonance stabilised. Five equivalent contributing structures, **28–32**, show each carbon atom to be equivalent and hence to carry one fifth of the negative charge.

cyclopentadienyl
anion

Pyrrole is isoelectronic with the cyclopentadienyl anion, but is electrically neutral because of the higher nuclear charge on nitrogen. The other consequence of the presence of nitrogen in the ring is the loss of radial symmetry, so that pyrrole does not have five equivalent canonical forms: it has one with no charge separation, **33**, and two pairs of equivalent forms in which there is charge separation, indicating electron density drift away from the nitrogen. These forms do not contribute equally; the order of importance is: **33** > **35,37** > **34,36**.

Resonance leads, then, to the establishment of partial negative charges on the carbons and a partial positive charge on the nitrogen. Of course the inductive effect of the nitrogen is, as usual, towards the hetero atom and away from carbon, so that the electronic distribution in pyrrole is a balance of two opposing effects, of which the mesomeric effect is probably the more significant. The lengths of the bonds in pyrrole are in accord with this exposition, thus the 3,4-bond is very much longer than the 2,3-/4,5-bonds, but appreciably shorter than a normal single bond betwen sp^2 hybridised carbons, in accord with contributions from the polarised structures **34–37**. It is because of this electronic drift away from nitrogen and towards the ring carbons that five-membered heterocycles of the pyrrole type are referred to as 'electron-rich', or sometimes 'π-excessive'.

It is most important to recognise that the nitrogen lone pair in pyrrole forms part of the aromatic six-electron system.

1.3.2 Structures of thiophene and furan

The structures of thiophene and furan are closely analogous to that discussed in detail for pyrrole above, except that the NH is replaced by S and O respectively. A consequence is that the hetero atom in each has one lone pair as part of the aromatic sextet, as in pyrrole, but also has a second lone pair which is not involved, and is located in an sp^2 hybrid orbital in the plane of the ring. Canonical forms exactly analogous to those (above) for pyrrole can be written for each, but the higher electronegativity of both sulfur and oxygen means that the polarised forms, with positive charges on the hetero atoms, make a smaller contribution. The decreased mesomeric electron drift away from the hetero atoms is insufficient, in these two cases, to overcome the inductive polarisation towards the hetero atom (the dipole moments of tetrahydrothiophene and tetrahydrofuran, 1.87D and 1.68D, respectively, both towards the hetero atom, are in any case larger) and the net effect is to give dipoles directed towards the hetero atoms in thiophene and furan.

The larger bonding radius of sulfur is one of the influences making thiophene more stable (more aromatic) than pyrrole or furan – the bonding angles are larger and angle strain is somewhat relieved, but in addition, a contribution to the stabilisation involving sulfur d orbital participation may be significant.

1.3.3 Structures of azoles

The 1,3- and 1,2-azoles, five-membered rings with two hetero atoms, present a fascinating combination of hetero atom types – in all cases, one hetero atom must be of the five-membered heterocycle (pyrrole, thiophene, furan) type and one of the imine type, as in pyridine; imidazole with two nitrogen atoms illustrates this best. Contributor **39** is a particularly favourable one.

1.4 Structures of bicyclic heteroaromatic compounds

Once the ideas of the structures of benzene, naphthalene, pyridine and pyrrole, as prototypes, have been assimilated it is straightforward to extrapolate to those systems which combine two (or more) of these types, thus quinoline is like naphthalene, only with one of the rings a pyridine, and indole is like pyrrole, but with a benzene ring attached.

Resonance representations must take account of the pattern established for benzene and the relevant heterocycle. Contributors in which both aromatic rings are disrupted make a very much smaller contribution and are shown in parentheses.

1.5 Tautomerism in heterocyclic systems[6]

A topic which has attracted an inordinately large research effort over the years is the determination of precise structure of heterocyclic molecules which are potentially tautomeric – the pyridinol/pyridone relationship (section 1.2.4) is one such situation. In principle, when an oxygen is located on a carbon α or γ to nitrogen, two tautomeric forms can exist; the same is true of amino groups.

Early attempts to use the results of chemical reactions to assess the form of a particular compound were misguided, since these can give entirely the wrong answer: the minor partner in such a tautomeric equilibrium may be the one which is the more reactive, so a major product may be actually derived from the minor component in the tautomeric equilibrium. Most secure evidence on these questions has come from comparisons of spectroscopic data for the compound in question with unambiguous models – often *N*- and *O*-methyl derivatives.

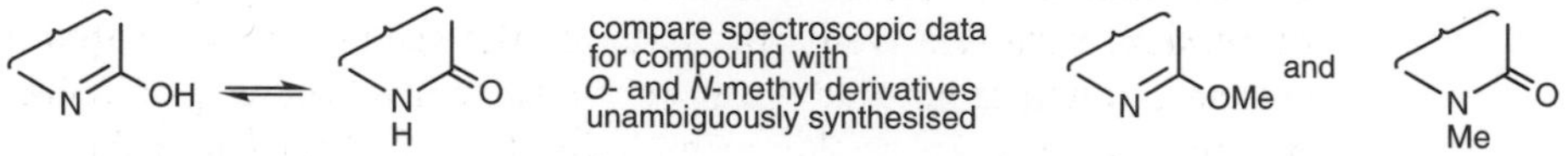

After all the effort that has been expended on this area, the picture which emerges is fairly straightforward: α and γ oxy-heterocycles generally prefer the carbonyl form; amino-heterocycles nearly always exist as amino tautomers. Sulfur analogues – potentially thiol or thione – tend to exist as thione in six-membered situtations, but as thiol in five-membered rings.

The establishment of tautomeric form is perhaps of most importance in connection with the purine and pyrimidine bases which form part of DNA and RNA, and, through H-bonding involving carbonyl oxygen, provide the mechanism for base pairing (cf. section 24.1).

1.6   Mesoionic systems[7]

There are a substantial number of heterocyclic substances for which no plausible, unpolarised canonical structure can be written: such systems are termed 'mesoionic'. Despite the presence of a nominal positive and negative charge in all resonance contributors to such compounds, they are not salt-like, are of course overall neutral, and behave like 'organic' substances, dissolving in the usual solvents. Examples of mesoionic structures occur throughout the text. Amongst the earliest mesoionic substances to be studied were the sydnones, for which several contributing structures can be drawn.

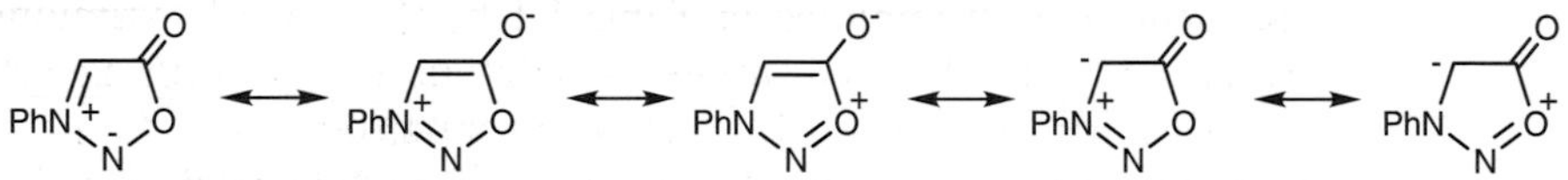

Mesoionic structures occur amongst six-membered systems too – two are illustrated below:

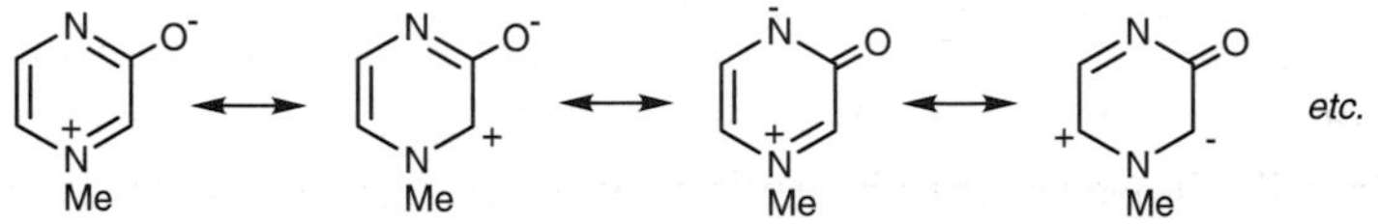

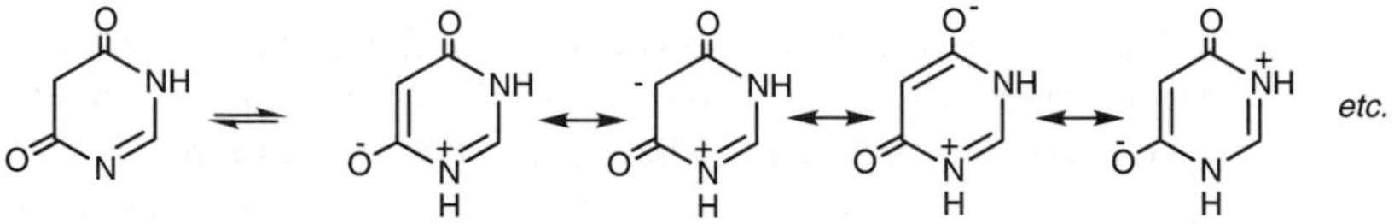

If there is any one feature which characterises mesionic compounds it is that their dipolar structures lead to reactions in which they serve as 1,3-dipoles in cycloadditions.

1.7   Some spectroscopic properties of some heteroaromatic systems

The use of spectroscopy is at the heart of chemical research and analysis, but a knowledge of the particular chemical shift of, say a proton on a pyridine, or the particular UV absorption maximum of, say, an indole, is only of direct relevance to those actually pursuing such research and analysis, and adds nothing to the understanding of heteroaromatic reactivity. Accordingly, we give here only a brief discussion, with relatively little data, of the spectroscopic properties of heterocyclic systems, anticipating that those who may be involved in particular research projects will turn to reviews[1] or the original literature for particular data.

The ultraviolet and infrared spectra of heteroaromatic systems are in accord with their aromatic character. Spectroscopic investigation, particularly ultraviolet/visible (UV/VIS) and nuclear magnetic resonance (NMR) spectroscopies, is particularly useful in the context of assessing the extent of such properties, in determining the position of tautomeric equilibria, and in testing for the existence of non-isolable intermediates.

1.7.1   Ultraviolet/visible (electronic) spectroscopy

The simple unsubstituted heterocyclic systems show a wide range of electronic absorption, from the simple 200 nm band of furan, for example, to the 340 nm maximum shown by pyridazine. As is true for benzenoid compounds, the presence of substituents which can conjugate causes profound changes in electronic absorption, but the many variations possible are outside the scope of this section.

The UV spectra of the monocyclic azines show two bands, each with fine structure: one occurs in the relatively narrow range of 240–260 nm and corresponds to the $\pi \to \pi^*$ transitions, analogous with the $\pi \to \pi^*$ transitions in the same region in benzene (see Table 1.1). The other band occurs at longer wavelengths, from 270 nm in pyridine to 340 nm in pyridazine and corresponds to the interaction of the hetero atom lone pair with aromatic π electrons, the $n \to \pi^*$ transitions, which of course cannot occur in benzene. The absorptions due to $n \to \pi^*$ transitions are very solvent dependent, as is exemplified in Table 1.1 by the case of pyrimidine. With pyridine, this band is only observed in hexane solution, for in alcoholic solution the shift to shorter wavelengths results in masking by the main $\pi \to \pi^*$ band. Protonation of the ring nitrogen naturally quenches the $n \to \pi^*$ band, by removing the hetero atom lone pair; protonation also has the effect of considerably increasing the intensity of the $\pi \to \pi^*$ band, without changing its position significantly, the observation of which can have considerable diagnostic utility.

The bicyclic azines have much more complex electronic absorption, and the $n \to \pi^*$ and $\pi \to \pi^*$ bands overlap; being much more intense, the latter mask the former. Broadly, however, the absorptions of the bicyclic azines resemble that of naphthalene (Table 1.2).

The UV spectra of the simple five-membered heteroaromatic systems all show just one medium-to-strong low-wavelength band with no fine structure. Their absorptions have no obvious similarity to that of benzene, and no detectable $n \to \pi^*$ absorption, not even in the azoles, which contain a pyridine-like nitrogen (Tables 1.3 and 1.4).

1.7.2   Nuclear magnetic resonance (NMR) spectroscopy[8]

The chemical shifts[9] of protons attached to, and in particular of the carbons in, heterocyclic systems, can be taken as relating to the electron-density at that position, with lower fields corresponding to electron-deficient carbons. For example, in the ^{1}H spectrum of pyridine, the lowest field signals are for the α-protons (Table 1.5), the next lowest is that for the γ-proton and the highest field signal corresponds to the β-

Table 1.1 Ultraviolet spectra of monocyclic azines (fine structure not given)

heterocycle (solvent)	$n \to \pi^*$ $\lambda_{max.}$ (nm)	ε	$\pi \to \pi^*$ $\lambda_{max.}$ (nm)	$\pi \to \pi^*$ $\lambda_{max.}$ (nm)	ε	ε
pyridine (hexane)	270	450	195	251	7500	2000
pyridine (ethanol)	-	-	-	257	-	2750
pyridinium (ethanol)	-	-	-	256	-	5300
pyridazine (hexane)	340	315	-	246	-	1400
pyrimidine (hexane)	298	326	-	243	-	2030
pyrazine (hexane)	328	1040	-	260	-	5600
pyrimidine (water)	271	410	-	243	-	3210
pyrimidinium (water)	-	-	-	242	-	5500
pyrylium (90% aq. $HClO_4$)	-	-	220	269	1400	8500
benzene (hexane)	-	-	204	254	7400	200

Table 1.2 Ultraviolet spectra of bicyclic azines (fine structure not given)

heterocycle	$\lambda_{max.}$ (nm)	$\lambda_{max.}$ (nm)	$\lambda_{max.}$ (nm)	ε	ε	ε
quinoline	313	270	226	2360	3880	35500
quinolinium	313	-	233	6350		34700
isoquinoline	317	266	217	3100	4030	37000
isoquinolinium	331	274	228	4170	1960	37500
quinolizinium	324	284	225	14500	2700	17000
naphthalene	312	275	220	250	5600	100000

protons, and this is echoed in the corresponding ^{13}C shifts (Table 1.6). A second generality relates to the inductive electron withdrawal by the hetero atom – for example it is the hydrogens on the α-carbons of pyridine which are at lower field than that at the γ-carbon, and it is the signals for protons at the α-positions of furan which are at lower field than those at the β-positions. Protons at the α-positions of pyrylium cations present the lowest field ^1H signals. In direct contrast, the chemical shifts for *C*-protons on electron-rich heterocycles, such as pyrrole, occur at much higher fields.

Coupling constants between 1,2-related (*ortho*) protons on heterocyclic systems vary considerably. Typical values round six-membered systems show smaller values closer to the hetero atom(s). In five-membered heterocycles, altogether smaller values are typically found, but again those involving a hydrogen closer to the hetero atom are smaller, except in thiophenes, where the larger size of the sulfur atom influences the coupling constant. The magnitude of such coupling constants reflects the degree of double bond character (bond fixation) in a particular C–C bond.

Table 1.3 Ultraviolet spectra of monocyclic five-membered heterocycles

heterocycle	$\lambda_{max.}$ (nm)	$\lambda_{max.}$ (nm)	ε	ε
pyrrole	210	-	5100	-
furan	200	-	10000	-
thiophene	235	-	4300	-
imidazole	206	-	3500	-
oxazole	205	-	3900	-
thiazole	235	-	3000	-
cyclopentadiene	200	239	10000	3400

The use of ^{15}N NMR spectroscopy is of obvious relevance to the study of nitrogen-containing heterocycles – it can for example be used to estimate the hybridisation of nitrogen atoms.[10]

Table 1.4 Ultraviolet spectra of bicyclic compounds with five-membered heterocyclic rings

heterocycle	$\lambda_{max.}$ (nm)	$\lambda_{max.}$ (nm)	$\lambda_{max.}$ (nm)	ε	ε	ε
indole	288	261	219	4900	6300	25000
benzo[*b*]thiophene	288	257	227	2000	5500	28000
benzo[*b*]furan	281		244	2600		11000
2-*t*-Bu-isoindole	223, 266	270, 277	289, 329	48000, 1800	1650, 1850	1250, 3900
isobenzofuran	215, 244, 249	254, 261, 313	319, 327, 334, 343	14800, 2500, 2350	2250, 1325, 5000	5000, 7400, 4575, 6150
indolizine	347	295	238	1950	3600	32000
benzimidazole	259	275		5620	5010	
benzothiazole	217, 251	285	295	18620, 5500	1700	1350
benzoxazole	231, 263	270	276	7940, 2400	3390	3240
2-methyl-2*H*-indazole	275	292	295	6310	6170	6030
2,1-benzisothiazole	203, 221	288sh, 298	315sh	14450, 16220	7590, 2880	3980
purine	263	-	-	7950	-	-

Table 1.5 ^{1}H chemical shifts (ppm) for heteroaromatic ring protons

heterocycle	δ_1	δ_2	δ_3	δ_4	δ_5	δ_6	δ_7	δ_8	others
pyridine	-	8.5	7.1	7.5	-	-	-	-	-
2-pyridone	-	-	6.6	7.3	6.2	7.3	-	-	-
quinoline	-	8.8	7.3	8.0	7.7	7.4	7.6	8.1	-
quinoline N-oxide	-	8.6	7.3	7.7	-	-	-	8.8	-
isoquinoline	9.1	-	8.5	7.5	7.7	7.6	7.5	7.9	-
isoquinoline N-oxide	8.8	-	8.1	-	-	-	-	-	-
pyridazine	-	-	9.2	7.7	-	-	-	-	-
pyrimidine	-	9.2	-	8.6	7.1		-	-	-
pyrimidine N-oxide	-	9.0	-	8.2	7.3	8.4	-	-	-
pyrazine	-	8.5	-	-	-	-	-	-	-
1,2,4-triazine	-	-	9.6	-	8.5	9.2	-	-	-
1,3,5-triazine	-	9.2	-	-	-	-	-	-	-
cinnoline	-	-	9.15	7.75	-	-	-	-	-
quinazoline	-	9.2	-	9.3	-	-	-	-	-
quinoxaline	-	9.7	-	-	-	-	-	-	-
phthalazine	9.4	-	-	-	-	-	-	-	-
pyrylium	-	9.6	8.5	9.3	-	-	-	-	in SO$_2$ (liq.)
pyrrole	-	6.6	6.2	-	-	-	-	-	-
thiophene	-	7.2	7.1	-	-	-	-	-	-
furan	-	7.4	6.3	-	-	-	-	-	-
indole	-	6.5	6.3	7.5	7.0	7.1	7.4	-	-
benzo[b]furan	-	7.5	6.7	7.5	7.1	7.2	7.4	-	-
benzo[b]thiophene	-	7.3	7.3	7.7	7.3	7.3	7.8	-	-
indolizine	6.3	6.6	7.1	-	7.8	6.3	6.5	7.2	-
imidazole	-	7.9	-	7.25	-	-	-	-	-
1-methylimidazole	-	7.5	-	7.1	6.9	-	-	-	-
pyrazole	-	-	7.6	6.3		-	-	-	-
1-methylpyrazole	-	-	7.5	6.2	7.4	-	-	-	3.8 (CH$_3$)
thiazole	-	8.9	-	8.0	7.4	-	-	-	-
oxazole	-	7.95	-	7.1	7.7	-	-	-	-
benzimidazole	-	7.4	-	7.0	6.9	-	-	-	-
benzoxazole	-	7.5	-	7.7	7.8	7.8	7.7	-	-
pyrazole	-	-	7.6	7.3	-	-	-	-	-
isothiazole	-	-	8.5	7.3	8.7	-	-	-	-
isoxazole	-	-	8.1	6.3	8.4	-	-	-	-
indazole	-	-	8.1	7.8	7.1	7.35	7.55		
1,2,3-triazole	-	-	-	7.75	-	-	-	-	-
1,2,4-triazole	-	-	7.9	-	8.85	-	-	-	-
tetrazole	-	-	-	9.5	-	-	-	-	-
purine	-	9.0	-	-	-	9.2	-	8.6	-
benzene	7.27	-	-	-	-	-	-	-	-

Table 1.5 (*continued*)

heterocycle	δ_1	δ_2	δ_3	δ_4	δ_5	δ_6	δ_7	δ_8	others
anisole	-	6.9	7.2	6.9	-	-	-	-	-
aniline	-	6.5	7.0	6.6	-	-	-	-	-
nitrobenzene	-	8.2	7.4	7.6	-	-	-	-	-
naphthalene	7.8	7.5	-	-	-	-	-	-	-

Table 1.6 ^{13}C chemical shifts (ppm) for heteroaromatic ring carbons

heterocycle	δ_1	δ_2	δ_3	δ_4	δ_5	δ_6	δ_7	δ_8	$\delta_{\text{ring junction}}$	$\delta_{\text{ring junction}}$	other
pyridine	-	150	124	136	-	-	-	-	-	-	-
1-*H*-pyridinium	-	143	129	148	-	-	-	-	-	-	-
pyridine *N*-oxide	-	139	126	126	-	-	-	-	-	-	-
1-Me-pyridinium	-	146	129	146	-	-	-	-	-	-	50 (CH$_3$)
2-pyridone	-	165	121	142	107	136	-	-	-	-	-
4-pyridone	-	140	116	176	-	-	-	-	-	-	-
quinoline	-	151	122	136	1289	127	130	131	129 (4*a*)	149 (8*a*)	-
isoquinoline	153	-	143	120	126	130	127	128	136 (4*a*)	129 (8*a*)	-
pyridazine	-	-	153	128	-	-	-	-	-	-	-
1-*H*-pyridazinium	-	-	152	138	-	-	-	-	-	-	-
pyrimidine	-	158	-	156	121	-	-	-	-	-	-
1-*H*-pyrimidinium	-	152	-	159	125	-	-	-	-	-	-
pyrazine	-	146	-	-	-	-	-	-	-	-	-
1-*H*-pyrazinium	-	143	-	-	-	-	-	-	-	-	-
cinnoline	-	-	146	125	128	132	132	130	127 (4*a*)	151 (8*a*)	-
quinazoline	-	161	-	156	127	128	134	129	135 (4*a*)	150 (8*a*)	-
quinoxaline	-	146	-	-	130	130	-	-	143 (4*a*)	-	-
phthalazine	152	-	-	-	127	133	-	-	126 (4*a*)	-	-
1,2,3-triazine	-	-	-	150	118	-	-	-	-	-	-
1,2,4-triazine	-	-	158	-	150	151	-	-	-	-	-
1,3,5-triazine	-	166	-	-	-	-	-	-	-	-	-
pyrylium (BF$_4^-$)	-	169	128	161			-	-	-	-	-
2-pyrone	-	162	117	143	106	152	-	-	-	-	-
2,6-Me$_2$-4-pyrone	-	166	114	180			-	-	-	-	20 (CH$_3$)
coumarin	-	161	117	144	129	124	132	117	119 (4*a*)	154 (8*a*)	-
chromone	-	156	113	177	125	126	134	118	125 (4*a*)	156 (8*a*)	-
pyrrole	-	117	108	-	-	-	-	-	-	-	-
thiophene	-	126	127	-	-	-	-	-	-	-	-
furan	-	144	110	-	-	-	-	-	-	-	-
indole	-	124	102	121	122	120	111	-	128 (3*a*)	136 (7*a*)	-
oxindole	-	179	36	124	122	128	110	-	125 (3*a*)	143 (7*a*)	-
benzo[*b*]furan	-	145	107	122	123	125	112	-	128 (3*a*)	155 (7*a*)	-

Table 1.6 (*continued*)

heterocycle	δ_1	δ_2	δ_3	δ_4	δ_5	δ_6	δ_7	δ_8	$\delta_{\text{ring junction}}$	$\delta_{\text{ring junction}}$	other
benzo[*b*]thiophene	-	126	124	124	124	124	123	-	140 (3*a*)	140 (7*a*)	-
indolizine	100	114	113	-	126	111	117	120	133 (8*a*)	-	-
imidazole	-	135	-	122		-	-	-	-	-	-
1-methylimidazole	-	138	-	130	120	-	-	-	-		33 (CH$_3$)
thiazole	-	154	-	143	120	-	-	-	-		-
oxazole	-	151	-	125	138	-	-	-	-		-
benzimidazole	-	144	-	110	123	122	119	-	-	-	-
benzothiazole	-	155	-	123	126	125	122	-	153 (3*a*)	134 (7*a*)	-
benzoxazole	-	153	-	121	125	124	111	-	140 (3*a*)	150 (7*a*)	-
pyrazole	-	-	135	106	135		-	-	-	-	-
isothiazole	-	-	157	123	148	-	-	-	-	-	-
isoxazole	-	-	150	105	159	-	-	-	-	-	-
indazole	-	-	133	120	120	126	110	-	123 (3*a*)	140 (7*a*)	-
3-methyl-1,2-benzisothiazole	-	163	-	-	-	-	-	-	152 (7*a*)	-	-
purine	-	152	-	155	131	146	-	146	-	-	-
uracil	-	151	-	142	100	164	-	-	-	-	-
benzene	129	-	-	-	-	-	-	-	-	-	-
anisole	160	114	130	121	-	-	-	-	-	-	-
aniline	149	114	129	116	-	-	-	-	-	-	-
nitrobenzene	149	124	130	135	-	-	-	-	-	-	-
naphthalene	128	126	-	-	-	-	-	-	133 (4*a*)	-	-

References

1 'Physical methods in heterocyclic chemistry', Vols 15, Ed. Katritzky, A. R., Academic Press, New York, **1960–72**; 'Comprehensive heterocyclic chemistry. The structure, reactions, synthesis, and uses of heterocyclic compounds', Eds. Katritzky, A. R. and Rees, C. W., Vols 1-8, Pergamon Press, Oxford, **1984**; 'Comprehensive heterocyclic chemistry II. A review of the literature 1982-1995', Eds. Katritzky, A. R., Rees, C. W., and Scriven, E. F. V., Vols 1-11, Pergamon Press, **1996**.

2 'Frontier orbitals and organic chemical reactions', Fleming, I., Wiley-Interscience, **1976**.

3 'Aromaticity of heterocycles', Cook, M. J., Katritzky, A. R., and Linda, P., *Adv. Heterocycl. Chem.*, **1974**, *17*, 257; 'Aromaticity of heterocycles: experimental realisation of DewarBreslow definition of aromaticity', Hosmane, R. A. and Liebman, J. F., *Tetrahedron Lett.*, **1991**, *32*, 3949; 'The relationship between bond type, bond order and bond lengths. A re-evaluation of the aromaticity of some heterocyclic molecules', Box, V. G. S., *Heterocycles*, **1991**, *32*, 2023; 'Heterocyclic aromaticity', Katritzky, A. R., Karelson, M., and Malhotra, N., *Heterocycles*, **1991**, *32*, 127; 'The concept of aromaticity in heterocyclic chemistry', Simkin, B. Ya., Minkin, V. I., and Glukhovtsev, M. N., *Adv. Heterocycl. Chem.*, **1993**, *56*, 303.

4 In solution at high dilution, or in the gas phase, hydroxypyridine tautomers are more important or even dominant: Beak, P., Covington, J. B., Smith, S. G., White, J. M., and Zeigler, J. M., *J. Org. Chem.*, **1980**, *45*, 1354.

5 Fringuelli, F., Marino, G., Taticchi, A., and Grandolini, G., *J. Chem. Soc., Perkin Trans. 2*, **1974**, 332.

6 'The tautomerism of heterocycles', Elguero, J., Marzin, C., Katritzky, A. R., and Linda, P., *Adv. Heterocycl. Chem.*, Supplement 1, **1976**; 'Energies and alkylations of tautomeric

heterocyclic compounds: old problems new answers', Beak, P., *Acc. Chem. Res.*, **1977**, *10*, 186; 'Prototropic tautomerism of heteroaromatic compounds', Katritzky, A. R., Karelson, M., and Harris, P. A., *Heterocycles*, **1991**, *32*, 329.

7 'Mesoionic compounds', Ollis, W. D. and Ramsden, C. A., *Adv. Heterocycl. Chem.*, **1976**, *19*, 1; 'Heterocyclic betaine derivatives of alternant hydrocarbons', Ramsden, C. A., *ibid.*, **1980**, *26*, 1; 'Mesoionic heterocycles (1976–1980)', Newton, C. G. and Ramsden, C. A., *Tetrahedron*, **1982**, *38*, 2965; 'Six-membered mesoionic heterocycles of the *m*-quinodimethane dianion type', Friedrichsen, W., Kappe, T., and Böttcher, A., *Heterocycles*, **1982**, *19*, 1083.

8 'Applications of nuclear magnetic resonance spectroscopy in organic chemistry', Jackman, L. M. and Sternhell, S., Pergamon Press, **1969**; 'Carbon-13 NMR spectroscopy', Breitmaier, E. and Voelter, W., VCH, **1990**.

9 Both proton and carbon chemical shifts are solvent dependent – the figures given in the Tables are a guide to the relative shift positions of proton and carbon signals in these heterocycles.

10 von Philipsborn, W. and Müller, R., *Angew. Chem., Int. Ed. Engl.*, **1986**, *25*, 383.

2 Reactivity of aromatic heterocycles

This chapter describes in general terms the types of reactivity found in the typical six- and five-membered aromatic heterocycles. In addition to discussions of classical substitution chemistry, considerable space is devoted to radical substitution, metallation and palladium-catalysed reactions, since these areas have become very important in heterocyclic manipulations. In order to gain a proper appreciation of their importance in the heterocyclic context we provide an introduction to these topics, since they are only poorly covered in general organic text-books. Emphasis on the typical chemistry of individual heterocyclic systems is to be found in the summary/revision chapters (4, 7, 10, 12, 16, and 20) and a more detailed examination of typical heterocyclic reactivity, and many more examples for particular heterocyclic systems are to be found in the chapters – 'Pyridines: reactions and synthesis' etc. For the advanced student, it is recommended that this present chapter should be read in its entirety before moving on to the later chapters, and that the introductory summary/revision chapters, like 'Typical reactivity of pyridines, quinolines and isoquinolines' should be read before the more detailed discussions.

2.1 Electrophilic addition at nitrogen

Heterocycles which contain an imine unit ($C=N$) as part of their ring structure – pyridines, quinolines, isoquinolines, 1,2- and 1,3-azoles, etc. – do not utilise the nitrogen lone pair in their aromatic π-system (cf. section 1.2) and therefore it is available for donation to electrophiles, just as in any simpler amine. In other words, such heterocycles are basic and will react with protons, or other electrophilic species, at nitrogen, by addition. In many instances the product salts, from such additions, are isolable.

For reversible additions, for example of a proton, the position of equilibrium depends on the pK_a of the heterocycle,[1] and this in turn is influenced by the substituents present on the ring: electron-releasing groups enhance the basicity and electron-withdrawing substituents reduce the basic strength. The pK_a of simple pyridines is of the order of 5, while those for 1,2- and 1,3-azoles depends on the character of the other heteroatom: pyrazole and imidazole, with two nitrogen atoms, have values of 2.5 and 7.1 respectively.

Related to basicity, but certainly not always mirroring it, is the N-nucleophilicity of imine-containing heterocycles. Here, the presence of substituents adjacent to the nitrogen can have a considerable effect on how easily reaction with alkyl halides takes place and indeed whether nitrogen attacks at carbon, forming N^+-alkyl salts,[2] or by deprotonation, bringing about a 1,2-dehydrohalogenation of the halide, the heterocycle then being converted into an N^+-hydrogen salt. The classical study of the slowing of *N*-alkylation by the introduction of steric interference at α-positions of pyridines showed one methyl to slow the rate by about threefold, whereas 2,6-dimethyl substitution slowed the rate between 12 and 40 times.[3] Taking this to an extreme, 2,6-di-*t*-butylpyridine will not react at all with iodomethane, even under high pressure; the very reactive methyl fluorosulfonate will *N*-methylate it, but only under high pressure.[4] The quantitative assessment of reactivity at nitrogen must always take into account both steric (especially at the α-positions) and electronic effects: 3-methylpyridine reacts faster (x 1.6) but 3-chloropyridine reacts slower (x 0.14) than pyridine. *Peri* substituents have a significant effect on the relative rates of reaction with iodomethane: for pyridine, isoquinoline (no *peri* hydrogen), quinoline, and 8-methylquinoline, rates are 50, 69, 8, and 0.008, respectively.

Other factors can influence the rate of quaternisation: all the diazines react with iodomethane more slowly than does pyridine. Pyridazine, much more weakly basic (pK_a 2.3) than pyridine, reacts with iodomethane faster than the other diazines, a result which is ascribed to the 'α effect', i.e. the increased nucleophilicity is deemed to be due to electron repulsion between the two immediately adjacent lone pairs.[5] Reaction rates for iodomethane with pyridazine, pyrimidine and pyrazine are respectively 0.25, 0.044, and 0.036 relative to the rate with pyridine.

2.2 Electrophilic substitution at carbon[6]

The study of aromatic heterocyclic reactivity can be said to have begun with the results of electrophilic substitution processes – these were traditionally the means for the introduction of substitutents onto heterocylic rings. To a considerable extent that methodology has been superseded, especially for the introduction of carbon substituents, by methods relying on the formation of heteroaryllithium nucleophiles (section 2.6) and on palladium-catalysed processes (section 2.7). Nonetheless the reaction of heterocycles with electrophilic reagents is still extremely useful in many cases, particularly for electron-rich, five-membered heterocycles.

2.2.1 Aromatic electrophilic substitution – mechanism

Electrophilic substitution of aromatic (and heteroaromatic) molecules proceeds *via* a two-step sequence, initial addition (of X^+) giving a positively charged intermediate (a σ-complex, or Wheland intermediate), then elimination (normally of H^+), of which the former is usually the slower (rate-determining) step. Under most circumstances such substitutions are irreversible and the product ratio is determined by kinetic control.

aromatic electrophilic
substitution

σ-complex   sp^3 hybridised

2.2.2 Six-membered heterocycles

An initial broad division must be made in considering heteroaromatic electrophilic substitution, into those heterocycles which are basic and those which are not, for in the case of the former the interaction of nitrogen lone pair with the electrophile (cf. section 2.1), or indeed with any other electrophilic species in the proposed reaction mixture (protons in a nitrating mixture, or aluminium chloride in a Friedel-Crafts combination) will take place far faster than any *C*-substitution, thus converting the substrate into a positively charged salt and therefore hugely reducing its susceptibility to attack by X^+ at carbon. It is worth recalling the rate reduction attendant upon the change from benzene to *N,N,N*-trimethylanilinium cation (PhN^+Me_3) where the electrophilic substitution rate goes down by a factor of 10^8 even though in this instance the charged atom is only attached to, and not a component of, the aromatic ring. Thus all heterocycles with a pyridine-type nitrogen (i.e. those containing $C=N$) do not easily undergo *C*-electrophilic substitution, unless (a) there are other substituents on the ring which 'activate' it for attack, or (b) the molecule has another, fused benzene ring in which substitution can take place, or (c) there is a second hetero atom in a five-membered ring, which can release electrons to the attacking electrophile. For example, simple pyridines do not undergo many useful electrophilic substitutions, but quinolines and isoquinolines undergo substitution in the benzene ring. It has been estimated that the intrinsic reactivity of a pyridine (i.e. not protonated) to electrophilic substitution is around 10^7 times less than that of benzene, that is to say, about the same as that of nitrobenzene.

When quinoline or isoquinoline undergo nitration in the benzene ring the actual species attacked is the *N*-protonated heterocycle, and even though substitution is taking place in the benzene ring, it must necessarily proceed through a doubly charged intermediate: this results in a much slower rate of substitution than for the obvious comparison, naphthalene – the 5- and 8-positions of quinolinium are attacked at about a 10^{10} slower rate than the 1-position of naphthalene, and it was estimated that the nitration of pyridinium cation is at least 10^5 slower still.[7] A study of the bromination of methylpyridines in acidic solution allowed an estimate of 10^{-13} for the partial rate factor for bromination of a pyridinium cation.[8]

'Activating' substitutents,[9] i.e. groups which can release electrons either inductively or mesomerically, make the electrophilic substitution of pyridine rings to which they are attached faster, for example 4-pyridone nitrates at the 3-position *via* the *O*-protonated salt.[10] In order to understand the activation, it is helpful to view the species attacked as a (protonated) phenol-like substrate. Electrophilic attack on neutral pyridones is best visualised as attack on an enamide. Dimethoxypyridines also undergo nitration *via* their cations, but the balance is often delicate, for example 2-aminopyridine brominates at C-5, in acidic solution, *via* the free base.[11]

Pyridines carrying activating substituents at C-2 are attacked at C-3/C-5, those with such groups at C-3 are attacked at C-2, and not at C-4, whilst those with substituents at C-4 undergo attack at C-3.

positions of electrophilic attack on substituted pyridines

Substituents which reduce the basicity of a pyridine nitrogen can also influence the susceptibility of the heterocycle to electrophilic susbtitution, in these cases by increasing the proportion of neutral (more reactive) pyridine present at equilibrium: 2,6-dichloropyridine nitrates at C-3, as the free base, and only 10^3 times more slowly than 1,3-dichlorobenzene. As a rule-of-thumb it has been suggested that (i) pyridines with a $pK_a > 1$ will nitrate as cations, slowly unless strongly activated, and at an α or β position depending on the position of the substituent, (ii) weakly basic pyridines, $pK_a < -2.5$, nitrate as free bases, and at an α or β position depending on the position of the substituent.[11]

Pyridines carrying strongly electron-withdrawing substituents, or heterocycles with additional heteroatoms, diazines for example, are so deactivated that electrophilic substitutions do not take place.

2.2.3   Five-membered heterocycles

For five-membered, electron-rich heterocycles the utility of electrophilic substitutions is much greater.[12] Heterocycles such as pyrrole, thiophene and furan undergo a range of electrophilic substitutions with great ease, at either type of ring position, but with a preference for attack adjacent to the hetero atom – at their α-positions. These substitutions are facilitated by electron-release from the hetero atom and, as a consequence, pyrroles are more reactive than furans which are in turn more reactive than thiophenes. Quantitative comparisons[13] of the relative reactivities of the three heterocycles vary from electrophile to electrophile, but for trifluoroacetylation, for example, the pyrrole:furan:thiophene ratio is: 5×10^7:1.5×10^2:1;[14] in formylation, furan is 12 times more reactive than thiophene,[15] and for acetylation, the value is 9.3.[16] In hydrogen exchange (deuteriodeprotonation) the partial rate factors for the α and β positions of N-methylpyrrole[17] are 3.9×10^{10} and 2.0×10^{10} respectively; for this same process, the values for furan are 1.6×10^8 and 3.2×10^4 and for thiophene, 3.9×10^8 and 1.0×10^5 respectively,[18] and in a study of thiophene, α:β ratios ranging from 100:1 to 1000:1 were found for different electrophiles.[19] Relative substrate reactivity parallels positional selectivity i.e. the α:β ratio decreases in the order furan > thiophene > pyrrole.[20] Nice illustrations of these relative reactivities are found in acylations of compounds containing two different systems linked together.[21]

The positional selectivity of attack on pyrroles can be completely altered by the presence of bulky groups on nitrogen: 1-(t-butyldimethylsilyl)pyrrole and 1-(tri-i-propylsilyl)pyrrole are attacked exclusively at their β-positions.[22] Extremely electrophilic reagents (hard electrophiles) such as trimethylsilyl triflate attack N-methylpyrroles exclusively at a β-position.[23]

Indoles are only slightly less reactive than pyrroles, electrophilic substitution taking place in the heterocyclic ring, at a β-position: in acetylation using a Vilsmeier combination (N,N-dimethylacetamide/phosgene), the rate ratio compared with pyrrole is 1:3.[24] In contrast to pyrrole there is a very large difference in reactivity betwen the two hetero-ring position in indoles: 2600:1, $\beta:\alpha$, in Vilsmeier acetylation. With reference to benzene, indole reacts at its β-position around 5×10^{13} times as fast.[25] Again, these differences can be illustrated conveniently using an example[26] which contains two types of system linked together.

The reactivity of an indole is very comparable to that of a phenol: typical of phenols is their ability to be substituted even by weak electrophiles, like benzenediazonium cations, and indeed indoles (and pyrroles) also undergo such couplings; depending on pH, indoles can undergo such processes *via* a small equilibrium concentration of anion formed by loss of N-proton (cf. section 2.5); of course this is an even more rapid process, shown to be 10^8 faster than for the neutral heterocycle.[27] The Mannich substitution (electrophile: $CH_2 = N^+Me_2$) of 5- and 6-hydroxyindoles, takes place *ortho* to the phenolic activating group on the benzene ring, and not at the indole β-position.[28] Comparisons of the rates of substitution of the pairs furan/benzo[*b*]furan and thiophene/benzo[*b*]thiophene showed the bicyclic systems to be less reactive than the monocyclic heterocycles, the exact degree of difference varying from electrophile to electrophile.[29]

Finally, in the 1,2- and 1,3-azoles there is a fascinating interplay of the propensities of an electron-rich five-membered heterocycle with an imine, basic nitrogen. This latter reduces the reactivity of the heterocycle towards electrophilic attack at carbon, both by inductive and by mesomeric withdrawal, and also by conversion into salt in acidic media. For example, depending on acidity, the nitration of pyrazole can proceed by attack on the pyrazolium cation,[30] or *via* the free base.[31] A study of acid-catalysed exchange showed the order: pyrazole > isoxazole > isothiazole, paralleling pyrrole > furan > thiophene, but each is much less reactive than the corresponding heterocycle without the azomethine nitrogen, but equally, that each is still more reactive than benzene, the partial rate factors for exchange at their 4-positions being 6.3×10^9, 2.0×10^4 and 4.0×10^3 respectively. Thiophene is 3×10^5 times more rapidly nitrated than 4-methylthiazoles;[32] the nitration of a 2-(thien-2-yl)thiazole illustrates the relative reactivities.[33]

2.3 Nucleophilic substitution at carbon[34]

2.3.1 Aromatic nucleophilic substitution – mechanism

Nucleophilic substitution of aromatic compounds proceeds *via* an **addition** (of Y^-) then **elimination** (of a negatively charged entity, most often Hal$^-$) two-step sequence, of which the former is usually rate-determining. It is the stabilisation (delocalisation of charge) of the negatively charged intermediates (Meisenheimer complexes) which is the key to such processes, for example in reactions of *ortho* and *para* chloronitrobenzenes the nitro group is involved in the charge dispersal.

2.3.2   Six-membered heterocycles

In the heterocyclic field, the displacement of good leaving groups, often halide, by a nucleophile is a very important general process, especially for six-membered electron-poor systems. In the chemistry of five-membered aromatic heterocycles, such processes only come into play in special situations such as where, as in benzene chemistry, the leaving group is activated by an *ortho* or *para* nitro group, or in the azoles, where the leaving group is attached to an imine unit.

Positions α and γ to an imine nitrogen are activated for the initial addition of a nucleophile by two factors: (i) inductive and mesomeric withdrawal of electrons by the nitrogen and (ii) inductive withdrawal of electrons by the halogen. The σ-adduct intermediate is also specially stabilised when attack is at α- and γ-positions, since in these intermediates the negative charge resides largely on the nitrogen: α and γ positions are much more reactive in nucleophilic displacements than β positions. A quantitative comparison for displacements of chloride with sodium methoxide in methanol showed the 2- and 4-chloropyridines to react at roughly the same rate as 4-chloronitrobenzene, with the γ-isomer somewhat more reactive than the α-halide.[35] It is notable that even 3-chloropyridine, where only inductive activation can operate, is appreciably more reactive than chlorobenzene.

Rates of reaction with MeO$^-$, relative to chlorobenzene, at 50 °C

2.8×10^8	9.1×10^4	7.4×10^9	7×10^{10}	4.5×10^5

The presence of a formal positive charge on the nitrogen, as in *N*-oxides and *N*-alkylpyridinium salts, has a further very considerable enhancing effect on the rate of nucleophilic substitutions, *N*-oxidation having a smaller effect than quaternisation – in the latter there is a full formal positive charge on the molecule but *N*-oxides are overall electrically neutral. In reactions with methoxide, the 2-, 3- and 4-chloropyridine *N*-oxides are 1.9×10^4, 1.1×10^5, and 1.1×10^3 times more reactive than the corresponding chloropyridines, and displacements of halide in the 2-, 3- and 4-chloro-1-methylpyridinium salts are 4.6×10^{12}, 2.9×10^8, and 5.7×10^9 times more rapid. Another significant point to emerge from these rate studies concerns the relative rate enhancements, at the three ring positions: the effect of the charge is much greater at an α than at a γ position such that in the salts the order is $2 > 4 > 3$, as opposed to both neutral pyridines, where the order of reactivity is $4 > 2 > 3$, and *N*-oxides, where the α-positions end up at about the same reactivity as the γ-position.[36] The utility of nitrite as a leaving group in heterocyclic chemistry is emphasised by a comparison of its relative reactivity to nucleophilic displacement: 4-nitropyridine is about 1100 times more reactive than 4-bromopyridine. A comparison of the rates of displacement of 4-methylsulfonylpyridine with its *N*-methyl quaternary salt showed a rise in rate by a factor of 7×10^8.[37] Although methoxide is not generally a good leaving group, when attached to a pyridinium salt it is only about 4 times less easily displaced than iodide, bromide and chloride; fluoride in the same situation is displaced about 250 times faster than the other halides.[38]

Turning to bicyclic systems, and a study of reaction with ethoxide, a small increase in the rate of reaction relative to pyridines was found for chloroquinolines at comparable positions.[39] In the bicyclic compounds, quaternisation again greatly increases the rate of nucleophilic substitution, having a larger effect ($\sim 10^7$) at C-2 than at C-4 ($\sim 10^5$).[40]

Relative rates for nucleophilic displacement with EtO⁻ at 20 °C

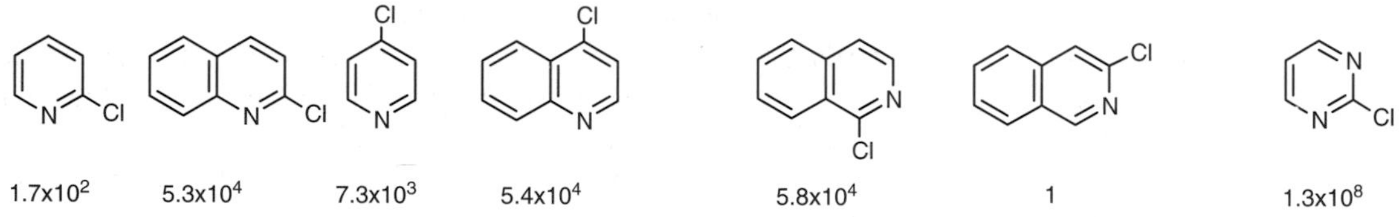

| 1.7x10² | 5.3x10⁴ | 7.3x10³ | 5.4x10⁴ | 5.8x10⁴ | 1 | 1.3x10⁸ |

Diazines with halogen α and γ to nitrogen are much more reactive than similar pyridines, for example 2-chloropyrimidine is $\sim 10^6$ times more reactive than 2-chloropyridine.

2.3.3   Vicarious nucleophilic substitution (VNS substitution)[41]

A process known as 'Vicarious Nucleophilic Substitution' (VNS) of hydrogen has been widely applied to carboaromatic and to heteroaromatic compounds. In general form the process requires the presence of a nitro group on the substrate which permits the addition of a carbon nucleophile, of the form C(X)(Y)(R), where X is a potential leaving group and Y is an anion-stabilising group which permits the formation of the carbanion in the first place. Most often X is a halogen and Y is arylsulfonyl; with these, a typical sequence is shown below. Following addition, *ortho* or *para* to the nitro group, elimination of HX takes place to form a conjugated, non-aromatic nitronate which on reprotonation returns the molecule to aromaticity and produces the substituted product. Excess of the base used to generate the initial carbanion must be used in order to drive the process forward by bringing about the irreversible elimination of HX.

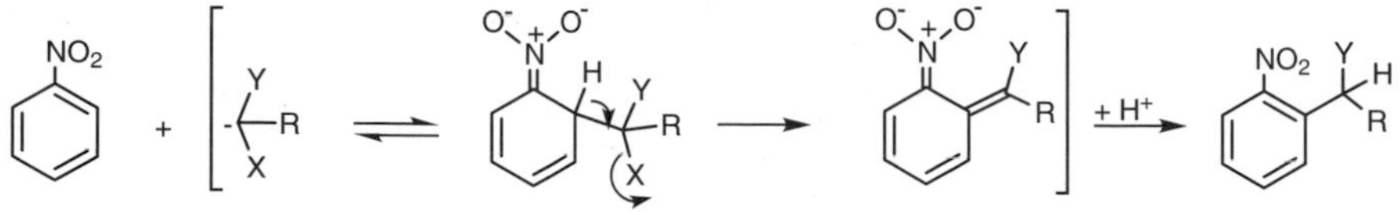

Examples of VNS processes will be found in several later chapters; given below are three typical sequences. The first example shows the operation of a VNS substitution in a five-membered heterocycle;[42] in the second example the anion-stabilising group (Y) (trifluomethanesulfonyl) also serves as the leaving group (X).[43] The third example is somewhat unusual in that the attacking nucleophile does not enter *ortho* or *para* to the nitro group: addition at C-2 in 6-nitroquinoxaline produces an anion stabilised by delocalisation involving both *N*-1 and the nitro group.[44]

2.4   Radical substitution at carbon[45]

Both electron-rich and electron-poor heterocyclic rings are susceptible to substitution of hydrogen by free radicals. Although electrically neutral, radicals exhibit varying degrees of nucleophilic or electrophilic character and this has a very significant effect on their reactivity towards different heterocyclic types. These electronic properties are a consequence of the interaction between the SOMO (**S**ingly **O**ccupied **M**olecular **O**rbital) of the radical and either the HOMO, or the LUMO, of the substrate, depending on their relative energies; these interactions are usefully compared with charge transfer interactions.

Nucleophilic radicals carry cation-stabilising groups on the radical carbon, allowing electron density to be transferred from the radical to an electron-deficient heterocycle; they react therefore only with electron-poor heterocycles and will not attack electron-rich systems: examples of such radicals are ·CH_2OH, alkyl·, and acyl·. Substitution by such a radical can be represented in the following general way:

Electrophilic radicals, conversely, are those which would form stabilised anions on gaining an electron, and therefore react readily with electron-rich systems: examples are ·CF_3 and ·$CH(CO_2Et)_2$. Substitution by such a radical can be represented in the following general way:

Aryl radicals can show both types of reactivity. A considerable effort (mainly older work) was devoted to substitutions by aryl radicals; they react with electron-rich and electron-poor systems at about the same rate but often with poor regioselectivity.[46]

2.4.1   Reactions of heterocycles with nucleophilic radicals
The Minisci reaction[47]

The reaction of nucleophilic radicals, under acidic conditions, with heterocycles containing an imine unit is by far the most important and synthetically useful radical substitution of heterocyclic compounds. Pyridines, quinolines, diazines, imidazoles, benzothiazoles, and purines are amongst the systems which have been shown to react with a wide range of nucleophilic radicals, selectively at positions α and γ to the nitrogen, with replacement of hydrogen. Acidic conditions are essential because N-protonation of the heterocycle both greatly increases its reactivity and promotes regioselectivity towards a nucleophilic radical, most of which hardly react at all with the neutral base. A particularly useful feature of the process is that it can be used to introduce acyl groups, directly, i.e. to effect the equivalent of a Friedel-Crafts substitution – impossible under normal conditions for such systems (cf. section 2.2.2). Tertiary radicals are more stable, but also more nucleophilic and therefore more reactive than methyl radicals in Minisci reactions. The majority of Minisci substitutions have been carried out in aqueous, or at least partially aqueous, media, making isolation of organic products particularly convenient.

Several methods have been employed to generate the required radical, many depending on the initial formation of oxy- or methyl radicals which then abstract hydrogen or iodine from suitable substrates; both these are illustrated by the typical examples shown below.[48] The re-aromatisation of the intermediate radical-cation is usually brought about by its reaction with excess of the oxidant used to form the initial radical.

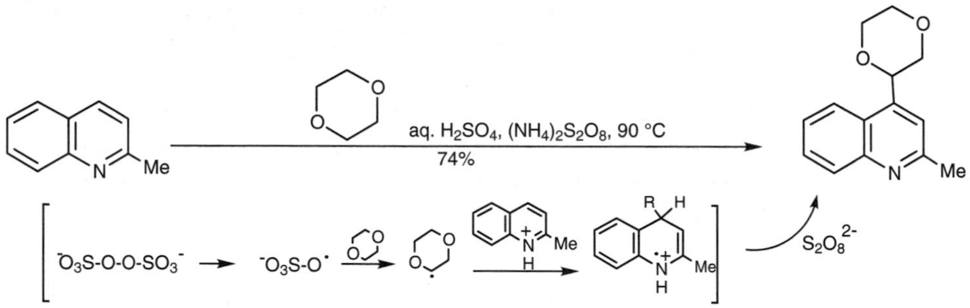

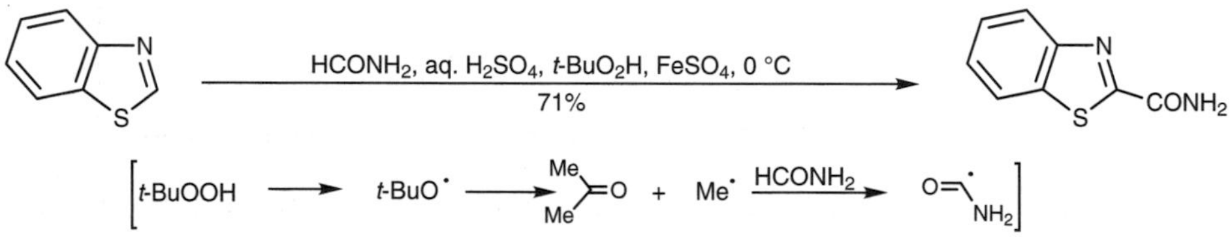

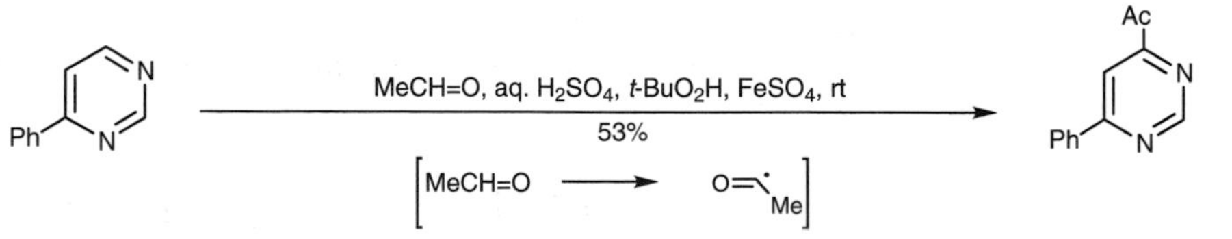

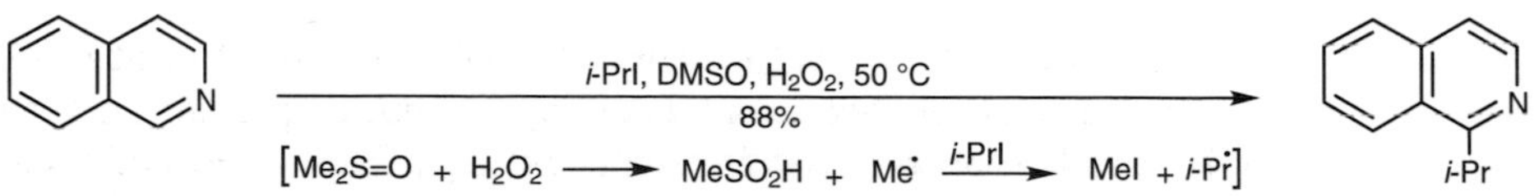

In contrast to the oxidative generation of radicals described above, reductions of alkyl iodides using tris(trimethylsilyl)silane also produces alkyl radicals under conditions suitable for Minisci-type substitution.[49] Carboxylic acids (α-keto acids) are also useful precursors for alkyl[50] (acyl[51]) radicals *via* silver-catalysed peroxide oxidation, or from their 1-hydroxypyridine-2-thione derivatives using Barton's method,[52] the latter in non-aqueous conditions.

$$RCO_2H \xrightarrow{AgNO_3 \text{ (cat.)}, (NH_4)_2S_2O_8} [R^\cdot] + CO_2$$

$$RCOCO_2H \xrightarrow{AgNO_3 \text{ (cat.)}, (NH_4)_2S_2O_8} [R\dot{C}O] + CO_2$$

$$Ag^+ + S_2O_8^{2-} \longrightarrow Ag^{2+}$$

1-Ad = 1-adamantyl

N,N-Dialkylformamides can be converted into either alkyl or acyl radicals, depending on the conditions.[53]

An instructive and useful process is the two-component coupling of an alkene with an electrophilic radical: the latter will of course not react with the protonated heterocycle, but after addition to the alkene a nucleophilic radical is generated which will react.[54]

(an electrophilic radical) (a nucleophilic radical)

When more than one reactive position is available in a heterocyclic substrate, as is often the case for pyridines for example, there are potential problems with regioselectivity or/and disubstitution (since the product of the first substitution is often as reactive as the starting material). Regioselectivity is dependent to a certain extent on the nature of the attacking radical and the solvent, but may be difficult to control satisfactorily.[55]

A point to note is that for optimum yields, radical substitutions are often not taken to full conversion (of starting heterocycle), but as product and starting material are often easily separated this is usually not a problem. Ways of avoiding disubstitution include control of pH (when the product is less basic than the starting material), or the use of a two-phase medium to allow extraction (removal) of a more lipophilic product out of the aqueous acidic reaction phase.

Very selective monosubstitution can also be achieved by the ingenious use of an N^+-methoxy-quaternary salt, in place of the usual protonic salt. Here, rearomatisation is the result of loss of methanol, leaving as a product a much less reactive, neutral pyridine.[56]

In addition to substitution of hydrogen, *ipso* replacement of nitro, sulfonyl, and acyl substituents can occur, and may compete with normal substitution.[57]

2.4.2 Reactions with electrophilic radicals

Although much less well developed than the Minisci reaction, substitution with electrophilic radicals can be used in some cases to achieve selective reaction in electron-rich heterocycles.[58]

2.5 Deprotonation of N-hydrogen[59]

Pyrroles, imidazoles, pyrazoles and benzo-fused derivatives which have a free *N*-hydrogen have pK_a values for the loss of the *N*-hydrogen as a proton in the region of 14–18. This is to say that they can be completely converted into anions by reaction with strong bases like sodium hydride or *n*-butyllithium. Even in the simplest of these examples, pyrrole itself, the acidity (pK_a 17.5) is very considerably greater than that of its saturated counterpart, pyrrolidine ($pK_a \sim 44$); similarly the acidity of indole (pK_a 16.2) is much greater than that of aniline (pK_a 30.7). One may rationalise this relatively increased acidity on the grounds that the charge is not localised, and this is illustrated by resonance forms which show the delocalisation of charge around the heterocycle. With the addition of electron-withdrawing substituents, or with the inclusion of extra heteroatoms, especially imine groups, the acidity is enhanced. A nice, though extreme, example is tetrazole for which the pK_a is 4.8, of the same order as a carboxylic acid.

pyrryl anion pK_a 17.5   tetrazolyl anion pK_a 4.8

2.6 Organometallic derivatives

The most important developments in heterocyclic chemistry in the last twenty or so years are probably in the area of organometallic chemistry, particularly transition-metal-catalysed reactions and the reactions of lithio-derivatives, reflecting development in these areas in organic chemistry as a whole. Even since the 3rd Edition of this book, significant further advances have been made, with improved preparations of boron, magnesium, and zinc compounds and with new ligands for palladium-catalysed reactions which considerably broaden their scope.

2.6.1   Lithium derivatives[60]

Lithio-heterocycles have proved to be the most useful organometallic derivatives: they react with the whole range of electrophiles in a manner exactly comparable to that of aryllithiums and can often be prepared by direct metallation (*C*-hydrogen deprotonation), as well as by halogen exchange between halo-heterocycle and alkyllithium. As well as reaction with carbon electrophiles, lithiated species are often the most convenient source of heterocyclic derivatives of less electropositive metals, such as zinc, boron, silicon, and tin (sections 2.6.2 and 2.6.3), which are now widely used in coupling reactions (section 2.7.2.2).

2.6.1.1   Direct lithiation (C-hydrogen deprotonation)

Many heterocyclic systems react directly with alkyllithiums or with lithium amides to give the lithio-heterocycle *via* abstraction of a proton. Although a 'free' anion is never formed, the ease of lithiation correlates well with *C*-hydrogen acidity and of course this, with the stability of the corresponding conjugate base (carbanion).[61] Lithiations by deprotonation are therefore directly related to base-catalysed proton exchange[62] using reagents such as sodium methoxide, at much higher temperatures, which historically provided the first indication that preparative deprotonations might be regioselective and thus of synthetic value. It must be remembered that kinetic and equilibrium acidities may be different; thermodynamic products are favoured by higher temperatures and by more polar solvents.

The detail of the mechanism of metallation is still under discussion; it may involve a four-centre transition state.

The main factor giving increased acidity of heterocyclic *C*-hydrogen relative to benzenoid *C*-hydrogen is the inductive effect of the heteroatom(s) thus metallation occurs at the carbon α to the heteroatom, where the inductive effect is felt most strongly, unless other factors, with varying degrees of importance, intervene. These include the following:

Mesomerism

Except in the case of side-chain anions (section 2.6.4), the 'anion' orbital is orthogonal to the π-system and so it is not mesomerically delocalised. However, electron density and therefore *C*-hydrogen acidity at ring carbons, is affected by resonance effects.

Coordination of the metal counterion to the heteroatom

Stronger coordination between the metal of the base and a heteroatom leads to enhanced acidity of the adjacent *C*-hydrogen due to increased inductive withdrawal of electron density – it is proportionately stronger, for example, for oxygen than for sulfur.

Lone pair interactions

Repulsion between the electrons in the orbital of the 'anion' and an adjacent heteroatom lone pair has a destabilising influence. This effect is thought to be important in pyridines and other azines.[63]

Polarisability of the heteroatom

More polarisable atoms such as sulfur are able to disperse charge more effectively.

Substituent effects

Directed metallation (DoM)[64] is extremely useful in heterocyclic chemistry, just as in carbocyclic chemistry. Metallation *ortho* to the directing group is promoted by either inductive effects (e.g. Cl, F), or chelation (e.g. $CH_2OH \rightarrow CH_2OLi$), or a combination of these, and may overcome the intrinsic regioselectivity of metallation of a particular heterocycle. When available, this is by far the most important additional factor influencing the regioselectivity of lithiation.

Lithiating agents

Lithiations are normally carried out with alkyllithiums or lithium amides. *n*-Butyllithium is the most widely used alkyllithium but *t*-butyllithium and occasionally *s*-butyllithium are used when more powerful reagents are required. Phenyllithium was used in older work but is uncommon now although it can be of value when a less reactive, more selective base is required.[65] A very powerful metallating reagent is formed from a mixture of *n*-butyllithium and potassium *t*-butoxide: this produces the potassium derivative of the heterocycle.

Lithium diisopropylamide (LiN(*i*-Pr)$_2$; LDA) is the most widely used lithium amide but lithium 2,2,6,6-tetramethylpiperidide (LiTMP) is rather more basic and less nucleophilic – it has found particular use in the metallation of diazines. Alkyllithiums are stronger bases than the lithium amides, but usually react at slower rates. Metallations with the lithium amides are reversible so for efficient conversion, the heterocyclic substrate must be more acidic (> 4 pK_a units) than the corresponding amine.

Solvents

Ether solvents – Et$_2$O and THF – are normally used. The more strongly coordinating THF increases the reactivity of the lithiating agent by increasing its dissociation. A mixture of ether, THF and pentane (Trapp's solvent) can be employed for very low temperature reactions (< 100 °C) (THF alone freezes at this temperature). To increase the reactivity of the reagents even further, ligands such as TMEDA (*N,N,N,N'*-tetramethylethylenediamine; Me$_2$N(CH$_2$)$_2$NMe$_2$) or HMPA ((Me$_2$N)$_3$PO) (**CAUTION**: carcinogen) are sometimes added – these strongly and specifically coordinate the metal cation. While these additives are undoubtedly beneficial in some cases, the efficacy of TMEDA has been questioned.[66]

2.6.1.2   *Halogen exchange*

Bromo- and iodo-heterocycles react rapidly with alkyllithiums, even at temperatures as low as –100 °C, to give the lithio-heterocycle. Where alternative exchanges are possible, the site of reaction is governed by the stability of the 'anion' formed, just as for direct lithiation by deprotonation. Exchange of fluorine is unknown and of chlorine, rare enough to assume that it is inert.

$$(\text{Het})\!-\!\text{Br} \quad + \quad \text{RLi} \quad \longrightarrow \quad (\text{Het})\!-\!\text{Li} \quad + \quad \text{RBr}$$

Mechanistically, the exchange process may involve a four-membered transition state, or may possibly proceed *via* an electron-transfer sequence, however direct nucleophilic attack, at least on iodine, has been demonstrated in the case of iodobenzene,[67] and cannot therefore be dismissed as a mechanism.

four-membered
transition state

direct nucleophilic
attack on halogen

Halogen exchange reagents

n-Butyllithium is the usual exchange reagent; the *n*-butyl bromide byproduct does not usually interfere with subsequent steps. When the presence of an alkyl bromide is undesirable, two equivalents of *t*-butyllithium can be employed – the initially formed *t*-butyl bromide is consumed by reaction with the second equivalent of alkyllithium, producing isobutene.

$$(\text{Het})\!-\!\text{Br} \quad + \quad 2t\text{-BuLi} \quad \longrightarrow \quad (\text{Het})\!-\!\text{Li} \quad \left[+ \quad t\text{-BuBr} \xrightarrow{t\text{-BuLi}} t\text{-BuH} \quad + \quad \text{Me}_2\text{C=CH}_2 \quad + \quad \text{LiBr} \right]$$

It is very important to differentiate between pure bases, such as lithium diisopropylamide, which act only by deprotonation, and alkyllithiums which can act as bases or take part in halogen exchange. When using alkyllithiums, exchange is favoured over deprotonation by the use of lower temperatures. The reaction of 3-iodo-1-phenylsulfonylindole with the two types is illustrative.[68]

2.6.1.3   Ring lithiation of five-membered heterocycles

α-hydrogens acidified
relative to β-hydrogens   X = NR, S, or O

The inductive effect of the heteroatom, which withdraws electrons to a greater extent from an adjacent carbon atom (α-positions), allows direct α-lithiation of practically all five-membered heterocycles. The relative 'acidities' of α-hydrogens in some different classes are illustrated in the table below.

Equilibrium pK_a values[#] for deprotonation of some five-membered heterocycles in THF[69]

35.6   33.0   Me 39.5   NMe₂ 37.1   Me 33.7   n-Pr 35.9   n-Pr 26.2   >28.3

33.2   32.4   Me 38.1   ~28

[#]Measured pK_a values vary according to solvent etc.

Despite the lower electronegativity of sulfur, and hence a weaker inductive effect, thiophene metallates about as readily as furan, probably in part because the higher polarisability of sulfur allows more efficient charge distribution;[70] d-orbital participation is thought to be relatively unimportant in the stabilisation of carbanionic centres adjacent to sulfur. The lithiation of 2-(2-furyl)thiophene, in either ring depending on conditions, is instructive;[71] preferential lithiation of the furan ring in the non-polar solvent is probably due to stronger coordination of lithium to the oxygen, thus increasing the inductive effect on the α-hydrogen in the furan ring.

The use of stronger bases can result in dimetallation.[72]

Directing groups can overcome the normal tendency for α-lithiation in five-membered heterocycles, as shown in the thiophene example below, however the use of lithium diisopropylamide does allow 'normal' α-lithiation.[73]

Lithiation of pyrroles is complicated by the presence of a much more acidic hydrogen on nitrogen, however 1-methylpyrrole lithiates, at C-2, albeit under slightly more vigorous conditions than for furan.[74] Removable protecting groups on the pyrrole nitrogen allow α-lithiation, t-butoxycarbonyl (Boc), is an example; it has additional advantages: not only is it easily hydrolytically removed, but it also withdraws electrons thus acidifying the α-hydrogen further, and finally, provides chelation assistance.[75]

Benzo[*b*]thiophenes and benzo[*b*]furans, and *N*-blocked indoles lithiate on the heterocyclic ring, α to the heteroatom.[76] Lithiation at the other hetero-ring position can be achieved *via* halogen exchange, but low temperatures must be maintained to prevent equilibration to the more stable 2-lithiated heterocycle.[68]

Benzene ring lithiated intermediates can be prepared by metal-halogen exchange, even, in the case of indoles, without protection of the NH, i.e. it is possible to produce an *N,C*-dilithiated species.[77]

The 1,3-azoles lithiate very readily, at C-2. One may understand this in terms of a combination of the acidifying effects seen at an α-position of pyridine (both inductive and mesomeric electron withdrawal, see section 2.6.1.4) with that at the α-positions of thiophene, furan, and pyrrole (inductive only). 2-Substituted-1,3-azoles generally lithiate at C-5.[78]

For imidazoles, it is usual for the *N*-hydrogen first to be masked,[79] and a variety of protecting groups have been used for that purpose, many of which provide additional stabilisation and an additional reason for regioselective α-lithiation by coordinating the lithium: trimethylsilylethoxymethyl ($Me_3Si(CH_2)_2OCH_2$; SEM) is one such group.[80]

It is a significant comment on the relative ease of α-lithiation in six- and five-membered systems that (*N*-protected) pyrazoles lithiate at C-5, i.e. in the pyrrole-like α-position, though, again chelation assistance from the *N*-protecting group also directs to C-5.[81]

One must be aware that hetero-ring cleavage[82] can occur in β-lithiated five-membered systems, because the heteroatom can act as a leaving group, if the temperature is allowed to rise.[83]

2.6.1.4 Ring lithiation of six-membered heterocycles

The preparation of lithiated derivatives of six-membered heterocycles like pyridines, quinolines and diazines must overcome the problem that they are susceptible to nucleophilic addition/substitution (section 2.3.2) by the lithium reagents. In contrast to the selective lithiation of five-membered rings, the direct metallation of pyridine is quite difficult and complex, but it can be achieved using the very strong base combination n-butyllithium/potassium t-butoxide. In relatively non-polar solvents (ether/hexane) kinetic 2-metallation predominates but in a polar solvent (THF/HMPA/hexane), or under equilibrating conditions, the 4-isomer is the major product. The pyridine α- and γ-positions, being more electron-deficient than a β-position, have the kinetically most acidic protons, and of the two former anions, location of negative charge at the γ-position is the more stable situation, perhaps due to unfavourable repulsion between the coplanar nitrogen lone pair and the α-'anion' only in the former. In non-polar solvents stronger coordination of the metal cation with the nitrogen lone pair will reduce this repulsive interaction and thus increase the relative stability of the α 'anion'.[84] As a corollory of this, pyridine can be selectively lithiated at C-2 when the lone pair is tied up as a complex with boron trifluoride.[85] This is consistent with much earlier studies of base-catalysed exchange when it was demonstrated that N-oxides and N^+-alkyl quaternary salts exchange more rapidly at an α position.[86]

All the isomerically pure lithio-pyridines can be prepared by halogen exchange, though 3-bromopyridine requires a lower temperature to discourage nucleophilic addition; bromopicolines can be similarly converted, without deprotonation at the methyl groups (cf. section 2.6.3.1).

Pyridines carrying groups which direct metallation *ortho*, using chelation and/or inductive influences, can be directly lithiated without risk of nucleophilic addition. When the group is at a 2-[87] or 4-position[88]-, lithiation must occur at a β-carbon; pyridines with *ortho*-directing groups located at a β position usually lithiate at C-4: this is true for example of chloro- and fluoropyridines;[89] 3-methoxymethoxy-,[90] 3-pivaloylamino,-[91] 3-trimethylsilylethoxymethoxy-[92], 3-t-butylaminosulfonyl-,[93] pyridines; pyridines carrying a 3-diethylaminocarbonyloxy or 3-diethylaminothiocarbonyloxy group;[94] and the adduct from 3-formylpyridine and $Me_2N(CH_2)_2NMeLi$,[95] however 3-ethoxypyridine metallates at C-2.[96]

Quinolines react like pyridines but are more susceptible to nucleophilic addition;[97] this is also an increased problem with pyrimidines, relative to pyridines, but nevertheless they can be lithiated by deprotonation or by halogen exchange at low temperatures, around −100 °C. The presence of 2- and/or 4-substituents adds some stability to lithiated pyrimidines.[98]

Pyrazines and pyridazines react in accord with the principles discussed above.[99]

2.6.2 Magnesium derivatives[100]

While Grignard reagents have been widely used in carboaromatic chemistry, the direct preparation of heterocyclic Grignard reagents by the standard method – halo-compound with magnesium – is often difficult, particularly for heterocycles containing a basic nitrogen. However, exchange of bromo- or iodo-heterocycles with alkyl Grignard reagents, preferably *i*-propylmagnesium halides or di-*i*-propylmagnesium, allows access to the magnesium derivatives of a wide range of heterocycles, from pyrroles to thiazoles and pyridines. The preparation of heteroaryl Grignards in this way has even been used in solid phase synthesis. Pyridyl sulfoxides will also undergo exchange to give pyridyl Grignard species.[101] While possibly not quite as reactive as their carboaromatic counterparts, heteroaryl Grignard reagents will react with a good range of electrophiles, sometimes requiring the assistance of a copper salt as catalyst.

2.6.3   Boron, silicon, and tin reagents

Caution: while very useful, many organotin compounds are toxic and should be handled with care. Trimethyltin derivatives in particular are highly toxic and whenever possible should be replaced by the slightly less reactive but much less toxic, tri-*n*-butyl analogues.

2.6.3.1   *Synthesis*

The most general preparative method for silanes,[102] stannanes, and boronic acids is the reaction of a heteroaryllithium with a chlorosilane, a chlorostannane, or with a borate ester,[103] respectively. 3-Diethylborylpyridine can be similarly prepared by reaction of the lithiopyridine with triethylborane, followed by cleavage of an ethyl group with iodine; this method does not work for electron-rich systems such as furan due to preferential cleavage of the heterocyclic group.[104] Transmetallation reactions can also be of use in specific cases.[105]

Boronic esters are also available via the palladium-catalysed boronation of halides.[106]

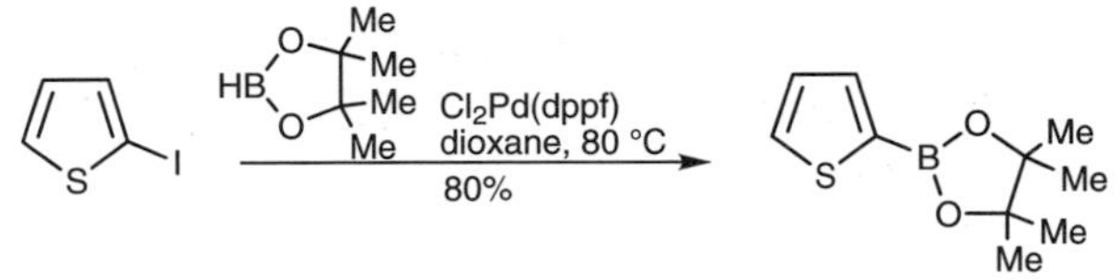

It is possible to directly silylate indoles and pyrroles *via* electrophilic substitution.[107]

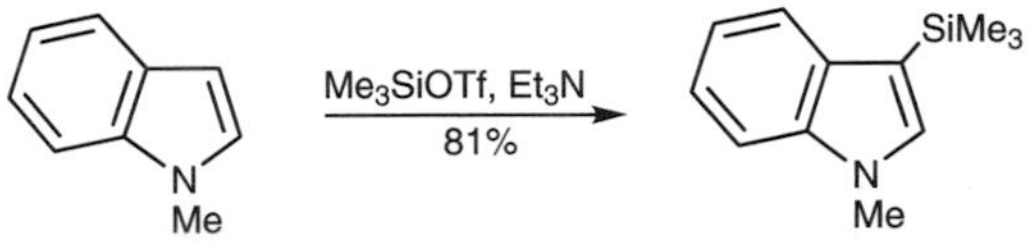

Useful alternative preparations of stannanes include palladium-catalysed decarboxylation of stannyl esters or coupling of halo compounds with hexaalkyldistannanes;[108] coupling with hexaalkyldisilanes requires rather more vigorous conditions.[109] Trialkylstannyl and trialkylsilyl anions are highly reactive and will displace halogen without the use of a catalyst.[110]

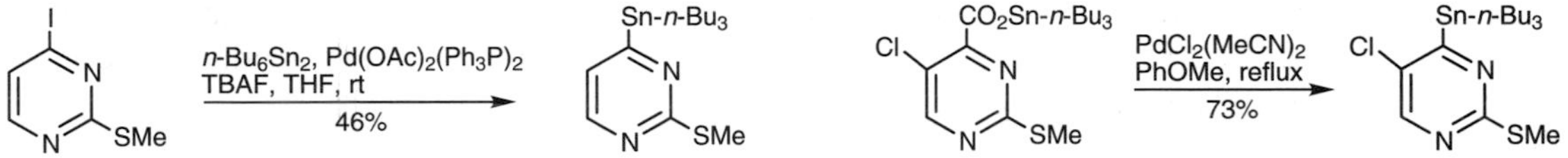

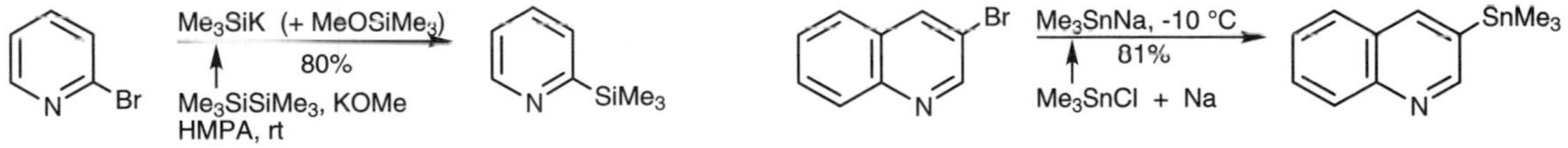

The relatively high stability of carbon-silicon/boron/tin bonds allows the 'metal' to be carried through many heterocyclic syntheses as an inert substitutent: some examples are shown below.[111]

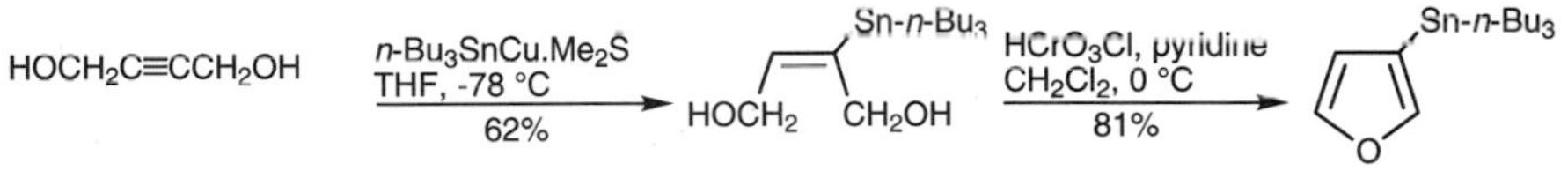

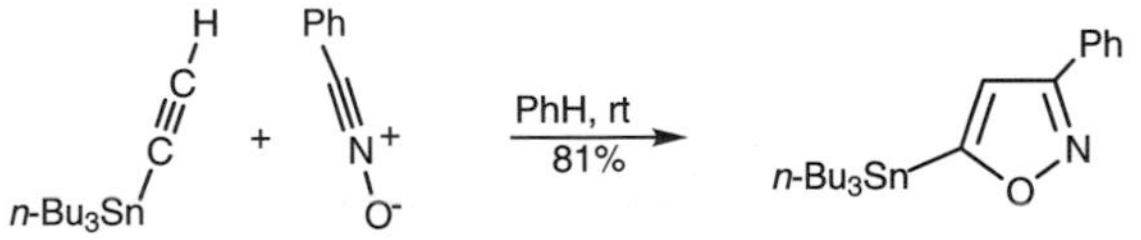

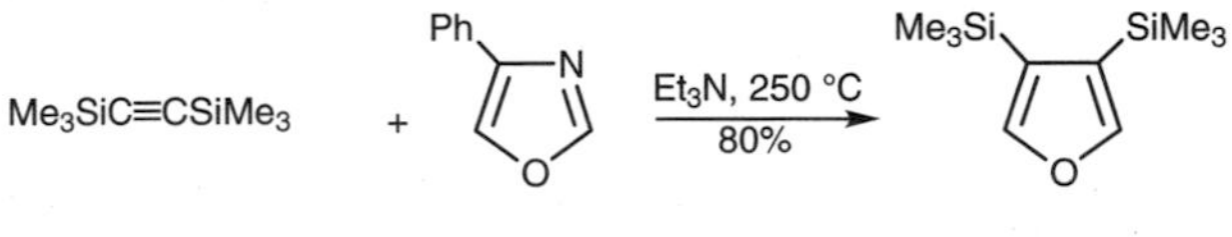

2.6.3.2 Reactions

The heteroaryl derivatives of boron, silicon, and tin, which show related patterns of reactivity, have found considerable application in synthesis. Unlike lithium compounds, they are generally fairly stable to air and water but will undergo a range of selective reactions under relatively mild conditions. Heteroaryl boronic acids and stannanes are particularly useful as the organometallic component in palladium-catalysed coupling reactions (section 2.7.2.2); heteroaromatic silanes such as 2-

(ethyldifluorosilyl)thiophene,[112] 2-(fluorodimethylsilyl)thiophene,[113] 2-trimethylsi-lylthiazole and 1-methyl-2-(trimethyl(methoxy)silyl)pyrrole[114] also participate in cross coupling reactions.

All three elements are susceptible to *ipso* replacement by electrophiles – such reactions have been studied extensively for arylsilanes and arylstannanes, where they occur *via* an electrophilic addition/silicon elimination mechanism analogous to other aromatic substitutions, but at a much faster rate than the corresponding replacement of hydrogen.[115] *Ipso* substitutions also take place on heterocycles and, in the case of electron-rich systems, probably *via* the same type of mechanism.

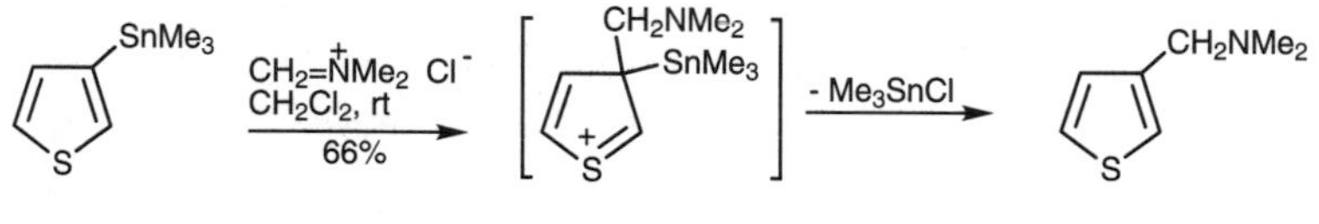

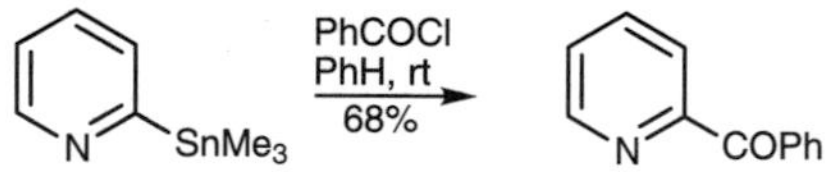

Most applications, however, have been in heterocycles containing an imine unit with the silicon (tin) directly attached;[116] such heterocycles undergo electrophilic attack reluctantly (section 2.2.2) so a mechanism involving coordination to nitrogen may be involved;[117] for example a 2-trimethylstannylpyridine will react readily with an acid chloride but its 3-isomer is inert under the same conditions, though palladium-catalysed coupling can be achieved with the 3- and 4-isomers under different conditions and *via* a different mechanism.[118] The oxazole example shown below illustrates prior interaction with the ring nitrogen.

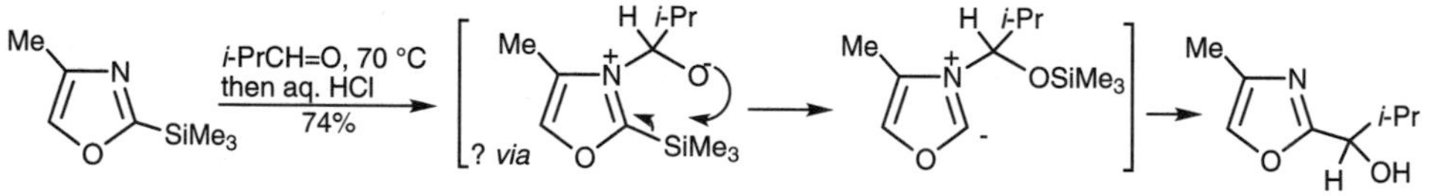

Silanes will also react with electrophiles with catalysis by fluoride or methoxide.[119] Here, an intermediate complex is formed *in situ* which reacts like a carbanion but under much milder conditions than would a lithio-derivative. This reaction can even be used to generate the equivalent of a CH_2-carbanion on a five-membered heterocyclic nitrogen.[120]

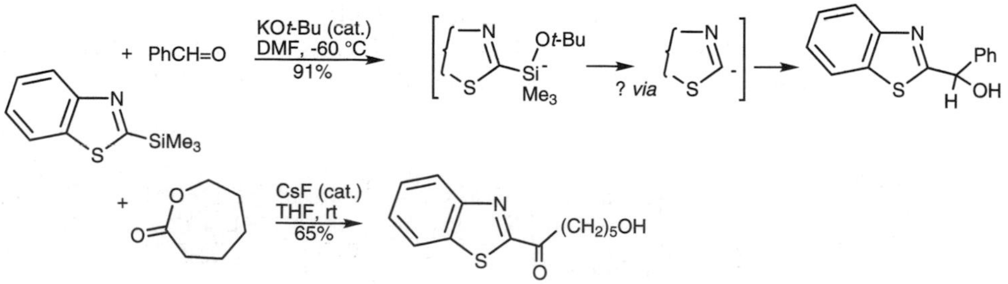

In addition to acting as a functional group, silanes can also be used as protecting groups for 'acidic' *C*-hydrogen, being removable at a later stage using fluoride or acid.[121] Stannanes are also valuable precursors for regiospecific formation of heteroaryllithiums *via* reaction with alkyllithiums.[111]

Although boronic acids are very reactive to *ipso* displacement by some electrophiles such as halonium ions, these reactions have found only occasional synthetic use. The *C*-boron bond can be cleaved by base, or acid, at rates depending on the corresponding carbanion stability or ease of protonation of the ring, respectively. When a relatively stable carbanion can be formed, such as in furan boronic acids containing electron-withdrawing groups, base-catalysed deboronation can be become an important unwanted side-reaction during palladium-catalysed boronic acid couplings.[122] Indeed, imidazole and oxazole 2-boronic acids have not yet been isolated, possibly due to their very ready deboronation.

2.6.4 Zinc reagents

Heteroarylzinc compounds are of particular use in palladium-catalysed couplings, being compatible with many functional groups. They have usually been prepared by exchange reactions[123] (*in situ*) of zinc halides with heteroaryllithiums but this method limits their usefulness. Efficient methods are now available for their direct preparation from either Rieke zinc[124] or commercial zinc dust[125] and the heteroaryl halide, in both electron-rich and electron-poor systems.

Directed *ortho* metallation can be carried out with a tetramethylpiperidyl zincate, but the only reported reactions of the resulting heteroarylzincates have been with iodine and benzaldehyde.[126]

2.6.5 Side-chain metallation of 6-membered heterocycles ('lateral metallation')[127]

Anions on alkyl side-chains and immediately adjacent to a heterocyclic ring are subject to varying degrees of stabilisation by interaction with the ring. The most favourable situation is where the side-chain is on an α or a γ carbon with respect to a C=N, as in the 2-, 6-, and 4-positions of a pyridine. Such anions are stabilised in much the same way as an enolate (conjugated enolate). We use the word 'enaminate' to describe this nitrogen-containing, enolate-like anion.

Quantitative measures for some methyl deprotonations are: 2-methylpyridine (pK_a 34), 3-methylpyridine (pK_a 37.7), 4-methylpyridine (pK_a 32.2), 4-methylquinoline (pK_a 27.5).[128] These values can be usefully compared with those typical for ketone α-deprotonation (19–20) and toluene side-chain deprotonation ($\sim$41). Thus strong bases can be used to convert methylpyridines quantitatively into side-chain anions, however the enolate-like stabilisation of the anion is sufficient that reactions can often be carried out using weaker bases under equilibrating conditions, i.e. under conditions where there is only a small percentage of anion present at any one time. It may be that under such conditions, side-chain deprotonation involves *N*-hydrogen-bonded or *N*-coordinated pyridines.

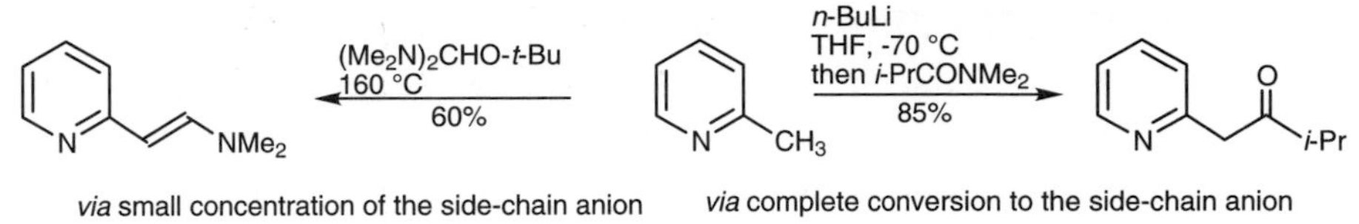

via small concentration of the side-chain anion   *via* complete conversion to the side-chain anion

An alternative means for effecting reaction at a side-chain depends on a prior electrophilic addition to the nitrogen: this acidifies further the side-chain hydrogens, then deprotonation generates an enamine or an enamide, each being nucleophilic at the side-chain carbon.

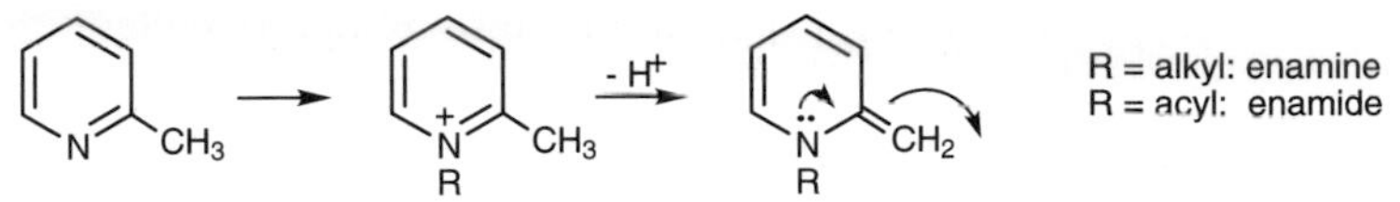

R = alkyl: enamine
R = acyl: enamide

One of the most elegant examples of this principle is the generation and use of *N*-dialkylboryl derivatives.[129]

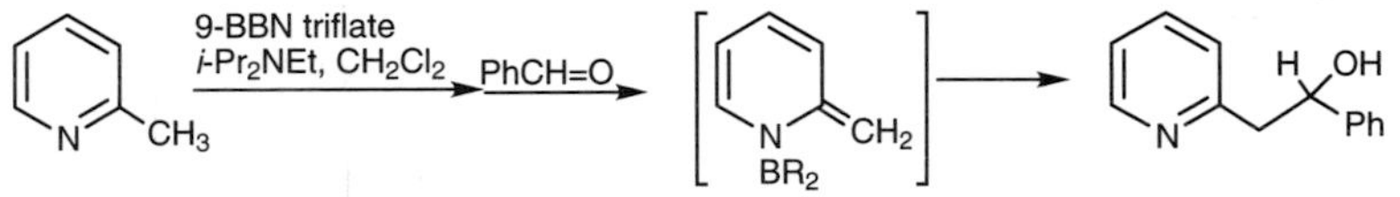

2.6.6   Side-chain metallation of five-membered heterocycles

The metalation of a side-chain on a simple five-membered heterocycle is much more difficult than in the six-membered series, because no enaminate stabilising resonance is available. Nonetheless it also is selective for an alkyl adjacent to the heteroatom, because the heteroatom acidifies by induction. Relatively more forcing conditions need to be applied, especially if an *N*-hydrogen is present,[130] but an elegant method has been developed for indoles, in which the first-formed *N*-anion is blocked with carbon dioxide, the lithium carboxylate thus formed then neatly also facilitating 2-methyl-lithiation by intramolecular chelation; this device has the further advantage that, following reaction of the side-chain anion with an electrophile, the *N*-protecting group is removed simply, during aqueous processing.[131]

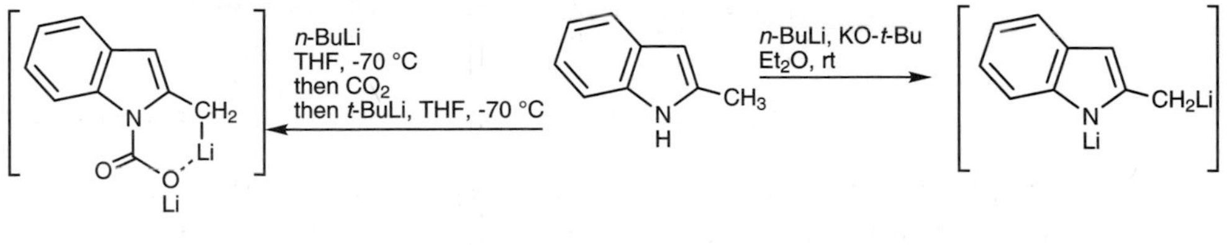

Side-chains at C-2 on 1,3-azoles are activated in a manner analogous to pyridine α-alkyl groups, and can be metallated, but more care is needed to avoid ring metallation.[132]

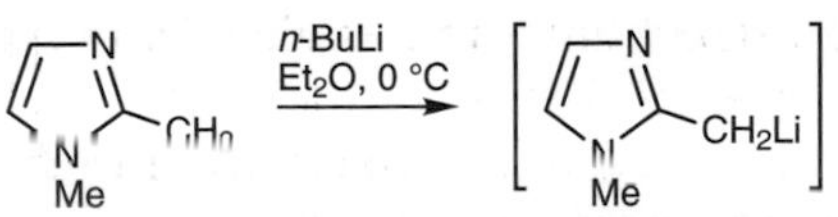

2.7   Palladium-catalysed reactions[133]

Transition metal-catalysed reactions are probably the most rapidly expanding area in organic chemistry at present and they have been used extensively in both the ring synthesis and the functionalisation of heterocycles. As well as completely new modes of reactivity, variants of older synthetic methods have been developed using the milder and more selective processes which attach to the use of transition metal catalysts. Palladium is by far the most important and widely used catalyst due to the very wide range of reaction types in which it can function. Nickel catalysts (mechanistically similar to palladium) have also been used, but for a narrower range of reactions.

In general, heterocyclic compounds undergo palladium-catalysed reactions in a way analogous to carbocycles; heterocyclic sulfur and nitrogen atoms seldom interfere with these (homogeneous) palladium catalysts, which must be contrasted with the well-known poisoning of hydrogenation catalysts such as palladium metal on carbon by sulfur- and nitrogen-containing molecules.

Palladium-catalysed processes typically utilise only 1–5 mol% of the catalyst and proceed through small concentrations of transient palladium species: there is a sequence of steps, each with an organopalladium intermediate, and it is important to become familiar with these basic organopalladium processes in order to rationalise the overall conversion. Concerted, rather than ionic, mechanisms are the rule so it is misleading to compare them too closely with apparently similar 'classical' organic mechanisms, however 'curly arrows' can be used as a memory aid (in the same way as one may use them for cycloaddition reactions), and this is the way in which palladium-catalysed reactions are 'explained' in the following discussion.

2.7.1   Basic organopalladium processes[134]

Note: For clarity, ligands which are not involved in the transformation under consideration are omitted from the following schemes, however it is important to understand that most organopalladium compounds normally exist as 4-coordinate, square-planar complexes:

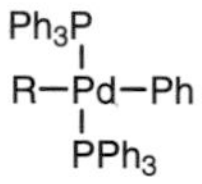

$$
\begin{array}{c}
Ph_3P \\
| \\
R-Pd-Ph \\
| \\
PPh_3
\end{array}
\quad \text{will be referred to as RPdPh} \qquad Pd(0) \text{ is usually} \quad Pd\big\langle\begin{array}{c}PPh_3\\PPh_3\end{array}
$$

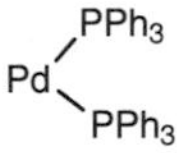

Despite an apparent similarity between RPdX and RMgX, their chemical properties are very different. The former are usually stable to air and water and unreactive to the usual electrophilic centres such as carbonyl, whereas RMgX do react with oxygen, water, and carbonyl compounds.

2.7.1.1   Concerted reactions

Oxidative addition

Aromatic and vinylic halides react with Pd(0) to give an organopalladium halide: aryl(or alkenyl)PdHal. This is formally similar to the formation of a Grignard reagent from magnesium metal, Mg(0), and a halide, but mechanistically, a concerted, direct 'insertion' of palladium into the carbon–halogen bond is believed to be involved. The ease of reaction: X = I > Br ~ OTf >> Cl >> F, explains why chloro and fluoro substituents can normally be tolerated, not interfering in palladium-catalysed processes. As a simple illustration, $Pd(PPh_3)_4$ reacts with iodobenzene at room temperature, but requires heating to 80 °C for a comparable insertion into bromobenzene. Although alkyl halides will undergo oxidative addition to Pd(0), the products are generally much less stable.

Palladium(0) exhibits a degree of nucleophilic character, thus electron-with-drawing substituents increase the reactivity of aryl halides in oxidative additions. This is exemplified in the heterocyclic context: the inductive effect of imine units allows 2-chloropyrimidine (it is slightly less reactive than bromobenzene), and even 3-chloropyridine[135] to react (even the moderate inductive effect at the β-position gives rise to a significantly higher rate of reaction relative to chlorobenzene) although a more reactive catalyst is required for the latter case (cf. section 2.7.2.2). However, the parallel with reactivity towards nucleophiles is not always exact. For example, 4-chloropyridine is more reactive than the 2-chloro isomer towards nucleophilic substitution, but the reverse is true in the palladium-catalysed couplings of these isomers.[136]

Reductive elimination

Organopalladium species with two organic units attached to the metal, R^1PdR^2, are generally unstable: extrusion of the metal, in a zero oxidation state, takes place, with the consequent linking of the two organic units. Because this is again a concerted process, stereochemistry in the organic moiety(ies) is conserved.

1,2-Insertion

Organopalladium halides add readily to double and triple bonds in a concerted, and therefore *syn,* manner (*via* a π-complex, not shown for clarity).

This process works best with electron-deficient alkenes such as ethyl acrylate, but will also take place with isolated, or even with electron-rich, alkenes. In reactions with acrylates, the palladium becomes attached to the carbon adjacent to the ester, i.e. the aromatic moiety becomes attached to the carbon β to the ester.

1,1-Insertion

Carbon monoxide, and isonitriles, will insert into a carbon-palladium bond.

β-Hydride elimination

When a *syn* β-hydrogen is present in an alkylpalladium species a rapid elimination of a palladium hydride occurs, generating an alkene. This reaction is much faster in RPdX than in R_2Pd and is the reason that attempted palladium-catalysed reactions of alkyl halides often fail.

Transmetallation

Palladium(II) compounds such as ArPdX and PdX_2 generally react readily with a wide variety of organometallic reagents, of varying nucleophilicity, such as R_4Sn, $RB(OH)_2$, RMgX, and RZnX, transfering the R group to palladium with overall displacement of X. The details of the reactions are not fully understood and probably vary from metal to metal, but a concerted transfer is probably the best means for their interpretation.

2.7.1.2   *Ionic reactions*

Addition to palladium-alkene π-complexes

Like those of Hg^{2+} and Br^+, Pd^{2+}-alkene complexes are very susceptible to attack by nucleophiles. In contrast to the reactions described in section 2.7.1.1 (1,2-insertion), this process exhibits *anti* stereospecificity.

Aromatic palladation

In reactions like aromatic mercuration, palladium(II) compounds will metallate aromatic rings *via* an electrophilic substitution, hence electron-rich systems are the most reactive.[137] *ortho*-Palladation assisted by electron-releasing chelating groups has been used frequently.[138]

2.7.2   Palladium-catalysed reactions in heterocyclic chemistry
2.7.2.1   Heck reactions[139]

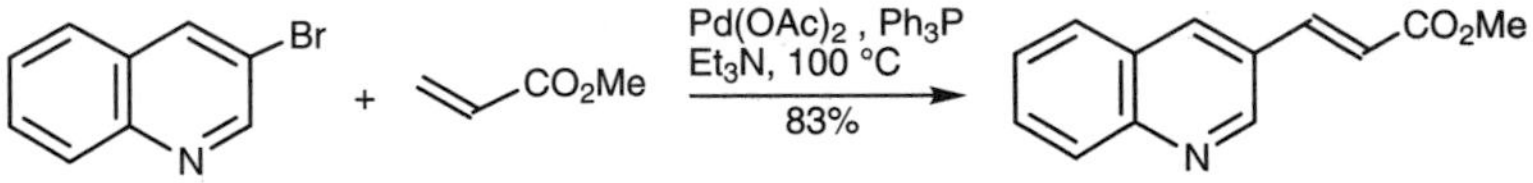

The standard Heck conditions shown in the example above[140] illustrate a common cause of confusion in understanding palladium-catalysed reactions, for while Pd(0) is actually involved in the catalytic cycle, palladium(II) acetate is generally used as an ingredient. This is just a matter of convenience because palladium acetate is stable and easily stored; it is reduced to Pd(0) by the phosphine (with a trace of water) or triethylamine *in situ* in a preliminary, initiating step.

The standard Heck reaction involves the reaction of an aryl halide with an alkene, commonly acrylate, to give a styrene (cinnamate) in the presence of a catalytic amount of palladium (often less than 1 mol %). The sequence involves (i) oxidative addition of the halide to Pd(0) followed by (ii) 1,2-insertion into the alkene; rotation then occurs to produce a species with hydrogen *syn* to palladium, then (iii) β-hydride elimination gives the styrene and regenerates Pd(0), which rejoins the catalytic cycle and can take part in a second oxidative addition, and so on.

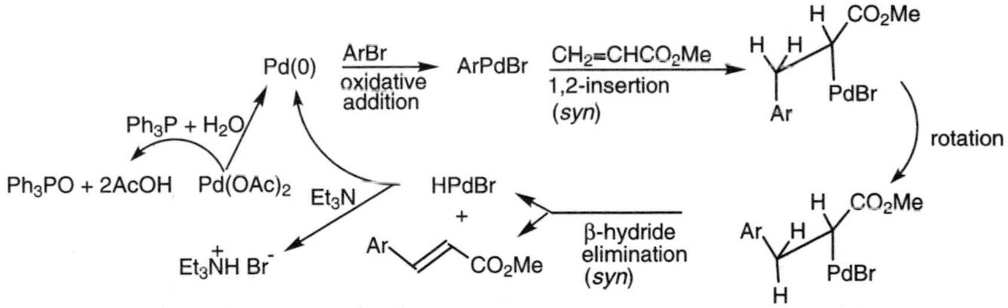

The electron-rich nature of heterocycles such as indoles, furans, and thiophenes allows a different type of Heck reaction to be carried out.[141] In this 'oxidative' modification the aryl palladium derivative is generated by electrophilic palladation with a palladium(II) reagent. This process is not catalytic in the standard way, but can be made so by the addition of a reoxidant selective for Pd(0); note, that the catalytic Pd(0) could not effect the first (electrophilic) ring palladation.[142]

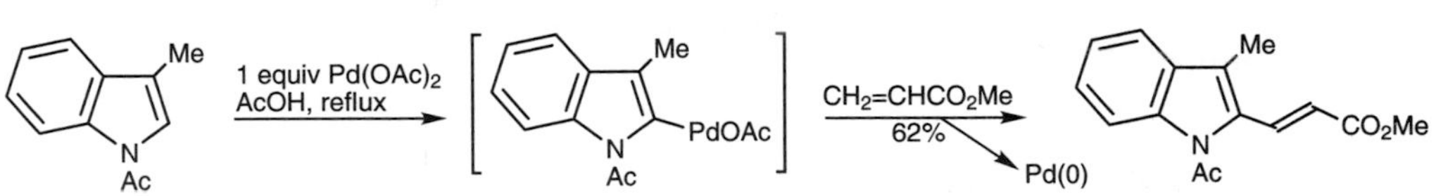

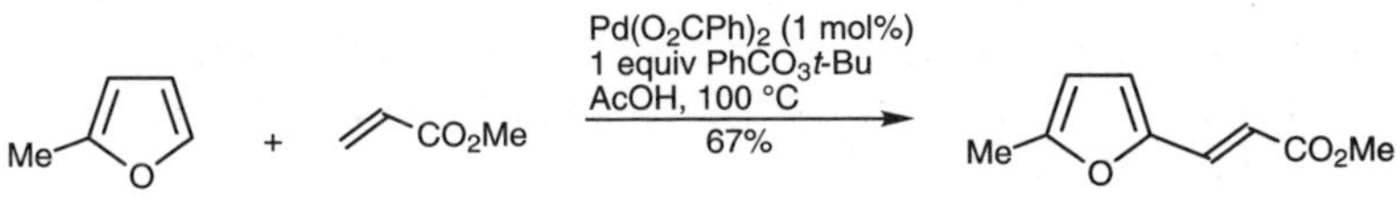

A different type of metallation, directed by acyl groups at either the pyridine 3- or 4-position, uses a catalytic ruthenium complex in the presence of an alkene and results in a reductive Heck-type substitution, as illustrated below. The mechanism involves insertion of the metal into a C–H bond. The process is non-polar and works equally well with electron-rich heterocycles, for example indole.[143]

2.7.2.2 Coupling reactions

Heteroaryl halides (or phenolic triflates) take part in palladium-catalysed couplings with a wide range of organometallic and anionic reagents; in contrast to the Heck reaction, the catalyst is often provided as preformed Pd(0), in a complex such as tetrakis(triphenylphosphine)palladium(0), Pd(Ph$_3$P)$_4$.[144]

These conversions also involve catalytic cycles: (i) oxidative addition is again the first step, but then (ii) transmetallation, and (iii) reductive elimination give product and regenerate Pd(0).

In the heterocyclic context there are examples in which RM is HetSnR$_3$ HetB(OH)$_2$, HetBEt$_2$, HetMgX, HetZnX, HetTiX$_3$, HetZrXL$_2$, M-≡-Het

The electron-withdrawing effect of typical azines makes chlorine substituents sufficiently reactive that they can participate in palladium-catalysed reactions, even at a pyridine β-position.[145] α-Activation can serve to allow regioselective reaction in the presence of a β-halogen (cf. section 2.7.1.1, oxidative addition) and this should be contrasted with lithiation by exchange which shows the opposite regioselectivity.

Where a heterocyclic organometallic reagent is required, Grignard and zinc derivatives are often satisfactory; complications sometimes attend the use of lithio derivatives. The use of boronic acids has become very popular on account of their clean reactions, general stability to air and water, and their compatibility with practically any functional group: furan, thiophene, indole and pyridine boronic acids have all been used.[146]

MeO$_2$C   Br   +   (HO)$_2$B

Pd(OAc)$_2$, o-Tol$_3$P
Et$_3$N, DMF, 100 °C
86%

Br   NHAc   +   (HO)$_2$B   CHO

Pd(PPh$_3$)$_4$
aq NaHCO$_3$
DME, reflux

CHO   NHAc

2NHCl
heat
71%

Some boronic acids may not be so stable, particularly diazole boronic acids, and in these cases tin derivatives can be used,[147] though they must be treated with caution as some are highly toxic.

Me   N   SnMe$_3$ +   Br   N

Pd(PPh$_3$)$_4$
PhI, reflux
100%

2.7.2.3   Carbonylation reactions

Acyl palladium species, formed by insertion of carbon monoxide into the usual aryl palladium halides, react readily with nucleophiles such as amines and alcohols to give amides and esters respectively; interception with hydride produces aldehydes.[148]

Br   N   S

Cl$_2$Pd(PPh$_3$)$_2$
CO (100 psi)
Et$_3$N, EtOH, 100 °C
95%

Pd(0)
oxidative
addition

BrPd

CO
1,1-insertion

Br Pd
EtOH

EtO$_2$C   N   S

2.7.2.4   Synthesis of benzo-fused heterocycles

Nucleophilic cyclisations onto palladium-complexed alkenes have been used to prepare indoles, benzofurans and other fused systems. The process can be made catalytic in some cases by the use of reoxidants such as benzoquinone or copper(II) salts.

NH$_2$

1 equiv PdCl$_2$(MeCN)$_2$
THF, Et$_3$N, rt
85%
or
0.1 equiv PdCl$_2$(MeCN)$_2$, LiCl
plus 1 equiv p-benzoquinone
86%

N
H
Me

Cl Cl
Pd

NH$_2$

CH$_2$PdCl
N
H

β-hydride
elimination

Pd (0)

CH$_2$
N
H

2.8   Oxidation and reduction[149] of heterocyclic rings

Generally speaking the electron-poor heterocycles are more resistant to oxidative degradation than are electron-rich systems – it is usually possible to oxidise alkyl side-chains attached to electron-poor heterocycles whilst leaving the ring intact; this is not generally true of electron-rich, five-membered systems.

The conversion of monocyclic heteroaromatic systems into reduced, or partially reduced derivatives is generally possible, especially in acidic solutions where it is a cation which is the actual species reduced. It follows that the six-membered types, which always have a basic nitrogen, are more easily reduced than the electron-rich, five-membered counterparts; heteroaromatic quaternary salts are likewise easily reduced.

2.9   Bioprocesses in heterocyclic chemistry[150]

The use of biological methods has a small but significant niche in synthetic heterocyclic chemistry, being used both on a research scale and for fine chemicals production. The processes may use isolated enzymes or whole microorganisms, the main reactions being oxidations of a heterocyclic nucleus or of side-chains. Some other reaction types are referred to later in the book, for example enzyme-catalysed base exchange in nucleosides and the deamination of adenosine.

A particular advantage of biological methods is their potential regio-, stereo- and enantioselectivity, which may not be attainable using chemical reagents. On the other hand, non-selective reactions have their uses: for example, the subjection of natural products to non-selective oxidations will generate a series of starting materials for the preparation of a wider range of analogues for biological evaluation.

The oxidation of pyridines to pyridones[151] and the selective oxidation of a side-chain in alkylpyridines and other azines, have been well studied.[152]

The enantioselective *cis* dihydroxylation[153] of benzothiophenes and benzofurans by *Pseudomonas putida* is analogous to well known conversions of simple benzenoid compounds,[154] but in the heterocyclic context, hydroxyl groups introduced at an α-carbon easily epimerise. Indole gives indoxyl probably via dehydration of an intermediate 2,3-diol.

The introduction of an amino acid side-chain onto 4-, 5-, 6-, and 7-azaindoles by an enzyme-catalysed alkylation with serine is an impressive demonstration of the power of biological methods.[155]

References

1. Gas-phase proton affinities (PAs) (cf. 'The reactivity of heteroaromatic compounds in the gas phase', Speranza, M., *Adv. Heterocycl. Chem.*, **1986**, *40*, 25) are rather similar for all bases; such measurements, though of considerable theoretical interest, are of limited value in considerations of solution chemistry.

2. 'The quaternisation of heterocyclic compounds', Duffin, G. F., *Adv. Heterocycl. Chem.*, **1964**, *3*, 1; 'Quaternisation of heteroaromatic compounds: quantitative aspects', Zoltewicz, J. A. and Deady, L. W., *Adv. Heterocycl. Chem.*, **1978**, *22*, 71; 'The quantitative analysis of steric effects in heteroaromatics', Gallo, R., Roussel, C., and Berg, U., *Adv. Heterocycl. Chem.*, **1988**, *43*, 173.

3. Brown, H. C. and Cahn, A., *J. Am. Chem. Soc.*, **1955**, *77*, 1715.

4. Okamoto, Y. and Lee, K. I., *J. Am. Chem. Soc.*, **1975**, *97*, 4015.

5. Zoltewicz, J. A. and Deady, L. W., *J. Am. Chem. Soc.*, **1972**, *94*, 2765.

6. 'Electrophilic substitution of heterocycles: quantitative aspects'; 'Part I, Electrophilic substitution reactions; Part II, Five-membered heterocyclic rings; Part III, Six-membered heterocyclic rings', Katritzky, A. R. and Taylor, R., *Adv. Heterocycl. Chem.*, **1990**, *47*, 1; 'Halogenation of heterocyclic compounds', Eisch, J. J., *Adv. Heterocycl. Chem.*, **1966**, *7*, 1; 'Halogenation of heterocycles: I. Five-membered rings', Grimmett, M. R., *ibid.*, **1993**, *57*, 291; 'II. Six- and seven-membered rings', *ibid.*, *58*, 271.

7. Austin, M. W. and Ridd, J. H., *J. Chem. Soc.*, **1963**, 4204.

8. Gilow, H. M. and Ridd, J. H., *J. Org. Chem.*, **1974**, *39*, 3481.

9. 'Substitution in the pyridine series: effect of substituents', Abramovitch, R. A. and Saha, J. G., *Adv. Heterocycl. Chem.*, **1966**, *6*, 229.

10. 'Mechanisms and rates of the electrophilic substitution reactions of heterocycles', Katritzky, A. R. and Fan, W.-Q., *Heterocycles*, **1992**, *34*, 2179.

11. 'Electrophilic substitution of heteroaromatic six-membered rings', Katritzky, A. R. and Johnson, C. D., *Angew. Chem., Int. Ed. Engl.*, **1967**, *6*, 608.

12. 'Electrophilic substitutions of five-membered rings', Marino, G., *Adv. Heterocycl. Chem.*, **1971**, *13*, 235.

13. Marino, G., *J. Heterocycl. Chem.*, **1972**, *9*, 817.

14. Clementi, S. and Marino, G., *Tetrahedron*, **1969**, *25*, 4599.

15. Clementi, S., Fringuelli, F., Linda, P., Marino, G., Savelli, G., and Taticchi, A., *J. Chem. Soc., Perkin Trans. 2*, **1973**, 2097.

16. Linda, P. and Marino, S., *Tetrahedron*, **1967**, *23*, 1739.

17. Quantitative comparisons must not ignore the considerable activating effect of a methyl group on an aromatic ring, whether attached to carbon or to nitrogen.

18. Bean, G. P., *J. Chem. Soc., Chem. Commun.*, **1971**, 421; Clementi, S., Forsythe, P. P., Johnson, C. D., and Katritzky, A. R., *J. Chem. Soc., Perkin Trans. 2*, **1973**, 1675; Clementi, S., Forsythe, P. P., Johnson, C. D., Katritzky, A. R., and Terem, B., *ibid.*, **1974**, 399.

19. Clementi, S., Linda, P., and Marino, G., *J. Chem. Soc. (B)*, **1970**, 1153.

20. Clementi, S. and Marino, G., *J. Chem. Soc., Perkin Trans. 2*, **1972**, 71.

21. Gol'dfarb, Y. L. and Danyushevskii, Y. L., *J. Gen. Chem. USSR (Engl. Transl.)*, **1961**, *31*, 3410; Boukou-Poba, J.-P., Farnier, M., and Guilard, R., *Can. J. Chem.*, **1981**, *59*, 2962.

22. Muchowski, J. M. and Naef, R., *Helv. Chim. Acta*, **1984**, *67*, 1168; Simchen, G. and Majchrzak, M. W., *Tetrahedron*, **1985**, *26*, 5035.

23. Frick, V. and Simchen, G., *Synthesis*, **1984**, 929; Majchrzak, M. W. and Simchen, G., *Tetrahedron*, **1986**, *42*, 1299.

24. Cipiciani, A., Clementi, S., Linda, P., Marino, G., and Savelli, G., *J. Chem. Soc., Perkin Trans. 2*, **1977**, 1284.

25. Laws, A. P. and Taylor, R., *J. Chem. Soc., Perkin Trans. 2*, **1987**, 591.
26. Holla, B. S. and Ambekar, S. Y., *Indian J. Chem., Sect. B*, **1976**, *14B*, 579.
27. Challis, B. C. and Rzepa, H. S., *J. Chem. Soc., Perkin Trans. 2*, **1975**, 1209; Butler, A. R., Pogorzelec, P., and Shepherd, P. R., *idem.*, **1977**, 1452.
28. Monti, S. A. and Johnson, W. O., *Tetrahedron*, **1970**, *26*, 3685.
29. Clementi, S., Linda, P., and Marino, G., *J. Chem. Soc., (B)*, **1971**, 79.
30. Austin, M. W., Blackborrow, J. R., Ridd, J. H., and Smith, B. V., *J. Chem. Soc.*, **1965**, 1051.
31. Austin, M. W., *Chem. Ind.*, **1982**, 57.
32. Poite, C., Roggero, J., Dou, H. J. M., Vernin, G., and Metzsger, J., *Bull. Soc. Chim. Fr.*, **1972**, 162.
33. Chauvin, P., Morel, J., Pastour, P., and Martinez, J., *Bull. Soc. Chim. Fr.*, **1974**, 2099.
34. 'Nucleophilic heteroaromatic substitution', Illuminati, G., *Adv. Heterocycl. Chem.*, **1964**, *3*, 285; 'Reactivity of azine, benzoazine, and azinoazine derivatives with simple nucleophiles', Shepherd, R. G. and Fedrick, J. L., *ibid.*, **1965**, *4*, 145; 'Formation of anionic σ-adducts from heteroaromatic compounds: structures, rates and equilibria', Illuminati, G., and Stegel, F., *ibid.*, **1983**, *34*, 306.
35. Liveris, M. and Miller, J., *J. Chem. Soc.*, **1963**, 3486; Miller, J. and Kai-Yan, W., *ibid.*, 3492.
36. Johnson, R. M., *J. Chem. Soc. (B)*, **1966**, 1058.
37. Barlin, G. B. and Benbow, J. A., *J. Chem. Soc., Perkin Trans. 2*, **1974**, 790.
38. O'Leary, M. H. and Stach, R. W., *J. Org. Chem.*, **1972**, *37*, 1491.
39. Chapman, N. B. and Russell-Hill, D. Q., *J. Chem. Soc.*, **1956**, 1563.
40. Barlin, G. B. and Benbow, J. A., *J. Chem. Soc., Perkin Trans. 2*, **1975**, 298.
41. 'Vicarious nucleophilic substitution of hydrogen', Makosza, M. and Winiarski, J., *Acc. Chem. Res.*, **1987**, *20*, 282; 'Applications of vicarious nucleophilic substitution in organic synthesis', Makosza, M. and Wojciechowski, K., *Liebigs Ann./Receuil*, **1997**,1805.
42. Wojciechowski, K., *Synth. Commun.*, **1997**, *27*, 135.
43. Wróbel, Z. and Makosza, M., *Org. Prep. Proc. Int.*, **1990**, 575.
44. Ostrowski, S. and Makosza, M., *Tetrahedron*, **1988**, *44*, 1721.
45. 'Radicals in organic synthesis: formation of carbon-carbon bonds', Giese, B., Pergamon Press, **1986**; 'Free radical substitution of heteroaromatic compounds', Norman, R. O. C., and Radda, G. K., *Adv. Heterocycl. Chem.*, **1963**, *2*, 131.
46. Klemm, L. H. and Dorsey, J., *J. Heterocycl. Chem.*, **1991**, *28*, 1153.
47. Minisci, F., Galli, R., Cecere, M., Malatesta, V., and Caronna, T.,*Tetrahedron Lett.*, **1968**, 5609; 'Substitutions by nucleophilic free radicals: a new general reaction of heteroaromatic bases', Minisci, F., Fontana, F., and Vismara, E., *J. Heterocycl. Chem.*, **1990**, *27*, 79; Minisci, F., Citterio, A., Vismara, E., and Giordano, C., *Tetrahedron*, **1985**, *41*, 4157; 'Advances in the synthesis of substituted pyridazines *via* introduction of carbon functional groups into the parent heterocycle', Heinisch, G., *Heterocycles*, **1987**, *26*, 481; 'Recent developments of free radical substitutions of heteroaromatic bases', Minisci, F., Vismara, E., and Fonatana, F., *Heterocycles*, **1989**, *28*, 489.
48. Buratti, W., Gardini, G. P., Minisci, F., Bertini, F., Galli, R., and Perchinunno, M., *Tetrahedron*, **1971**, 3655; Minisci, F., Gardini, G. P., Galli, R., and Bertini, F., *Tetrahedron Lett.*, **1970**, 15; Sakamoto, T., Sakasai, T., and Yamanaka, H., *Chem. Pharm. Bull.*, **1980**, *28*, 571; Minisci, F., Vismara, E., and Fonatana, F., *J. Org. Chem.*, **1989**, *54*, 5224.
49. Togo, H., Hayashi, K., and Yokoyama, M., *Chem. Lett.*, **1993**, 641.
50. Fontana, F., Minisci, F., Nogueira-Barbosa, M. C., and Vismara, E., *Tetrahedron*, **1990**, *46*, 2525.
51. Fontana, F., Minisci, F., Nogueira-Barbosa, M. C., and Vismara, E., *J. Org. Chem.*, **1991**, *56*, 2866.
52. Barton, D. H. R., Garcia, B., Togo, H., and Zard, S. Z., *Tetrahedron Lett.*, **1986**, *27*, 1327; Barton D. H. R., Chern, C.-Y., and Jaszberenyi, J. Cs., *ibid.*, **1992**, *33*, 5013.
53. Gardini, G. P., Minisci, F., Palla, G., Arnone, A., and Galli, R., *Tetrahedron Lett.*, **1971**, 59; Citterio, A., Gentile, A., Minisci, F., Serravalle, M., and Ventura, S., *J. Org. Chem.*, **1984**, *49*, 3364.
54. Citterio, A., Gentile, A., and Minisci, F., *Tetrahedron Lett.*, **1982**, *23*, 5587.
55. Minisci, F., Vismara, E., Fontana, F., Morini, G., Serravalle, M., and Giordano, G., *J. Org. Chem.*, **1987**, *52*, 730.

56. Katz, R. B., Mistry, J., and Mitchell, M. B., *Synth. Commun.*, **1989**, *19*, 317.
57. 'Radical *ipso* attack and *ipso* substitution in aromatic compounds', Tiecco, M., *Acc. Chem. Res.*, **1980**, *13*, 51.
58. Tordeaux, M., Langlois, B., and Wakselman, C., *J. Chem. Soc., Perkin Trans. 1*, **1990**, 2293; Cho, I.-S. and Muchowski, J. M., *Synthesis*, **1991**, 567.
59. 'Basicity and acidity of azoles', Catalan, J., Abboud, J. L. M., and Elguero, J., *Adv. Heterocycl. Chem.*, **1987**, *41*, 187.
60. 'Preparative polar organometallic chemistry', Brandsma, L. and Verkruijsse, H. D., Vol. 1, Springer-Verlag, **1987**; 'The chemistry of organolithium compounds', Wakefield, B. J., Pergamon Press, **1974**; 'Generation and reactions of sp²-carbanionic centers in the vicinity of heterocyclic nitrogen atoms', Rewcastle, G. W. and Katritzky, A. R., *Adv. Heterocycl. Chem.*, **1993**, *56*, 155; 'Ring and lateral metalation of heteroaromatic substrates using strong base systems' , FMC Lithium Link, Spring **1993**, FMC Lithium Division, 449, North Cox Road, Gastonia, NC 28054, USA.
61. von Ragué Schleyer, P., Chandrasekhar, J., Kos, A. J., Clark, T., and Spitznagel, G. W., *J. Chem. Soc., Chem. Commun.*, **1981**, 882.
62. 'Base catalysed hydrogen exchange', Elvidge, J. A., Jones, J. R., O'Brien, C., Evans, E. A., and Sheppard, H. C., *Adv. Heterocycl. Chem.*, **1974**, *16*, 1.
63. Zoltewicz, J. A., Grahe, G., and Smith, C. L., *J. Am. Chem. Soc.*, **1969**, *91*, 5501; Meot-Ner (Mautner), M , and Kafafi, S. A., *ibid.*, **1988**, *110*, 6297.
64. 'Heteroatom-facilitated lithiations', Gschwend, H. W. and Rodriguez, H. R., *Org. React.*, **1979**, *26*, 1; 'Directed metallation of π-deficient azaaromatics: strategies of functionalisation of pyridines, quinolines, and diazines', Quéguiner, G., Marsais, F., Snieckus, V., and Epsztajn, J., *Adv. Heterocycl. Chem.*, **1991**, *52*, 187; 'Metallation and metal-assisted bond formation in π-electron deficient heterocycles', Undheim, K. and Benneche, T., *Heterocycles*, **1990**, *30*, 1155; 'Directed *ortho* metallation. Tertiary amide and *O*-carbamate directors in synthetic strategies for polysubstituted aromatics', Snieckus, V., *Chem. Rev.*, **1990**, *90*, 879.
65. Mallet, M., *J. Organomet. Chem.*, **1991**, *406*, 49.
66. 'Is *N,N,N',N'*-tetramethylethylenediamine a good ligand for lithium?', Coolum, D. B., *Acc. Chem. Res.*, **1992**, *25*, 448.
67. Reich, H. J., Phillips, N. H., and Reich, I. L., *J. Am. Chem. Soc.*, **1985**, *107*, 4101.
68. Saulnier, M. G. and Gribble, G. W., *J. Org. Chem.*, **1982**, *47*, 757.
69. Fraser, R. R., Mansour, T. S., and Savard, S., *Can. J. Chem.*, **1985**, *63*, 3505.
70. von Ragué Schleyer, P., Clark, T., Kos, A. J., Spitznagel, G. W., Rohde, C., Arad, D., Houk, K. N., and Rondan, N. G., *J. Am. Chem. Soc.*, **1984**, *106*, 6467.
71. Carpita, A., Rossi, R., and Veracini, C. A., *Tetrahedron*, **1985**, *41*, 1919.
72. Feringa, B.L., Hulst, R., Rikers, R., and Brandsma, L., *Synthesis*, **1988**, 316.
73. Carpenter, A. J. and Chadwick, D. J., *Tetrahedron Lett.*, **1985**, *26*, 1777.
74. Gronowitz, S. and Kada, R., *J. Heterocycl. Chem.*, **1984**, *21*, 1041.
75. Hasan, I., Marinelli, E. R., Lin, L.-C. C., Fowler, F. W., and Levy, A. B., *J. Org. Chem.*, **1981**, *46*, 157.
76. *e.g.* for benzo[*b*]thiophen and benzo[*b*]furan: Kerdesky, F. A. J. and Basha, A., *Tetrahedron Lett.*, **1991**, *32*, 2003.
77. Moyer, M. P., Shiurba, J. F., and Rapoport, H., *J. Org. Chem.*, **1986**, *51*, 5106.
78. 'Metallation and metal–halogen exchange reactions of imidazoles', Iddon, B., *Heterocycles*, **1985**, *23*, 417.
79. See however Katritzky, A. R., Slawinski, J. J., Brunner, F., and Gorun, S., *J. Chem. Soc., Perkin Trans. 1*, **1989**, 1139.
80. Lipshutz, B. H., Huff, B., and Hagen, W., *Tetrahedron Lett.*, **1988**, *29*, 3411.
81. Katritzky, A. R., Rewcastle, G. W., and Fan, W.-Q., *J. Org. Chem.*, **1988**, *53*, 5685; Heinisch, G., Holzer, W., and Pock S., *J. Chem. Soc., Perkin Trans. 1*, **1990**, 1829.
82. 'Ring-opening of five-membered heteroaromatic anions', Gilchrist, T. L., *Adv. Heterocycl. Chem.*, **1987**, *41*, 41.
83. Dickinson, R. P. and Iddon, B., *J. Chem. Soc. (C)*, **1971**, 3447.
84. Verbeek, J., George, A. V. E., de Jong, R. L. P., and Brandsma, L., *J. Chem. Soc., Chem. Commun.*, **1984**, 257; Verbeek, J. and Brandsma, L., *J. Org. Chem.*, **1984**, *49*, 3857.
85. Kessar, S. V., Singh, P., Singh, K. N., and Dutt, M., *J. Chem. Soc., Chem. Commun.*, **1991**, 570.

86. Zoltewicz, J. A. and Helmick, L. S., *J. Am. Chem. Soc.*, **1970**, *92*, 7547; Zoltewicz, J. A. and Cantwell, V. W., *J. Org. Chem.*, **1973**, *38*, 829; Zoltewicz, J. A. and Sale, A. A., *J. Am. Chem. Soc.*, **1973**, *95*, 3928.

87. Mallet, M., *J. Organomet. Chem.*, **1991**, *406*, 49.

88. Miah, M. A. J. and Snieckus, V., *J. Org. Chem.*, **1985**, *50*, 5436.

89. Gribble, G. W. and Saulnier, M. G., *Tetrahedron Lett.*, **1980**, *21*, 4137.

90. Ronald, R. C. and Winkle, M. R., *Tetrahedron*, **1983**, *39*, 2031.

91. Estel, L., Linard, F., Marsais, F., Godard, A., and Quéguiner, G., *J. Heterocycl. Chem.*, **1989**, *26*, 105.

92. Sengupta, S. and Snieckus, V., *Tetrahedron Lett.*, **1990**, *31*, 4267.

93. Alo, B. I., Familoni, O. B., Marsais, F., and Quéguiner, G., *J. Heterocycl. Chem.*, **1992**, *29*, 61.

94. Beaulieu, F. and Snieckus, V., *Synthesis*, **1992**, 112; Tsukazaki, M. and Snieckus, V., *Heterocycles*, **1992**, *33*, 533.

95. Kelly, T. R., Xu, W., and Sundaresan, J., *Tetrahedron Lett.*, **1993**, *34*, 6173.

96. Marsais, F., Le Nard, G., and Quéguiner, G., *Synthesis*, **1982**, 235.

97. Godard, A., Jaquelin, J.-M., and Quéguiner, G., *J. Organomet. Chem.*, **1988**, *354*, 273.

98. Kress, T. J., *J. Org. Chem.*, **1979**, *44*, 2081; Frissen, A. E., Marcelis, A. T. M., Buurman, D. G., Pollmann, C. A. M., and van der Plas, H. C., *Tetrahedron*, **1989**, *45*, 5611; Turck, A., Plé, N., Majovic, L., and Quéguiner, G., *J. Heterocycl. Chem.*, **1990**, *27*, 1377.

99. Turk, A., Mojovic, L., and Quéguiner, G., *Synthesis*, **1988**, 881; Ward, J. S. and Merritt, L., *J. Heterocycl. Chem.*, **1991**, *28*, 765.

100. Turner, R. M., Ley, S. V., and Lindell, S. D., *Synlett*, **1993**, 748; Boymond, L., Rottlander, M., Cahiez, G., and Knochel, P., *Angew. Chem., Int. Ed. Engl.*, **1998**, *37*, 1701; Berillon, L., Lepetre, A., Turck, A., Plé, N., Quéguiner, G., Cahiez, G., and Knochel, P., *Synlett*, **1998**, 1359; Trecourt, F., Breton, G., Bonnet, V., Mongin, F., Marsais, F., and Quéguiner, G., *Tetrahedron Lett.*, **1999**, *40*, 4339; Abarbi, M., Dehmel, F., and Knochel, P., *ibid*, 7449.

101. Furukawa, N., Shibutani, T., Matsumura, K., Fujihara, H., and Oae, S., *Tetrahedron Lett.*, **1986**, *27*, 3899

102. 'Preparation of aryl- and heteroaryltrimethylsilanes', Häbich, D. and Effenberger, F., *Synthesis*, **1979**, 841.

103. Florentin, D. and Roques, B. P., and Fournie-Zalaski, M. C., *Bull. Soc. Chim. Fr.*, **1976**, 1999.

104. Terashima, M., Kakimi, H., Ishikura, M., and Kamata, K., *Chem. Pharm. Bull.*, **1983**, *31*, 4573.

105. Song, Z. Z., Zhan, Z. Y., Mak, T. C. W., and Wong, H. N. C., *Angew. Chem., Int. Ed. Engl.*, **1993**, *32*, 432; Zheng, Q., Yang, Y., and Martin, A. R., *Tetrahedron Lett.*, **1993**, *34*, 2235.

106. Ishiyama, T., Murata, M., and Miyaura, N., *J. Org. Chem.*, **1995**, *60*, 7508; Murata, M., Watanabe, S., and Masuda, Y., *ibid.*, **1997**, *62*, 6458.

107. Frick, U. and Simchen, G., *Synthesis*, **1984**, 929.

108. Majeed, A. J., Antonsen, Ø., Benneche, T., and Undheim, K., *Tetrahedron*, **1989**, *45*, 993.

109. Shippey, M. A. and Dervan, P. B., *J. Org. Chem.*, **1977**, *42*, 2654; Babin, P., Bennetau, B., Theurig, M., and Dunogues, J., *J. Organometal. Chem.*, **1993**, *446*, 135.

110. Yamamoto, Y. and Yanagi, A., *Chem. Pharm. Bull.*, **1982**, *30*, 1731.

111. Fleming, I. and Taddei, M., *Synthesis*, **1985**, 898; Kondo, Y., Uchiyama, D., Sakamaoto, T., and Yamanaka, H., *Tetrahedron Lett.*, **1989**, *30*, 4249.

112. Hatanaka, Y., Fukushima, S., and Hiyama, T., *Heterocycles*, **1990**, *30*, 303.

113. Kang, S.-K., Yamaguchi, T., Hong, R.-K., Kim, T.-H., and Pyun, S.-J., *Tetrahedron*, **1997**, *53*, 3027.

114. Ito, H., Sensui, H.-o., Arimoto, K., Miura, K., and Hosomi, A., *Chem. Lett.*, **1997**, 639.

115. 'Silicon reagents in organic synthesis', Colvin, E., Academic Press, **1988**; 'Tin in organic synthesis', Pereyre, M., Quintard, J.-P., and Rahm, A., Butterworths, **1987**; 'Non-conventional electrophilic aromatic substitutions and related reactions', Hartshorn, S. R., *Chem. Soc. Rev.*, **1974**, *3*, 167; 'Unusual electrophilic substitution in the aromatic series *via* organosilicon intermediates', Bennetau, B. and Dunogues, J., *Synlett*, **1993**, 171; 'Tin for organic synthesis. VI. The new role for organotin reagents in organic synthesis', Neumann, W. P., *J. Organometal. Chem.*, **1992**, *437*, 23; Cooper, M. S.,

Fairhurst, R. A., Heaney, H., Papageorgiou, G., and Wilkins, R. F., *Tetrahedron*, **1989**, *45*, 1155.

116. Pinkerton, F. H. and Thames, S. F., *J. Heterocycl. Chem.*, **1969**, *6*, 433; *ibid.*, **1971**, *8*, 257; *ibid.*, **1972**, *9*, 67; Dondoni, A., Dall'Occo, T., Galliani, G., Mastellari, A., and Medici, A., *Tetrahedron Lett.*, **1984**, *25*, 3637.

117. Jutzi, P. and Gilge, U., *J. Heterocycl. Chem.*, **1983**, *20*, 1011; Dondoni, A., Fantin, G., Fogognolo, M., Medici, A., and Pedrini, P., *J. Org. Chem.*, **1987**, *52*, 3413.

118. Yamamoto, Y. and Yanagi, A., *Chem. Pharm. Bull.*, **1982**, *30*, 2003.

119. Effenberger, F. and Schöllkopf, K., *Angew. Chem., Int. Ed. Engl.*, **1981**, *20*, 266; Ricci, A., Fiorenza, M., Grifagni, M. A., and Bartolini, G., *Tetrahedron Lett.*, **1982**, *23*, 5079.

120. Shimizu, S. and Ogata, M., *J. Org. Chem.*, **1986**, *51*, 3897.

121. Dondoni, A., Mastellari, A. R., Medici, A., Negrini, E., and Pedrini, P., *Synthesis*, **1986**, 757.

122. Florentin, D., Fournié-Zaluski, M. C., Callanquin, M., and Roques, B. P., *J. Heterocycl. Chem.*, **1976**, *13*, 1265; Brandão, M. A., de Oliveira, A. B., and Snieckus, V., *Tetrahedron Lett.*, **1993**, *34*, 2437.

123. Bell, A. S., Roberts, D. A., and Ruddock, K. S., *Synthesis*, **1987**, 843; Sakamoto, T., Kondo, Y., Takazawa, N., and Yamanaka, H., *Heterocycles*, **1993**, *36*, 941; Browder, C. C., Mitchell, M. O., Smith, R. L., and el-Sulayman, G., *Tetrahedron Lett.*, **1993**, *34*, 6245.

124. Sakamoto, T., Kondo, Y., Takazawa, N., and Yamanaka, H., *Tetrahedron Lett.*, **1993**, *34*, 5955; Sakamoto, T., Kondo, Y., Murata, N., and Yamanaka, H., *Tetrahedron*, **1993**, *49*, 9713.

125. Prasad, A. S. B., Stevenson, T. M., Citineni, J. R., Nyzam, V., and Knochel, P., *Tetrahedron*, **1997**, *53*, 7237.

126. Kondo, Y., Shilai, M., Uchiyama, N., and Sakamoto, T., *J. Am. Chem. Soc.*, **1999**, *121*, 3539.

127. 'Lateral lithiation reactions promoted by heteroatomic substitutents', Clark, R. D. and Jahinger, A., *Org. React.*, **1995**, *47*, 1.

128. Fraser, R. R., Mansour, T. S., and Savard, S., *J. Org. Chem.*, **1985**, *50*, 3232.

129. Hamana, H. and Sugasawa, T., *Chem. Lett.*, **1984**, 1591.

130. Naruse, Y., Ito, Y., and Inagaki, S., *J. Org. Chem.*, **1991**, *56*, 2256.

131. Katritzky, A. R. and Akutagawa, K., *J. Am. Chem. Soc.*, **1986**, *108*, 6808.

132. Noyce, D. S., Stowe, G. T., and Wong, W., *J. Org. Chem.*, **1974**, *39*, 2301.

133. 'Carbon–carbon bond formation in heterocycles using Ni- and Pd catalysed reactions', Kalinin, V. N., *Synthesis*, **1992**, 413; 'Transition metals in the synthesis and functionalisation of indoles', Hegedus, L. S., *Angew. Chem., Int. Ed. Engl.*, **1988**, *27*, 1113; 'Synthesis of condensed heteroaromatic compounds using palladium-catalysed reactions', Sakamoto, T., Kondo, Y., and Yamanaka, H., *Heterocycles*, **1988**, *27*, 2225; 'Organometallics in coupling reactions in π-deficient azaheterocycles', Undheim, K. and Benneche, T., *Adv. Heterocycl. Chem.*, **1995**, *62*, 305; 'Connection between metalation of azines and diazines and cross-coupling strategies for the synthesis of natural and biologically active molecules', Godard, A., Marsais, F., Plé, N., Trécourt, F., Turck, A., and Quéguiner, G., *Heterocycles*, **1995**, *40*, 1055.

134. 'Palladium reagents in organic synthesis', Heck, R. F., Academic Press, **1985**; 'Principles and applications of organotransition metal chemistry', Collman, J. P., Hegedus, L. S., Norton, J. R., and Finke, R. G., University Science Books, **1987**.

135. Ali, N. M., McKillop, A., Mitchell, M. B., Rebelo, R. A., and Wallbank, P. J., *Tetrahedron*, **1992**, *40*, 8117.

136. Lohse, O., Thevenin, P., and Waldvogel, E., *Synlett*, **1999**, 45.

137. Stock, L. M., Tse, K., Vorvick, L. J., and Walstrum, S. A., *J. Org. Chem.*, **1981**, *46*, 1757.

138. 'Cyclopalladated complexes in organic synthesis', Ryabov, A. D., *Synthesis*, **1985**, 233; 'Mechanisms of intramolecular activation of C–H bonds in transition metal complexes', *idem*, *Chem. Rev.*, **1990**, *90*, 403.

139. 'Palladium catalysed vinylation of organic halides', Heck, R. F., *Org. React.*, **1982**, *27*, 345.

140. Frank, W. C., Kim, Y. C., and Heck, R. F., *J. Org. Chem.*, **1978**, *43*, 2947.

141. Itahara, T., Ikeda, M., and Sakakibara, T., *J. Chem. Soc., Perkin Trans. 1*, **1983**, 1361.

142. Tsuji, J. and Nagashima, H., *Tetrahedron*, **1984**, *46*, 2699.

143. Grigg, R. and Savic, V., *Tetrahedron Lett.*, **1997**, *38*, 5737.

144. Arcadi, A., Burini, A., Cacchi, S., Delmastro, M., Marinelli, F. and Pietrani, B., *Synlett*, **1990**, 47; Sakamoto, T., Kondo, Y., Watanabe, R., and Yamanaka, H., *Chem. Pharm. Bull.*, **1986**, *34*, 2719; Sakamoto, T., Katoh, E., Kondo, Y., and Yamanaka, H., *Heterocycles*, **1988**, *27*, 1353.
145. Ali, N. M., McKillop, A., Mitchell, M. B., Rehelo, R. A., and Wallbank, P. J., *Tetrahedron*, **1992**, *37*, 8117.
146. Thompson, W. J. and Gaudino, J., *J. Org. Chem.*, **1984**, *49*, 5237; Yang, Y., Hörnfeldt, A.-B., and Gronowitz, S., *J. Heterocycl. Chem.*, **1989**, *26*, 865.
147. Dondoni, A., Fantin, G., Fogagnolo, M., Medici, A., and Pedrini, P., *Synthesis*, **1987**, 693; Bailey, T. R., *Tetrahedron Lett.*, **1986**, *27*, 4407.
148. Head, R. A. and Ibbotson, A., *Tetrahedron Lett.*, **1984**, *25*, 5939; Baillargean, V. P. and Stille, J. K., *J. Am. Chem. Soc.*, **1983**, *105*, 7175.
149. 'The reduction of nitrogen heterocycles with complex metal hydrides', Lyle, R. E. and Anderson, P. S., *Adv. Heterocycl. Chem.*, **1966**, *6*, 46; 'The reduction of nitrogen heterocycles with complex metal hydrides', Keay, J. G., *ibid.*, **1986**, *39*, 1.
150. 'Biocatalysis', Petersen, M. and Kiener, A., *Green Chem.*, **1999**, *1*, 99; 'Biotransformations for fine chemical production', Meyer, H. P., Kiener, A., Imwinkelried, R., and Shaw, N., *Chimia*, **1997**, *51*, 287; 'Biotechnological processes in the fine chemicals industry', Birch, O. M., Brass, J. M., Kiener, A., Robins, K., Schmidhalter, D., Shaw, N. M., and Zimmermann, T., *Chim. Oggi*, **1995**, *13*, 9; 'Biosynthesis of functionalised aromatic *N*-heterocycles', Kiener, A., *Chemtech*, **1995**, *25*, 31.
151. Kiener, A., Glocker, R., and Heinzmann, K., *J. Chem. Soc., Perkin Trans. 1*, **1993**, 1201.
152. Kiener, A., *Angew. Chem., Int. Ed. Engl.*, **1992**, *31*, 774.
153. Boyd, D. R., Sharma, N. D., Boyle, R., McMurray, B. T., Evans, T. A., Malone, J. F., Dalton, H., Chima, J., and Sheldrake, G. N., *J. Chem. Soc., Chem. Commun.*, **1993**, 49; Boyd, D. R., Sharma, N. D., Brannigan, I. N., Haughey, S. A., Malone, J. F., Clarke, D. A., and Dalton, H., *Chem. Commun.*, **1996**, 2361.
154. 'Enzymatic dihydroxylation of aromatics in enantioselective synthesis: expanding asymmetric methodology', Hudlicky, T., Gonzalez, D., and Gibson, D. T., *Aldrichimica Acta*, **1999**, *32*, 35.
155. Sloan, M. J. and Phillips, R. S., *Bioorg. Med. Chem. Lett.*, **1992**, *2*, 1053.

3 Synthesis of aromatic heterocycles

The preparation of benzenoid compounds nearly always begins with an appropriately substituted, and often readily available, benzene derivative – only on very rare occasions is it necessary to start from compounds lacking the ring, and to form it during the synthesis. The preparation of heteroaromatic compounds presents a very different picture, for it involves ring synthesis[1] more often than not. Of course when first considering a suitable route to a desired target, it is always important to give thought to the possibility of utilising a commercially available compound which contains the heterocyclic nucleus and which could be modified by manipulation, introduction and/or elimination of substituents[2] – a synthesis of tryptophan (section 17.12) for example would start from indole – however if there is no obvious route, a ring synthesis has to be designed which leads to a heterocylic intermediate appropriately substituted for further elaboration into the desired target.

This chapter shows how just a few general principles allow one to understand the methods, at first sight apparently diverse, which are used in the construction of the heterocyclic ring of an aromatic heterocyclic compound from precursors which do not have that ring. It discusses the principles, and analyses the types of reaction frequently used in constructing an aromatic heterocycle, and also the way in which appropriate functional groups are placed, in the reactants, in order to achieve the desired ring synthesis.

3.1 Reaction types most frequently used in heterocyclic ring synthesis

By far the most frequently used process is the addition of a nucleophile to a carbonyl carbon (or the more reactive carbon of an *O*-protonated carbonyl). When the reaction leads to C–C bond formation, then the nucleophile is the β-carbon of an enol or enolate anion, or of an enamine, and the reaction is aldol in type:

When the process leads to C–heteroatom bond formation, then the nucleophile is an appropriate heteroatom, either anionic (-X⁻) or neutral (-XH):

In all cases, subsequent loss of water produces a double bond, either a C–C or a C–heteroatom double bond. Simple examples are the formation of an aldol condensation product, and the formation of an imine or enamine, respectively.

9. Huisgen, R., Gotthardt, H., Bayer, H. O., and Schaefer, F. C., *Chem. Ber.*, **1970**, *103*, 2611; Potts, K. T. and McKeough, D., *J. Am. Chem. Soc.*, **1974**, *96*, 4268.
10. 'Synthesis of heterocycles through nitrenes', Kametani, T., Ebetino, F. F., Yamanaka, T., and Nyu, Y., *Heterocycles*, **1974**, *2*, 209.
11. MacKenzie, A. R., Moody, C. J., and Rees, C. W., *J. Chem. Soc., Chem. Commun.*, **1983**, 1372.
12. 'Recent advances in the chemistry of carbazoles', Joule, J. A., *Adv. Heterocycl. Chem.*, **1984**, *35*, 84; 'Phosphite-reduction of aromatic nitro-compounds as a route to heterocycles', Cadogan, J. I. G., *Synthesis*, **1969**, 11.
13. 'Heterocyclic *ortho*-quinodimethanes', Collier, S. J. and Storr, R. C., *Prog. Heterocycl. Chem.*, **1998**, *10*, 25.
14. Carly, P. R., Cappelle, S. L., Compernolle, F., and Hoornaert, G. J., *Tetrahedron*, **1996**, *52*, 11889.
15. Munzel, N. and Schweig, A., *Chem. Ber.*, **1988**, *121*, 791.
16. Trahanovsky, W. S., Cassady, T. J., and Woods, T. L., *J. Am. Chem. Soc.*, **1981**, *103*, 6691.
17. White, L. A., O'Neill, P. M., Park, B. K., and Storr, R. C., *Tetrahedron Lett.*, **1995**, *37*, 5983.
18. Herrera, A., Martinez, R., González, B., Illescas, B., Martin, N., and Seoane, C., *Tetrahedron Lett.*, **1997**, *38*, 4873.
19. Mertzanos, G. E., Stephanidou-Stephanatou, J., Tsoleridis, C. A., and Alexandrou, N. E., *Tetrahedron Lett.*, **1992**, *33*, 4499; Alexandrou, N. E., Mertzanos, Stephanidou-Stephanatou, J., Tsoleridis, C. A., and Zachariou, P., *ibid.*, **1995**, *36*, 6777; G. E., Pindur, U., Gonzalez, E., and Mehrabani, F., *J. Chem. Soc., Perkin Trans. 1*, **1997**, 1861.
20. Kinsman, A. C. and Snieckus, V., *Tetrahedron Lett.*, **1999**, *40*, 2453.
21. Liu, G.-B., Mori, H., and Katsumura, S., *Chem. Commun.*, **1996**, 2251.
22. Leusink, F. R., ten Have, R., van der Berg, K. J, and van Leusen, A. M., *J. Chem. Soc., Chem. Commun.*, **1992**, 1401.
23. Magnus, P., Gallagher, T., Brown, P., and Pappalardo, P., *Acc. Chem. Res.*, **1984**, *17*, 25.
24. Ko, C.-W. and Chou, T., *Tetrahedron Lett.*, **1997**, *38*, 5315; Tomé, A. C., Cavaleiro, J. A. S., and Storr, R. C., *Tetrahedron*, **1996**, *52*, 1723; Chen, H.-C and Chou, T.-s, *Tetrahedron*, **1998**, *54*, 12609.

4 Typical reactivity of pyridines, quinolines and isoquinolines

pyridine   quinoline   isoquinoline

Before detailed descriptions of the chemistry of the heterocyclic systems covered in this book, and at intervals during the book, we provide six highly condensed and simplified discussions of the types of reaction, ease of such reactions, and regiochemistry of such reactions for groups of related heterocyles. In this chapter the group comprises pyridine, as **the** prototype electron-poor six-membered heterocycle and its benzo-fused analogues, quinoline and isoquinoline. As in each of these summary chapters, reactions are shown in brief and either as the simplest possible example, or in general terms.

Typical pyridine reactivities

nucleophilic substitution at α-position

electrophilic addition at nitrogen produces pyridinium salts

radical substitution in acid solution at α-position

pyridine N-oxide

catalytic reduction of ring easy (in acid solution)

electrophilic substitution at β-position but very difficult

The formal replacement of a CH in benzene, by N, leads to far-reaching changes in typical reactivity: pyridines are much less susceptible to electrophilic substitution than benzene, and much more susceptible to nucleophilic attack. However, pyridine undergoes a range of simple electrophilic additions, some reversible, some forming isolable products, each involving donation of the nitrogen lone pair to an electrophile, and thence the formation of 'pyridinium' salts which, of course, do not have a counterpart in benzene chemistry at all. It is essential to understand that the ready donation of the pyridine lone pair in this way does not destroy the aromatic sextet (compare with pyrrole, chapter 12) – pyridinium salts are still aromatic, though of course much more polarised than neutral pyridines (see section 1.2.3).

Electrophilic substitution of aromatic compounds proceeds *via* a two-step sequence – addition (of X^+) then elimination (of H^+), of which the former is usually the slower (rate-determining) step. Qualitative predictions of relative rates of substitution at different ring positions can be made by inspecting the structures of the σ-complexes (Wheland intermediates) thus formed, on the assumption that their relative stabilities reflect the relative energies of the transition states which lead to them.

σ-complex   sp³ hybridised

Electrophilic substitution at carbon, in simple pyridines at least, is very difficult in contrast to the reactions of benzene – Friedel-Crafts acylations, for example, do not occur at all with pyridines. This unreactivity can be traced to two factors:

very slow rate of reaction due to:
(i) low concentration of neutral pyridine
(ii) intrinsically low reactivity of pyridine

very slow rate of reaction due to: energy barrier to formation of doubly charged intermediate

(1) Exposure of a pyridine to a medium containing electrophilic species immediately converts the heterocycle into a pyridinium cation with the electrophile (or a proton from the medium, or a Lewis acid) attached to the nitrogen. The extent of conversion depends on the nature and concentration of the electrophile (or protons) and the basicity of the particular pyridine, and is usually nearly complete. Obviously, the positively charged pyridinium cation is many orders of magnitude less easily attacked at carbon by the would-be electrophile than the original neutral heterocycle. The electrophile, therefore, has Hobson's choice – it must either attack an already-positively charged species, or seek out a neutral pyridine from the very low concentration of uncharged heterocyclic molecules.

(2) The carbons of a pyridine are, in any case, electron-poor, particularly at the α- and γ-positions: formation of a σ-complex between a pyridine and an electrophile is intrinsically disfavoured. The least disfavoured, i.e. best option, is attack at a β-position – resonance contributors to the cation thus produced, do not include one with the particularly unfavourable sextet, positively-charged nitrogen situation (shown in parentheses for the α- and γ-intermediates). The situation has a direct counterpart in benzene chemistry where a consideration of possible intermediates for electrophilic substitution of nitrobenzene provides a rationalisation of the observed *meta* selectivity.

intermediate for β-attack by X⁺

intermediate for α-attack by X⁺

intermediate for γ-attack by X⁺

Substituents exert an influence on the ease of electrophilic attack, just as in benzene chemistry. Strongly electron-withdrawing substituents simply render the pyridine even more inert, however activating groups – amino and oxy, and even alkyl – allow substitution to take place, even though by way of the protonated heterocycle i.e. *via* a dicationic intermediate. The presence of halogen substituents, which have a base-weakening effect and are only weakly deactivating, can allow substitution to take place in a different way – by allowing an appreciably larger concentration of the un-protonated pyridine to be present.

Pyridine rings are resistant to oxidative destruction, as are benzene rings. In terms of reduction, however, the heterocyclic system is much more easily catalytically reduced, especially in acidic solution. Similarly, *N*-alkyl- and *N*-arylpyridinium salts can be easily reduced both with hydrogen over a catalyst, and by nucleophilic chemical reducing agents.

Nucleophilic substitution of aromatic compounds proceeds *via* an addition (of Y^-) then elimination (of a negatively charged entity, most often Hal^-) two-step sequence of which the former is usually rate-determining (the $S_N(AE)$ mechanism: **S**ubstitution **N**ucleophilic **A**ddition **E**limination). Rates of substitution at different ring positions can be assessed by inspecting the structures of the negatively charged intermediates (Meisenheimer complexes) thus formed, on the assumption that their relative stabilities (degree of delocalisation of negative charge) reflect the relative energies of the transition states which lead to them. For example 2- and 4-halonitrobenzenes react in this way because the anionic adduct derives stabilisation by delocalisation of the charge onto the nitro group(s).

Hal Y⁻ → [Hal, Y] – Hal⁻ → Y

Meisenheimer complex   sp³ hybridised

The electron-deficiency of the carbons in pyridines, particularly α- and γ-carbons, makes nucleophilic addition and, especially nucleophilic displacement of halide (and other good leaving groups), a very important feature of pyridine chemistry.

Y⁻ / N Hal → N Y

nucleophilic displacements at α- and γ-positions are a very important aspect of pyridine chemistry

Such substitutions follow the same mechanistic route as the displacement of halide from 2- and 4-halo-nitro-benzenes, i.e. the nucleophile first adds and then the halide departs. By analogy with the benzenoid situation, the addition is facilitated by (i) the electron-deficiency at α- and γ-carbons, further increased by the halogen substituent, and (ii) the ability of the heteroatom to accommodate negative charge in the intermediate thus produced. Once again, a comparison of the three possible intermediates makes it immediately plain that this latter is not available for attack at a β-position, and thus β nucleophilic displacements are very much slower – for practical purposes they do not occur.

intermediate for α-attack by Y⁻
on α-halopyridine

intermediate for γ-attack by Y⁻
on γ-halopyridine

intermediate for β-attack by Y⁻
on β-halopyridine

It is useful to compare the reactivity of α- and γ-halopyridines with the reaction of acid halides and β-halo-α,β-unsaturated ketones, respectively, both of which also interact easily with nucleophiles and also by an addition/elimination sequence resulting in overall displacement of the halide by the nucleophile.

The generation of metallated aromatics has become extremely important for the introduction of substituents, especially carbon substituents, by subsequent reaction with an electrophile. It is very important, in the light of the discussion above on the ease of nucleophilic addition and substitution, to realise that iodine and bromine at all positions of a pyridine can be exchanged at low temperature *without* nucleophilic displacement or addition, with formation of the pyridyllithium.

the use of lithiopyridines as nucleophiles is a very important method for the introduction of electrophilic species

In the absence of an α- or γ-halogen, pyridines are less reactive and, of course, do not have a substituent suitable for leaving as an anion to complete a nucleophilic substitution. Nucleophilic additions do however take place, but the resultant dihydropyridine adduct requires an oxidant – to remove 'hydride' – to complete an overall substitution. Such reactions, for example with amide or with organometallic reagents, are selective for an α-position, possibly because the nucleophile is delivered *via* a complex involving interaction of the ring nitrogen with the metal cation associated with the reactant. The addition of organometallic or hydride reagents to N^+-acylpyridinium salts is an extremely useful process: the product, dihydropyridines are stable because the nitrogen is an amide, most often a urethane.

Radical substitution of pyridines, in acid solution, is now a preparatively useful process. For efficient reaction, the radicals must be 'nucleophilic', like ·CH₂OH, alkyl·, and acyl· – aminocarbonylation provides an example.

Pyridines carrying oxygen at an α- or γ-position exist as tautomers having carbonyl groups – pyridones. Nonetheless, there is considerable parallelism between their reactions and those of phenols: pyridones are activated towards electrophilic substitution, attack taking place *ortho* and *para* to the oxygen, and they readily form anions, by loss of the N-hydrogen, which are analogous in structure and reactivity to phenolates, though in the heterocyclic system the anion can react at either oxygen or nitrogen, depending on conditions.

Where pyridones differ from phenols is in their interaction with phosphorus and sulfur halides, where transformation of the oxygen substituent into halide occurs. Here, the pyridones react in an amide-like fashion, the inorganic reagent reacting first at the amide-like oxygen.

The special properties associated with pyridine α- and γ-positions show again in the reactions of alkylpyridines: the protons on alkyl groups at those positions are particularly acidified because the 'enaminate' anions formed are delocalised. The ability to form side-chain anions provides an extremely useful means for the manipulation of α- and γ-side-chains.

Pyridinium salts show the properties which have been discussed above, but in extreme: they are highly resistant to electrophilic substitution but, conversely, nucleophiles add very easily. The hydrogens of α- and γ-alkyl side-chains on pyridinium salts are further acidified compared with the uncharged alkylpyridine.

Pyridine N-oxide chemistry, which clearly has no parallel in benzenoid chemistry, is an extremely important and useful aspect of the chemistry of heterocycles of the pyridine series. The structure of these derivatives means that they are both more susceptible to electrophilic substitution *and* react more easily with nucleophiles – an extraordinary concept when first encountered. On the one hand, the formally negatively charged oxygen can release electrons to stabilise an intermediate from electrophilic attack and, on the other, the positively charged ring nitrogen can act as an electron sink to encourage nucleophilic addition.

There are a number of very useful processes in which the *N*-oxide function allows the introduction of substituents usually mainly at an α position and in the process, the oxide function is removed; reaction with thionyl chloride is an example.

Quinoline and isoquinoline, the two possible structures in which a benzene ring is annelated to a pyridine ring, represent an opportunity to examine the effect of fusing one aromatic ring to another. Clearly, both the effect the benzene ring has on the reactivity of the pyridine ring, and *vice versa*, as well as comparisons with the chemistry of naphthalene must be considered. Thus the regioselectivity of electrophilic substitution, which in naphthalene favours an α-position, is mirrored in quinoline/isoquinoline chemistry by substitution at 5- and 8-positions. It should be noted that such substitutions usually involve attack on the species formed by electrophilic addition (often protonation) at the nitrogen, which has the effect of discouraging (preventing) attack on the heterocyclic ring.

Just as for naphthalene, the regiochemistry of attack is readily interpreted by looking at possible intermediates: those for attack at C-5/8 allow delocalisation of charge without disruption of the pyridinium ring aromatic resonance, while those for attack at C-6/7 would necessitate disrupting that resonance in order to allow delocalisation of charge.

So, just as quinoline and isoquinoline are reactive towards electrophiles in their benzene ring, so they are reactive to nucleophiles in the pyridine ring, especially (see above) at the positions α and γ to the nitrogen and, further, are more reactive in this sense than pyridines. This is consistent with the structures of the intermediates for, in these, a full and complete, aromatic benzene ring is retained. Since the resonance stabilisation of the bicyclic aromatic is considerably less than twice that of either benzene or pyridine, the loss in resonance stabilisation in proceeding from the bicyclic system to the intermediate is considerably less than in going from pyridine to an intermediate adduct. There is an obvious analogy: the rate of electrophilic substitution of naphthalene is greater than that of benzene for, in forming a σ-complex from the former, less resonance energy is sacrificed.

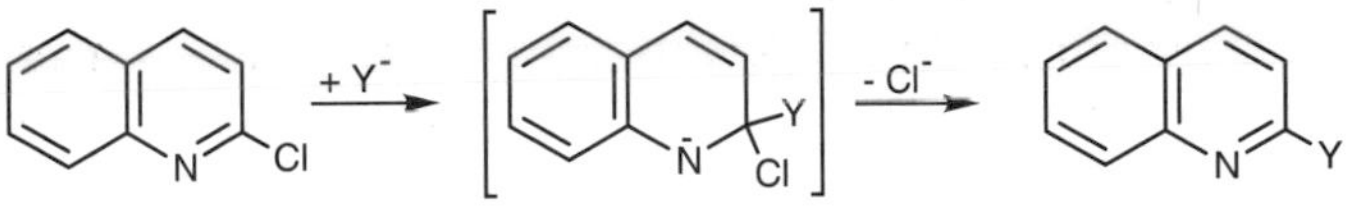

A significant difference in this typical behaviour applies to the isoquinoline 3-position – the special reactivity which the discussion above has developed for positions α to pyridine nitrogen, and which also applies to the isoquinoline 1-position, does not apply at C-3. In the context of nucleophilic displacements, for example, an intermediate for reaction of a 3-halo-isoquinoline cannot achieve delocalisation of negative charge onto the nitrogen unless the aromaticity of the benzene ring is disrupted. Therefore, such intermediates are considerably less stabilised and reactivity considerably tempered.

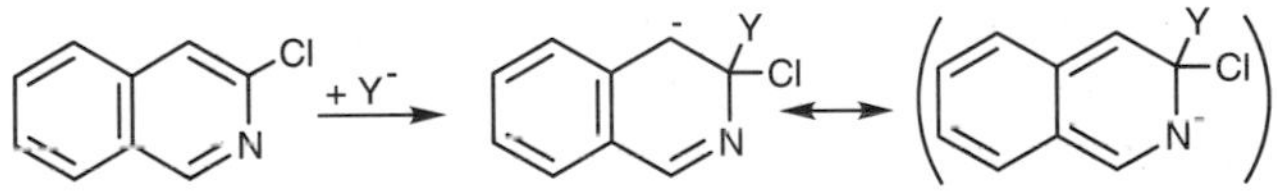

The displacement of halogen at all positions of the pyridine and isoquinoline nucleus is achievable using metal, usually palladium(0), catalysis (see section 2.7 for a detailed discussion). Couplings with alkenes (Heck reactions), with alkynes, with alkenyl- or aryltin or -boron species are complemented by couplings in the opposite sense using pyridinyl/quinolinyl/isoquinolinyl metal (often tin) reagents, with alkenyl or arylhalides or triflates. This extremely useful methodology allows transformations in one step which would formerly have required such an extensive sequence of steps that they might not even have been attempted: two examples are shown below.

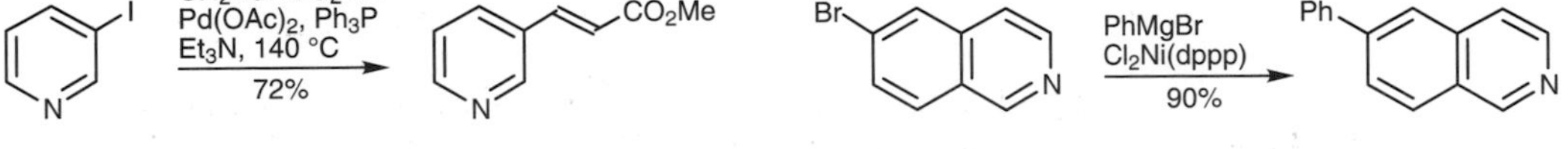

A great variety of methods is available for the ring synthesis of pyridines: the most obvious approach is to construct a 1,5-dicarbonyl compound, preferably also having further unsaturation and allow it to react with ammonia, addition of which at each carbonyl group, with losses of water, producing the pyridine. 1,4-Dihydropyridines, which can easily be dehydrogenated to the fully aromatic system, result from the interaction of aldehydes with two mol equivalents of 1,3-diketones (or 1,3-keto-esters, *etc.*) and ammonia; aldol and Michael reactions and addition of ammonia at the termini, produces the heterocycle.

Nearly all quinoline syntheses begin from an arylamine: that shown generally below – the acid-catalysed interaction with a 1,3-diketone – involves addition of the amine nitrogen to one of the carbonyl groups and a ring closure onto the aromatic ring having the character of an electrophilic substitution. Another much-used route utilises the aldol-type interaction of an *ortho*-aminoaraldehyde (or -ketone) with a ketone have an α methylene.

The amides of 2-(aryl)ethanamines can be made to ring close producing 3,4-dihydroisoquinolines (which can be easily dehydrogenated to the aromatic systems) using reagents such as phosphoryl chloride; again, the ring-closure step is an intramolecular electrophilic substitution of the aromatic ring.

5 Pyridines: reactions and synthesis

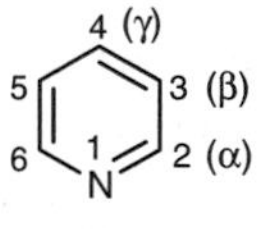

pyridine

Pyridine and its simple derivatives are stable and relatively unreactive liquids, with strong penetrating odours that are unpleasant to some people. They are much used as solvents and bases, especially pyridine itself, in reactions such as N- and O-tosylation and -acylation. Pyridine and the monomethylpyridines (picolines) are completely miscible with water.

Pyridine was first isolated, like pyrrole, from bone pyrolysates: the name is constructed from the Greek for fire, 'pyr', and the suffix '$idine$', which was at the time being used for all aromatic bases – phenetidine, toluidine, etc. Pyridine and its simple alkyl derivatives were for a long time produced by isolation from coal tar, in which they occur in quantity. In recent years this source has been displaced by synthetic processes: pyridine itself, for example, can be produced on a commercial scale in 60–70% yields by the gas-phase high-temperature interaction of crotonaldehyde, formaldehyde, steam, air and ammonia over a silica-alumina catalyst. Processes for the manufacture of alkylpyridines involve reaction of acetylenes and nitriles over a cobalt catalyst.

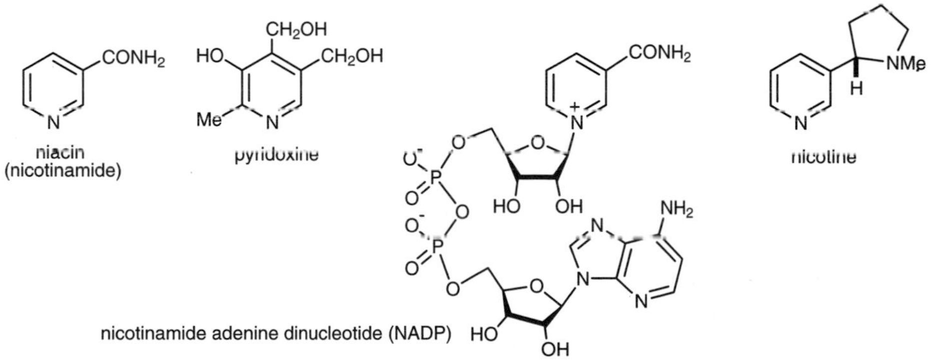

niacin
(nicotinamide)

pyridoxine

nicotinamide adenine dinucleotide (NADP)

nicotine

The pyridine ring plays a key role in several biological processes, most notably in the oxidation/reduction coenzyme nicotine adenine dinucleotide (NADP); the vitamin niacin (or the corresponding acid) is required for its biosynthesis. Pyridoxine (vitamin B_6) plays a key role as the coenzyme in transaminases. Nicotine, a highly toxic alkaloid, is the major active component in tobacco, and the most addictive drug known.[1]

Isoniazide   Sulphapyridine   Prialdoxime   Amlodipine

Many synthetic pyridine derivatives are important as therapeutic agents, for example Isoniazide is a major antituberculosis agent, Sulphapyridine is one of the sulfonamide antibacterials, Prialdoxime is an antidote for poisoning by organophosphates, and Amlodipine is one of several antihypertensive 1,4-dihydropyridines. Some herbicides (Paraquat)[2] and fungicides (Davicil) are also pyridine derivatives. Nemertelline (for a synthesis see section 5.15.2.4) is a neurotoxin from a marine worm; epibatidine, isolated from a South American frog, shows promise as an analgetic agent (for a synthesis see section 13.18.3.6).

Paraquat   Davicil   epibatidine   nemertelline

5.1 Reactions with electrophilic reagents

5.1.1 Addition to nitrogen

In reactions which involve bond formation using the lone pair of electrons on the ring nitrogen, such as protonation and quaternisation, pyridines behave just like tertiary aliphatic or aromatic amines. When a pyridine reacts as a base or a nucleophile it forms a pyridinium cation in which the aromatic sextet is retained and the nitrogen acquires a formal positive charge.

5.1.1.1 Protonation of nitrogen

Pyridines form crystalline, frequently hygroscopic, salts with most protic acids. Pyridine itself, with pK_a 5.2 in water, is a much weaker base than saturated aliphatic amines which have pK_a values mostly between 9 and 11. Since the gas-phase proton affinity of pyridine is actually very similar to those of aliphatic amines, the observed solution values reflect relatively strong solvation of aliphatic ammonium cations;[3] this difference may in turn be related to the mesomerically delocalised charge in pyridinium ions and the consequent reduced requirement for external stabilisation via solvation.

Electron-releasing substituents generally increase the basic strength; 2-methyl-(pK_a 5.97), 3-methyl (5.68) and 4-methylpyridine (6.02) illustrate this. The basicities of pyridines carrying groups which can interact mesomerically as well as inductively vary in more complex ways, for example 2-methoxypyridine (3.3) is a weaker, but 4-methoxypyridine (6.6) a stronger base than pyridine; the effect of inductive withdrawal of electrons by the electronegative oxygen is felt more strongly when it is closer to the nitrogen, i.e. at C-2.

Large 2- and 6-substituents impede solvation of the protonated form: 2,6-di-*t*-butylpyridine is less basic than pyridine by one pK_a unit and 2,6-di(tri-*i*-propylsilyl)pyridine will not dissolve even in 6N hydrochloric acid.[4]

5.1.1.2   Nitration at nitrogen (see also section 5.1.2.2)

This occurs readily by reaction of pyridines with nitronium salts, such as nitronium tetrafluoroborate.[5] Protic nitrating agents such as nitric acid of course lead exclusively to *N*-protonation.

1-Nitro-2,6-dimethylpyridinium tetrafluoroborate is one of several *N*-nitropyridinium salts which can be used as non-acidic nitrating agents with good substrate and positional selectivity. The 2,6-disubstitution serves to sterically inhibit resonance overlap between nitro group and ring and consequently increase reactivity as a nitronium ion donor, however the balance between this advantageous effect and hindering approach of the aromatic substrate is illustrated by the lack of transfer nitration reactivity in 2,6-dihalo-analogues.[6]

5.1.1.3   Amination of nitrogen

The introduction of nitrogen at a different oxidation level can be achieved with hydroxylamine *O*-sulfate.[7]

5.1.1.4   Oxidation of nitrogen

In common with other tertiary amines, pyridines react smoothly with percarboxylic acids to give *N*-oxides, which have their own rich chemistry (section 5.14).

5.1.1.5   Sulfonation at nitrogen

Pyridine reacts[8] with sulfur trioxide to give the commercially available, crystalline, zwitterionic pyridinium-1-sulfonate, usually known as the pyridine sulfur trioxide complex. This compound is hydrolysed in hot water to sulfuric acid and pyridine (for its reaction with hydroxide see section 5.13.4), but more usefully it can serve as a mild sulfonating agent (for examples see sections 13.1.3 and 15.1.3) and as an activating agent for dimethylsulfoxide in Moffat oxidations.

When pyridine is treated with thionyl chloride a synthetically useful dichloride salt is formed, which can, for example, be transformed into pyridine-4-sulfonic acid. The reaction is believed to involve initial attack by sulfur at nitrogen, followed by nucleophilic addition of a second pyridine at C-4 (cf. section 5.13.3).[9]

5.1.1.6 Halogenation at nitrogen

Pyridines react easily with halogens and interhalogens[10] to give crystalline compounds, largely undissociated when dissolved in solvents such as carbon tetrachloride. Structurally they are best formulated as resonance hybrids related to trihalide anions. 1-Fluoropyridinium triflate is also crystalline and serves as an electrophilic fluorinating agent.[11]

These salts must be distinguished from pyridinium tribromide, obtained by treating pyridine hydrobromide with bromine, which does not contain an N-halogen bond, but does include a trihalide anion. The stable, crystalline, commercially available salt can be used as a source of molecular bromine especially where small accurately known quantities are required.

5.1.1.7 Acylation at nitrogen

Carboxylic, and arylsulfonic acid halides react rapidly with pyridines generating 1-acyl- and 1-arylsulfonylpyridinium salts in solution, and in suitable cases some of these can even be isolated as crystalline solids.[12] The solutions, generally in excess pyridine, are commonly used for the preparation of esters and sulfonates from alcohols and of amides and sulfonamides from amines. 4-Dimethylaminopyridine[13] (DMAP) is widely used (in catalytic quantities) to activate anhydrides in a similar manner. The salt derived from DMAP and t-butyl chloroformate is stable even in aqueous solution at room temperature.[14]

4-dimethylaminopyridine (DMAP)

5.1.1.8 Alkylation at nitrogen

Alkyl halides and sulfates react readily with pyridines giving quaternary pyridinium salts. As with aliphatic tertiary amines, increasing substitution around the nitrogen, or around the halogen-bearing carbon, causes an increase in the alternative, competing, elimination process which gives alkene and N-proto-pyridinium salt, thus 2,4,6-trimethylpyridine (collidine) is useful as a base in dehydrohalogenation reactions.

5.1.1.9 Reaction with metal centres

The normal behaviour of pyridines in the presence of metal cations is complexation involving donation of the nitrogen lone pair to the metal centre. This means that for simple pyridines, formation of π-complexes like benzene-chromium carbonyl complexes, does not take place. However, if the nitrogen lone pair is hindered, then η^6-complexes can be formed.[15]

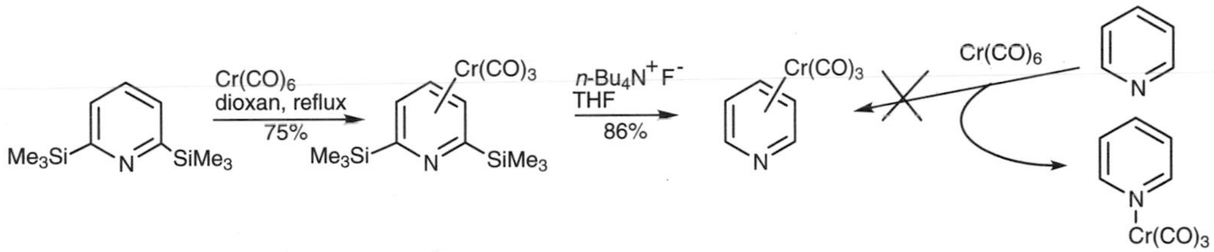

5.1.2 Substitution at carbon

In most cases, electrophilic substitution of pyridines occurs very much less readily than for the correspondingly substituted benzene. The main reason is that the electrophilic reagent, or a proton in the reaction medium, adds preferentially to the pyridine nitrogen, generating a pyridinium cation, which is naturally very resistant to a further attack by an electrophile. When it does occur then, electrophilic substitution at carbon must involve either highly unfavoured attack on a pyridinium cation or relatively easier attack but on a very low equilibrium concentration of uncharged free pyridine base.

Some of the typical electrophilic substitution reactions do not occur at all – Friedel-Crafts alkylation and acylation are examples – but it is worth recalling that these also fail with nitrobenzene. Milder reagents, such as Mannich reactants, diazonium ions and nitrous acid, which in any case require activated benzenes for success, naturally fail with pyridines.

5.1.2.1 Proton exchange

H–D exchange *via* an electrophilic addition process, such as operates for benzene, does not take place with pyridine. A special mechanism allows selective exchange at the two α-positions in DCl–D$_2$O or even in water at 200 °C, the key species being an ylide formed by 2/6-deprotonation of the 1H-pyridinium cation (see also section 5.12).[16]

5.1.2.2 Nitration

Pyridine itself can be converted into 3-nitropyridine only inefficiently by direct nitration even with vigorous conditions,[17] as shown below, however a couple of ring methyl groups facilitate electrophilic substitution sufficiently to allow nitration to compete with side-chain oxidation.[18] Steric or/and inductive inhibition of *N*-nitration allows *C*-substitution using nitronium tetrafluoroborate, an example is nitration of 2,6-dichloropyridine.[6]

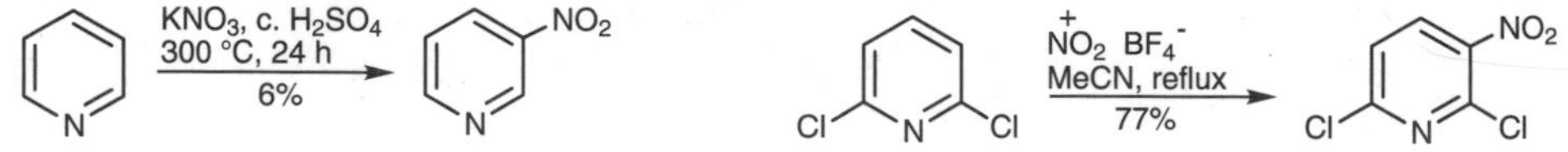

3-Nitropyridine itself, and some of its substituted derivatives, can now be prepared efficiently by reaction with dinitrogen pentoxide as shown below. The initially formed *N*-nitropyridinium nitrate suffers addition of a nucleophile – sulfur dioxide when this is used as solvent or co-solvent, or sulfite, added subsequently – forming a 1,2-dihydropyridine. Transfer of the nitro group to a 3- or 5-position, via a [1,5]-sigmatropic migration, is then followed by elimination of the nucleophile regenerating the aromatic system.[19]

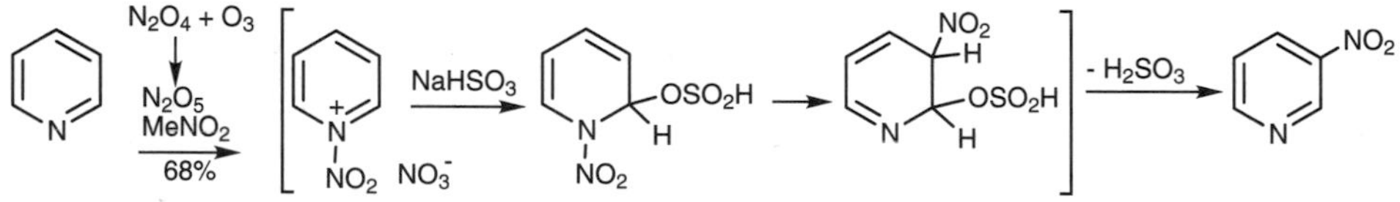

Both collidine and its quaternary salt are nitrated at similar rates under the same conditions, showing that the former reacts *via* its *N*-protonic salt.[20]

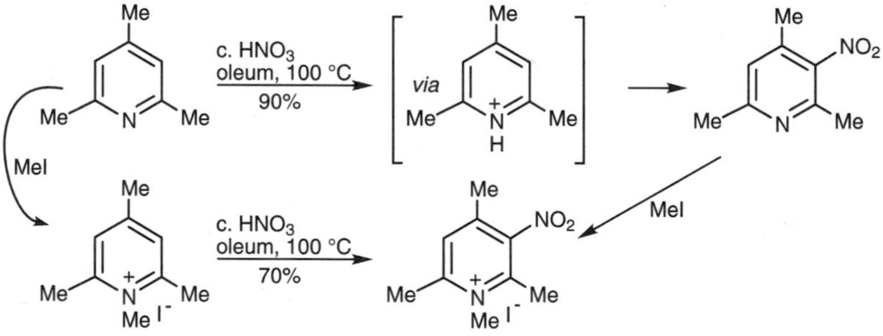

5.1.2.3 *Sulfonation*

Pyridine is very resistant to sulfonation using concentrated sulfuric acid or oleum, only very low yields of the 3-sulfonic acid being produced after prolonged reaction periods at 320 °C. However, addition of mercuric sulfate in catalytic quantities allows smooth sulfonation at a somewhat lower temperature. The role of the catalyst is not established; one possibility is that *C*-mercuration is the first step (cf. section 5.1.2.5).[21]

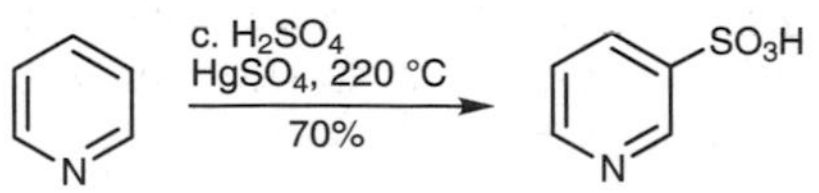

The *C*-sulfonation of 2,6-di-*t*-butylpyridine[22] is a good guide to the intrinsic reactivity of a pyridine ring, for in this situation the bulky alkyl groups effectively prevent addition of sulfur trioxide to the ring nitrogen allowing progress to a 'normal' electrophilic *C*-substitution intermediate, at about the same rate as for sulfonation of nitrobenzene. A maximum conversion of 50% is all that is achieved because for every *C*-substitution a proton is produced which 'consumes' a molecule of starting material by *N*-protonation.

5.1.2.4 Halogenation

3-Bromopyridine is produced in good yield by the action of bromine in oleum.[23] The process is thought to involve pyridinium-1-sulfonate as the reactive species, since no bromination occurs in 95% sulfuric acid. 3-Chloropyridine can be produced by chlorination at 200 °C or at 100 °C in the presence of aluminium chloride.[24]

2-Bromo- and 2-chloropyridines can be made extremely efficiently by reaction of pyridine with the halogen, at 0–5 °C in the presence of palladium(II) chloride.[25]

5.1.2.5 Acetoxymercuration

The salt formed by the interaction of pyridine with mercuric acetate at room temperature can be rearranged to 3-acetoxymercuripyridine by heating to only 180 °C.[26] This process, where again there is *C*-attack by a relatively weakly electrophilic reagent, like that described for mercuric sulfate-catalysed sulfonation, may involve attack on an equilibrium concentration of free pyridine.

5.1.2.6 Substitution in pyridines carrying activating nitrogen and oxygen substituents

See sections 5.10.2.1 and 5.10.3.1

5.2 Reactions with oxidising agents

The pyridine ring is generally resistant to oxidising agents, vigorous conditions being required, thus pyridine itself is oxidised by neutral aqueous potassium permanganate at about the same rate as benzene (sealed tube, 100 °C), to give carbon dioxide. In acidic solution pyridine is more resistant, but in alkaline media more rapidly oxidised, than benzene.

In most situations, carbon substituents can be oxidised with survival of the ring, thus alkylpyridines can be converted into pyridine carboxylic acids with a variety of reagents.[27] Some selectivity can be achieved: only α- and γ-groups are attacked by selenium dioxide; the oxidation can be halted at the aldehyde oxidation level.[28]

5.3 Reactions with nucleophilic reagents

Just as electrophilic substitution is the characteristic reaction of benzene and electron-rich heteroaromatic compounds (pyrrole, furan etc.), so substitution reactions with nucleophiles can be looked on as characteristic of pyridines.

It is important to realise that nucleophilic substitution of hydrogen differs in an important way from electrophilic substitution: whereas the last step in electrophilic substitution is loss of proton, an easy process, the last step in nucleophilic substitution of hydrogen has to be a hydride transfer, which is less straightforward and generally needs the presence of an oxidising agent as hydride acceptor. Nucleophilic substitution of an atom or group which is a good anionic leaving group however is an easy and straightforward process.

5.3.1 Nucleophilic substitution with 'hydride' transfer[29]

5.3.1.1 Alkylation and arylation

Reaction with alkyl- or aryllithiums proceeds in two discrete steps: addition to give a dihydropyridine N-lithio-salt which can then be converted into the substituted aromatic pyridine by oxidation (e.g. by air), disproportionation, or elimination of lithium hydride.[30] The N-lithio-salts can be observed spectroscopically and in some cases isolated as solids.[31] Attack is nearly always at an α-position; reaction with 3-substituted-pyridines usually takes place at both available α-positions, but predominantly at C-2.[32] This regioselectivity may be associated with relief of strain when the 2-position rehybridises to sp^3 during addition.

From the preparative viewpoint nucleophilic alkylations can be greatly facilitated by the device of prior quaternisation of the pyridine in such a way that the N-substituent can be subsequently removed – these processes are dealt with in section 5.13.2.

5.3.1.2 Amination

Amination of pyridines and related heterocycles, generally at a position α to the nitrogen, is called the Chichibabin reaction,[33] the pyridine reacting with sodamide with the evolution of hydrogen. The 'hydride' transfer and production of hydrogen probably involve interaction of aminopyridine product, acting as an acid, with the anionic intermediate. The preference for α-substitution may be associated with an intramolecular delivery of the nucleophile, perhaps guided by complexation of ring nitrogen with metal cation.

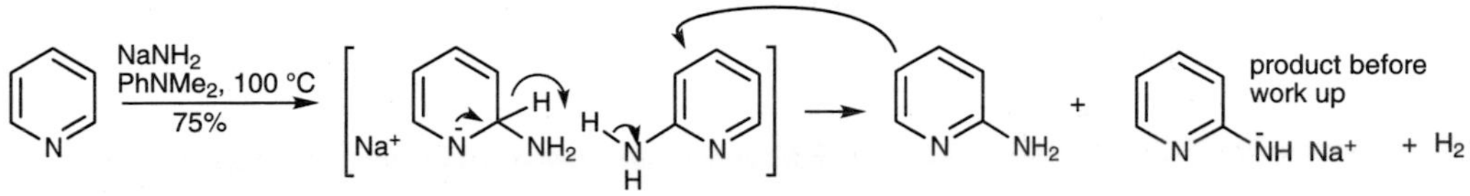

More vigorous conditions are required for the amination of 2- or 4-alkylpyridines since proton abstraction from the side-chain by the amide occurs first and ring attack must therefore involve a dianionic intermediate.[34] Amination of 3-alkylpyridines is regioselective for the 2-position.[35]

Vicarious nucleophilic substitution (section 2.3.3) permits the introduction of amino groups *ortho* to nitro groups by reaction with methoxyamine as illustrated below.[36]

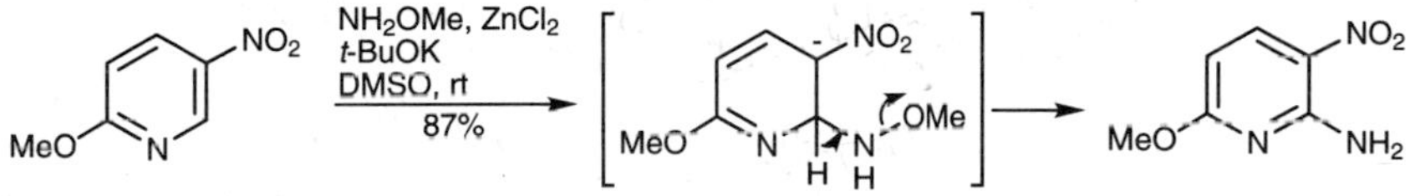

5.3.1.3 Hydroxylation

Hydroxide ion, being a much weaker nucleophile than amide, attacks pyridine only at very high temperatures to produce a low yield of 2-pyridone,[37] which can be usefully contrasted with the much more efficient reaction of hydroxide with quinoline and isoquinoline (section 6.3.1.3) and with pyridinium salts (section 5.13.4).

5.3.2 Nucleophilic substitution with displacement of good leaving groups

Halogen, and also, though with fewer examples, nitro,[38] alkoxysulfonyl,[39] and methoxy[40] substituents at α- or γ-positions, but not at β-positions, are relatively easily displaced by a wide range of nucleophiles *via* an addition-elimination mechanism facilitated by (a) electron withdrawal by the substituent and (b) the good leaving ability of the substituent. γ-Halopyridines are more reactive than the α-isomers; β-halopyridines are very much less reactive, being much closer to, but still somewhat more reactive than halobenzenes. Fluorides are more reactive than the other halides.[41]

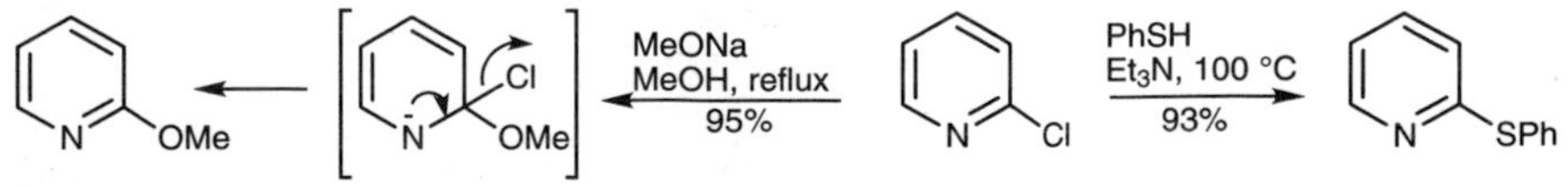

Replacement of halide by reaction with ammonia can be achieved at considerably lower temperatures under 6–8 kbar pressure.[42] The inclusion of Aliquat is an alternative means for improving the efficiency of such nucleophilic displacements.[43]

In some, apparently straightforward, displacements, more detailed mechanistic study reveals the operation of alternative mechanisms. For example the reaction of either 3- or 4-bromopyridine with secondary amines in the presence of sodamide/ sodium t-butoxide, produces the same mixture of 3- and 4-dialkylaminopyridines; this proceeds *via* an elimination process ($S_N(EA)$ – Substitution Nucleophilic Elimination Addition) and the intermediacy of 3,4-didehydropyridine (3,4-pyridyne).[44] That no 2-aminated pyridine is produced shows a greater difficulty in generating 2,3-pyridyne, it can however be formed by reaction of 3-bromo-2-chloropyridines with butyllithium[45] or via the reaction of 3-trimethylsilyl-2-trifluoromethanesulfonyloxypyridine with fluoride.[46]

5.4 Reactions with bases

5.4.1 Deprotonation of C-hydrogen

When pyridine is heated to 165 °C in MeONa–MeOD, H–D exchange occurs at all positions *via* small concentrations of deprotonated species, at the relative rates $\alpha : \beta : \gamma$, 1 : 9.3 : 12.[47] However, using the combination n-butyllithium/potassium t-butoxide, efficient formation of 2-pyridylpotassium or 4-pyridylpotassium has been achieved.[48] Some pyridines have been selectively lithiated at C-2 *via* complexes with hexafluoroacetone;[49] complexation removes the lone pair (cf. section 5.5.1) and additionally provides inductive and chelation effects to assist the regioselective metallation. In practice, simple lithiopyridines are generally prepared by metal–halogen exchange, however the presence of chlorine or fluorine, or other substituents which direct *ortho* metallation, allows direct lithiation (section 5.5.1).

5.5 Reactions of C-metallated pyridines

5.5.1 Lithium and magnesium derivatives

Lithium derivatives are easily prepared and behave as typical organometallic nucleophiles,[50-52] thus for example, 3-bromopyridine undergoes efficient exchange with *n*-butyllithium in ether at $-78\,°C$. In the more basic tetrahydrofuran as solvent, and at this temperature, the alkyllithium becomes more nucleophilic and only addition occurs, although the exchange can be carried out in tetrahydrofuran at lower temperatures.[53] Lithiopyridines can also be prepared from halopyridines, including chloropyridines, via exchange with lithium naphthalenide.[54] 2-Bromo-6-methylpyridine can be converted into its lithio derivative without deprotonation of the methyl.[55]

Direct regioselective α lithiation of pyridine, 2-methoxypyridine, or 2-methylthio-pyridine, can be carried out using a complex base consisting of a mixture of butyllithium and the lithium salt of 2-dimethylaminoethanol. The process may be more complex than simple deprotonation, possibly involving a radical anion intermediate.[56]

Metal/halogen exchange with 2,5-dibromopyridine leads exclusively and efficiently to 2-bromo-5-lithiopyridine in a thermodynamically controlled process;[57] it has been suggested that the 2-pyridyl anion is destabilised by electrostatic repulsion between nitrogen lone pair and the adjacent anion;[46] this same factor is probably important in the greater difficulty found in generating 2,3-pyridyne (see section 5.3.2). The example below illustrates the use of the 'Weinreb amide' of formic acid as a formyl transfer reagent.[58]

Monolithiation of 2,6-dibromopyridine is best achieved by 'inverse addition' – dibromide to *n*-butyllithium, or by using dichloromethane as solvent – probably a unique application of this solvent to lithiation.[59] A normal lithiation of 2,5-dibromopyridine, but at $-90\,°C$, produces clean 2-substituted product with a hindered silicon electrophile.[60]

Halo-, particularly chloro-, or better, fluoropyridines, but even bromopyridines[61] undergo lithiation by deprotonation *ortho* to the halogen best using lithium di-*i*-propylamide, 3-halopyridines reacting mainly at C-4 and 2- and 4-halopyridines necessarily lithiating at a β-position.[62] In the lithiation of methoxypyridines using mesityllithium, the 3-isomer metallated at C-2.[63] 3-Methoxymethoxypyridine,[64] 3-di-*i*-propylaminocarbonyl-[65] and 3-*t*-butylcarbonylamino-[66] -pyridines all lithiate at C-4. Lithiation assisted by the dimethyloxazoline group requires lithium 2,2,6,6-tetramethylpiperidide, otherwise C-4-addition of alkyllithium or Grignard occurs; subsequent aerial oxidation produces 4-alkylated derivatives efficiently.[67] A directed magnesiation required a much higher temperature, as shown below.[68]

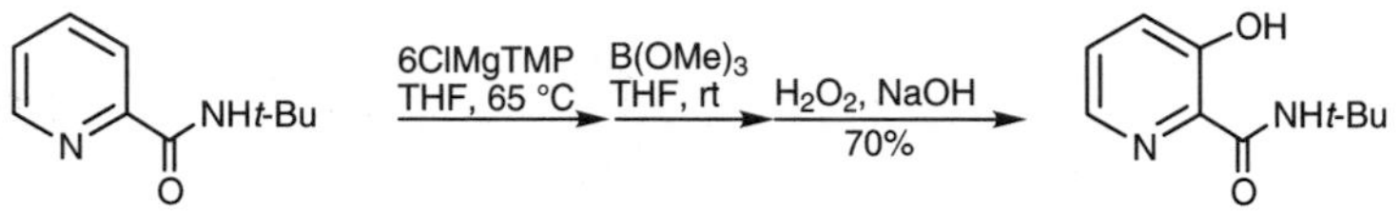

Lithiation of 2- and 4-*t*-butoxycarbonylaminopyridines can only take place at C-3; a neat sequence involving first, ring lithiation to allow introduction a methyl group and secondly side-chain lithiation (section 5.11) at the introduced methyl group provided a route to azaindoles (section 17.17.7), as illustrated below for the synthesis of 5-azaindole (pyrrolo[3,2-*c*]pyridine).[69]

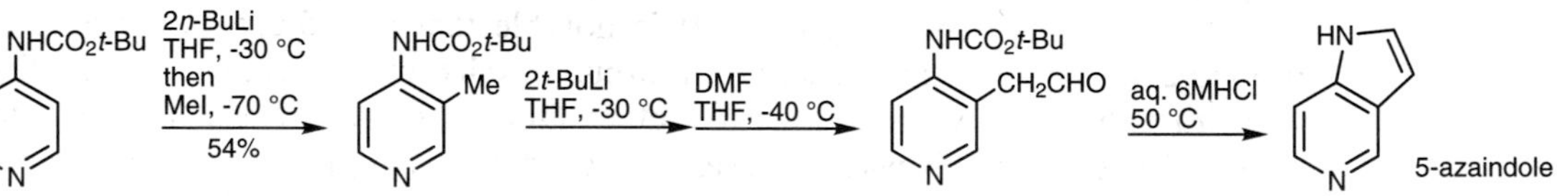

Lithiated pyridines react with the normal range of electrophilic species, for example they are acylated by tertiary amides.[70]

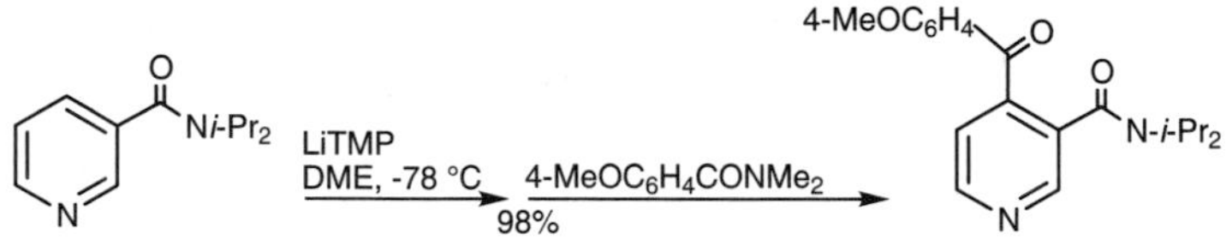

The use of halogen to direct lithiation can be combined with the ability to subsequently displace the halogen with a nucleophile.[71]

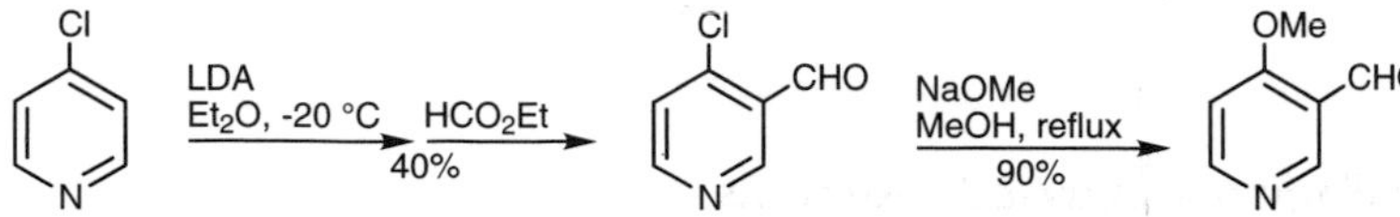

The combination of metal–halogen exchange with the presence of a directing substitutent can permit regioselective exchange;[72] two 1,3-related directing groups causes lithiation between the two groups.[73]

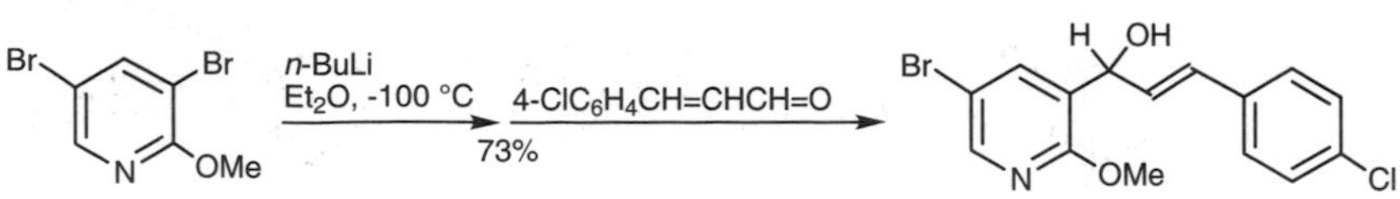

Bromine and iodine also direct lithiations, but isomerisation ('halogen dance', see section 14.5.1) can be a problem. The sequence below shows how advantage was taken of the isomerisation to the more stable lithio derivative i.e. that in which the formally negatively charged ring carbon is located between two halogen-bearing carbon atoms.[74]

The selective 2-lithiation of pyridine *N*-oxides can be achieved in favourable circumstances: one instructive example is the regioselective 6-lithiation of 2-pivaloylaminopyridine *N*-oxide, i.e. adjacent to the *N*-oxide group, and not adjacent to the *ortho*-directing 2-substitutent. The regioselective C-2-lithiation of 3,4-dimethoxypyridine *N*-oxide also shows the influence of the *N*-oxide functionality.[75]

Pyridyl Grignard reagents are readily prepared by exchange of bromine or iodine using *i*-propyl Grignard reagents.[76] It is notable that in 2,5-dibromopyridine, the exchange follows the same pattern as in lithium exchange that is, selective reaction at C-5; other dibromopyridines also give clean mono-exchange. Formation of pyridyl Grignard species in this way will even tolerate functional groups such as esters and nitriles, provided the temperature is kept low. While they are probably not quite as versatile as lithium compounds, the pyridinyl Grignards have obvious advantages in some circumstances.

5.5.2 Palladium-catalysed reactions

Halopyridines or pyridinyl triflates[77,78] take part in palladium-catalysed reactions – Heck,[79] carbonylation,[80] and coupling reactions, for example with alkynes,[81] or in Suzuki reactions with arylboronic acids,[78,82] and cyclopropylboronic acids.[83]

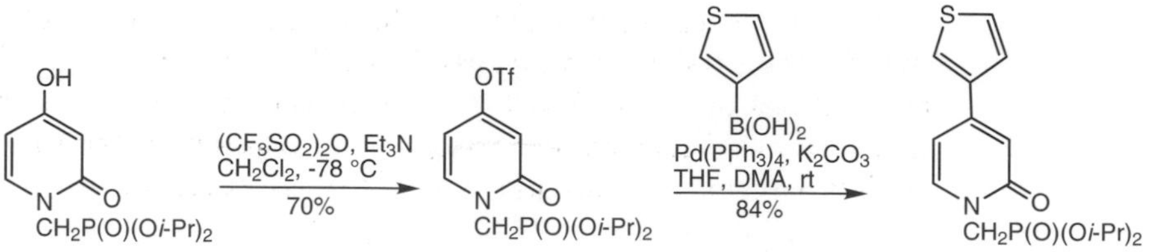

Couplings requiring pyridyl organometallic species are best achieved with boron, zinc, or tin compounds; the last are available either by reaction of pyridyl halides with sodium trialkylstannate or, in the opposite sense, by the reaction of a pyridyllithium with chlorotrimethylstannane.[84] Pyridyltin compounds have been coupled for example with haloarenes[85] and with halopyridine N-oxides.[86]

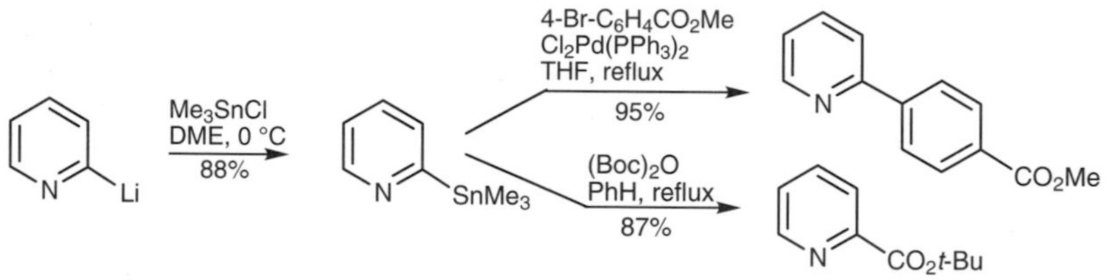

Pyridinyltin reagents also provide a means for the effective acylation of a pyridine, unachievable by conventional Friedel-Crafts processes, as discussed earlier, and illustrated above by 2-*t*-butoxycarbonylation.[87]

Palladium(0) catalysis also provides an excellent means for the overall displacement of halide with nitrogen[88] or sulfur,[89] taking place equally well at all three pyridine ring positions, *i.e.* not relying on the increased susceptibility to nucleophilic displacement at α- and γ-positions (section 5.3.2).

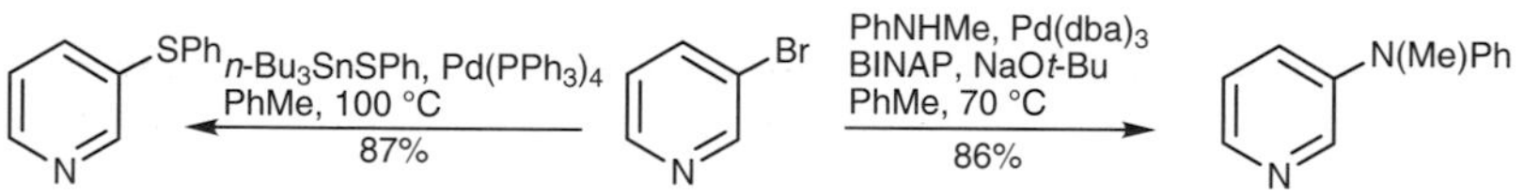

5.6   Reactions with radical reagents; reactions of pyridyl radicals

5.6.1   Halogenation

At temperatures where bromine (500 °C) and chlorine (270 °C) are appreciably dissociated into atoms, 2- and 2,6-dihalopyridines are obtained *via* radical substitution.[90]

5.6.2   Carbon radicals

This same preference for α-attack is demonstrated by phenyl radical attack, but the exact proportions of products depend on the method of generation of the radicals.[91] Greater selectivity for phenylation at the 2- and 4-positions is found in pyridinium salts.[92]

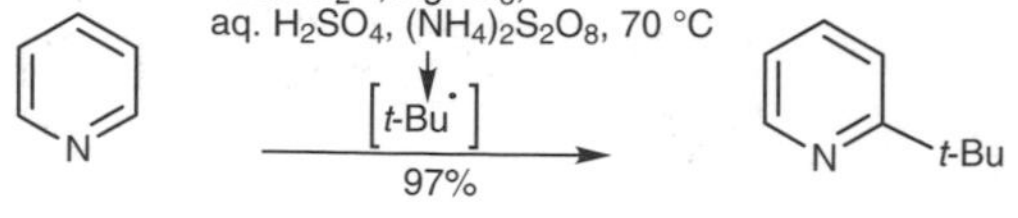

Of more preparative value are the reactions of nucleophilic radicals, such as HOCH$_2$· and R$_2$NCO· which can be easily generated under mild conditions. These substitutions are carried out on the pyridine protonic salt, which provides both increased reactivity and selectivity for an α-position; the process is known as the Minisci reaction (cf. section 2.4.1).[93] It is accelerated by electron-withdrawing substituents on the ring.

5.6.3 Dimerisation

Both sodium and nickel bring about 'oxidative' dimerisations,[94] despite the apparently reducing conditions, the former giving 4,4'-bipyridine and the latter 2,2'-bipyridine.[95] Each reaction is considered to involve the same anion-radical resulting from transfer of an electron from metal to heterocycle, and the species has been observed by ESR spectroscopy when generated by single electron transfer (SET) from lithium diisopropylamide.[96] In the case of nickel, the 2,2'-mode of dimerisation may be favoured by chelation to the metal surface. Bipyridyls are important for the preparation of Paraquat-type weedkillers.

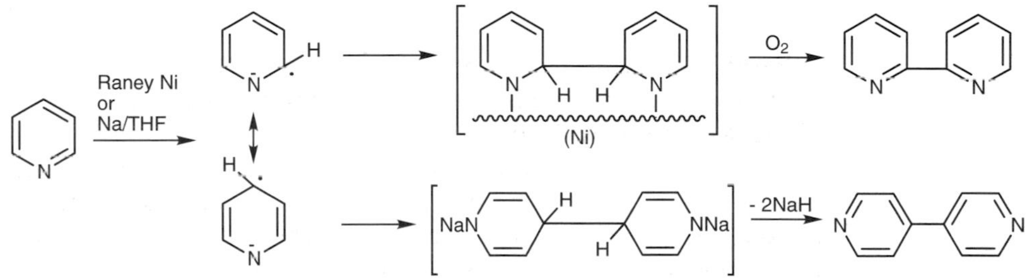

Intermediate, reduced dimers can be trapped under milder conditions,[97] and reduced monomers when the pyridine carries a 4-substituent.[98]

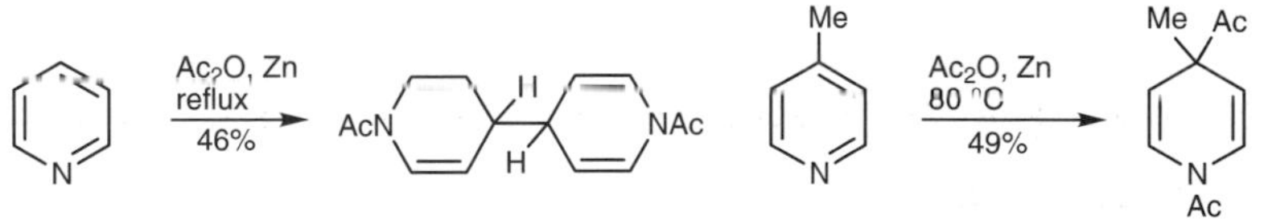

5.6.4 Pyridyl radicals

Irradiation of iodopyridines generates pyridinyl radicals which will effect radical substitution of aromatic compounds.[99] Pyridinyl radicals, like aryl radicals, can also be generated from halopyridines using tin hydrides and participate in typical radical reactions, as in the cyclisation shown below.[100]

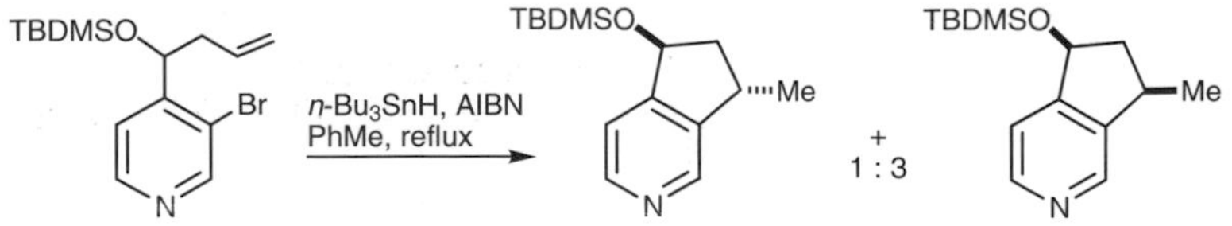

5.7 Reactions with reducing agents

Pyridines are much more easily reduced than benzenes, for example catalytic reduction proceeds easily at atmospheric temperature and pressure, usually in weakly acidic solution but also in dilute alkali over nickel.[101]

Of the hydride reagents, sodium borohydride is without effect on pyridines, though it does reduce pyridinium salts (section 5.13.1), lithium aluminium hydride effects the addition of one hydride equivalent to pyridine,[102] but lithium triethylborohydride reduces to piperidine efficiently.[103]

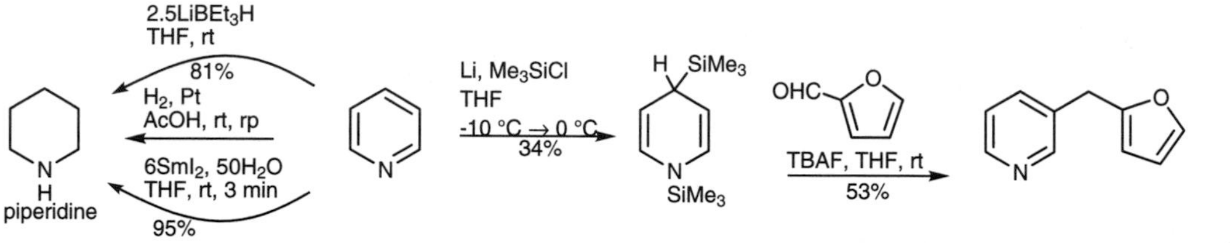

The combination lithium/chlorotrimethylsilane produces a 1,4-dihydro doubly-silylated product, the enamine character in which can be utilised for the introduction of 3-alkyl groups *via* reaction with aldehydes.[104]

Sodium in liquid ammonia, in the presence of ethanol, affords the 1,4-dihydropyridine[105] and 4-pyridones are reduced to 2,3-dihydro-derivatives.[106] Metal/acid combinations, which in other contexts do bring about reduction of iminium groups, are without effect on pyridines. Samarium(II) iodide in the presence of water smoothly reduces pyridine to piperidine.[107]

Trimethylsilane in the presence of palladium gives 1,4-dihydro-1-trimethylsilylpyridine, together with silylated dimer;[108] titanium-catalysed hydrosilylation produces a tetrahydro-derivative cleanly.[109]

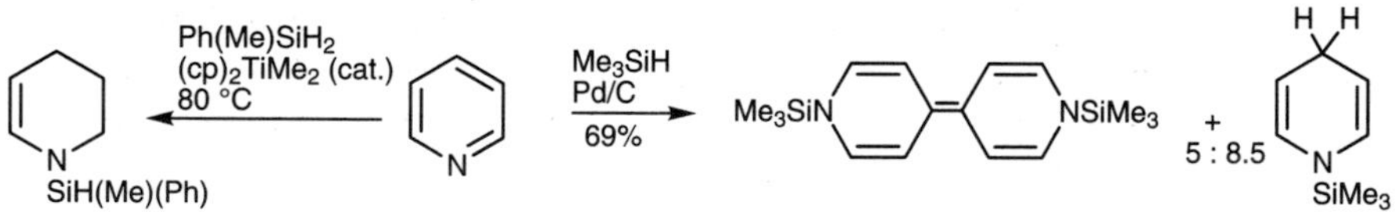

5.8 Electrocyclic reactions (ground state)

There are no reports of thermal electrocyclic reactions involving simple pyridines; 2-pyridones however participate as 4π components in Diels-Alder additions, especially under high pressure.[110]

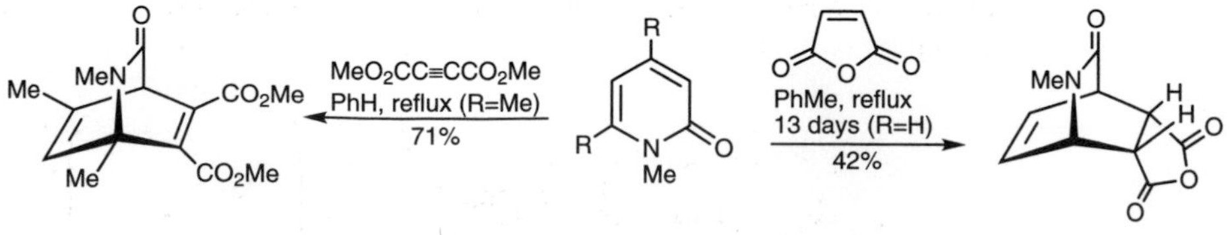

N-Tosyl-2-pyridones with a 3-alkoxy or 3-arylthio substituent, undergo cycloaddition with electron-deficient alkenes under milder conditions, as illustrated below.[111]

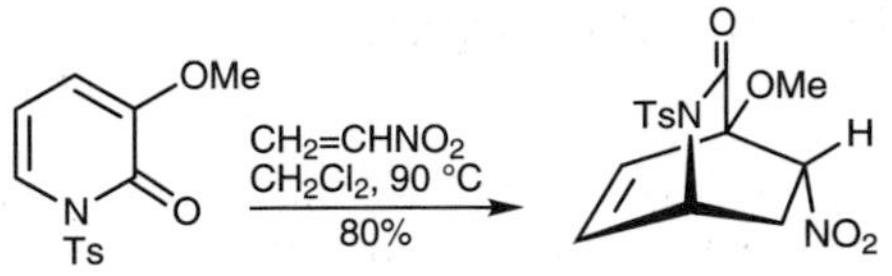

The quaternary salts of 3-hydroxypyridines are converted by mild base into zwitterionic, organic-solvent-soluble species for which no neutral canonical form can be drawn. These 3-oxidopyridiniums undergo a number of dipolar cycloaddition reactions, especially across the 2,6-positions.[112]

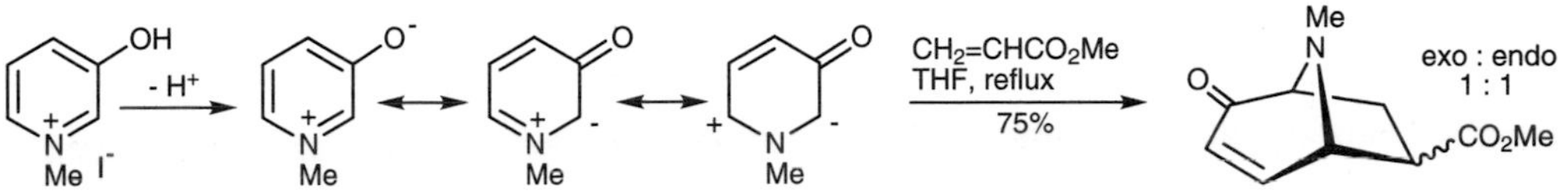

5.9   Photochemical reactions

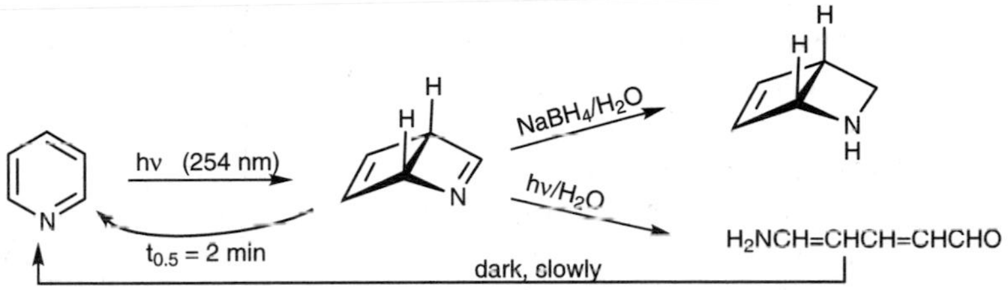

Ultraviolet irradiation of pyridines can produce highly strained species which may lead to isomerised pyridines or can be trapped. From pyridines[113] and from 2-pyridones[114] 2-azabicyclo[2.2.0]hexadienes and -hexenones are obtained; in the case of pyridines these are usually unstable and revert thermally to the aromatic heterocycle, but 2-alkylpyridines with an electron-withdrawing group on the alkyl substituent give stable products by base-catalysed proton shift.[115] Pyridone-derived bicycles are relatively stable, 4-alkoxy- and -acyloxypyridones are converted in particularly good yields.

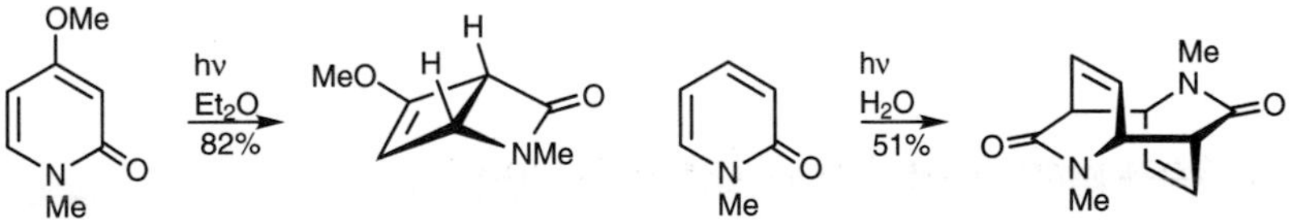

Irradiation of *N*-methyl-2-pyridone in aqueous solution produces a mixture of regio- and stereoisomeric dimers such as the one shown above;[116] such 4π plus 4π photo-cycloadditions of 2-pyridones[117] have also been conducted between two tethered pyridones as illustrated below[118] and between a side-chain 1,3-diene and a 2-pyridone.[119]

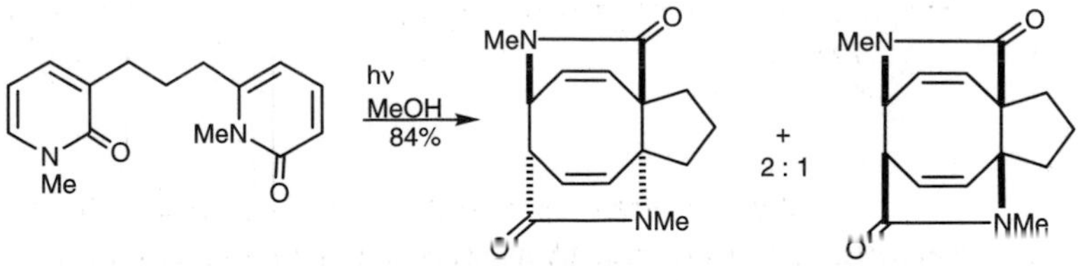

Photocatalysed 2π plus 2π cycloadditions between a pair of tethered 4-pyridones[120] as shown below, can also generate complex rings systems spectacularly easily. Photocatalysed 2π plus 2π cycloaddition between the 5,6-bond of 2-pyridones and an alkene tethered to nitrogen are also known.[121]

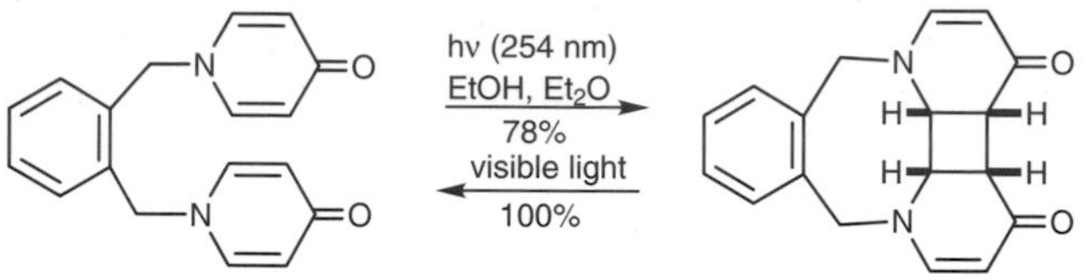

The photoreactions of pyridinium salts in water give 6-azabicyclo[3.1.0]hex-3-en-2-ols or the corresponding ethers, which can undergo regio- and stereoselective ring openings of the aziridine by attack of nucleophiles under acidic conditions. These products are useful starting materials for the synthesis of aminopentanol-derived natural products.[122] At a higher oxidation level, comparable irradiation of 3-methopyridinium salts in neutral solution produces bicyclic aziridines as final products; the sequence shown below shows the first photo-intermediate as an azabenzvalene – an alternative interpretation of its structure.[123]

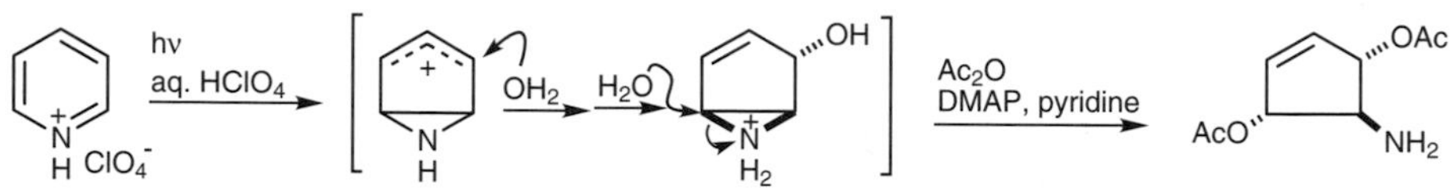

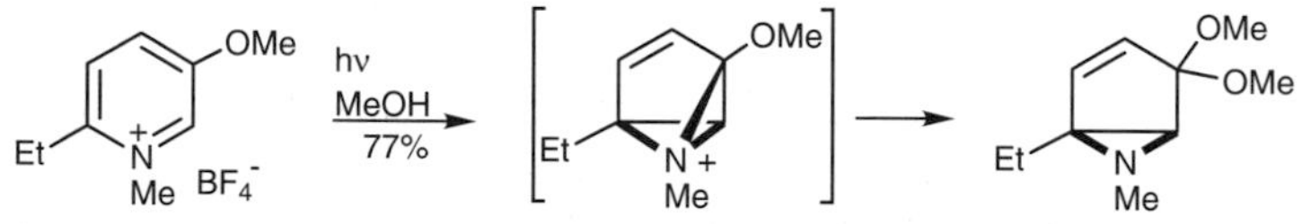

On photolysis of pyridine *N*-oxides in alkaline solution, ring opening produces cyano-dienolates.[124]

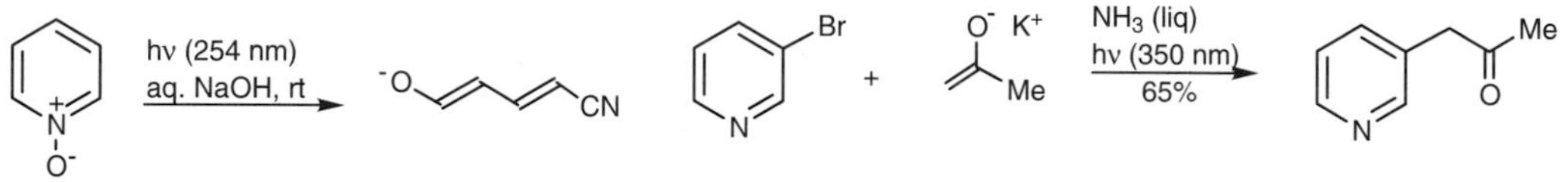

The displacement of bromine, in the relative order $2 > 3 > 4$, by an enolate or related anion under irradiation, known as an $S_{RN}1$ process (**S**ubstitution **R**adical **N**ucleophilic, unimolecular), involves photostimulated transfer of an electron from the enolate to the heterocycle, loss of bromide to generate a pyridyl radical which then combines with a second mol of enolate, generating the radical anion of product, transfer of an electron from which sustains the chain process.[125] The equivalent photo-catalysed displacement of bromide by hydroxide gives 3-hydroxypyridine.[126]

5.10   Oxy- and aminopyridines

5.10.1   Structure

The three oxy-pyridines are subject to tautomerism involving hydrogen interchange between oxygen and nitrogen, but again with a significant difference between α- and γ- on the one hand and β-isomers on the other.

Under all normal conditions, α- and γ-isomers exist almost entirely in the carbonyl tautomeric form, and are accordingly known as pyridones; the hydroxy tautomers are detected in significant amounts only in very dilute solutions in non-polar solvents like petrol, or in the gas phase where, for the α-isomer, 2-hydroxypyridine is actually the dominant tautomer by 2.5:1.[127] The polarised pyridone form is favoured by solvation.[128] 3-Hydroxypyridine exists in equilibrium with a corresponding zwitterionic tautomer, the exact ratio depending on solvent.

All three aminopyridines exist in the amino form; the α- and γ-isomers are polarised in a sense opposite to that in the pyridones.

5.10.2   Reactions of pyridones
5.10.2.1   Electrophilic addition and substitution

3-Hydroxypyridine protonates on nitrogen, with a typical pyridine pK_a of 5.2; the pyridones however are much less basic, and both, like amides, protonate on oxygen.[129] 2,6-Dimethyl-4-pyridones produce isolable 4-hydroxypyridinium bromides on reaction with t-butyl bromide.[130]

An apparent exception to this pattern is the reaction of 4-pyridone with acid chlorides producing N-acyl derivatives. 1-Acetyl-4-pyridone subsequently equilibrates in solution affording a mixture with 4-acetoxypyridine.[131]

Electrophilic substitution at carbon can be effected much more readily with the three oxy-pyridines than with pyridine itself, and it occurs *ortho* and *para* to the oxygen function, as indicated below. Acid catalysed exchange of 4-pyridone in deuterium oxide, for example, gives 3,5-dideuterio-4-pyridone, *via* C-protonation of the neutral pyridone.[132]

positions of electrophilic substitution of oxypyridines

Substitutions in acidic solutions usually proceed *via* attack on the free pyridone,[133] but in very strong acid, where there is almost complete protonation, 4-pyridone undergoes a slower nitration, *via* the *O*-protonated salt, but with the same regioselectivity.[134]

N-Methyl-2-pyridone undergoes electrophilic palladation at C-5, allowing a subsequent direct coupling *via* a modified Heck reaction (cf. section 2.7.2.1).[135]

Some electrophilic substitutions of 3-hydroxypyridine take place at C-2 and some at C-6 thus nitration gives 3-hydroxy-2-nitropyridine[136] and Mannich condensation also takes place at C-2,[137] but iodination goes at C-6[138] (complimentarily, 2-methoxypyridine brominates at C-5).[139]

5.10.2.2 *Deprotonation and reaction of salts*

N-Unsubstituted pyridones are acidic, with pK_a values of about 11 for deprotonation giving mesomeric anions. These ambient anions can be alkylated on either oxygen or nitrogen, producing alkoxypyridines or *N*-alkylpyridones, respectively, the relative proportions depending on the reaction conditions;[140] *N*-alkylation is usually predominant for primary halides; *O*-alkylation for secondary halides.[141] A clean method for the synthesis of *N*-alkylated 4-pyridones is to convert the pyridone first into the *O*-trimethylsilyl ether[142] which can then be reacted selectively at nitrogen, subsequent removal of the silicon giving the *N*-alkylpyridone.[102] 2-Pyridone is sufficiently acidic to take part in Mitsunobu reactions with alcohols though again, mixtures of *O*- and *N*-alkylation products result.[143]

Alkylation of the sodium salt of 2-pyridone with chloromethyltrimethylsilane allows subsequent introduction of further groups on to the nitrogen substituent.[144]

Aqueous sodium hydroxide at 100 °C causes exchange of the α-protons in 1-methyl-4-pyridone[145] and synthetically useful metallation at an α-position can be effected with *n*-butyllithium;[146] 1-methyl-2-pyridone, in contrast, metallates on the methyl,[147] but 2-pyridones, protected by carboxylation at nitrogen, lithiate at C-4.[148] The metallated *N*-methylpyridones tend to dimerise in the sense that they add to free pyridone in a Michael fashion. Metallation then condensation at side-chain methyl in a pyridone is also known.[149]

5.10.2.3   Replacement of oxygen

The conversion of the carbonyl group in pyridones into a leaving group has a very important place in the chemistry of these compounds, the most frequently encountered examples involving reaction with phosphoryl chloride and/or phosphorus pentachloride leading to the chloropyridine, *via* an assumed dichlorophosphate intermediate as indicated below. Conversion into halo derivative can also be conveniently achieved with *N*-bromosuccinimide and triphenylphosphine in refluxing dioxane.[150] Similarly, treatment with a secondary amine and phosphorus pentaoxide, or of 2- or 4-trimethylsilyloxypyridines with secondary amines produces aminopyridines.[142]

The usual way to remove oxygen completely from a pyridone is by conversion, as described, into halogen followed by catalytic hydrogenolysis.[151] Alternatively, reaction of the pyridone salt with 5-chloro-1-phenyltetrazole then hydrogenolysis of the resulting ether can be used.[152]

5.10.2.4   Thio-2-pyridone

Thio-2-pyridone[153] can be converted efficiently into 2-acylthiopyridines by reaction with an acid chloride in the presence of triethylamine; the combination of an acid, triphenylphosphine, and 2,2'-pyridyldisulfide also produces such thioesters.[154] These 2-acylthiopyridines react smoothly with Grignard reagents giving ketones, the thiopyridone anion being the leaving group. 2-Acylthiopyridines have also been used as acyl-transfer reagents to nitrogen, in peptide synthesis,[155] and to oxygen in medium-sized lactone construction.[156]

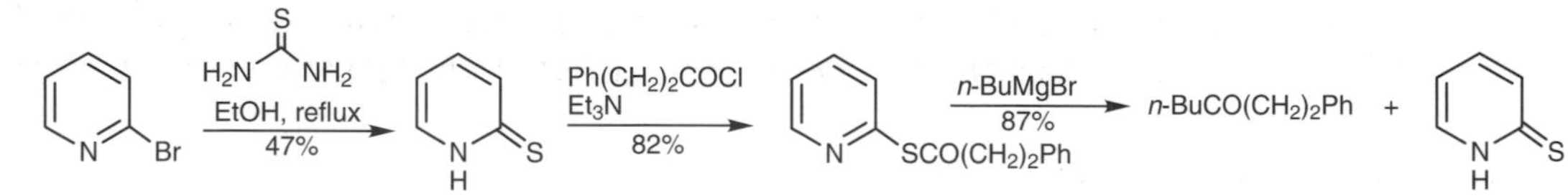

5.10.3 Reactions of aminopyridines

5.10.3.1 Electrophilic addition and substitution

The three aminopyridines are all more basic than pyridine itself and form crystalline salts by protonation at the ring nitrogen. The α- and γ-isomers are monobasic only, because charge delocalisation over both nitrogen atoms, in the manner of an amidinium cation, prevents the addition of a second proton. The effect of the delocalisation is strongest in 4-aminopyridine (pK_a 9.1) and much weaker in 2-aminopyridine (pK_a 7.2). Delocalisation is not possible for the β-isomer which thus can form a di-cation in strong acid (pK_as 6.6 and −1.5).[157]

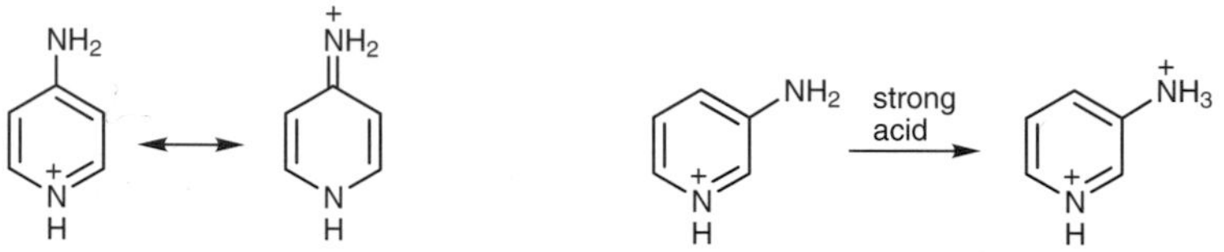

Whereas alkylation, irreversible at room temperature, gives the product of kinetically controlled attack at the most nucleophilic nitrogen, the ring nitrogen,[158] acetylation gives the product of reaction at a side-chain amino group. The acetylaminopyridine which is isolated probably results from side-chain deprotonation of an N-acylpyridinium salt followed by side-chain N-acylation, with loss of the ring-N-acetyl during aqueous work up as shown below.

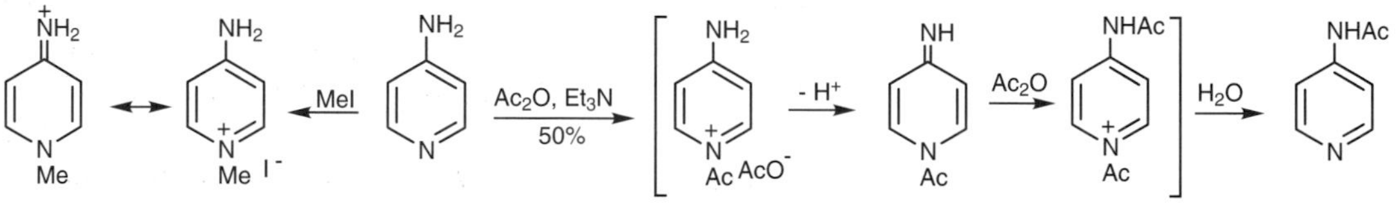

As in benzene chemistry, electron-releasing groups facilitate electrophilic substitution, so that, for example, 2-aminopyridine undergoes 5-bromination in acetic acid even at room temperature; this product can then can be nitrated, at room temperature, forming 2-amino-5-bromo-3-nitropyridine.[159] Chlorination of 3-aminopyridine affords 3-amino-2-chloropyridine.[160] Nitration of aminopyridines in acid solution is also relatively easy, with selective attack of 2- and 4-isomers at β-positions. A study of dialkylaminopyridines showed nitration to take place by attack on the salts.[161]

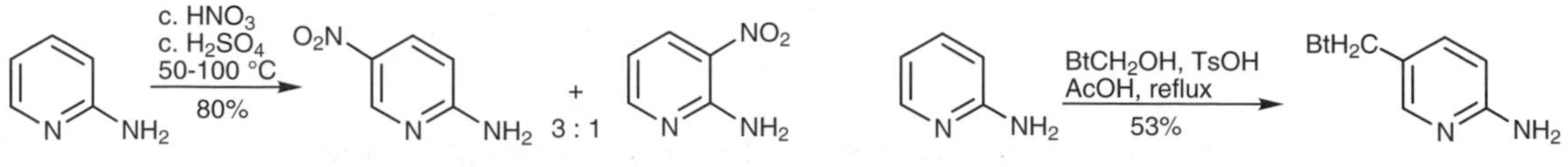

Whereas previously some C-alkylations of aminopyridines had been reported under very vigorous conditions, now a much milder alkylation results from reaction of 2-aminopyridines with 1-hydroxymethylbenzotriazole (section 26.3) in the presence of acid. One must assume that this C-substitution is made possible by a reversible ring-*N*-alkylation.[162] The products of such alkylations display all the characteristics of benzotriazole derivatives for further manipulation (section 26.3).

5.10.3.2 Reactions of the amino group

β-Aminopyridines give normal diazonium salts on reaction with nitrous acid, but with α- and γ-isomers, unless precautions are taken, the corresponding pyridones are then produced *via* easy hydrolysis,[163] water addition at the diazonium-bearing carbon being rapid.[164] With care however, this same susceptibility to nucleophilic displacement can be harnessed in effecting Sandmeyer-type reactions, without the use of copper, of either 2- or 4-aminopyridines.[163,165,166]

5.11 Alkylpyridines

The main feature of the reactivity of alkylpyridines is deprotonation of the alkyl group at the carbon adjacent to the ring.[167] Measurements of side-chain-exchange in methanolic sodium methoxide, 4:2:3, 1800:130:1,[168] and of pK_a values in tetrahydrofuran[169] each have the γ-isomer more acidic than the α-isomer, both being much more acidic than the β-isomer, though the actual carbanion produced in competitive situations can depend on both the counterion and the solvent. Alkyllithiums selectively deprotonate an α-methyl where amide bases produce the more stable γ-anion.[170] The much greater ease of deprotonation[171] of the α- and γ-isomers is related to mesomeric stabilisation of the anion involving the ring nitrogen, not available to the β-isomer for which there is only inductive facilitation, but deprotonation can be effected at a β-methyl under suitable conditions;[172] the difference in acidity between 2- and 3-methyl groups allows selective reaction at the former.[173]

The 'enaminate' anions produced by deprotonating α- and γ-alkylpyridines can participate in a wide range of reactions,[174] being closely analogous to enolate anions. Similar side-chain carbanion formation is seen in *ortho*- but not *meta*-nitrotoluene. Side-chain metallation of 2-*t*-butylcarbonylamino-4-methylpyridine proceeds at room temperature.[175]

In the quaternary salts of alkylpyridines, the side-chain hydrogens are considerably more acidic and condensations can be brought about under quite mild conditions, the reactive species being an enamine;[176] side-chain deprotonation of N-oxides can also be achieved, though it can be complicated by ring deprotonation at C-2.[177]

A further consequence of the stabilisation of carbanionic centres at pyridine α- and γ-positions is the facility with which vinylpyridines,[178] and alkynylpyridines, add nucleophiles, in Michael-like processes (mercury-catalysed hydration goes in the opposite sense[179]). Complimentarily, pyridin-2-yl- and 4-ylethyl esters, sulfides or sulfones can serve as protecting groups, being readily and mildly removed by pyridine nitrogen quaternisation (iodomethane), causing elimination of the vinylpyridinium salt.[180]

In considering reactions of side-chain halides, it is significant that calculations, supported by mass spectroscopic measurements, showed that pyridyl-2-cations are stabilised significantly by overlap with the coplanar nitrogen lone pair.[181]

5.12   Pyridine aldehydes, ketones, carboxylic acids and esters

These compounds all closely resemble the corresponding benzene compounds in their reactivity because the carbonyl group cannot interact mesomerically with the ring nitrogen. The pyridine 2- (picolinic), 3- (nicotinic), and 4- (isonicotinic) acids exist almost entirely in their zwitterionic forms in aqueous solution; they are slightly stronger acids than benzoic acid. Decarboxylation of picolinic acids is relatively easy and results in the transient formation of the same type of ylide which is responsible for specific proton α-exchange of pyridine in acid solution (see section 5.1.2.1).[182] This transient ylide can be trapped by aromatic or aliphatic aldehydes in a reaction known as the Hammick reaction.[183] As implied by this mechanism, quaternary salts of picolinic acids also undergo easy decarboxylation.[184] The process can also be carried out by heating a silyl ester of picolinic acid in the presence of a carbonyl electrophile.[185]

5.13 Quaternary pyridinium salts

The main features of the reactivity of pyridinium salts are (i) the greatly enhanced susceptibility to nucleophilic addition and displacement at the α- and γ-positions, sometimes followed by ring opening and (ii) the easy deprotonation of α- and γ-alkyl groups (see also section 5.11).

5.13.1 Reduction and oxidation

The oxidation of pyridinium salts[186] to pyridones by alkaline ferricyanide is presumed to involve a very small concentration of hydroxide adduct. 3-Substituted pyridinium ions are transformed into mixtures of 2- and 6-pyridones, for example oxidation of 1,3-dimethylpyridinium iodide gives a 9:1 ratio of 2- and 6-pyridones.

Catalytic reduction of pyridinium salts to piperidines is particularly easy; they are also susceptible to hydride addition by complex metal hydrides[187] or formate,[188] and lithium/ammonia reduction.[189] In the reduction with sodium borohydride in protic media the main product is a tetrahydro-derivative with the double bond at the allylic, 3,4-position. These cyclic allylamines are formed by initial hydride addition at C-2, followed by enamine β-protonation and a second hydride addition. Some fully reduced material is always produced and its relative percentage increases with increasing N-substitutent bulk, consistent with a competing sequence having initial attack at C-4, generating a dienamine which can then undergo two successive proton-then-hydride addition steps. When 3-substituted pyridinium salts are reduced with sodium borohydride, 3-substituted-1,2,5,6-tetrahydropyridines result. Care must be taken to destroy amine-borane which can be present at the end of such reductions.[190] When 1,4-dihydro-1-methylpyridine and 1,2-dihydro-1-methylpyridine are equilibrated using strong base, the former predominates to the extent of approximately 9:1.[191]

N-Acyl, particularly *N*-alkoxy- or *N*-aryloxycarbonylpyridiniums can be reductively trapped as dihydro-derivatives by borohydride;[192] no further reduction occurs because the immediate product is an enamide and not an enamine and therefore does not protonate under the conditions of the reduction.[193] The 1,2-dihydro-isomers, which can be produced essentially exclusively by reduction at $-70\,^{\circ}$C in methanol, serve as dienes in Diels-Alder reactions. Irradiation causes conversion into 2-azabicyclo[2.2.0]hexenes; removal of the carbamate and *N*-alkylation gives derivatives which are synthons for unstable *N*-alkyldihydropyridines, and convertible into the latter thermally.[194]

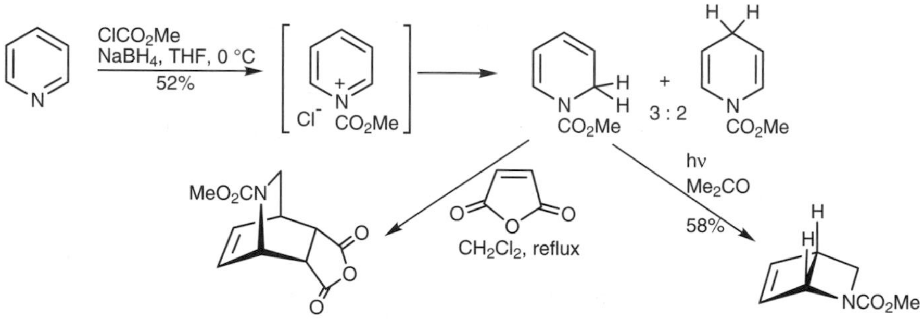

The easy specific reduction of 3-acylpyridinium salts giving stable 3-acyl-1,4-dihydropyridines using sodium dithionite is often quoted, because of its perceived relevance to nicotinamide coenzyme activity; the mechanism involves addition of sulfur at C-4 as its first step, as shown below.[195] 1,4-Dihydropyridines are normally air-sensitive, easily rearomatised molecules; the stability of 3-acyl-1,4-dihydropyridines is related to the conjugation between ring nitrogen and side-chain carbonyl group (see also Hantzsch synthesis, section 5.15.1.2). However, even simple pyridinium salts, provided the *N*-substituent is larger than propyl, or for example benzyl, can be reduced to 1,4-dihydropyridines with sodium dithionite.[196]

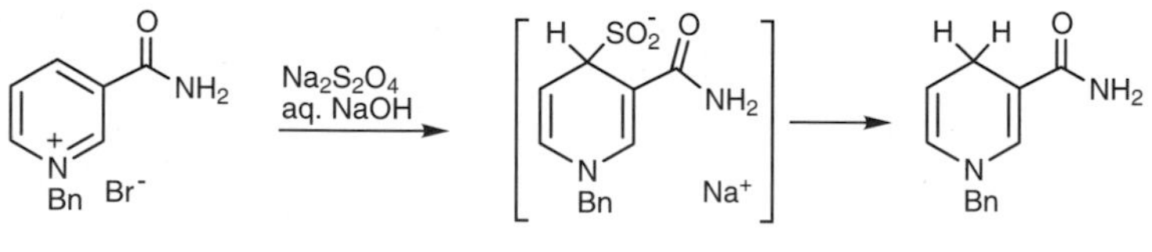

5.13.2   Organometallic addition

Organometallic reagents add very readily to *N*-alkyl-, *N*-aryl- and with important synthetic significance, *N*-acylpyridinium salts. In the simplest cases, addition is to an α-carbon; the resulting 2-substituted-1,2-dihydropyridine can be handled and spectroscopically identified, with care, but more importantly can be easily oxidised to a 2-substituted pyridinium salt.[197]

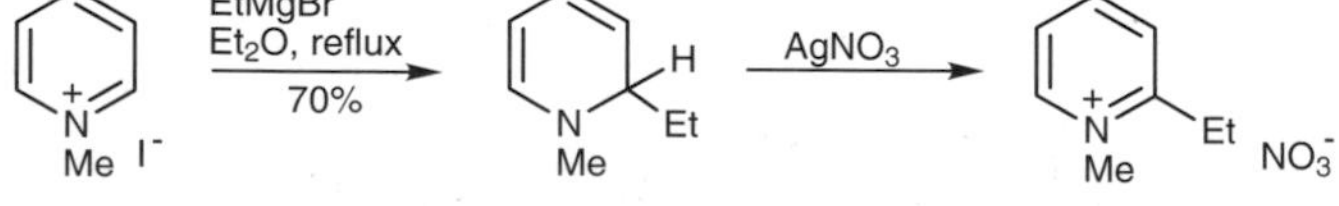

The great significance of the later discovery, that exactly comparable additions to *N*-acylpyridinium cations, generated and reacted *in situ*, is that the dihydropyridines which result can be further manipulated if required and that during rearomatisation the *N*-substituent can be easily removed to give a substituted pyridine. It is worth noting the contrast to the use of *N*-acylpyridinium salts for reaction with alcohol, amine nucleophiles (section 5.1.1.7) when attack is at the carbonyl carbon; the use of an *N*-alkoxycarbonyl pyridinium salt in the present context aids this discrimination.

Generally, organometallic addition to *N*-alkoxy- or *N*-aryloxycarbonylpyridinium salts[192] takes place at both 2- and 4-positions,[198] however higher selectivity for the 4-position can be achieved using copper reagents.[199] Indole as the neutral molecule, reacts with *N*-benzoylpyridinium chloride at C-4,[200] but its anion will add to *N*-methylpyridinium salts having acyl groups at C-3 either at C-6 or at C-4 depending on the solvent.[201] High selectivity for the 2-position is found in the addition of phenyl,[202] alkenyl and alkynyl organometallics,[203] including ethoxycarbonyl-methyl[204] and alkynyl[205] tin reagents.

Examples of the further manipulation of dihydropyridines produced by the methods described above include introduction of substituents at a β-position, by acylation of the enamide,[150] and at an α-position, *via* 2-lithiation, each of which is illustrated below.[150]

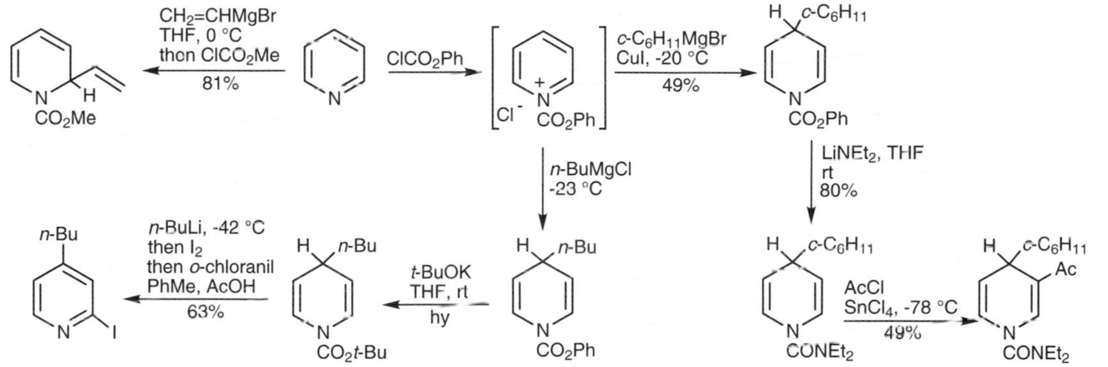

Silylation at nitrogen with *t*-butyldimethylsilyl triflate, generates pyridinium salts which, because of the size of the *N*-substitutent, react with Grignard reagents exclusively at C-4;[206] montmorillonite-catalysed addition of silyl enol ethers to pyridines has a comparable effect in producing 1-trimethylsilyl-1,4-dihydropyridines carrying an acylalkyl substituent at C-4.[207]

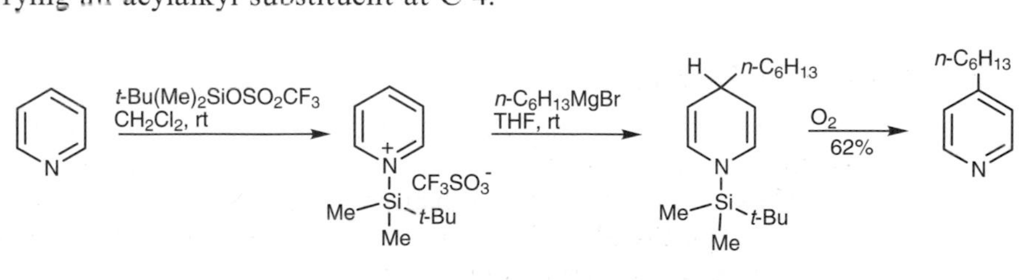

4-Substituents direct attack to an α-carbon;[208,209] the use of a removable 4-blocking group – trimethyltin in the example below - can be made the means for the production of 2-substituted isomers.[210]

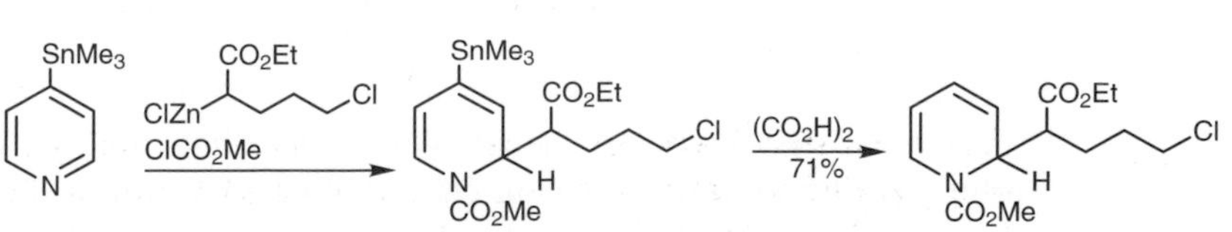

The use of chiral chloroformates such as that derived from *trans*-2-(α-cumyl)cyclohexanol allows diastereoselective additions to 4-methoxypyridine. The introduction of a tri-*i*-propylsilyl group at C-3 greatly enhances the diastereoselectivity. The products of these reactions are multifunctional chiral piperidines which have found use in the asymmetric synthesis of natural products.[211]

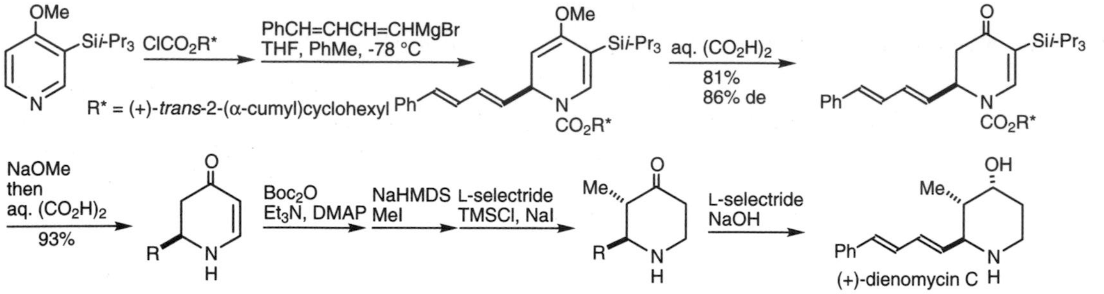

Some nucleophiles add to *N*-fluoropyridinium salts to give dihydropyridines in which elimination of fluoride occurs *in situ* to give the 2-substituted pyridine, thus avoiding the need for a dehydrogenation step. The main disadvantages of this method are that the preparation of the pyridinium salts require the use of elemental fluorine and that some carbanions give only modest yields due to competitive reactions such as C-fluorination. However, silyl enol ethers do react efficiently; stabilised heteronucleophiles (phenolate, azide) can also be used. Addition to *N*-fluoropyridinium salts shows a strong preference for attack at an α-position.[212]

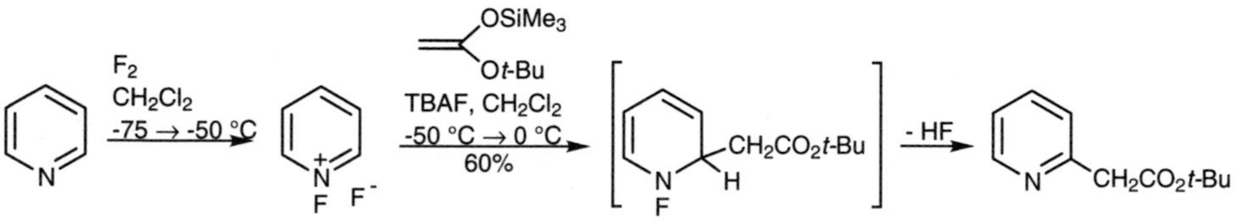

In a similar way, pyridine phosphonium salts and phosphonates can be prepared by reaction of trivalent phosphorus compounds with the more accessible *N*-trifluoromethanesulfonyl pyridines, when trifluoromethansulfinate is the leaving group from nitrogen; attack is normally at C-4 as illustrated below.[213]

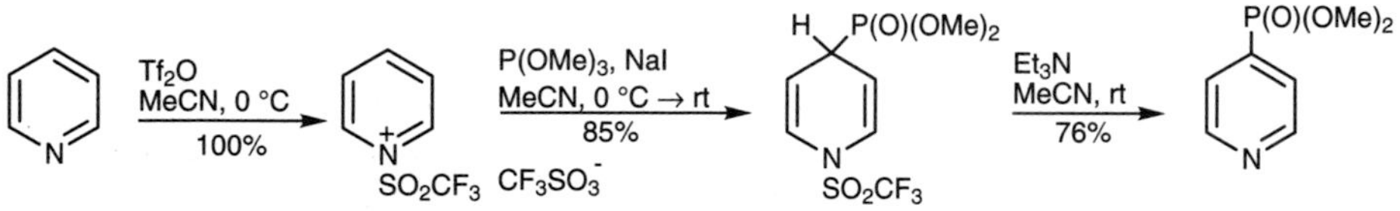

5.13.3   Other nucleophilic additions

There are a variety of examples of other nucleophiles adding to *N*-alkylpyridinium salts. A study[214] of reversible additions to 3-cyano-1-methylpyridinium iodide showed α-attack to be kinetically favoured but the γ-adduct to be the more thermodynamically stable. Similarly, in thermodynamically-controlled processes, 1-methyl-3-nitropyridinium gives products resulting from addition at C-4 in which again there is stabilising conjugation between ring nitrogen and 3-substituent.[215] Products of γ-addition, even in 1-methyl- or -phenylpyridinium iodides, lacking a conjugating 3-substituent, can be trapped via attack by added oxidant as illustrated.[216]

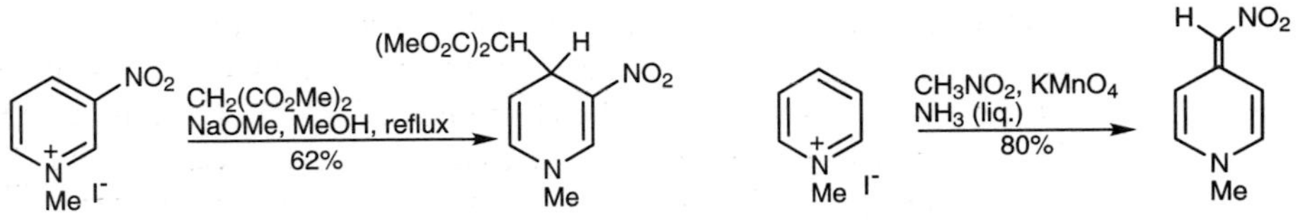

NO2
Me I-
CH2(CO2Me)2
NaOMe, MeOH, reflux
62%
(MeO2C)2CH
H
NO2
Me
Me I-
CH3NO2, KMnO4
NH3 (liq.)
80%
H
NO2
Me

5.13.4 Nucleophilic addition followed by ring opening[217]

There are many examples of pyridinium salts, particularly, but not exclusively, those with powerful electron-withdrawing *N*-substituents, adding a nucleophile at C-2 and then undergoing a ring opening. Perhaps the classic example is addition of hydroxide to the pyridine sulfur trioxide complex, which produces the sodium salt of glutaconaldehyde as shown below.[218]

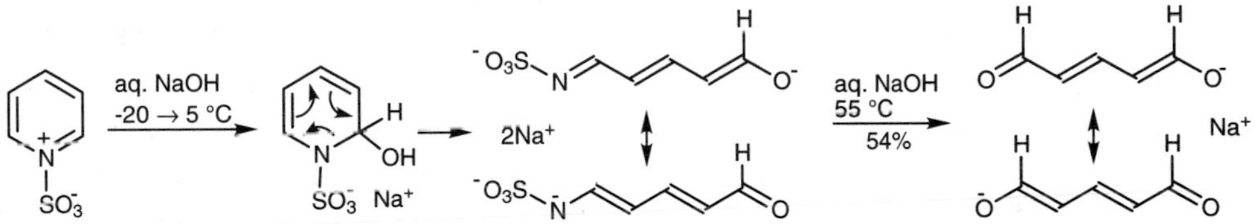

SO3-
aq. NaOH
-20 → 5 °C
H
OH
SO3- Na+
2Na+
-O3S
aq. NaOH
55 °C
54%
Na+

Another well known example is a synthesis of azulene which utilises the bis dimethylamine derivative of glutaconaldehyde produced with loss of 2,4-dinitroaniline from 1-(2,4-dinitrophenyl)pyridinium chloride (Zincke's salt).[219]

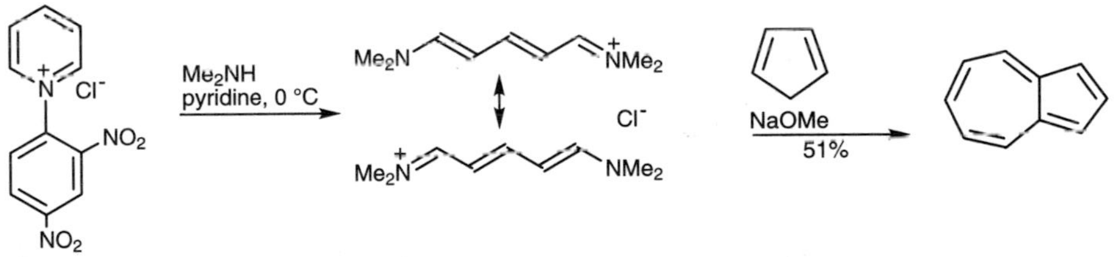

Cl-
NO2
NO2
Me2NH
pyridine, 0 °C
Me2N
NMe2
Cl-
NaOMe
51%

The reaction of such pyridinium salts with primary amines, including amino acid esters is a useful synthesis of chirally-*N*-substituted pyridinium salts.[220]

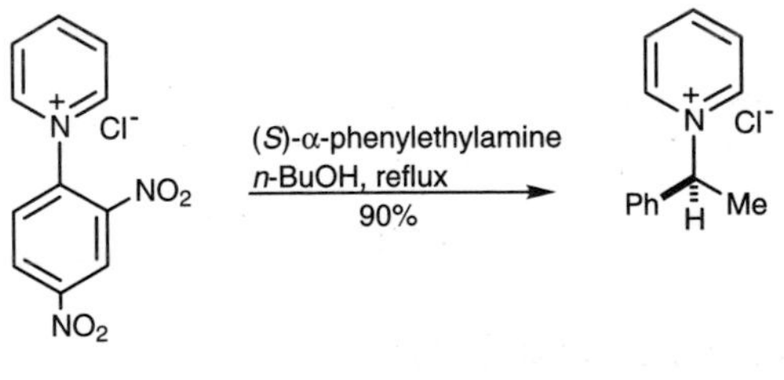

Cl-
NO2
NO2
(S)-α-phenylethylamine
n-BuOH, reflux
90%
Cl-
Ph
H
Me

As a final example of nucleophilic addition then ring opening, it has even been possible to isolate the ring-opened 'hydrate' of pyridine, by reaction with benzaldehyde and benzoyl chloride, as shown below.[221]

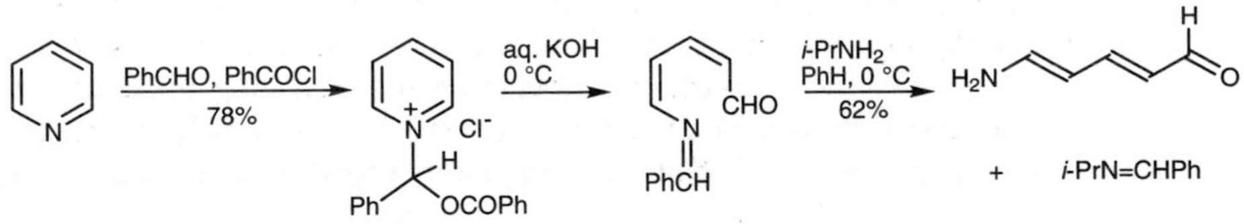

PhCHO, PhCOCl
78%
Cl-
H
Ph
OCOPh
aq. KOH
0 °C
CHO
PhCH
i-PrNH2
PhH, 0 °C
62%
H2N
H
+ i-PrN=CHPh

5.13.5   Cyclisations involving an α-position or an α-substituent

It is often possible to achieve cyclisation of pyridinium salts, in which the ring closure involves an α-substituent or the electrophilic nature of the α-position (see also section 5.1.1.8) and gives a neutral product – sections 25.1.2, 25.2.1, 25.2.2, and 25.2.3 give examples.

5.13.6   N-Dealkylation

The conversion of N-alkyl- or -arylpyridinium salts into the corresponding pyridine, i.e. the removal of the N-substitutent, is generally not an easy process, however triphenylphosphine[222] or simply heating the iodide salt[223] can work for metho-salts. 1-Triphenylmethyl-4-dimethylaminopyridinium chloride[224] and 1-trialkylsilylpyridinium triflates[225] are isolable and relatively stable salts; O-tritylations and O-silylations involving transfer of trityl or trialkylsilyl from the positively charged nitrogen in such salts are usually carried out without isolation using mixtures of 4-dimethylaminopyridine (DMAP) with chlorotriphenylmethane or, for example, chloro-t-butyldimethylsilane.[226]

Pyridinium salts corresponding to 2,4,6-trisubstituted pyridines, which must be prepared by reacting a primary amine with 2,4,6-trisubstituted pyrylium perchlorate (see section 8.1.2.2) are attacked by a variety of nucleophiles with transfer of the N-substituent to the attacking reagent and as such are convenient alkylating agents,[227] and, recalling that the precursor to the pyridinium salt is the primary amine, the sequence also represents the overall transformation of a primary amine into a variety of derivatives.

5.14   Pyridine N-oxides[228]

The reactions of pyridine N-oxides are of great interest,[229] differing significantly from those of both neutral pyridines and pyridinium salts.

A striking difference between pyridines and their N-oxides is the much greater susceptibility of the latter to electrophilic nitration. This can be understood in terms of mesomeric release from the oxide oxygen, and is parallel to electron release by oxygen and hence increased reactivity towards electrophilic substitution in phenols and phenoxides. One can find support for this rationalisation by a comparison of the dipole moments of trimethylamine and its N-oxide, on the one hand, and pyridine and its N-oxide, on the other: the difference of 2.03 D for the latter pair is much smaller than the 4.37 D found for the former. The smaller difference signals significant contributions from those canonical forms in which the oxygen is neutral and the ring negatively charged. Clearly, however, the situation is subtle, as those contributors carrying formal positive charges on α- and γ-carbons suggest a polarisation in the opposite sense and thus an increased susceptibility to nucleophilic attack too, compared with the neutral pyridine, and this is indeed found to be the case. Summarising: the *N*-oxide function in pyridine *N*-oxides serves to facilitate, on demand, both electrophilic and nucleophilic addition to the α- and γ-positions.

Many methods are available for the removal of oxygen from *N*-oxides, samarium iodide, chromous chloride, stannous chloride with low-valent titanium, ammonium formate with palladium, and catalytic hydrogenation all do the job at room temperature.[230] The most frequently used methods have involved oxygen transfer to trivalent phosphorus[228] or divalent sulfur.[231]

5.14.1 Electrophilic addition and substitution

Pyridine *N*-oxides protonate and are alkylated at oxygen; stable salts can be isolated in some cases.[232] Hot aqueous sodium hydroxide treatment of alkoxypyridinium salts produces aldehydes corresponding to the alkoxy substituent.[233]

Electrophilic nitration and bromination of pyridine *N*-oxides can be controlled to give 4-substituted products[234] by way of attack on the free *N*-oxide.[235] Under conditions where the *N*-oxide is *O*-protonated, substitution follows the typical pyridine/pyridinium reactivity pattern thus, in fuming sulfuric acid, bromination shows β-regioselectivity,[236] mercuration, however, takes place at the α-position.[237]

5.14.2 Nucleophilic addition and substitution

The *N*-oxide function enhances the rate of nucleophilic displacement of halogen from α- and γ-positions. The relative rates 4 > 2 > 3 found for pyridines are echoed for the *N*-oxides, but interestingly altered to 2 > 4 > 3 in methiodides.[238]

Grignard reagents add to pyridine *N*-oxide forming adducts, which can be characterised from a low temperature reaction, but which at room temperature undergo disrotatory ring opening, the isolated product being an acyclic, unsaturated oxime. Heating with acetic anhydride brings about rearomatisation, *via* electrocyclic ring closure rendered irreversible by the loss of acetic acid.[239]

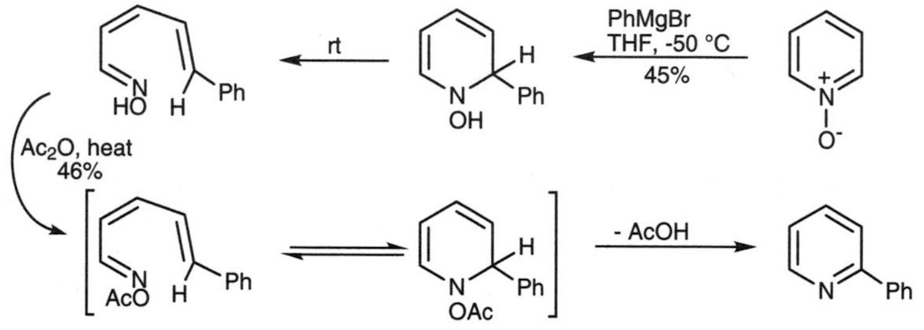

Comparable addition/ring openings can be observed with 1-alkoxypyridiniums,[240,241] however prior acylation at the *N*-oxide oxygen before addition of alkyl or aryl Grignard or acetylide leads through to 2-substituted-pyridines.[241,242]

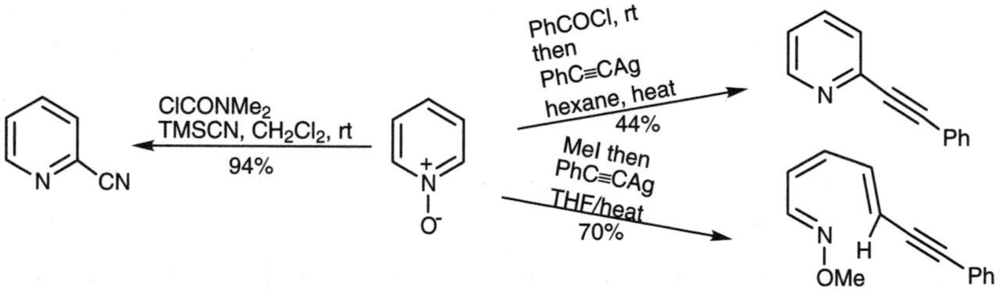

Very clean conversions of pyridine *N*-oxide into 2-cyanopyridine depend on prior conversion of oxide into silyloxy or carbamate,[243] and displace earlier methods which utilised *N*-alkoxypyridinium salts.[244]

5.14.3 Rearrangements

A range of synthetically useful rearrangements convert pyridine *N*-oxides into variously substituted pyridines in which an α-(γ-)position, or an α-substitutent has been modified.

2-Methylpyridine *N*-oxides react with hot acetic anhydride and produce 2-acetoxymethylpyridines; using trifluoroacetic anhydride permits reaction at room temperature with fewer by-products.[245] Repetition of the sequence affords 2-aldehydes after hydrolysis.[246] The course[247] of the rearrangement would seem to be most simply explained by invoking an electrocyclic sequence, as shown below.

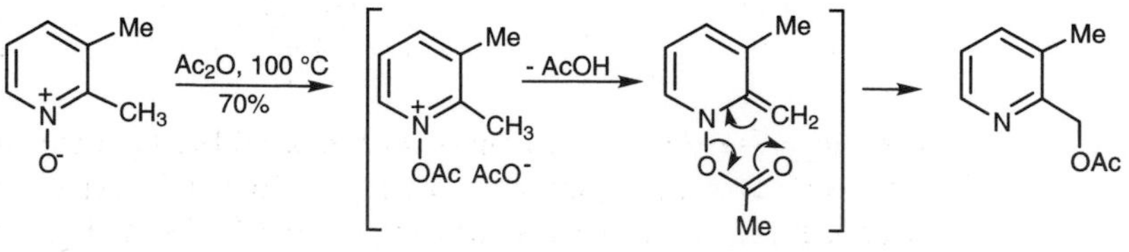

In the absence of a 2-substituent, reaction with thionyl chloride or with acetic anhydride leads to the formation of 2- and 4-chloro- or 2-acetoxypyridines. Mechanistically, electrophilic addition to oxide is followed by nucleophilic addition to an α- or γ-position, the process being completed by an elimination.[228] The sequence below shows an example and also illustrates other aspects of *N*-oxide chemistry discussed above.[248]

5.15   Synthesis of pyridines

5.15.1   Ring synthesis

There are very many ways of achieving the synthesis of a pyridine ring; the following section describes the main general methods.

5.15.1.1   *From 1,5-dicarbonyl compounds and ammonia*

Ammonia reacts with 1,5-dicarbonyl compounds to give 1,4-dihydropyridines which are easily dehydrogenated to pyridines. With unsaturated 1,5-dicarbonyl compounds, or their equivalents (e.g. pyrylium ions) ammonia reacts to give pyridines directly.

1,5-Diketones are accessible *via* a number routes, for example by Michael addition of enolate to enone (or precursor Mannich base[249]), by ozonolysis of a cyclopentene precursor, or by reaction of silyl enol ethers with 3-methoxyallylic alcohols.[250] They react with ammonia, with loss of two mol equivalents of water to produce a cyclic bis-enamine, i.e. a 1,4-dihydropyridine, which is generally unstable but can be easily and efficiently dehydrogenated to the aromatic heterocycle.

The oxidative final step can be neatly avoided by the use of hydroxylamine,[251] instead of ammonia, when a final 1,4-loss of water produces the aromatic heterocycle. In an extension of this concept, the construction of a 1,5-diketone equivalent by tandem Michael addition of dimethylhydrazone anion to an enone, then acylation, has loss of dimethylamine as the final aromatisation step.[252]

It follows, that the use of an unsaturated 1,5-dicarbonyl compound will also afford aromatic pyridine directly; a number of methods are available for the assembly of the unsaturated diketone, including the use of pyrylium ions or 2-pyrones[253] (see chapter 8) as synthons, or the alkylation of an enolate with a 3,3-bis(methylthio)-enone.[254]

When one of the carbonyl carbons is at the oxidation level of acid (as in a 2-pyrone, section 8.2) then the product, reflecting this oxidation level, is a 2-pyridone.[255] Similarly, 4-pyrones (section 8.2) react with ammonia or primary amines to give 4-pyridones[256] and similarly the bis-enamines which can be obtained directly from ketones by condensation on both sides of the carbonyl group with dimethylformamide dimethylacetal, produce 4-pyridones on reaction with primary amines.[257] When one of the the 'carbonyl' units is actually a nitrile, then an aminopyridine results.[258]

5.15.1.2 *From an aldehyde, two equivalents of a 1,3-dicarbonyl compound, and ammonia*

Symmetrical 1,4-dihydropyridines, which can be easily dehydrogenated, are produced from the interaction of ammonia, an aldehyde, and two equivalents of a 1,3-dicarbonyl compound which must have a central methylene.

The Hantzsch synthesis[259]

The product from the classical Hantzsch synthesis is necessarily a symmetrically substituted 1,4-dihydropyridine since two mol equivalents of the one dicarbonyl component are utilised, the aldehyde carbonyl carbon becoming the pyridine C-4. The precise sequence of intermediate steps is not known for certain, and may indeed vary from case to case, for example the ammonia may become involved early or late, but a reasonable sequence would be aldol condensation followed by Michael addition generating, *in situ*, a 1,5-dicarbonyl compound.

The 1,4-dihydropyridines produced in this approach, carrying conjugating substituents at each β-position, are stable, and can be easily isolated before dehydrogenation; classically the oxidation has been achieved with nitric acid, or nitrous acid, but other oxidants such as cerium(IV) ammonium nitrate, copper(II) nitrate on montmorillonite, and manganese dioxide on bentonite also all achieve this objective smoothly.[260]

Unsymmetrical 1,4-dihydropyridines can be produced by conducting the Hantzsch synthesis in two stages, i.e. by making the (presumed) aldol condensation product separately, then reacting with ammonia and a different 1,3-dicarbonyl component, or an enaminoketone, in a second step.[261]

5.15.1.3 *From 1,3-dicarbonyl compounds and 3-amino-enones or -nitriles*

Pyridines are formed from the interaction between a 1,3-dicarbonyl compound and a 3-amino-enone or 3-aminoacrylate; 3-cyano-2-pyridones result if cyanoacetamide is used instead of an amino-enone.

This approach, in its various forms, is probably the most versatile and useful since it allows the construction of unsymmetrically substituted pyridines from relatively simple precursors. Again, in this pyridine ring construction, intermediates are not isolated and it is difficult to be sure of the exact sequence of events.

3-Amino-enones or 3-amino-acrylates can be prepared by the straightforward reaction of ammonia with a 1,3-diketone or a 1,3-keto-ester. The simplest 1,3-dicarbonyl compound, malondialdehyde, is too unstable to be useful, but its readily available acetal enol ether can be used instead, as shown below.[262] Vinamidinium $(R_2NCH=CR'CH=N^+R_2)$ salts will serve as synthons for substituted malondialdehydes or unsaturated keto-aldehydes in these syntheses.[263]

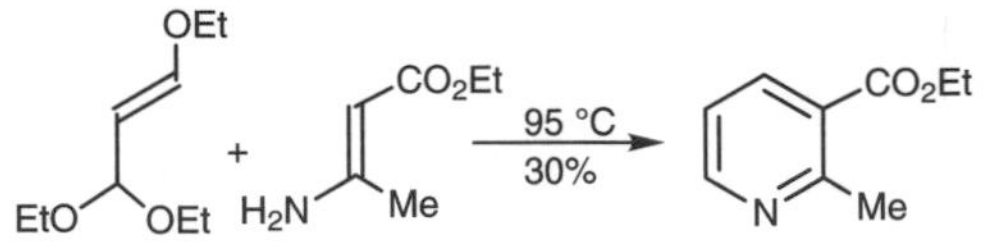

The Guareschi synthesis

The variation which makes use of cyanoacetamide as the nitrogen-containing component leads to 3-cyano-2-pyridones, from which the carbonyl group and/or the cyano group can be subsequently removed.

Providing the two carbonyl groups are sufficiently different in reactivity, only one of the two possible isomeric pyridine/pyridone products is formed *via* reaction of the more electrophilic carbonyl group with the central carbon of the 3-amino-enone, 3-aminoacrylate, or cyanoacetamide.[264,265]

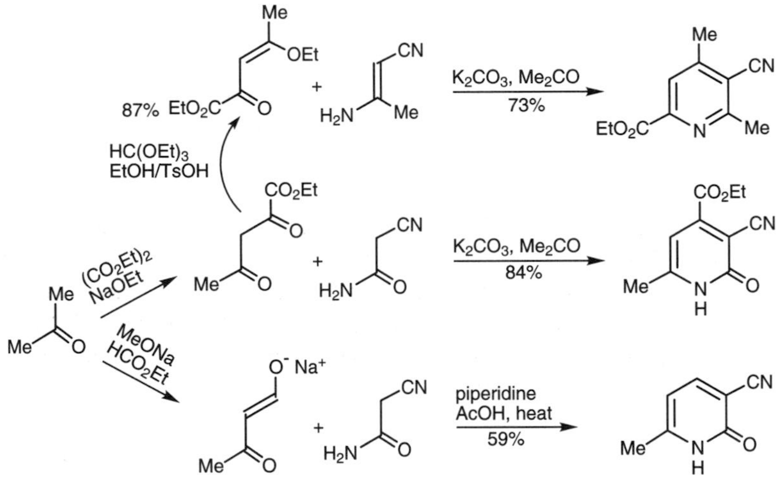

Variations include the use of yne-ones, when conjugate addition of the cyanoacetamide controls the regiochemistry of reaction,[266] and 3-alkoxy-enones (i.e. the enol ethers of 1,3-diketones) when comparably, the initial Michael-type interaction dictates the regiochemistry.[264,267] Using $H_2NCOCH_2C(NH_2)=N^+H_2\ Cl^-$ instead of cyanoacetamide gives 2-aminopyridine-3-carboxamides[268] and using $H_2NCOCH_2NO_2$ instead of cyanoacetamide produces 3-nitro-2-pyridones.[269]

Successful ring closure to produce pyridines and pyridones can also be carried out with starting materials at a lower oxidation level, with *in situ* dehydrogenation by air or added oxygen, *i.e.* instead of using a 1,3-dicarbonyl component, an α,β-unsaturated ketone/aldehyde is employed, as illustrated below.[270]

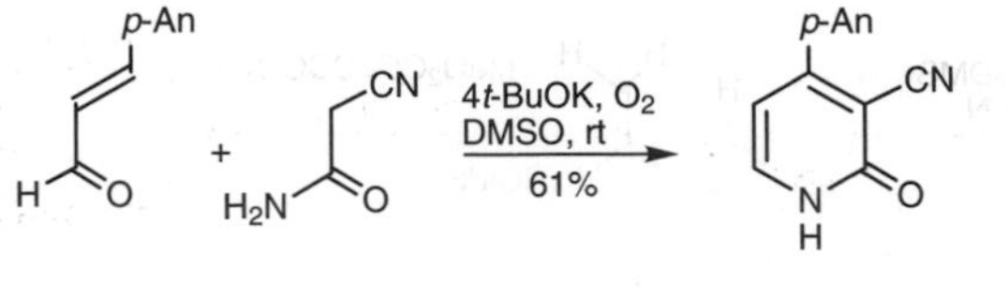

5.15.1.4　By cycloadditions

Various electrocyclic additions, with subsequent extrusion of a small molecule have been used to construct pyridines: addition to oxazoles is one of these.

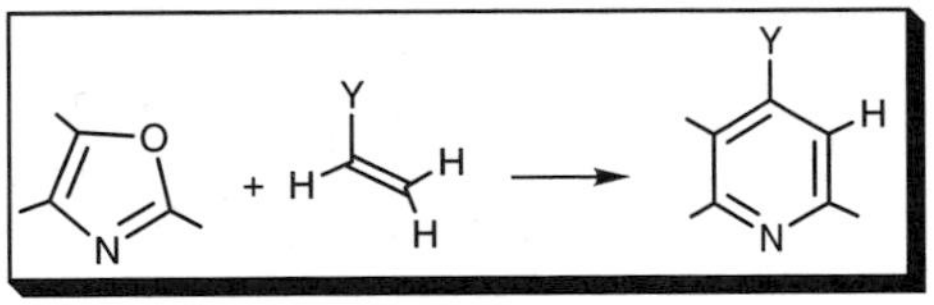

A number of 6π cycloadditions, some with inverse electron-demand, have been developed into useful means for the construction of pyridines. Historically, the first of these was the addition of a dienophile to an oxazole; sometimes the oxazole oxygen is retained (giving 3-hydroxypyridines) and sometimes it is lost, as illustrated below.[271]

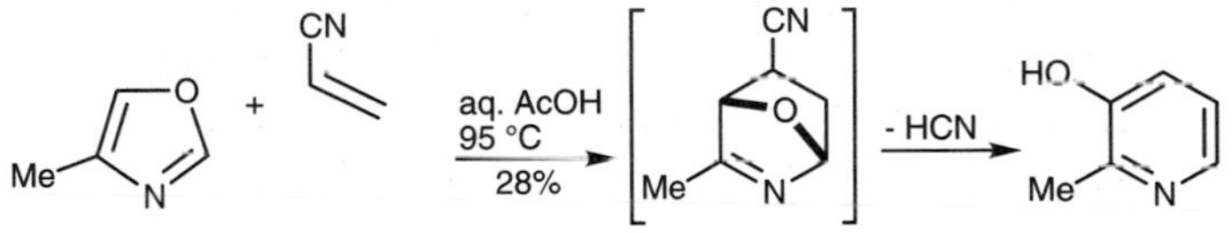

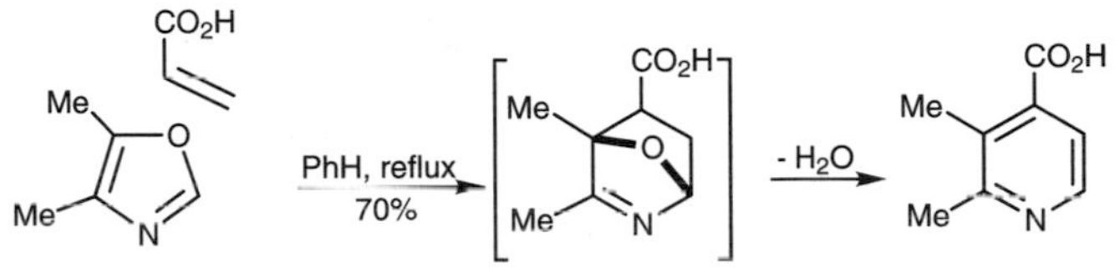

1,2,3-[272] and 1,2,4-Triazines, acting as inverse electron-demand azadienes, add to enamines and thus, following extrusion of nitrogen and loss of amine, a pyridine is produced (see section 25.2.1).[273] 1,2,4-Triazines will also react with other dienophiles: reaction with ethynyltributyltin for example gives 4-stannylpyridines;[274] norbornadiene is useful as an acetylene equivalent;[275] oxazinones can also be used as the 'diene' component.[276]

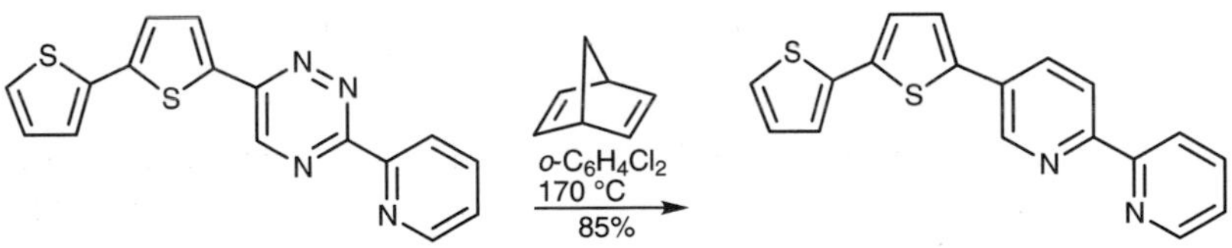

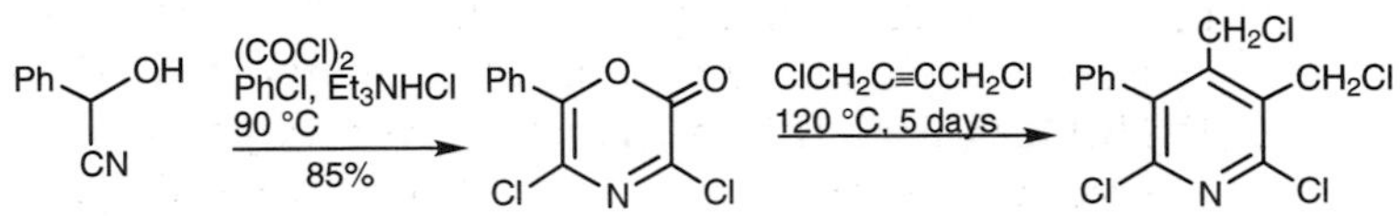

The O,O'-bis-t-butyldimethylsilyl derivative of an imide serves as an azadiene in reaction with dienophiles; 2-pyridones are the result, following desilylation.[277]

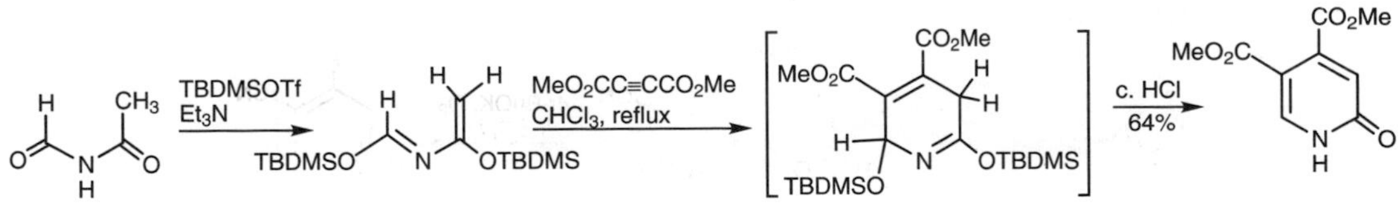

5.15.1.5 By thermal electrocyclisations
From 1-aza-1,3,5-trienes

Electrocyclisation of 1-aza-1,3,5-trienes generates dihydropyridines which can be oxidised to pyridines. If an oxime or hydrazine derivative is used, elimination of water or an amine *in situ* gives the pyridine directly. This method is particularly useful for fusion of pyridines to other ring systems and is illustrated by the examples below.[278]

X = H, OR, NR$_2$

5.15.1.6 From furans

2-Furfurylamines, can be converted *via* ring-opening ring-closure sequences, for example through 2,5-dimethoxy-2,5-dihydrofurans, into 3-hydroxypyridines.

Ring-opening and reclosure processes using furans include several significant methods for the construction of pyridines. 2,5-Dihydro-2,5-dimethoxyfurans (see section 15.1.4) carrying as side-chain an aminoalkyl group, give rise to 3-hydroxypyridines.[279]

Furfurylamines react with formaldehyde, directly,[280] (or with an aromatic aldehyde *via* 5-lithiation after *N*-protection[281]) to give 3-hydroxy-6-substituted pyridines.

5.15.1.7 Miscellaneous methods

Many alkylpyridines are manufactured commercially by chemically complex processes which often produce them as mixtures. A good example is the extraordinary *Chichibabin synthesis*, in which paraldehyde and ammonium hydroxide react together at 230 °C under pressure to afford 52% of 5-ethyl-2-methylpyridine; so here, four mol equivalents of acetaldehyde and one of ammonia combine.[282] Also of commercial significance is the cobalt-catalysed interaction of a nitrile and acetylene.[283]

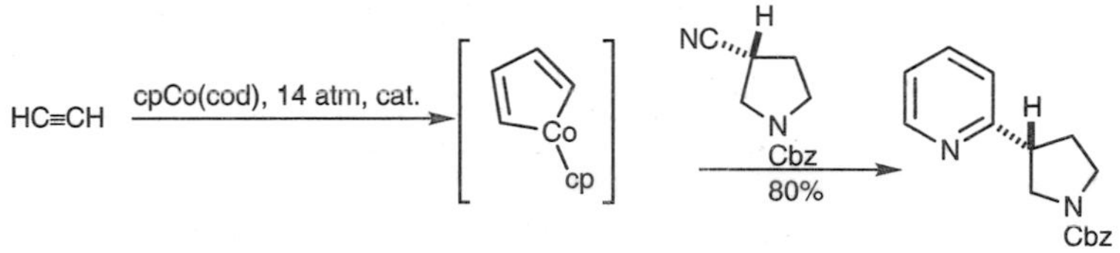

5.15.2 Examples of notable syntheses of pyridine compounds

5.15.2.1 Fusarinic acid

Fusarinic acid is a mould metabolite with antibiotic and antihypertensive activity. Two syntheses of this substance employ cycloadditions, the earlier[284] as a means to produce a 1,5-diketone, and the second[285] to generate a 1-dimethylamino-1,4-dihydropyridine.

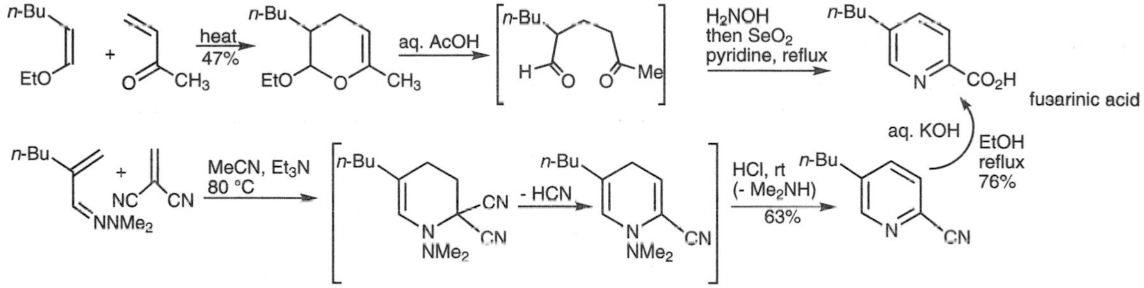

5.15.2.2 Pyridoxine

Pyridoxine, vitamin B_6, has been synthesised by several routes, including one which utilises a Guareschi ring synthesis, as shown below.[286]

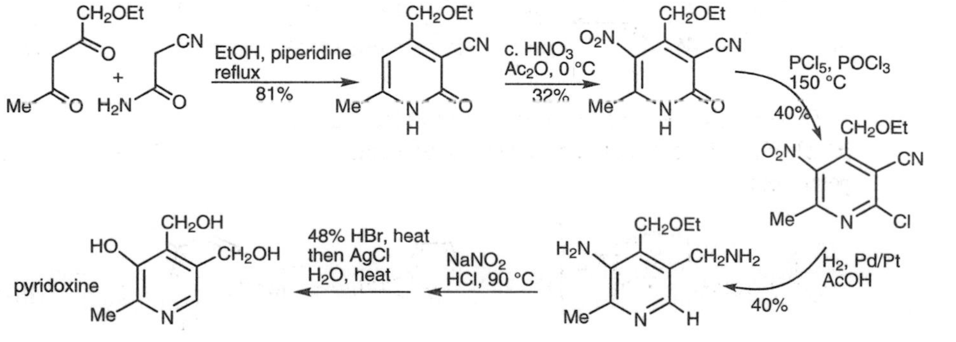

5.15.2.3 2-Methoxy-4-methyl-5-nitropyridine

2-Methoxy-4-methyl-5-nitropyridine is an intermediate used in a synthesis of porphobilinogen (section 13.18.3.1).

5.15.2.4 Nemertelline

The total synthesis of nemertelline, a hoploemertin worm toxin, illustrates the use of metallation and palladium-catalysed couplings.[287]

Exercises for chapter 5

Straightforward revision exercises (consult chapters 4 and 5)

(a) In what way does pyridine react with electrophilic reagents such as acids and alkyl halides?

(b) What factors make it much more difficult to bring about electrophilic substitution of pyridine than benzene?

(c) How do pyridines compare with benzenes with regard to (i) oxidative destruction of the ring and (ii) reduction of the ring?

(d) Give two examples of pyridines reacting with nucleophilic reagents with substitution of a hydrogen.

(e) What are the relative reactivities of bromobenzene, 2-bromopyridine, 3-bromopyridine towards replacement of the halide with ethoxide on treatment with NaOEt?

(f) How could one generate 2-lithiopyridine?

(g) What would result from treatment of 3-chloropyridine with LDA at low temperature?

(h) Draw the main tautomeric forms of 2-hydroxypyridine (2-pyridone), 3-hydroxypyridine and 2-aminopyridine.

(i) How could one convert 4-pyridone cleanly into 1-ethyl-4-pyridone?

(j) What would be the result of treating a 1 : 1 mixture of 2- and 3-methylpyridines with 0.5 equivalents of LDA and then 0.5 equivalents of MeI?

(k) Draw the structure of the product(s) you would expect to be formed if pyridine were reacted successively with methyl chloroformate and then phenyllithium.

(l) In pyridine N-oxides, both electrophilic substitution and nucleophilic displacement of halide from C-4 go more rapidly than in pyridine – explain.

(m) Describe two important methods for the synthesis of pyridines from precursors which do not contain the ring.

(n) What compounds would result from the following reagent combinations: (i) H_2NCOCH_2CN (cyanoacetamide) with $MeCOCH_2COMe$; (ii) $MeC(NH_2)=CHCO_2Et$ (ethyl 3-aminocrotonate) with $MeCOCH_2COMe$; (iii) $PhCH=O$, $MeCOCH_2COMe$, and NH_3?

More advanced exercises

1. Suggest a structure for the products: (i) $C_7H_8N_2O_3$ produced by treating 3-ethoxypyridine with f. HNO_3/c. H_2SO_4 at $100\,°C$, (ii) $C_6H_4BrNO_2$ produced by reaction of 4-methylpyridine first with Br_2/H_2SO_4/oleum then with hot $KMnO_4$.

2. Deduce a structure for the product $C_9H_{15}N_3$ produced by reacting pyridine with the potassium salt of $Me_2N(CH_2)_2NH_2$.

3. Deduce structures for the product formed by (i) reacting 2-chloropyridine with (a) hydrazine → $C_5H_7N_3$, (b) water → C_5H_5NO; (ii) 4-nitropyridine heated with water at $60\,°C$ → C_5H_5NO.

4. Deduce structures for the products formed in turn by reacting 4-chloropyridine with (i) sodium methoxide → C_6H_7NO, A, this with iodomethane → $C_7H_{10}INO$, then this heated at $185\,°C$ → C_6H_7NO, isomeric with A.

5. Treatment of 4-bromopyridine with $NaNH_2$ in NH_3 (liq) gives two products (isomers, $C_5H_6N_2$) but reaction with sodium methoxide gives a single product, C_6H_7NO. What are the products and why is there a difference?

6. Write structures for the products to be expected in the following sequences: (i) 4-diisopropylaminocarbonyl pyridine with LDA then with benzophenone, then with hot acid → $C_{19}H_{13}NO_2$; (ii) 2-chloropyridine with LDA then iodine → C_5H_3ClNI; (iii) 3-fluoropyridine with LDA then with acetone → $C_8H_{10}FNO$; (iv) 2-bromopyridine with butyllithium at $-78\,°C$ then chlorotrimethylstannane → $C_8H_{13}NSn$.

7. A crystalline solid $C_9H_{11}BrN_2O_3$ is formed when 2-methyl-5-nitropyridine is reacted with bromoacetone, subsequent treatment with $NaHCO_3$ affords $C_9H_8N_2O_2$ – deduce the structures and write out a mechanism.

8. When the salt, $C_9H_{13}IN^+$ I^- produced by reacting pyridine with 1,4-diiodobutane is then treated with Bu_3SnH in the presence of AIBN, a new salt, $C_9H_{12}N^+$ I^- is formed, which had ^1H NMR signals for four aromatic protons. Suggest structures for the two salts and a mechanism of formation of the latter.

9. Deduce a structure for the product, $C_6H_{11}NO_3$, produced by exposing 4-methyl-2-pyridone to the following sequence: (i) irradiation at 310 nm, (ii) O_3/MeOH/–78 °C then $NaBH_4$.

10. Write structures for the compounds produced at each stage in the following sequence: 4-methylpyridine reacted with $NaNH_2 \rightarrow C_6H_8N_2$, this then with $NaNO_2/H_2SO_4$ at $0\,^\circ C \rightarrow$ rt $\rightarrow C_6H_7NO$, then this with sodium methoxide and iodomethane $\rightarrow C_7H_9NO$ and finally this with $KOEt/(CO_2Et)_2 \rightarrow C_{11}H_{13}NO_4$.

11. Nitration of aniline is not generally possible, yet nitration of 2- and 4-aminopyridines can be achieved easily – why?

12. When 3-hydroxypyridine is reacted with 5-bromopent-1-ene a crystalline salt, $C_{10}H_{14}NBrO$ is formed. Treatment of the salt with mild base gave a dipolar substance $C_{10}H_{13}NO$ which on heating provided a neutral, non-aromatic isomer. Deduce the structures of these compounds.

13. Give an explanation for the relatively easy decarboxylation of pyridine-2-acetic acid; what is the organic product?

14. Suggest a structure for the product, $C_{16}H_{22}N_2O_5$ resulting from the interaction of 4-vinylpyridine with diethyl acetamidomalonate ($AcNHCH(CO_2Et)_2$) and base.

15. Write structures for the products of reacting (i) 2,3-dimethylpyridine with butyllithium then diphenyldisulfide $\rightarrow C_{13}H_{13}NS$; (ii) 2,3-dimethylpyridine with NBS then with PhSH $\rightarrow C_{13}H_{13}NS$ isomeric with the product in (i).

16. Write structures for the isomeric compounds $C_7H_6N_2O$ (formed in a ratio of 4:3) when 3-cyanopyridine methiodide is reacted with alkaline potassium ferricyanide.

17. Predict the sites at which deuterium would be found when 1-butylpyridinium iodide is reduced with $NaBD_4$ in EtOH forming (mainly) 1-butyl-1,2,5,6-tetrahydropyridine.

18. Deduce structures for the final product, and intermediate, in the following sequence: pyridine with methyl chloroformate and sodium borohydride gave $C_7H_9NO_2$, then this irradiated gave an isomer which had NMR signals for only two alkene protons – what are the compounds?

19. When pyridine N-oxide is heated with c. H_2SO_4 and c. HNO_3 a product $C_5H_4N_2O_3$ is formed; separate reactions of this with PCl_3 then H_2/Pd-C produces $C_5H_4N_2O_2$ and $C_5H_6N_2$ sequentially. What are the three products?

20. Write a structure for the cyclic product, $C_{18}H_{21}NO_4$, from the reaction of ammonia, phenylacetaldehyde ($PhCH_2CH{=}O$), and two mol equivalents of methyl acetoacetate. How might it be converted into a pyridine?

21. 2,3-Dihydrofuran reacts with acrolein to give $C_7H_{10}O_2$; reaction of this with aq. H_2NOH/HCl gave a pyridine, C_7H_9NO: deduce structures.

22. What pyridines or pyridones would be produced from the following combinations of reactants: (a) H_2NCOCH_2CN (cyanoacetamide) with (i) $EtCOCH_2$-CO_2Et; (ii) 2-acetylcyclohexanone; (iii) ethyl propiolate; (b) $MeC(NH_2){=}CH$-CO_2Et (ethyl 3-aminocrotonate) with (i) but-3-yne-2-one; (ii) $MeCOC(CO_2Et){=}CHOEt$.

23. When the sodium salt of formyl acetone ($MeCOCH{=}CHO^- Na^+$) is treated with ammonia a pyridine, C_8H_9NO, is formed. Deduce a structure and explain the regiochemistry of reaction.

References

1. 'Drugs against drugs', Gaskell, D., *Chem. Brit.*, **1998** (December), 27.
2. 'The bipyridinium herbicides', Summers, L. A., Academic Press, London, **1980**.

3. Arnett, E. M., Chawla, B., Bell, L., Taagepera, M., Hehre, W. J., and Taft, R. W., *J. Am. Chem. Soc.*, **1977**, *99*, 5729.
4. Corey, E. J. and Zheng, G. Z., *Tetrahedron Lett.*, **1998**, *39*, 6151.
5. Olah, G. A., Narang, S. C., Olah, J. A., Pearson, R. L., Cupas, C. A., *J. Am. Chem. Soc.*, **1980**, *102*, 3507.
6. Duffy, J. L. and Laali, K. K., *J. Org. Chem.*, **1991**, *56*, 3006.
7. Gösl, R. and Meuwsen, A., *Org. Synth., Coll. Vol. V*, **1973**, 43.
8. Baumgarten, P., *Chem. Ber.*, **1926**, 59, 1166.
9. Evans, R. F., Brown, H. C., and van der Plas, H. C., *Org. Synth., Coll. Vol. V*, **1973**, 977.
10. Popov, A. I. and Rygg, R. H., *J. Am. Chem. Soc.*, **1957**, *79*, 4622.
11. Umemoto, T., Tomita, K., and Kawada, K., *Org. Synth.*, **1990**, *69*, 129.
12. King, J. A. and Bryant, G. L., *J. Org. Chem.*, **1992**, *57*, 5136.
13. '4-Dialkylaminopyridines as highly active acylation catalysts', Höfle, G, Steglich, W., and Vorbrüggen, H., *Angew. Chem., Int. Ed. Engl.*, **1979**, *17*, 569; '4-Dialkylaminopyridines: super acylation and alkylation catalysts', Scriven, E. F. V., *Chem. Soc. Rev.*, **1983**, *12*, 129.
14. Guibé-Jampel, E., and Wakselman, M., *J. Chem. Soc., Chem. Commun.*, **1971**, 267.
15. Davies, S. G. and Shipton, M. R., *J. Chem. Soc., Perkin Trans. 1*, **1991**, 501.
16. Zoltewicz, J. A. and Smith, C. L., *J. Am. Chem. Soc.*, **1967**, *89*, 3358; Zoltewicz, J. A. and Cross, R. E., *J. Chem. Soc., Perkin Trans. 2*, **1974**, 1363 and 1368; Werstuik, N. H., and Ju, C., *Can. J. Chem.*, **1989**, *67*, 5; Zoltewicz, J. A., and Meyer, J. D., *Tetrahedron Lett.*, **1968**, 421.
17. Den Hertog, H. J. and Overhoff, J., *Recl. Trav. Chim. Pays-Bas*, **1930**, *49*, 552.
18. Brown, E. V. and Neil, R. H., *J. Org. Chem.*, **1961**, *26*, 3546.
19. Bakke, J. M. and Ranes, E., *Synthesis*, **1997**, 281; Suzuki, H., Iwaya, M., and Mori, T., *Tetrahedron Lett*, **1997**, *38*, 5647; Bakke, J. M. and Ranes, E., *J. Chem. Soc., Perkin Trans. 1*, **1997**, 1919; Bakke, J. M., Svensen, H., and Ranes, E., *J. Chem. Soc., Perkin Trans. 2*, **1998**, 2477; Bakke, J. M., Ranes, E., Riha, J., and Svensen, II., *Acta Cherm. Scand.*, **1999**, *53*, 141; Bakke, J. M. and Riha, J., *ibid.*, 356.
20. Johnson, C. D., Katritzky, A. R., Ridgewell, B. J., and Viney, M., *J. Chem. Soc., B*, **1967**, 1204.
21. McElvain, S. M. and Goese, M. A., *J. Am. Chem. Soc.*, **1943**, *65*, 2233.
22. '2,6-Di-*t*-butylpyridine – an unusual base', Kanner, B., *Heterocycles*, **1982**, *18*, 411.
23. den Hertog, H. J., den Does, L. V., Laandheer, C. A., *Recl. Trav. Chim. Pays-Bas*, **1962**, *91*, 864.
24. Pearson, D. E., Hargreave, W. W., Chow, J. K. T., and Suthers, B. R., *J. Org. Chem.*, **1961**, *26*, 789.
25. Paraskewas, S., *Synthesis*, **1980**, 378.
26. McCleland, N. P. and Wilson, R. H., *J. Chem. Soc.*, **1932**, 1263.
27. Bartok, W., Rosenfeld, D. D., and Schriesheim, A., *J. Org. Chem.*, **1963**, *28*, 410; Black, G., Depp, E., and Corson, B. B., *J. Org. Chem.*, **1949**, *14*, 14.
28. Jerchel, D., Heider, J., and Wagner, H., *Justus Liebigs Ann. Chem.*, **1958**, *613*, 153.
29. 'Nucleophilic aromatic substitution of hydrogen', Chupakhin, O. N., Charushin, V. N., and van der Plas, H. C., Academic Press, San Diego, **1994**.
30. Evans, J. C. W. and Allen, C. F. H., *Org. Synth., Coll. Vol. II*, **1943**, 517.
31. 'Formation of anionic σ-adducts from heterocyclic compounds: structures, rates and equilibria', Illuminati, G., and Stegel, F., *Adv. Heterocycl. Chem.*, **1983**, 34, 305.
32. Abramovitch, R. A. and Giam, C.-S., *Canad. J. Chem.*, **1964**, *42*, 1627; Abramovitch, R. A. and Poulton, G. A., *J. Chem. Soc., B*, **1969**, 901.
33. 'Amination of heterocyclic bases by alkali amides', Leffler, M. T., *Org. Reactions*, **1942**, *1*, 91; 'Advances in the Chichibabin reaction', McGill, C. K. and Rappa, A., *Adv. Heterocycl. Chem.*, **1988**, *44*, 2; 'Advances in the amination of nitrogen heterocycles', Vorbrüggen, H., *Adv. Heterocycl. Chem.*, **1990**, *49*, 117.
34. Viscardi, G., Savarino, P. Quagliotto, P., Barni, E., and Bottam M., *J. Heterocycl. Chem.*, **1996**, *33*, 1195.
35. Abramovitch, R. A., Helmer, F., and Saha, J. G., *Chem. Ind.*, **1964**, 659; Ban, Y. and Wakamatsu, T., *ibid.*, 710.
36. Seko, S. and Miyake, K., *Chem. Commun.*, **1998**, 1519.
37. Chichibabin, A. E., *Chem. Ber.*, **1923**, *56*, 1879.

38. Katada, M., *J. Pharm. Soc. Jpn.*, **1947**, *67*, 59 (*Chem. Abs.*, **1951**, *45*, 9537); den Hertog, H. J., Broekman, F. W., and Combé, W. P., *Recl. Trav. Chim. Pays-Bas*, **1951**, *70*, 105.
39. Hanessian, S. and Kagotani, M., *Synthesis*, **1987**, 409.
40. Yamanaka, H. and Ohba, S., *Heterocycles*, **1990**, *31*, 895.
41. DuPriest, M. T., Schmidt, C. L., Kuzmich, D., and Williams, S. B., *J. Org. Chem.*, **1986**, *51*, 2021.
42. Hashimoto, S., Otani, S., Okamoto, T., and Matsumoto, K., *Heterocycles*, **1988**, *27*, 319.
43. Loupy, A., Philippon, N., Pigeon, P., and Galons, H., *Heterocycles*, **1991**, *32*, 1947.
44. Jamart-Gregoire, B., Leger, C., and Caubere, P., *Tetrahedron Lett.*, **1990**, *31*, 7599; 'Hetarynes', den Hertog, H. J. and van der Plas, H. C., *Adv. Heterocycl. Chem.*, **1965**, *4*, 121; 'Hetarynes', Reinecke, M. G., *Tetrahedron*, **1982**, *38*, 427.
45. Mallet, M. and Queguiner, G., *C. R. Acad. Sci., Ser. C*, **1972**, *274*, 719; Walters, M. A., Carter, P. H., and Banerjee, S., *Synth. Commun.*, **1992**, 22, 2829.
46. Walters, M. A. and Shay, J. J., *Synth. Commun.*, **1997**, *27*, 3573.
47. Zoltewicz, J. A., Grahe, G., and Smith, C. L., *J. Am. Chem. Soc.*, **1969**, *91*, 5501.
48. Verbeek, J., George, A. V. E., de Jong, R. L. P., and Brandsma, L., *J. Chem. Soc., Chem. Commun.*, **1984**, 257; Verbeek, J. and Brandsma, L., *J. Org. Chem.*, **1984**, *49*, 3857.
49. Taylor, S. L., Lee, D. Y., and Martin, J. C., *J. Org. Chem.*, **1983**, *48*, 4157.
50. French, H. E. and Sears, K., *J. Am. Chem. Soc.*, **1951**, *73*, 469.
51. Ishikura, M., Kamada, M., Oda, I., and Tereshima, M., *Heterocycles*, **1985**, *23*, 117.
52. Ishikura, M., Ohta, T., and Terashima, M., *Chem. Pharm. Bull.*, **1985**, *33*, 4755.
53. Bell, A. S., Roberts, D. A., and Ruddock, K. S., *Synthesis*, **1987**, 843
54. Kondo, Y., Murata, N., and Sakamoto, T., *Heterocycles*, **1994**, *37*, 1467.
55. Ishikura, M., Mano, T., Oda, I., and Terashima, M., *Heterocycles*, **1984**, *22*, 2471.
56. Gros, P., Fort, Y., and Caubere, P., *J. Chem. Soc., Perkin Trans. 1*, **1997**, 3597; Gros, P. and Fort, Y, *ibid.*, **1998**, 3515.
57. Parham, W. E. and Piccirilli, R. M., *J. Org. Chem.*, **1977**, *42*, 257.
58. Lipshutz, B. H., Pfeiffer, S. S., and Chrisman, W., *Tetrahedron Lett.*, **1999**, *40*, 7889.
59. Cai, D., Hughes, D. L., and Verhoeven, T. R., *Tetrahedron Lett.*, **1996**, *37*, 2537; Paterson, M. A. and Mitchell, J. R., *J. Org. Chem.*, **1997**, *62*, 8237.
60. Corey, E. J. and Zheng, G. Z., *Tetrahedron Lett.*, **1998**, *39*, 6151.
61. Gu, Y. G. and Baybunt, E. K., *Tetrahedron Lett.*, **1996**, *37*, 2563.
62. Gribble, G. W. and Saulnier, M. G., *Heterocycles*, **1993**, *35*, 151.
63. Commins, D. L. and LaMunyon, D. H., *Tetrahedron Lett.*, **1988**, *29*, 773.
64. Ronald, R. C. and Winkle, M. R., *Tetrahedron*, **1983**, *39*, 2031.
65. Epsztajn, J, Berski, Z., Brzezinski, J. Z., and Józwick, A., *Tetrahedron Lett.*, **1980**, *21*, 4739.
66. Güngör, T., Marsais, F., and Quéguiner, G., *Synthesis*, **1982**, 499.
67. Meyers, A. I. and Gabel, R. A., *Heterocycles*, **1978**, *11*, 133.
68. Schlecker, W., Huth, A., and Ottow, E., *J. Org. Chem.*, **1995**, *60*, 8414.
69. Hands, D., Bishop, B., Cameron, M., Edwards, J. S., Cottrell, I. F., and Wright, S. H. R., *Synthesis*, **1996**, 877.
70. Watanabe, M., Shinoda, E., Shimizu, Y., Furukawa, S., Iwao, M., and Kuraishi, T., *Tetrahedron*, **1987**, *43*, 5281.
71. Marsais, F., Trécourt, F., Bréant, P. and Quéguiner, G., *J. Heterocycl. Chem.*, **1988**, *25*, 81.
72. Bargar, T. M., Wilson, T., and Daniel, J. K., *J. Heterocycl. Chem.*, **1985**, *22*, 1583.
73. Epsztajn, J., Bieniek, A., and Kowalska, J. A., *Tetrahedron*, **1991**, *47*, 1697.
74. Guillier, F., Nivoliers, F., Cochennee, A., Godard, A., Marsais, F., and Queguiner, G., *Synth. Commun.*, **1996**, *26*, 4421.
75. Mongin, O., Rocca, P., Thomas-dit-Dumont, L., Trécourt, F., Marsais, F., Godard, A., and Quéguiner, G., *J. Chem. Soc., Perkin Trans. 1*, **1995**, 2503.
76. Berillon, L., Lepretre, A., Turck, A., Plé, N., Quéguiner, G., Cahiez, G., and Knochel, P., *Synlett*, **1998**, 1359; Trecourt, F., Breton, G., Bonnet, V., Mongin, F., Marsais, F., and Quéguiner, G., *Tetrahedron Lett.*, **1999**, *40*, 4339; Abarbi, M., Dehmel, F., and Knochel, P., *Tetrahedron Lett.*, **1999**, *40*, 7449.
77. Okita, T. and Isobe, M., *Synlett*, **1994**, 589; Godard, A., Rovera, J.-C., Marsais, F., Plé, N., and Quéguiner, G., *Tetrahedron*, **1992**, *48*, 4123.
78. Fu, J., Chen, Y., and Castelhano, A. L., *Synlett*, **1998**, 1408.
79. Edo, K., Sakamoto, T., and Yamanaka, H., *Chem. Pharm. Bull.*, **1979**, *27*, 193.

80. Head, R. A. and Ibbotson, A., *Tetrahedron Lett.*, **1984**, *25*, 5939.
81. Sakamoto, T., Shiraiwa, M., Kondo, Y., and Yamanaka, H., *Synthesis*, **1983**, 312; Sakamoto, T., Nagata, H., Kondo, Y., Sato, K., and Yamanaka, H., *Chem. Pharm. Bull.*, **1984**, *32*, 4866.
82. Zhang, H., Kwong, F. Y., Tian, Y., and Chan, K. S., *J. Org. Chem.*, **1998**, *63*, 6886; Lohse, O., Thevenin, P., Waldvogel, E., *Synlett*, **1999**, 45.
83. Ma, H., Wang, X., and Deng, M., *Synth. Commun.*, **1999**, *29*, 2477.
84. Yamamoto, T. and Yanagi, A., *Heterocycles*, **1981**, *16*, 1161.
85. Bailey, T. R., *Tetrahedron Lett.*, **1986**, *27*, 4407.
86. Yamamoto, Y., Tanaka, T., Yagi, M., and Inamoto, M., *Heterocycles*, **1996**, *42*, 189.
87. Yamamoto, Y. and Yanagi, A., *Chem. Pharm. Bull.*, **1982**, *30*, 2003; Yamamoto, Y., Ouchi, H., and Tanako, T., *ibid.*, **1995**, *43*, 916.
88. Wagaw, S. and Buchwald, S. L., *J. Org. Chem.*, **1996**, *61*, 7240.
89. Jixiang, C. and Crisp, G. T., *Synth. Commun.*, **1992**, *22*, 683.
90. Wibaut, J. P. and Nicolaï, J. R., *Recl. Trav. Chim. Pays-Bas*, **1939**, *58*, 709; McElvain, S. M., and Goese, M. A., *J. Am. Chem. Soc.*, **1943**, *65*, 2227.
91. Elofson, R. M., Gadallah, F. F., and Schutz, K. F., *J. Org. Chem.*, **1971**, *36*, 1526; Gurczynski, M. and Tomasik, P., *Org. Prep. Proc. Int.*, **1991**, *23*, 438.
92. Bonnier, J. M. and Court, J., *Bull. Soc. Chim. Fr.*, **1972**, 1834; Minisci, F., Vismara, E., Fontana, F., Morini, G., Serravelle, M., and Giordano, C., *J. Org. Chem.*, **1986**, *51*, 4411.
93. 'Recent developments of free radical substitutions of heteroaromatic bases', Minisci, F., Vismara, E., and Fontana, F., *Heterocycles*, **1989**, *28*, 489.
94. 'The action of metal catalysts on pyridines', Badger, G. M. and Sasse, W. H. F., *Adv. Heterocycl. Chem.*, **1963**, 2, 179; 'The bipyridines', Summers, L. A., *Adv. Heterocycl. Chem.*, **1984**, *35*, 281.
95. Sasse, W. H. F., *Org. Synth.*, *Coll. Vol. V*, **1973**, 102.
96. Newkome, G. R. and Hager, D. C., *J. Org. Chem.*, **1982**, *47*, 599.
97. Nielsen, A. T., Moore, D. W., Muha, G. M., and Berry, K. H., *J. Org. Chem.*, **1964**, *29*, 2175; Frank, R. L. and Smith, P. V., *Org. Synth.*, *Coll. Vol. III*, **1955**, 410.
98. Atlanti, P. M., Biellmann, J. F., and Moron, J., *Tetrahedron*, **1973**, *29*, 391.
99. Ryang, H.-S. and Sakurai, H., *J. Chem. Soc., Chem. Commun.*, **1972**, 594; Ohkura, K., Terashima, M., Kanaoka, Y., and Seki, K. *Chem. Pharm. Bull.*, **1993**, *41*, 1920.
100. Jones, K. and Fiumana, A., *Tetrahedron Lett.*, **1996**, *38*, 8049.
101. Lunn, G. and Sansome, E. B., *J. Org. Chem.*, **1986**, *51*, 513.
102. Giam, C. S., and Abbott, S. D., *J. Am. Chem. Soc.*, **1971**, *93*, 1294.
103. Blough, B. E. and Carroll, F. I., *Tetrahedron Lett.*, **1993**, *34*, 7239.
104. Tsuge, O., Kanemasa, S., Naritomi, T., and Tanaka, J., *Bull. Chem. Soc. Jpn.*, **1987**, *60*, 1497.
105. Birch, A. J. and Karakhanov, E. A., *J. Chem. Soc., Chem. Commun.*, **1975**, 480.
106. Guerry, P. and Neier, R., *Synthesis*, **1984**, 485.
107. Kamochi, Y. and Kudo, T., *Heterocycles*, **1993**, *36*, 2383.
108. Cook, N. C. and Lyons, J. E., *J. Am. Chem. Soc.*, **1966**, *88*, 3396.
109. Hao, L., Harrod, J. F., Lebuis, A.-M., Mu, Y., Shu, R., Samuel, E., and Woo, H.-G., *Angew. Chem., Int. Ed. Engl.*, **1998**, *37*, 3126.
110. Heap, U., *Tetrahedron*, **1975**, 31, 77; Tomisawa, H. and Hongo, H., *Chem. Pharm. Bull.*, **1970**, 18, 925; Matsumoto, K., Ikemi-Kono, Y., Uchida, T., and Acheson, R. M., *J. Chem. Soc., Chem. Commun.*, **1979**, 1091; 'Diels-Alder cycloadditions of 2-pyrones and 2-pyridones', Afarinkia, K., Viader, V., Nelson, T. D., and Posner, G. H., *Tetrahedron*, **1992**, *48*, 9111.
111. Posner, G. H., Vinader, V., and Afarinkia, K., *J. Org. Chem.*, **1992**, *57*, 4088.
112. 'Synthetic applications of heteroaromatic betaines with six-membered rings', Dennis, N., Katritzky, A. R., and Takeuchi, Y., *Angew. Chem., Int. Ed. Engl.*, **1976**, *15*, 1; 'Cycloaddition reactions of heteroaromatic six-membered rings', Katritzky, A. R. and Dennis, N., *Chem. Rev.*, **1989**, *89*, 827; Lomenzo, S. A., Enmon, J. L., Troyer, M. C., and Trudell, M. L., *Synth. Commun.*, **1995**, *25*, 3681.
113. Joussot-Dubien, J. and Houdard, J., *Tetrahedron Lett.*, **1967**, 4389; Wilzbach, K. E. and Rausch, D. J., *J. Am. Chem. Soc.*, **1970**, *92*, 2178;
114. De Selms, R. C. and Schleigh, W. R., *Tetrahedron Lett.*, **1972**, 3563; Kaneko, C., Shiba, K., Fujii, H., and Momose, Y., *J. Chem. Soc., Chem. Commun.*, **1980**, 1177

115. Ogata, Y. and Takagi, K., *J. Org. Chem.*, **1978**, *43*, 944.
116. Nakamura, Y., Kato, T., and Morita, Y., *J. Chem. Soc., Perkin Trans. 1*, **1982**, 1187.
117. First reported by Taylor, E. C. and Paudler, W. W., *Tetrahedron Lett.*, **1960**, *25*, 1.
118. Sieburth, S. McN., Hiel, G., Lin, C.-H., and Kuan, D. P., *J. Org. Chem.*, **1994**, *59*, 80.
119. Sieburth, S. McN. and Zhang, F., *Tetrahedron Lett.*, **1999**, *40*, 3527.
120. Johnson, B. L., Kitahara, Y., Weakley, T. J. R., and Keana, J. F. W., *Tetrahedron Lett.*, **1993**, *34*, 5555.
121. Somekawa, K., Okuhira, H., Sendayama, M., Suishu, T., and Shimo, T., *J. Org. Chem.*, **1992**, *57*, 5708.
122. Kaplan, L., Pavlik, J. W., and Wilzbach, K. E., *J. Am. Chem. Soc.*, **1972**, *94*, 3283; Ling, R., Yoshida, M., and Mariano, P. S., *J. Org. Chem.*, **1996**, *61*, 4439; Ling, R. and Mariano, P. S., *ibid.*, **1998**, *63*, 6072.
123. Penkett, C. S. and Simpson, I. D., *Tetrahedron*, **1999**, *55*, 6183.
124. Buchardt, O., Christensen, J. J., Nielsen, P. E., Koganty, R. R., Finsen, L., Lohse, C., and Becher, J., *Acta Chem. Scand., Ser. B*, **1980**, 34, 31; Lohse, C., Hagedorn, L., Albini, A., and Fasani, E., *Tetrahedron*, **1988**, *44*, 2591.
125. Komin, A. P. and Wolfe, J. F., *J. Org. Chem.*, **1977**, *42*, 2481; Rossi, R. A., de Rossi, R. H. and López, A. F., *J. Org. Chem.*, **1976**, 41, 3371.
126. van der Stegen, G. H. D., Poziomek, E. J., Kronenberg, M. E., and Haavinga, E., *Tetrahedron Lett.*, **1966**, 6371.
127. Beak, P., Fry, F. S., Lee, J., and Steele, F., *J. Am. Chem. Soc.*, **1976**, *98*, 171; Stefaniak, L., *Tetrahedron*, **1976**, *32*, 1065; Beak, P., Covington, J. B., Smith, S. G., White, J. M., and Zeigler, J. M., *J. Org. Chem.*, **1980**, *45*, 1354.
128. Bensaude, O., Chevrier, M., and Dubois, J.-E., *Tetrahedron Lett.*, **1978**, 2221.
129. Schoffner, J. P., Bauer, L., and Bell, C. L., *J. Heterocycl. Chem.*, **1970**, *7*, 479 and 487.
130. Cioffi, E. A. and Bailey, W. F., *Tetrahedron Lett.*, **1998**, *39*, 2679.
131. Flemming, I. and Philippides, D., *J. Chem. Soc. C*, **1970**, 2426; Effenberger, F., Mück, A. O., and Bessey, E., *Chem. Ber.*, **1980**, *113*, 2086.
132. Bellingham, P., Johnson, C. D., and Katritzky, A. R., *Chem. Ind.*, **1965**, 1384.
133. Burton, A. G., Halls, P. J., and Katritzky, A. R., *J. Chem. Soc., Perkin Trans. 2*, **1972**, 1953.
134. Brignell, P. J., Katritzky, A. R., and Tarhan, H. O., *J. Chem. Soc., B*, **1968**, 1477.
135. Itahara, T. and Ouseto, F., *Synthesis*, **1984**, 488.
136. De Selms, R. C., *J. Org. Chem.*, **1968**, *33*, 478.
137. Stempel, A. and Buzzi, E. C., *J. Am. Chem. Soc.*, **1949**, *71*, 2969.
138. Edgar, K. J. and Falling, S. N., *J. Org. Chem.*, **1990**, *55*, 5287.
139. Windscheif, P.-M. and Vögtle, F., *Synthesis*, **1994**, 87.
140. Kornblum, N. and Coffey, G. P., *J. Org. Chem.*, **1966**, *31*, 3449; Hopkins, G. C., Jonak, J. P., Minnemeyer, H. J., and Tieckelmann, H., *J. Org. Chem.*, **1967**, *32*, 4040; Dou, H. J.-M, Hassanaly, P., and Metzger, J., *J. Heterocycl. Chem.*, **1977**, *14*, 321.
141. Liu, H., Ko, S.-B., Josien, H., and Curran, D. P., *Tetrahedron Lett.*, **1995**, *36*, 845; Sato, T., Yoshimatsu, K., and Otera, J, *Synlett*, **1995**, 845.
142. Vorbrüggen, H. and Krolikiewicz, K., *Chem. Ber.*, **1984**, *117*, 1523.
143. Comins, D. L. and Jianhua, G., *Tetrahedron Lett.*, **1994**, *35*, 2819.
144. Katritzky, A. R. and Sengupta, S., *Tetrahedron Lett.*, **1987**, *28*, 5419.
145. Beak, P. and Bonham, J., *J. Am. Chem. Soc.*, **1965**, *87*, 3365.
146. Meghani, P. and Joule, J. A., *J. Chem. Soc., Perkin Trans. 1*, **1988**, 1.
147. Katritzky, A. R., Grzeskowiak, N. E., and Winwood, D., *J. Mol. Sci.*, **1983**, *1*, 71.
148. Katritzky, A. R., Fan, W.-Q., Koziol, A. E., and Palenik, G. J., *Tetrahedron*, **1987**, *43*, 2343.
149. Adams, R. and Schrecker, A. W., *J. Am. Chem. Soc.*, **1949**, *71*, 1186.
150. Sugimoto, O., Mori, M., and Tanji, K., *Tetrahedron Lett.*, **1999**, *40*, 7477.
151. Isler, O., Gutmann, H., Straub, U., Fust, B., Böhni, E., and Studer, A., *Helv. Chim. Acta*, **1955**, *38*, 1033.
152. Lowe, J. A., Ewing, F. E., and Drozda, S. E., *Synth. Commun.*, **1989**, *19*, 3027.
153. Philips. M. A. and Shapiro, H., *J. Chem. Soc.*, **1942**, 584.
154. Araki, M., Sakata, S., Takei, H., and Mukaiyama, T., *Bull Chem. Soc. Jpn.*, **1974**, *47*, 1777; Nicolaou, K. C., Claremon, D. A., and Papahatjis, D. P., *Tetrahedron Lett.*, **1981**, *22*, 4647.
155. Lloyd, K. and Young, G. T., *J. Chem. Soc., C*, **1971**, 2890.

156. Corey, E. J. and Nicolaou, K. C., *J. Am. Chem. Soc.*, **1974**, *96*, 5614.
157. Girault, G., Coustal, S., and Rumpf, P., *Bull Soc. Chim. France*, **1972**, 2787; Forsythe, P., Frampton, R., Johnson, C. D., and Katritzky, A. R., *J. Chem. Soc., Perkin Trans. 2*, **1972**, 671.
158. Frampton, R., Johnson, C. D., and Katritzky, A. R., *Justus Liebigs Ann. Chem.*, **1971**, *749*, 12.
159. Fox, B. A. and Threlfall, T. L., *Org. Synth., Coll. Vol. V*, **1973**, 346.
160. v. Schickh, O., Binz, A., and Schule, A., *Chem. Ber.*, **1936**, 69, 2593.
161. Burton, A. G., Frampton, R. D., Johnson, C. D., and Katritzky, A. R., *J. Chem. Soc., Perkin Trans. 2*, **1972**, 1940.
162. Katritzky, A. R., El-Zemity, S., and Lang, H., *J. Chem. Soc., Perkin Trans. 1*, **1995**, 3129.
163. Kalatzis, E., *J. Chem. Soc., B*, **1967**, 273 and 277.
164. Bunnett, J. F. and Singh, P., *J. Org. Chem.*, **1981**, *46*, 4567.
165. Allen, C. F. H. and Thirtle, J. R., *Org. Synth., Coll. Vol. III*, **1955**, 136; Windscheif, P.-M. and Vgtle, *Synthesis*, **1994**, 87.
166. Yoneda, N. and Fukuhara, T., *Tetrahedron*, **1996**, *52*, 23; Coudret, C., *Synth. Commun.*, **1996**, *26*, 3543; .
167. 'Methylpyridines and other methylazines as precursors of bicycles and polycycles', Abu-Shanab, F. A., Wakefield, B. J., and Elnagdi, M. H., *Adv. Heterocycl. Chem.*, **1997**, *68*, 181.
168. White, W. N. and Lazdins, D., *J. Org. Chem.*, **1969**, *34*, 2756.
169. Fraser, R. R., Mansour, T. S., and Savard, S., *J. Org. Chem.*, **1985**, *50*, 3232.
170. Levine, R., Dimmig, D. A., and Kadunce, W. M., *J. Org. Chem.*, **1974**, *39*, 3834; Kaiser, E. M., Bartling, G. J., Thomas, W. R., Nichols, S. B., and Nash, D. R., *J. Org. Chem.*, **1973**, *38*, 71.
171. Beumel, O. F., Smith, W. N., and Rybalka, B., *Synthesis*, **1974**, 43.
172. Kaiser, E. M. and Petty, J. D., *Synthesis*, **1975**, 705; Davis, M. L., Wakefield, B. J., and Wardell, J. A., *Tetrahedron*, **1992**, *48*, 939.
173. Lochte, H. L. and Cheavers, T. H., *J. Am. Chem. Soc.*, **1957**, 79, 1667; Ghera, E., David, Y. B., and Rapoport, H., *J. Org. Chem.*, **1981**, *46*, 2059.
174. Williams, J. L. R., Adel, R. E., Carlson, J. M., Reynolds, G. A., Borden, D. G., and Ford, J. A., *J. Org. Chem.*, **1963**, *28*, 387; Cassity, R. P., Taylor, L. T., and Wolfe, J. F., *J. Org. Chem.*, **1978**, *43*, 2286; Konakahara, T. and Takagi, Y., *Heterocycles*, **1980**, *14*, 393; Konakahara, T. and Takagi, Y., *Synthesis*, **1979**, 192; Woodward, R. B. and Kornfield, E. C., *Org. Synth., Coll. Vol. III*, **1973**, 413; Feucr, H. and Lawrence, J. P., *J. Am. Chem. Soc.*, **1969**, *91*, 1856; Markovac, A., Stevens, C. L., Ash, A. B., and Hackley, B. E., *J. Org. Chem.*, **1970**, *35*, 841; Bredereck, H., Simchen, G., and Wahl, R., *Chem. Ber.*, **1968**, *101*, 4048; Pasquinet, E., Rocca, P., Godard, A., Marsais, F., and Quguiner, G., *J. Chem. Soc., Perkin Trans. 1*, **1998**, 3807.
175. Ihle, N. C. and Krause, A. E., *J. Org. Chem.*, **1996**, *61*, 4810.
176. Jerchel, D. and Heck, H. E., *Justus Liebigs Ann. Chem.*, **1958**, *613*, 171.
177. Abramovitch, R. A., Coutts, R. T., and Smith, E. M., *J. Org. Chem.*, **1972**, *37*, 3584; Abramovitch, R. A., Smith, E. M., Knaus, E. E., and Saha, M., *J. Org. Chem.*, **1972**, *37*, 1690.
178. Doering, W. E. and Weil, R. A. N., *J. Am. Chem. Soc.*, **1947**, *69*, 2461; Leonard, F. and Pschannen, W., *J. Med. Chem.*, **1966**, *9*, 140.
179. Sakamoto, T., Kondo, Y., Shiraiwa, M., and Yamanaka, H., *Synthesis*, **1984**, 245.
180. Katritzky, A. R., Khan, G. R., and Schwarz, O. A., *Tetrahedron Lett.*, **1984**, *25*, 1223.
181. Gozzo, F. C. and Eberlin, M. N., *J. Org. Chem.*, **1999**, *64*, 2188.
182. Moser, R. J. and Brown, E. V., *J. Org. Chem.*, **1972**, *37*, 3938.
183. Brown, E. V. and Shambhu, M. B., *J. Org. Chem.*, **1971**, *36*, 2002; Rapoport, H. and Volcheck, E. J., *J. Am. Chem. Soc.*, **1956**, *78*, 2451
184. Quast, H. and Schmitt, E., *Justus Liebigs Ann. Chem.*, **1970**, 732, 43; Katritzky, A. R. and Faid-Allah, H. M., *Synthesis*, **1983**, 149.
185. Effenberger, F. and Knig, J., *Tetrahedron*, **1988**, *44*, 3281; Bohn, B., Heinrich, N., and Vorbrggen, H., *Heterocycles*, **1994**, *37*, 1731.
186. Prill, E. A. and McElvain, S. M., *Org. Synth., Coll. Vol II*, **1943**, 419; 'Oxidative transformation of heterocyclic iminium salts', Weber, H., *Adv. Heterocycl. Chem.*, **1987**, *41*, 275.

187. Anderson, P. S., Kruger, W. E., and Lyle, R. E., *Tetrahedron Lett.*, **1965**, 4011; Holik, M. and Ferles, M., *Coll. Czech. Chem. Commun.*, **1967**, *32*, 3067.
188. Cervinka, O. and Kriz, O., *Coll. Czech. Chem. Commun.*, **1965**, *30*, 1700.
189. de Koning, A. J., Budzelaar, P. H. M., Brandsma, L., de Bie, M. J. A., and Boersma, J., *Tetrahedron Lett.*, **1980**, *21*, 2105.
190. Gessner, W. and Brossi, A., *Synth. Commun.*, **1985**, *15*, 911.
191. Fowler, F. W., *J. Am. Chem. Soc.*, **1972**, *94*, 5926.
192. 'N-Dienyl amides and lactams. Preparation and Diels-Alder reactivity', Smith, M. B., *Org. Prep. Proc. Int.*, **1990**, *22*, 315.
193. Fowler, F. W., *J. Org. Chem.*, **1972**, *37*, 1321.
194. Beeken, P., Bonfiglio, J. N., Hassan, I., Piwinski,. J. J., Weinstein, B., Zollo, K. A., and Fowler, F. W., *J. Am. Chem. Soc.*, **1979**, *101*, 6677; Comins, D. L. and Mantlo, N. B., *J. Org. Chem.*, **1986**, *51*, 5456.
195. Caughey, W. S. and Schellenberg, K. A., *J. Org. Chem.*, **1966**, *31*, 1978; Biellman, J.-F. and Callot, H. J., *Bull. Chem. Soc. Fr.*, **1968**, 1154; Blankenhorn, G. and Moore, E. G., *J. Am. Chem. Soc.*, **1980**, *102*, 1092.
196. Wong, Y. S., Marazano, C., Grecco, D., and Das, B. C., *Tetrahedron Lett.*, **1994**, *35*, 707.
197. Thiessen, L. M., Lepoivre, J. A., and Alderweireldt, F. C., *Tetrahedron Lett.*, **1974**, 59.
198. Comins, D. L. and Abdullah, A. H., *J. Org. Chem.*, **1982**, *47*, 4315.
199. Piers, E. and Soucy, M., *Canad. J. Chem.*, **1974**, *52*, 3563; Akiba, K., Iseki, Y., and Wada, M., *Tetrahedron Lett.,* **1982**, *23,* 429; Comins, D. L. and Mantlo, N. B., *Tetrahedron Lett.*, **1983**, *24*, 3683; Comins, D. L., Smith, R. K., and Stroud, E. D., *Heterocycles*, **1984**, *22*, 339; Shiao, M.-J., Shih, L.-H., Chia, W.-L., and Chau, T.-Y., *Heterocycles*, **1991**, *32*, 2111.
200. von Dobeneck, H. and Goltzsche, W., *Chem. Ber.*, **1962**, *95*, 1484.
201. Lavilla, R., Gotsens, T., and Bosch, J., *Synthesis*, **1991**, 842.
202. Lyle, R. E., Marshall, J. L., and Comins, D. L., *Tetrahedron Lett.*, **1977**, 1015.
203. Yamaguchi, R., Nakazono, Y., and Kawanisi, M., *Tetrahedron Lett.*, **1983**, *24*, 1801
204. Dhar, T. G. and Gluchowski, C., *Tetrahedron Lett.*, **1994**, *35*, 989.
205. Itoh, T., Hasegawa, H., Nagata, K., and Ohsawa, A., *Synlett*, **1994**, 557.
206. Akiba, K., Iseki, Y., and Wada, M., *Tetrahedron Lett.*, **1982**, *23*, 3935.
207. Onaka, M., Ohno, R., and Izumi, Y., *Tetrahedron Lett.*, **1989**, *30*, 747.
208. Fraenkel, G., Cooper, J. W., and Fink, C. M., *Angew. Chem., Int. Ed. Engl.*, **1970**, *7*, 523.
209. Comins, D. L., Weglarz, M. A., and O'Connor, S., *Tetrahedron Lett.*, **1988**, *29*, 1751.
210. Commins, D. L. and Brown, J. D., *Tetrahedron Lett.*, **1986**, *27*, 2219.
211. Comins, D. L. and Joseph, S. P., *Adv. Nitrogen Heterocyles*, **1996**, *2*, 251; Comins, D. L., Joseph, S. P., Hong, H., Al-awar, R. S., Foti, C. J., Zhang, Y., Chen, X., LaMunyon, D. H., and Weltzien, M., *Pure Appl. Chem.*, **1997**, *69*, 477; Comins, D. L. and Green, G. M., *Tetrahedron Lett.*, **1999**, *40*, 217.
212. Kiselyov, A. S. and Strekowski, L., *J. Org. Chem.*, **1993**, *58*, 4476; *idem*, *J. Heterocycl. Chem.*, **1993**, *30*, 1361.
213. Haase, M., Goerls, H., and Anders, E., *Synthesis*, **1998**, 195.
214. Damji, S. W. H. and Fyfe, C. A., *J. Org. Chem.*, **1979**, *44*, 1757.
215. Severin, T., Lerche, H., and Btz, D., *Chem. Ber.*, **1969**, *102*, 2163.
216. Buurman, D. J. and van der Plas, H. C., *Org. Prep. Proc. Int.*, **1988**, *20*, 591.
217. 'Pyridine ring nucleophilic recyclisations', Kost, A. N., Gromov, S. P., and Sagitullin, R. S., *Tetrahedron*, **1981**, *37*, 3423; 'Synthesis and reactions of glutaconaldehyde and 5-amino-2,4-pentadienals', Becher, J., *Synthesis*, **1980**, 589.
218. Becher, J., *Org. Synth.*, **1979**, *59*, 79.
219. Hafner, K. and Meinhardt, K.-P., *Org. Synth.*, **1984**, *62*, 134.
220. Genisson, Y., Marazano, C., Mehmandoust, M., Grecco, D., and Das, B. C., *Synlett*, **1992**, 431.
221. Reinehr, D. and Winkler, T., *Angew. Chem., Int. Ed. Engl.*, **1981**, *20*, 881.
222. Kutney, J. P. and Greenhouse, R., *Synth. Commun.*, **1975**, 119; Berg, U., Gallo, R., and Metzger, J., *J. Org. Chem.*, **1976**, *41*, 2621.
223. Aumann, D. and Deady, L. W., *J. Chem. Soc., Chem. Commun.*, **1973**, 32.
224. Bhatia, A. V., Chaudhury, S. K., and Hernandez, O., *Org. Synth.*, **1997**, *75*, 184.
225. Olah, G. A. and Klumpp, D. A., *Synthesis*, **1997**, 744.
226. Chaudhury, S. K. and Hernandez, O., *Tetrahedron Lett.*, **1979**, 95; *ibid.*, 99.

227. 'Pyrylium mediated transformations of primary amino groups into other functional groups'. Katritzky, A. R. and Marson, C. M., *Angew. Chem., Int. Ed. Engl.*, **1984**, *23*, 420.

228. 'Heterocyclic *N*-oxides', Katritzky, A.R. and Lagowski, J. M., Methuen, London, **1967**; 'Aromatic amine oxides', Ochiai, E., Am. Elsevier, New York, **1967**; 'Heterocyclic *N*-oxides', Albini, A. and Pietra, S., CRC Press Wolfe Publishing, London, **1991**; 'Heterocyclic *N*-oxides and *N*-imides', Katritzky, A. R. and Lam, J. N., *Heterocycles*, **1992**, *33*, 1011.

229. 'Rearrangements of *t*-amine oxides', Oae, S. and Ogino, K., *Heterocycles*, **1977**, *6*, 583.

230. Zhang, Y. and Lin, R., *Synth Commun.*, **1987**, *17*, 329; Akita, Y., Misu, K., Watanabe, T., and Ohto, A., *Chem. Pharm. Bull.*, **1976**, *24*, 1839; Malinowski, M. and Kaczmarek, L., *Synthesis*, **1987**, 1013; Balicki, R., Kaczmarek, L., and Malinowski, M., *Synth. Commun.*, **1989**, 897; Balicki, R., *Synthesis*, **1989**, 645.

231. Olah, G. A., Arvanaghi, M., and Vankar, Y. D., *Synthesis*, **1980**, 660.

232. Reichardt, C., *Chem. Ber.*, **1966**, *99*, 1769.

233. Manning, R. F. and Schaefer, F. M., *Tetrahedron Lett.*, **1975**, 213.

234. Taylor, E. C. and Crovetti, A. J., *Org. Synth., Coll. Vol. IV*, **1963**, 654; Saito, H. and Hamana, M., *Heterocycles*, **1979**, *12*, 475.

235. Johnson, C. D., Katritzky, A. R., Shakir, N., and Viney, M., *J. Chem. Soc. (B)*, **1967**, 1213.

236. van Ammers, M., den Hertog, H. J., and Haasc, B., *Tetrahedron*, **1962**, *18*, 227.

237. Van Ammers, M. and Den Hertog, H. J., *Recl. Trav. Chim. Pays Bas*, **1962**, *81*, 124.

238. Liveris, M. and Miller, J., *J. Chem. Soc*, **1963**, 3486; Johnson, R. M., *J. Chem. Soc. C*, **1966**, 1058.

239. Van Bergen, T. J. and Kellogg, R. M., *J. Org. Chem.*, **1971**, *36*, 1705; Schiess, P. and Ringele, P., *Tetrahedron Lett.*, **1972**, 311; Schiess, P., Monnier, C., Ringele, P., and Sendi, E., *Helv. Chim. Acta*, **1974**, *57*, 1676.

240. Schnckenburger, J. and Heber, D., *Chem. Ber.*, **1974**, *107*, 3408.

241. Nishiwaki, N., Minakata, S., Komatsu, M., and Ohshiro, Y., *Chem. Lett.*, **1989**, 773.

242. Webb, T. R., *Tetrahedron Lett.*, **1985**, *26*, 3190.

243. Fife, W. K., *J. Org. Chem.*, **1983**, *48*, 1375; Vorbrggen, H. and Krolikiewicz, K., *Synthesis*, **1983**, 316.

244. Feely, W. E., Evanega, G., and Beavers, E. M., *Org. Synth., Coll. Vol. V*, **1973**, 269.

245. Fontenas, C., Bejan, E., Ait Haddou, H., and Balavoine, G. G. A,, *Synth. Commun.*, **1995**, *25*, 629.

246. Ginsberg, S. and Wilson, I. B., *J. Am. Chem. Soc.*, **1957**, *79*, 481.

247. Bodalski, R and Katritzky, A. R., *J. Chem. Soc. (B)*, **1968**, 831; Koenig, T., *J. Am. Chem. Soc.*, **1966**, *88*, 4045.

248. Walters, M. A. and Shay, J. J., *Tetrahedron Lett.*, **1995**, *36*, 7575.

249. Gill, N. S., James, K. B., Lions, F., and Potts, K. T., *J. Am. Chem. Soc.*, **1952**, *74*, 4923; Keuper, R., Risch, N., Flrke, U., Haupt, H.-J., *Liebigs Ann.*, **1996**, 705; Keuper, R., Risch, N., , *ibid*, 717.

250. Duhamel, P., Hennequin, L., Poirier, J. M., Tavel, G., and Vottero, C., *Tetrahedron*, **1986**, *42*, 4777.

251. Knoevenagel, E., *Justus Liebigs Ann. Chem.*, **1894**, *281*, 25; Stobbe, H. and Vollard, H., *Chem. Ber.*, **1902**, *35*, 3973; Stobbe, H., *ibid.*, 3978.

252. Kelly, T. R. and Liu, H., *J. Am. Chem. Soc.*, **1985**, *107*, 4998.

253. Katritzky, A. R., Murugan, R., and Sakizadeh, K., *J. Heterocycl. Chem.*, **1984**, *21*, 1465.

254. Potts, K. T., Cipullo, M. J., Ralli, P., and Theodoridis, G., *J. Am. Chem. Soc.*, **1981**, *103*, 3584 and 3585; Potts, K. T., Ralli, P., Theodoridis, G., and Winslow, P., *Org. Synth.*, **1986**, *64*, 189.

255. Wiley, R. H., and Slaymaker, S. C., *J. Am. Chem. Soc.*, **1956**, *78*, 2393; Cook, P. D., Day, R. T., and Robins, R. K., *J. Heterocycl. Chem.*, **1977**, *14*, 1295.

256. Bickel, A. F., *J. Am. Chem. Soc.*, **1947**, *69*, 1805; Campbell, K. N., Ackermann, J. F., and Campbell, B. K., *J. Org. Chem.*, **1950**, *15*, 221.

257. Abdulla, R. F., Fuhr, K. H., and Williams, J. C., *J. Org. Chem.*, **1979**, *44*, 1349.

258. Kurihara, H. and Mishima, H., *J. Heterocycl. Chem.*, **1977**, *14*, 1077; Johnson, F., Panella, J. P., Carlson, A. A, and Hunneman, D. H., *J. Org, Chem.*, **1962**, *27*, 2473.

259. '4-Aryldihydropyridines, a new class of highly active calcium antagonists', Bossart, F., Meyer, H., and Wehinger, E., *Angew. Chem., Int. Ed. Engl.* **1981**, *20*, 762

260. Pfister, J. R., *Synthesis*, **1990**, 689; Maquestian, A., Mayence, A., and Eynde, J.-J. V., *Tetrahedron Lett.*, **1991**, *32*, 3839; Alvarez, C., Delgado, F., Garc'a, O., Medina, S., and Mrquez, C., *Synth. Commun.*, **1991**, *21*, 619.

261. Satoh, Y., Ichihashi, M., and Okumura, K., *Chem. Pharm. Bull.*, **1992**, *40*, 912.

262. Baumgarten, P. and Dornow, A., *Chem. Ber.*, **1939**, *72*, 563.

263. Petrich, S. A., Hicks, F. A., Wilkinson, D. R., Tarrant, J. G., Bruno, S. M., Vargas, M., Hosein, K. N., Gupton, J. T., and Sikorski, J. A., *Tetrahedron*, **1995**, *51*, 1575.

264. Henecke, H., *Chem. Ber.*, **1949**, *82*, 36.

265. Mariella, R. P., *Org. Synth.*, *Coll. Vol. IV*, **1963**, 210.

266. Bohlmann, F. and Rahtz, D., *Chem. Ber.*, **1957**, *90*, 2265.

267. Bottorff, E. M., Jones, R. G., Kornfield, E. C., and Mann, M. J., *J. Am. Chem. Soc.*, **1951**, *73*, 4380.

268. Dornow, A. and Neuse, E., *Chem. Ber.*, **1951**, *84*, 296.

269. Wai, J. S. *et al.*, *J. Med. Chem.*, **1993**, *36*, 249.

270. Matsui, M., Oji, A., Kiramatsu, K., Shibata, K., and Muramatsu, H., *J. Chem. Soc., Perkin Trans 2*, **1992**, 201; Jain, R., Roschangar, F., and Ciufolini, M. A., *Tetrahedron Lett.*, **1995**, *36*, 3307.

271. Naito, T., Yoshikawa, T., Ishikawa, F., Isoda, S., Omura, Y., and Takamura, I., *Chem. Pharm. Bull.*, **1965**, *13*, 869; Kondrat'eva, G. Ya. and Huan, C.-H., *Dokl. Akad. Nauk, SSSR*, **1965**, *164*, 816 (*Chem. Abs.*, **1966**, *64*, 2079).

272. Okatani, T., Koyama, J., Suzata, Y., and Tagahara, K., *Heterocycles*, **1988**, *27*, 2213.

273. Boger, D. L. and Panek, J. S., *J. Org. Chem.*, **1981**, *46*, 2179.

274. Sauer, J. and Heldmann, D. K., *Tetrahedron Lett.*, **1998**, *39*, 2549.

275. Pfller, O. C. and Sauer, J., *Tetrahedron Lett.*, **1998**, *39*, 8821.

276. Carly, P. R., Cappelle, S. L., Compernolle, F., and Hoornaert, G. J., *Tetrahedron*, **1996**, *52*, 11889; Meerpoel, L. and Hoornaert, G., *Tetrahedron Lett.*, **1989**, *30*, 3183.

277. Sainte, F., Serckx-Poncin, B., Hesbain-Frisque, A.-M., and Ghosez, L., *J. Am. Chem. Soc.*, **1982**, *104*, 1428.

278. Germain, A. L., Gilchrist, T. L., and Kemmitt, P. D., *Heterocycles*, **1994**, *37*, 697; Gilchrist, T. L. and Healy, M. A. M., *Tetrahedron*, **1993**, *49*, 2543; Hibino, S., Sugino, E., Kuwada, T., Ogura, N., Sat, K., and Choshi, T., *J. Org. Chem.*, **1992**, *57*, 5917.

279. Clauson-Kaas, N. *et al.*, *Acta Chem. Scand.*, **1955**, *9*, 1, 9, 14, 23, and 30; Clauson-Kaas, N., Petersen, J. B., Sorensen, G. O., Olsen, G., and Jansen, G., *Acta Chem. Scand.*, **1965**, *19*, 1146.

280. Clauson-Kaas, N. and Meister, M., *Acta Chem. Scand.*, **1967**, *21*, 1104

281. Barrett, A. G. M. and Lebold, S. A., *Tetrahedron Lett.*, **1987**, *28*, 5791.

282. Frank, R. C., Pilgrim, F. J., and Riener, E. F., *Org. Synth.*, *Coll. Vol. IV*, **1963**, 451.

283. Chelucci, G., Faloni, M., and Giacomelli, G., *Synthesis*, **1990**, 1121; 'Organocobalt-catalysed synthesis of pyridines', Bnnemann, H. and Brijoux, W., *Adv. Heterocycl. Chem.*, **1990**, *48*, 177.

284. Chumakov, Yu I. and Sherstyuk, V. P., *Tetrahedron Lett.*, **1965**, 129.

285. Waldner, A., *Synth. Commun.*, **1989**, *19*, 2371.

286. Harris, S. A. and Folkers, K., *J. Am. Chem. Soc.*, **1939**, *61*, 1245.

287. Cruskie, M. P., Zoltewicz, J. A., and Abboud, K. A., *J. Org. Chem.*, **1995**, *60*, 7491.

6 Quinolines and isoquinolines: reactions and synthesis

quinoline   isoquinoline

Quinoline and isoquinoline are stable; the former is a high-boiling liquid the latter a low-melting solid, each with a sweetish odour. Both bases have been known for a long time: quinoline was first isolated from coal tar in 1834, isoquinoline from the same source in 1885. Shortly after the isolation of quinoline from coal tar it was also recognised as a pyrolytic degradation product of cinchonamine, an alkaloid closely related to quinine, from which the name quinoline is derived; the word quinine, in turn, derives from *quina*, a Spanish version of a local South American name for the bark of quinine-containing *Cinchona* species. Several synthetic antimalarial drugs are based on the quinoline nucleus, chloroquine is an example. Ciprofloxacin is one of several 4-quinolone-based antibiotics in use.

quinine   Chloroquine   methoxatin   Ciprofloxacin

Quinolines play a relatively minor role in fundamental metabolism, methoxatin, an enzyme cofactor of methylotrophic bacteria, being one of the small number of examples. There are also comparitively few quinoline-containing secondary metabolites, in contrast to isoquinoline, which occurs, mainly at the 1,2,3,4-tetrahydro-level, in a large number of alkaloids – the opium poppy alkaloids papaverine and, in more-elaborated form, morphine are examples.[1] Emetine, with two tetrahydroisoquinoline units, is a medicinally important amoebicide.

papaverine   morphine   emetine

Quinoline compounds provided the first photographic film sensitizers: the cyanine dye[?] ethyl red extended photography from the blue into the green and then in 1904, with pinacyanol, into the red. Subsequently, thousands of sensitizing dyes have been made and tested and quinoline-based dyes replaced by other, more efficient systems.

ethyl red   pinacyanol

6.1   Reactions with electrophilic reagents

6.1.1   Addition to nitrogen

All the reactions noted in this category for pyridine (section 5.1.1), which involve donation of the nitrogen lone pair to electrophiles, also occur with quinoline and isoquinoline and little further comment is necessary, for example the respective pK_a values, 4.94 and 5.4, show them to be of similar basicity to pyridine. Each, like pyridine, forms N-oxides and quaternary salts.

6.1.2   Substitution at carbon

6.1.2.1   *Proton exchange*

Benzene ring C-protonation, and thence exchange, *via* N-protonated quinoline, requires strong sulfuric acid and occurs fastest at C-8, then at C-5 and C-6; comparable exchange in isoquinoline takes place somewhat faster at C-5 than at C-8.[3] At lower acid strengths each system undergoes exchange α to nitrogen, at C-2 for quinoline and C-1 for isoquinoline. These processes involve a zwitterion produced by deprotonation of the N-protonated heterocycle.

6.1.2.2   *Nitration (see also 6.3.1.2)*

The positional selectivity for proton exchange is partly mirrored in nitrations, quinoline gives approximately equal amounts of 5- and 8-nitroquinolines whereas isoquinoline produces almost exclusively the 5-nitro-isomer;[4] mechanistically the substitutions involve nitronium ion attack on the N-protonated heterocycles.

6.1.2.3   *Sulfonation*

Sulfonation of quinoline gives largely the 8-sulfonic acid whereas isoquinoline affords the 5-acid.[5] Reactions at higher temperatures produce other isomers, under thermodynamic control, for example both quinoline 8-sulfonic acid and quinoline 5-sulfonic acid are isomerised to the 6-acid.[6]

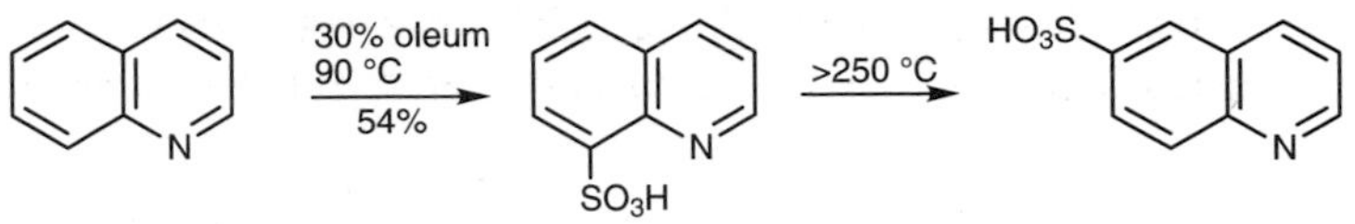

6.1.2.4 Halogenation

Ring substitution of quinoline and isoquinoline by halogens is rather complex, products depending on the conditions used.[7] In concentrated sulfuric acid quinoline gives a mixture of 5- and 8-bromo-derivatives; comparably, isoquinoline is efficiently converted into the 5-bromo derivative in the presence of aluminium chloride;[8] both processes involve halogen attack on a salt.

Introduction of halogen to the hetero-rings occurs under remarkably mild conditions in which the nitrogen lone pair initiates a sequence by interaction with an electrophile. Thus treatment of quinoline and isoquinoline hydrochlorides with bromine produces 3-bromoquinoline and 4-bromoisoquinoline respectively as illustrated below for the latter.[9]

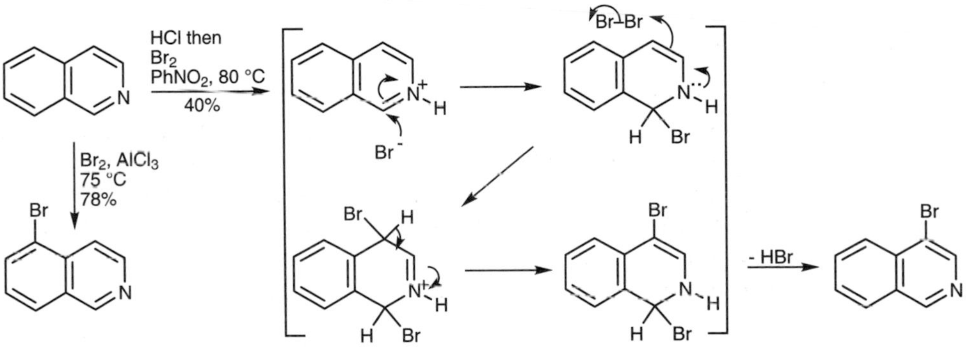

6.1.2.5 Acylation and alkylation

There are no generally useful processes for the introduction of carbon substituents by electrophilic substitution of quinolines or isoquinolines except for a few examples in which a ring has a strong electron-releasing substituent.

6.2 Reactions with oxidising agents

It requires vigorous conditions to degrade a ring in quinoline and isoquinoline: examples of attack at both rings are known, though degradation of the benzene ring, generating pyridine diacids, should be considered usual;[10] ozonolysis can be employed to produce pyridine dialdehydes,[11] or after subsequent hydrogen peroxide treatment, diacids.[12] Electrolytic oxidation of quinoline is the optimal way to convert quinoline to pyridine-2,3-dicarboxylic acid ('quinolinic acid')[13]; alkaline potassium permanganate converts isoquinoline into a mixture of pyridine-3,4-dicarboxylic acid ('cinchomeronic acid') and phthalic acid.[14]

6.3 Reactions with nucleophilic reagents

6.3.1 Nucleophilic substitution with hydride transfer

Reactions of this type occur fastest at C-2 in quinoline and at C-1 in isoquinolines.

6.3.1.1 Alkylation and arylation

The immediate products of addition of alkyl and aryl Grignard reagents and alkyl- and aryllithiums are dihydroquinolines and -isoquinolines and can be characterised as such, but can be oxidised to afford the *C*-substituted, rearomatised heterocycle; illustrated below is a 2-arylation of quinoline.[15]

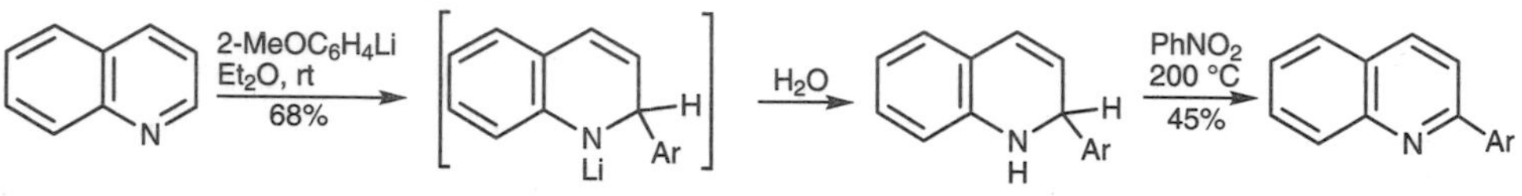

Vicarious nucleophilic substitution (section 2.3.3) allows the introduction of substituents into nitroquinolines: cyanomethyl and phenylsulfonylmethyl groups, for example, can be introduced *ortho* to the nitro group, in 5-nitroquinolines at C-6 and in 6-nitroquinolines at C-5.[16]

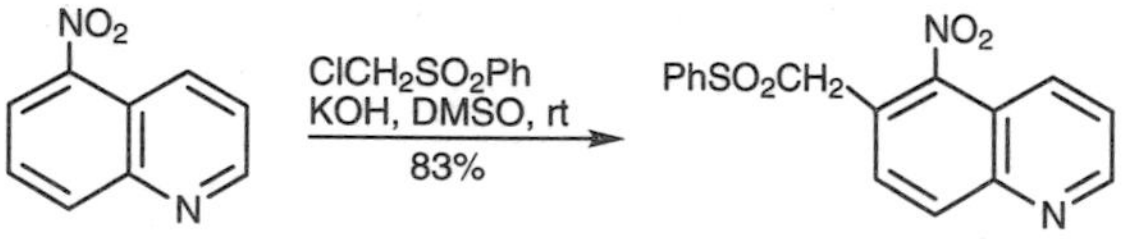

There are considerable possibilities inherent in addition of organolithiums to osmium complexes of quinoline for the synthesis of 5-substituted quinolines.[17]

6.3.1.2 Amination and nitration

Sodium amide reacts rapidly and completely with quinoline and isoquinoline, even at -45 °C, to give dihydro-adducts with initial amide attack at C-2 (main) and C-4 (minor) in quinoline and C-1 in isoquinoline. The quinoline 2-adduct rearranges to the more stable 4-aminated adduct at higher temperatures.[18] Oxidative trapping of the quinoline adducts provides 2- or 4-aminoquinoline;[19] isoquinoline reacts with potassium amide in liquid ammonia at room temperature to give 1-aminoisoquino-line.[20]

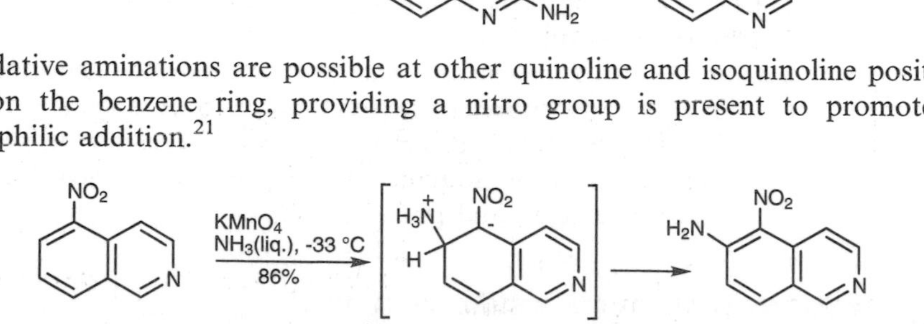

Oxidative aminations are possible at other quinoline and isoquinoline positions, even on the benzene ring, providing a nitro group is present to promote the nucleophilic addition.[21]

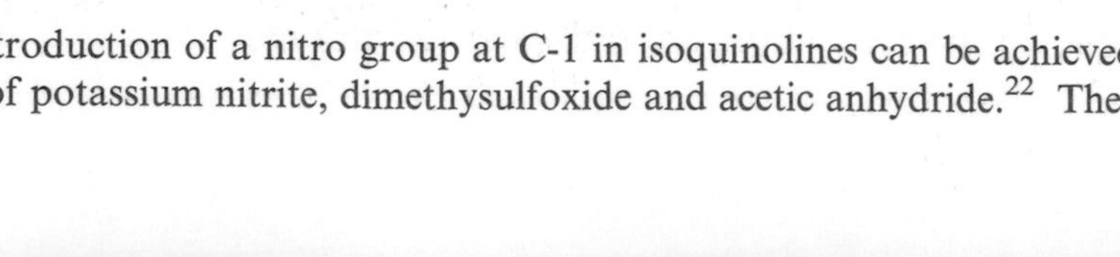

The introduction of a nitro group at C-1 in isoquinolines can be achieved using a mixture of potassium nitrite, dimethysulfoxide and acetic anhydride.[22] The key step is the nucleophilic addition of nitrite to the heterocycle previously quaternised by reaction through nitrogen with a complex of dimethylsulfoxide and the anhydride.

6.3.1.3 Hydroxylation

Both quinoline and isoquinoline can be directly hydroxylated with potassium hydroxide at high temperature with the evolution of hydrogen.[23] 2-Quinolone ('carbostyril') and 1-isoquinolone ('isocarbostyril') are the isolated products.

6.3.2 Nucleophilic substitution with displacement of halide

The main principle here is that halogen on the homocyclic rings of quinoline and isoquinoline, and at the quinoline-3- and the isoquinoline 4-positions behave as would halobenzenes. In contrast, 2- and 4-haloquinolines and 1-haloisoquinolines have the same susceptibility as α- and γ-halopyridines (see section 5.3.2). 3-Haloisoquinolines are intermediate in their reactivity to nucleophiles.[24]

An apparent exception to the relative unreactivity of 3-haloisoquinolines is provided by the reaction of 3-bromoisoquinoline with sodium amide. Here, a different mechanism, known by the acronym ANRORC (*A*ddition of *N*ucleophile, *R*ing *O*pening and *R*ing *C*losure), leads to the product, apparently of direct displacement, but in which a switching of the ring nitrogen, to become the substituent nitrogen, has occurred.[25]

6.4 Reactions with bases

6.4.1 Deprotonation of C-hydrogen

Direct lithiation, i.e. *C*-deprotonation of quinolines[26] requires an adjacent substituent such as chlorine, fluorine, or alkoxy – probably the first ever strong base *C*-deprotonation of a six-membered heterocycle was the 3-lithiation of 2-ethoxyquinoline.[27] Both 4- and 2-dimethylaminocarbonyloxyquinolines lithiate at C-3. 4-Pivaloylaminoquinoline lithiates at the *peri* position, C-5. 3-Substituted quinolines lithiate at C-4, not at C-2.

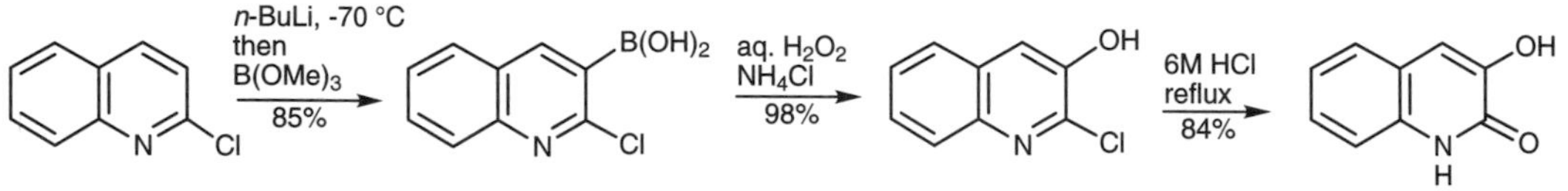

2-Lithiation of 1-substituted 4-quinolones[28] and 3-lithiation of 2-quinolone[29] provides derivatives with the usual nucleophilic propensity, as illustrated below.

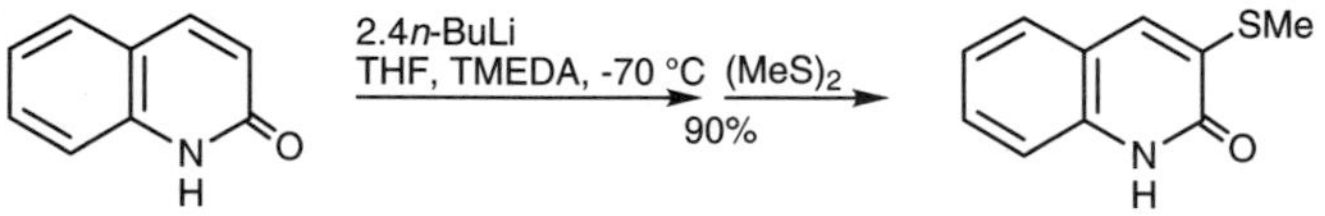

6.5 Reactions of C-metallated quinolines and isoquinolines

6.5.1 Lithium derivatives

The preparation of lithioquinolines and -isoquinolines *via* metal–halogen exchange is complicated by competing nucleophilic addition. Low temperatures do allow metal–halogen exchange at both pyridine[30] and benzene ring positions[31] in quinolines, and the isoquinoline-1-[30] and 4-positions,[32] subsequent reaction with electrophiles generating *C*-substituted products. It seems that for benzene ring lithiation two mol equivalents of butyllithium are necessary to allow one butyllithium to associate with the ring nitrogen as suggested below.

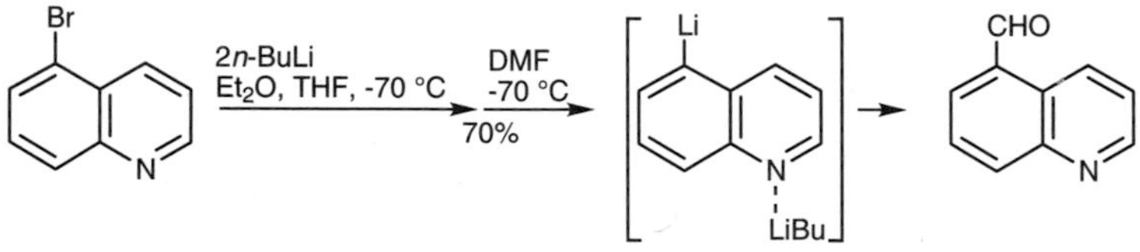

6.5.2 Zinc derivatives

Direct formation of quinolinylzinc reagents is illustrated below.[33]

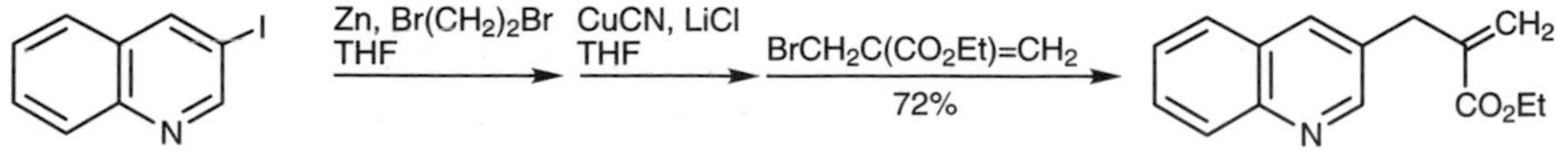

6.5.3 Palladium- and nickel-catalysed reactions

The coupling of various haloquinolines and -isoquinolines has been effected with palladium[34] or nickel reagents[35] – a couple of examples are shown below. Even chlorine, at the quinoline 2-position, will take part in such processes.[36] A nice distinction is made between chlorine at the isoquinoline 1- and 3-positions: 1,3-dichloroisoquinoline couples only at the 1-position with arylboronic acids using palladium catalysis (*cf.* section 6.3.2).[37]

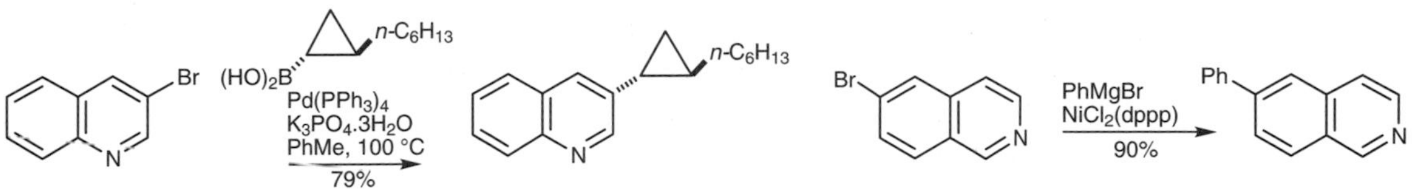

6.6   Reactions with radical reagents

Phenyl radicals generated by the decomposition of dibenzoyl peroxide attack quinoline and isoquinoline with formation of mixtures of all the isomeric phenyl-substituted products. Much more discriminating substitutions can be achieved with more nucleophilic radicals in acid solution (cf. section 2.4.1) .

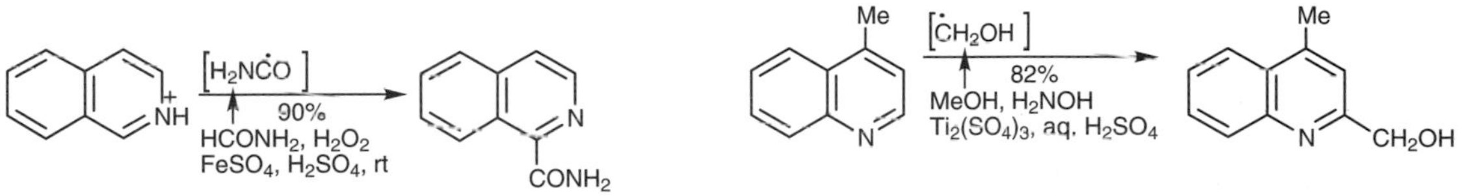

6.7   Reactions with reducing agents

Selective reduction of either the pyridine or the benzene rings in quinolines and isoquinoline can be achieved: the heterocyclic ring is reduced to the tetrahydro level by sodium cyanoborohydride in acid solution,[38] by sodium borohydride in the presence of nickel(II) chloride,[39] by zinc borohydride,[40] or, traditionally, by room temperature and room pressure catalytic hydrogenation in methanol. However, in strong acid solution it is the benzene ring which is selectively saturated;[41] longer reaction times can then lead to decahydro-derivatives.

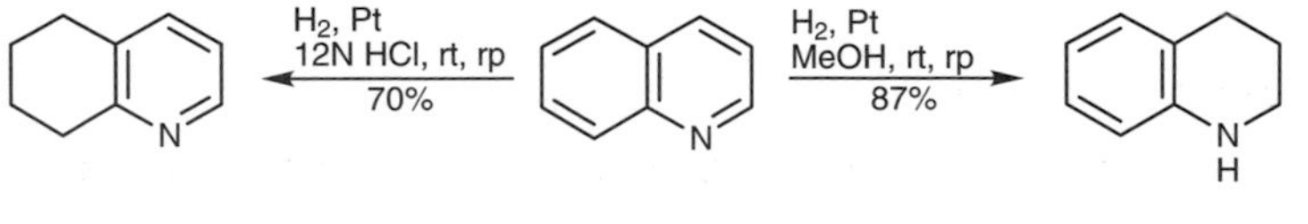

Lithium in liquid ammonia conditions can produce 1,4-dihydroquinoline[42] and 3,4-dihydroisoquinoline[43] under certain conditions. Conversely, lithium aluminium hydride reduces generating 1,2-dihydroquinoline[44] and 1,2-dihydroisoquinoline.[45] These dihydro-heterocycles[46] can be easily oxidised back to the fully aromatic systems, or disproportionate,[47] especially in acid solution, to give a mixture of tetrahydro- and rearomatised compounds as shown below. Stable dihydro derivatives can be obtained by trapping following reduction, as a urethane by reaction with a chloroformate.[48]

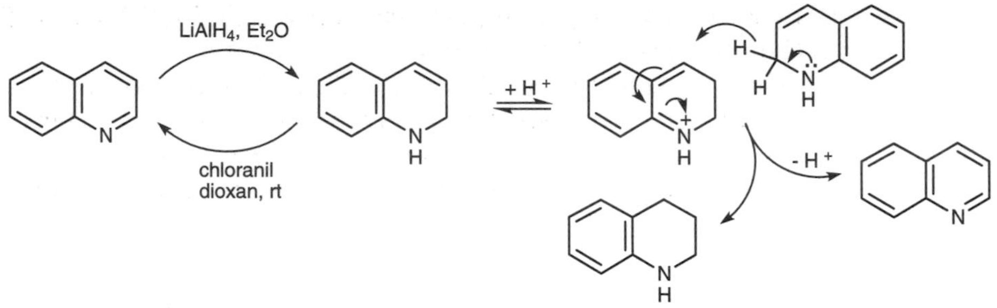

Quaternary salts of quinoline and isoquinoline are particularly easily reduced in the heterocyclic ring, either catalytically or with a borohydride in protic solution.

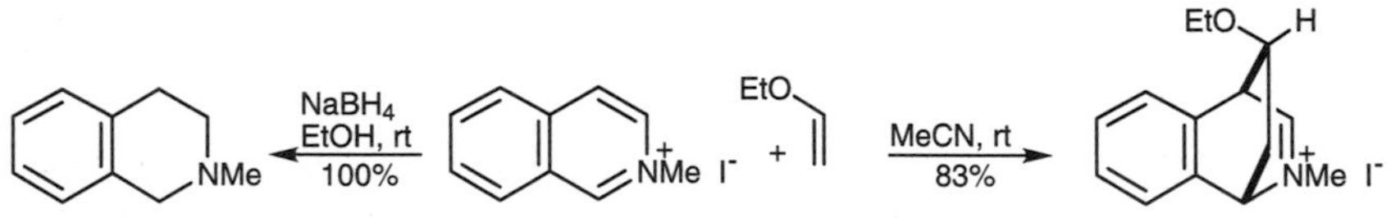

1-Methylquinolinium iodide is reduced by tributylstannane to give mainly the 1,2-dihydro-isomer, which isomerises at room temperature to the 1,4-dihydro-isomer; reduction with this reagent but with concurrent irradiation, produced exclusively 1,4-dihydro-1-methylquinoline, quantitatively.[49]

6.8 Electrocyclic reactions (ground state)

The tendency for relatively easy nucleophilic addition to the pyridinium ring in isoquinolinium salts is echoed in the cycloaddition (shown above) of electron-rich dienophiles such as ethoxyethene, which is reversed on refluxing in acetonitrile.[50] The comparable reaction of 4-oxygenated 2-arylisoquinolinium salts is a neat method for the synthesis of heavily functionalised tetralins.[51]

6.9 Photochemical reactions

Of a comparitively small range of photochemical reactions described for quinolines and isoquinolines perhaps the most intriguing are some hetero-ring rearrangements of quaternary derivatives, which can be illustrated by the ring expansions of their N-oxides.[52] As with 2-pyridones, 2-quinolones undergo $2+2$ photo-dimerisation.[53]

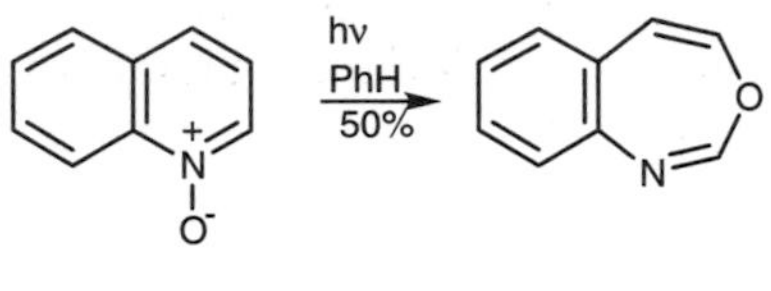

6.10 Oxyquinolines and -isoquinolines

Quinolinols and isoquinolinols in which the oxygen is at any position other than 2- and 4- for quinolines and 1- and 3- for isoquinolines are true phenols i.e. have an hydroxyl group, though they exist in equilibrium with variable concentrations of zwitterionic structures with the nitrogen protonated and the oxygen deprotonated. They show the typical reactivity of naphthols.[54] 8-Quinolinol has long been used in analysis as a chelating agent, especially for Zn(II), Mg(II), and Al(III) cations; the Cu(II) chelate is used as a fungicide.

2-Quinolone (strictly 2(1H)-quinolinone) and 4-quinolone and 1-isoquinolone are completely in the carbonyl tautomeric form[55] for all practical purposes – the hydroxyl tautomers lack a favourable polarised resonance contribution, as illustrated below for 1-isoquinolone. There is considerable interest in quinolones as antibacterial agents.[56]

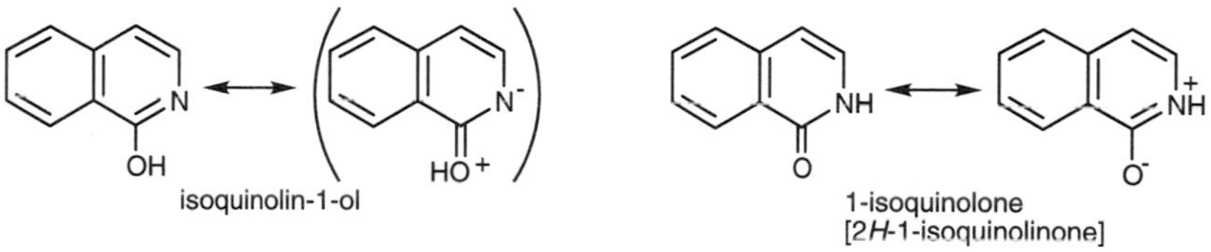

isoquinolin-1-ol

1-isoquinolone
[2H-1-isoquinolinone]

In 3-oxy-isoquinoline there is an interesting and instructive situation: here the two tautomers are of comparable stability. 3-Isoquinolinol is dominant in dry ether solution, 3-isoquinolone is dominant in aqueous solution. A colourless ether solution of 3-isoquinolinol turns yellow on addition of a little methanol because of the production of some of the carbonyl tautomer. The similar stabilities is the consequence of the balancing of two opposing tendencies: the presence of an amide unit in 3-isoquinolone forces the benzene ring into a less favoured quinoid structure, conversely, the complete benzene ring in isoquinolinol necessarily means loss of the amide unit and its contribution to stability. One may contrast this with 1-isoquinolone which has an amide, as well as a complete benzene, unit.[57]

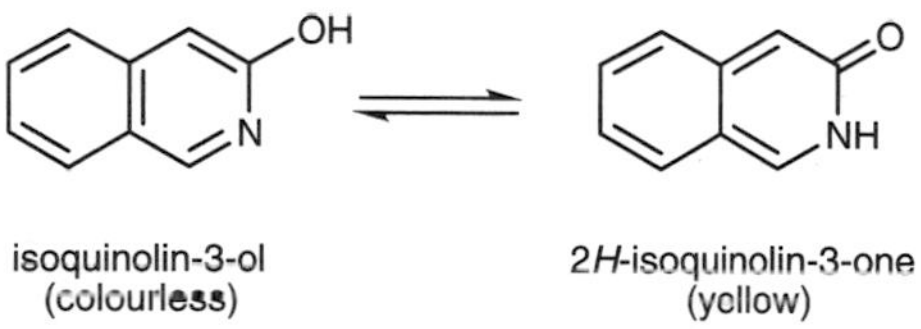

isoquinolin-3-ol
(colourless)

2H-isoquinolin-3-one
(yellow)

The position of electrophilic substitution of quinolones and isoquinolones depends upon the pH of the reaction medium. Each type protonates on carbonyl oxygen so reactions in strongly acidic media involve attack on this cation: the contrast can be illustrated by the nitration of 4-quinolone at different acid strengths.[58] The balance between benzene ring and unprotonated heterocyclic ring selectivity is small, for example 2-quinolone chlorinates preferentially, as a neutral molecule, at C-6, and only secondly at C-3.

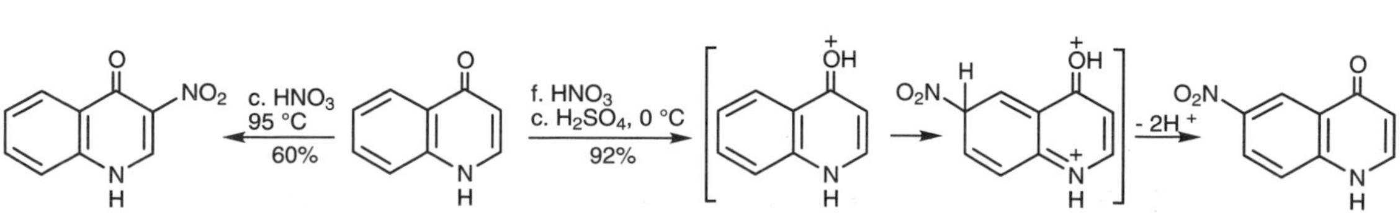

Strong acid-catalysed H-exchange of 2-quinolone proceeds fastest at C-6 and C-8; of 1-isoquinolone at C-4, then 5–7.[59] This is echoed in various electrophilic substitutions, for example formylation.[60]

heterocycles. *N*-Sulfonyl analogues of Reissert adducts easily eliminate arylsulfinate, thus providing a method for the introduction of a cyano group.[78]

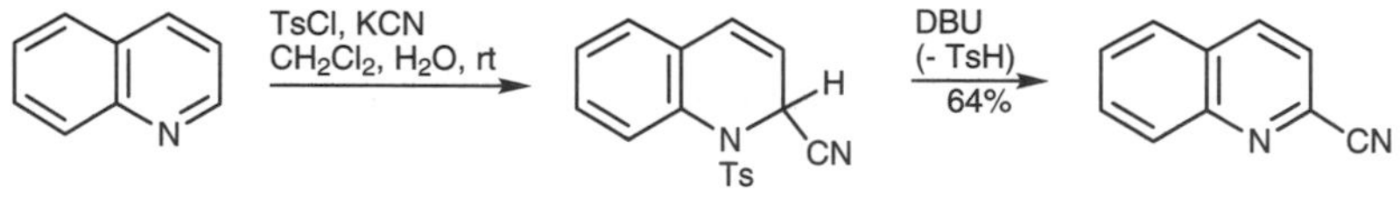

Allylsilanes will also trap *N*-acylisoquinolinium[79] and *N*-acylquinolinium[80] salts, and silyl alkynes will add,[81] also with silver ion catalysis.

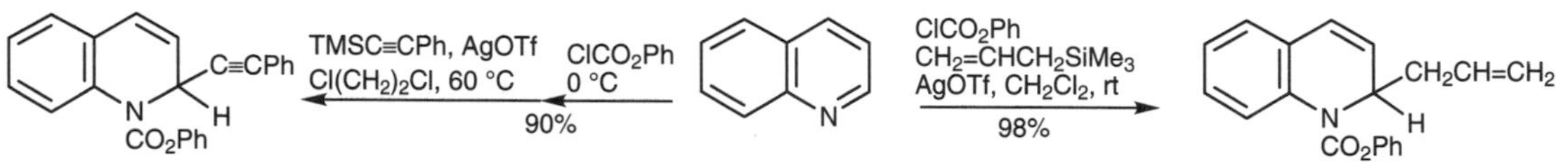

The Zincke salt of isoquinoline[82] is easily transformed into chiral isoquinolinium salts on reaction with chiral amines.[83] Nucleophilic addition of Grignard reagents to these salts shows good stereoselectivity.[84]

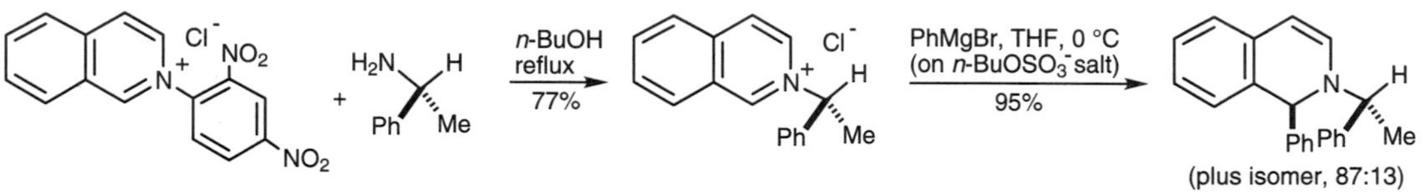

6.15 Quinoline and isoquinoline *N*-oxides

N-Oxide chemistry in these bicyclic systems largely parallels the processes described for pyridine *N*-oxide, with the additional possibility of benzene ring electrophilic susbstitution, for example mixed acid nitration of quinoline *N*-oxide takes place at C-5 and C-8 via the *O*-protonated species, but at C-4 at lower acid strength;[85] nitration of isoquinoline *N*-oxide takes place at C-5.[86]

Diethyl cyanophosphonate converts quinoline and isoquinoline *N*-oxides into the 1- and 2-cyano heterocycles in high yields in a process which must have *O*-phosphorylation as a first step, and in which the elimination of diethylphosphate may proceed via a cyclic transition state;[87] trimethylsilyl cyanide and diazabicycloundecene effect the same transformation.[88] Chloroformates and an alcohol convert the *N*-oxides into ethers, as illustrated below for isoquinoline *N*-oxide.[89]

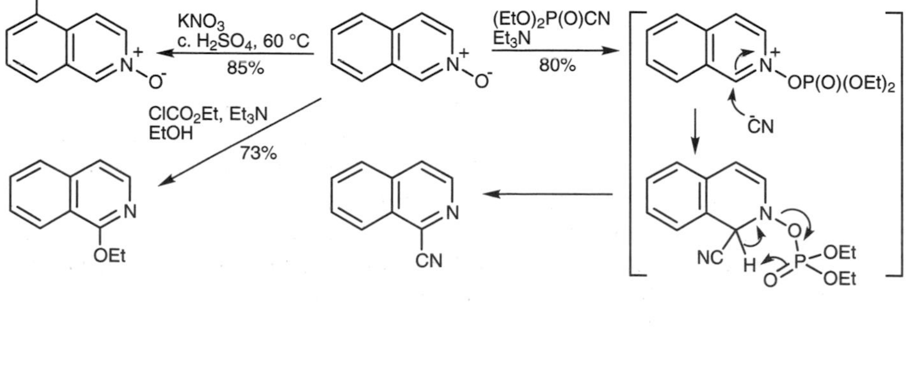

6.16   Synthesis of quinolines and isoquinolines

6.16.1   Ring syntheses

Three of the more generally important approaches to quinoline and three to isoquinoline compounds from non-heterocyclic precursors, are summarised in this section.

6.16.1.1   Quinolines from arylamines and 1,3-dicarbonyl compounds

Anilines react with 1,3-dicarbonyl compounds to give intermediates which can be cyclised with acid.

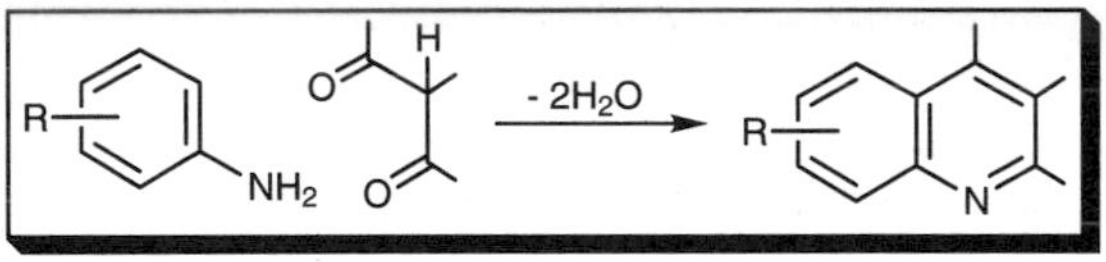

The Combes synthesis

Condensation of a 1,3-dicarbonyl compound with an arylamine gives a high yield of a β-amino-enone, which can then be cyclised with concentrated acid.[90] Mechanistically, the cyclisation step can be viewed as an electrophilic substitution by the O-protonated amino-enone, as shown, followed by loss of water to give the aromatic quinoline.

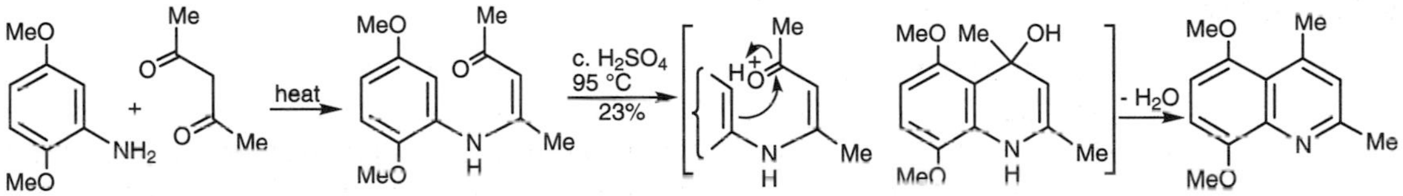

An example of this approach which provides 4-unsubstituted 3-nitroquinolines, is shown below.[91]

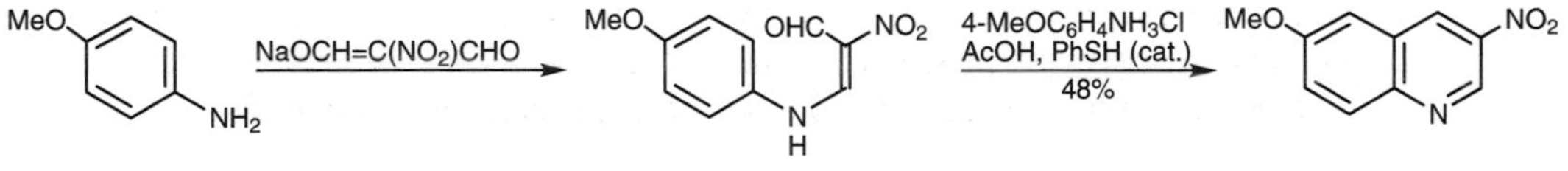

The Conrad–Limpach–Knorr synthesis

This closely related synthesis uses β-ketoesters and leads to quinolones.[92] Anilines and β-ketoesters can react at lower temperatures to give the kinetic product, a β-aminoacrylate, cyclisation of which gives a 4-quinolone. At higher temperatures, β-ketoester anilides are formed and cyclisation of these affords 2-quinolones. β-Aminoacrylates, for cyclisation to 4-quinolones, are also available *via* the addition of anilines to acetylenic esters[93] or by reaction with diethyl ethoxymethylenemalonate $(EtOCH = C(CO_2Et)_2 \rightarrow ArNHCH = C(CO_2Et)_2)$.[94]

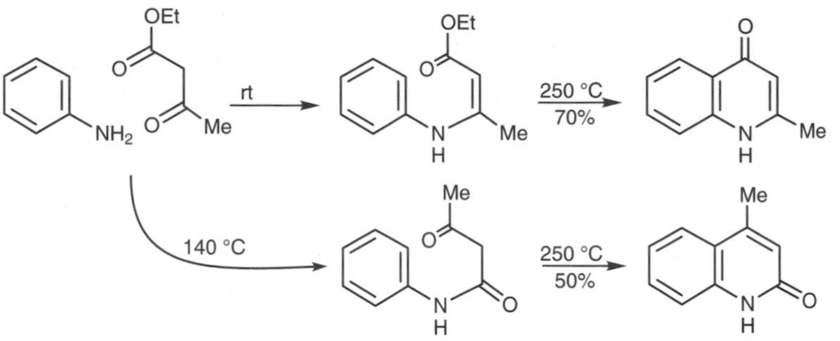

Cyclisations where the benzene ring carries an electron-withdrawing group can be effected using the variant shown below – the substrate is simply heated strongly – the mechanism of ring closure probably does not involve electrophilic attack on the benzene ring but rather is better viewed as the electrocyclic cyclisation of a 1,3,5-3-azatriene.[95]

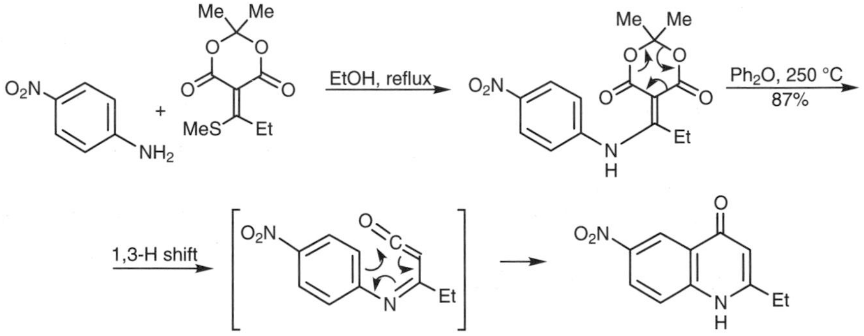

6.16.1.2 *Quinolines from arylamines and α,β-unsaturated carbonyl compounds*

Arylamines react with an α,β-unsaturated carbonyl compound in the presence of an oxidising agent to give quinolines. When glycerol is used as an *in situ* source of acrolein, quinolines carrying no substituents on the heterocyclic ring are produced.

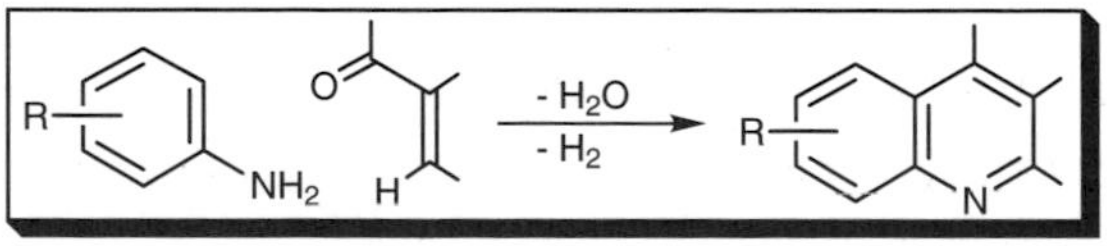

The Skraup synthesis[96]

In this extraordinary reaction, quinoline is produced when aniline, concentrated sulfuric acid, glycerol and a mild oxidising agent are heated together.[97] The reaction has been shown to proceed by dehydration of the glycerol to acrolein to which aniline then adds in a conjugate fashion. Acid-catalysed cyclisation produces a 1,2-dihydroquinoline finally dehydrogenated by the oxidising agent – the corresponding nitrobenzene or arsenic acid have been used classically, though with the inclusion of a little sodium iodide, the sulfuric acid can serve as oxidant.[12] The Skraup synthesis is the best for the ring synthesis of quinolines unsubstituted on the hetero-ring.[98]

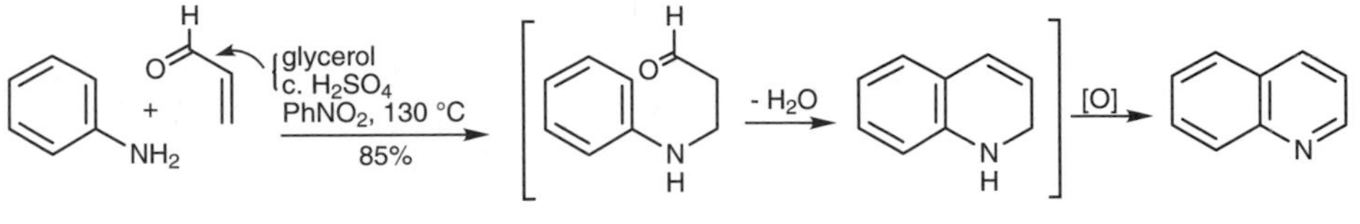

The use of substituted carbonyl components confirms the mechanism, showing that interaction of the aniline amino group with the carbonyl group is not the first step.

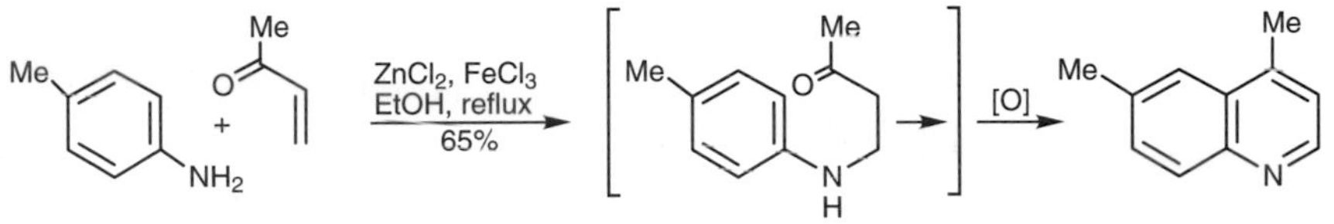

Skraup syntheses sometimes become very vigorous and care must be taken to control their potential violence; preforming the Michael adduct and using an alternative oxidant (*p*-chloranil was the best) has been shown to be advantageous in terms of yield and as a better means for controlling the reaction.[99]

Orientation of ring closure in Skraup syntheses

In principle, *meta*-substituted arylamines could give rise to both 5- and 7-substituted quinolines. In practice, electron-donating substituents direct ring closure *para*, thus producing 7-substituted quinolines; *meta*-halo-arylamines produce mainly the 7-halo isomer. Arylamines with a strong electron-withdrawing *meta* substituent give rise mainly to the 5-substituted quinoline.

6.16.1.3   *Quinolines from ortho-acylarylamines and carbonyl compounds*

ortho-Acylarylamines react with ketones having an α-methylene to give quinolines.

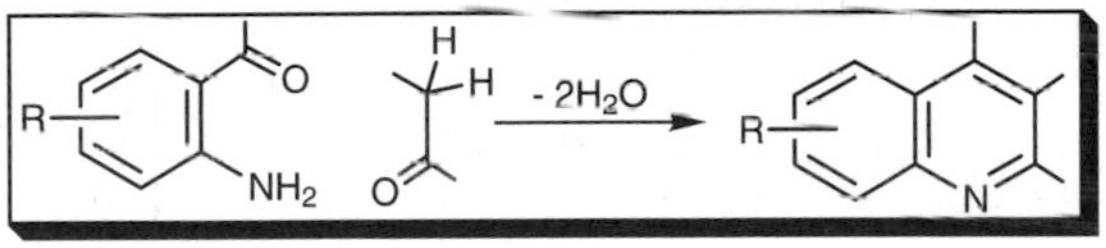

The Friedländer synthesis[100]

ortho-Acylarylamines[101] condense with a ketone or aldehyde (which must contain an α-methylene group) by base or acid catalysis to yield quinolines. The orientation of condensation depends on the orientation of enolate or enol formation.[102] The use of an oxime ether, as synthon for the α-methylene ketone, has been shown to be advantageous.[103]

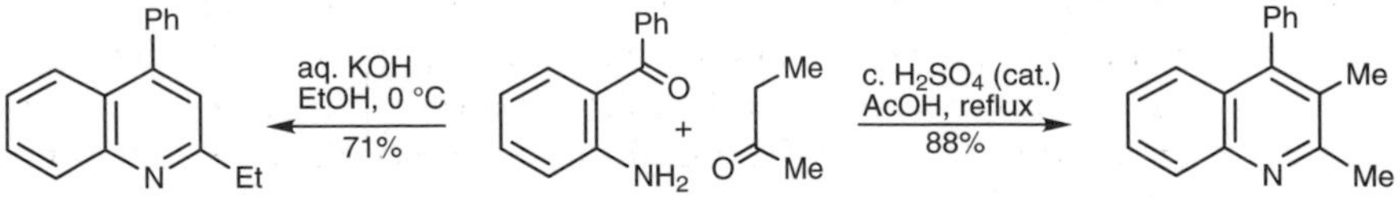

The potential of *ortho,N*-dilithiated *N*-protected arylamines for the preparation of *ortho*-acylarylamines starting components has been described,[104] and this idea taken further to provide a one-pot modification as illustrated below.

The Pfitzinger synthesis

ortho-Aminoaraldehydes are sometimes difficult of access; in this modification, isatins (sections 17.14.3 and 17.17.4), which are easy to synthesise, are hydrolysed to *ortho*-aminoarylglyoxylates, which react with ketones affording quinoline-4-carboxylic acids.[105] The carboxylic acid group can be removed, if required, by pyrolysis with calcium oxide.

6.16.1.4   *Isoquinolines from arylaldehydes and aminoacetal*

Aromatic aldehydes react with aminoacetal (2,2-diethoxyethanamine) to generate imines which can be cyclised with acid to isoquinolines carrying no substituents on the heterocyclic ring.

The Pomeranz-Fritsch synthesis

The Pomeranz-Fritsch synthesis[106] is normally carried out in two stages. Firstly, an aryl aldehyde is condensed with aminoacetal to form an aryl aldimine. This stage proceeds in high yield under mild conditions. Secondly the aldimine is cyclised by treatment with strong acid; hydrolysis of the imine competes and reduces the efficiency of this step and for this reason trifluoroacetic acid with boron trifluoride is a useful reagent.[107]

   The second step is similar to those in the Combes and Skraup syntheses, in that the acid initially protonates, causing elimination of ethanol and the production of a species which can attack the aromatic ring as an electrophile. Final elimination of a second mole of alcohol completes the process.

The electrophilic nature of the cyclisation step explains why the process works best for araldimines carrying electron-donating substituents (especially when these are oriented *para* to the point of closure leading to 7-substituted isoquinolines) and least well for systems deactivated by electron-withdrawing groups.

The problem of imine hydrolysis can be avoided by cyclising at a lower oxidation level, with tosyl on nitrogen for subsequent elimination as toluenesulfinic acid. The ring closure substrates can be obtained by reduction and tosylation of imine condensation products[108] or by benzylating the sodium salt of 2-tosylaminoethanal acetal.[109]   Alternatively, a benzylic alcohol can be reacted with N-sulfonyl-aminoacetals in a Mitsunobu reaction.[110] Cyclisation of benzylaminoethanal acetals using chlorosulfonic acid gives the aromatic isoquinoline directly.[111]

Isoquinolines substituted at C-1 are not easily formed by the standard Pomeranz-Fritsch procedure. The first step would require formation of a ketimine from aminoacetal and a aromatic ketone, which would proceed much less well than for an aryl aldehyde. A variation, which overcomes this difficulty, has a benzylamine condensing with glyoxal diethyl acetal; the resulting, isomeric imine, can be cyclised with acid.[112]

6.16.1.5   Isoquinolines from arylethanamides

The amide or imine from reaction of 2-arylethanamines with an acid derivative or with an aldehyde, can be ring-closed to a 3,4-dihydro- or 1,2,3,4-tetrahydroisoquinoline respectively. Subsequent dehydrogenation produces the aromatic heterocycle.

The Bischler-Napieralski synthesis[113]

In the classical process, a 2-arylethanamine reacts with a carboxylic acid chloride or anhydride to form an amide, which can be cyclised, with loss of water, to a 3,4-

dihydroisoquinoline, then readily dehydrogenated to the isoquinoline using palladium, sulfur, or diphenyl disulfide. Common cyclisation agents are phosphorus pentaoxide, phosphoryl chloride and phosphorus pentachloride. The electrophilic intermediate is very probably an imino chloride,[114] or imino phosphate; the former have been isolated and treated with Lewis acids when they are converted into isonitrilium salts, which cyclise efficiently to 3,4-dihydroisoquinolines.[115]

Here, once again, the cyclising step involves electrophilic attack on the aromatic ring so the method works best for activated rings, and *meta*-substituted substrates give exclusively 6-substituted isoquinolines.

Urethanes can be cyclised to 1-isoquinolones using trifluoromethanesulfonic anhydride and 4-dimethylaminopyridine; it may be that this reagent combination is well suited to the standard formation of 3,4-dihydroisoquinolines too.[116] The cyclisation of urethanes with a combination of phosphorus pentaoxide and phosphoryl chloride is not restricted to substrates with activated benzene rings.[117]

Pictet-Gams modification

By conducting the Bischler-Napieralski sequence with a potentially unsaturated arylethanamine, a fully aromatic isoquinoline can be obtained directly. The amide of a β-methoxy- or β-hydroxy-β-arylethanamine is heated with the usual type of cyclisation catalyst. It is not clear whether dehydration to an unsaturated amide or to an oxazolidine[118] is an initial stage in the overall sequence.

6.16.1.6   *Isoquinolines from activated arylethanamines and aldehydes*
The Pictet-Spengler synthesis[119]

Arylethanamines react with aldehydes easily and in good yields to give imines. 1,2,3,4-Tetrahydroisoquinolines result from their cyclisation with acid catalysis. Note that the lower oxidation level imine, *versus* amide, leads to a tetrahydro- not a dihydroisoquinoline. After protonation of the imine, a Mannich-type electrophile is generated; since these are intrinsically less electrophilic than the intermediates in Bischler-Napieralski closure, a strong activating substituent must be present, and appropriately sited on the aromatic ring, for efficient ring closure.

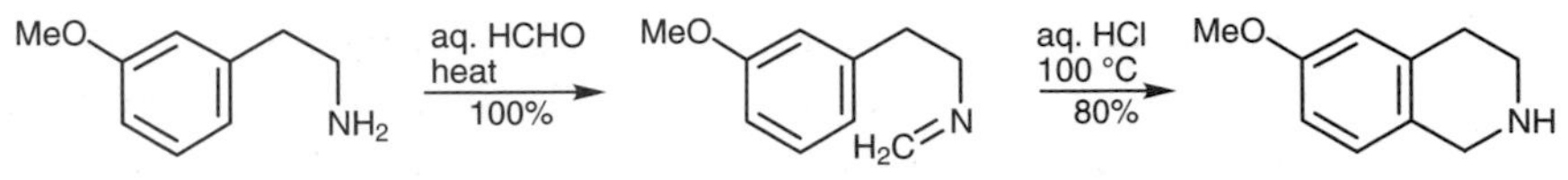

Highly activated hydroxylated aromatic rings permit Pictet-Spengler ring closure under very mild, 'physiological' conditions.[120]

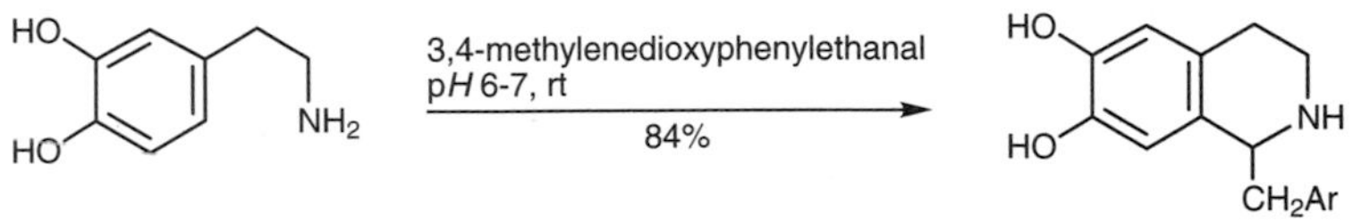

Routine dehydrogenation easily converts the tetrahydroisoquinolines produced by this route into fully aromatic species. Perhaps of more interest is their selective conversion into 3,4-dihydroisoquinolines using potassium permanganate and a crown ether.[121]

6.16.1.7 Newer methods

A number of recently described routes take rather different approaches to the synthesis of quinolines and isoquinolines, for example ozonolyses of indenes, as shown below provides homophthalaldehydes which are at exactly the right oxidation level for aromatic pyridine ring closure with ammonia.[122] Another method for the generation of equivalent species depends on the side-chain lithiation of ortho-methylaraldehyde cyclohexylamine imines, then acylation with a Weinreb amide.[123]

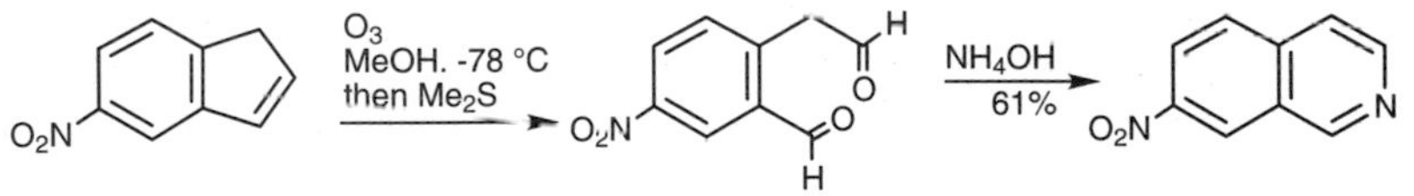

A route[124] in which a synthon for such a dialdehyde is central depends on ortho lithiation of an aryl bromide for conversion to ortho bromoaryl aldehyde, then palladium-catalysed replacement of the halide with an alkyne, subsequent reaction with ammonia producing the isoquinoline. The sequence below shows how this type of approach can be used to produce naphthyridine mono-N-oxides by reaction of the alkynyl-aldehyde with hydroxylamine instead of ammonia.[125]

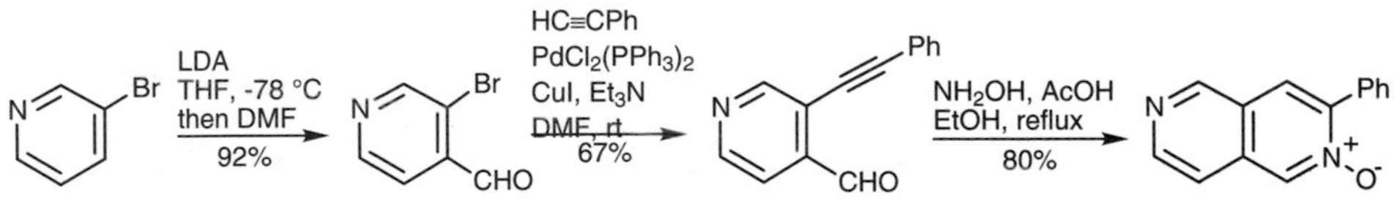

An isoquinoline ring disconnection, with considerable potential, could not be brought to practical fruition until a 1,2-monoazabisylide was synthesised; it involves Wittig and aza-Wittig condensations on phthalic dicarboxaldehyde to generate isoquinoline.[126]

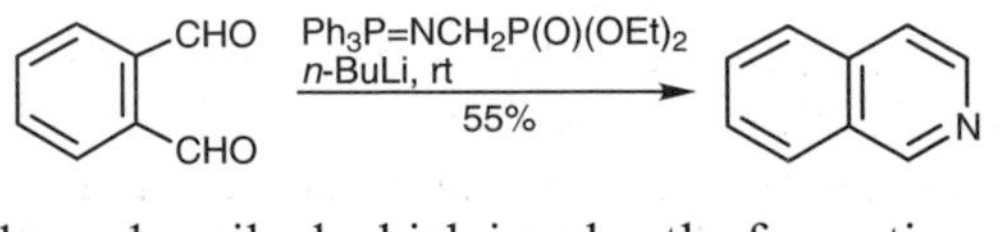

Some routes have been described which involve the formation of two C–C bonds in the same pot to produce quinolines and isoquinolines. For the production of 2-arylquinolines a Schiff base is reacted with an enol ether in presence of ytterbium(III) triflate.[127] 2-Chloro-3-formylquinolines result from a practically simple process involving the interaction of acetanilide with phosphoryl chloride and dimethylfor-mamide,[128] and for the synthesis of isoquinolines, an arylacetate is treated with a nitrile and trifluoromethanesulfonic anhydride.[129]

In a 2-quinolone synthesis[130] a *trans* cinnamate must isomerise during the ring closing process.

ortho-Iodoaraldehyde imines react directly with alkynes, using palladium(0) catalysis, generating isoquinolines in which the original nitrogen substituent has been lost.[131] An unusual quinoline synthesis has a bond making between the aromatic ring and the nitrogen of the future heterocycle as a key step. Reaction of alkyl 2-arylethylketone oximes with tetra-*n*-butylammonium perrhenate produces quinolines; depending on the aromatic substituents the mechanism involves either direct cyclising attack *ortho*, or *ipso* attack followed by rearrangement.[132]

A radical-based quinoline ring synthesis was developed[133] with the total synthesis[134] of the anti-cancer alkaloid camptothecin in mind. The sequence[135] below shows the construction of the ring skeleton of camptothecin and suggests a mechanism for the process.

6.16.2   Examples of notable syntheses of quinoline and isoquinoline compounds

6.16.2.1   Chloroquine[136]

Chloroquine is a synthetic antimalarial.

6.16.2.2   Papaverine[137]

Papaverine is an alkaloid from opium; it is a smooth muscle relaxant and thus useful as a coronary vasodilator – the synthesis illustrates the Pictet-Gams variation.

6.16.2.3   Methoxatin[138]

Methoxatin is an enzyme cofactor of bacteria which metabolise methanol. This total synthesis is a particularly instructive one since it includes an isatin synthesis (section 17.17.4), a quinoline synthesis, and an indole synthesis.

9. Kress, T. J. and Costantino, S. M., *J. Heterocycl. Chem.*, **1973**, *10*, 409.
10. Hoogerwerff, S. and van Dorp, W. A., *Chem. Ber.*, **1880**, *12*, 747.
11. Quéguiner, G. and Pastour, P., *Bull. Soc. Chim. Fr.*, **1968**, 4117.
12. O'Murcha, C., *Synthesis*, **1989**, 880.
13. Cochran, J. C. and Little, W. F., *J. Org. Chem.*, **1961**, *26*, 808.
14. Hoogerwerff, S. and van Dorp, W. A., *Recl. Trav. Chim. Pays-Bas*, **1885**, *4*, 285.
15. Geissman, T. A., Schlatter, M. J., Webb, I. D., and Roberts, J. D., *J. Org. Chem.*, **1946**, *11*, 741.
16. Makosza, M., Kinowski, A., Danikiewicz, W., and Mudryk, B., *Justus Liebigs Ann. Chem.*, **1986**, 69.
17. Bergman, B., Holmquist, R., Smith, R., Rosenberg, E., Ciurash, J., Hardcastle, K., and Visi, M., *J. Am. Chem. Soc.*, **1998**, *120*, 12818.
18. Zoltewicz, J., Helmick, L. S., Oestreich, T. M., King, R. W., and Kandetzki, P. E., *J. Org. Chem.*, **1973**, *38*, 1947.
19. Tondys, H., van der Plas, H. C., and Wozniak, M., *J. Heterocycl. Chem.*, **1985**, *22*, 353.
20. Bergstrom, F. W., *Justus Liebigs Ann. Chem.*, **1935**, *515*, 34; Ewing, G. W. and Steck, E. A., *J. Am. Chem. Soc.*, **1946**, *68*, 2181.
21. Wozniak, M., Baranski, A., Nowak, K., and van der Plas, H. C., *J. Org. Chem.*, **1987**, *52*, 5643; Wozniak, M., Baranski, A., Nowak, K., and Poradowska, H., *Justus Liebigs Ann. Chem.*, **1990**, 653; Wozniak, M. and Nowak, K., *Justus Liebigs Ann. Chem.*, **1994**, 355.
22. Baik, W., Yun, S., Rhee, J. U., and Russell, G. A., *J. Chem. Soc., Perkin Trans. 1*, **1996**, 1777.
23. Vanderwalle, J. J. M., de Ruiter, E., Reimlinger, H., and Lenaers, R. A., *Chem. Ber.*, **1975**, *108*, 3898.
24. Simchen, G. and Krämer, W., *Chem. Ber.*, **1969**, *102*, 3666.
25. Sanders, G. M., van Dijk, M., and den Hertog, H. J., *Recl. Trav. Chim. Pays-Bas*, **1974**, *93*, 198.
26. Verbeek, J., George, A. V. E., de Jong, R. C. P., and Brandsma, L., *J. Chem. Soc., Chem. Commun.*, **1984**, 257; Godard, A., Jaquelin, J.-M., and Quéguiner, G., *J. Organomet. Chem.*, **1988**, *354*, 273; Jacquelin, J. M., Robin, Y., Godard, A., and Quéguiner, G., *Can. J. Chem.*, **1988**, *66*, 1135; Marsais, F., Godard, A., and Quéginer, G., *J. Heterocycl. Chem.*, **1989**, *26*, 1589.
27. Gilman, H. and Beel, J. A., *J. Am. Chem. Soc.*, **1951**, *73*, 32.
28. Alvarez, M., Salas, M., Rigat, Ll., de Veciana, A., and Joule, J. A., *J. Chem. Soc., Perkin Trans. 1*, **1992**, 351.
29. Fernández, M., de la Cuesta, E., and Avendaño, C., *Synthesis*, **1995**, 1362.
30. Gilman, H. and Soddy, T. S., *J. Org. Chem.*, **1957**, *22*, 565; *ibid.*, **1958**, *23*, 1584; Pinder, R. M. and Burger, A., *J. Med. Chem.*, **1968**, 11, 267; Ishikura, M., Mano, T., Oda, I., and Terashima, M., *Heterocycles*, **1984**, *22*, 2471.
31. Wommack, J. B., Barbee, T. G., Thoennes, D. J., McDonald, M. A., and Pearson, D. E., *J. Heterocycl. Chem.*, **1969**, *6*, 243.
32. Baradarani, M. M., Dalton, L., Heatley, F., and Joule, J. A., *J. Chem. Soc., Perkin Trans. 1*, **1985**, 1503.
33. Prasad, A. S., Stevenson, T. M., Citineni, J. R., Nyzam, V., and Knochel, P., *Tetrahedron*, **1997**, *53*, 7237.
34. Edo, K., Sakamoto, T., and Yamanaka, H., *Chem. Pharm. Bull.*, **1979**, *27*, 193; Sakamoto, T., Shiraiwa, M., Kondo, Y., and Yamanaka, H., *Synthesis*, **1983**, 312; Ma, H., Wang, X., and Deng, M., *Synth. Commun.*, **1999**, *29*, 2477.
35. Pridgen, L. N., *J. Heterocycl. Chem.*, **1980**, *17*, 1289; Tamao, K., Kodama, S., Nakajima, I., Kumada, M., Minato, A., and Suzuki, K., *Tetrahedron*, **1982**, *38*, 3347.
36. Ciufolini, M. A., Mitchell, J. W., and Roschanger, F., *Tetrahedron Lett.*, **1996**, *37*, 8281.
37. Ford, A., Sinn, E., and Woodward, S., *J. Chem. Soc., Perkin Trans. 1*, **1997**, 927.
38. Girard, G. R., Bondinelli, W. E., Hillegass, L. M., Holden, K. G., Pendleton, R. G., and Vzinskas, I. *J. Med. Chem.*, **1989**, *32*, 1566.
39. Nose, A. and Kudo, T., *Chem. Pharm. Bull.*, **1984**, *32*, 2421.
40. Ranu, B. C., Jana, U., and Sarkar, A., *Synth. Commun.*, **1998**, *28*, 485.
41. Vierhapper, F. W. and Eliel, E. L., *J. Org. Chem.*, **1975**, *40*, 2729; Patrick, G. L, *J. Chem. Soc., Perkin Trans. 1*, **1995**, 1273.
42. Birch, A. J. and Lehman, P. G., *Tetrahedron Lett.*, **1974**, 2395.
43. Hückel, W. and Graner, G., *Chem. Ber.*, **1957**, *90*, 2017.

44. Braude, E. A., Hannah, J., and Linstead, R., *J. Chem. Soc.*, **1960**, 3249.
45. Jackman, L. M. and Packham, D. I., *Chem. Ind. (London)*, **1955**, 360.
46. '1,2-Dihydroisoquinolines', Dyke, S. F., *Adv. Heterocycl. Chem.*, **1972**, *14*, 279.
47. Muren, J. F. and Weissman, A., *J. Med. Chem.*, **1971**, *14*, 49.
48. Minter, D. E. and Stotter, P. L., *J. Org. Chem.* **1981**, *46*, 3965.
49. Fukuzumi, S. and Noura, S., *J. Chem. Soc., Chem. Commun.*, **1994**, 287.
50. Day, F. H., Bradsher, C. K., and Chen, T.-K., *J. Org. Chem.*, **1975**, *40*, 1195.
51. Nicolas, T. E. and Franck, R. W., *J. Org. Chem.*, **1995**, *60*, 6904.
52. Albini, A., Bettinetti, G. F., and Minoli, G., *Tetrahedron Lett.*, **1979**, 3761; *idem, Org. Synth.*, **1983**, *61*, 98.
53. Buchardt, O., *Acta Chem. Scand.*, **1964**, *18*, 1389.
54. Woodward, R. B. and Doering, W. E., *J. Am. Chem. Soc.*, **1945**, *67*, 860.
55. Pfister-Guillouzo, G., Guimon, C., Frank, J., Ellison, J. and Katritzky, A. R., *Justus Libeigs Ann. Chem.*, **1981**, 366.
56. 'The quinolones', Ed. Andriole, V. T., Academic Press, London, **1988**.
57. '3(2*H*)-Isoquinolones and their saturated derivatives', Hazai, L., *Adv. Heterocycl. Chem.*, **1991**, *52*, 155.
58. Adams, A. and Hey, D. H., *J. Chem. Soc.*, **1949**, 255; Schofield, K. and Swain, T., *ibid.*, 1367.
59. Kawazoe, Y. and Yoshioka, Y., *Chem. Pharm. Bull.*, **1968**, *16*, 715.
60. Horning, D. E., Lacasse, G., and Muchowski, J. M., *Can. J. Chem.*, **1971**, *49*, 2785.
61. Gabriel, S. and Coman, J., *Chem. Ber.*, **1900**, *33*, 980.
62. Kanishi, K., Onari, Y., Goto, S., and Takahashi, K., *Chem. Lett.*, **1975**, 717.
63. Ogata, Y., Kawasaki, A., and Hirata, H., *J. Chem. Soc., Perkin Trans. 2*, **1972**, 1120.
64. Kaslow, C. E. and Stayner, R. D., *J. Am. Chem. Soc.*, **1945**, *67*, 1716.
65. Wislicenus, W. and Kleisinger, E., *Chem. Ber.*, **1909**, *42*, 1140.
66. Burger, A. and Modlin, L. R., *J. Am. Chem. Soc.*, **1940**, *62*, 1079.
67. Dyson, P. and Hammick, D. Ll., *J. Chem. Soc.*, **1937**, 1724.
68. Quast, H. and Schmitt, E., *Justus Liebigs Ann. Chem.*, **1970**, *732*, 43.
69. Gensler, W. J. and Shamasundar, K. T., *J. Org. Chem.*, **1975**, 40, 123.
70. For '-one' production by ferricyanide oxidation: Bunting, J. W., Lee-Young, P. A., and Norris, D. J., *J. Org. Chem.*, **1978**, 43, 1132; by O_2 oxidation: Ruchirawat, S., Sunkul, S., Thebtaranonth, Y., and Thirasasna, N., *Tetrahedron Lett.*, **1977**, 2335.
71. Schultz, O.-E. and Amschler, U., *Justus Liebigs Ann. Chem.*, **1970**, *740*, 192; see also Bunting, J. W. and Meathrel, W. G., *Can. J. Chem.*, **1974**, *52*, 303.
72. Diaba, F., Lewis, I., Grignon-Dubois, M., and Navarre, S., *J. Org. Chem.*, **1996**, *61*, 4830.
73. Katayama, H., Ohkoshi, M., and Yasue, M., *Chem. Pharm. Bull.*, **1980**, *28*, 2226.
74. Weinstock, J. and Boekelheide, V., *Org. Synth., Coll. Vol. IV*, **1963**, 641.
75. 'Reissert compounds', Popp, F. D., *Adv. Heterocycl. Chem.*, **1968**, *9*, 1; 'Developments in the chemistry of Reissert compounds (1968–1978)', *idem*, **1979**, *24*, 187.
76. Ezquerra, J. and Alvarez-Builla, J., *J. Chem. Soc., Chem. Commun.*, **1984**, 54.
77. Chênevert, R., Lemieux, E., and Voyer, N., *Synth. Commun.*, **1983**, *13*, 1095.
78. Boger, D. L., Brotherton, C. E., Panek, J. S., and Yohannes, D., *J. Org. Chem.*, **1984**, *49*, 4056.
79. Yamaguchi, R., Mochizuki, K., Kozima, S., and Takaya, H., *Chem. Lett.*, **1994**, 1809.
80. Yamaguchi, R., Hatano, B., Nakayasu, T., and Kozima, S., *Tetrahedron Lett.*, **1997**, *38*, 403.
81. Yamaguchi, R., Omoto, Y., Miyake, M., and Fujita, K., *Chem. Lett.*, **1998**, 547.
82. Zincke, T. H. and Weisspfenning, G., *Justus Liebig's Ann. Chem.*, **1913**, *396*, 103.
83. Barbier, D., Marazano, C., Das, B. C., and Potier, P., *J. Org. Chem.*, **1996**, *61*, 9596.
84. Barbier, D., Marazano, C., Riche, C., Das, B. C., and Potier, P., *J. Org. Chem.*, **1998**, *63*, 1767.
85. Ochiai, E., *J. Org. Chem.*, **1953**, *18*, 534; Yokoyama, A., Ohwada, T., Saito, S., and Shudo, K., *Chem. Pharm. Bull.*, **1997**, *45*, 279.
86. Ochiai, E. and Ikehara, M., *J. Pharm. Soc. Japan*, **1953**, *73*, 666.
87. Harusawa, S., Hamada, Y., and Shiorii, T., *Heterocycles*, **1981**, *15*, 981.
88. Miyashita, A., Matsuda, H., Iijima, C., and Higashino, T., *Heterocycles*, **1992**, *33*, 211.
89. Hayashida, M., Honda, H., and Hamana, M., *Heterocycles*, **1990**, *31*, 1025.

7 Typical reactivity of pyrylium and benzopyrylium ions, pyrones and benzopyrones

The pyrylium cation presents an intriguing dichotomy – it is both 'aromatic', and therefore, the beginning student would be tempted to understand, 'stable', yet it is very reactive – the tropylium cation and the cyclopentadienyl anion can also be described in this way. However, all is relative, and that pyrylium cations react rapidly with nucleophiles to produce adducts which are not aromatic, is merely an expression of their relative stability – if they were not 'aromatic' it is doubtful whether such cations could exist at all. Pyrylium perchlorate is surprisingly stable – it does not decompose below 275 °C but, nonetheless, it will react with water, even at room temperature, producing a non-aromatic product.

Typical pyrylium reactivities

The properties of pyrylium cations are best compared with those of pyridinium cations: the system does not undergo electrophilic substitution nor, indeed, are benzopyrylium cations substituted in the benzene ring. This is a considerable contrast with the chemistry of quinolinium and isoquinolinium cations and is a comment on the stronger deactivating effect of the positively charged oxygen.

Pyrylium ions readily add nucleophilic reagents, at an α-position, generating 1,2-dihydropyrans which then often ring open. Virtually all the known reactions of pyrylium salts fall into this general category. Often, the initial product of ring opening also subsequently and spontaneously takes part in an alternative ring closure, generating a benzenoid aromatic system (if Y contains active hydrogen attached to carbon) or a pyridine (if Y is an amine nitrogen).

Resonance contributors to the pyrylium cation show that there is greater positive charge at the α- and γ-positions, but nearly all of the known nucleophilic additions take place at an α-position. It is relevant to recall here the greater influence of the heteroatom positive charge on pyridine α-positions versus the γ-position. Pyrylium is more reactive in such nucleophilic additions than pyridinium – oxygen tolerates a positive charge less well than nitrogen. It is worth pointing out again the analogy with carbonyl chemistry – the nucleophilic additions which characterise pyrylium systems are nothing more nor less than those which occur frequently in acid-catalysed (*O*-protonated) chemistry of carbonyl groups.

Turning to benzopyrylium systems, one finds exactly comparable behaviour – a readiness to add nucleophiles, adjacent to the positively charged oxygen, in the heterocyclic ring. The interaction of the two isomeric bicycles with ammonia is instructive: one can be converted into an isoquinoline, the other cannot be converted into a quinoline for, although in the last case the addition can and does take place, in the subsequent ring-opened species, no low energy mechanism is available to allow the nitrogen to become attached to the benzene ring.

Pyrones, which are the oxygen equivalent of pyridones, are simply α- and γ-hydroxypyrylium salts from which an *O*-proton has been removed. There is little to recommend that 2- and 4-pyrones be viewed as aromatic: they are perhaps best seen as cyclic unsaturated lactones and cyclic β-oxy-α,β-unsaturated ketones, respectively, for example 2-pyrones are hydrolysed by alkali just like simpler esters (lactones). It is instructive that whereas the pyrones are converted into pyridones by reaction with amines or ammonia, the reverse is not the case – pyridones are not transformed into pyrones by water, or hydroxide. Some electrophilic *C*-substitutions are known for pyrones and benzopyrones, the carbonyl oxygen guiding the electrophile *ortho* or *para*, however there is a tendency for electrophilic addition to a double bond of the heterocyclic ring, again reflecting their non-aromatic nature. Easy Diels-Alder additions to 2-pyrones are further evidence for diene, rather than aromatic, character.

The cyclisation of an unsaturated 1,5-dicarbonyl compound produces pyrylium salts, providing of course that the acidic medium chosen is suitable – it must not contain nucleophilic species which would add to the salt, once formed. Acid-catalysed ring closure of 1,3,5-triketones produces 4-pyrones.

Benzopyrylium salts are formed when phenols react with 1,3-dicarbonyl compounds under acidic, dehydrating conditions. The comparable use of 1,3-keto-esters leads benzopyrones.

8.1.8 Alkylpyryliums[26]

Hydrogens on alkyl groups at the α- and γ-positions of pyrylium salts are, as might be expected, quite acidic: reaction at a γ-methyl is somewhat faster than at an α-methyl.[27] Condensations with aromatic aldehydes (illustrated below),[28] triethyl orthoformate,[29] and dimethylformamide[30] have been described.

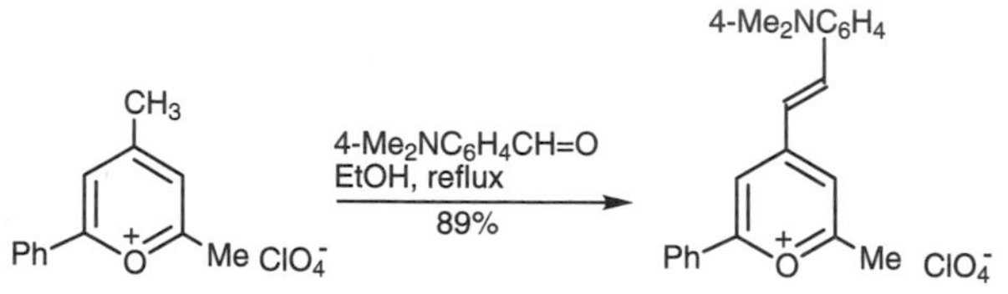

8.2 2-Pyrones and 4-pyrones (2H-pyran-2-ones and 4H-pyran-4-ones; α-pyrones and γ-pyrones)

8.2.1 Structure of pyrones

2- and 4-Hydroxypyrylium salts are quite strongly acidic and are therefore much better known as their conjugate bases, the 2- and 4- (α- and γ-) -pyrones. The simple 4-pyrones are quite stable crystalline substances, whereas the 2-pyrones are much less stable, 2-pyrone itself, which has the smell of fresh-mown hay, polymerising slowly on standing. There are relatively few simple pyrone natural products in great contrast with the widespread occurrence and importance of their benzo-derivatives (section 9.2), the coumarins and chromones, in nature.

8.2.2 Reactions of pyrones

8.2.2.1 Electrophilic addition and substitution

4-Pyrone is a weak base, pK_a −0.3 which is protonated on the carbonyl oxygen to afford often crystalline 4-hydroxypyrylium salts. The reaction of 2,6-dimethyl-4-pyrone with t-butylbromide in hot chloroform provides a neat way to form the corresponding 4-hydroxy-2,6-dimethylpyrylium bromide.[31] 2-Pyrones are much weaker bases and though they are likewise protonated on carbonyl oxygen in solution in strong acids, salts cannot be isolated. This difference is mirrored in reactions with alkylating agents: the former give 4-methoxypyrylium salts with dimethyl sulfate,[32] whereas 2-pyrones require Meerwein salts, MeO^+ BF_4^-, for carbonyl-O-methylation. Acid-catalysed exchange in 4-pyrone, presumably via C-protonation of a concentration of neutral molecule, takes place at the 3/5-positions.[33]

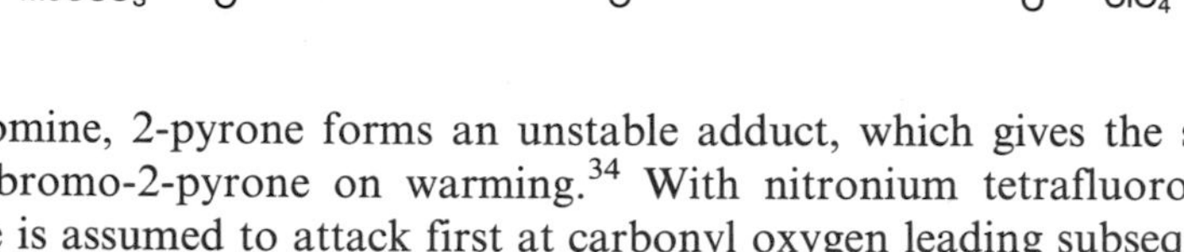

With bromine, 2-pyrone forms an unstable adduct, which gives the substitution product 3-bromo-2-pyrone on warming.[34] With nitronium tetrafluoroborate, the electrophile is assumed to attack first at carbonyl oxygen leading subsequently to 5-nitro-2-pyrone.[35] Simple examples of electrophilic substitution of 4-pyrones are remarkably rare, however bis-dimethylaminomethylation of the parent heterocycle takes place under quite mild conditions.[36]

8.2.2.2 Attack by nucleophilic reagents

2-Pyrones are in many ways best viewed as unsaturated lactones, and as such they are easily hydrolysed by aqueous alkali; 4-pyrones, too, easily undergo ring-opening with base, though for these vinylogous lactones, initial attack is at C-2, in a Michael fashion.[37]

2-Pyrones can in principle add nucleophilic reactants at either C-2 (carbonyl carbon), C-4, or C-6: their reactions with cyanide anion,[38] and ammonia/amines are examples of the latter, whereas the addition of Grignard nucleophiles occurs at carbonyl carbon.

4-Pyrones also add Grignard nucleophiles at the carbonyl carbon, C-4; dehydration of the immediate tertiary alcohol product with mineral acid provides an important access to 4-mono-substituted pyrylium salts.[39] More vigorous conditions lead to the reaction of both 2- and 4-pyrones with two mol equivalents of organometallic reagent and the formation of 2,2-disubstituted-2*H*- and 4,4-disubstituted-4*H*-pyrans respectively.[40] Perhaps surprisingly, hydride (lithium aluminium hydride) addition to 4,6-dimethyl-2-pyrone takes place, in contrast, at C-6.[41]

Ammonia and primary aliphatic and aromatic amines convert 4-pyrones into 4-pyridones:[42] this must involve attack at an α-position, then ring opening and reclosure; in some cases ring opened products of reaction with two mols of the amine have been isolated, though such structures are not necessarily intermediates on the route to pyridones.[43] The transformation can also be achieved by first, hydrolytic ring opening using barium hydroxide (see above), and then reaction of the barium salt with ammonium chloride.[44]

8.2.2.6 Side-chain reactions

4-Pyrones[60] and 2-pyrones[61] condense with aromatic aldehydes at 2- and 6-methyl groups respectively and 2,6-dimethyl-4-pyrone has been lithiated at a methyl and thereby substituted as illustrated.[62]

8.2.2.7 2,4-Dioxygenated pyrones

2,4-Dioxygenated pyrones exist as the 4-hydroxy tautomers. Such molecules are easily substituted by electrophiles, at the position between the two oxygens (C-3)[63] and can be side-chain deprotonated using two mol equivalents of strong base.[64]

8.3 Synthesis of pyryliums[1,8a]

Pyrylium rings are assembled by the cyclisation of a 1,5-dicarbonyl precursor, separately synthesised or generated *in situ*.

8.3.1 From 1,5-dicarbonyl compounds

1,5-Dicarbonyl compounds can be cyclised, with dehydration and in the presence of an oxidising agent.

Mono-enolisation of a 1,5-diketone, then the formation of a cyclic hemiacetal, and its dehydration, produces dienol ethers (4*H*-4-pyrans) which require only hydride abstraction to arrive at the pyrylium oxidation level. The diketones are often prepared *in situ* by the reaction of an aldehyde with two mols of a ketone (compare Hantzsch synthesis, section 5.15.1.2) or of a ketone with a previously prepared conjugated ketone – a 'chalcone' in the case of aromatic ketones/aldehydes. It is the excess chalcone which serves as the hydride acceptor in this approach.

Early work utilised acetic anhydride as solvent with the incorporation of an oxidising agent (hydride acceptor), often iron(III) chloride (though it is believed that it is the acylium cation which is the hydride acceptor); latterly the incorporation of

2,3-dichloro-5,6-dicyano-1,4-benzoquinone,[65] 2,6-dimethylpyrylium or most often, the triphenylmethyl cation[66] have proved efficient. In some cases the 4H-pyran is isolated then oxidised in a separate step.[67]

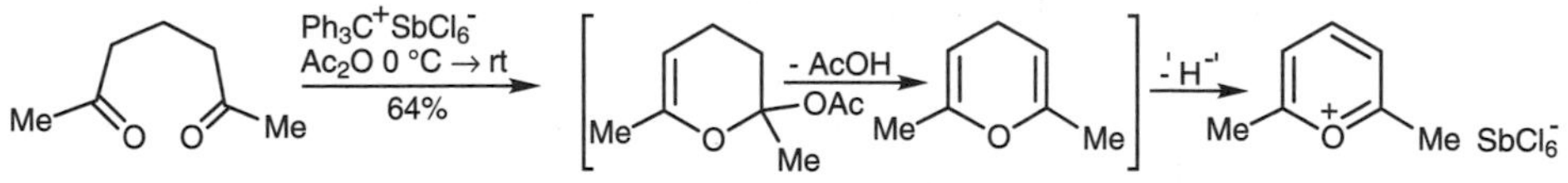

If an unsaturated dicarbonyl precursor is available, no oxidant needs to be added: a synthesis of the perchlorate of pyrylium itself, shown below, falls into this category: careful acid treatment of either glutaconaldehyde, or of its sodium salt, produces the parent salt[13,68] (CAUTION: potentially explosive).

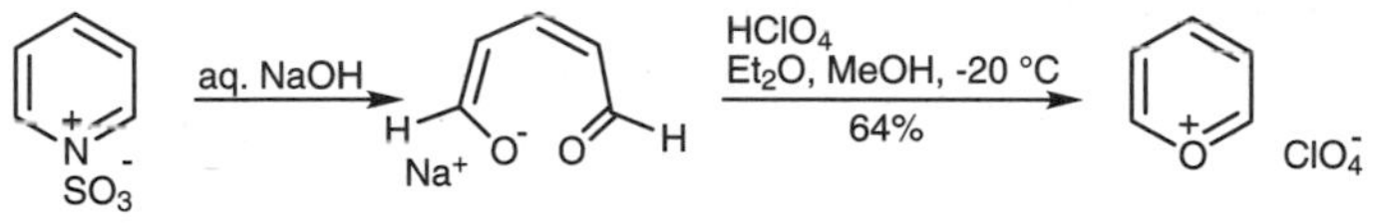

8.3.2 Alkene acylation

Alkenes can be diacylated with an acid chloride or anhydride generating an unsaturated 1,5-dicarbonyl compound which then cyclises with loss of water.

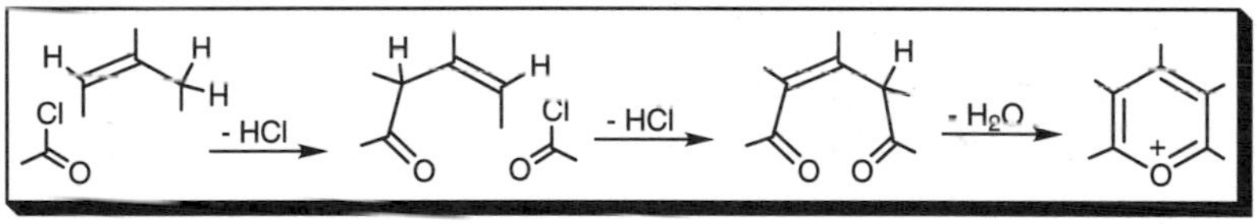

The aliphatic version of the classical aromatic Friedel-Crafts acylation produces, by loss of proton, a non-conjugated enone which can then undergo a second acylation thus generating an unsaturated 1,5-diketone. Clearly, if the alkene is not symmetrical, two isomeric diketones are formed.[69] Under the conditions of these acylations, the unsaturated diketone cyclises, loses water and forms a pyrylium salt. The formation of 2,4,6-trimethylpyrylium, best as its much more stable and non-hygroscopic carboxymethanesulfonate,[70] illustrates the process.

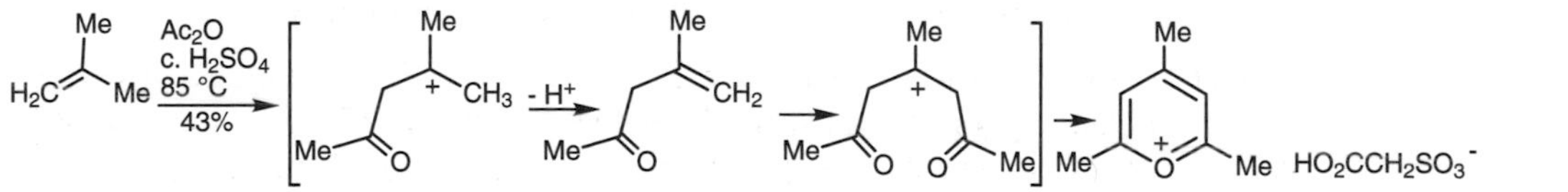

Common variations are the use of an alcohol, which dehydrates in situ,[71] or of a halide which dehydrohalogenates[72] to give the alkene.

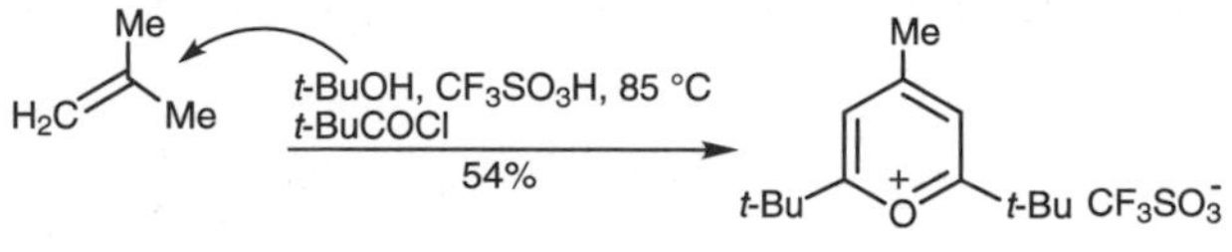

8.3.3 From 1,3-dicarbonyl compounds and ketones

The acid-catalysed condensation of a ketone with a 1,3-dicarbonyl compound, with dehydration in situ produces pyrylium salts.

Aldol condensation between a 1,3-dicarbonyl component and a ketone carrying an α-methylene under acidic, dehydrating conditions, produces pyrylium salts.[73] It is likely that the initial condensation is followed by a dehydration before the cyclic hemiacetal formation and loss of a second water molecule. The use of the bis-acetal of malondialdehyde, as synthon for the 1,3-dicarbonyl component is one of the few ways available for preparing α-unsubstituted pyryliums.[1]

Successful variations on this theme include the use, as synthons for the 1,3-dicarbonyl component, of β-chloro-α,β-unsaturated ketones,[74] or of conjugated alkynyl aldehydes.[75]

8.4   Synthesis of 2-pyrones

8.4.1   From 1,3-keto(aldehydo)-acids and carbonyl compounds

The classical general method for constructing 2-pyrones is that based on the cyclising condensation of a 1,3-keto(aldehydo)-acid with a second component which provides the other two ring carbons.

The long known synthesis of coumalic acid from treatment of malic acid with hot sulfuric acid illustrates this route: decarbonylation produces formylacetic acid, *in situ*, which serves as both 1,3-aldehydo-acid component and the second component.[76] Decarboxylation of coumalic acid is still used to access 2-pyrone itself.[77]

Conjugate addition of enolates to alkynyl-ketones[78] and to alkynyl-esters[79] are further variations on the synthetic theme.

8.4.2 Other methods

2-Pyrone itself can be prepared via Prins alkylation of but-3-enoic acid with subsequent lactonisation giving 5,6-dihydro-2-pyrone which *via* allylic bromination and then dehydrobromination is converted into 2-pyrone as shown below.[80] Alternative manipulation[81] of the dihydropyrone affords a convenient synthesis of a separable mixture of the important 3- and 5-bromo-2-pyrones (see section 8.2.2.4).

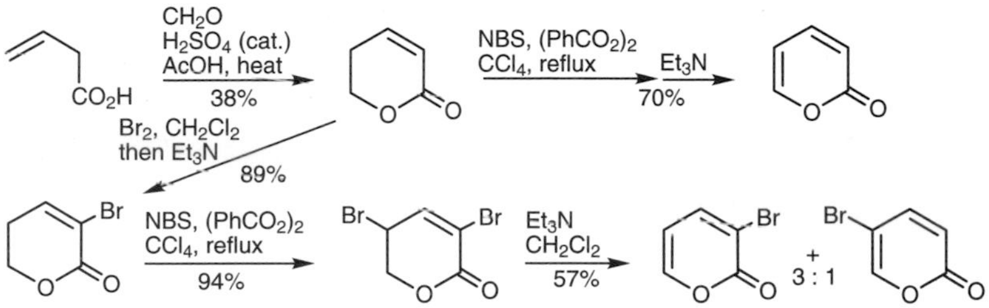

Formation of the 5,6-bond is also involved in the Claisen condensation between diethyl oxalate and an α,β-unsaturated ester at its γ position, to generate an intermediate in which ring closure via the ketone enol produces a 2-pyrone.[82]

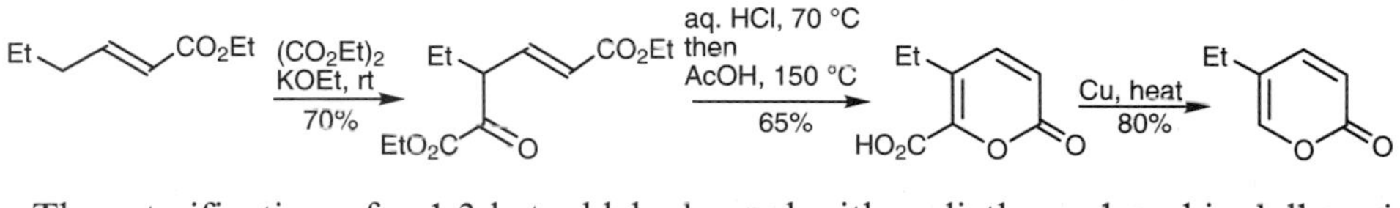

The esterification of a 1,3-ketoaldehyde enol with a diethoxyphosphinylalkanoic acid, forming the ester linkage of the final molecule first, allows ring closure *via* an intramolecular Horner-Emmons reaction.[83]

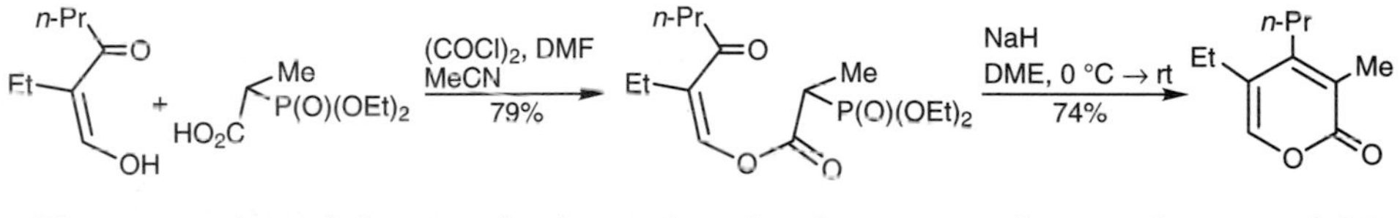

The conversion of glucosamine into a 3-amino-2-pyrone points up the potential for conversion of sugars into six-membered oxygen heterocycles.[84]

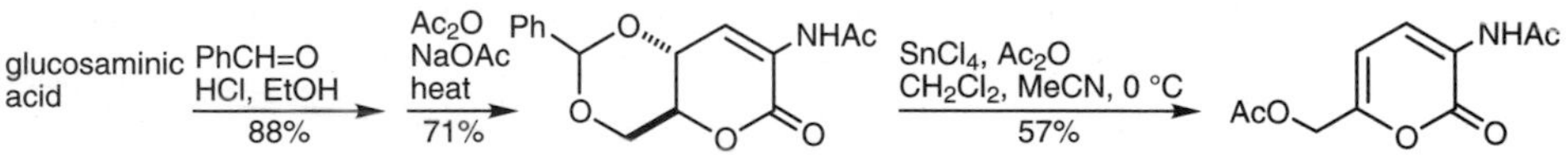

The inverse electron-demand cycloaddition of electron-rich or strained alkynes with 1,3,4-oxadiazin-6-ones leads to 2-pyrones because the adducts lose nitrogen (rather than carbon dioxide).[85] The example below shows the use of ethynyltribu-tyltin giving a mixture of regioisomers; the stannylated pyrones can be utilised in the usual ways, for example for the introduction of halogen.[86]

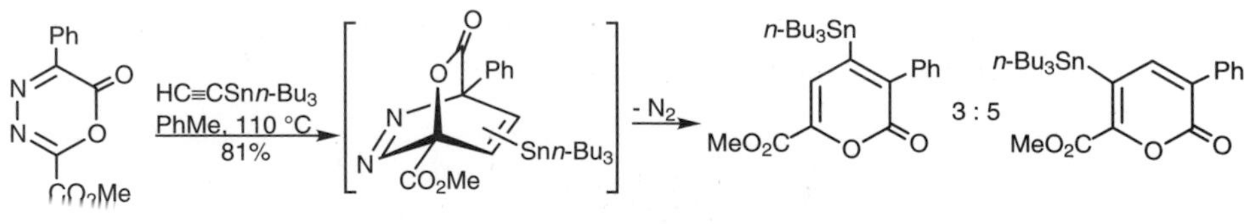

The palladium-catalysed coupling of alkynes with a 3-iodo-α,β-unsaturated ester, or with the enol triflate of a β-keto-ester as illustrated below, must surely be one of the shortest and most direct routes to 2-pyrones.[87] The cycloaddition (non-concerted) of ketenes with silyl enol ethers of α,β-unsaturated esters also provides a simple, direct route to usefully functionalised 2-pyrones.[88]

8.5 Synthesis of 4-pyrones

4-Pyrones result from the acid-catalysed closure of 1,3,5-tricarbonyl precursors.

The construction of a 4-pyrone is essentially the construction of a 1,3,5-tricarbonyl compound since such compounds easily form cyclic hemiacetals then requiring only dehydration. Strong acid has usually been used for this purpose, but where stereochemically sensitive centres are close, the reagent from triphenylphosphine and carbon tetrachloride has been employed.[89]

Several methods are available for the assembly of such precursors: the synthesis of chelidonic acid (4-pyrone-2,6-dicarboxylic acid)[90] represents the obvious approach of bringing about two Claisen condensations, one on each side of a ketone carbonyl group. Chelidonic acid can be decarboxylated to produce 4-pyrone itself.[91]

A variety of symmetrically substituted 4-pyrones can be made very simply by heating an alkanoic acid with polyphosphoric acid;[92] presumably a series of Claisen-type condensations, with a decarboxylation, lead to the assembly of the requisite acyclic, tricarbonyl precursor.

The Claisen condensation of a 1,3-diketone, *via* its dianion, with an ester,[93] or of a ketone enolate with an alkyne ester[94] also give the desired tricarbonyl arrays.

Another strategy to bring about acylation at the less acidic carbon of a β-keto ester, is to condense, firstly at the central methylene, with a formate equivalent; this has the added advantage that the added carbon can then provide the fifth carbon of the target heterocycle.[95]

α-Unsubstituted 4-pyrones have similarly been constructed *via* the enolate of methoxymethylene ketones.[96]

Dehydroacetic acid[97] was first synthesised in 1866;[98] it is formed very simply from ethyl acetoacetate by a Claisen condensation between two molecules, followed by the usual cyclisation and finally loss of ethanol. In a modern version, β-keto-acids can be self-condensed using carbonyl diimidazole as the condensing agent.[99]

The acylation of the enamine of a cyclic ketone with diketene leads directly to bicyclic 4-pyrones, as indicated below.[100]

Exercises for chapter 8

Straightforward revision exercises (consult chapters 7 and 8)

(a) Specify three nucleophiles which add easily to pyrylium salts and draw the structures of the products produced thereby.

(b) Certain derivatives of six-membered oxygen heterocycles undergo 4 + 2 cycloaddition reactions: draw out three examples.

(c) Draw a mechanism for the transformation of 2-pyrone into 1-methyl-2-pyridone on reaction with methylamine.

(d) What steps must take place to achieve the conversion of a saturated 1,5-diketone into a pyrylium salt?

(e) Describe how 5,6-dihydro-2-pyrone can be utilised to prepare either 2-pyrone, or 3- and 5-bromo-2-pyrones.

(f) 1,3,5-Tricarbonyl compounds are easily converted into 4-pyrones. Describe two ways to produce a 1,3,5-trione or a synthon thereof.

49. Diels, O. and Alder, K., *Justus Liebigs Ann. Chem.*, **1931**, *490*, 257; Goldstein, M. J. and Thayer, G. L., *J. Am. Chem. Soc.*, **1965**, *87*, 1925; Shimo, T., Kataoka, K., Maeda, A., and Somekawa, K., *J. Heterocycl. Chem.*, **1992**, *29*, 811.
50. Salomon, R. G., and Burns, J. R., and Dominic, W. J., *J. Org. Chem.*, **1976**, *41*, 2918.
51. Markó, I. E., Seres, P., Swarbrick, T. M., Staton, I., and Adams, H., *Tetrahedron Lett.*, **1992**, *33*, 5649; Markó, I. E., Evans, G. R., Seres, P., Chellé, I., and Janousek, Z., *Pure Appl. Chem*, **1996**, *68*, 113.
52. Posner, G. H., Nelson, T. D., Kinter, C. M., and Afarinkia, K., *Tetrahedron Lett.*, **1991**, *32*, 5295.
53. Afarinkia, K. and Posner, G. H., *Tetrahedron Lett.*, **1992**, *51*, 7839.
54. Posner, G. H., Nelson, T. D., Kinter, C. M., and Johnson, N., *J. Org. Chem.*, **1992**, *57*, 4083.
55. Afarinka, K., Daly, N. T., Gomez-Farnos, S., and Joshi, S., *Tetrahedron Lett.*, **1997**, *38*, 2369; Posner, G., Hutchings, R. H., and Woodard, B. T., *Synlett*, **1997**, 432.
56. Hong, B.-C. and Sun, S.-S., *Chem. Commun.*, **1996**, 937.
57. Liu, Z. and Meinwald, J., *J. Org. Chem.*, **1996**, *61*, 6693.
58. Van Allen, J. A., Chie Chang, S., and Reynolds, G. A., *J. Heterocycl. Chem.*, **1974**, *11*, 195.
59. Corey, E. J. and Streith, J., *J. Am. Chem. Soc.*, **1964**, *86*, 950; Pirkle, W. H. and McKendry, L. H., *J. Am. Chem. Soc.*, **1969**, *91*, 1179; Chapman, O. L., McKintosh, C. L., and Pacansky, J., *J. Am. Chem. Soc.*, **1973**, *95*, 614.
60. Woods, L. L., *J. Am. Chem. Soc.*, **1958**, *80*, 1440.
61. Adam, W., Saha-Möller, C. R., Veit, M., and Welke, B., *Synthesis*, **1994**, 1133.
62. West, F. G., Fisher, P. V., and Willoughby, C. A., *J. Org. Chem.*, **1990**, *55*, 5936; West, F. G., Amann, C. M., and Fisher, P. V. *Tetrahedron Lett.*, **1994**, *35*, 9653.
63. De March, P., Moreno-Mañas, M., Pleixats, R., and Roca, J. C., *J. Heterocycl. Chem.*, **1984**, *21*, 1369.
64. Groutas, W. C., Huang, T. L., Stanga, M. A., Brubaker, M. J., and Moi, M. K., *J. Heterocycl. Chem.*, **1985**, *22*, 433; Poulton, G. A. and Cyr, T. P., *Canad. J. Chem.*, **1980**, *58*, 2158.
65. Carretto, J. and Simalty, M., *Tetrahedron Lett.*, **1973**, 3445.
66. Rundel, W., *Chem. Ber.*, **1969**, *102*, 374; Farcasiu, D., Vasilescu, A., and Balaban, A. T., *Tetrahedron*, **1971**, *27*, 681; Farcasiu, D., *Tetrahedron*, **1969**, *25*, 1209.
67. Undheim, K. and Ostensen, E. T., *Acta Chem. Scand.*, **1973**, *27*, 1385.
68. Klager, F. and Träger, H., *Chem. Ber.*, **1953**, *86*, 1327.
69. Balaban, A. T. and Nenitzescu, C. D., *Justus Liebigs Ann. Chem.*, **1959**, *625*, 74; idem, *J. Chem. Soc.*, **1961**, 3553; Praill, P. F. G. and Whitear, A. L., *ibid.*, 3573.
70. Dinculescu, A. and Balaban, A. T., *Org. Prep. Proced. Int.*, **1982**, *14*, 39.
71. Anderson, A. G. and Stang, P. J., *J. Org. Chem.*, **1976**, *41*, 3034.
72. Balaban, A. T., *Org. Prep. Proced. Int.*, **1977**, *9*, 125.
73. Schroth, W. and Fischer, G. W., *Chem. Ber.*, **1969**, *102*, 1214; Dorofeenko, G. N., Shdanow, Ju. A., Shungijetu, G. I., and Kriwon, W. S. W., *Tetrahedron*, **1966**, *22*, 1821;
74. Schroth, W., Fischer, G. W., and Rottmann, J., *Chem. Ber.*, **1969**, *102*, 1202.
75. Stetter, H. and Reischl, A., *Chem. Ber.*, **1960**, *93*, 1253.
76. Wiley, R. H. and Smith, N. R., *Org. Synth., Coll. Vol. IV*, **1963**, 201.
77. Zimmerman, H. E., Grunewald, G. L., and Paufler, R. M., *Org. Synth., Coll. Vol. V*, **1973**, 982.
78. Anker, R. M. and Cook, A. H., *J. Chem. Soc.*, **1945**, 311.
79. El-Kholy, I., Rafla, F. K., and Soliman, G., *J. Chem. Soc.*, **1959**, 2588.
80. Nakagawa, M., Saegusa, J., Tonozuka, M., Obi, M., Kiuchi, M., Hino, T., and Ban, Y., *Org, Synth., Coll. Vol. VI*, **1988**, 462.
81. Posner, G., Afarinkia, K., and Dai, H. *Org. Synth.*, **1994**, *73*, 231.
82. Fried, J. and Elderfield, R. C., *J. Org. Chem.*, **1941**, *6*, 566.
83. Stetter, H. and Kogelnik, H.-J., *Synthesis*, **1986**, 140.
84. Nin, A. P., de Lederkremer, R. M., and Vanela, O., *Tetrahedron*, **1996**, *52*, 12911.
85. Steglich, W., Buschmann, E., Gansen, G., and Wilschowitz, L., *Synthesis*, **1977**, 252.
86. Sauer, J., Heldmann, D. K., Range, K.-J., and Zabel, M., *Tetrahedron*, **1998**, *54*, 12807.
87. Larock, R. C., Han, X., and Doty, M. J., *Tetrahedron Lett.*, **1998**, *39*, 5713.
88. Ito, T., Aoyama, T., and Shioiri, T., *Tetrahedron Lett.*, **1993**, *34*, 6583.
89. Arimoto, H., Nishiyama, S., and Yamamura, S., *Tetrahedron Lett.*, **1990**, *31*, 5491.

90. Riegel, E. R. and Zwilgmeyer, F. Z., *Org. Synth., Coll. Vol. II*, **1943**, 126.
91. De Souza, C., Hajikarimian, Y., and Sheldrake, P. W., *Synth. Commun.*, **1992**, *22*, 755.
92. Mullock, E. B. and Suschitzky, H., *J. Chem. Soc., C*, **1967**, 828.
93. Miles, M. L., Harris, T. M., and Hauser, C. R., *Org. Synth., Coll. Vol. V*, **1973**, 718; Miles, M. L. and Hauser, C. R., *ibid.*, 721.
94. Soliman, G. and El-Kholy, I. E.-S., *J. Chem. Soc.*, **1954**, 1755.
95. McCombie, S. W., Metz, W. A., Nazareno, D., Shankar, B. B., and Tagat, J. *J. Org, Chem.*, **1991**, *56*, 4963.
96. Morgan, T. A. and Ganem, B., *Tetrahedron Lett.*, **1980**, *21*, 2773; Koreeda, M. and Akagi, H., *ibid.*, 1197.
97. Arndt, F., *Org. Synth., Coll. Vol. III*, **1955**, 231; 'Dehydroacetic acid, triacetic acid lactone, and related pyrones', Moreno-Mañas, M. and Pleixats, R., *Adv. Heterocycl. Chem.*, **1993**, *53*, 1.
98. Oppenheim, A. and Precht, H., *Chem. Ber.*, **1866**, *9*, 324.
99. Ohta, S., Tsujimura, A., and Okamoto, M., *Chem. Pharm. Bull.*, **1981**, *29*, 2762.
100. Hünig, S., Benzing, E., and Hübner, K., *Chem. Ber.*, **1961**, *94*, 486.

9 Benzopyryliums, benzopyrones: reactions and synthesis

chromylium
(benzo[b]pyrylium)
[1-benzopyrylium]

coumarin
[2H-1-benzopyran-2-one]

chromone
[4H-1-benzopyran-4-one]

isochromylium
(benzo[c]pyrylium)
[2-benzopyrylium]

isocoumarin
[1H-2-benzopyran-1-one]

[3H-2-benzopyran-3-one]

1-Benzopyryliums, coumarins, and chromones are very widely distributed throughout the plant kingdom where many secondary metabolites contain them. Not the least of these are the anthocyanins[1] and flavones[2] which, grouped together, are known as the flavonoids,[3] and make up the majority of the flower pigments. In addition, many flavone and coumarin[4] derivatives have marked toxic and other physiological properties in animals, though they play no part in the normal metabolism of animals. The isomeric 2-benzopyrylium[5] system does not occur naturally and only a few isocoumarins[6] occur as natural products and as a consequence much less work on these has been described.

Chemotherapeutically valuable compounds in this group are a series of coumarins, of which Acenocoumarol is one, which are valuable as anticoagulants, and Intal, which is used in the treatment of bronchial asthma. One of the earliest optical brighteners was 7-diethylamino-4-methylcoumarin.[7]

Acenocoumarol

Intal

Processes initiated by nucleophilic additions to the positively charged heterocyclic ring are the main, almost the only, types of reaction known for benzopyryliums. The absence of examples of electrophilic substitution in the benzene ring is to be contrasted with the many examples of substitution in quinolinium and isoquinolinium salts, emphasising the greater electron-withdrawing and thus deactivating effect of positively charged oxygen.

Coumarins, chromones, and isocoumarins react with both nucleophiles and electrophiles in much the same way as do quinolones and isoquinolones.

9.1 Reactions of benzopyryliums

Much more work has been done on 1-benzopyryliums than on 2-benzopyryliums, because of their relevance to the flavylium (2-phenyl-1-benzopyrylium) nucleus which occurs widely in the anthocyanins, and much of that work has been conducted on flavylium itself. As with pyrylium salts, benzopyrylium salts usually add nucleophiles at the carbon adjacent to the oxygen.

9.1.1   Reactions with electrophilic reagents

No simple examples are known of electrophilic or radical substitution of either heterocyclic or homocyclic rings of benzopyrylium salts; flavylium[8] and 1-phenyl-2-benzopyrylium[5] salts nitrate in the substituent benzene ring. Having said this, the cyclisation of coumarin-4-propanoic acid may represent Friedel-Crafts type intramolecular attack on the carbonyl-*O*-protonated form i.e. on a 2-hydroxy-1-benzopyrylium system, at C-3.[9]

9.1.2   Reactions with oxidising agents

Oxidative general breakdown of flavylium salts was utilised in early structural work on the natural compounds. Baeyer-Villiger oxidation is such a process whereby the two 'halves' of the molecule can be separately examined (after ester hydrolysis of the product).[10] Flavylium salts can be oxidised to flavones using thallium(III) nitrate[11] and benzopyrylium itself can be converted into coumarin with manganese dioxide.[12]

9.1.3   Reactions with nucleophilic reagents

9.1.3.1   *Water and alcohols*

Water and alcohols add readily at C-2, and sometimes at C-4, generating chromenols or chromenol ethers.[13] It is difficult to obtain 2*H*-chromenols pure since they are always in equilibrium with ring-opened chalcones.[14]

Controlled conditions are required for the production of simple adducts, for under more vigorous alkaline treatment, ring opening then carbon–carbon bond cleavage *via* a retro-aldol mechanism takes place and such processes, which are essentially the reverse of a route used for the synthesis of 1-benzopyryliums (section 9.3.1) were utilised in early structural work on anthocyanin flower pigments.

9.1.3.2   *Ammonia and amines*

Ammonia and amines add to benzopyryliums, and simple adducts from secondary amines have been isolated.[15]

It is important to realise that 1-benzopyrylium salts cannot be converted into quinolines or quinolinium salts by reaction with ammonia or primary amines, whereas 2-benzopyrylium salts are converted, efficiently, into isoquinolines or isoquinolinium salts respectively.[16]

9.1.3.3 Carbon nucleophiles

Organometallic carbon nucleophiles add to flavylium salts[17] as do activated aromatics like phenol,[18] and enolates such as those from cyanoacetate, nitromethane,[19] dimedone,[20] all very efficiently, at C-4. Cyanide and azide add to 2-benzopyryliums at C-1.[21]

Silyl enol ethers, or allylsilanes will add at C-2 to benzopyrylium salts generated by *O*-silylation of chromones; in the case of silyl ethers of α,β-unsaturated ketones, cyclisation of the initial adduct is observed.[22]

9.1.4 Reactions with reducing agents

Catalytic hydrogenation of flavylium salts is generally straightforward and results in the saturation of the heterocyclic ring. Lithium aluminium hydride reduces flavylium salts generating 4*H*-chromenes,[23] unless there is a 3-methoxyl, when 2*H*-chromenes are the products.[24] 2-Benzopyryliums add hydride at C-1.[25]

9.1.5 Alkylbenzopyryliums

Alkyl groups oriented α or γ to the positively charged oxygen in benzopyryliums have acidified hydrogens which allow aldol-type condensations.[5,26]

9.1.6 1-Benzopyrylium pigments; anthocyanins and anthocyanidins

The anthocyanidins are polyhydroxyflavylium salts. They occur in a large proportion of the red to blue flower pigments and in fruit skins, for example grapes and therefore in red wines made therefrom.[27] Anthocyanidins are generally bound to sugars, and these glycosides are known as anthocyanins. As an example, cyanin (isolated as its chloride) is an anthocyanin which occurs in the petals of the red rose (*Rosa gallica*), the poppy (*Papaver rhoeas*), and very many other flowers. Another example is malvin chloride which has been isolated from many species, including *Primula viscosa*, a mauvy-red alpine primula.

In the living cell these compounds exist in more complex bound forms, interacting with other molecules, for example flavones,[28] and the actual observed colour will depend on these interactions. However it is interesting that even *in vitro*, simple pH changes bring about extreme changes in the electronic absorption of these molecules. For example cyanidin is red in acidic solution, violet at intermediate pH and blue in weakly alkaline solution, the deep colours being the result of extensive resonance delocalisation in each of the structures.

9.2 Benzopyrones (chromones, coumarins, and isocoumarins)

9.2.1 Reactions with electrophilic reagents

9.2.1.1 Addition to carbonyl oxygen

Addition to carbonyl oxygen of a proton produces a hydroxybenzopyrylium salt; chromones undergo this protonation more easily than the coumarins, for example passage of hydrogen chloride through a mixture of chromone and coumarin in ether solution leads to the precipitation of only chromone hydrochloride (i.e. 4-hydroxy-1-benzopyrylium chloride).[29] *O*-Alkylation requires the more powerful alkylating agents.[5,30] *O*-Silylation of benzopyrones is easy (section 9.1.3.3).

9.2.1.2 C-Substitution

C-Substitution of coumarins and chromones has been observed in both rings: in strongly acidic media, in which presumably it is an hydroxybenzopyrylium cation which is attacked, substitution takes place at C-6, for example nitration.[31] This can be contrasted with the dimethylaminomethylation of chromone[32], iodination of flavones[33] or the chloromethylation of coumarin[34] where hetero-ring substitution takes place, presumably *via* the non-protonated (complexed) heterocycle (CAUTION: CH₂O/HCl also produces some ClCH₂OCH₂Cl, a carcinogen).

Bromine in the presence of an excesss of aluminium chloride (the 'swamping catalyst' effect) converts coumarin into 6-bromocoumarin.[35] Reaction of coumarin with bromine alone results in simple addition to the double bond in the heterocyclic ring; 3-bromocoumarin can be obtained by then eliminating hydrogen bromide.[36] Copper(II) halides with alumina in refluxing chlorobenzene is an alternative method for 3-halogenation of coumarins.[37] Chromone can be efficiently brominated at C-6 using dibromoisocyanuric acid (DBI);[38] treatment of chromone with bromine in carbon disulfide results in addition, elimination of hydrogen bromide on warming giving 3-bromochromone.[39]

9.2.2 Reactions with oxidising agents

Non-phenolic coumarins are relatively stable to oxidative conditions. Various oxidative methods were used extensively in structure determinations of natural flavones.[40]

Flavones and isoflavones (3-arylchromones) are quantitatively converted into 2,3-epoxides by exposure to dimethyl dioxirane; such intermediates have obvious synthetic potential, flavone oxides, for example, being quantitatively converted by acid into 3-hydroxyflavones, which are naturally occurring.[41]

9.2.3 Reactions with nucleophilic reagents

9.2.3.1 Hydroxide

Coumarins (and isocoumarins) are quantitatively hydrolysed to give yellow solutions of the salts of the corresponding *cis* cinnamic acids (coumarinic acids) which cannot be isolated since acidification brings about immediate relactonisation; prolonged alkali treatment leads to isomerisation and the formation of the *trans* acid (coumaric acid) salt.

Cold sodium hydroxide comparably reversibly converts chromones into the salts of the corresponding ring-opened phenols, *via* initial attack at C-2, more vigorous alkaline treatment leading to reverse-Claisen degradation of the 1,3-diketo-side-chain.

9.2.3.2 Ammonia and amines

Ammonia and amines do not convert coumarins into 2-quinolones nor chromones into 4-quinolones, but isocoumarins do produce isoquinolones.[42] Ring-opened products from chromones and secondary amines can be obtained where again the nucleophile has attacked at C-2.

The interaction of 3-iodochromone with imidazole leads to substitution at the 2-position, presumably via an addition/elimination sequence as indicated.[43]

9.2.3.3 Carbon nucleophiles

Grignard reagents react with chromones at carbonyl carbon; the resulting chromenols can be converted by acid into the corresponding 4-substituted 1-benzopyrylium salts.[26]

Coumarins, and isocoumarins,[16] react with Grignard reagents, as do esters, and can give mixtures of products, resulting from ring opening of the initial carbonyl adduct; the reaction of coumarin with methylmagnesium iodide illustrates this.[44]

By conversion into a benzopyrylium salt with a leaving group, nucleophiles can be introduced at the chromone 4-position: treatment with acetic anhydride presumably forms a 4-acetoxybenzopyrylium.[45]

In efficient reactions, coumarin can be made to react with electron-rich aromatics using phosphoryl chloride, alone, or with zinc chloride.[46]

9.2.3.4 Organometallic derivatives

Flavone has been lithiated at C-3.[47]

Both 3-bromochromone and 4-bromocoumarin have been successfully used in coupling reactions using palladium(0) methodology.[48]

9.2.3.5   Reactions with reducing agents

Both coumarin and chromone are converted by diborane then alkaline hydrogen peroxide into 3-hydroxychroman.[49] Catalytic reduction of coumarin or chromone saturates the C–C double bond.[50] For both systems, hydride reagents can of course react either at carbonyl carbon or at the conjugate position and mixtures therefore tend to be produced. Zinc amalgam in acidic solution converts benzopyrones into 4-unsubstituted benzopyrylium salts.[51]

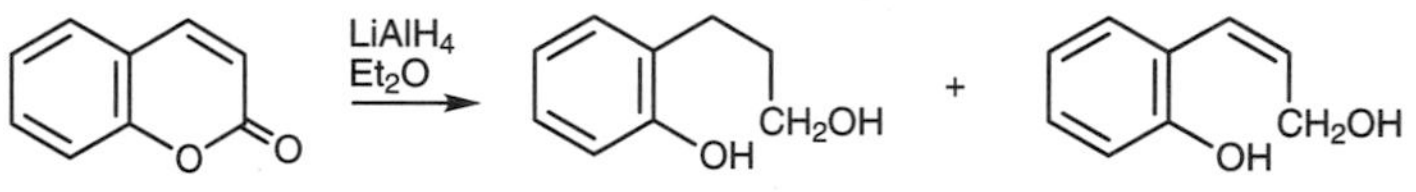

9.2.3.6   Reactions with dienophiles; cycloadditions

Coumarins, but not apparently, chromones, serve as dienophiles in Diels-Alder reactions, though under relatively forcing conditions.[52]

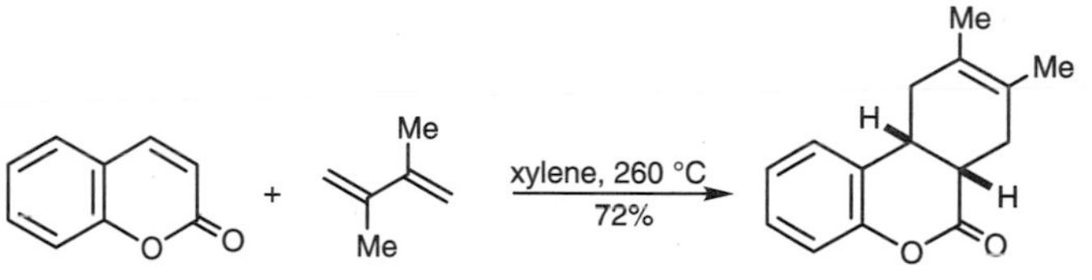

It can be taken as a measure of the low intrinsic aromaticity associated with fused pyrone rings, that 3-acylchromones undergo hetero Diels-Alder additions with enol ethers,[53] and ketene acetals,[54] 3-formylchromone reacting the most readily.

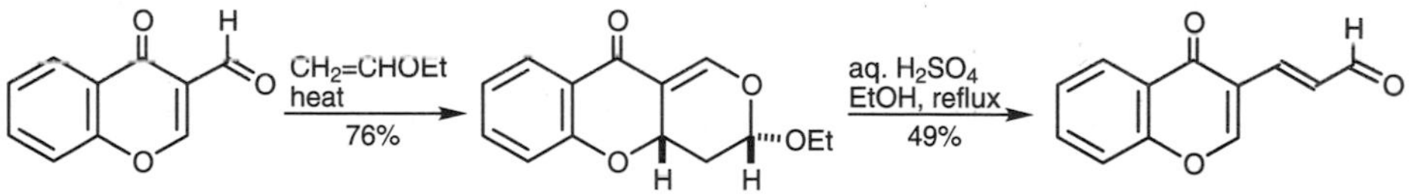

Chromone-3-esters, on the other hand, serve as dienophiles under Lewis acid catalysis.[55]

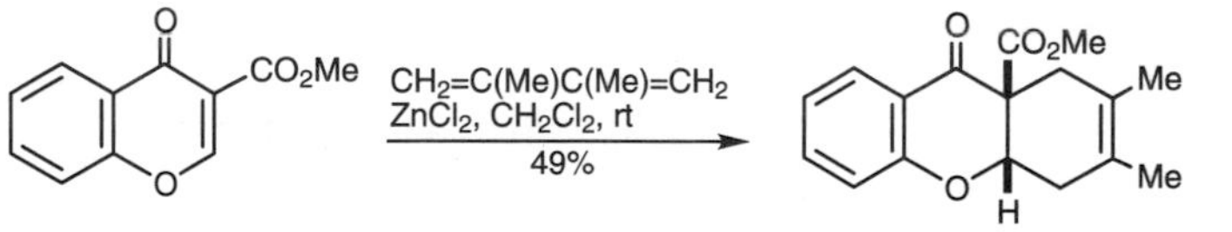

2-Benzopyran-3-ones, generated by cyclising dehydration of an *ortho* formylarylacetic acid take part in intramolecular Diels-Alder additions as shown below.[56]

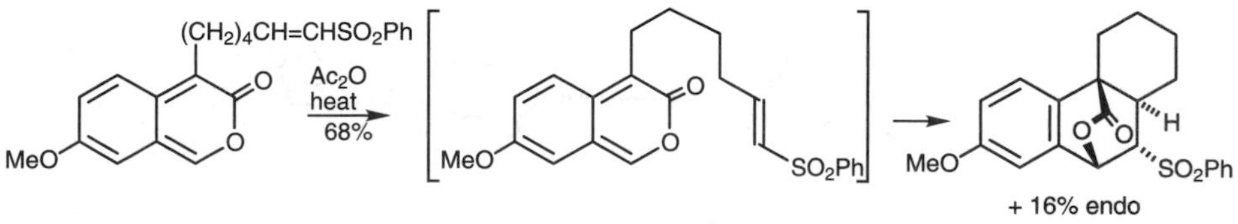

Decomposition of aryl diazoketones with an appropriately tethered alkene allows the intramolecular cycloaddition to the 4-oxidoisochromylium salts thus formed, as illustrated below.[57]

9.2.3.7 Photochemical reactions

Coumarin has been studied extensively in this context; in the absence of a sensitiser it gives a *syn* head-to-head dimer; in the presence of benzophenone, as sensitiser, the *anti* isomer is formed;[58] the *syn* head-to-tail dimer is obtained by irradiation in acetic acid.[59] Cyclobutane-containing products are obtained in modest yields by sensitiser-promoted cycloadditions of coumarins and 3-acyloxycoumarins with alkenes, ketene diethylacetal, and cyclopentene.[60]

9.2.3.8 Alkylcoumarins and alkylchromones

Methyl groups at C-2, but not at C-3, of chromones undergo condensations with aldehydes, because only the former can be deprotonated to give conjugated enolates.[61]

The 4-position of coumarins is the only one at which alkyl substituents have enhanced acidity in their hydrogens,[62] and this is considerably less than that of the methyl groups of 2-methylchromones.[63]

9.2.3.9 Flavone pigments

The naturally occurring flavones are yellow and are very widely distributed in plants. They accumulate in almost any part of a plant, from the roots to the flower petals.

Unlike the anthocyanins, which are too reactive and short-lived, the much more stable flavones have, from time immemorial, been used as dyes, for they impart various shades of yellow to wool. As an example, in the more recent past the inner bark of one of the North American oaks, *Quercus velutina*, was a commercial material known as quercitron bark and much used in dyeing: it contains quercetrin.

quercetrin   quercetin

The corresponding aglycone, quercetin, is one of the most widely occurring flavones, found, for example, in *Chrysanthemum* and *Rhododendron* species, horse chestnuts, lemons, onions, and hops.

9.3 Synthesis of benzopryliums, chromones, coumarins and isocoumarins

There are three important ways of putting together 1-benzopryliums, coumarins, and chromones; all begin with phenols. The isomeric 2-benzopryrylium and isocoumarin nuclei require the construction of an *ortho*-carboxy- or *ortho*-formyl-arylacetaldehyde (homophthalaldehyde).

Subject to the restrictions set out below, phenols react with 1,3-dicarbonyl compounds to produce 1-benzopryliums or coumarins depending on the oxidation level of the 1,3-dicarbonyl component.

ortho-Hydroxybenzaldehydes react with carbonyl compounds having an α-methylene, to give 1-benzopryliums or coumarins depending on the nature of the aliphatic unit.

ortho-Hydroxyaryl alkyl ketones react with esters to give chromones.

9.3.1 Ring synthesis of 1-benzopryliums[1b]

9.3.1.1 From phenols and 1,3-dicarbonyl compounds

The simplest reaction, that between a diketone and a phenol, works best with resorcinol, for the second hydroxyl facilitates the cyclising electrophilic attack. This synthesis can give mixtures with unsymmetrical diketones, and it is therefore well suited to the synthesis of 1-benzopryliums with identical groups at C-2 and C-4,[64] however diketones in which the two carbonyl groups are appreciably different in reactivity can produce high yields of single products.[65]

Acetylenic ketones, synthons for 1,3-keto-aldehydes, also take part regioselectively in condensations,[66] as do chalcones, though of course an oxidant must be incorporated in this latter case.[67]

For hetero-ring-unsubstituted targets, the bis-acetal of malondialdehyde can be employed; in this variant a heterocyclic acetal-ether is first obtained, from which two mol equivalents of ethanol must then be eliminated.[68]

9.3.1.2 From ortho-hydroxyaraldehydes and ketones

Salicylaldehydes can be condensed, by base or acid catalysis, with ketones which have an α-methylene. When base catalysis is used, the intermediate hydroxy-chalcones can be isolated,[8] but overall yields are often better when the whole sequence is carried out in one step, using acid.[69] It is important to note that because this route does not rely upon an electrophilic cyclisation on the benzene ring, benzene-ring-unsubstituted 1-benzopyryliums can be produced.

9.3.2 Ring synthesis of coumarins

9.3.2.1 From phenols and 1,3-ketoesters

The Pechmann synthesis[70]

Phenols react with β-ketoesters, including cyclic keto-esters,[71] to give coumarins under acid-catalysed conditions – concentrated sulfuric acid,[72] hydrogen fluoride,[73] or a cation exchange resin have been used.[74]

The Pechmann synthesis works best with the more nucleophilic aromatics such as resorcinols: electrophilic attack on the benzene ring *ortho* to phenolic oxygen by the protonated ketone carbonyl is the probable first step, though aryl acetoacetates, prepared from a phenol and diketene, also undergo ring closure to give coumarins.[75] The greater electrophilicity of the ketonic carbonyl determines the orientation of combination. The production of hetero-ring-unsubstituted coumarins can be achieved by condensing with formylacetic acid, generated *in situ* by the decarbonylation of malic acid.

Coumarins can be obtained directly, in a one-pot procedure, from phenols and a propiolate using palladium(0) catalysis. The catalytic cycle is considered to involve a formyloxypalladium hydride reacting with the phenol to produce an aryloxypalladium hydride which adds to the alkyne.[76]

9.3.2.2   From ortho-hydroxyaraldehydes and anhydrides (esters)

The simplest synthesis of coumarins is a special case of the Perkin condensation i.e. the condensation of an aromatic aldehyde with an anhydride. *ortho*-Hydroxy-*trans*-cinnamic acids cannot be intermediates since they do not isomerise under the conditions of the reaction; nor can *O*-acetylsalicylaldehyde be the immediate precursor of the coumarin, since it is not cyclised by sodium acetate on its own.[77]

The general approach can be enlarged and conditions for condensation made milder by the use of further-activated esters, thus condensation with methyl nitroacetate produces 3-nitrocoumarins,[78] condensations with Wittig ylides[79] allow *ortho*-hydroxyaryl ketones to be used[80] and the use of diethyl malonate (or malonic acid[81]), malononitrile or substituted acetonitriles in a Knoevenagel condensation, produces coumarins with a 3-ester[82] 3-cyano or 3-alkyl or -aryl substituent.[83] A 3-ester can be removed by hydrolysis and decarboxylation.[84]

ortho-Hydroxy-*trans*-cinnamates can be converted into coumarins and with concomitant introduction of an aryl or alkenyl substituent into the 4-position; the sequence depends on a Heck reaction at the double bond.[85]

9.3.5 Ring synthesis of 2-benzopyryliums

The first synthesis[105] of the 2-benzopyrylium cation provided the pattern for subsequent routes in which it is the aim to produce a homophthaldehyde, or diketone analogues, for acid-catalysed closure.

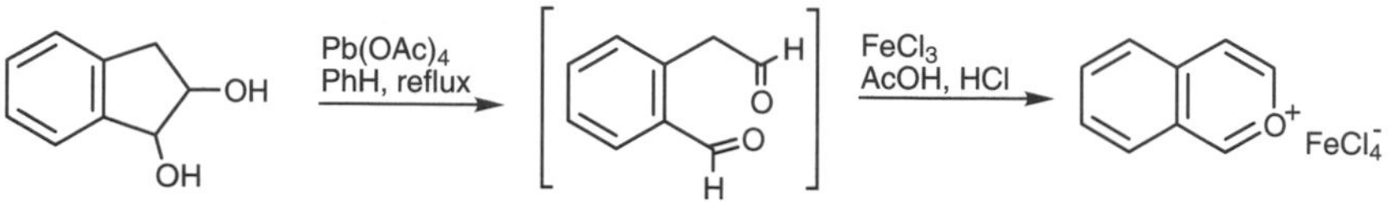

Most of the 2-benzopyrylium salts which have been synthesised subsequently have been 1,3-disubstituted and their precursors have been prepared by Friedel-Crafts acylation of activated benzyl ketones.[6,106]

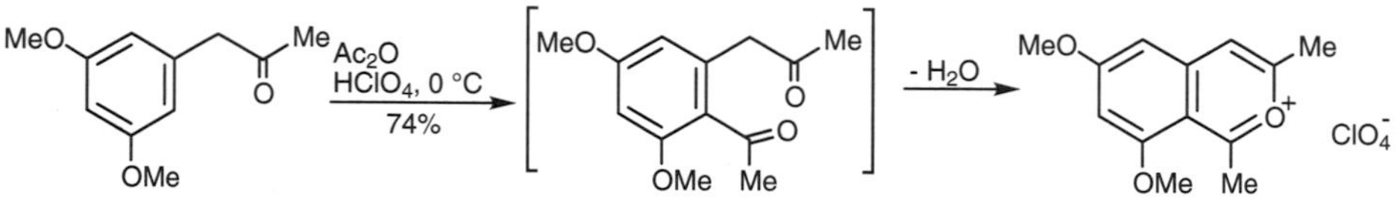

9.3.6 Ring synthesis of isocoumarins

One approach to isocoumarins is comparable to that above for 2-benzopyryliums, only the aromatic aldehyde needing to be changed to acid.[107]

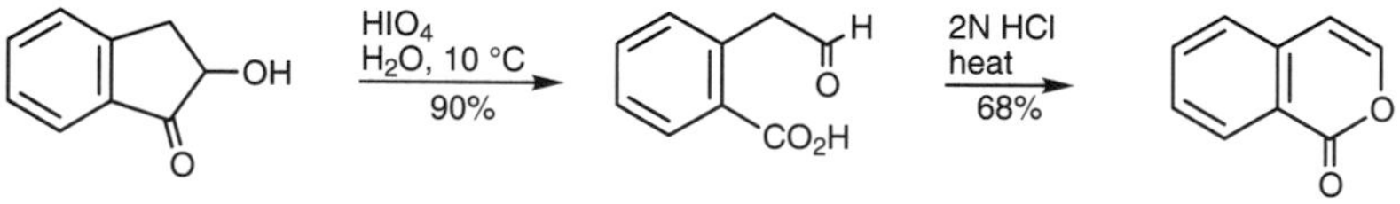

The direct introduction of the two-carbon unit of the heterocyclic ring, *ortho* to an existing carboxylic acid (ester) can be achieved in two ways: *ortho*-bromobenzoates can be coupled with π-(2-methoxyallyl)nickel bromide for the introduction of acetonyl,[108] or thallation of benzoic acids, *ortho* to the carboxyl, can be followed by palladium-catalysed coupling with alkenes.[109] Benzoates carrying an *ortho* acetylenic substituent can be ring closed using mercuric acetate, as shown below.[110]

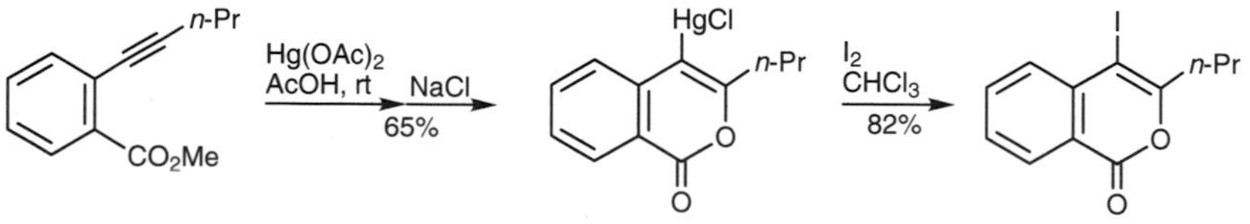

The most general route so far described involves coupling an alkyne with *ortho*-iodobenzoic acid or with methyl *ortho*-iodobenzoate.[111] Both mono- and disubstituted alkynes will serve allowing considerable flexibility for the construction of substituted isocoumarins.

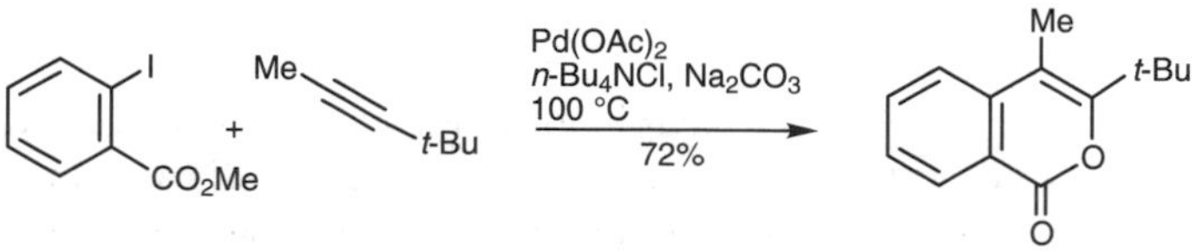

In a related process an *ortho* bromo ester is coupled with 1-tri-*n*-butylstannyl-2-ethoxyethene then acid used to close the ring; the example below shows how this sequence is applied to a pyridine ester.[112]

9.3.7 Notable examples of benzopyrylium and benzopyrone syntheses

9.3.7.1 *Pelargonidin chloride*

The first synthesis of pelargonidin chloride used methyl ethers as protecting groups for the phenolic hydroxyls during the Grignard addition step.[113]

9.3.7.2 *Apigenin*

The scheme below shows two contrasting routes to apigenin. The modern use of excess of a very strong base, and the reaction of the resulting 'polyanion' obviated the need for phenolic protection in one synthesis.[114]

An elegant and flexible strategy for the assembly of a synthon for the *ortho*-hydroxyaryl-1,3-diketone required for a chromone synthesis depends on the use of an isoxazole as surrogate for the 1,3-diketone unit (section 22.8). An isoxazole is produced by the cycloaddition of an arylnitrile oxide to tri-*n*-butylstannylacetylene (section 22.13.1.2), the product coupled with an aryl halide and then the N–O bond hydrogenolytically cleaved (section 22.8).[115]

Exercises for chapter 9

Straightforward revision exercises (consult chapters 7, 8, and 9)

(a) At which position(s) do benzopyrylium ions react with nucleophiles, for example water?

(b) What is the typical structure of an anthocyanin flower pigment? What is the typical structure of a flavone flower pigment?

(c) At which atom do coumarins and chromones protonate?

(d) At which positions do coumarins and chromones undergo electrophilic substitution?

(e) Describe a cycloaddition reaction in which (i) a coumarin and (ii) a chromone take part.

(f) How could one construct a 1-benzopyrylium salt from a phenol?

(g) How can *ortho*-hydroxyaryl aldehydes be used to prepare coumarins?

(h) How can *ortho*-hydroxyaryl ketones be used to prepare chromones?

More advanced exercises

1. When salicylaldehyde and 2,3-dimethyl-1-benzopyrylium chloride are heated together in acid, a condensation product $C_{18}H_{15}O_2^+$ Cl$^-$ is formed. Treatment of the salt with a weak base (pyridine) generates a neutral compound, $C_{18}H_{14}O_2$. Suggest structures for these two products.

2. When ethyl 2-methylchromone-3-carboxylate is treated with NaOH, then HCl, a product $C_{11}H_8O_4$ is produced which does not contain a carboxylic acid group but does dissolve in dilute alkali: suggest a structure and the means whereby it could be formed.

3. Deduce the structures of intermediate and final product in the sequence: salicylaldehyde/MeOCH$_2$CO$_2$Na/Ac$_2$O/heat $\rightarrow C_{10}H_8O_3$, this then with 1 mol equivalent of PhMgBr $\rightarrow C_{16}H_{14}O_3$ and finally this with HCl $\rightarrow C_{16}H_{13}O_2^+$ Cl$^-$.

4. Predict the structure of the major product from the interaction of resorcinol (1,3-dihydroxybenzene) and (i) PhCOCH$_2$COMe in AcOH/HCl; (ii) methyl 2-oxocyclopentanecarboxylate/H$_2$SO$_4$.

References

1. (a) 'The chemistry of plant pigments', Chichester, C. O., Ed., Academic Press, NY, **1972**; (b) 'The chemistry of anthocyanins, anthocyanidins, and related flavylium salts', Iacobucci, G. A. and Sweeny, J. E., *Tetrahedron*, **1983**, *39*, 3005.

2. 'Naturlich vorkommende chromone', Schmid, H., *Fortschr. Chem. Org. Naturst.*, **1954**, *11*, 124.

3. 'The flavonoids', Harborne, J. B., Mabry, T. J., and Mabry, H., Eds., Chapman and Hall, London, **1975**; 'Flavonoids: chemistry and biochemistry', Morita, N. and Arisawa, M., *Heterocycles*, **1976**, *4*, 373; 'The flavonoids: advances in research since 1980', Harborne, J. B. and Mabry, T. J., Eds., Chapman and Hall, London & NY, **1988**.

4. 'Synthesis of coumarins with 3,4-fused ring systems and their physiological activity', Darbarwar, M. and Sundaramurthy, V., *Synthesis*, **1982**, 337; 'Naturally occurring coumarins', Dean, F. M., *Fortschr. Chem. Org. Naturst.*, **1952**, *9*, 225; 'Naturally occurring plant coumarins', Murray, R. D. H., *Fortschr. Chem. Org. Naturst.*, **1978**, *35*,

199; 'The natural coumarins: occurrence, chemistry, and biochemistry', Murray, R. D. H., Méndez, J., and Brown, S. A., Chichester, Wiley, **1982**.

5. A large proportion of work on benzo[c]pyryliums is by Russian and Hungarian workers and is described in relatively inaccessible journals, however it is well reviewed as: 'Benzo[c]pyrylium salts: syntheses, reactions and physical properties', Kuznetsov, E. V., Shcherbakova, I. V., and Balaban, A. T., *Adv. Heterocycl. Chem.*, **1990**, *50*, 157.

6. 'Isocoumarins. Developments since 1950', Barry, R. D., *Chem. Rev.*, **1964**, *64*, 229.

7. 'Heterocycles as structural units in new optical brighteners', Dorlans, A., Schellhammer, C.-W., and Schroeder, J., *Angew. Chem., Int. Ed. Engl.*, **1975**, *14*, 665.

8. Le Fèvre, R. J. W., *J. Chem. Soc.*, **1929**, 2771.

9. Holker, J. S. E. and Underwood, J. G., *Chem. Ind. (London)*, **1964**, 1865.

10. Jurd, L., *Tetrahedron*, **1966**, *22*, 2913; *ibid.*, **1968**, *24*, 4449.

11. Meyer-Dayan, M., Bodo, B., Deschamps-Valley, C., and Molho, D., *Tetrahedron Lett.*, **1978**, 3359.

12. Degani, I. and Fochi, R., *Ann. Chim. (Rome)*, **1968**, *58*, 251.

13. Hill, D. W. and Melhuish, R. R., *J. Chem. Soc.*, **1935**, 1161.

14. Jurd, L., *Tetrahedron*, **1969**, *25*, 2367.

15. Sutton, R., *J. Org. Chem.*, **1972**, *37*, 1069.

16. Dimroth, K. and Odenwälder, H., *Chem. Ber.*, **1971**, *104*, 2984.

17. Lowenbein, A., *Chem. Ber.*, **1924**, *57*, 1517.

18. Pomilio, A. B., Müller, O., Schilling, G., and Weinges, K., *Justus Liebigs Ann. Chem.*, **1977**, 597.

19. Kröhnke, F. and Dickoré, K., *Chem. Ber.*, **1959**, *92*, 46.

20. Jurd, L., *Tetrahedron*, **1965**, *21*, 3707.

21. Shcherbakova, I. V., Kuznetsov, E. V., Yudilevich, I. A., Kompan, O. E., Balaban, A. T., Abolin, A. H., Polyakov, A. V., and Struchkov, Yu. T., *Tetrahedron*, **1988**, *44*, 6217; Le Roux, J.-P., Desbene, P.-L., and Cherton, J.-C., *J. Heterocycl.Chem.*, **1981**, *18*, 847.

22. Lee, Y. G., Ishimaru, K., Iwasaki, H., Okhata, K., and Akiba, K., *J. Org. Chem.*, **1991**, *56*, 2058.

23. Marathe, K. G., Philbin, E. M., and Wheeler, T. S., *Chem. Ind. (London)*, **1962**, 1793.

24. Clark-Lewis, J. W. and Baig, M. I., *Aust. J. Chem.*, **1971**, *24*, 2581.

25. Müller, A., Lempert-Stréter, M., and Karczag-Wilhelms, A., *J. Org. Chem.*, **1954**, *19*, 1533.

26. Heilbron, I. M. and Zaki, A., *J. Chem. Soc.*, **1926**, 1902.

27. 'A curious brew', Allen, M., *Chem. Brit.*, **1996** (May), 35.

28. 'Structure, stability and color variation of natural anthocyanins', Goto, T., *Fortschr. Chem. Org. Naturst.*, **1987**, *52*, 113; Goto, T. and Kondo, T., 'Structure and molecular stacking of anthocyanins - flower colour variation', *Angew. Chem., Int. Ed. Engl.*, **1991**, *30*, 17; 'The chemistry of rose pigments', Eugster, C. H. and Märki-Fischer, E., *ibid.*, 654; 'Nature's palette', Haslam, E., *Chem. Brit.*, **1993** (Oct.), 875.

29. Wittig, G., Baugert, F., and Richter, H. E., *Justus Liebigs Ann. Chem.*, **1925**, *446*, 155.

30. Meerwein, H., Hinz, G., Hofmann, P., Kroenigard, E., and Pfeil, E., *J. Prakt. Chem.*, **1937**, *147*, 257.

31. Clayton, A., *J. Chem. Soc.*, **1910**, 2106; Joshi, P. P., Ingle, T. R., and Bhide, B. V., *J. Ind. Chem. Soc.*, **1959**, *36*, 59.

32. Wiley, P. F., *J. Am. Chem. Soc.*, **1952**, *74*, 4326.

33. Zhang, F. J. and Li, Y. L., *Synthesis*, **1993**, 565.

34. Dean, F. M. and Murray, S., *J. Chem. Soc., Perkin Trans. 1*, **1975**, 1706.

35. Pearson, D. E., Stamper, W. E., and Suthers, B. R., *J. Org. Chem.*, **1963**, *28*, 3147.

36. Fuson, R. C., Kneisley, J. W. and Kaiser, E. W., *Org. Synth., Coll. Vol. III*, **1955**, 209; Perkin, W. H., *Justus Liebigs Ann. Chem.*, **1871**, *157*, 115.

37. Thapliyal, P. C., Singh, P. K., and Khauna, R. N., *Synth. Commun.*, **1993**, *23*, 2821.

38. Ellis, G. P. and Thomas, I. L., *J. Chem. Soc., Perkin Trans. 1*, **1973**, 2781.

39. Arndt, F., *Chem. Ber.*, **1925**, *58*, 1612.

40. For the use of singlet oxygen see Matsuura, T., *Tetrahedron*, **1977**, *33*, 2869.

41. Adam, W., Golsch, D., Hadjiarapoglou, L., and Patonay, T., *J. Org. Chem.*, **1991**, *56*, 7292; Adam, W., Hadjiarapoglou, L., and Levai, A., *Synthesis*, **1992**, 436.

42. Muller, E., *Chem. Ber.*, **1909**, *42*, 423.

43. Sugita, Y. and Yokoe, I., *Heterocycles*, **1996**, *43*, 2503.

10 Typical reactivity of the diazines: pyridazine, pyrimidine and pyrazine

pyridazine   pyrimidine   pyrazine

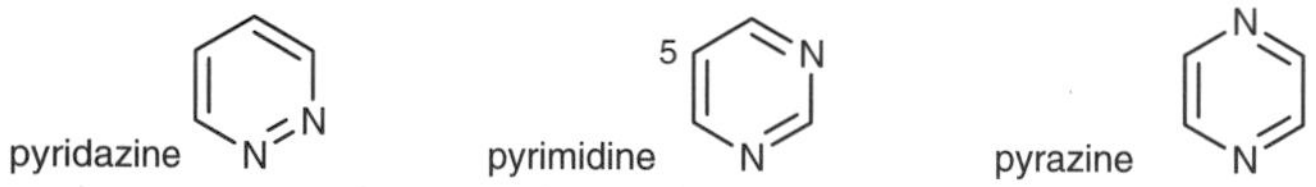

The diazines – pyridazine, pyrimidine and pyrazine – contain two imine nitrogen atoms, so the lessons learnt with regard to pyridine (chapter 5) are, in these heterocycles, exaggerated. Two heteroatoms withdraw electron density from the ring carbons even more than in pyridine, so the unsubstituted diazines are even more resistant to electrophilic substitution than is pyridine. A corollary of course, developed below, is that this same increased electron deficiency at carbon makes the diazines more easily attacked by nucleophiles than pyridines. The availability of nitrogen lone pair(s) is also reduced: each of the diazines is appreciably less basic than pyridine, reflecting the destabilising influence of the second nitrogen on the N-protocation. Nevertheless, diazines will form salts and will react with alkyl halides and with peracids to give N-alkyl quaternary salts and N-oxides, respectively. Generally speaking, such electrophilic additions take place at one nitrogen only, because the presence of the positive charge in the products renders the second nitrogen extremely unreactive towards a second electrophilic addition.

Typical reactivity of diazines exemplified by pyrimidine

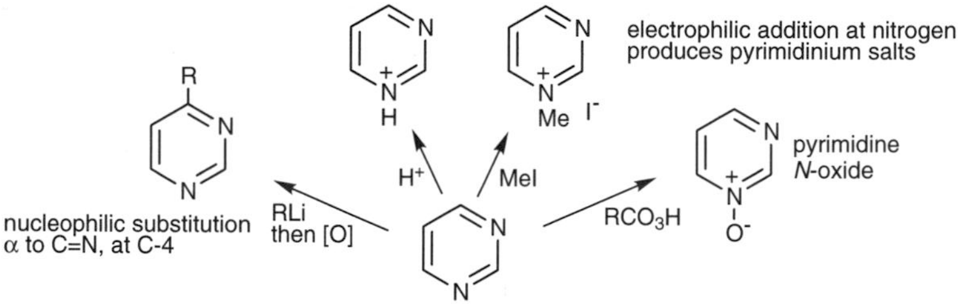

A very characteristic feature of the chemistry of diazines, which is associated with their strongly electron-poor nature, is that they add nucleophilic reagents easily. Without halide to be displaced, such adducts require an oxidation to complete an overall substitution. However, halo-diazines, where the halide is α or γ to a nitrogen, undergo very easy nucleophilic displacements, the intermediates being particularly well stabilised.

All positions on each of the diazines, with the sole exception of the 5-position of a pyrimidine, are α and/or γ to an imine ring nitrogen and, in considering nucleophilic addition/substitution, it must be remembered that there is also an additional nitrogen withdrawing electron density. As a consequence, all the monohalodiazines are more reactive than either 2- or 4-halopyridines. The 2- and 4-halopyrimidines are particularly reactive because the anionic intermediates (shown below for attack on 2-chloropyrimidine) derive direct mesomeric stabilisation from both nitrogen atoms.

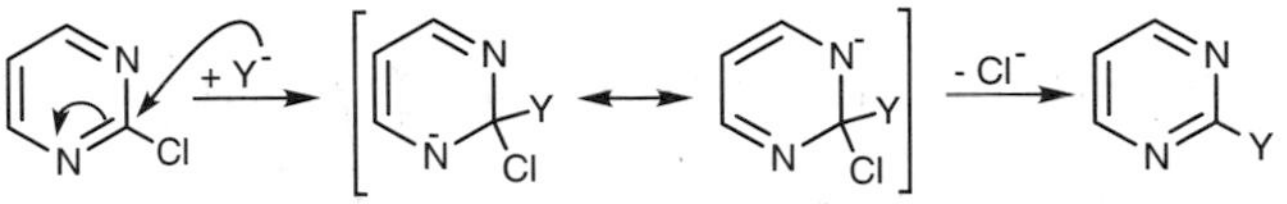

Despite this particularly strong propensity for nucleophilic addition, *C*-lithiation of diazines can be achieved by either metal–halogen exchange or, by deprotonation *ortho* to chloro or alkoxyl substituents, though very low temperatures must be utilised in order to avoid nucleophilic addition of the reagent.

In line with their susceptibility to nucleophilic addition, diazines also undergo Minisci radical substitution with ease. Considerable use has been made in diazine chemistry of palladium(0)-catalysed coupling processes, one of which is illustrated below (see section 2.7 for a detailed discussion).

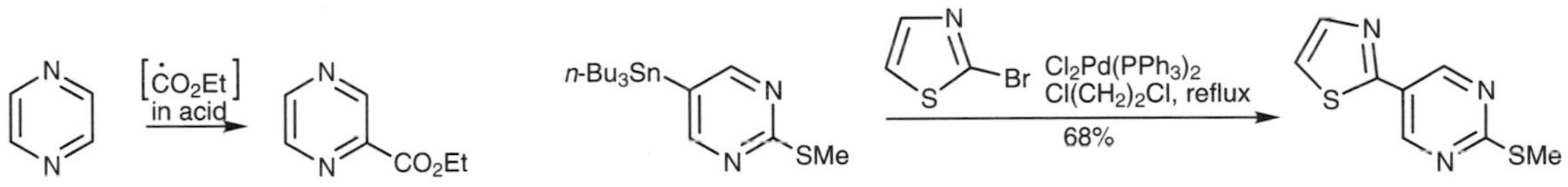

Further examples of the enhancement of those facets of pyridine chemistry associated with the imine electron withdrawal, include a general stability towards oxidative degradation but, on the other hand, a tendency to undergo rather easy reduction of the ring.

Although there is always debate about quantitative measures of aromaticity, it is agreed that the diazines are less resonance stabilised than pyridines – they are 'less aromatic'. Thus, Diels-Alder additions are known for all three systems, with the heterocycle acting as a diene; initial adducts lose a small molecule – hydrogen cyanide in the pyrimidine example shown – to afford a final stable product.

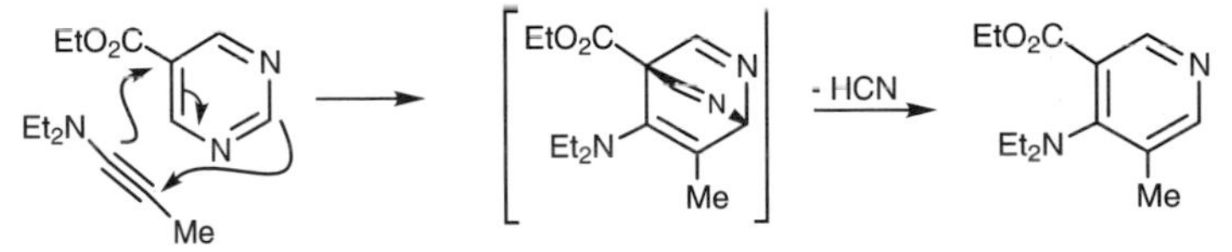

N-Oxides, just as in the pyridine series, show a remarkable duality of effect – they encourage both electrophilic substitutions and nucleophilic displacements. The sequence below shows pyridazine *N*-oxide undergoing first, electrophilic nitration, then, the product, nucleophilic displacement, with nitrite as leaving group.

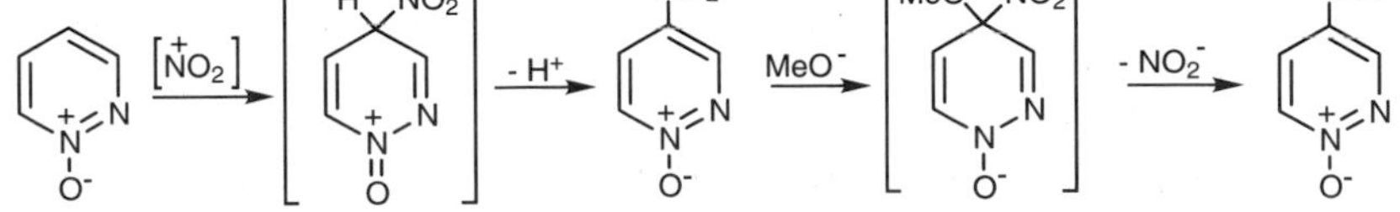

N-Oxide chemistry in six-membered heterocycles provides considerable scope for synthetic manipulations. One of the very useful transformations is the introduction of halide α to a nitrogen on reaction with phosphorus or sulfur halides, the conversion being initiated by oxygen attack on the phosphorus (sulfur). The power of

11 The diazines: pyridazine, pyrimidine, and pyrazine: reactions and synthesis

pyridazine   pyrimidine   pyrazine

cinnoline   phthalazine   quinazoline   quinoxaline

The three diazines, pyridazine,[1] pyrimidine,[2] and pyrazine[3] are stable, colourless compounds which are soluble in water. The three parent heterocycles, unlike pyridine, are expensive and not readily available and so are seldom used as starting materials for the synthesis of their derivatives. There are only four ways in which a benzene ring can be fused to a diazine: cinnoline, phthalazine, quinazoline and quinoxaline are the bicyclic systems thus generated.

One striking aspect of the physical properties of the diazine trio is the high boiling point of pyridazine (207 °C), 80–90 °C higher than that of pyrimidine (123 °C), pyrazine (118 °C), or indeed other azines, including 1,3,5-triazine, all of which also boil in the range 114–124 °C. The high boiling point of pyridazine is attributed to the polarisability of the N–N unit which results in extensive dipolar association in the liquid.

The most important naturally occuring diazines are the pyrimidine bases uracil, thymine and cytosine, which are constituents of the nucleic acids.[4] Following from this, several pyrimidine nucleoside analogues have been developed as anti-viral agents, for example Idoxuridine is used in the treatment of *Herpes* infections of the eye, and AZT is the most widely used anti-AIDS drug; 3-TC (Lamivudine) is used to treat both hepatitis B and AIDS, and d4T (Stavudine) is a fourth drug approved for treatment of HIV infection and AIDS. The pyrimidine ring also occurs in the vitamin thiamin (section 21.11). The nucleic acid pyrimidines are often drawn horizontally transposed from the representations used in this chapter, i.e. with N-3 to the 'North-West', mainly to draw attention to their structural similarity to the pyrimidine ring of the nucleic acid purines (chapter 24), which are traditionally drawn with the pyrimidine ring on the left.

uridine   thymidine   cytidine

Idoxuridine   AZT   3-TC (Lamivudine)   d4T (Stavudine)

The pyrazine ring system is found in the fungal metabolite aspergillic acid and in dihydro-form in the luciferins of several beetles, including the firefly, *Cypridina hilgendorfii*, and is responsible for the chemiluminescence[5] of this ostracod. Quite simple methoxypyrazines are very important components of the aromas of many fruits and vegetables, such as peas and Capsicum peppers, and also of wines.[6] Although present in very small amounts, they are extremely odorous and can be detected at concentrations as low as 0.00001 ppm. Related compounds, probably formed by the pyrolysis of amino acids during the process of cooking, are also important in the aroma of roasted meats. Several polyalkylpyrazines are insect pheromones, for example 2-ethyl-3,6-dimethylpyrazine is the major component of the trail pheremone of the South American leaf-cutting ant.

Cypridina luciferin   aspergillic acid   R = Me, *i*-Bu, *etc.* food aroma components   ant trail pheremone

There are relatively few naturally occuring pyridazines, for example some fungal metabolites from *Streptomyces* species, consisting mainly of reduced systems, and the quaternary salt pyridazinomycin. Divicine, present as a glucoside in broad beans and related plants, is the toxic agent responsible for 'favism', a serious hemolytic reaction in genetically susceptible people (of Mediterranean origin). Piperazine (hexahydro-pyrazine) is used in the treatment of intestinal nematode (worm) infections.

pyridazinomycin   divicine   piperazine

Derivatives of all three heterocyclic systems have been widely investigated for use in synthetic drugs (see also above); amongst the most commonly used compounds are the antibacterial Trimethoprim, the antimalarial Pyrimethamine and the anti-hypertensive agent Hydralazine (containing a phthalazine nucleus).

Trimethoprim   Pyrimethamine   Hydralazine   Pyridate   3-hydroxy-6(1H)-pyrazinone

A number of pyridazines are important as selective plant growth regulators and are used as herbicides, e.g. Pyridate; 3-hydroxy-6(1H)-pyrazinone, is used as a lawn weedkiller.

11.1 Reactions with electrophilic reagents

11.1.1 Addition at nitrogen[7]

11.1.1.1 Protonation

The diazines, pyridazine (pK_a 2.3), pyrimidine (1.3), and pyrazine (0.65) are essentially monobasic substances, and considerably weaker, as bases, than pyridine (5.2). This reduction in basicity is believed to be largely a consequence of destabilisation of the mono-protonated cations due to inductive withdrawal by the second nitrogen atom. Secondary effects, however, determine the order of basicity for the three systems: lone pair repulsion between the two adjacent nitrogen atoms in pyridazine means that protonation occurs more readily than if inductive effects, only, were operating. In the case of pyrazine, mesomeric interaction between the protonated and neutral nitrogen atoms probably destabilises the cation.

N,N'-Diprotonation is very much more difficult and has only been observed in very strongly acidic media. Of the trio, pyridazine ($pK_{a(2)}$ −7.1) is the most difficult from which to generate a dication, probably due to the high energy associated with the juxtaposition of two immediately adjacent positively charged atoms, but pyrimidine ($pK_{a(2)}$ −6.3) and pyrazine ($pK_{a(2)}$ −6.6) are only marginally easier to doubly protonate.

Substituents can affect basicity (and nucleophilicity) both inductively and mesomerically, but care is needed in the interpretation of pK_a changes, for example it is important to be sure which of the two nitrogens of the substituted azine is protonated (see also section 11.1.1.2).

11.1.1.2 Alkylation

The diazines react with alkyl halides to give mono-quaternary salts, though somewhat less readily than comparable pyridines. Dialkylation cannot be achieved with simple alkyl halides, however the very much more reactive trialkyloxonium tetrafluoroborates do convert all three systems into di-quaternary salts.[8]

Pyridazine is the most reactive in alkylation reactions and this again has its origin in the lone pair/lone pair interaction between the nitrogen atoms. This phenomenon is known as the 'α effect' and is also responsible, for example, for the relatively higher reactivity of hydrogen peroxide as a nucleophile, compared with water.

Unsymmetrically substituted diazines can give rise to two isomeric quaternary salts. Substituents influence the orientation mainly by steric and inductive, rather than mesomeric effects. For example, 3-methylpyridazine alkylates mainly at N-1, even though N-2 is the more electron-rich site. Again, quaternisation of 3-methoxy-6-methylpyridazine takes place adjacent to the methyl substituent, at N-1, although mesomeric release would have been expected to favour attack at N-2.[9]

11.1.1.3 Oxidation

All three systems react with peracids,[10] giving N-oxides, but care must be taken with pyrimidines[11] due to the relative instability of the products under the acidic conditions. Pyrazines[10] form N,N'-dioxides the most easily, but pyridazine[12] requires forcing conditions and pyrimidines, apart from some examples in which further activation is present, give poor yields.[13]

The regiochemistry of N-oxidation of substituted azines is governed by the same factors as alkylation (section 11.1.1.2), for example 3-methylpyridazine gives the 1-oxide as main (3:1) product,[14] but the pattern is not a simple one, for 4-methylpyrimidine N-oxidises principally (3.5:1) at the nitrogen adjacent to the methyl.[15] The acidity of the medium can also influence the regiochemistry of oxidation, for example 3-cyanopyridazine reacts at N-1 with peracetic acid, but under strongly acidic conditions, in which the heterocycle is mainly present as its N-1-protonic salt, oxidation, apparently involving attack on this salt, occurs at N-2.[16]

11.1.2 Substitution at carbon

Recalling the resistance of pyridines to electrophilic substitution, it is not surprising to find that introduction of a second azomethine nitrogen, in any of the three possible orientations, greatly increases this resistance: no nitration or sulfonation of a diazine or simple alkyldiazine has been reported, though some halogenations are known. It is to be noted that C-5 in pyrimidine is the only position, in all three diazines, which is not in an α- or γ-relationship to a ring nitrogen, and is therefore equivalent to a β-position in pyridine. Diazines carrying electron-releasing (activating) substituents undergo electrophilic substitution much more easily (sections 11.10.2.1 and 11.11).

11.1.2.1 Halogenation

Chlorination of 2-methylpyrazine occurs under such mild conditions that it is almost certain that an addition/elimination sequence is involved, rather than a classical aromatic electrophilic substitution.[17] Halogenation of pyrimidines may well also involve such processes.[18]

11.2 Reactions with oxidising agents

The diazines are generally resistant to oxidative attack at ring carbons, though alkaline oxidising agents can bring about degradation *via* intermediates produced by initial nucleophilic addition (section 11.3). Alkyl substituents[19] and fused aromatic rings[20] can be oxidised to carboxylic acid residues, leaving the heterocyclic ring untouched. An oxygen can be introduced into pyrimidines at vacant C-2 and/or C-4 positions using various bacteria.[21] Dimethyldioxirane converts *N,N*-dialkylated uracils into 5,6-diols probably via 5,6-epoxides.[22]

11.3 Reactions with nucleophilic reagents

The diazines are very susceptible to nucleophilic addition: pyrimidine, for example, is decomposed when heated with aqueous alkali by a process which involves hydroxide addition as a first step, and it is converted into pyrazole by reaction with hydrazine.

11.3.1 With replacement of hydrogen

11.3.1.1 Alkylation and arylation

The diazines readily add alkyl- and aryllithiums, and Grignard reagents, to give dihydro-adducts which can be aromatised by oxidation with reagents such as potassium permanganate or 2,3-dichloro-5,6-dicyano-1,4-benzoquinone. In reactions with organolithiums, pyrimidines react at C-4,[13] and pyridazines at C-3, but Grignard reagents add to pyridazines at C-4.[23]

An important point is that in diazines carrying chlorine or methylthio substituents, attack does not take place at the halogen- or methylthio-bearing carbon; halogen-[24] and methylthio-containing[25] products are therefore obtained.

11.3.1.2 Amination

The Chichibabin reaction can be carried out under the usual conditions in a few cases,[26] but is much less general than for pyridines. This may be a consequence of the lower aromaticity of the diazines for, although the initial addition is quite easy, the subsequent loss of hydride (rearomatisation) is difficult. However, high yields of 4-aminopyridazine, 4-aminopyrimidine and 2-aminopyrazine can be obtained by oxidation of the dihydro-adduct *in situ* with potassium permanganate.[27]

11.3.2 With replacement of good leaving groups

All the halodiazines, apart from 5-halopyrimidines, react readily with 'soft' nucleophiles such as amines, thiolates, and malonate anions, with substitution of the halide. All cases are more reactive than 2-halopyridines: the relative reactivity can be summarised:

and is illustrated by the following examples:[28]

Nucleophilic displacement of halogen with ammonia[29] and amines[30] can be accelerated by carrying out the displacements in acid solution, when the protonated heterocycle is more reactive than the neutral heterocycle.[31] Halogen can also be easily removed hydrogenolytically, for example treatment of 2,4-dichloropyrimidine, readily available from uracil, with hydrogen in the presence of palladium, or with hydrogen iodide, gives pyrimidine itself.[32]

The difference in reactivity between 2- and 4-halopyrimidines is relatively small and a discussion of the selectivity in nucleophilic displacement reactions of 2,4-dichloropyrimidine (an important synthetic intermediate) is instructive.

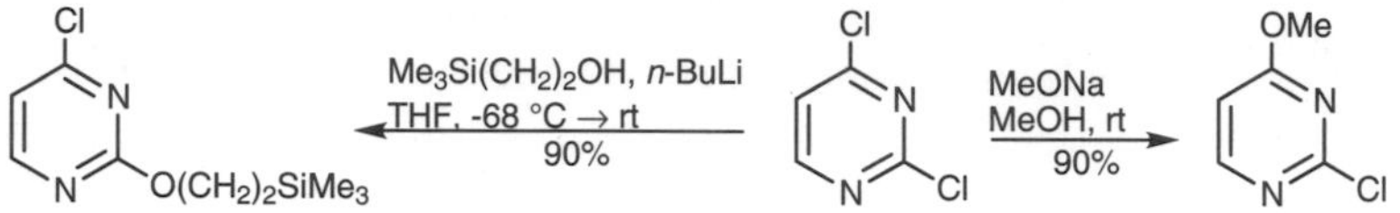

Reaction with sodium methoxide in methanol is highly selective for the 4-chloro substitutent[33] whereas, lithium 2-(trimethylsilyl)ethoxide is equally selective, but for the 2-chloro substituent.[34] The former is the normal situation for nucleophilic displacements[35] – 4-chloro > 2-chloro – the second case is the exception where strong co-ordination of lithium in a non-polar solvent to the more basic nitrogen, N-1, leads to activation, and possibly also internal attack, at C-2. Under acidic conditions, an approximately 1:1 mixture of the two methoxy products is formed. Here, hydrogen bonding to the proton on N-1 provides the mechanism for encouraging attack at C-2. Selectivity with other nucleophiles is dependent on the nature of the nucleophile and on reaction conditions.

These reactions are also sensitive to the presence of other substituents in the ring, either by electronic or steric effects and this sometimes leads to a reversal of the typical selectivity[36] as can changes in the nucleophile, for example tri-n-butylstannyllithium attacks 2,4-dichloropyrimidine at C-2.[37] Selective reductions are also possible.[38]

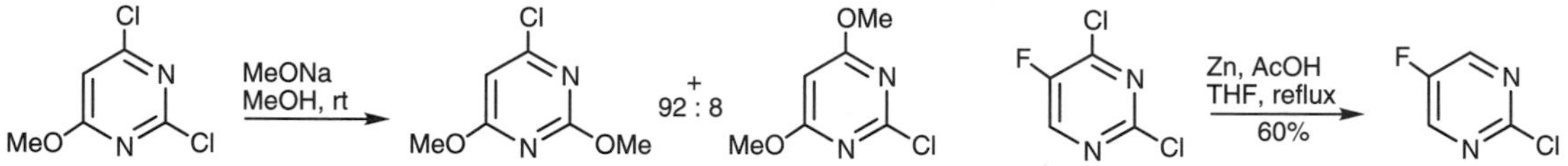

A device which is also used in pyridine and purine chemistry is the initial replacement of halogen with a tertiary amine, the resulting salt now having a better leaving group, as shown below.[39]

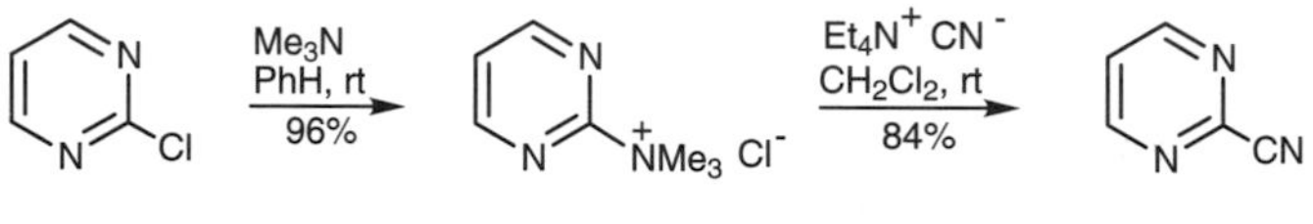

Halopyrimidines with other electron-donating substituents in the ring tend to be much less reactive to nucleophilic substitution: in one example this was overcome by use of the very nucleophilic O,N-dimethylhydroxylamine, followed by hydrogenolysis to reveal the amine.[40]

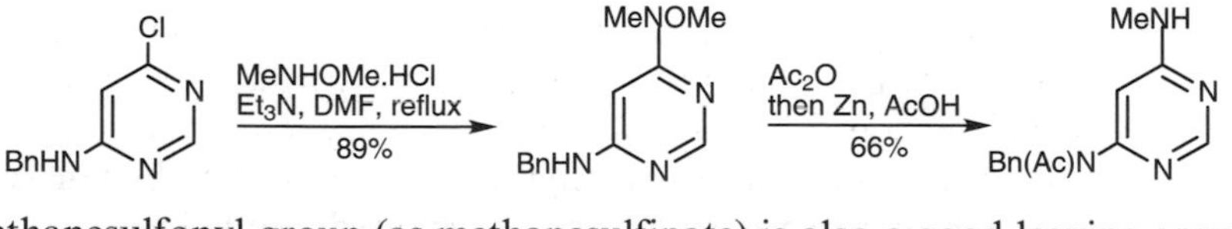

A methanesulfonyl group (as methanesulfinate) is also a good leaving group in all of the diazines,[41] generally better than chloro, sometimes considerably so, for example 3-methanesulfonylpyridazine reacts 90 times faster with methoxide than does 3-chloropyridazine. Sulfinates can be used to catalyse displacements of chlorine via the intermediacy of the sulfone.[42]

Even methoxy groups can be displaced by carbanions.[43]

Monsubstitution of 2,6-diiodopyridazine is easy, further manipulation via various palladium-catalysed couplings (see also 11.5.2) providing a good route to 2,6-disubstituted pyridazines.[44]

A highly regioselective VNS substitution (section 2.3.3) can be carried out on pyridazines bearing a 3-substitutent. Here, formation of a dicyanomethylene ylide only at N-1, due to hindrance of N-2, results in a specific activation of the hindered 4-position.[45]

11.4   Reactions with bases

11.4.1   Deprotonation of C-hydrogen

All three diazines undergo H/D exchange at all ring positions with MeONa/MeOD at 164 °C;[46] the transient carbanions which allow the exchange are formed somewhat faster than for pyridines, and again this is probably due to the acidifying, additional inductive withdrawal provided by the second nitrogen.

11.4.2   Metallation

The three parent diazines have been metallated adjacent to nitrogen (for pyrimidine at C-4, not C-2) using the non-nucleophilic lithium tetramethylpiperidide, but the resulting heteroaryllithiums are very unstable, readily forming dimeric compounds by self addition. Moderate to good yields of trapped products can however be obtained either by using very short lithiation times (pyridazine and pyrazine) or by *in situ* trapping where the electrophile is added *before* the metallating agent.[47] 4-Lithiopyridazine has been prepared by transmetallation of the corresponding tri-*n*-butylstannane using *n*-butyllithium. Lithiation of diazines with directing groups (chloro, fluoro, methoxy, methylthio, and various carboxamides) is straightforward[48] and such derivatives have been widely used (see below).

11.5   Reactions of C-metallated diazines

11.5.1   Lithium derivatives

Typically, lithium tetramethylpiperidide has been used for the kinetically-controlled metallation of substituted diazines;[49] in some cases the use of a somewhat higher temperature allows equilibration to a thermodynamic anion.[50]

Lithiodiazines can also be prepared by halogen exchange with alkyllithiums, but very low temperatures must be used in order to avoid nucleophilic addition to the ring.[51] The examples below show how 5-bromopyrimidine can be lithiated at C-4, using LDA, or alternatively can be made to undergo exchange, using *n*-butyllithium.[52] Note, also, that in some cases, reactions are carried out by adding the electrophile to the pyrimidine *before* lithiation, a practice which incidentally illustrates that metal-halogen exchange with *n*-butyllithium is faster than the addition of *n*-butyllithium to a carbonyl compound.[53]

Lithiopyrimidines, -pyrazines, and -pyridazines have been converted by exchange with zinc chloride into the more stable zinc compounds[54] for use in palladium-catalysed couplings (section 11.5.2). Magnesium derivatives have been prepared by reaction of 5-bromopyrimidines with *n*-butylmagnesium bromide and cerium compounds (which give better results than lithiopyrimidines in reactions with enolisable ketones) can be prepared from either bromo- or lithiopyrimidine.[55]

11.5.2   Palladium-catalysed reactions

Palladium- (and nickel-) -catalysed coupling reactions proceed normally on halodiazines and the equivalent triflates,[56] the most significant feature being the enhanced reactivity relative to chlorobenzenes of chlorine at position α and γ to a nitrogen, just as in pyridine chemistry. In some particularly activated caes, this extra activation is sufficient to overcome the normally higher reactivity of bromine, but not of iodine.[57] In the examples shown below the 4-chlorine has the edge over the bromine, although the reagent needed careful selection for good selectivity. The same selectivity in reactivity of pyrimidines for nucleophilic substitution (4 > 2) applies in palladium-catalysed reactions.[58]

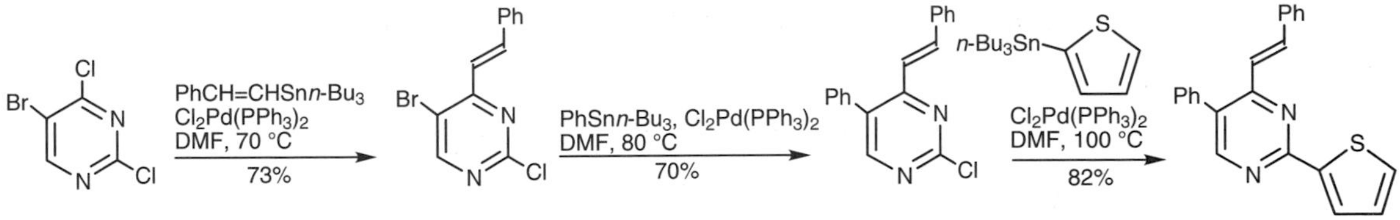

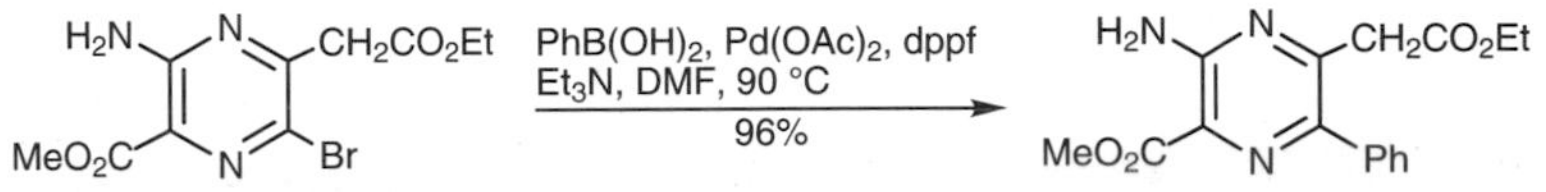

In diazine chemistry, tin compounds have normally been used[59] when an organometallic derivative of the heterocycle is required for a coupling reaction; they have the particular advantage that they can be prepared without the use of an organolithium intermcdiate. Boronic acids have been used occasionally,[60] but as a general rule they are difficult to prepare at positions α to azine nitrogens – a major disadvantage in diazine chemistry. Zinc compounds have advantages in some cases. A selection of cxamples involving palladium(0)-catalysed couplings is shown below.

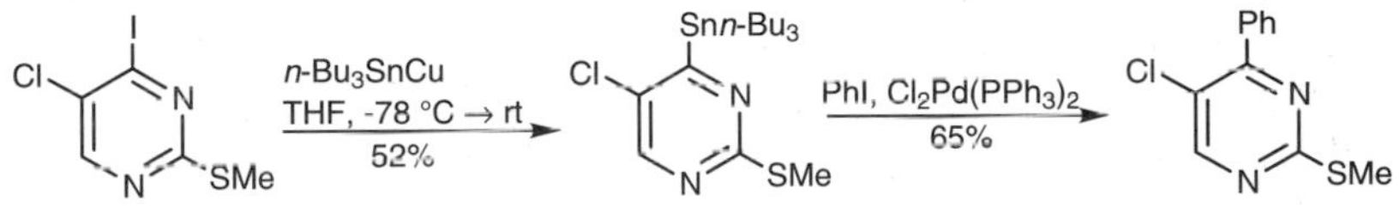

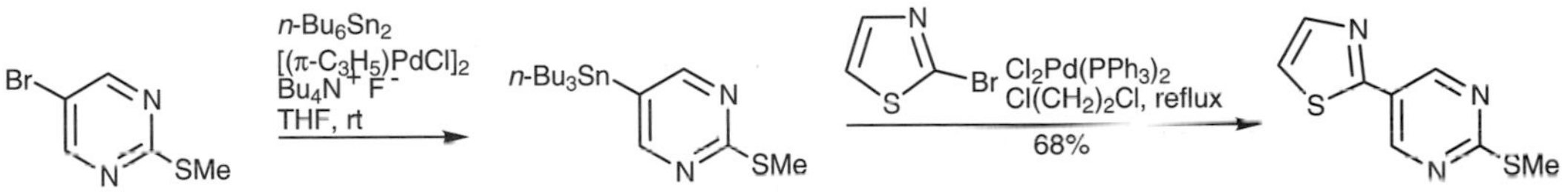

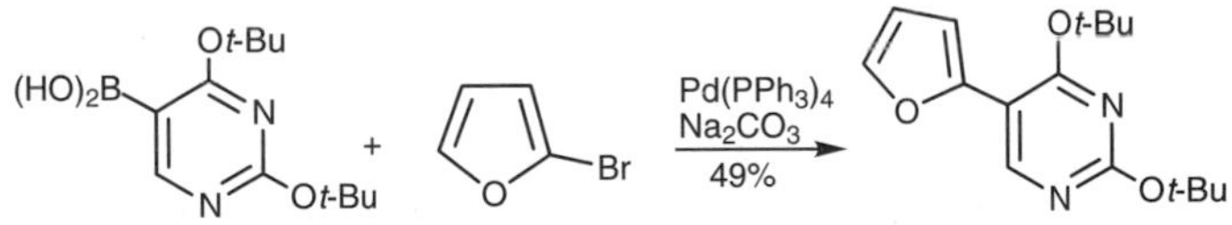

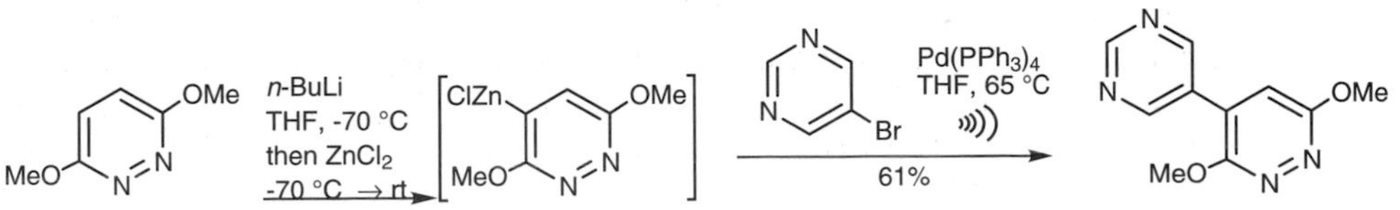

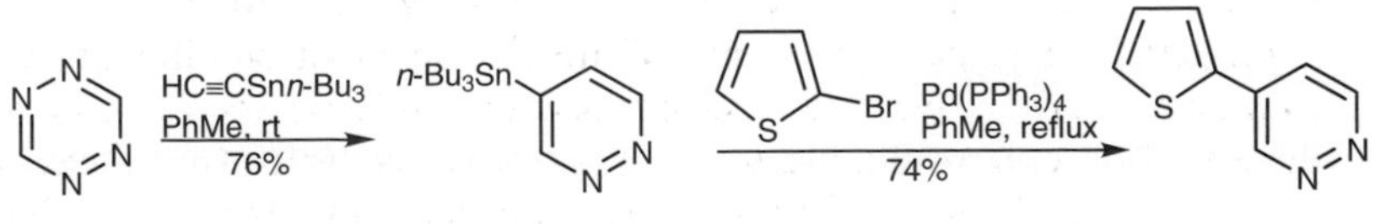

11.6   Reactions with reducing agents

Due to their lower aromaticity, the diazines are more easily reduced than pyridines. Pyrazine and pyridazine can be reduced to hexahydro derivatives with sodium in hot ethanol; under these conditions pyridazine has a tendency for subsequent reductive cleavage of the N–N bond.

Partial reductions of quaternary salts to dihydro compounds can be achieved with borohydride but such processes are much less well studied than in pyridinium salt chemistry.[61] 1,4-Dihydropyrazines have been produced with either silicon[62] or amide[63] substitution at the nitrogen atoms, and all the diazines can be reduced to tetrahydro-derivatives with carbamate protection on nitrogen, which aids in stabilisation and thus allows isolation.[64]

11.7   Reactions with radical reagents

Radicals add readily to diazines under Minisci conditions.[65] Additions to pyrimidine often show little selectivity, C-2 *versus* C-4, however a selective Minisci reaction on 5-bromopyrimidine provided a convenient synthesis of the 4-benzoyl derivative on a large scale;[66] attack at C-5 does not take place.[67] Radical attack on pyridazines shows selectivity for C-4,[68] even when C-3 is unsubstituted. Pyrazines[69] can of course substitute in only one type of position.

Substitutions of halide by an $S_{RN}1$ mechanism (section 5.9) have been carried out, but addition/elimination mechanisms compete in the more reactive halides.[70]

11.8   Electrocyclic reactions

All the diazines, providing they also have electron-withdrawing substituents, undergo Diels-Alder additions with dienophiles. Intramolecular reactions occur the most readily; these do not even require the presence of activating substituents. The immediate products of such process usually lose nitrogen (pyridazine adducts) or hydrogen cyanide (adducts from pyrimidines and pyrazines) to generate benzene and pyridine products,[71] respectively, as illustrated below.[72] Singlet oxygen has been added across the 2,5-positions of pyrazines.[73]

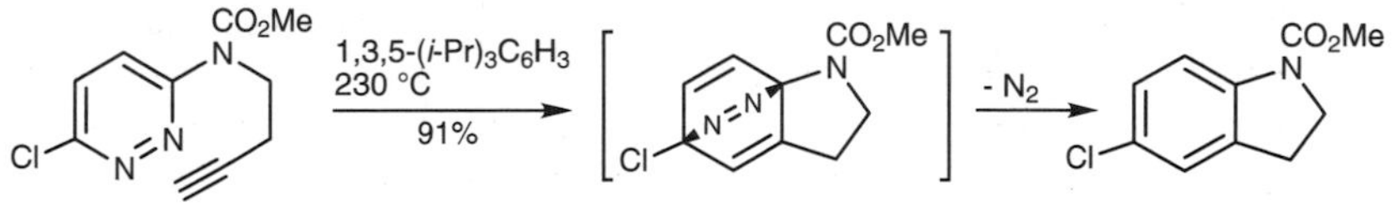

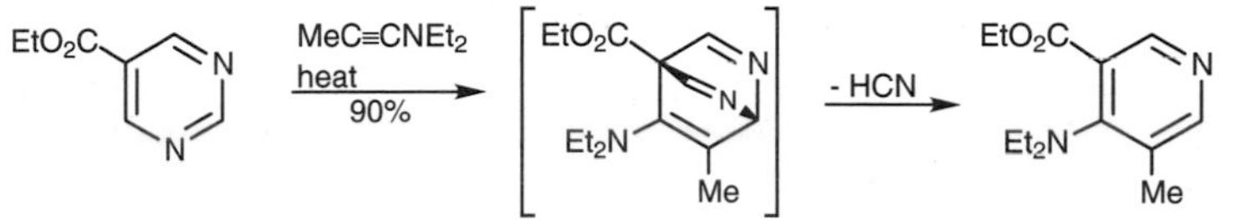

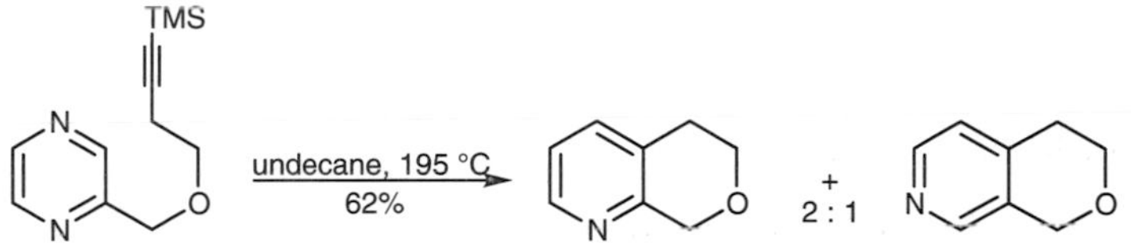

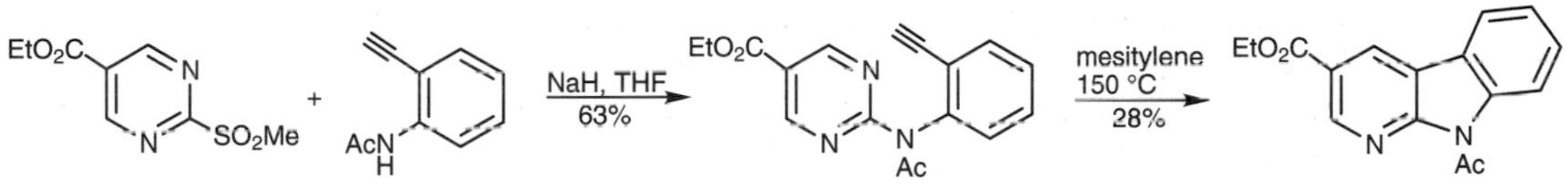

All three diazines undergo dipolar addition to the imine unit with benzonitrile N-oxide, the initial mono-adducts then undergoing further transformations.[74]

11.9   Diazine *N*-oxides[75]

Although pyridazine and pyrazine N-oxides can be readily prepared by oxidation of the parent heterocycles, pyrimidine N-oxides are more difficult to obtain in this way but they can conveniently be prepared by ring synthesis.[76]

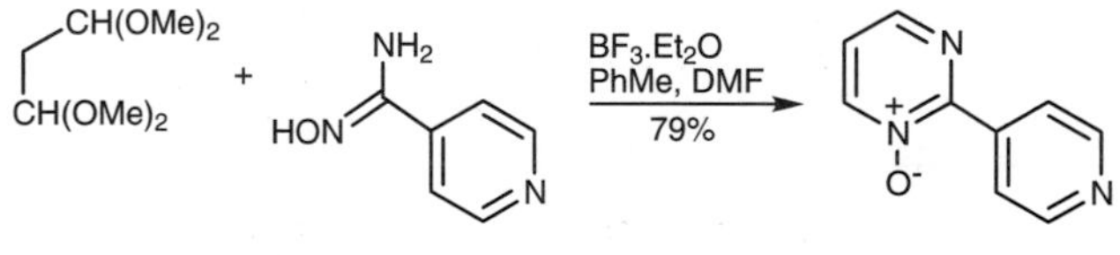

Pyridazine and pyrazine N-oxides behave like their pyridine counterparts in electrophilic substitution,[77] and nucleophilic displacement reactions involving loss of the oxygen. It is interesting that displacement of nitro β to the N-oxide function occurs about as readily as that of a γ nitro group, but certainly, displacements on N-oxides proceed faster[78] than for the corresponding base.

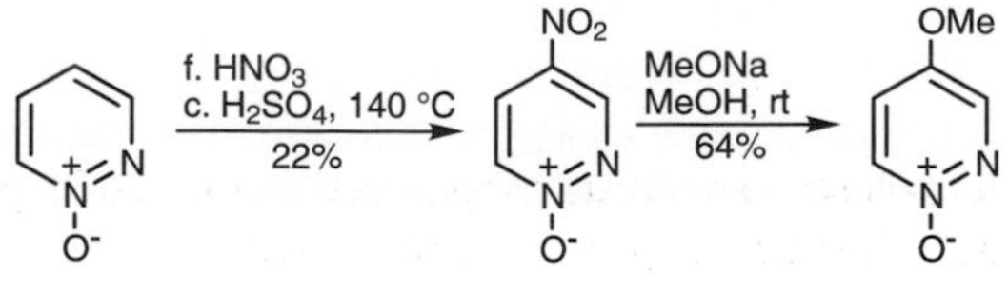

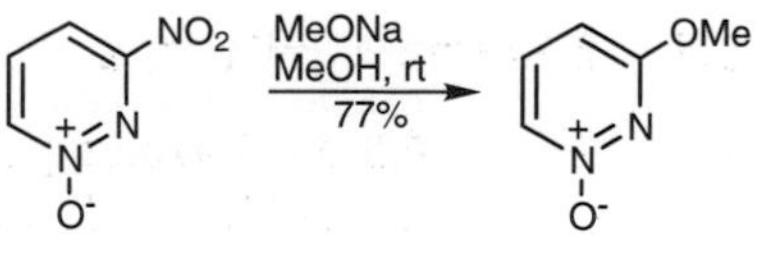

Nucleophilic substitution by halide, cyanide, carbon nucleophiles such as enamines, and acetate (by reaction with acetic anhydride), with concomitant loss of the oxide function occurs smoothly in all three systems,[79] though the site of introduction of the nucleophile is not always that (α to the N-oxide) predicted by analogy with pyridine chemistry, as illustrated by a couple of the examples below.[13]

The N-oxide grouping can also serve as an activating substituent to allow regioselective lithiation[80] or for the further acidification (section 11.12) of side-chain methyl groups for condensations with, for example, aromatic aldehydes or amyl nitrite.[81]

11.10   Oxydiazines

By far the most important naturally occuring diazines are the pyrimidones uracil, thymine, and cytosine, which as the nucleosides uridine, thymidine and cytidine, are components of the nucleic acids. As a consequence, a great deal of synthetic chemistry has been directed towards these types of compound in the search for anti-viral and anti-tumour agents.[82] Among other well known pyrimidones are the barbiturate sedatives.[83]

uracil   thymine   cytosine   barbiturates $R^1=R^2=H$ = barbituric acid

11.10.1   Structure of oxydiazines

With the exception of 5-hydroxypyrimidine, which is analogous to 3-hydroxypyridine, all the mono-oxygenated diazines exist predominantly as carbonyl tautomers and are thus categorised as diazinones.

The dioxydiazines present a more complicated picture, for in some cases, where both oxygens are α or γ to a nitrogen, and both might be expected to exist in carbonyl form, one actually takes up the hydroxy form: a well known example is 'maleic hydrazide'. One can rationalise the preference easily in this case, as resulting from the removal of the unfavourable interaction between two adjacent, partially positive nitrogen atoms in the dicarbonyl form. On the other hand, uracil exists as the dione and most of its reactions[84] can be interpreted on this basis. Barbituric acid adopts a tricarbonyl tautomeric form.

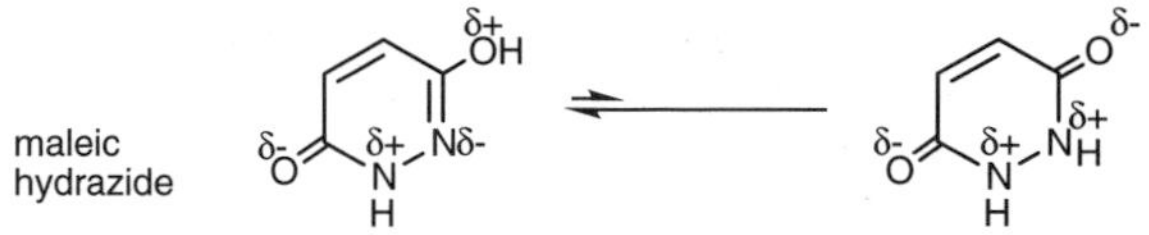

11.10.2   Reactions of oxydiazines

For many synthetic transformations, it is convenient to utilise halo- or alkoxydiazines, in lieu of the (oxidation level) equivalent -ones; often this device facilitates solubility; a final hydrolysis re-converts to the 'one'.

11.10.2.1   Reactions with electrophilic reagents

The deactivating effect of two ring nitrogens cannot always be overcome by a single oxygen substituent: 3-pyridazinone can be neither nitrated nor halogenated, or again, of the singly oxygenated pyrimidines, only 2(1H)-pyrimidone can be nitrated;[85] pyrazinones seem to be the most reactive towards electrophilic substitution.

5-Hydroxypyrimidine, the only phenolic diazine, is unstable even to dilute acid and no electrophilic substitutions have been reported.

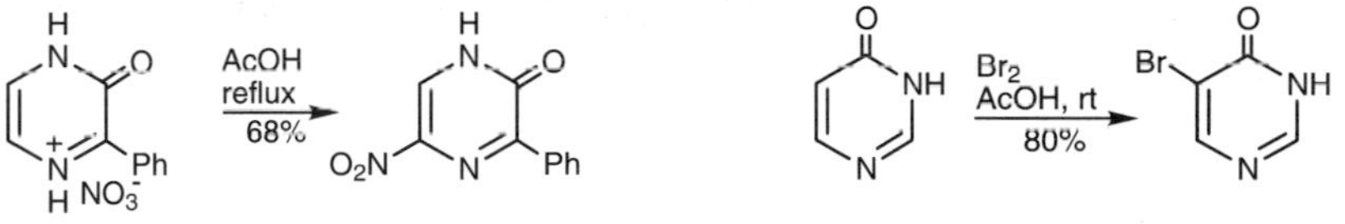

Uracils undergo a range of electrophilic substitution reactions such as halogenation, phenylsulfenylation,[86] mercuration,[87] and hydroxy- and chloromethylation.[88] Bromination of uracils has been shown to proceed *via* the bromohydrin adduct, and similarly of 2-pyrimidone, *via* the bromohydrin-hydrate;[89] iodine with tetrabutylammonium peroxydisulfate allows iodination.[90]

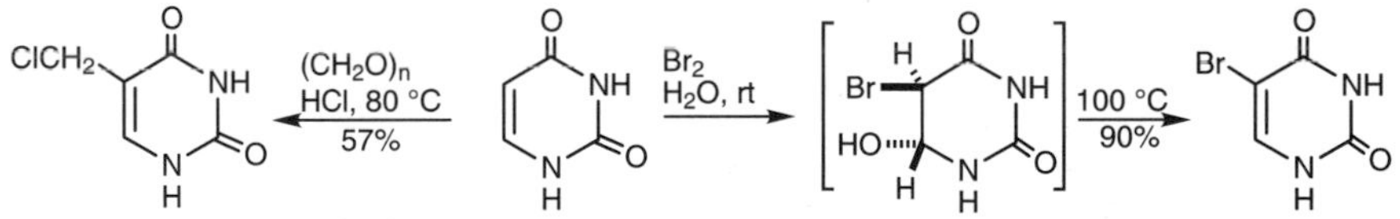

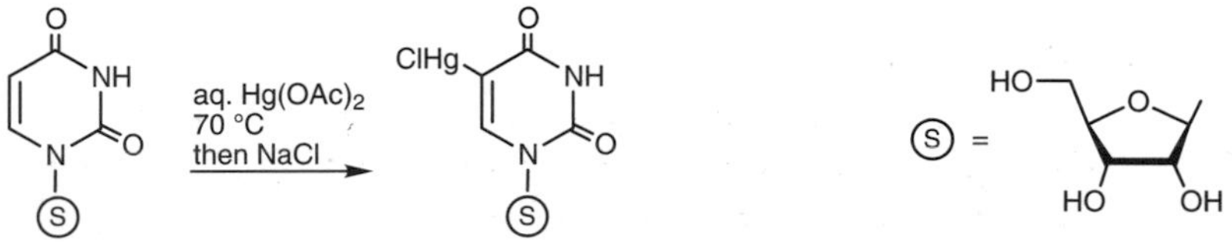

Uracil derivatives can be nitrated at C-5 under conditions which allow retention of a sugar residue at N-1. Nitration at N-3 can also be achieved: N-3-nitro compounds react with amines via an ANRORC mechanism, with displacement of nitramide and incorporation of the amine as a substituted N-3. This sequence has been utilised to prepared [15]N N-3-labelled pyrimidines and is illustrated below.[91]

11.10.2.2 Reactions with nucleophilic reagents

Diazinones are quite susceptible to nucleophilic attack, reaction taking place generally *via* Michael-type adducts rather than by attack at a carbonyl group, though there are exceptions[92] to this generalisation. Grignard reagents add to give dihydro-compounds and good leaving groups can be displaced.[93]

The reaction of cyanide with a protected 5-bromouridine[94] is instructive: under mild conditions a *cine*-substituted product is obtained *via* a Michael addition followed by β-elimination of bromide, but at higher temperatures, conversion of the 6- into the 5-cyano isomer is observed, i.e. the product of apparent, direct displacement of bromide is obtained. The higher temperature product arises *via* an isomerisation involving another Michael addition then elimination of the 6-cyano group.

In a related reaction with the anion of phenylacetonitrile, the initial addition is followed by an internal alkylation, generating a cyclopropane.[95]

The conversion of 1,3-dimethyluracil into a mixture of *N,N′*-dimethylurea and the disodium salt of formylacetic acid begins with the addition of hydroxide at C-6.[96]

The propensity for uracils to add nucleophiles can be put to synthetic use by reaction with double nucleophiles such as ureas, when a sequence of addition, ring opening and reclosure can achieve (at first sight) extraordinary transformations.[97]

11.10.2.3 Reactions with bases

N-Deprotonation

Like pyridones, oxydiazines are readily deprotonated under mild conditions to give ambident anions which can be alkylated, conveniently by phase-transfer methods, alkylation usually occuring at nitrogen.[98] 3-Pyridazinones alkylate cleanly on N-2 under phase-transfer conditions[99] but the regiochemistry of uracil alkylation is sometimes difficult to control (see also below). Carbon substitution can also be effected in some cases *via* delocalised *N*-anions, as in the reaction of 6-methyluracil with formaldehyde,[100] or with diazonium salts.[101]

O-Alkylation is also possible and is particularly important in ribosides where it occurs intramolecularly and can be used to control the stereochemistry of substitution in the sugar residue as illustrated in the following sequence for replacement of the 3′-hydroxyl with azide and with overall retention of configuration.[102]

Alternative methods for *N*-alkylation include heating with trimethyl phosphate[103] and the alkylation of *O*-silylated derivatives,[104] which is an important method for unambiguous *N*-alkylation especially ribosylation of uracils;[105] ribosylation is subject to the same stereochemical difficulties as in purine chemistry (for further discussion see section 24.2.1.2).

Stereospecific ribosylation of uracils and other pyrimdine bases has been carried out by attachement to the 5-hydroxymethyl substitutent of the sugar, followed by internal delivery to C-2.[106]

C-Metallation

C-Lithiation of uridine derivatives has been thoroughly studied as a means for the introduction of functional groups at C-5 and C-6. Chelating groups at C-5′ (hydroxyl or methoxymethoxy) favour 6-metallation,[107] as do equilibrating conditions, indicating that this is the most stable lithio-derivative. Kinetic lithiation, at C-5, can be achieved when weakly chelating silyloxy groups are used as protecting groups for the sugar.[108] It is remarkable that protection of the N-H group is not necessary and this is illustrated again in the 6-lithiation of uracil carrying an ethoxymethyl substitutent on N-1.[109]

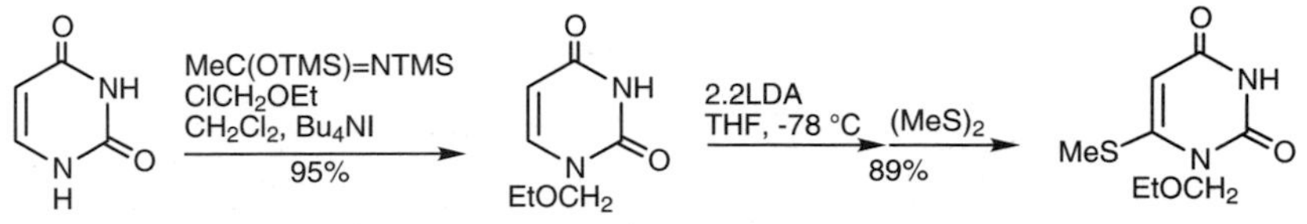

NH-Protection is also unnecessary for the side-chain metallation of 6-methylpyrimidin-2-one.[110]

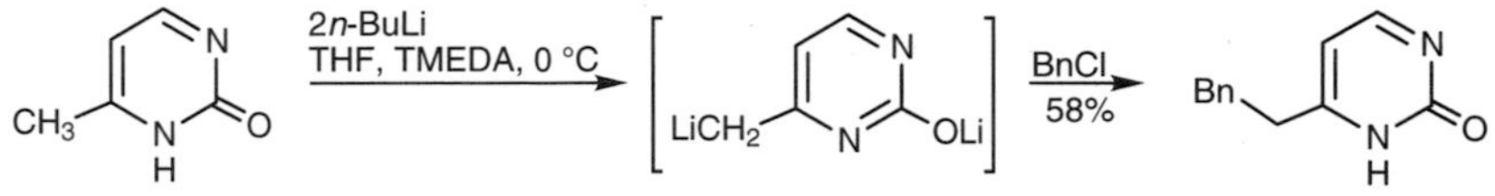

Zinc derivatives of uracils can be prepared directly by reaction of the appropriate halide with zinc dust. They react with a limited range of electrophiles but are particularly useful for palladium-catalysed couplings[111] (see also 11.10.2.5).

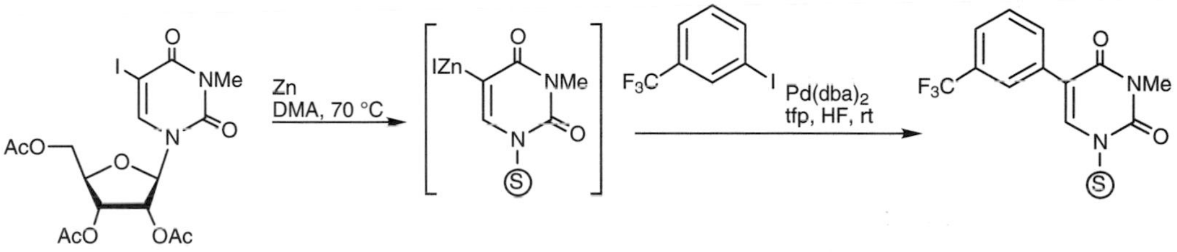

11.10.2.4 Replacement of oxygen

Oxydiazines, with the oxygen α to nitrogen, can be converted into halo-[29,112] and thio-compounds[113] using the same reagents used for 2- and 4-pyridones, including N-bromosuccinimide with triphenylphosphine.[114] The reactions of O-silylated pyrazinones with phosphorus(III) bromide or phosphorus(V) chloride are also efficient.[115]

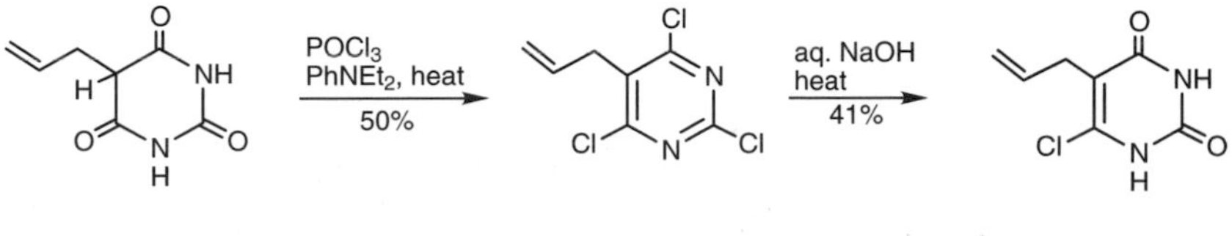

Diazinones can also now be converted directly into aminodiazines, without the (classical) intermediacy of an isolated halo-derivative by various processes including the use of 1,2,4-triazole, as illustrated below.[116]

5-Hydroxypyrimidine-2,4-diones react as ketones at the 5-position and undergo Wittig condensation, the double bond thus formed isomerising back into the stabler position in the ring.[117] Barbituric acid and C-5-derivatives can be converted into uracils by first forming a 6-mesylate and then catalytic hydrogenolysis.[118]

11.10.2.5   Transition metal-catalysed reactions

Halodiazinones undergo palladium-catalysed couplings with boronic acids and stannanes, but the reactions appear to be less consistent than with other systems. Temporary protection as silyl derivatives,[119] or the use of additives such as silver oxide[120] are beneficial in some cases, but it is often preferable to carry out transformations on alkoxydiazines, followed by hydrolysis. Direct coupling with organocopper reagents has also been described.[121]

Stannyluridines, prepared via lithiation, have been used in coupling reactions, but again the use of the corresponding dialkoxypyrimidine is usually to be preferred when an organometallic derivative of the heterocycle is required.[122]

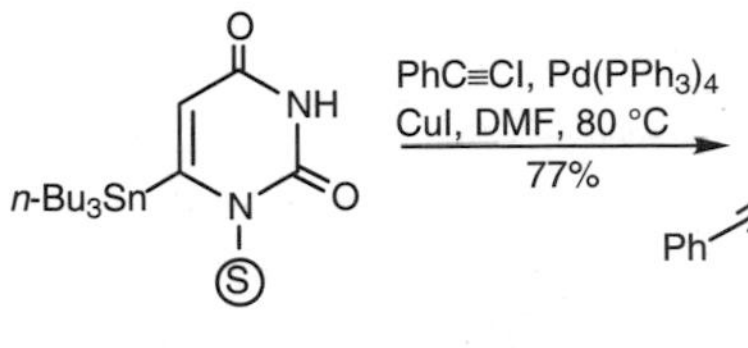

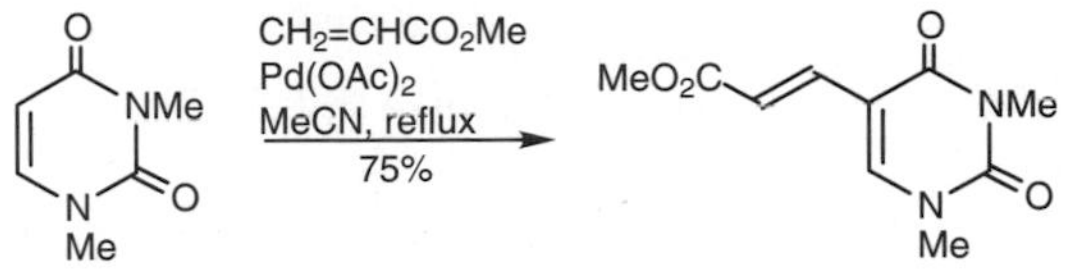

Heck reactions can be carried out on halo- or the readily available mercuri-derivatives, the latter requiring the use of one equivalent of palladium.[123] In addition, due to their susceptibility to electrophilic substitution, 'oxidative' Heck couplings (proceeding via *in situ* palladation) have found use in uracil chemistry, as shown above.[124]

11.10.2.6   Electrocyclic reactions

Mesoionic oxidopyraziniums undergo cycloadditions[125] similar to those known for oxidopyridiniums (section 5.8) and oxidopyryliums (section 8.1.7).

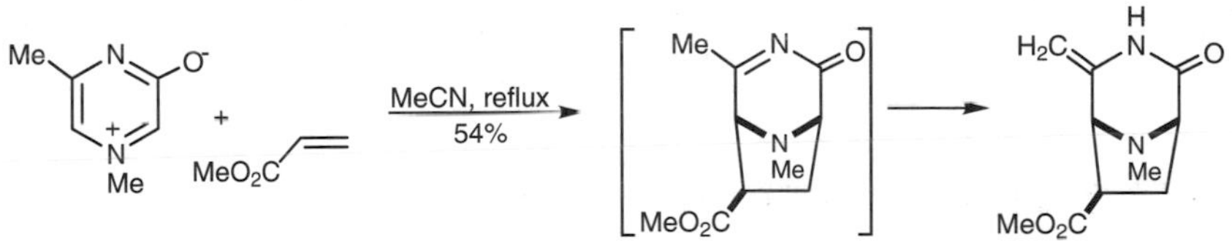

Heterodienophiles have also been studied: diethyl azodicarboxylate adds across the 2,6-positions of a pyridazin-3-one, and singlet oxygen across the 2,5-positions of pyrazinones.[126] The immediate cycloadduct is isolable when acryloyl cyanide is used as the heterodiene component in reaction with a pyrimidine-2,4 dione.[127]

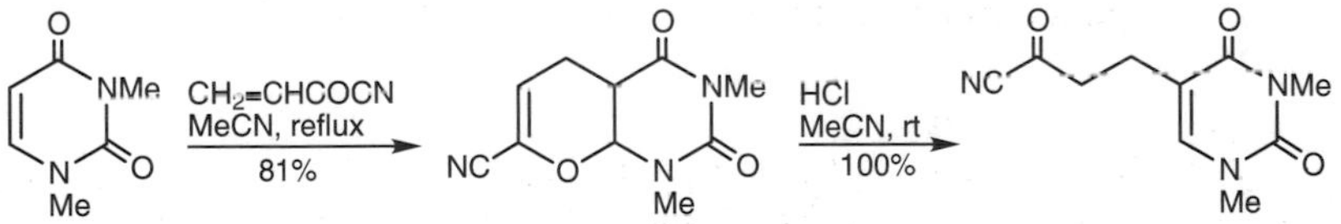

Because of possible relevance to mutagenesis, considerable study has been devoted to study of the photochemical transformations of oxypyrimidines; uracil, for example, takes part in a 2 + 2 cycloaddition with itself,[128] or with vinylene carbonate (1,3-dioxol-2-one).[129] Uracils undergo radical additions;[130] these too are of possible relevance to mutagenesis mechanisms.

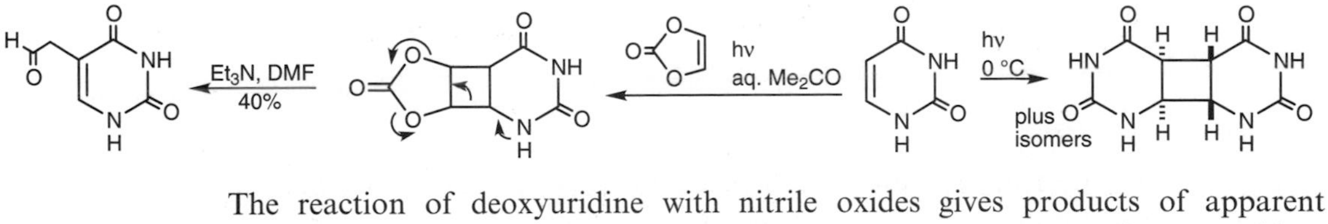

The reaction of deoxyuridine with nitrile oxides gives products of apparent electrophilic substitution, but these probably arise by ring opening of a cycloadduct (*cf.* 11.8).[131]

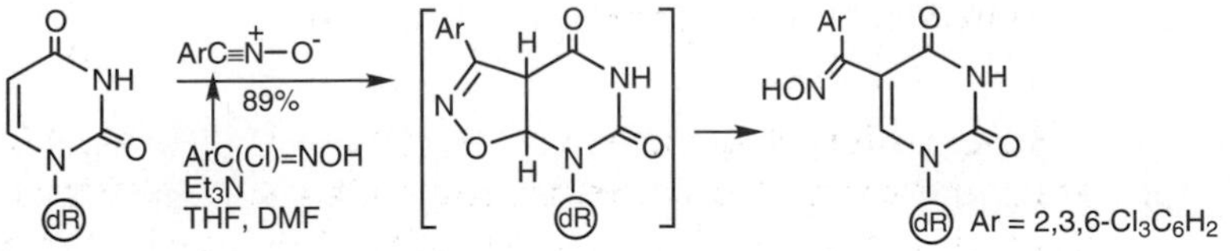

2-Pyrimidones[132] and 4-pyrimidones[133] form bicyclic systems and pyrazine isomerises into pyrimidine, on exposure to light.[134]

24%, at equilibrium, separable from starting material

11.11 Aminodiazines

Aminodiazines exist in the amino form. They are stronger bases than the corresponding unsubstituted systems and always protonate on one of the ring nitrogen atoms: where two isomeric cations are possible, the order of preference for protonation is of a ring nitrogen which is $\gamma > \alpha > \beta$ to the amino group, as can be seen in the two examples below. A corollary of this is that those aminodiazines which contain a γ-aminoazine system are the strongest bases.

The alkali-promoted rearrangement of quaternary salts derived from 2-aminopyrimidine provides the simplest example of the Dimroth rearrangement.[135] The larger the substituent on the positively charged ring nitrogen the more rapidly the rearrangement proceeds, no doubt as a result of the consequent relief in strain between the substituent and the adjacent amino group.

All of the aminodiazines react with nitrous acid to give the corresponding diazinones,[29] by way of highly reactive diazonium salts; even 5-aminopyrimidine does not give a stable diazonium salt, though a low yield of 2-chloropyrimidine can be obtained by diazotisation of 2-aminopyrimidine in concentrated hydrochloric acid.[136]

One amino group is sufficient in most cases to allow easy electrophilic substitution, halogenation[137] for example, and two amino groups activate the ring to attack even by weaker electrophilic reagents – for example by thiocyanogen.[138] Diaminopyrimidines will couple with diazonium salts[139] which provides a means for the introduction of a third nitrogen substituent.

Amino-oxy-pyrimidines,[140] and amino-dioxy-pyrimidines[141] can be *C*-nitrosated, and such 5,6-dinitrogen-substituted pyrimidines, after reduction to 5,6-diaminopyrimidines, are important intermediates for the synthesis of purines (an example is shown below; see also section 24.13.1.1) and pteridines (section 11.13.4.6).

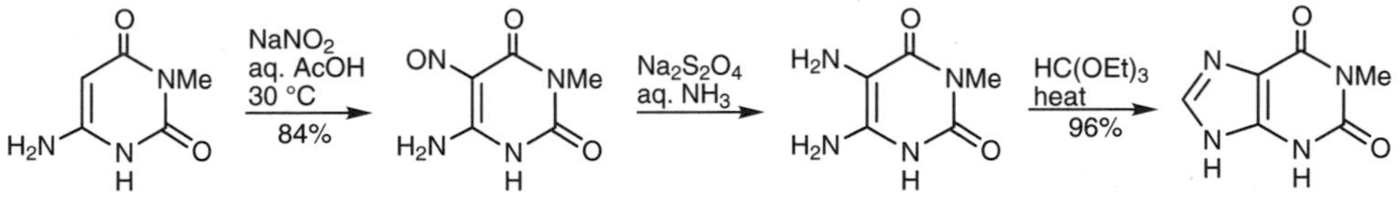

11.12 Alkyldiazines

All alkyldiazines, with the exception of 5-alkylpyrimidines, undergo condensations which involve deprotonation of the alkyl group,[142] in the same way as α- and γ-picolines. The intermediate anions are stabilised by mesomerism involving one, or in the case of 2- and 4-alkylpyrimidines, both nitrogens.

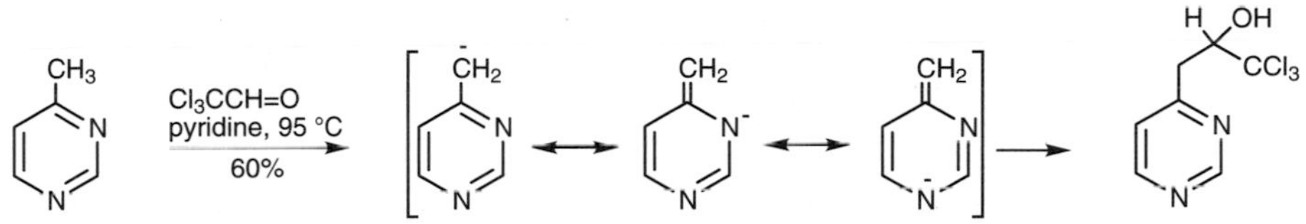

In pyrimidines, a 4-alkyl is deprotonated more readily that a 2-alkyl group;[143] here again one sees the greater stability associated with a γ-quinonoid resonating ion.

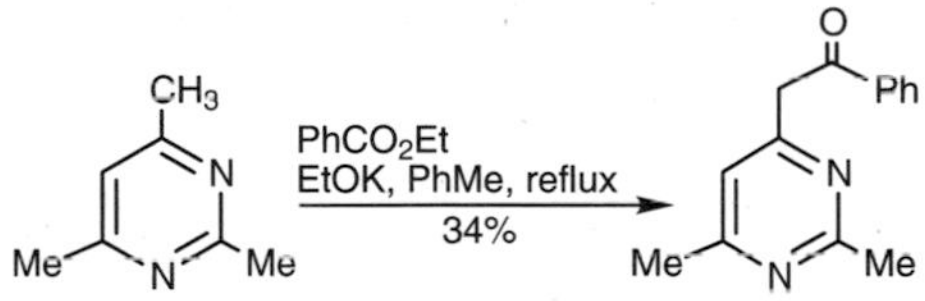

Side-chain radical halogenation selects a pyrimidine 5-methyl over a pyrimidine 4-methyl; the reverse selectivity can be achieved by halogenation in acid solution – presumably an *N*-protonated, side-chain deprotonated species, i.e. the enamine tautomer, is involved.[144]

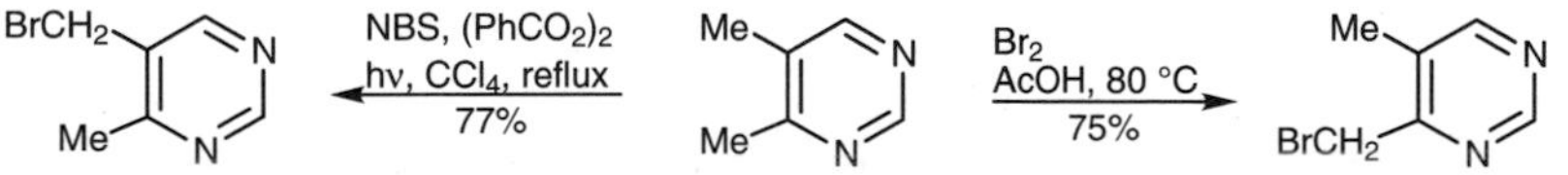

11.13 Quaternary azinium salts

The already high susceptibility of the diazines to nucleophilic addition is greatly increased by quaternisation. Addition of organometallic reagents to *N*-acyl quaternary salts has been achieved in some cases but is much more restricted than is the case with pyridines (section 5.13). Thus, allylstannanes[145] and -silanes[146] and silyl enol ethers have been added to diazine salts (hydride also traps such salts (section 11.6)). Pyridazines give good yields of monoadducts with attack mainly α to the acylated nitrogen, but the regioselectivity of silyl ether addition[147] is sensitive to substituents. Pyrazine gives mainly double addition products and pyrimidine produces only the double adduct. Reissert adducts have been described for pyridazine and pyrimidine.[148]

11.14 Synthesis of diazines

Routes for the ring synthesis of the isomeric diazines are, as one would expect, quite different one from the other, and must therefore be dealt with separately.

11.14.1 Ring synthesis of pyridazines
11.14.1.1 From a 1,4-dicarbonyl compound and a hydrazine

By far the most common method for the synthesis of pyridazines involves a 1,4-dicarbonyl compound reacting with hydrazine; unless the four-carbon component is unsaturated, a final oxidative step is needed to give an aromatic pyridazine.

The most useful procedure makes use of a 1,4-keto-ester giving a dihydropyridazinone which can be easily dehydrogenated to the fully aromatic heterocycle, often by C-bromination then dehydrobromination;[149] sodium 3-nitrobenzoate as an oxidant is a good alternative if complications would attend the use of halogenation. 6-Arylpyridazinones have been produced by this route in a number of ways: using an α-amino-nitrile as a masked ketone in the four-carbon component,[150] or by reaction of an acetophenone with glyoxylic acid and then hydrazine.[151] Friedel-Crafts acylation using succinic anhydride is an alternative route to 1,4-keto-acids, reaction with hydrazine again giving 6-arylpyridazinones.[152] Alkylation of an enamine with a phenacyl bromide produces 1-aryl-1,4-diketones allowing synthesis of 3-arylpyridazines.[153]

Maleic anhydride and hydrazine give the hydroxypyridazinone directly,[154] the additional unsaturation in the 1,4-dicarbonyl component meaning that an oxidative step is not required; conversion of 3-hydroxypyridazin-6-one into 3,6-dichloropyridazine makes this useful intermediate very easily available.

Saturated 1,4-diketones can suffer in this approach from the disadvantage that they can react with hydrazine in two ways, giving mixtures of the desired dihydropyridazine and an *N*-aminopyrrole; this complication does not arise when unsaturated 1,4-diketones are employed.[155] Synthons for unsaturated 1,4-diketones are available as cyclic acetals from the oxidation of furans (section 15.1.4), and react with hydrazines to give the fully aromatic pyridazines directly.[156]

11.14.1.2 By cycloaddition of a 1,2,4,5-tetrazine with an alkyne

Cycloaddition of a 1,2,4,5-tetrazine with an alkyne (or its equivalent), with elimination of nitrogen gives pyridazines.

This process works best when the tetrazine has electron-withdrawing substituents, but a wide range of substituents can be incorporated on the acetylene, including nitro, trimethylsilyl, and trimethyltin, affording routes to substituted pyridazines[157] not easily available by other methods.

The addition of ketone and aldehyde enolates to tetrazines, though not a concerted process, has the same overall effect.[158]

11.14.1.3 By other cycloaddition-based processes

11.14.1.3.1 From halohydrazones

Di- and trihalohydrazones react with enol ethers or enamines in the presence of base to give pyridazines via the intermediacy of an azadiene. The final pyridazine may be formed directly in the reaction mixture or with intermediate di- or tetrahydro-intermediate isolated and further treated with base.[159]

11.14.1.3.2 From thiophene S,S-dioxides

Thiophene *S,S*-dioxides with bulky 3,4-disubstitution undergo Diels-Alder additions with *N*-phenyltriazolinedione to give 1:2 adducts which on hydrolysis are converted into 4,5-disubstituted pyridazines.[160]

11.14.1.3.3 From halocyclopropenes

Di, tri-, and tetrahalocyclopropenes undergo cycloaddition with diazoalkanes to give unstable pyrazolines, which readily rearrange to pyridazines with loss of hydrogen halide. Tetrachlorocyclopropene is commercially available but many of the bromo compounds are easily prepared in two steps from vinyl bromides by addition of dibromocarbene, followed by reaction with methyllithium, the last step being carried out *in situ* for the cycloaddition step.[161]

11.14.2 Ring synthesis of pyrimidines

11.14.2.1 From a 1,3-dicarbonyl compound and an N–C–N fragment

The most general pyrimidine ring synthesis involves the combination of a 1,3-dicarbonyl component with an N–C–N fragment such as a urea, an amidine, or a guanidine.

The choice of N–C–N component – amidine,[162] guanidine,[163] or a urea[164] (thiourea[165]) – governs the substitution at C-2 in the product heterocycle. Although not formally 'N–C–N' components, formamide,[166] or an orthoester plus ammonia[167] can serve instead in this type of approach. The dicarbonyl component can be generated *in situ*, for example formylacetic acid (by decarbonylation of malic acid), or a synthon used (1,1,3,3-tetramethoxypropane for malondialdehyde), or a nitrile can serve as a carbonyl equivalent, the resulting heterocycle now carrying an amino substituent, as shown in the examples below.[168] The use of 2-bromo-1,1,3,3-tetramethoxypropane provides a route to 5-bromopyrimidine[169] and methanetricarboxaldehyde reacts with amidines to give 5-formylpyrimidines.[170]

Other synthons for 1,3-dicarbonyl compounds which have been successfully applied include β-chloro-α,β-unsaturated ketones and aldehydes,[171] β-amino-α,β-unsaturated ketones,[172] vinylamidinium salts,[173] and propiolic acid, reaction of which with urea gives uracil directly in about 50% yield.[174] 1,3-Ketoesters with formamidine produce 4-pyrimidones.[175] 2-Aminomalondialdehyde leads to 5-aminopyrimidines.[176] In analogy, pyrimidines fused to other rings, for example as in quinazolines, can be made from *ortho*-aminonitriles[177] and in general, from β-enaminoesters.[178]

Barbituric acid and barbiturates can be synthesised by reacting a malonate with a urea,[179] or a bis primary amide of a substituted malonic acid with diethyl carbonate.[180]

11.14.2.2 By cycloaddition of a 1,3,5-triazine with an alkyne

Cycloaddition of a 1,3,5-triazine with an alkyne (or its equivalent) gives pyrimidines after loss of hydrogen cyanide.

The formation of pyrimidines[181] *via* aza-Diels-Alder reactions is similar to the preparation of pyridazines from tetrazines (see also section 25.2.1).

11.14.2.3 From 3-ethoxyacryloyl isocyanate and primary amines

Primary amines add to the isocyanate group in 3-ethoxyacryloyl isocyanate; ring closure then gives pyrimidines *via* intramolecular displacement of the ethoxy group.

Uracils can be prepared *via* reaction of primary amines with 3-ethoxyacryloyl isocyanate;[182] this method is particularly suitable for complex amines and has found much use in recent years in the synthesis of, for example, carbocyclic nucleoside analogues as potential anti-viral agents.[183] The immediate product of amine/isocyanate interaction can be cyclised under either acidic or basic conditions and the method can also be applied to thiouracil synthesis.

Condensation of ketones with two mol equivalents of a nitrile in the presence of trifluoromethanesulfonic acid anhydride is a useful method for the production of a limited range of pyrimidines, where the substituents at C-2 and C-4 are identical.[184]

11.14.3 Ring synthesis of pyrazines

Pyrazine is not easily made in the laboratory. Commercially, the high temperature cyclodehydrogenation of precursors such as *N*-hydroxyethyl ethane-1,2-diamine is used.

11.14.3.1 From the self condensation of a 2-aminoketone

Symmetrical pyrazines result from the spontaneous self condensation of two mol equivalents of a 2-aminoketone, or 2-aminoaldehyde, followed by an oxidation.

2-Amino-carbonyl compounds, which are stable only as their salts, are usually prepared *in situ* by the reduction of 2-diazo-, -oximino- or -azido-ketones. The dihydropyrazines produced by this strategy are very easily aromatised, for example by air oxidation, and often distillation alone is sufficient to bring about disproportionation.[185]

α-Amino-esters are more stable than α-amino-ketones but nonetheless easily self-condense to give heterocycles, known as 2,5-diketopiperazines. These compounds are resistant to oxidation but can be used to prepare aromatic pyrazines after first converting them into dichloro- or dialkoxy-dihydropyrazines.[186]

11.14.3.2 From 1,2-dicarbonyl compounds and 1,2-diamines

1,2-Dicarbonyl compounds undergo double condensation with 1,2-diamines; an oxidation is then required.

This method is well suited to the formation of symmetrical pyrazines,[187] if both diketone and diamine are unsymmetrical, two isomeric pyrazines are formed.

11.14.4.5   2,5-Dimethyl-3-n-propylpyrazine

Alkylpyrazines can be produced by an ingenious sequence involving an electrocyclic ring closure of a 1,4-diazatriene, aromatisation being completed by loss of the oxygen from the original oxime hydroxyl group.[196]

11.15   Pteridines

Pyrazino[2,3-*d*]pyrimidines are known as 'pteridines',[197] because the first examples of the ring system, as natural products, were found in pigments, like xanthopterin (yellow), in the wings of butterflies (*Lepidoptera*). The pteridine ring system has subsequently been found in coenzymes which use tetrahydrofolic acid (derived from the vitamin folic acid) and in the cofactor of the oxomolybdoenzymes[198] and comparable tungsten enzymes. It is also present in the anti-cancer drug Methotrexate.

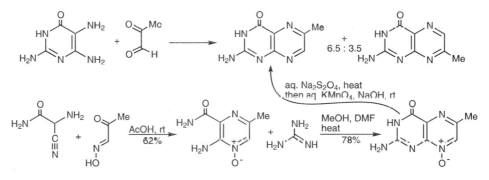

The synthesis of the pteridine ring system has been approached by two obvious routes: one is the fusion of the pyrazine ring onto a preformed 4,5-diaminopyrimidine, and the second, the elaboration of the pyrimidine ring on a preformed pyrazine. The first of these, the *Isay synthesis*, suffers from the disadvantage that condensation of the heterocyclic 1,2-diamine with an unsymmetrical 1,2-dicarbonyl compound usually leads to a mixture of two 5/6-substituted isomers.[199] It was to avoid this difficulty that the alternative strategy, the *Taylor synthesis*, now widely used, starting with a pyrazine, was developed.[200] This approach has the further advantage that because the pyrazine ring is presynthesised, using 2-cyanoglycinamide,[201] it eventually produces, regioselectively, 6-substituted pteridines – substitution at the 6-position is the common pattern in natural pteridines.

Exercises for chapter 11

Straightforward revision exercises (consult chapters 10 and 11)

(a) Why is it difficult to form diprotonic salts from diazines?

(b) How do uridine, thymidine, and cytidine differ?

(c) Are the diazines more or less reactive towards *C*-electrophilic substitution than pyridine?

(d) What factor assists and what factor mediates against nucleophilic displacement of hydrogen in diazines?

(e) Which is the only chlorodiazine which does not undergo easy nucleophilic displacement, and why?

(f) What precaution is usually necessary in order to lithiate a diazine?

42. Miyashita, A., Suzuki, Y., Ohta, K., Iwamoto, K., amd Higashito, T., *Heterocycles*, **1998**, *47*, 407.
43. Yamanake, H. and Ohba, S., *Heterocycles*, **1990**, *31*, 895.
44. Draper, T. L. and Bailey, T. R., *J. Org. Chem.*, **1995**, *60*, 748.
45. Itoh, T., Matsuya, Y., Nagata, K., Okada, M., and Ohsawa, A., *J. Chem. Soc., Chem. Commun.*, **1995**, 2067.
46. Zoltewicz, J. A., Grahe, G., and Smith, C. L., *J. Am. Chem. Soc.*, **1969**, *91*, 5501.
47. Plé, N., Turck, A., Couture, K., and Quéguiner, G., *J. Org. Chem.*, **1995**, *60*, 3781.
48. 'Directed metallation of π-deficient azaaromatics: strategies of functionalisation of pyridines, quinolines, and diazines', Quéguiner, G., Marsais, F., Snieckus, V., and Epsztajn, J., *Adv. Heterocycl. Chem.*, **1991**, *52*, 187.
49. Wada, A., Yamamoto, J., and Kanatomo, S., *Heterocycles*, **1987**, *26*, 585; Turck, A., Majovic, L., and Quéguiner, G., *Synthesis*, **1988**, 881; Plé, N., Turck, A., Bardin, F., and Quéguiner, G., *J. Heterocycl. Chem.*, **1992**, *29*, 467; Turck, A., Plé, N., Trohay, D., Ndzi, B., and Quéguiner, G., *ibid.*, **1992**, *29*, 699; Mattson, R. J. and Sloan, C. P., *J. Org. Chem.*, **1990**, *55*, 3410; Plé, N., Turck, A., Heynderickx, A., and Quéguiner, G., *J. Heterocycl. Chem.*, **1994**, *31*, 1311; Turck, A., Plé, N., Mojovic, L., Ndzi, B., Quéguiner, G., Haider, N., Schuller, H., and Heinisch, G., *J. Heterocycl. Chem.*, **1995**, *32*, 841; Nakamura, H., Aizawa, M., and Murai, A., *Synlett*, **1996**, 1015.
50. Turck, A., Plé, N., Mojovic, L., and Quéguiner, G., *J. Heterocycl. Chem.*, **1990**, *27*, 1377.
51. Gronowitz, S. and Röe, J., *Acta Chem. Scand.*, **1965**, *19*, 1741; Frissen, A. E., Marcelis, A. T. M., Buurman, D. G., Pollmann, C. A. M., and van der Plas, H. C., *Tetrahedron*, **1989**, *45*, 5611.
52. Kress, T. J., *J. Org. Chem.*, **1979**, *44*, 2081; Rho, T. and Abah, Y. F., *Synth. Commun.*, **1994**, *24*, 253.
53. Taylor, H. M., Jones, C. D., Davenport, J. D., Hirsch, K. S., Kress, T. J., and Weaver, D., *J. Med. Chem.*, **1987**, *30*, 1359.
54. Turck, A., Plé, N., Lepretre-Graquere, A., and Quéguiner, G., *Heterocycles*, **1998**, *49*, 205.
55. Zheng, J. and Undheim, K., *Acta Chem. Scand.*, **1989**, *43*, 816; Shimura, A., Momotake, A., Togo, H., and Yokoyama, M., *Synthesis*, **1999**, 495.
56. Toussaint, D., Suffert, J., and Wermuth, C. G., *Heterocycles*, **1994**, *38*, 1273.
57. Kondo, Y., Watanabe, R., Sakamoto, T., and Yamanaka, H., *Chem. Pharm. Bull.*, **1989**, *37*, 2814; Thompson, W. J., Jones, J. H., Lyle, P. A., and Thies, J. E., *J. Org. Chem.*, **1988**, *53*, 2052; Solberg, J. and Undheim, K., *Acta Chem. Scand.*, **1989**, *38*, 1273.
58. Benneche, T., *Acta Chem. Scand.*, **1990**, *44*, 927.
59. Majeed, A. J., Antonsen, Ø., Benneche, T., and Undheim, K., *Tetrahedron*, **1989**, *45*, 993; Sandosham, J. and Undheim, K., *Acta Chem. Scand.*, **1989**, *43*, 684; Heldmann, D. K. and Sauer, J., *Tetrahedron Lett.*, **1997**, *38*, 5791.
60. Peters, D., Hörnfeldt, A.-B., and Gronowitz, S., *J. Heterocycl. Chem.*, **1990**, *27*, 2165; Peters, D., Hörnfeldt, A.-B., Gronowitz, S., and Johanssen, N. G., *ibid.*, **1991**, *28*, 529.
61. 'Recent advances in the chemistry of dihydroazines', Weis, A. L., *Adv. Heterocycl. Chem.*, **1985**, *38*, 1.
62. Becker, H. P. and Neumann, W. P., *J. Organomet. Chem.*, **1972**, *37*, 57; Bessenbacher, C., Kaim, W., and Stahl, T., *Chem. Ber.*, **1989**, *122*, 933.
63. Gottlieb, R., and Pfleiderer, W., *Justus Liebigs Ann. Chem.*, **1981**, 1451.
64. Russell, J. R., Garner, C. D., and Joule, J. A., *J. Chem. Soc., Perkin Trans. 1*, **1992**, 409.
65. 'Advances in the synthesis of substituted pyridazines *via* introduction of carbon functional groups into the parent heterocycle', Heinisch, G., *Heterocycles*, **1987**, *26*, 481.
66. Phillips, O. A., Keshava Murthy, K. S., Fiakpui, C. Y., and Knaus, E. E., *Can. J. Chem.*, **1999**, *77*, 216.
67. Togo, H., Ishigami, S., and Yokoyama, M., *Chem. Lett.*, **1992**, 1673.
68. Samaritoni, J. G. and Babbitt, G., *J. Heterocycl. Chem.*, **1991**, *28*, 583.
69. Houminer, Y., Southwick, E. W., and Williams, D. L., *J. Org. Chem.*, **1989**, *54*, 640.
70. Carver, D. R., Greenwood, T. D., Hubbard, J. S., Komin, A. P., Sachdeva, Y. P., and Wolfe, J. F., *J. Org. Chem.*, **1983**, *48*, 1180.
71. 'Hetero Diels-Alder methodology in organic synthesis', Ch. 10, Boger, D. L. and Weinreb, S. M., Academic press, **1987**; Stolle, W. A. W., Frissen, A. E., Marcelis, A. T. M., and van der Plas, H. C., *J. Org. Chem.*, **1992**, *57*, 3000.

72. Boger, D. L. and Coleman, R. S., *J. Org. Chem.*, **1984**, *49*, 2240; Neunhoffer, H. and Werner, G., *Justus Liebigs Ann. Chem.*, **1974**, 1190; Biedrzycki, M., de Bie, D. A., and van der Plas, H. C., *Tetrahedron*, **1989**, *45*, 6211.

73. Dawson, I. M., Pappin, A. J., Peck, C. J., and Sammes, P. G., *J. Chem. Soc., Perkin Trans. 1*, **1989**, 453.

74. Grassi, G., Risitano, F., and Foti, F., *Tetrahedron*, **1995**, *51*, 11855.

75. 'Pyrimidine *N*-oxides: synthesis, structure and chemical properties', Yamanaka, H., Sakamoto, T., and Niitsuma, S., *Heterocycles*, **1990**, *31*, 923.

76. Kocevar, M., Mlakar, B., Perdih, M., Petric, A., Polanc, S., and Vercek, B., *Tetrahedron Lett.*, **1992**, *33*, 2195.

77. Itai, T. and Natsume, S., *Chem. Pharm. Bull.*, **1963**, *11*, 83.

78. Sako, S., *Chem. Pharm. Bull.*, **1963**, *11*, 261; Klein, B., O'Donnell, E., and Gordon, J. M., *J. Org. Chem.*, **1964**, *29*, 2623.

79. Ogata, M., *Chem. Pharm. Bull.*, **1963**, *11*, 1522; Iwao, M. and Kuraishi, T., *J. Heterocycl. Chem.*, **1978**, *15*, 1425; Sato, N., *J. Heterocycl. Chem.*, **1989**, *26*, 817; Koelsch, C. F. and Gumprecht, W. H., *J. Org. Chem.*, **1958**, *23*, 1603.

80. Aoyagi, Y., Maeda, A., Inoue, M., Shiraishi, M., Sakakibara, Y., Fukui, Y., and Ohta, A., *Heterocycles*, **1991**, *32*, 735.

81. Itai, T., Sako, S., and Okusa, G., *Chem. Pharm. Bull.*, **1963**, *11*, 1146; Ogata, M., *ibid.*, 1517.

82. 'AIDS-Driven nucleoside chemistry', Huryn, P. M. and Okabe, M., *Chem. Rev.*, **1992**, *92*, 1745.

83. 'Recent progress in barbituric acid chemistry', Bojarski, J. T., Mokrosz, J. L., Bartón, H. J., and Paluchowska, *Adv. Heterocycl. Chem.*, **1985**, *38*, 229.

84. 'Uracils: versatile starting materials in heterocyclic synthesis', Wamhoff, H., Dzenis, J., and Hirota, K., *Adv. Heterocycl. Chem.*, **1992**, *55*, 129.

85. Johnson, C. D., Katritzky, A. R., Kingsland, M., and Scriven, E. F. V., *J. Chem. Soc., (B)*, **1971**, 1.

86. Lee, C. H. and Kim, Y. H., *Tetrahedron Lett.*, **1991**, *32*, 2401.

87. Bergstrom, D. E. and Ruth, J. L., *J. Carbohyd. Nucleosides, Nucleotides*, **1977**, *4*, 257.

88. Skinner, W. A., Schelstraete, M. G. M., and Baker, B. R., *J. Org. Chem.*, **1960**, *25*, 149; Delia, T. J., Scovill, J. P., Munslow, W. D., and Burckhalter, J. H., *J. Med. Chem.*, **1976**, *19*, 344.

89. Wang, S. Y., *J. Org. Chem.*, **1959**, *24*, 11; Tee, O. S. and Banerjee, S., *Can. J. Chem.*, **1974**, *52*, 451.

90. Whang, J. P., Yang, S. G., and Kim, Y. H., *Chem. Commun.*, **1997**, 1355.

91. Ariza, X., Bou, V., and Vilarrasa, J., *J. Am. Chem. Sooc.*, **1995**, *117*, 3665; Giziewicz, J., Wnuk, S. F., and Robins, M. J., *J. Org. Chem.*, **1999**, *64*, 2149.

92. Saito, I., Sugiyama, H., and Matsuura, T., *J. Am. Chem. Soc.*, **1983**, *105*, 956.

93. Fateen, A. K., Moustafa, A. H., Kaddah, A. M., and Shams, N. A., *Synthesis*, **1980**, 457; Bischofberger, N., *Tetrahedron Lett.*, **1989**, *30*, 1621.

94. Inoue, H. and Ueda, T., *Chem. Pharm. Bull.*, **1978**, *26*, 2657.

95. Hirota, K., Sajiki, H., Maki, Y., Inoue, H., and Ueda, T., *J. Chem. Soc., Chem. Commun.*, **1989**, 1659.

96. Lovett, E. G. and Lipkin, D., *J. Org. Chem.*, **1977**, *42*, 2574.

97. Hirota, K., Watanabe, K. A., and Fox, J. J., *J. Org. Chem.*, **1978**, *43*, 1193; Hirota, K., Kitade, Y., Sagiki, H., and Maki, Y., *Heterocycles*, **1984**, *22*, 2259.

98. Heddayatulla, M., *J. Heterocycl. Chem.*, **1981**, *18*, 339; Tanabe, T., Yamauchi, K., and Kinoshita, M., *Bull. Chem. Soc. Jpn.*, **1977**, *50*, 3021.

99. Yamada, T. and Ohki, M., *Synthesis*, **1981**, 631.

100. Cline, R. E., Fink, R. M., and Fink, K., *J. Am. Chem. Soc.*, **1959**, *81*, 2521.

101. Ottenheijm, H. C. J., van Nispen, S. P. J. M., and Sinnige, M. J., *Tetrahedron Lett.*, **1976**, 1899.

102. Hiebl, J., Zbiral, E., Balzarini, J., and De Clercq, E., *J. Med. Chem.*, **1992**, *35*, 3016; Verheyden, J. P. H., Wagner, D., and Moffatt, J. G., *J. Org. Chem.*, **1971**, *36*, 250.

103. Yamauchi, M. and Kinoshita, M., *J. Chem. Soc., Perkin Trans. 1*, **1973**, 391.

104. Müller, C. E., *Tetrahedron Lett.*, **1991**, *32*, 6539.

105. 'Nucleoside synthesis, organosilicon methods', Lukevics, E. and Zablacka, A., Ellis Horwood, London, **1991**; Matsuda, A., Kurasawa, Y., and Watanabe, K., *Synthesis*, **1981**, 748.

175. Butters, M., *J. Heterocycl. Chem.*, **1992**, *29*, 1369.
176. Reichardt, C. and Schagerer, K., *Justus Liebigs Ann. Chem.*, **1982**, 530.
177. Ch. II in 'o-Aminonitriles', Taylor, E. C. and McKillop, A., *Adv. Org. Chem.*, **1970**, *7*, 79
178. 'Heterocyclic β-enaminoesters, versatile synthons in heterocyclic synthesis', Wamhoff, H., *Adv. Heterocycl. Chem.*, **1985**, *38*, 299.
179. Dickey, J. B. and Gray, A. R., *Org. Synth., Coll. Vol. II*, **1943**, 60.
180. Shimo, K. and Wakamatsu, S., *J. Org. Chem.*, **1959**, *24*, 19.
181. Boger, D. L., Schumacher, J., Mullican, M. D., Patel, M., and Panek, J. S., *J. Org. Chem.*, **1982**, *47*, 2673; Boger, D. L. and Menezes, R. F., *ibid.*, **1992**, *57*, 4331.
182. Shaw, G. and Warrener, R. N., *J. Chem. Soc.*, **1958**, 157.
183. Hronowski, L. J. J. and Szarek, W. A., *Can. J. Chem.*, **1985**, *63*, 2787.
184. Martínez, A. G., Fernández, A. H., Jiménez, F. M., Fraile, A. G., Subramanian, L. R., and Hanack, M., *J. Org. Chem.*,, **1992**, *57*, 1627; Herrera, A., Martinez, R., González, B., Illescas, B., Martin, N., and Seoane, C., *Tetrahedron Lett.*, **1997**, *38*, 4873.
185. Birkofer, L., *Chem. Ber.*, **1947**, *80*, 83.
186. Blake, K. W., Porter, A. E. A., and Sammes, P. G., *J. Chem. Soc., Perkin Trans. 1*, **1972**, 2494.
187. Flament, I. and Stoll, M., *Helv. Chim. Acta*, **1967**, *50*, 1754.
188. Rothkopf, H. W., Wöhrle, D., Müller, R., and Kossmehl, G., *Chem. Ber.*, **1975**, *108*, 875.
189. Bradbury, R. H., Griffiths, D., and Rivett, J. E., *Heterocycles*, **1990**, *31*, 1647.
190. Muehlmann, F. L. and Day, A. R., *J. Am. Chem. Soc.*, **1956**, *7δ*, 242.
191. Weijlard, J., Tishler, M., and Erickson, A. E., *J. Am. Chem. Soc.*, **1945**, *67*, 802.
192. Nakayama, J., Konishi, T., Ishii, A., and Hoshino, M., *Bull. Chem. Soc. Jpn.*, **1989**, *62*, 2608.
193. Todd., A. R. and Bergel, F., *J. Chem. Soc.*, **1937**, 364.
194. Herdewijn, P., De Clerq, E., Balzarini, J., and Vandehaeghe, H., *J. Med. Chem.*, **1985**, *28*, 550.
195. Jones, K., Keenan, M., and Hibbert, F., *Synlett*, **1996**, 509; Keenan, M., Jones, K., and Hibbert, F., *Chem. Commun.*, **1997**, 323.
196. Büchi, G. and Galindo, J., *J. Org. Chem.*, **1991**, *56*, 2605.
197. 'Pteridine Chemistry', Pfleiderer, W. and Taylor, E. C., Eds., Pergammon Press, London, **1964**; 'Pteridines. Properties, reactivities and biological significance', Pfleiderer, W., *J. Heterocycl. Chem.*, **1992**, *29*, 583.
198. For reviews see *J. Biol. Inorg. Chem.*, **1997**, *2*, 772, 773, 782, 786, 790, 797, 804, 810, and 817.
199. Waring, P. and Armarego, W. L. F., *Aust. J. Chem.*, **1985**, *38*, 629.
200. Taylor, E. C., Perlman, K. L., Sword, I. P., Séquin-Frey, M., and Jacobi, P. A., *J. Am. Chem. Soc.*, **1973**, *29*, 3610.
201. Cook, A. H., Heilbron, I., and Smith, E., *J. Chem. Soc.*, **1949**, 1440.

12 Typical reactivity of pyrroles, thiophenes, and furans

In this chapter are gathered the most important generalisations which can be made, and the general lessons which can be learned about the reactivity, and relative reactivities, one with the other, of the prototypical five-membered aromatic heterocycles: pyrroles, thiophenes and furans.

Typical reactivities of pyrrole, furan, and thiophene

substitution at α-position via α-lithiated intermediates

reductive removal of sulfur in thiophenes

formation of cycloadducts on reaction with dienophiles

substitution on pyrrole nitrogen via pyrryl anion

pyrryl anion

electrophilic substitution easy; preferred at α-position; also easy at β-position

The chemistry of pyrrole, thiophene and furan is dominated by a readiness to undergo electrophilic substitution, preferentially at an α-position but, with only slightly less alacrity, also at a β-position, should the α positions be blocked. For the beginning student of heterocyclic chemistry it is worth re-emphasising the stark contrast between the five- and six-membered heterocycles – the former react much more readily with electrophiles than does benzene and the latter much less readily.

intermediate for α-attack by X⁺

intermediate for β-attack by X⁺

Positional selectivity in these five-membered systems, indeed their high reactivity to electrophilic attack, are well explained by a consideration of the Wheland intermediates (and by implication, the transition states which lead to them) for electrophilic substitution. Intermediate cations from both α- and β-attack are stabilised (shown for attack on pyrrole). The delocalisation, involving donation of electron density from the hetero atom, is greater in the intermediate from α-attack, as illustrated by the number of low energy resonance contributors which can be drawn. Note that the C–C double bond in the intermediate for β-attack is not, and cannot become involved in delocalisation of the charge.

There is a simple parallelism between the reaction of a pyrrole with an electrophile and the comparable reaction of an aniline, and indeed pyrrole is in the same range of reactivity towards electrophiles as is aniline.

compare

The five-membered heterocycles do not react with electrophiles at the hetero atom; perhaps this surprises the heterocyclic newcomer most obviously with respect to pyrrole, for here, it might have been anticipated, the nitrogen lone pair would be easily donated to an incoming electrophile, as it certainly would be in reactions of its saturated counterpart, pyrrolidine. The difference is that in pyrrole, electrophilic addition at the nitrogen would lead to a substantial loss of resonance stabilisation – the molecule would be converted into a cyclic butadiene, with an attached nitrogen carrying a positive charge localised on that nitrogen atom. The analogy with aniline falls down for, of course, anilines do react easily with simple electrophiles (e.g. protons) at nitrogen. The key difference is that, although some stabilisation in terms of overlap between the aniline nitrogen lone pair and benzenoid π-system is lost, the majority of the stabilisation energy, associated with the six-electron aromatic π-system, is retained when aniline nitrogen donates its lone pair of electrons to a proton (electrophile).

Of the trio – pyrrole, furan and thiophene – the first is by far the most susceptible to electrophilic attack: this susceptibility is linked to the greater electron-releasing ability of neutral trivalent nitrogen, and the concomitant greater stability of a positive charge on tetravalent nitrogen. This finds its simplest expression in the relative basicities of saturated amines, sulfides and ethers, respectively, which are seen to parallel nicely the relative order of reactivity of pyrrole, furan and thiophene towards electrophilic attack at carbon, but involving major assistance by donation from the hetero atom, i.e. the development of positive charge on the hetero atom.

In qualitative terms, the much greater reactivity of pyrrole is illustrated by its rapid reaction with weak electrophiles like benzenediazonium cation and nitrous acid, neither of which react with furan or thiophene. It is relevant to note that N,N-dimethylaniline reacts rapidly with these reactants, where anisole does not.

Substituents ranged on five-membered rings have directing effects comparable to those which they exert on a benzene ring. Alkyl groups, for example, direct *ortho* and *para*, and nitro groups direct *meta* although, strictly, these two terms cannot be applied to the five-membered situation. The very strong tendency for α electrophilic substitution is however the dominating influence in most instances, and products resulting from attack following guidance from the substituent are generally minor products in mixtures where the dominant substitution is at an available α-position. The influence of substituents is felt least in furans.

An aspect of the chemistry of furans is the occurrence of a number of 2,5-additions initiated by electrophilic attack: a Wheland intermediate is formed normally but then adds a nucleophile, when a sufficiently reactive one is present, instead of then losing a proton. Conditions can, however, usually be chosen to allow the formation of a 'normal' α-substitution product if desired. The occurrence of such processes in the case of furan is generally considered to be associated with its lower aromatic resonance stabilisation energy – there is less to regain by loss of a proton and the consequent return to an aromatic furan.

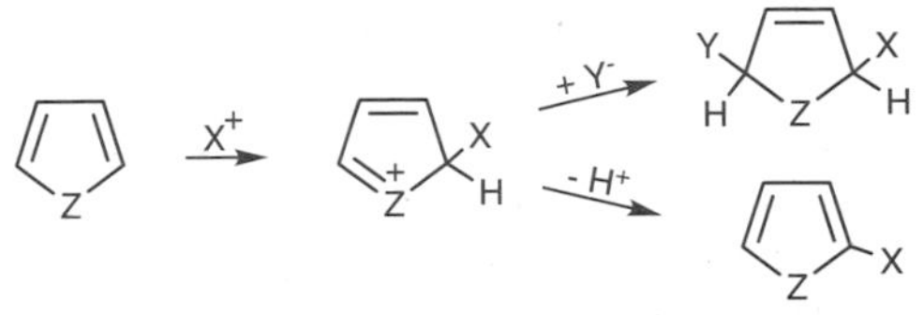

formation of 2,5-adducts sometimes for Z = O

Z = NR or S

The lower aromaticity of furans also manifests itself in a much greater tendency to undergo cycloadditions, as a 4-π, diene component in Diels-Alder reactions. That is to say, furans are much more like dienes, and less like a six-electron aromatic system, than are pyrroles and thiophenes. However, the last two systems can be made to undergo cycloadditions by increasing the pressure or, in the case of pyrroles, by 'reducing the aromaticity' by the device of inserting an electron-withdrawing group onto the nitrogen.

In direct contrast with electron-deficient heterocycles like pyridines and the diazines, the five-membered systems do not undergo nucleophilic substitutions, except in situations (especially in furan and thiophene chemistry) where halide is situated *ortho* or *para* to a nitro group. In the manipulation of the five-membered heterocycles of this group, extensive use has been made of the various palladium(0)-catalysed couplings available, as illustrated below (see section 2.7 for a detailed discussion).

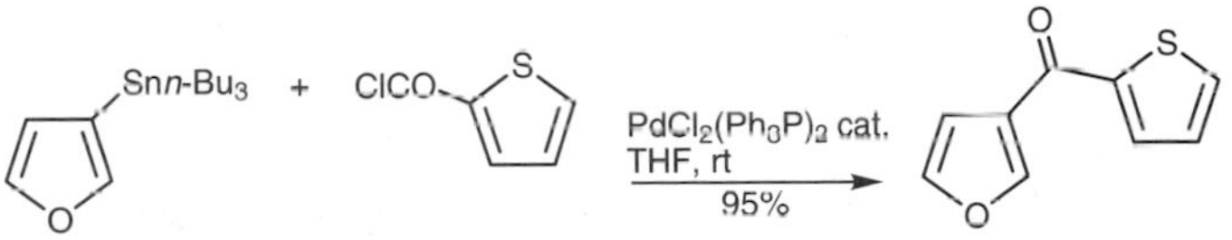

Deprotonations are extremely important: furan and thiophene are deprotonated by strong bases, such as *n*-butyllithium or lithium diisopropylamide, at their α-positions, because here the heteroatom can exert its greatest acidifying influence by inductive withdrawal of electrons, to give anions which can then be made to react with the whole range of electrophiles affording α-substituted furans and thiophenes. This methodology compliments the use of electrophilic substitutions to introduce groups, also regioselectively α, but has the advantage that even weak electrophiles can be utilised. The employment of metallated *N*-substituted (blocked) pyrroles is an equally valid strategy for producing α-substituted pyrroles. Pyrroles which have an *N*-hydrogen are deprotonated at the nitrogen, and the pyrryl anion thus generated is nucleophilic at the hetero atom, providing a means for the introduction of electrophilic groups on nitrogen.

The potential for interaction of the hetero atom (electron donation) with positive charge on a side-chain, especially at an α-position, has a number of effects: amongst the most important is the enhanced reactivity of side-chain derivatives carrying leaving groups.

Similarly, carbonyl groups attached to five-membered heterocycles have somewhat reduced reactivity, as implied by the resonance contributor shown.

Generally speaking, the five-membered heterocycles are far less stable to oxidative conditions than benzenoid aromatic substances, with thiophenes bearing the closest similarity – in many ways thiophenes, of the trio, are the most like carboaromatic compounds. Hydrogenation of thiophenes, particularly over nickel as catalyst, leads to saturation and removal of the hetero atom. Some controlled chemical reductions of pyrroles and furans are known, which give dihydro-products.

The ring synthesis of five-membered heterocycles has been investigated extensively and many and subtle methods have been devised. Each system can be prepared from 1,4-dicarbonyl compounds, for furans by acid catalysed cyclising dehydration, and for pyrroles and thiophenes by interaction with ammonia or primary amine, or a source of sulfur, respectively.

As illustrations of the variety of methods available, the three processes below show (i) the addition of isonitrile anions to α,β-unsaturated nitro compounds, with loss of nitrous acid to aromatise, (ii) the interaction of thioglycolates with 1,3-dicarbonyl compounds, for the synthesis of thiophene-2-esters, and (iii) the cycloaddition/cycloreversion preparation of furans from oxazoles.

13   Pyrroles: reactions and synthesis

pyrrole
[1H-pyrrole]

Pyrrole[1] and the simple alkyl pyrroles are colourless liquids, with relatively weak odours rather like that of aniline, which, also like the anilines, darken by autoxidation. Pyrrole itself is readily available commercially, and is manufactured by alumina-catalysed gas-phase interaction of furan and ammonia. Pyrrole was first isolated from coal tar in 1834 and then in 1857 from the pyrolysate of bone by a process which is similar to an early laboratory method for the preparation of pyrrole – the pyrolysis of the ammonium salt of the sugar acid, mucic acid. The word pyrrole is derived from the Greek for red, which refers to the bright red colour which pyrrole imparts to a pinewood shaving moistened with concentrated hydrochloric acid.

The early impetus for the study of pyrroles came from degradative work relating to the structures of two pigments central to life processes, the blood respiratory pigment haem, and chlorophyll, the green photosynthetic pigment of plants.[2] Such degradations led to the formation of mixtures of alkylpyrroles. Chlorophyll and haem are synthesised in the living cell from porphobilinogen, the only aromatic pyrrole to play a function – a vitally important function – in fundamental metabolism.[3,4]

porphobilinogen

Ultimately, all life on earth depends on the incorporation of atmospheric carbon dioxide into carbohydrates. The energy for this highly endergonic process is sunlight, and the whole is called photosynthesis. The very first step in the complex sequence is the absorption of a photon by pigments, of which the most important in multicellular plants is chlorophyll-a. This photonic energy is then used chemically to achieve a crucial carbon-carbon bonding reaction to carbon dioxide, in which ultimately oxygen is liberated. Thus, formation of the by-product of this process, molecular oxygen, allowed the evolution of aerobic organisms of which man is one.

chlorophyll-a

haem

13.1.2 Nitration

Nitrating mixtures suitable for benzenoid compounds cause complete decomposition of pyrrole, but reaction occurs smoothly with acetyl nitrate at low temperature, giving mainly 2-nitropyrrole. This nitrating agent is formed by mixing fuming nitric acid with acetic anhydride to form acetyl nitrate and acetic acid, thus removing the strong mineral acid. In the nitration of pyrrole with this reagent it has been shown that C-2 is 1.3×10^5 and C-3 is 3×10^4 times more reactive than benzene.[14]

N-Substitution of pyrroles gives rise to increased proportions of β-substitution, even methyl causing the $\beta:\alpha$ ratio to change to 1:3, the much larger *t*-butyl actually reverses the relative positional reactivities, with a $\beta:\alpha$ ratio of 4:1,[15] and the intrinsic α-reactivity can be effectively completely blocked with a very large substituent such as a triisopropylsilyl (TIPS) group, especially useful since it can be subsequently easily removed.[16]

13.1.3 Sulfonation and reactions with other sulfur electrophiles

For sulfonation, a mild reagent of low acidity must be used: the pyridine-sulfur trioxide compound smoothly converts pyrrole into the 2-sulfonate.[17]

Sulfinylation of pyrrole[18] and thiocyanation of pyrrole[19] or of 1-phenylsulfonyl-pyrrole[20] also provide means for the electrophilic introduction of sulfur, but at lower oxidation levels.

Acid catalyses rearrangement of sulfur substituents from the α-position (kinetically-controlled substitution) to give β-substituted pyrroles[20,21] (see also section 13.1.5); the scheme above shows a reasonable mechanism for this transposition, though work on acid-catalysed rearrangement of arylthioindoles revealed a more complex sequence.[22]

13.1.4 Halogenation

Pyrrole halogenates so readily that unless controlled conditions are used, stable tetrahalopyrroles are the only isolable products.[23] Attempts to monohalogenate simple alkylpyrroles fail, probably because of side-chain halogenation and the generation of extremely reactive pyrrylalkyl halides (section 13.12).

2-Bromo- and 2-chloropyrroles, unstable compounds, can be prepared by direct halogenation of pyrrole.[24] Using 1,3-dibromo-5,5-dimethylhydantoin as brominating agent, both 2-bromo- and 2,5-dibromopyrroles can be obtained, the products stabilised by immediate conversion to their N-t-butoxycarbonyl derivatives.[25] Conversely, bromination of N-Boc-pyrrole with N-bromosuccinimide gives the 2,5-dibromo derivative.[26]

N-Triisopropylsilylpyrrole monobrominates and monoiodinates cleanly and nearly exclusively at C-3, and with two mol equivalents of N-bromosuccinimide dibrominates, at C-3 and C-4.[16,27]

13.1.5 Acylation

Direct acetylation of pyrrole with acetic anhydride at 200 °C leads to 2-acetylpyrrole as main product together with some 3-acetylpyrrole, but no N-acetylpyrrole.[28] N-Acetylpyrrole can be obtained in high yield by heating pyrrole with N-acetylimidazole.[29] Alkyl substitution facilitates C-acylation, so that 2,3,4-trimethylpyrrole yields the 5-acetyl derivative even on refluxing in acetic acid. The more reactive trifluoroacetic anhydride and trichloroacetyl chloride react with pyrrole efficiently, even at room temperature, to give 2-substituted products, alcoholysis or hydrolysis of which provides a clean route to pyrrole-2-esters or -acids.[30] Strong electron-withdrawing (*meta*-directing) substituents at a pyrrole α-position tend to override the intrinsic pyrrole regioselectivity and further substitution takes place mainly at C-4 as illustrated, not the remaining α-position.[31]

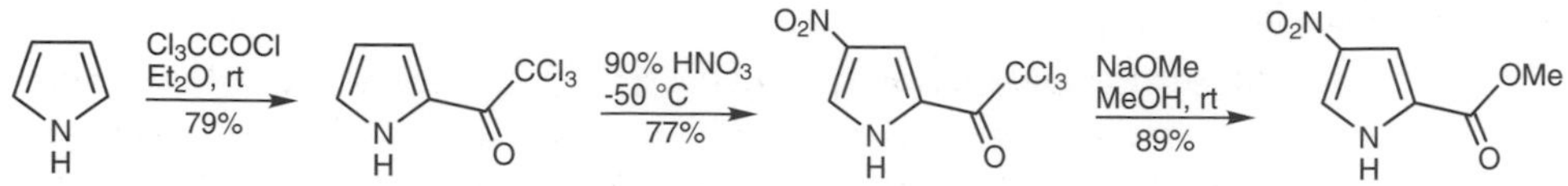

Vilsmeier[32,33] acylation of pyrroles, formylation with dimethylformamide/phosphoryl chloride in particular, is a generally applicable process.[34] As shown below, the actual electrophilic species is an N,N-dialkyl chloromethyleneiminium cation.[35] Here again, the presence of a large pyrrole-N-substituent perturbs the intrinsic α-selectivity, formylation of N-tritylpyrrole favouring the β-position by 2.8:1 and trifluoroacetylation of this pyrrole giving only the 3-ketone;[36] the use of bulky N-silyl-substituents allows β-acylation with subsequent removal of the N-substituent.[37] The electrophilic species produced by the combination of dimethylformamide with pyrophosphoryl chloride is more bulky and leads to increased proportions of β-attack on N-substituted pyrroles.[38] The iminium salt intermediates under Vilsmeier conditions, before hydrolysis, can be neatly utilised for further Friedel Crafts substitution. The substituent is strongly *meta* directing, thus leading to 2,4-diacylated pyrroles.[39] Where a cyclic secondary amide is used, hydrolysis does not take place and the isolated product is a cyclic imine.[40]

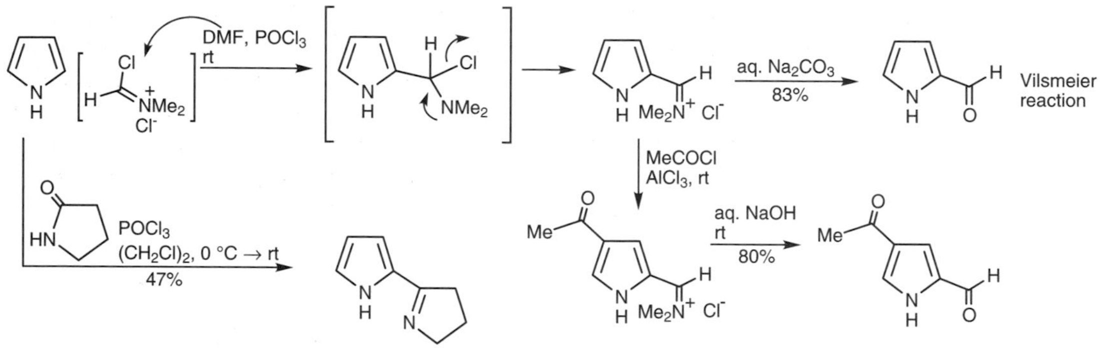

Acylation of 1-phenylsulfonyl pyrrole, with its deactivating N-substituent, requires more forcing conditions in the form of a Lewis acid as catalyst, the regioselectivity of attack depending both on choice of catalyst and on the particular acylating agent as illustrated below.[41] The use of weaker Lewis acid catalysts leads to a greater proportion of α-substitution. Regioselectivity of Friedel-Crafts acylations, depending on the strength of the Lewis acids employed, also extends to pyrroles with electron-withdrawing/stabilising groups, like esters, on carbon.[42] Lewis acid catalysed acylation of 3-acylpyrroles, easily obtained by hydrolysis of 1-phenylsulfonyl-3-acylpyrroles, proceeds smoothly to give 2,4-diacylpyrroles;[43] Vilsmeier formylation of methyl pyrrolyl-2-carboxylate takes place at C-5.[44] Oxidation of 3-acetyl-1-phenylsulfonylpyrrole[45] or hydrolysis and detritylation of 3-trifluoroacetyl-1-tritylpyrrole are each efficient routes to pyrrole-3-carboxylic acid.

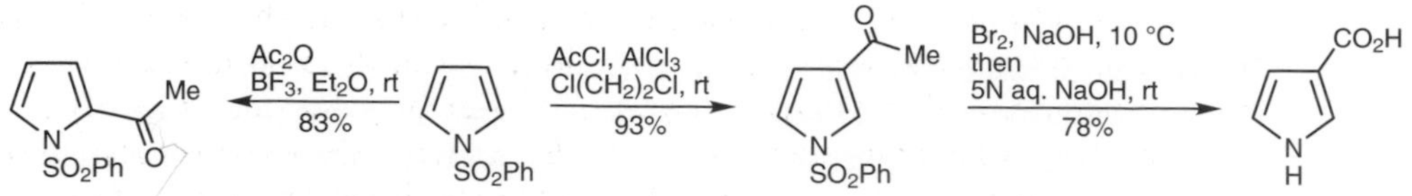

13.1.6 Alkylation

Mono-*C*-alkylation of pyrroles cannot be achieved by direct reaction with simple alkyl halides either alone or with a Lewis acid catalyst, for example pyrrole does not react with methyl iodide below 100 °C; above about 150 °C a series of reactions occurs leading to a complex mixture made up mostly of polymeric material together with some poly-methylated pyrroles. The more reactive allyl bromide reacts with pyrrole at room temperature, but mixtures of mono- to tetrallylpyrroles together with oligomers and polymers are obtained. Alkylations with conjugated enones carrying a leaving group at the *β*-position proceed smoothly, usefully producing mono-alkenylated pyrroles.[46]

13.1.7 Condensation with aldehydes and ketones

Condensations of pyrroles with aldehydes and ketones occur easily by acid catalysis but the resulting pyrrolylcarbinols cannot usually be isolated, for under the reaction conditions proton-catalysed loss of water produces 2-alkylidenepyrrolium cations which are themselves highly reactive electrophiles. Thus, in the case of pyrrole itself, reaction with aliphatic aldehydes in acid inevitably leads to resins, probably linear polymers. Reductive trapping of these cationic intermediates produces alkylated pyrroles; all free positions react and as the example shows, acyl and alkoxycarbonyl substituents are unaffected.[47] A mechanistically related process is the clean 4-chloromethylation of pyrroles carrying acyl groups at C-2.[48]

a 2-alkylidenepyrrolium cation

Syntheses of dipyrromethanes have usually involved pyrroles with electron-withdrawing substituents and only one free *α*-position, the dipyrromethane resulting from attack by a second mol equivalent of the pyrrole on the 2-alkylidenepyrrolium intermediate.[49]

a dipyrromethane

However, conditions have been established for the production and isolation of bis(pyrrol-2-yl)methane itself from treatment of pyrrole with aqueous formalin in acetic acid;[50] from reaction with formalin in the presence of potassium carbonate a bis-hydroxymethylation product is obtained.[51] This reacts with pyrrole in dilute acid to give tripyrrane and from this, as the scheme shows, reaction with 2,5-bis(hydroxymethyl)pyrrole gives porphyrinogen which can be oxidised to porphyrin.

Acetone, reacting in a comparable manner, gives a cyclic tetramer directly and in high yield, perhaps because the geminal methyl groups tend to force the pyrrole rings into a coplanar conformation, greatly increasing the chances of cyclisation of a linear tetrapyrrolic precursor.[52]

Condensations with aromatic aldehydes carrying appropriate electron-releasing substituents produce cations which are sufficiently stabilised by mesomerism to be isolated. Such cations are coloured: the reaction with *p*-dimethylaminobenzaldehyde is the basis for the classical Ehrlich test, deep red/violet colours being produced by pyrroles (and also by furans and indoles) which have a free nuclear position.

Analogous condensations with a pyrrole aldehyde lead to mesomeric dipyrromethene cations, which play an important part in porphyrin synthesis. Under appropriate conditions one can combine four mol equivalents of pyrrole and four of an aromatic aldehyde to produce a tetra-aryl substituted porphyrin in one pot.[53]

a dipyrromethene cation

13.1.8 Condensation with imines and iminium ions

The imine and iminium functional groupings are, of course, the nitrogen equivalents of carbonyl and *O*-protonated carbonyl groups, and their reactivity is analogous. The Mannich reaction of pyrrole produces dialkylaminomethyl derivatives, the iminium electrophile being generated *in situ* from formaldehyde, dialkylamine, and acetic acid.[54] There are only a few examples of the reactions of imines themselves with pyrroles; the condensation of 1-pyrroline with pyrrole as reactant and solvent is one such example.[55]

The mineral acid-catalysed polymerisation of pyrrole involves a series of Mannich reactions, but under controlled conditions pyrrole can be converted into an isolable trimer, which is probably an intermediate in the polymerisation. The key to understanding the formation of the observed trimer is that the less stable, therefore more reactive *β*-protonated pyrrolium cation is the electrophile which initiates the sequence attacking a second mol equivalent of the heterocycle. The dimer, an enamine, is too reactive to be isolable, however the trimer, relatively protected as its salt, reacts further only slowly.[56]

pyrrole trimer
2:1 *trans* : *cis*

13.1.9 Diazo-coupling[57]

The high reactivity of pyrroles is illustrated by their ready reaction with benzenediazonium salts. Pyrrole itself gives a mono-azo derivative by reacting as a neutral species below p*H* 8, but by way of the pyrryl anion (section 13.4), and 10^8 times faster, in solutions above p*H* 10. In more strongly alkaline conditions 2,5-bisdiazo derivatives are formed.

13.2 Reactions with oxidising agents[58]

Simple pyrroles are generally easily attacked by strong chemical oxidising agents, frequently with complete breakdown. When the ring does survive, maleimide derivatives are the commonest products, even when there was originally a 2- or 5-alkyl substituent. This kind of oxidative degradation played an important part in early porphyrin structure determination, in which chromium trioxide in aqueous sulfuric acid or fuming nitric acid were usually used as oxidising agents. Hydrogen peroxide is a more selective reagent and can convert pyrrole itself into a tautomeric mixture of pyrrolin-2-ones in good yield (section 13.17.1).

Pyrroles which have a ketone or ester substituent are more resistant to ring degradation and high yielding side-chain oxidation can be achieved using cerium(IV) ammonium nitrate with selectivity for an α-alkyl.[59]

13.3 Reactions with nucleophilic reagents

Pyrrole and its derivatives do not react with nucleophilic reagents by addition or by substitution, except in the same type of situation which allows nucleophilic substitution in benzene chemistry: the two examples below are illustrative.[60]

A key step in a synthesis of Ketorolac involves an intramolecular nucleophilic displacement of a methanesulfonyl group activated by a 5-ketone.[61]

13.4 Reactions with bases

13.4.1 Deprotonation of N-hydrogen

Pyrrole N-hydrogen is much more acidic (pK_a 17.5) than that of a comparable saturated amine, say pyrrolidine ($pK_a \sim 44$), or aniline (pK_a 30.7), and of the same order as that of 2,4-dinitroaniline. Any very strong base will effect complete conversion of an N-unsubstituted pyrrole into the corresponding pyrryl anion, perhaps the most convenient being commercial *n*-butyllithium in hexane exemplified below by the preparation of 1-triisopropylsilylpyrrole, however reactions at nitrogen can proceed *via* smaller, equilibrium concentrations of pyrryl anion as in the formation of 1-chloropyrrole (in solution) by treatment with sodium hypochlorite[62] or the preparation of 1-*t*-butoxycarbonylpyrrole.[63]

13.4.2 Deprotonation of *C*-hydrogen

The *C*-deprotonation of pyrroles requires the absence of the much more acidic *N*-hydrogen i.e. the presence of an *N*-substituent, either alkyl[64] or, if required, a removable group like phenylsulfonyl,[65] carboxylate,[66] trimethylsilylethoxymethyl,[67] or *t*-butylaminocarbonyl.[68] Even in the absence of chelation assistance to lithiation, which is certainly an additional feature in each of the latter examples, metallation proceeds at the α-position. Deprotonation of *N*-methylpyrrole proceeds further, amazingly easily, to a dilithio derivative, either 2,4- or 2,5-dilithio-1-methylpyrrole depending on the exact conditions.[69] Lithiation of 1-*t*-butoxycarbonyl-3-*n*-hexylpyrrole occurs at C-5, avoiding both steric and electronic discouragement of the alternative C-2 deprotonation.[70]

13.5 Reactions of *N*-metallated pyrroles

13.5.1 Lithium, sodium, potassium, magnesium, and zinc derivatives

N-Metallated pyrroles can react with electrophiles to give either *N*- or *C*-substituted pyrroles: generally speaking the more ionic the metal–nitrogen bond and/or the better the solvating power of the solvent, the greater is the percentage of attack at nitrogen.[71] Based on these principles, several methods are available for efficient *N*-alkylation of pyrroles including the use of potassium hydroxide in dimethylsulfoxide,[72] or in benzene with 18-crown-6,[73] thallous ethoxide,[74] using phase-transfer methodology,[75] or of course by reaction of the pyrryl anion generated using *n*-butyllithium. The thallium salt acylates[76] and the potassium salt arylsulfonylates[77] efficiently on nitrogen. *N*-Acylpyrroles can be reduced to *N*-alkylpyrroles using borane.[78]

Pyrryl Grignard reagents, obtained in solution by treating an *N*-unsubstituted pyrrole with alkyl Grignard, tend to react at carbon with alkylating and acylating agents, but sometimes give mixtures of 2- and 3-substituted products with the former predominating,[79] *via* neutral, non-aromatic intermediates. Clean α-acylation can be achieved for example with bromacetates[80] or, as exemplified below, using 2-acylthiopyridines (section 5.10.2.4) as acylating agents.[81]

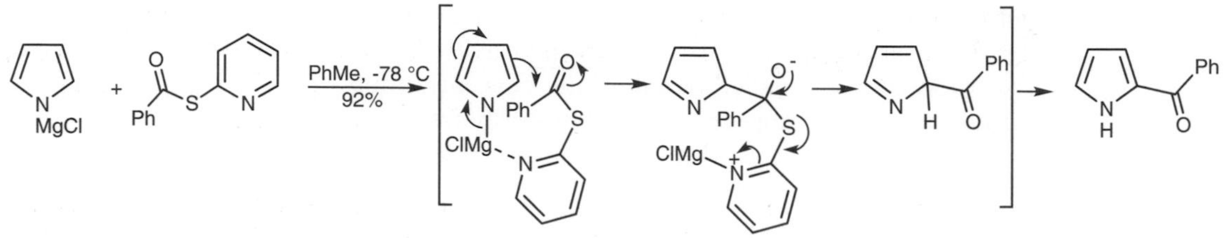

N-Arylation of pyrroles can be achieved by conversion of 1-lithiopyrroles into the corresponding zinc compounds and then reaction with aryl bromides using palladium(0) catalysis[82] or by direct reaction of the pyrrole with an aryl halide in the presence of base and the palladium catalyst.[83]

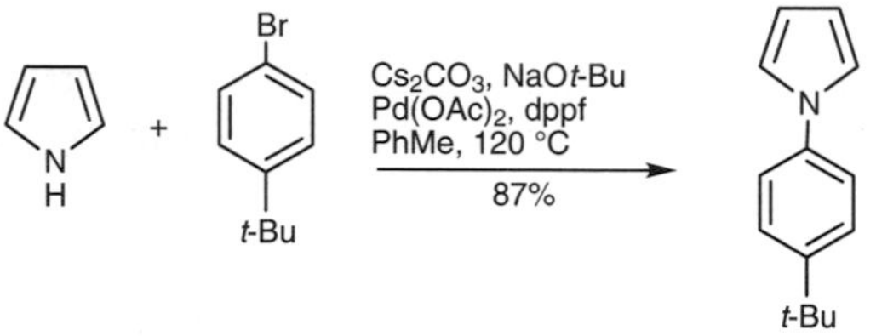

13.6   Reactions of C-metallated pyrroles

13.6.1   Lithium derivatives

Reactions of the species produced by the lithiation of N-substituted pyrroles are efficient for the introduction of groups to the 2-position, either by reaction with electrophiles[64-68] or by coupling processes based on boron or palladium chemistry.[84]

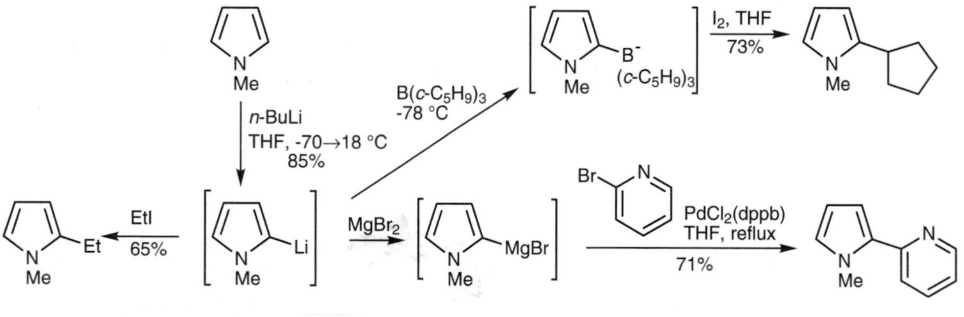

Some examples where removable N-blocking groups have been used in the synthesis of 2-substituted pyrroles, via lithiation, are shown below.

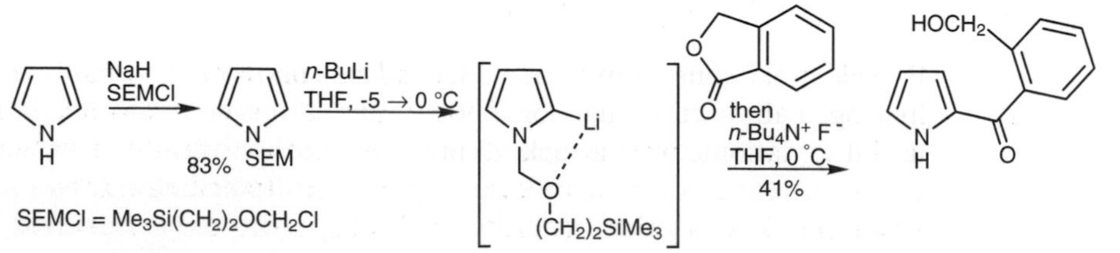

SEMCl = Me$_3$Si(CH$_2$)$_2$OCH$_2$Cl

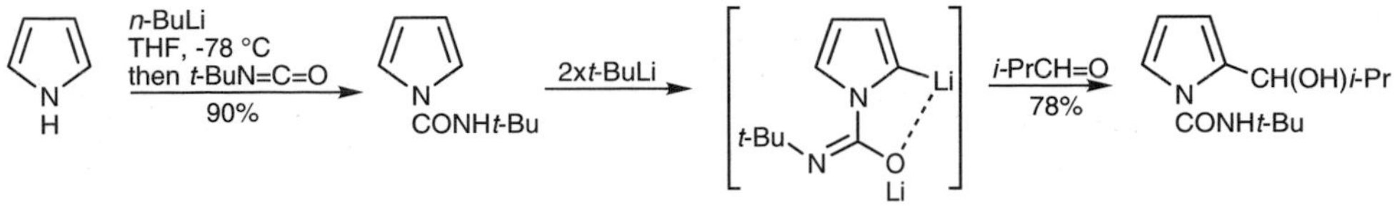

Metal/halogen exchange on 2-bromo-1-t-butoxycarbonylpyrrole and on its 2,5-dibromo conterpart proceed normally and the mono- or dilithiated species thus produced react with the usual range of electrophiles, as illustrated below.[25,26]

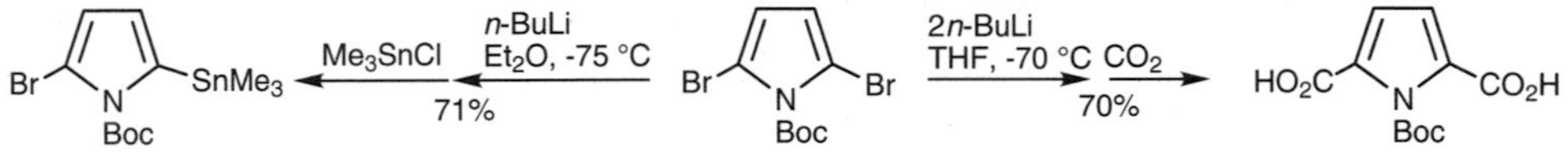

Metal/halogen exchange using 3-bromo-N-triisopropylsilylpyrrole very usefully allows the introduction of groups to the pyrrole β-position and can complement direct electrophilic substitution of N-triisopropylsilylpyrrole (see sections 13.1.2 and 13.1.4).

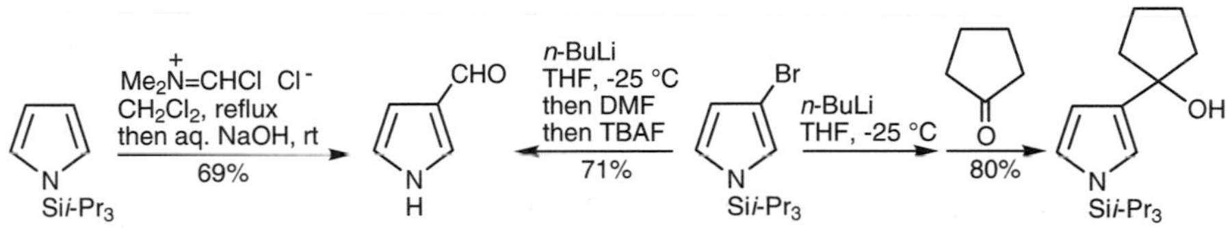

13.6.2 Palladium-catalysed reactions

Pyrrolylstannanes and boronic acids can be synthesised and utilised in the standard manner. The examples below show palladium(0)-catalysed coupling to an aromatic halide.[85]

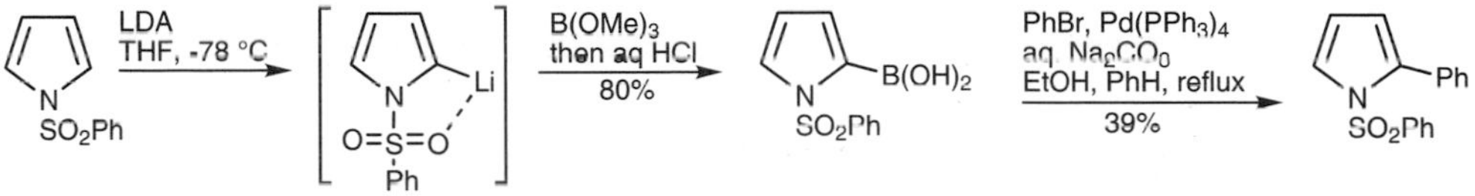

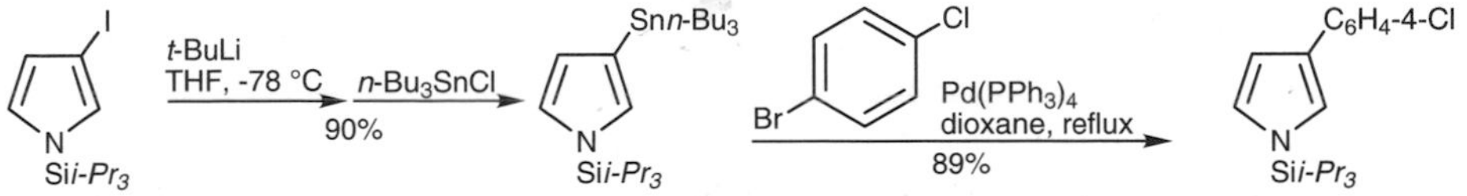

13.7 Reactions with radical reagents

Pyrrole itself tends to give tars under radical conditions, probably by way of initial N-hydrogen abstraction, but some N-substituted derivatives will undergo preparatively useful arylations, with attack taking place predominantly at an α-position.[86] More efficient routes to arylpyrroles depend on transition metal-mediated coupling processes (see section 2.7.2.2). N-Methylpyrrole is attacked by electrophilic benzoyloxy radicals at its α-positions.[87]

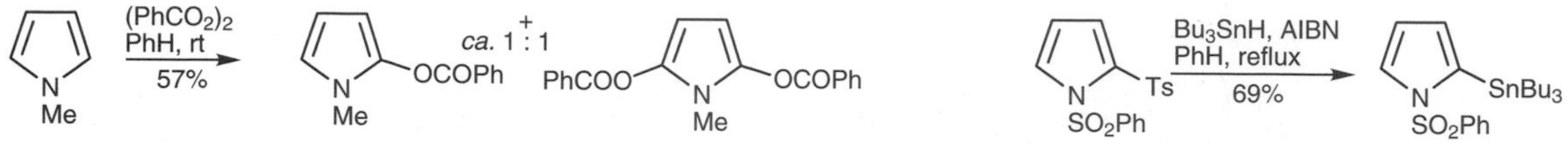

Radical substitution of hydrogen[88] or of toluenesulfonyl[89] at an α-position are processes which will no doubt be developed further in future.

13.8 Reactions with reducing agents

Simple pyrroles are not reduced by hydride reducing agents, diborane, or alkali metal/ethanol or /liquid ammonia combinations, but are reduced in acidic media, in which the species under attack is the protonated pyrrole. The products are 2,5-dihydropyrroles, accompanied by some of the pyrrolidine as by-product.[90] Reduction[91] of pyrroles to pyrrolidines can be effected catalytically over a range of catalysts, is especially easy if the nitrogen carries an electron-withdrawing group, and is not complicated by carbon–heteroatom hydrogenolysis and ring opening as is the case for furans.

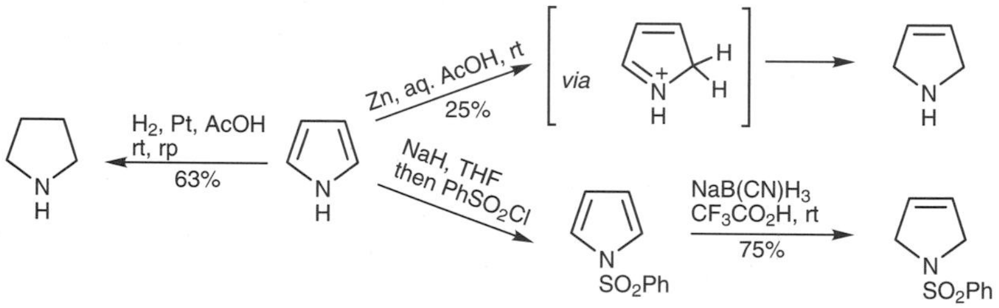

Birch reduction of pyrrole carboxylic esters and tertiary amides gives dihydro derivatives; the presence of an electron-withdrawing group on the nitrogen serves both to remove the acidic N-hydrogen and also to reduce the electron density on the ring. Quenching with an alkyl halide produces alkylated dihydropyrroles.[92]

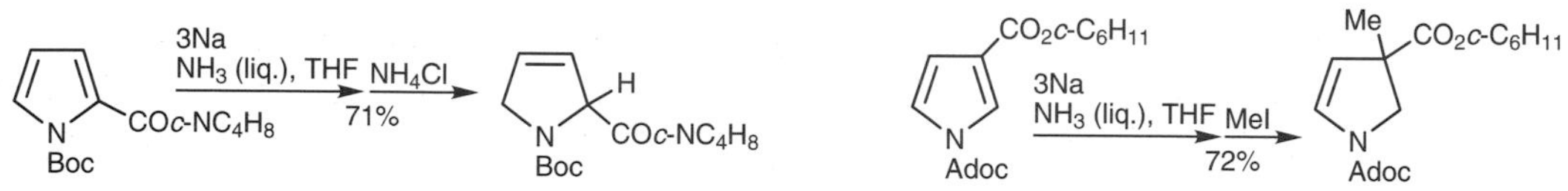

13.9 Electrocyclic reactions (ground state)

Simple pyrroles do not react as 4π components in cycloadditions – exposure of pyrrole to benzyne for example leads only to 2-phenylpyrrole, in low yield.[93] However N-substitution, particularly with an electron-withdrawing group, does allow such reactions to occur,[94] thus adducts with arynes were obtained using 1-trimethylsilylpyrrole.[95] Whereas pyrrole itself reacts with dimethyl acetylenedicarboxylate only by α-substitution, even at 15 kbar,[96] 1-acetyl- and 1-alkoxycarbonylpyrroles give cycloadducts,[97] addition being much accelerated by high pressure or by aluminium chloride catalysis.[98] The most popular N-substituted pyrrole in this context has been its N-t-butoxycarbonyl (Boc) derivative: the reactions shown below illustrate this.[99]

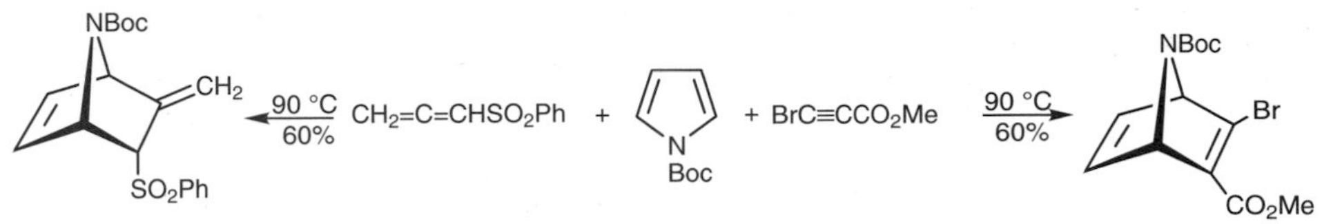

A completely different device to encourage pyrroles to react as dienes is their conversion into osmium complexes;[100] in this way even traditional dienophiles will react under mild conditions as shown below; adducts can be subsequently obtained by oxidative destruction of the metal complex.[101]

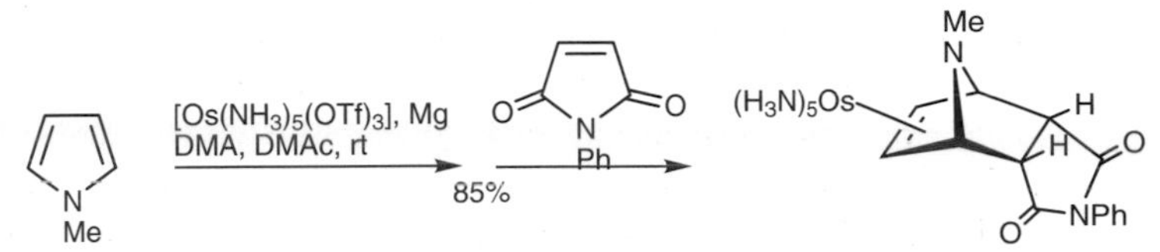

A process which has proved valuable in synthesis is the addition of singlet oxygen to *N*-alkyl- and especially *N*-acylpyrroles[102] producing 2,3-dioxa-7-azabicyclo[2.2.1]-heptanes which react with nucleophiles, such as silyl enol ethers, mediated by tin(II) chloride, generating 2-substituted pyrroles which can be used, as shown, for the synthesis of indoles.

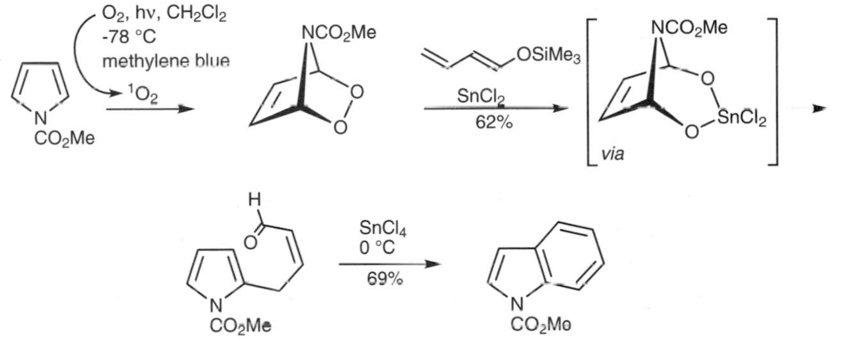

Vinylpyrroles take part in Diels-Alder processes as 4-π components;[103] this reactivity is best controlled by the presence of a phenylsulfonyl group on the pyrrole nitrogen as illustrated below, the presumed initial product easily isomerising in the reaction conditions to reform an aromatic pyrrole.[104]

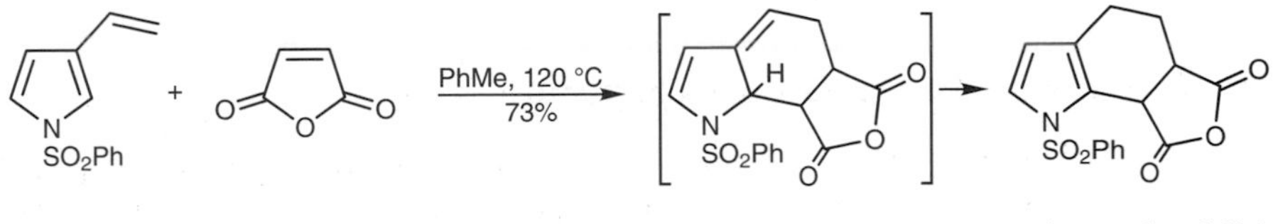

Intermolecular examples of pyrroles serving as 2π components in cycloadditions are very rare, however in an intramolecular sense tricyclic 6-azaindoles have been produced effectively where the 4π component is a 1,2,4-triazene (section 25.2.1).[105]

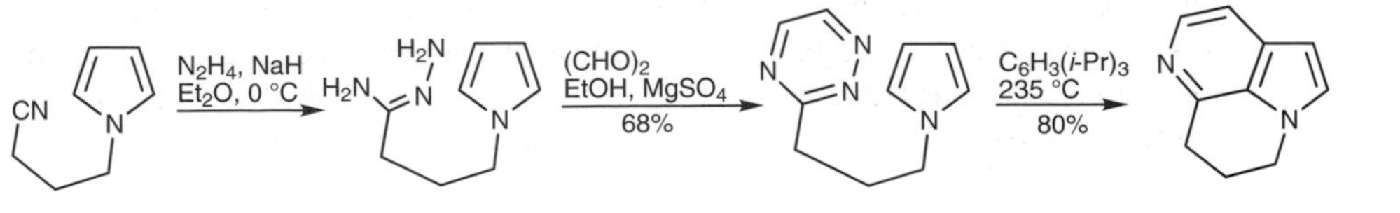

13.10   Reactions with carbenes and carbenoids

The reaction of pyrrole with dichlorocarbene proceeds in part *via* a dichlorocyclo-propane intermediate, ring expansion of which leads to 3-chloropyridine.[106,107] There are relatively few (section 14.1.2) reported isolable cyclopropane-containing adducts from pyrroles – 1-methoxycarbonylpyrrole[108] or *N*-acylpyrroles.[109] 1-Methylpyrrole with ethoxycarbonylcarbene gives only substitution products.[110]

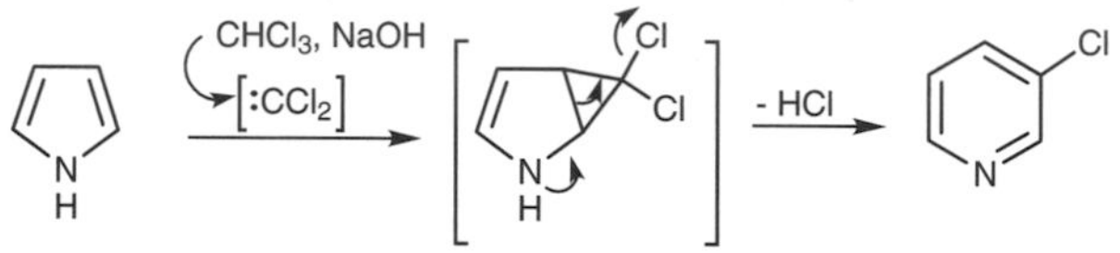

More useful regimes involve the interaction of pyrroles with electron-withdrawing groups on nitrogen with carbenoids generated from diazoalkanes and rhodium(II) compounds; in the example shown below, addition of a vinyl carbene produces a cyclopropanated intermediate which undergoes a Cope rearrangement neatly producing an 8-azabicyclo[3.2.1]octadiene – the ring skeleton of cocaine.[111]

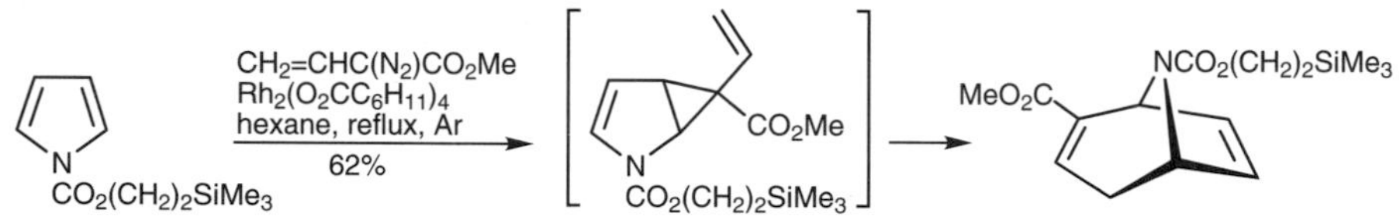

13.11   Photochemical reactions[112]

The photo-catalysed rearrangement of 2- to 3-cyanopyrroles is considered to involve a 1,3-shift in an initially-formed bicyclic aziridine.[113]

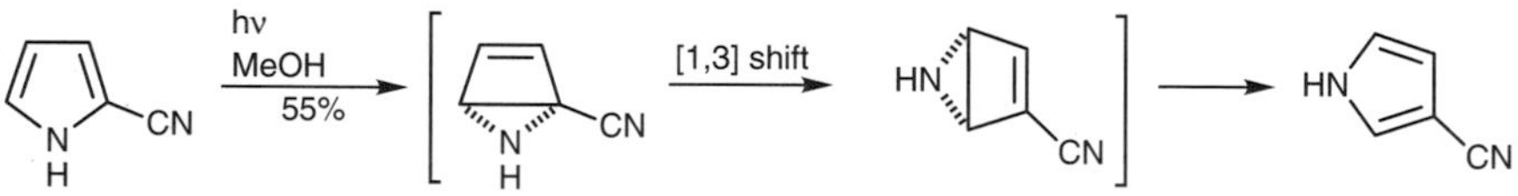

13.12   Pyrryl-C-X compounds

Pyrroles of this type, where X is halogen, alcohol, alkoxy, or amine, and especially protonated alcohol or alkoxy, or quaternised amine, easily lose X generating very reactive electrophilic species. Thus ketones can be reduced to alkane, *via* the loss of oxygen from the initially formed alcohol (cf. section 13.1.7), and quaternary ammonium salts, typified by 2-dimethylaminomethylpyrrole metho-salts, react with nucleophiles by loss of trimethylamine in an elimination/addition sequence of considerable synthetic utility.[114]

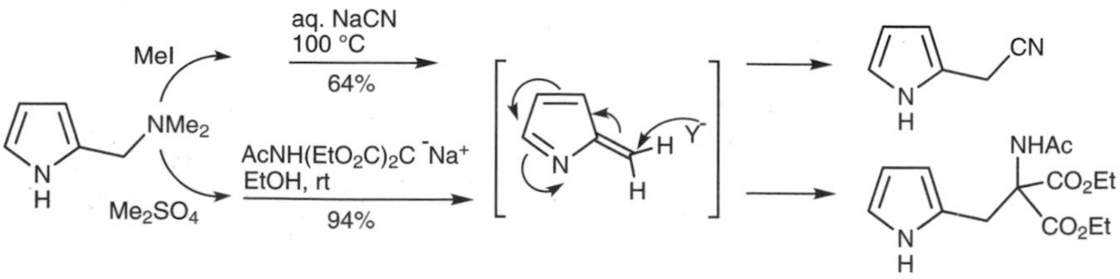

13.13 Pyrrole aldehydes and ketones

These are stable compounds which do not polymerise or autoxidise. For the most part, pyrrole aldehydes and ketones are typical aryl ketones, though less reactive – such ketones can be viewed as vinylogous amides. They can be reduced to alkylpyrroles by the Wolff-Kishner method, or by sodium borohydride *via* elimination from the initial alcoholic product.[115] Treatment of acylated 1-phenylsulfonylpyrroles with *t*-butylamine-borane effects conversion to the corresponding alkyl derivatives.[116]

β- and *α*-Acylpyrroles can be equilibrated one with the other using acid; for *N*-alkyl-*C*-acylpyrroles, the equilibrium lies completely on the side of the 3-isomer.[117]

13.14 Pyrrole carboxylic acids

The main feature within this group is the ease with which loss of the carboxyl group occurs. Simply heating[118] pyrrole acids causes easy loss of carbon dioxide in what is essentially *ipso* displacement of carbon dioxide by proton.[119] This facility is of considerable relevance to pyrrole synthesis since many of the ring-forming routes (e.g. see sections 13.18.1.2 and 13.18.1.3) produce pyrrole esters, in which the ester function may not be required ultimately.

Displacement of carboxyl groups by other electrophiles such as halogens[120] or under nitrating conditions, or with aryl diazonium cations occurs more readily than at a carbon carrying hydrogen.

13.15 Pyrrole carboxylic acid esters

The electrophilic substitution of these stable compounds has been much studied; the *meta*-directing effect of the ester overcomes the normally dominant tendency for α-substitution.[121]

An ester group can also activate side-chain alkyl for halogenation, and such pyrrolylalkyl halides have been used extensively in synthesis.[122] Cerium(IV) triflate in methanol can be used for the analogous introduction of methoxide onto an alkyl side chain.[123]

The rates of alkaline hydrolysis of α- and β-esters are markedly different, the former being faster than the latter, possibly because of stabilisation of the intermediate by intramolecular hydrogen bonding involving the ring hetero-atom.[124]

13.16 Halopyrroles

Simple 2-halopyrroles are very unstable compounds whereas 3-halopyrroles are relatively stable, as indeed are 2-halopyrrole ketones and esters. Chemical manipulation of halopyrroles is best achieved with an electron-withdrawing substituent on nitrogen. Pyrrole halides have typical aryl halide reactivity being inert to nucleophilic displacement but undergoing exchange with *n*-butyllithium and palladium-catalysed couplings.[125] Pyrrole halides undergo catalytic hydrogenolysis, which has allowed the use of halide as a blocking substituent.

13.17 Oxy- and aminopyrroles

13.17.1 2-Oxypyrroles

2-Oxypyrroles exist in the hydroxyl form, if at all, only as a minor component of the tautomeric mixture which favours 3-pyrrolin-2-one over 4-pyrrolin-2-one by 9:1.[126]

After *N*-protection, silylation produces 2-silyloxypyrroles which react with aldehydes to give substituted 3-pyrrolin-2-ones.[127]

13.17.2   3-Oxypyrroles

3-Oxypyrroles exist largely in the carbonyl form, unless flanked by an ester group at C-2 which favours the hydroxyl tautomer by intramolecular hydrogen bonding.[128]

13.17.3   Aminopyrroles

Aminopyrroles have been very little studied because they are relatively unstable and difficult to prepare.[129] Simple 2-aminopyrroles can be prepared and stored in acidic solution.[130]

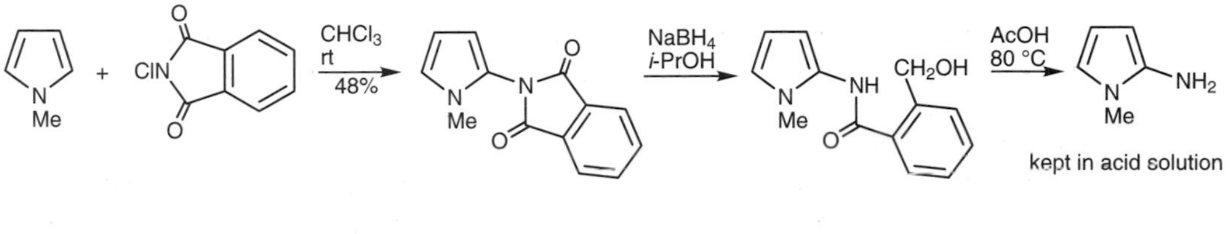

13.18   Synthesis of pyrroles[7,131]

13.18.1   Ring synthesis

13.18.1.1   From 1,4-dicarbonyl compounds and ammonia or primary amines[132]

1,4-Dicarbonyl compounds react with ammonia or primary amines to give pyrroles.

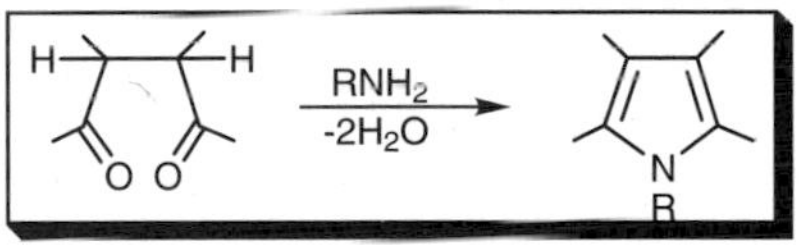

Paal-Knorr synthesis[133]

Pyrroles are formed by the reaction of ammonia or a primary amine with a 1,4-dicarbonyl compound[134] (see also 15.14.1.1). An alternative to the use of ammonia for the synthesis of N-unsubstituted pyrroles by this method employs hexamethyldisilazide with alumina.[135] Successive nucleophilic additions of the amine nitrogen to the two carbonyl carbon atoms and the loss of two mol equivalents of water represent the net course of the synthesis; a reasonable sequence[136] for this is shown below using the synthesis of 2,5-dimethylpyrrole[137] as an example.

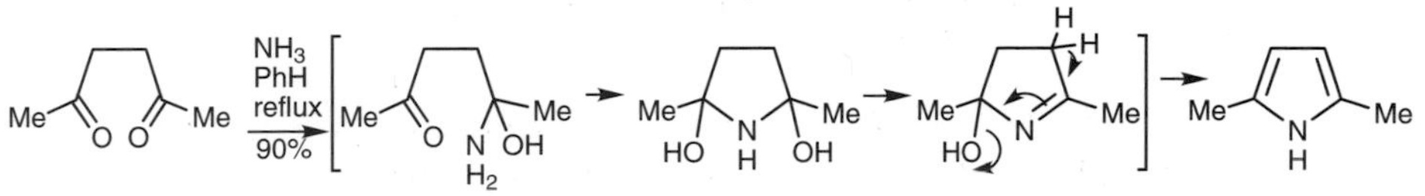

The best synthon for unstable succindialdehyde, for the ring synthesis of C-unsubstituted pyrroles, is 2,5-dimethoxytetrahydrofuran (section 15.1.4),[138] or 1,4-dichloro-1,4-dimethoxybutane obtainable from it.[139] 2,5-Dimethoxytetrahydrofuran will react with aliphatic and aromatic amines, amino esters, arylsulfonamides, trimethylsilylethoxycarbonylhydrazine,[140] or primary amides to give the corresponding N-substituted pyrroles.[141]

A still useful synthesis of *N*-substituted pyrroles, which consists of dry distillation of the alkylammonium salt of mucic or saccharic acid,[142] probably also proceeds by way of a 1,4-dicarbonyl intermediate. The overall process involves loss of four mol equivalents of water and two of carbon dioxide, and may proceed as shown.

13.18.1.2 From α-aminocarbonyl compounds and activated ketones

α-Aminoketones react with carbonyl compounds which have an α-methylene grouping, preferably further activated, for example by ester, as in the illustration.

Knorr synthesis

This widely used general approach to pyrroles, utilizes two components: one, the α-aminocarbonyl component, supplies the nitrogen and C-2 and C-3, and the second component supplies C-4 and C-5 and must possess a methylene group α to carbonyl. The Knorr synthesis works well only if the methylene group of the second component is further activated (e.g. as in acetoacetic ester) to enable the desired condensation leading to pyrrole to compete effectively with the self-condensation of the α-aminocarbonyl component. The synthesis of 4-methylpyrrole-3-carboxylic acid and therefrom, 3-methylpyrrole illustrates the process.

Since free α-aminocarbonyl compounds self-condense very readily producing dihydropyrazines (section 11.13.3.1), they have traditionally been prepared and used in the form of their salts, to be liberated for reaction by the base present in the reaction mixture. Alternatively, carbonyl-protected amines, such as aminoacetal ($H_2NCH_2CH(OEt)_2$), have been used, in this case with the enol ether of a 1,3-diketone as synthon for the activated carbonyl component.[143]

A way of avoiding the difficulty of handling α-aminocarbonyl compounds is to prepare them in the presence of the second component, with which they are to react. Zinc–acetic acid or sodium dithionite[144] can be used to reduce oximino groups to amino while leaving ketone and ester groups untouched.

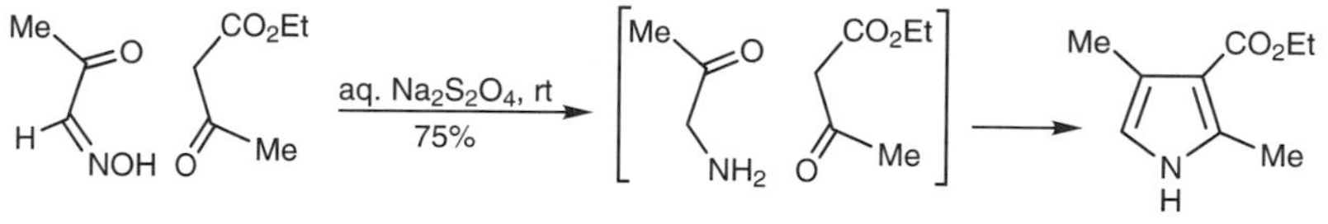

In the classical synthesis, which gives this route its name, the α-aminocarbonyl component is simply an amino-derivative of the other carbonyl component, and it is even possible to generate the oximino precursor of the amine *in situ*.[145]

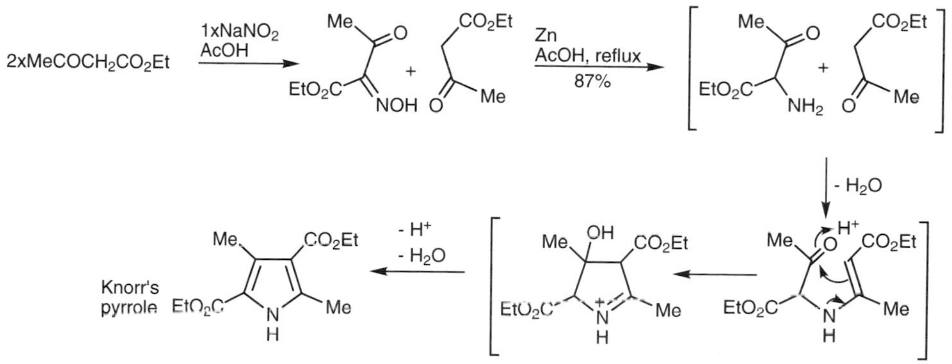

It is believed that in the mechanism, shown for Knorr's pyrrole, an N–C-2 bond is the first formed, which implies that the nitrogen becomes attached to the more electrophilic of the two carbonyl groups of the other component. Similarly, the C-3–C-4 bond is made to the more electrophilic carbonyl group of the original α-aminocarbonyl component, where there is a choice. There are many elegant examples of the use of this approach; an interesting example in which two pyrrole rings are formed using a phenylhydrazone as precursor of the α-aminocarbonyl component is shown below.[146]

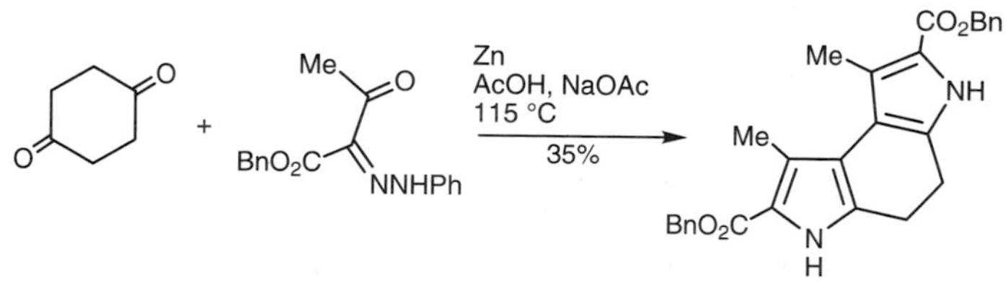

Modern alternatives for the assembly of the α-aminocarbonyl component include the reaction of a 2-bromoketone with sodium diformamide producing an α-formamido-ketone,[147] and the reaction of a Weinreb amide of an N-protected α-amino acid with a Grignard reagent, then release of the N-protection in the presence of the second component, as illustrated below.[148] Hydride reduction of the Weinreb amide of an N-protected α-amino acid gives N-protected α-amino-aldehydes for use in this approach.[149]

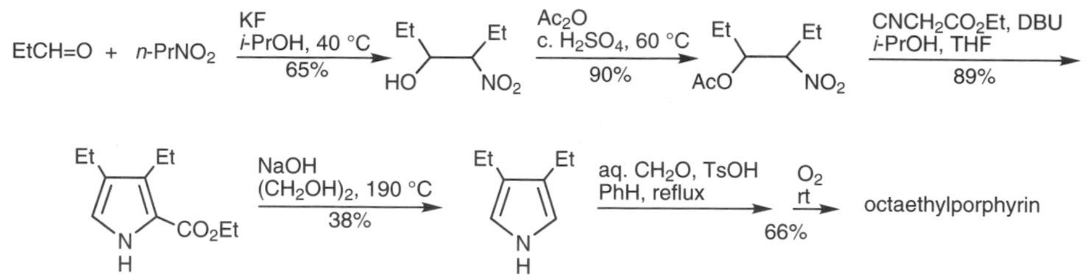

13.18.3.4 Octaethylhemiporphycene[174,175]

All of the non-natural isomers (porphycenes) of the porphyrin ring system comprising permutations of four pyrrole rings, four methines, and having an 18 π-electron main conjugation pathway, have been synthesised. The scheme below shows the use of a MacDonald condensation[176] to assemble a tetrapyrrole and then the use of the McMurray reaction to construct the macrocycle.[177]

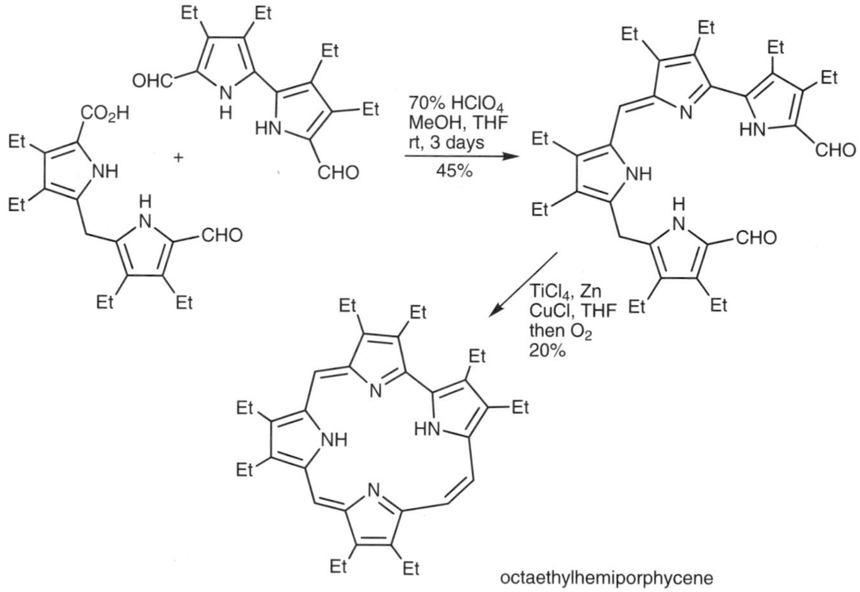

octaethylhemiporphycene

13.18.3.5 Benzo[1,2-b:4,3-b']dipyrroles

Several ingenious approaches[178] have been described for the elaboration of the pyrrolo-indole unit (strictly a benzo[1,2-b:4,3-b']dipyrrole) three of which are present in the potent anti-tumour agent CC-1065;[179] the approach shown here employs the method described in section 13.18.1.4 for the construction of the pyrrole nuclei.[180]

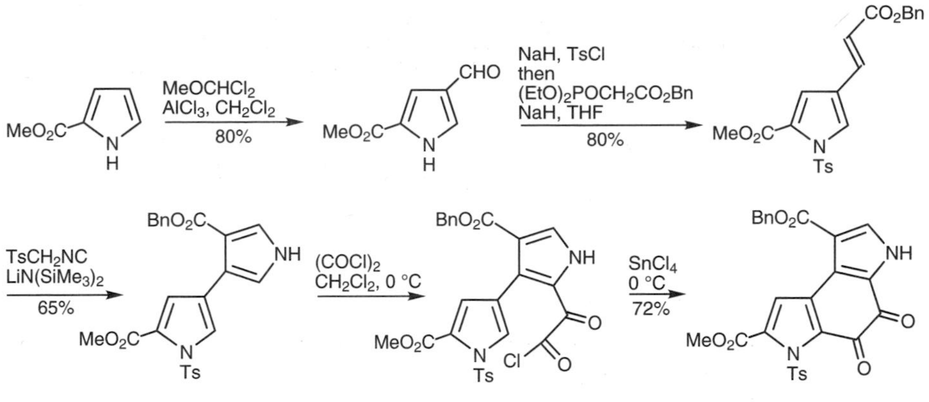

13.18.3.6 Epibatidine[181]

Cycloaddition of *N*-Boc-pyrrole with ethynyl *p*-tolyl sulfone generated the bicyclic system; selective reduction of a double bond then conjugate addition of 5-lithio-2-methoxypyridine produced an intermediate, with the required stereochemistry, requiring only straightforward manipulations to produce epibatidine.

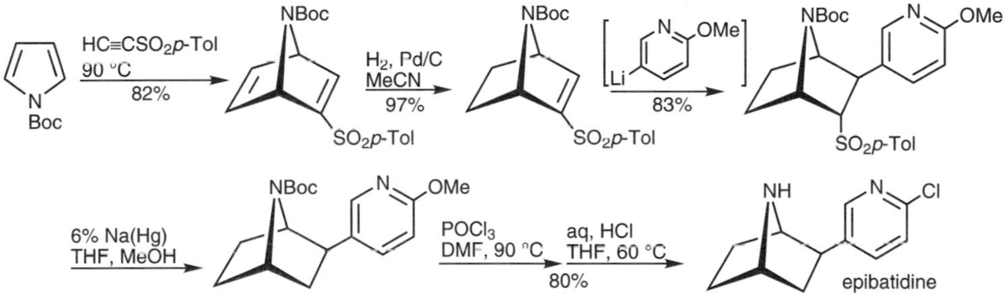

Exercises for chapter 13

Straightforward revision exercises (consult chapters 12 and 13)

(a) Why does pyrrole not form salts by protonation on nitrogen?

(b) Starting from pyrrole, how would one prepare, cleanly, 2-bromopyrrole, 3-bromopyrrole, 2-formylpyrrole, 3-nitropyrrole? (more than one step necessary in some cases)

(c) What would be the structures of the products from the following reactions: (i) pyrrole with CH_2O/pyrrolidine/AcOH; (ii) pyrrole with NaH/MeI; (iii) 1-tri-*i*-propylsilylpyrrole with LDA then $Me_3CCH=O$?

(e) How could one produce a 3-lithiated pyrrole?

(f) Give two ways in which pyrrole could be encouraged to react as a diene in Diels-Alder type processes.

(g) How could pyrrole be converted into pyrrol-2-yl-CH_2CN in two steps?

(h) By what mechanism are pyrrole carboxylic acids readily decarboxylated on heating?

(i) Which ring synthesis method and what reactants would be appropriate for the synthesis of a pyrrole, unsubstituted on carbon but carrying $CH(Me)(CO_2Me)$ on nitrogen?

(j) With what compound would ethyl acetoacetate ($MeCOCH_2CO_2Et$) need to be reacted to produce ethyl 2-methyl-4,5-diphenylpyrrole-3-carboxylate?

(k) With what compound would TosMIC ($TsCH_2NC$) need to be reacted to produce methyl 4-ethylpyrrol-3-carboxylate?

(l) With what reactants would 3-nitrohex-3-ene need to be treated to produce ethyl 3,4-diethylpyrrole-2-carboxylate?

More advanced exercises

1. Two isomeric mono-nitro-derivatives, $C_5H_6N_2O_2$, are formed in a ratio of 6:1, by treating 2-methylpyrrole with Ac_2O/HNO_3. What are their structures and which would you predict to be the major product?

2. Write structures for the products of the following sequences: (i) pyrrole treated with $Cl_3CCO.Cl$, then the product with Br_2, then this product with $MeONa/MeOH \rightarrow C_6H_6BrNO_2$, (ii) pyrrole treated with $DMF/POCl_3$, then with $MeCOCl/AlCl_3$, then finally with aq. $NaOH \rightarrow C_7H_7NO_2$, (iii) 2-chloropyrrole treated with $DMF/POCl_3$, then aq. $NaOH$, then the product with $LiAlH_4 \rightarrow C_5H_6ClN$.

3. Write structures for the products formed by the reaction of pyrrole with $POCl_3$ in combination with (i) N,N-dimethylbenzamide; (ii) pyrrole-2-carboxylic acid N,N-dimethylamide; (iii) 2-pyrrolidone $\rightarrow C_8H_{10}N_2$, in each case followed by aq. $NaOH$.

4. Treatment of 2-methylpyrrole with HCl produces a dimer, not a trimer as does pyrrole itself (section 13.1.8). Suggest a structure for the dimer, $C_{10}H_{14}N_2$, and explain the non-formation of a trimer.

5. Treatment of 2,5-dimethylpyrrole with Zn/HCl gave a mixture of two isomeric products $C_6H_{11}N$: suggest structures.

6. (i) Heating 1-methoxycarbonylpyrrole with diethyl acetylenedicarboxylate at 160 °C produced diethyl 1-methoxycarbonylpyrrole-3,4-dicarboxylate; suggest a mechanism and a key intermediate; (ii) deduce the structure of the product, $C_{11}H_{12}N_2O_2$, resulting from successive treatment of 1-methoxycarbonylpyrrole with singlet oxygen then a mixture of 1-methylpyrrole and $SnCl_2$.

7. Deduce structures for the products formed at each stage by treating pyrrole successively with (i) $Me_2NH/HCHO/AcOH$, (ii) CH_3I, (iii) piperidine in hot $EtOH \rightarrow C_{10}H_{16}N_2$.

8. From a precursor which does not contain a pyrrole ring how might one synthesise (i) 1-propylpyrrole; (ii) 1-(thien-2-yl)pyrrole; (iii) 1-phenylsulfonyl-pyrrole?

9. Reaction of $MeCOCH_2CO_2Et$ with HNO_2, then a combination of Zn/AcOH and pentane-2,4-dione gave a pyrrole, $C_{11}H_{15}NO_3$. Deduce the structure of the pyrrole, write out a sequence for its formation, and suggest a route whereby it could then be converted into 2,4-dimethyl-3-ethylpyrrole.

10. How might one prepare (i) diethyl 4-methylpyrrole-2,3-dicarboxylate, (ii) ethyl 2,4,5-trimethylpyrrole-3-carboxylate; (iii) ethyl 4-amino-2-methylpyrrole-3-carboxylate; (iv) ethyl 3,4,5-trimethylpyrrole-2-carboxylate?

References

1. 'Pyrrole. From Dippel to Du Pont', Anderson, H. J., *J. Chem. Ed.*, **1995**, *72*, 875; 'The chemistry of pyrroles', Jones, R. A. and Bean, G. P., Academic Press, New York, **1977**;

'Physicochemical properties of pyrroles', Jones, R. A., *Adv. Heterocycl. Chem.*, **1970**, *11*, 383.

2. 'The pyrrole pigments', Smith, K. M. in 'Rodd's Chemistry of Carbon Compounds', **1977**, Vol. IVB, and supplement, **1997**, chs. 12.
3. 'Nature's pathways to the pigments of life', Battersby, A. R., *Nat. Prod. Rep.*, **1987**, *4*, 77.
4. 'The colours of life. An introduction to the chemistry of porphyrins and related compounds', Milgram, L. R., Oxford University Press, **1997**.
5. 'Cytochromes', Lemberg, R. and Barrett, J., Academic Press, London and New York, **1973**.
6. 'Vitamin B_{12}', Eds. Zagalak, B. and Friedrich, W., Walter de Gruyter, Berlin and New York, **1979**.
7. 'The synthesis of 3-substituted pyrroles from pyrrole', Anderson, H. J. and Loader, C. E., *Synthesis*, **1985**, 353.
8. Bean, G. P., *J. Chem. Soc., Chem. Commun.*, **1971**, 421; Muir, D. M. and Whiting, M. C., *J. Chem. Soc., Perkin Trans. 2*, **1975**, 1316.
9. Nguyen, V. Q. and Turecek, F., *J. Mass Spectrom.*, **1996**, *31*, 1173.
10. Chiang, Y., Hinman, R. L., Theodoropulos, S., and Whipple, E. B., *Tetrahedron*, **1967**, *23*, 745.
11. Gassner, R., Krumbholz, E., and Steuber, F. W., *Justus Liebigs Ann. Chem.*, **1981**, 789.
12. Findlay, S. P., *J. Org. Chem.*, **1956**, *21*, 644; Garrido, D. O., A., Buldain, G., and Frydman, B., *J. Org. Chem.*, **1984**, *49*, 2619.
13. Breukelman, S. P., Leach, S. E., Meakins, G. D., and Tirel, M. D., *J. Chem. Soc., Perkin Trans. 1*, **1984**, 2801.
14. Cooksey, A. R., Morgan, K. J., and Morrey, D. P., *Tetrahedron*, **1970**, *26*, 5101.
15. Doddi, G., Mencarelli, P., Razzini, A., and Stegel, F., *J. Org. Chem.*, **1979**, *44*, 2321.
16. Muchowski, J. M. and Naef, R., *Helv. Chim. Acta*, **1984**, *67*, 1168; Bray, B. I., Mathies, P. H., Naef, R., Solas, D. R., Tidwell, T. T., Artis, D. R., and Muchowski, J. M., *J. Org. Chem.*, **1990**, *55*, 6317.
17. Terent'ev, A. P. and Shadkhina, M. A., *Compt. rend. acad. sci. U.R.S.S.*, **1947**, *55*, 227 (*Chem. Abstr.*, **1947**, *41*, 5873).
18. Carmona, O., Greenhouse, R., Landeros, R., and Muchowski, J. M., *J. Org. Chem.*, **1980**, *45*, 5336.
19. Nair, V., George, T. G., Nair, L. G., and Panicker, S. B., *Tetrahedron Lett.*, **1999**, *40*, 1195.
20. Kakushima, M. and Frenette, R., *J. Org. Chem.*, **1984**, *49*, 2025.
21. Ortiz, C. and Greenhouse, R., *Tetrahedron Lett.*, **1985**, *26*, 2831.
22. Hamel, P., Girard, Y., and Atkinson, J. G., *J. Org. Chem.*, **1992**, *57*, 2694.
23. Treibs, A. and Kolm, H. G., *Justus Liebigs Ann. Chem.*, **1958**, *614*, 176.
24. Cordell, G. A., *J. Org. Chem.*, **1975**, *40*, 3161; Gilow, H. M. and Burton, D. E., *ibid.*, **1981**, *46*, 2221.
25. Chen, W. and Cava, M. P., *Tetrahedron Lett.*, **1987**, *28*, 6025; Chen, W., Stephenson, E. K., Cava, M. P., and Jackson, Y. A., *Org. Synth.*, **1992**, *70*, 151.
26. Martina, S., Enkelmann, V., Wegner, G., and Schlüter, A.-D., *Synthesis*, **1991**, 613.
27. Kozikowski, A. P. and Cheng, X.-M., *J. Org. Chem.*, **1984**, *49*, 3239; Shum, P. W. and Kozikowski, A. P., *Tetrahedron Lett.*, **1990**, *31*, 6785.
28. Anderson, A. G. and Exner, M. M., *J. Org. Chem.*, **1977**, *42*, 3952.
29. Reddy, G. S., *Chem. Ind.*, **1965**, 1426.
30. Harbuck, J. W. and Rapoport, H., *J. Org. Chem.*, **1972**, *37*, 3618; Bailey, D. M., Johnson, R. E., and Albertson, N. F., *Org. Synth., Coll. Vol. VI*, **1988**, 618; Chadwick, D. J., Meakins, G. D., and Rhodes, C. A., *J. Chem. Res., (S)*, **1980**, 42.
31. Bélanger, P. *Tetrahedron Lett.*, **1979**, 2505.
32. Vilsmeier, A. and Haack, A., *Chem. Ber.*, **1927**, *60*, 119.
33. 'The Vilsmeier reaction of fully conjugated carbocycles and heterocycles', Jones, G. and Stanforth, S. P., *Org. React.*, **1997**, *49*, 1.
34. Smith, G. F., *J. Chem. Soc.*, **1954**, 3842; Silverstein, R. M., Ryskiewicz, E. E., and Willard, C., *Org. Synth., Coll. Vol. IV*, **1963**, 831; de Groot, J. A., Gorter-La Roy, G. M., van Koeveringe, J. A., and Lugtenburg, J., *Org. Prep. Proc. Int.*, **1981**, *13*, 97.
35. Jugie, G., Smith, J. A. S., and Martin, G. J., *J. Chem. Soc., Perkin Trans. 2*, **1975**, 925.
36. Chadwick, D. J. and Hodgson, S. T., *J. Chem. Soc., Perkin Trans. 1*, **1983**, 93.
37. Simchen, G. and Majchrzak, M. W., *Tetrahedron Lett.*, **1985**, *26*, 5035.

38. Downie, I. M., Earle, M. J., Heaney, H., and Shuhaibar, K. F., *Tetrahedron*, **1993**, *49*, 4015.
39. Anderson, H. J., Loader, C. E., and Foster, A., *Can. J. Chem.*, **1980**, *58*, 2527.
40. Rapoport, H. and Castagnoli, N., *J. Am. Chem. Soc.*, **1962**, *84*, 2178; Oishi, T., Hirama, M., Sita, L. R., and Masamune, S., *Synthesis*, **1991**, 789.
41. Rokach, J., Hamel, P., Kakushima, M., and Smith, G. M., *Tetrahedron Lett.*, **1981**, *22*, 4901;Kakushima, M., Hamel, P., Frenette, R., and Rokach, J., *J. Org. Chem.*, **1983**, *48*, 3214; Anderson, H. J., Loader, C. E., Xu, R. X., Le, N., Gogan, N. J., McDonald, R., and Edwards, L. G., *Can. J. Chem.*, **1985**, *63*, 896; Xiao, D., Schreier, J. A., Cook, J. H., Seybold, P. G., and Ketcha, D. M., *Tetrahedron Lett.*, **1996**, *37*, 1523; Nicolaou, I. and Demopoulos, V. J., *J. Heterocycl. Chem.*, **1998**, *35*, 1345.
42. Tani, M., Ariyasu, T., Nishiysama, C., Hagiwara, H., Watanabe, T., Yokoyama, Y., and Murakami, Y., *Chem. Pharm. Bull.*, **1996**, *44*, 48.
43. Cadamuro, S., Degani, I., Dughera, S., Fochi, R., Gatti, A., and Piscopo, L., *J. Chem. Soc., Perkin Trans. 1*, **1993**, 273.
44. Hong, F., Zaidi, J., Pang, Y.-P., Cusack, B., and Richelson, E., *J. Chem. Soc., Perkin Trans. 1*, **1997**, 2997.
45. Cativiela, C and Garcia, J. I., *Org. Prep. Proc. Intl.*, **1986**, *18*, 283.
46. Hayakawa, K., Yodo, M., Ohsuki, S., and Kanematsu, K., *J. Am. Chem. Soc.*, **1984**, *106*, 6735.
47. Gregorovich, B. V., Liang, K. S. Y., Clugston, D. M., and MacDonald, S. F., *Can. J. Chem.*, **1968**, *46*, 3291.
48. Barker, P. L. and Bahia, C., *Tetrahedron*, **1990**, *46*, 2691.
49. Fischer; Schubert, *Hoppe-Seyler's Z. Physiol. Chem.*, **1926**, *155*; 88.
50. Wang, Q. M. and Bruce, D. W., *Synlett*, **1995**, 1267.
51. Taniguchi, S., Hasegawa, H., Nishimura, M., and Takahashi, M., *Synlett*, **1999**, 73.
52. Rothemund, P. and Gage, C. L., *J. Am. Chem. Soc.*, **1955**, *77*, 3340; Corwin, A. H., Chivvis, A. B., and Storm, C. B., *J. Org. Chem.*, **1964**, *29*, 3702.
53. Gonsalvez, A. M. d'A. R., Varejão, J. M. T. B., and Pereira, M. M., *J. Heterocycl. Chem.*, **1991**, *28*, 635.
54. Hanck, A. and Kutscher, W., *Z. Physiol. Chem.*, **1964**, *338*, 272.
55. Fuhlhage, D. W. and VanderWerf, C. A. *J. Am. Chem. Soc.*, **1958**, *80*, 6249.
56. 'The acid-catalysed polymerisation of pyrroles and indoles', Smith, G. F., *Adv. Heterocycl. Chem.*, **1963**, *2*, 287; Zhao, Y., Beddoes, R. L. and Joule, J. A., *J. Chem. Res., (S)*, **1997**, 42–43; (*M*), **1997**, 0401–0429.
57. Fischer, O. and Hepp, E., *Chem. Ber.*, **1886**, *19*, 2252; Butler, A. R., Pogorzelec, P., and Shepherd, P. T., *J. Chem. Soc., Perkin Trans. 2*, **1977**, 1452.
58. 'The oxidation of monocyclic pyrroles', Gardini, G. P., *Adv. Heterocycl. Chem.*, **1973**, *15*, 67.
59. Thyrann, T. and Lightner, D. A., *Tetrahedron Lett.* **1995**, *36*, 4345; *ibid.*, **1996**, *37*, 315; Moreno-Vargas, A. J., Robina, I., Fernández-Bolaños, J. G., and Fuentes, J., *ibid.*, **1998**, *39*, 9271.
60. Doddi, G., Mercarelli, P., and Stegel, F., *J. Chem. Soc., Chem. Commun.*, **1975**, 273; Di Lorenzo, A. D., Mercarelli, P., and Stegel, F., *J. Org. Chem.*, **1986**, *51*, 2125.
61. Franco, F., Greenhouse, R., and Muchowski, J. M., *J. Org. Chem.*, **1982**, *47*, 1682.
62. De Rosa, M., *J. Org. Chem.*, **1982**, *47*, 1008.
63. Grehn, L. and Ragnarsson, U., *Angew. Chem., Int. Ed. Engl.*, **1984**, *23*, 296.
64. Brittain, J. M., Jones, R. A., Arques, J. S., and Saliente, T. A., *Synth. Commun.*, **1982**, 231.
65. Hasan, I, Marinelli, E. R., Lin, L.-C. C., Fowler, F. W., and Levy, A. B., *J. Org. Chem.*, **1981**, *46*, 157; Grieb, J. G. and Ketcha, D. M., *Synth. Commun.*, **1995**, *25*, 2145.
66. A. R. Katritzky, A. R. and Akutagawa, K., *Org. Prep. Proc. Int.*, **1988**, *20*, 585.
67. Edwards, M. P., Doherty, A. M., Ley, S. V., and Organ, H. M., *Tetrahedron*, **1986**, *42*, 3723; Muchowski, J. M. and Solas, D. R., *J. Org. Chem.*, **1984**, *49*, 203.
68. Gharpure, M., Stoller, A., Bellamy, F., Firnau, G., and Snieckus, V., *Synthesis*, **1991**, 1079.
69. Chadwick, D. J., *J. Chem. Soc., Chem. Commun.*, **1974**, 790; Chadwick, D. J. and Willbe, C., *J. Chem. Soc., Perkin Trans. 1*, **1977**, 887; Chadwick, D. J., McKnight, M. V., and Ngochindo, R., *ibid.*, **1982**, 1343.

70. Groenedaal, L., M. E. Van Loo, M. E., Vekemans, J. A. J. M., and Meijer, E. W., *Synth. Commun.*, **1995**, *25*, 1589.
71. Hobbs, C. F., McMillin, C. K., Papadopoulos, E. P., and VanderWerf, C. A., *J. Am. Chem. Soc.*, **1962**, *84*, 43.
72. Heaney, H. and Ley, S. V., *J. Chem. Soc., Perkin Trans. 1*, **1973**, 499.
73. Santaniello, E., Farachi, C., and Ponti, F., *Synthesis*, **1979**, 617.
74. Candy, C. F. and Jones, R. A., *J. Org. Chem.*, **1971**, *36*, 3993.
75. Jonczyk, A. and Makosza, M., *Rocz. Chem.*, **1975**, *49*, 1203; Wang, N.-C., Teo, K.-E., and Anderson, H. J., *Can. J. Chem.*, **1977**, *55*, 4112; Hamaide, T., *Synth. Commun.*, **1990**, 2913.
76. Candy, C. F., Jones, R. A., and Wright, P. H., *J. Chem. Soc., C*, **1970**, 2563.
77. Papadopoulos, E. P. and Haidar, N. F., *Tetrahedron Lett.*, **1968**, 1721.
78. D'Silva, C. and Iqbal, R., *Synthesis*, **1996**, 457.
79. Skell, P. S. and Bean, G. P., *J. Am. Chem. Soc.*, **1962**, *84*, 4655; Bean, G. P., *J. Heterocycl. Chem.*, **1965**, *2*, 473.
80. Schloemer, G. C., Greenhouse, R., and Muchowski, J. M., *J. Org. Chem.*, **1994**, *59*, 5230.
81. Nicolau, C., Claremon, D. A., and Papahatjis, D. P., *Tetrahedron Lett.*, **1981**, *22*, 4647.
82. Fillippini, L., Gusmeroli, M., and Riva, R., *Tetrahedron Lett.*, **1992**, *33*, 1755.
83. Mann, G., Hartwig, J. F., Driver, M. S., and Fernández-Rivas, C., *J. Am. Chem. Soc.*, **1998**, *120*, 827.
84. Sotoyama, T., Hara, S., and Suzuki, A., *Bull. Chem. Soc. Jpn.*, **1979**, *52*, 1865; Marinelli, E. R. and Levy, A. B., *Tetrahedron Lett.*, **1979**, 2313; Minato, A., Tamao, K., Hayashi, T., Suzuki, K., and Kumada, M., *Tetrahedron Lett.*, **1981**, *22*, 5319.
85. Alvarez, A., Guzmán, A., Ruiz, A., Velarde, E., and Muchowski, J. M., *J. Org. Chem.*, **1992**, *57*, 1653.
86. Rapoport, H. and Look, M., *J. Am. Chem. Soc.*, **1953**, *75*, 4605; Sacki, S., Hayashi, T., and Hamana, M., *Chem. Pharm. Bull.*, **1984**, *32*, 2154; Jones, K., Ho, T. C. T., and Wilkinson, J., *Tetrahedron Lett.*, **1995**, *36*, 6734.
87. Aiura, M. and Kanaoka, Y., *Chem. Pharm. Bull.*, **1975**, *23*, 2835.
88. Aldabbagh, F., Bowman, W. R., and Mann, E., *Tetrahedron Lett.*, **1997**, *38*, 7937.
89. Aboutayab, K., Cadick, S., Jenkins, K., Joshi, S., and Khan, S., *Tetrahedron*, **1996**, *52*, 11329.
90. Hudson, C. B. and Robertson, A. V., *Tetrahedron Lett.*, **1967**, 4015; Ketcha, D. M., Carpenter, K. P., and Zhou, Q., *J. Org. Chem.*, **1991**, *56*, 1318.
91. Andrews, L. H. and McElvain, S. M., *J. Am. Chem. Soc.*, **1929**, *51*, 887; Kaiser, H.-P. and Muchowski, J. M., *J. Org. Chem.*, **1984**, *49*, 4203; Jefford, C. W., Sienkiewicz, K., and Thornton, S. R., *Helv. Chim. Acta*, **1995**, *78*, 1511; Hext, N. M., Hansen, J., Blake, A. J., Hibbs, D. E., Hursthouse, M. B., Shishkin, O. V., and Mascal, M., *J. Org. Chem.*, **1998**, *63*, 6016.
92. Donohoe, T. J. and Guyo, P. M., *J. Org. Chem.*, **1996**, *61*, 7664; Donohoe, T. J., Guyo, P. M., Beddoes, R. L., and Helliwell, M., *J. Chem. Soc., Perkin Trans. 1*, **1998**, 667; Donohoe, T. J., Guyo, P. M., Helliwell, M., *Tetrahedron Lett.*, **1999**, *40*, 435.
93. Wittig, G. and Reichel, B., *Chem. Ber.*, **1963**, *96*, 2851.
94. 'Chemistry of 7-azabicyclo[2.2.1]hepta-2,5-dienes, 7-azabicyclo[2.2.1]hept-2-enes and 7-azabicyclo[2.2.1]heptanes', Chen, Z. and Trudell, M. L., *Chem. Rev.*, **1996**, *96*, 1179.
95. Anderson, P. S., Christy, M. E., Lundell, G. F., and Ponticello, G. S., *Tetrahedron Lett.*, **1975**, 2553.
96. Kotsuki, H., Mori, Y., Nishizawa, H., Ochi, M., and Matsuoka, K., *Heterocycles*, **1982**, *19*, 1915.
97. Kitzing, R., Fuchs, R., Joyeux, M., and Prinzbach, H., *Helv. Chim. Acta*, **1968**, *51*, 888.
98. Bansal, R. C., McCulloch, A. W., and McInnes, A. G., *Can. J. Chem.*, **1969**, *47*, 2391; *ibid.* **1970**, *48*, 1472.
99. Zhang, C. and Trudell, M. L., *J. Org. Chem.*, **1996**, *61*, 7189; Pavri, N. P. and Trudell, M. L., *Tetrahedron Lett.*, **1997**, *38*, 7993.
100. Hodges, L. M. and Harman, W. D., *Adv. Nitrogen Heterocycles*, **1998**, *3*, 1.
101. Gonzalez, J., Koontz, J. I., Hodges, L. M., Nilsson, K. R., Neely, L. K., Myers, W. H., Sabat, M., and Harman, W. D., *J. Am. Chem. Soc.*, **1995**, *117*, 3405.
102. Lightner, D. A., Bisacchi, G. S., and Norris, R. D., *J. Am. Chem. Soc.*, **1976**, *98*, 802; Natsume, M. and Muratake, H., *Tetrahedron Lett.*, **1979**, 3477.

103. 'Cycloaddition reactions with vinyl heterocycles', Sepúlveda-Arques, J., Abarca-González, B., and Medio-Simón, *Adv. Heterocycl. Chem.*, **1995**, *63*, 339.
104. Xiao, D. and Ketcha, D. M., *J. Heterocycl. Chem.*, **1995**, *32*, 499.
105. Li, J.-H. and Snyder, J. K., *J. Org. Chem.*, **1993**, *58*, 516.
106. Jones, R. C. and Rees, C. W., *J. Chem. Soc. (C)*, **1969**, 2249; Gambacorta, Nicoletti, R., Cerrini, S., Fedeli, W., and Gavuzzo, E., *Tetrahedron Lett.*, **1978**, 2439.
107. de Angelis, F., Gambacorta, A., and Nicoletti, R., *Synthesis, 1976*, 798.
108. Fowler, F. W., *Angew. Chem., Int. Ed. Engl.*, **1971**, *10*, 135; *idem, J. Chem. Soc., Chem. Commun.*, **1969**, 1359.
109. Voigt, J., Noltemeyer, M., and Reiser, O., *Synlett*, **1997**, 202.
110. Maryanoff, B. E., *J. Org. Chem.*, **1979**, *44*, 4410; Pomeranz, M. and Rooney, P., *J. Org. Chem.*, **1988**, *53*, 4374.
111. H. M. L. Davies, E. Saikali, and W. B. Young, *J. Org. Chem.*, **1991**, *56*, 5696.
112. 'Photochemical reactions involving pyrroles, Parts I and II', D'Auria, M., *Heterocycles*, **1996**, *43*, 1305 amp; 1529.
113. Hiraoka, H., *J. Chem. Soc., Chem. Commun.*, **1970**, 1306; Barltrop, J. A., Day, A. C., and Ward, R. W., *ibid.*, **1978**, 131.
114. Herz, W. and Tocker, S., *J. Am. Chem. Soc.*, **1955**, *77*, 6353; Herz, W., Dittmer, K., and Cristol, S. J., *ibid., 1948*, *70*, 504.
115. Greenhouse, R., Ramirez, C., and Muchowski, J. M., *J. Org. Chem.*, **1985**, *50*, 2961; Schumacher, D. P. andHall, S. S., *ibid.*, **1981**, *46*, 5060.
116. Ketcha, D. M., Carpenter, K. P., Atkinson, S. T., and Rajagopalan, H. R., *Synth. Commun.*, **1990**, *20*, 1647.
117. Carson, J. R. and Davis, N. M., *J. Org. Chem.*, **1981**, *46*, 839.
118. For example Badger, G. M., Harris, R. L. N., and Jones, R. A., *Aust. J. Chem.*, **1964**, *17*, 1022 and Lancaster, R. E. and VanderWerf, C. A., *J. Org. Chem.*, **1958**, *23*, 1208.
119. Dunn, G. E. and Lee, G. K. J., *Can. J. Chem.*, **1971**, *49*, 1032.
120. Joh, Y., *Seikagaku*, **1961**, *33*, 787 (*Chem. Abs.*, **1962**, *57*, 11144).
121. Anderson, H. J. and Lee, S.-F., *Can. J. Chem.*, **1965**, *43*, 409.
122. Norris, R. D. and Lightner, D. A., *J. Heterocycl. Chem.*, **1979**, *16*, 263.
123. Thyrann, T. and Lightner, D. A., *Tetrahedron Lett.*, **1996**, *37*, 315.
124. Khan, M. K. A. and Morgan, K. J., *Tetrahedron*, **1965**, *21*, 2197; Williams, A. and Salvadori, G., *J. Chem. Soc., Perkin Trans. 2*, **1972**, 883.
125. Muratake, H., Tonegawa, M., and Natsume, M., *Chem. Pharm. Bull.*, **1996**, *44*, 1631.
126. Bocchi, V., Chierichi, L., Gardini, G. P., and Mondelli, R., *Tetrahedron*, **1970**, *26*, 4073.
127. Casiraghi, G., Rassu, G., Spanu, P., and Pinna, L., *J. Org. Chem.*, **1992**, *57*, 3760; *idem, Tetrahedron Lett.*, **1994**, *35*, 2423; 'Furan-, pyrrole-, and thiophene-based siloxydienes for synthesis of densely functionalised homochiral compounds', Casiraghi, G. and Rassu, G., *Synthesis*, **1995**, 607.
128. Momose, T., Tanaka, T., Yokota, T., Nagamoto, N., and Yamada, K., *Chem. Pharm. Bull.*, **1979**, *27*, 1448.
129. 'Synthesis of amino derivatives of five-membered heterocycles by Thorpe-Ziegler cyclisation', Granik, V. G., Kadushkin, A. V., and Liebscher, J., *Adv. Heterocycl. Chem.*, **1998**, *72*, 79.
130. De Rosa, M., Issac, R. P., Houghton, G., *Tetrahedron Lett.*, **1995**, *36*, 9261.
131. 'Recent synthetic methods for pyrroles and pyrrolenines (2*H*- or 3*H*-pyrroles)', Patterson, J. M., *Synthesis*, **1976**, 281.
132. Bishop, W. S., *J. Am. Chem. Soc.*, **1945**, *67*, 2261.
133. Knorr, L., *Chem. Ber.*, **1884**, *17*, 1635.
134. For one of several methods for the synthesis of 1,4-dicarbonyl compounds see: Wedler, C. and Schick, H., *Synthesis*, **1992**, 543.
135. Rousseau, B., Nydegger, E., Gossauer, A., Bennau-Skalmowski, B., and Vorbrüggen, H., *Synthesis*, **1996**, 1336.
136. Amarnath, V., Anthony, D. C., Amarnath, K., Valentine, W. M., Wetteran, L. A., and Graham, D. G., *J. Org. Chem.*, **1991**, *56*, 6924.
137. Young, D. M. and Allen, C. F. H., *Org. Synth., Coll. Vol. II*, **1943**, 219.
138. Elming, N. and Clauson-Kaas, N., *Acta Chem. Scand.*, **1952**, *6*, 867.
139. Lee, S. D., Brook, M. A., and Chan, T. H., *Tetrahedron Lett.*, **1983**, 1569; Chan T. H. and Lee, S. D., *J. Org. Chem.*, **1983**, *48*, 3059.
140. McLeod, M., Boudreault, N., and Leblanc, Y., *J. Org. Chem.*, **1996**, *61*, 1180.

141. Josey, A. D., *Org. Synth., Coll. Vol. V*, **1973**, 716; Jefford, C. W., Thornton, S. R., and Sienkiewicz, K., *Tetrahedron Lett.* **1994**, *35*, 3905; Fang, V., Leysend, D., and Ottenheijm, H. C. J., *Synth. Commun.*, **1995**, *25*, 1857.
142. McElvain, S. M. and Bolliger, K. M., *Org. Synth., Coll. Vol. I*, **1932**, 473.
143. Okada, E., Masuda, R., Hojo, M., and Yoshida, R., *Heterocycles*, **1992**, *34*, 1435.
144. Treibs, A., Schmidt, R., and Zinsmeister, R., *Chem. Ber.*, **1957**, *90*, 79.
145. Fischer, H., *Org. Synth., Coll. Vol. II*, **1943**, 202.
146. Quizon-Colquitt, D. M. and Lash, T. D., *J. Heterocycl. Chem.*, **1993**, *30*, 477.
147. Yinglin, H. and Hongwer, H., *Synthesis,*, **1990**, 615.
148. Hamby, J. M. and Hodges, J. C., *Heterocycles*, **1993**, *35*, 843.
149. Konieczny, M. T. and Cushman, M., *Tetrahedron Lett.*, **1992**, *33*, 6939.
150. Terang, N., Mehta, B. K., Ha, H., and Junjappa, H., *Tetrahedron*, **1998**, *54*, 12973.
151. Kolar, P. and Tisler, M., *Synth. Commun.*, **1994**, *24*, 1887.
152. Roomi M. W. and MacDonald, S. F., *Can. J. Chem.*, **1970**, *48*, 1689.
153. Katritzky, A. R., Cheng, D., and Musgrave, R. P., *Heterocycles*, **1997**, *44*, 67.
154. Hoppe, D., *Angew. Chem., Int. Ed. Engl.*, **1974**, *13*, 789; van Leusen, A. M., Siderius, H., Hoogenboom, B. E., and van Leusen, D., *Tetrahedron Lett.*, **1972**, 5337; Possel O. and van Leusen, A. M., *Heterocycles*, **1977**, *7*, 77; Parvi, N. P. and Trudell, M. L., *J. Org. Chem.*, **1997**, *62*, 2649; for a related process see Houwing H. A. and van Leusen, A. M., *J. Heterocycl. Chem.*, **1981**, *18*, 1127.
155. Ono, T., Muratani, E., and Ogawa, T., *J. Heterocycl. Chem.*, **1991**, *28*, 2053.
156. Barton, D. H. R., Kervagoret, J., Zard, S. Z., *Tetrahedron*, **1990**, *46*, 7587.
157. Boëlle, J., Schneider, R., Gérardin, P., and Loubinoux, B., *Synthesis*, **1997**, 1451.
158. Lash, T. D., Belletini, J. R., Bastian, J. A., and Couch, K. B., *Synthesis*, **1994**, 170.
159. Ono, N., Hironaga, H., Ono, K., Kaneko, S., Murashima, T., Ueda, T., Tsukamura, C., and Ogawa, T., *J. Chem. Soc., Perkin Trans. 1*, **1996**, 417.
160. Abel, Y., Haake, E., Haake, G., Schmidt, W., Struve, D., Walter, A., and Montforts, F.-P., *Helv. Chim. Acta*, **1998**, *81*, 1978.
161. Mataka, S., Takahashi, K., Tsuda, Y., and Tashiro, M., *Synthesis*, **1982**, 157; Walizei G. H. and Breitmaier, E., *ibid.*, **1989**, 337; Hombrecher H. K. and Horter, G., *ibid.*, **1990**, 389.
162. Gupton, J. T., Petrich, S. A., Hicks, F. A., Wilkinson, D. R., Vargas, M., Hosein, K. N., and Sikorski, J. A., *Heterocycles*, **1998**, *47*, 689.
163. Terry, W. G., Jackson, A. H., Kenner, G. W., and Kornis, G., *J. Chem. Soc.*, **1965**, 4389.
164. Lash, T. D. and Hoehner, M. C., *J. Heterocycl. Chem.*, **1991**, *28*, 1671.
165. Bayer, H. O., Gotthard, H., and Huisgen, R., *Chem. Ber.*, **1970**, *103*, 2356; Huisgen, R., Gotthard, H., Bayer, H. O., and Schafer, F. C., *ibid.*, 2611; Padwa, A., Burgess, E. M., Gingrich, H. L. and Roush, D. M., *J. Org. Chem.*, **1982**, *47*, 786.
166. Enders, D., Maassen, R., and Han, S.-H., *Liebigs Ann.*, **1996**, 1565.
167. Shiraishi, H., Nishitani, T., Sakaguchi, S., and Ishii, Y., *J. Org. Chem.*, **1998**, *63*, 6234.
168. Knight, D. W., Redfern, A. L., and Gilmore, J., *Chem. Commun.*, **1998**, 2207.
169. Bonnand, B. and Bigg, D. C. H., *Synthesis*, **1994**, 465.
170. Ogawa, H., Aoyama, T., and Shioiri, T., *Heterocycles*, **1996**, *42*, 75.
171. Frydman, B., Reil, S., Despuy, M. E., and Rapoport, H., *J. Am. Chem. Soc.*, **1969**, *91*, 2338.
172. Paine, J. B., Kirshner, W. B., Maskowitz, D. W., and Dolphin, D., *J. Org. Chem.*, **1976**, *41*, 3857; Wang, C.-B. and Chang, C. K., *Synthesis*, **1979**, 548.
173. Sessler, J. L., Mozaffari, A., and Johnson, M. R., *Org. Synth.*, **1992**, *70*, 68.
174. 'Novel porphyrinoid macrocycles and their metal complexes', E. Vogel, *J. Heterocycl. Chem.*, **1996**, *33*, 1461.
175. 'Expanded, contracted, and isomeric porphyrins', Sessler, J. L. and Weghorn, S. J., Pergamon, Oxford, **1997**.
176. G. P. Arsenault, E. Bullock, and S. F. MacDonald, *J. Am. Chem. Soc.*, **1960**, *82*, 4384.
177. E. Vogel, M. Bröring, S. J. Weghorn, P. Scholz, R. Deponte, J. Lex, H. Schmickler, K. Schaffner, S. E. Braslavsky, M. Müller, S. Pörting, C. J. Fowler, and J. C. Sessler, *Angew. Chem., Int. Ed. Engl.*, **1997**, *36*, 1651.
178. 'CC-1065 and the duocarmicins: synthetic studies', Boger, D. C., Boyce, C. W., Garbaccio, R. M., and Goldberg, J. A., *Chem. Rev.*, **1997**, *97*, 787.
179. 'The chemistry, mode of action and biological properties of CC 1065', Reynolds, V. L., McGovern, J. P., and Hurley, L. H., *J. Antiobiotics*, **1986**, *39*, 319.

180. Carter, P., Fitzjohn, S., Halazy, S., and Magnus, P., *J. Am. Chem., Soc.*, **1987**, *109*, 2711.
181. Giblin, G. M. P., Jones, C. D., and Simpkins, N. S., *J., Chem. Soc., Perkin Trans. 1*, **1998**, 3689.

14 Thiophenes: reactions and synthesis

thiophene

The simple thiophenes[1] are stable liquids which closely resemble the corresponding benzene compounds in boiling points and even in smell. They occur in coal tar distillates – the discovery of thiophene in coal tar benzene provides one of the classic anecdotes of organic chemistry. In the early days, colour reactions were of great value in diagnosis: an important one for benzene involved the production of a blue colour on heating with isatin (section 14.1.1.7) and concentrated sulfuric acid. In 1882, during a lecture-demonstration by Viktor Meyer before an undergraduate audience, this test failed, no doubt to the delight of everybody except the professor, and especially except the professor's lecture assistant. An inquiry revealed that the lecture assistant had run out of commercial benzene and had provided a sample of benzene which he had prepared by decarboxylation of pure benzoic acid. It was thus clear that commercial benzene contained an impurity and that it was this, not benzene, which was responsible for the colour reaction. In subsequent investigations, Meyer isolated the impurity *via* its sulfonic acid derivative and showed it to be the first representative of a then new ring system, which was named thiophene from *theion*, the Greek word for sulfur, and another Greek word *phaino* which means shining, a root first used in phenic acid (phenol) because of its occurrence in coal tar, a by-product of the manufacture of illuminating gas.

Aromatic thiophenes play no part in animal metabolism; biotin, one of the vitamins, is a tetrahydrothiophene, however aromatic thiophenes do occur in some plants, in association with polyacetylenes with which they are biogenetically linked. Banminth (Pyrantel), a valuable anthelminth used in animal husbandry, is one of the few thiophene compounds in chemotherapy.

14.1 Reactions with electrophilic reagents

14.1.1 Substitution at carbon

14.1.1.1 Protonation

Thiophene is stable to all but very strongly acidic conditions so many reagent combinations which lead to acid-catalysed decomposition or polymerisation of furans and pyrroles, can be applied successfully to thiophenes.

Measurements of acid-catalysed exchange, or of protonolysis of other groups, for example silicon,[2] or mercury,[3] show the rate of proton attack at C-2 to be about 1000 times faster than at C-3.[4] The pK_a for 2,5-di-t-butylthiophene forming a salt by protonation at C-2, is -10.2.[5]

Reactions of protonated thiophenes

The action of hot phosphoric acid on thiophene leads to a trimer;[6] its structure suggests that, in contrast with pyrrole (section 13.1.8), the electrophile involved in the first C–C bonding step is the α-protonated cation.

14.1.1.2 Nitration

Nitration of thiophene needs to be conducted in the absence of nitrous acid which can lead to an explosive reaction;[7] the use of acetyl nitrate[8] or nitronium tetrafluoroborate[9] are satisfactory. Invariably the major 2-nitro-product is accompanied by approximately 10% of the 3-isomer.[10] Further nitration of either 2- or 3-nitrothiophenes[11] also leads to mixtures – equal amounts of 2,4- and 2,5-dinitrothiophenes from the 2-isomer, and mainly the former from 3-nitrothiophene.[12] Similar, predictable isomer mixtures are produced in other nitrations of substituted thiophenes, for example 2-methylthiophene gives rise to 2-methyl-5- and 2-methyl-3-nitrothiophenes,[13] and 3-methylthiophene gives 4-methyl-2-nitro- and 3-methyl-2-nitrothiophenes,[14] in each case in ratios of 4:1.

14.1.1.3 Sulfonation

As discussed in the introduction, the production of thiophene-2-sulfonic acid by sulfuric acid sulfonation of the heterocycle has been long known;[15] use of the pyridine-sulfur trioxide complex is probably the best method.[16] 2-Chlorosulfonation[17] and 2-thiocyanation[18] are similarly efficient.

14.1.1.4 Halogenation

Halogenation of thiophene occurs very readily at room temperature and is rapid even at $-30\,°C$ in the dark; tetrasubstitution occurs easily.[19] The rate of halogenation of thiophene, at 25 °C, is about 10^8 times that of benzene.[20] Both 2,5-dibromo- and 2,5-dichlorothiophenes[21] and 2-bromo-[22] and 2-iodothiophene[23] can be produced cleanly under various controlled conditions. Controlled bromination of 3-bromothiophene produces 2,3-dibromothiophene.[24]

2,3,5-Tribromination of thiophene goes smoothly in 48% hydrobromic acid solution.[25] Since it has long been known that treatment of polyhalogenothiophenes with zinc and acid brings about selective removal of α-halogen, this compound can be used to access 3-bromothiophene[26] just as 3,4-dibromothiophene can be obtained by reduction of the tetrabromide.[27] One interpretation of the selective reductive removal is that it involves first, electron transfer to the bromine, then subsequently, transient 'anions', thus halogen can be selectively removed from that position where such an anion is best stabilised – normally an α position (section 14.4.1). The use of sodium borohydride, respectively with and without palladium(0) catalysis, converts 2,3,5-tribromothiophene into 2,3-dibromothiophene and 2,4-dibromothiophene.[28]

Monoiodination of 2-substituted thiophenes, whether the substituent is activating or deactivating, proceeds efficiently at the remaining α-position using iodine with iodobenzene diacetate.[29] 3-Alkylthiophenes can be monobrominated or monoiodinated at C-2 using N-bromosuccinimide[30] or iodine with mercury(II) oxide,[31] respectively.

14.1.1.5 Acylation

The Friedel-Crafts acylation of thiophenes is a much-used reaction and generally gives good yields under controlled conditions, despite the fact that aluminium chloride reacts with thiophene to generate tars: this problem can be avoided by adding catalyst to the thiophene and the acylating agent;[32] tin tetrachloride has been used most frequently. Acylation with anhydrides, catalysed by phosphoric acid[33] is an efficient method. Reaction with acetyl p-toluenesulfonate, in the absence of any catalyst produces 2-acetylthiophene in high yield.[34] Vilsmeier formylation of thiophene leads efficiently to 2-formylthiophene;[35] 2-formylation results when 3-phenylthiophene is subjected to Vilsmeier conditions.[36]

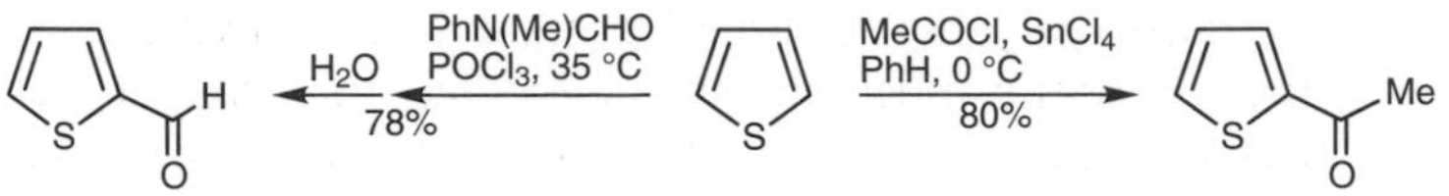

In acylations, almost exclusive α-substitution is observed, but where both α-positions are substituted, β-substitution occurs easily. This is nicely illustrated by the synthesis of the isomeric bicyclic ketones shown below.[37]

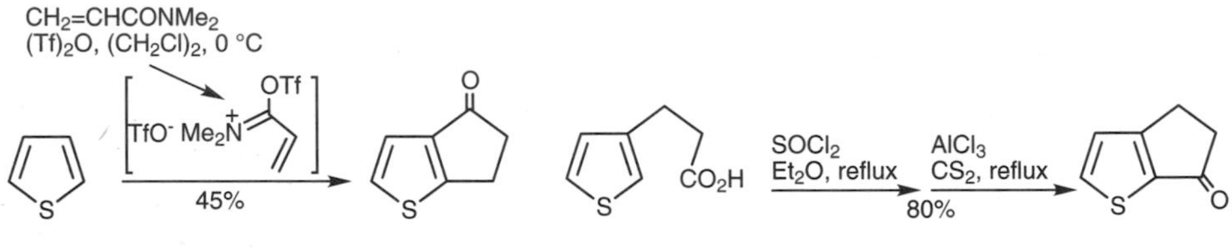

14.1.1.6   Alkylation

Alkylation occurs readily, but is rarely of preparative use; the efficient 2,5-bis-t-butylation of thiophene is one such example.[38]

14.1.1.7   Condensation with aldehydes and ketones

Acid-catalysed reaction of thiophene with aldehydes and ketones is not a viable route to hydroxyalkylthiophenes, for these are unstable under the reaction conditions. Chloroalkylation can however be achieved[39] and with the use of zinc chloride, even thiophenes carrying electron-withdrawing groups react.[40] Care is needed in choosing conditions; there is a tendency for formation of either di-2-thienylmethanes[41] or 2,5-bis(chloromethyl)thiophene.[42]

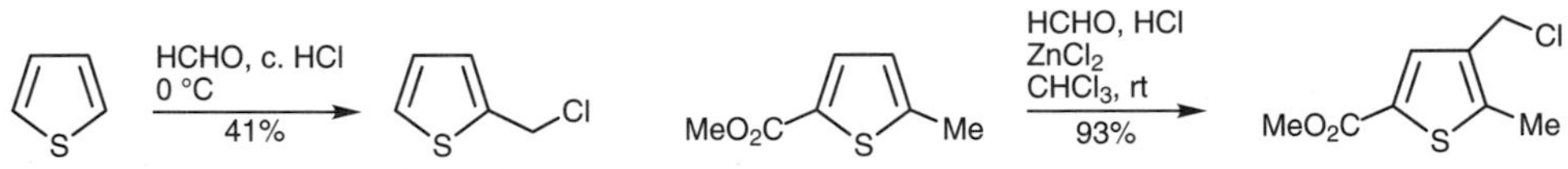

A reaction of special historical interest, mentioned in the introduction to this chapter, is the condensation of thiophene with isatin in concentrated sulfuric acid, to give the deep blue indophenine[43] as a mixture of geometrical isomers.[44]

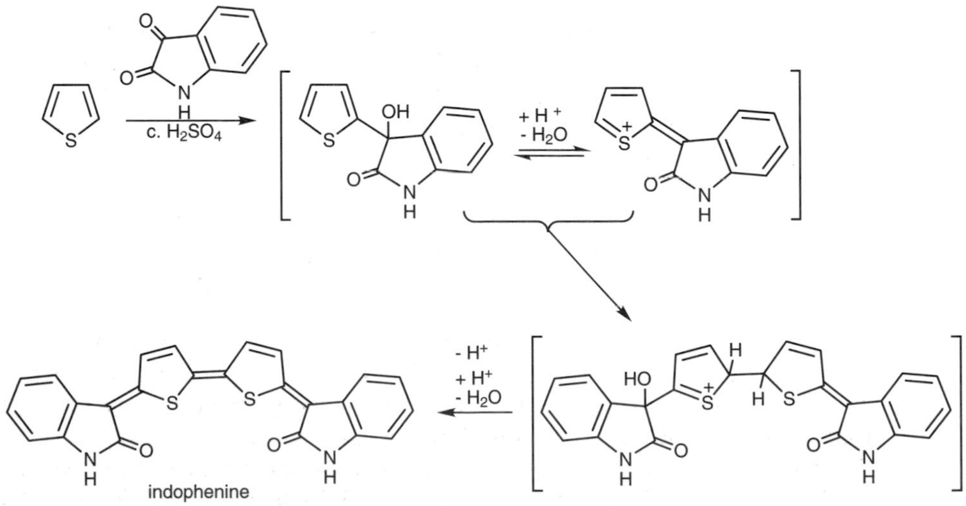

indophenine

Hydroxyalkylation at the 5-position of 2-formylthiophene results from exposure of the thiophene aldehyde and a second aldehyde, to samarium(II) iodide; in the example shown below the other aldehyde is 1-methylpyrrole-2-aldehyde.[45]

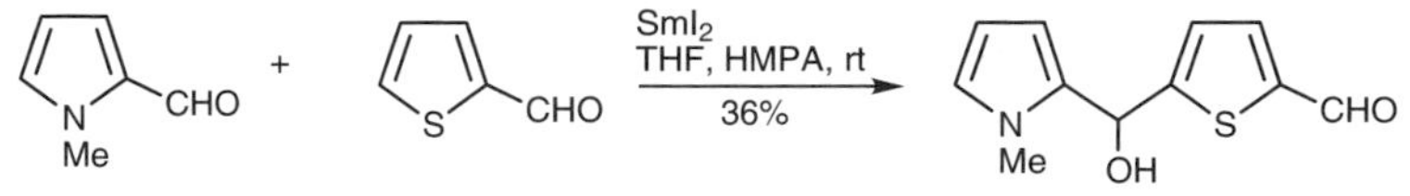

14.1.1.8 Condensation with imines and iminium ions

Aminomethylation of thiophene[46] was reported long before the more common Mannich reaction – dimethylaminomethylation, which, although it can be achieved under routine conditions with methoxythiophenes,[47] requires the use of $Me_2N^+ = CH_2\ Cl^-$ ('Eschenmoser's salt' is the iodide) for thiophene and alkylthiophenes.[48]

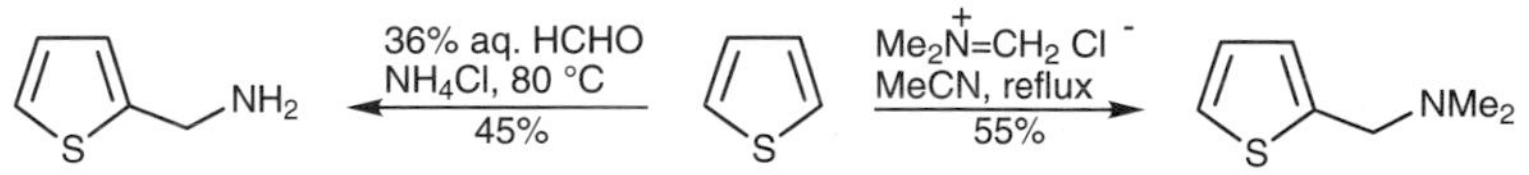

Another device for bringing thiophenes into reaction with Mannich intermediates is to utilise thiophene boronic acids, as illustrated below; primary aromatic amines can be used as the amine component.[49]

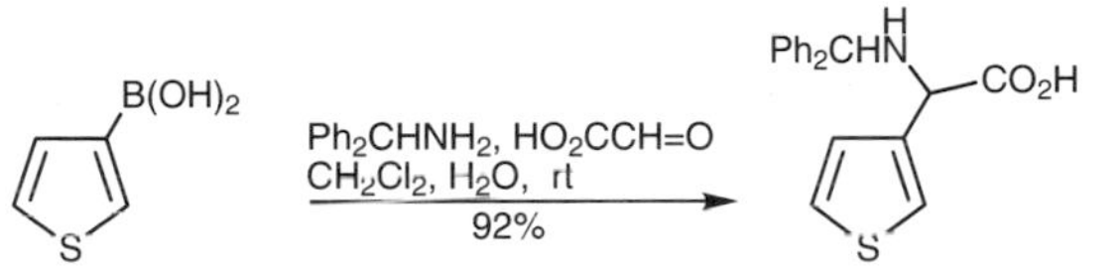

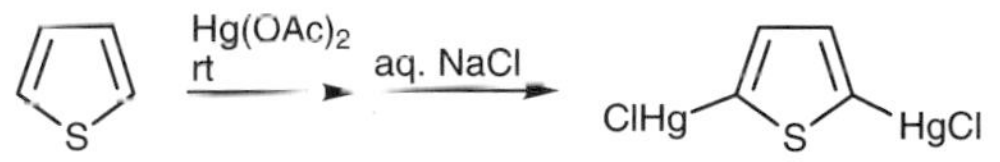

14.1.1.9 Mercuration

Mercuration of thiophenes occurs with great ease; mercuric acetate is more reactive than the chloride;[50] tetrasubstitution and easy replacement of the metal with halogen can also be achieved straightforwardly.[51]

14.1.2 Addition at sulfur

In reactions not possible with the second row element-containing pyrrole and furan, thiophene sulfur can add electrophilic species. Thiophenium salts[52] though not formed efficiently from thiophene itself, are produced in high yields with polyalkyl-substituted thiophenes.[53] The sulfur in such salts is probably tetrahedral,[54] i.e. the sulfur is sp[3] hybridised.

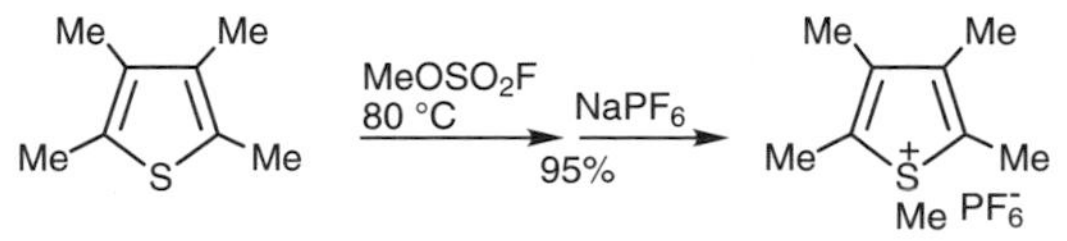

Even thiophene itself will react with carbenes, at sulfur, to produce isolable thiophenium ylides, and in these, the sulfur is definitely tetrahedral.[55] The rearrangement[56] of thiophenium bis(methoxycarbonyl)methylide to the 2-substituted thiophene provides a rationalisation for the reaction of thiophene with ethyl diazoacetate[57] which produces what appears to be the product of carbene addition to the 2,3-double bond; perhaps this goes *via* initial attack at sulfur followed by S→C-2

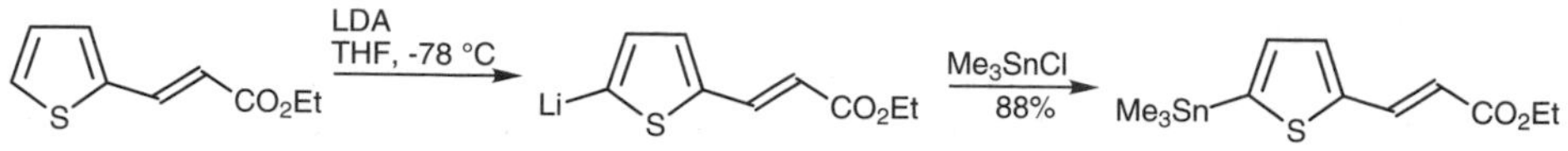

The formation of arynes has often been achieved by base-induced dehydrohalo-genation but for the formation of 3,4-didehydrothiophene a fluoride-induced process can be used, following *ipso* electrophilic displacement of one of the silicons to generate the appropriate precursor.[81]

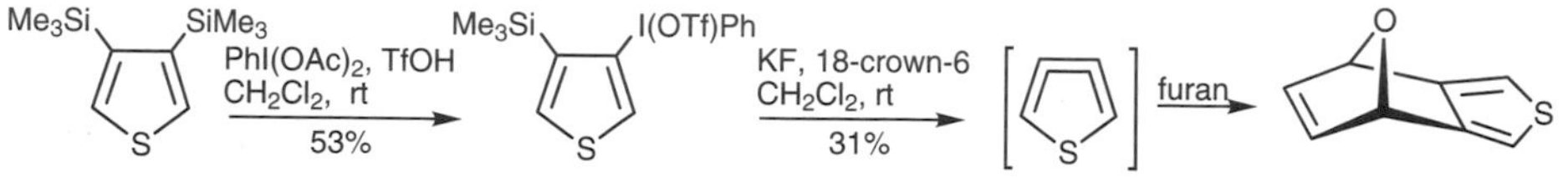

14.5   Reactions of C-metallated thiophenes

14.5.1   Lithium and magnesium derivatives

2-Bromo- and 2-iodothiophenes readily form thienyl Grignard reagents[82] though 3-iodothiophene requires the use of Rieke magnesium.[73] Bromine and iodine at either α- or β-positions undergo exchange with alkyllithiums giving lithiated thiophenes. The reaction of 2,3-dibromothiophene with *n*-butyllithium produces 3-bromothien-2-yllithium.[83]

The use of thienyl Grignard reagents, and more recently lithiated thiophenes, has been extensive and can be illustrated by citing formation of oxythiophenes, either by reaction of the former with *t*-butyl perbenzoate[84] or the latter directly with bis(trimethylsilyl) peroxide[85] or *via* the boronic acid,[86] the synthesis of thiophene carboxylic acids by reaction of the organometallic with carbon dioxide,[87] the synthesis of ketones, by reaction with a nitrile,[88] or alcohols by reaction with aldehydes,[73] by the reaction of 2-lithiothiophene with *N*-tosylaziridine,[89] and by syntheses of thieno[3,2-*b*]thiophene[90] and of dithieno[3,2-*b*:2',3'-*d*]thiophene.[91] Some of these are illustrated below.

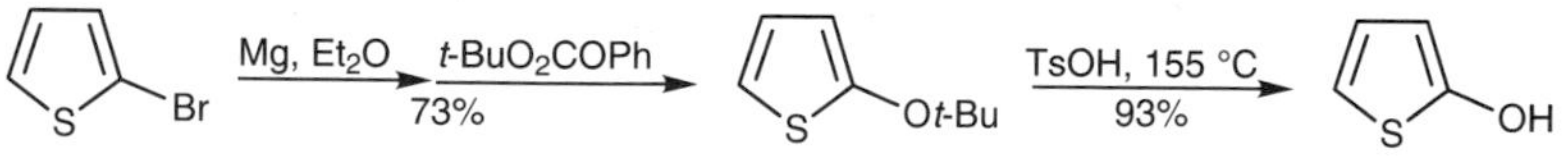

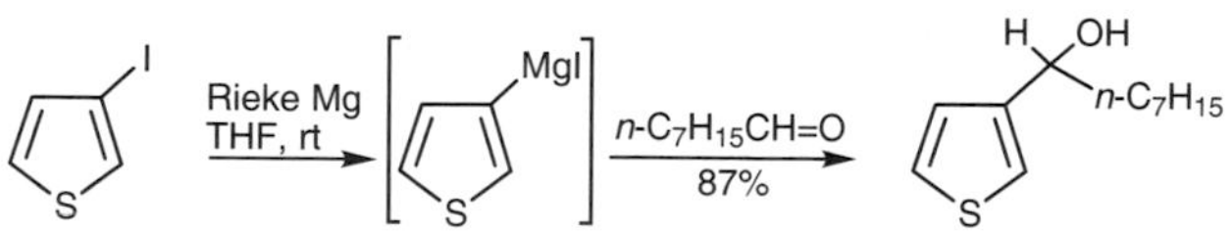

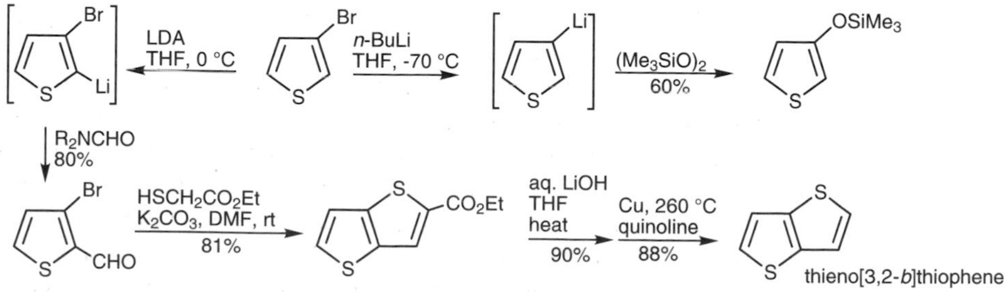

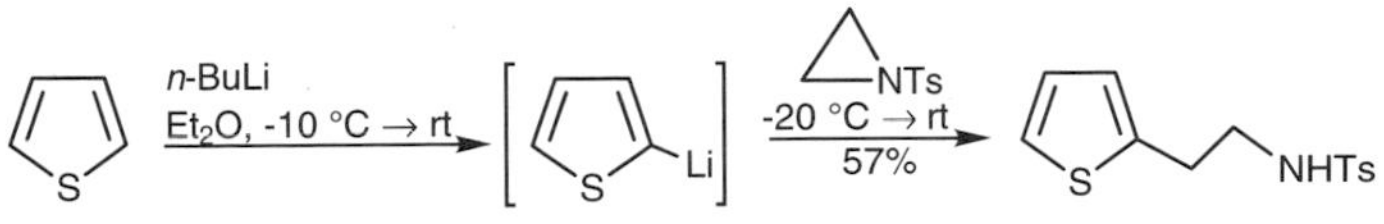

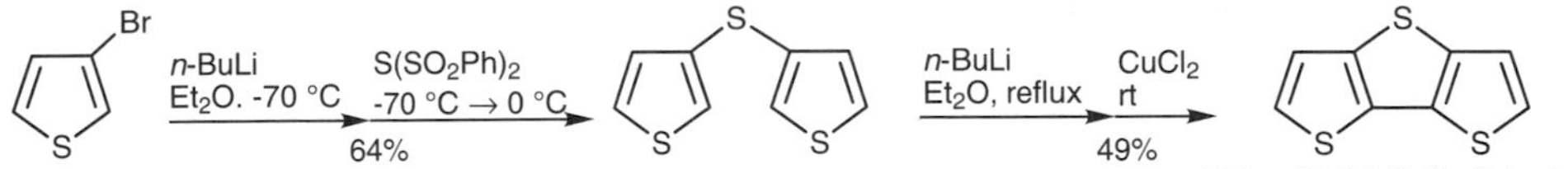

There are two complications which can arise in the formation and the use of lithiated thiophenes: the occurence of a 'Base Catalysed Halogen Dance',[92] and the isomerisation or ring opening of 3-lithiated thiophenes. As an example of the first of these, and one in which the phenomenon is put to good use, consider the transformation of 2-bromothiophene into 3-bromothiophene by reaction with sodamide in ammonia.[93] The final result is governed, in a set of equilibrations, by the stability of the final anion: the system settles to an anion in which the charge is both adjacent to halogen and at an α-position.

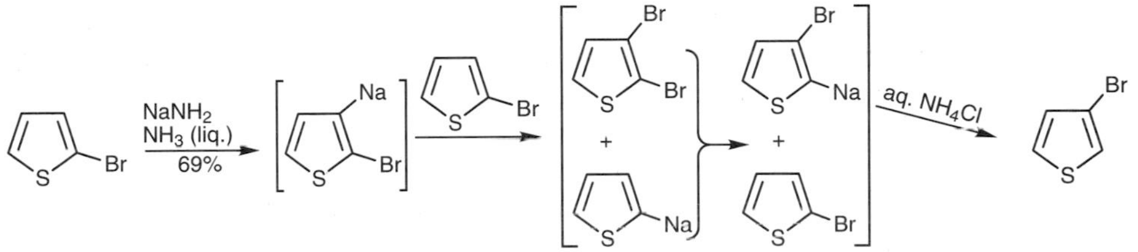

3-Lithiothiophene can be utilised straightforwardly at low temperature but if the temperature is increased, ring-opening can occur. The ring opening can be used to advantage in the synthesis of Z-enynes by trapping with an alkyl halide, as illustrated below.[94]

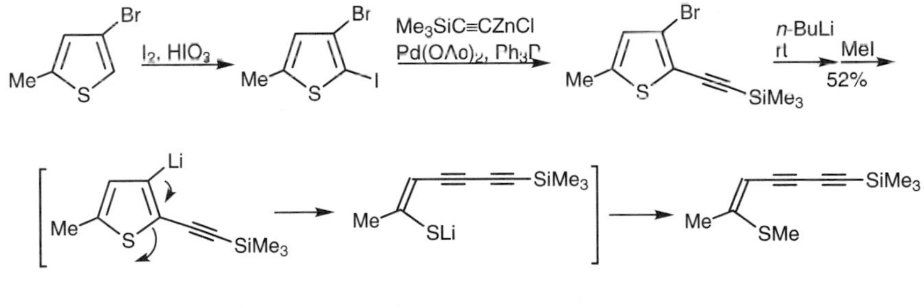

14.5.2 Palladium-catalysed reactions (also nickel- and copper-catalysed reactions)

There are by now many examples involving palladium(0)-catalysed couplings using halothiophenes, thiophene boronic acids, thienylstannanes, and thienylzinc reagents.[95] In substrates where there is halogen at both thiophene α- and β-positions, the former enters into reaction. Representative examples, some of which are illustrated below, include the production of arylthiophenes using thiophene-2- and -3-boronic acids[96] and thiophene stannanes,[24] and directly from the heterocycle with an aryl halide leading to 2-substitution;[97] the formation of thienyl-alkynes,[81,98] thienyl-alkenes from thienyl halides and from thienyl stannanes,[99] the formation of fused ring systems from bromo-amino-thiophenes or formylthiophene boronic

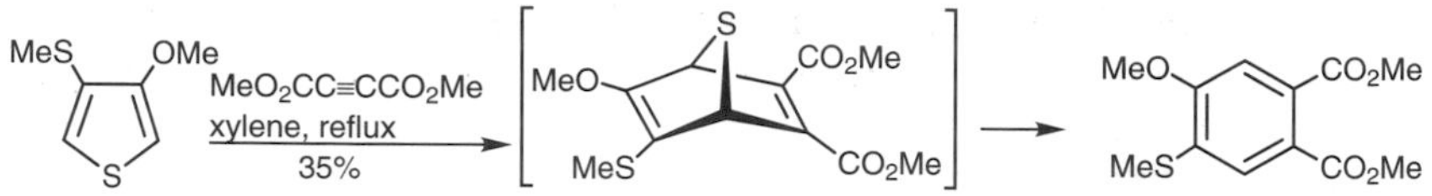

The strong tendency for thiophene-*S,S*-dioxides to undergo cycloaddition processes (section 14.1.2) is echoed, to a lesser degree by thiophene *S*-oxides. Thus, when thiophenes are oxidised with *meta*-chloroperbenzoic acid and boron trifluoride (without which *S,S*-dioxides are formed), in the presence of a dienophile, adducts from 2 + 4 addition can be isolated.[119] Thiophenes 2,5- or 3,4-disubstituted with bulky groups can be converted into isolable *S*-oxides[120] which undergo cycloadditions as shown below.[121]

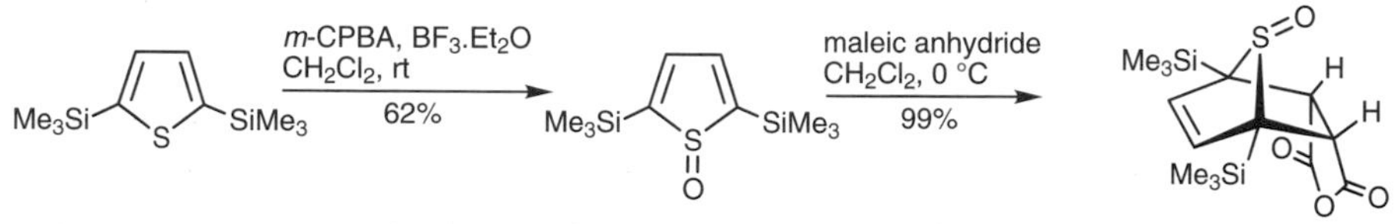

14.9   Photochemical reactions

The classic photochemical reaction involving thiophenes is the isomerisation of 2-arylthiophenes to 3-arylthiophenes;[122] the aromatic substituent remains attached to the same carbon and the net effect has been shown to involve interchange of C-2 and C-3, with C-4 and C-5 remaining in the same relative positions; scrambling of deuterium labelling is however observed and the detailed mechanism for the rearrangement is still a matter for discussion.

There are an appreciable number of examples in which photochemical ring closure of a 1-thienyl-2-aryl (or heteroaryl) ethene, carried out in the presence of an oxidant (often oxygen) to trap/aromatise a cyclised intermediate, leads to polycyclic products; an example[123,124] is shown below.

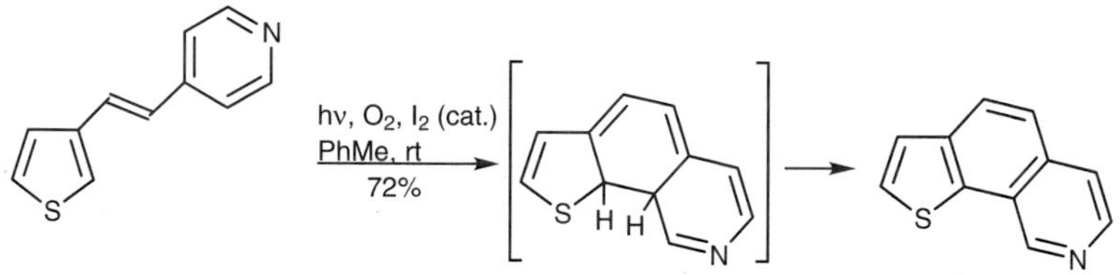

14.10   Thiophene-C-X compounds: thenyl derivatives

The unit – thiophene linked to a carbon – is termed thenyl, hence thenyl chloride is the product of chloromethylation (section 14.1.1.7); thenyl bromides are usually made by side-chain radical substitution,[125] substitution at an α-methyl being preferred over a β-methyl.[126]

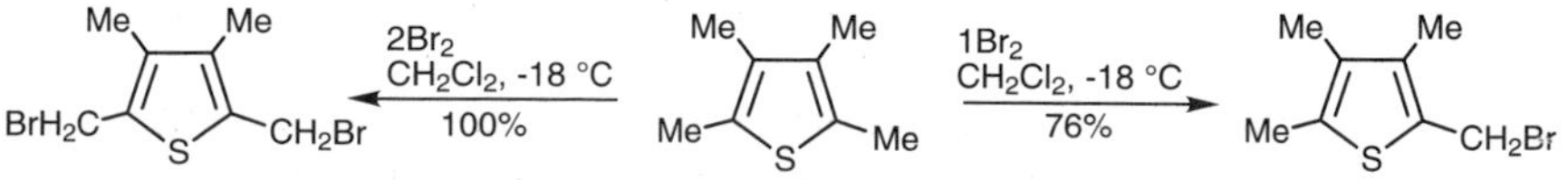

Relatively straightforward benzene-analogue reactivity is found with thenyl halides, alcohols (conveniently preparable by reducing aldehydes) and amines, from for example, reduction of oximes. One exception is that 2-thenyl Grignard reagents usually react to give 3-substituted derivatives, presumably *via* a non-aromatic intermediate.[127]

14.11   Thiophene aldehydes and ketones, and carboxylic acids and esters

Here, the parallels with benzenoid counterparts continue, for these compounds have no special properties – their reactivities are those typical of benzenoid aldehydes, ketones, acids, and esters. For example, in contrast to the easy decarboxylation of α-acids observed for pyrrole and furan, thiophene-2-acids do not easily lose carbon dioxide nevertheless, high temperature decarboxylations are of preparative value (see also 14.13.1.2).[128]

Just as in benzene chemistry, Wolff-Kischner or Clemmensen reduction of ketones is a much-used route to alkylthiophenes, hypochlorite oxidation of acetylthiophenes a good route to thiophene acids, Beckmann rearrangement of thiophene oximes is a useful route to acylaminothiophenes and hence aminothiophenes, and esters and acids are interconvertible without complications.

14.12   Oxy- and aminothiophenes

14.12.1   Oxythiophenes

These compounds are much more difficult to handle and much less accessible than phenols. Neither 2-hydroxythiophene nor its 4-thiolen-2-one tautomer are detectable, the compound existing as the conjugated enone isomer.[129] Thiophene can be converted directly into its 2-oxygenated derivative.[130]

The inclusion of alkyl groups both stabilise the oxy-compounds and the double bond to which they are attached. In these more stable compounds alternative tautomers are found, thus 5-methyl-2-hydroxythiophene exists as a mixture (actually separable by fractional distillation!) of the two enone tautomers.[131]

β-Hydroxythiophenes are even more unstable than α-hydroxythiophenes; 3-hydroxy-2-methylthiophene exist as a mixture of hydroxyl and carbonyl tautomeric forms with the former predominating.[132]

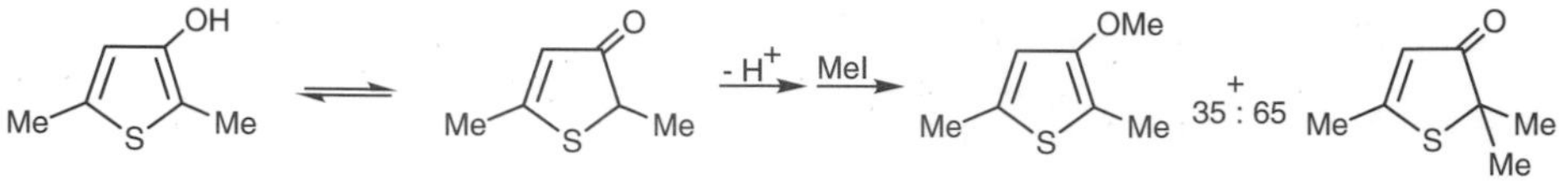

The acidities of the thiolenones are comparable with those of phenols, with pK_as of about 10. Oxythiophene anions can react at oxygen or carbon and products from reaction of electrophiles at both centres can be obtained.[133] Silylation generates 2-silyloxy derivatives which react with aldehydes in the presence of boron trifluoride as shown below.[134]

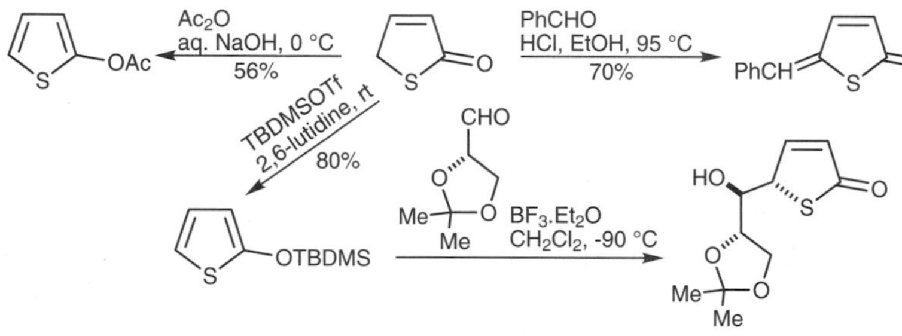

14.12.2 Aminothiophenes

Here again, these thiophene derivatives are much less stable than their benzenoid counterparts, unless the ring is provided with other substitution.[135] The unsubstituted aminothiophenes (thiophenamines) can be obtained by reduction of the nitrothiophenes,[136] but in such a way as to isolate them as salts – usually hexachlorostannates – or *via* Beckmann rearrangements[137] or Hofmann degradation,[138] as acyl derivatives which are stable. Many substituted amines have been prepared by nucleophilic displacement of halogen in nitro-halo-thiophenes. In so far as it can be studied, in simple cases, and certainly in substituted thiophenamines, the amino form is the only detectable tautomer.[139]

14.13   Synthesis of thiophenes[140]

Thiophene is manufactured by the gas-phase interaction of C_4 hydrocarbons and elementary sulfur at 600 °C. Using *n*-butane the sulfur first effects dehydrogenation and then interacts with the unsaturated hydrocarbon by addition, further dehydrogenation generating the aromatic system.

14.13.1   Ring synthesis
14.13.1.1   *From 1,4-dicarbonyl compounds and a source of sulfur*

1,4-Dicarbonyl compounds can be reacted with a source of sulfur to give thiophenes.

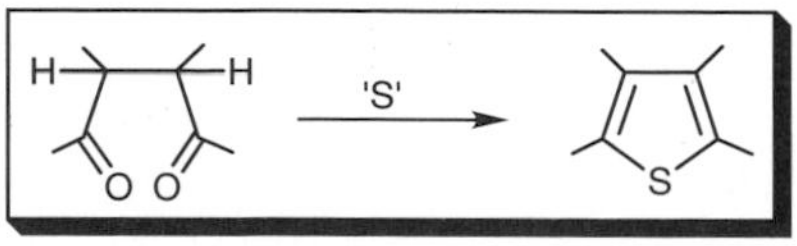

The reaction of a 1,4-dicarbonyl compound (see also 15.14.1.1) with a source of sulfur, traditionally phosphorus sulfides, latterly Lawesson's reagent (LR),[141] or bis(trimethylsilyl)sulfide,[142] gives thiophenes, presumably, but not necessarily, *via* the bis(thioketone).

When the process is employed with 1,4-dicarboxylic acids, a reduction must occur at some stage, for thiophenes, and not 2-/5-oxygenated thiophenes result.[143]

Much use has been made of conjugated diynes, also at the oxidation level of 1,4-dicarbonyl compounds, which react smoothly with hydrosulfide or sulfide, under mild conditions, to give 3,4-unsubstituted thiophenes. Unsymmetrical 2,5-disubstituted thiophenes can be produced in this way too.[144] Since nearly all naturally-occurring thiophenes are found in plant genera, and co-occur with polyynes, this laboratory ring synthesis may be mechanistically related to their biosynthesis.

14.13.1.2   From thiodiacetates and 1,2-dicarbonyl compounds

1,2-Dicarbonyl compounds condense with thiodiacetates (or thiobismethyleneketones) to give thiophene-2,5-diacids (-diketones).

The Hinsberg synthesis

Two consecutive aldol condensations between a 1,2-dicarbonyl compound and diethyl thiodiacetate give thiophenes. The immediate product is an ester-acid, produced[145] by a Stobbe-type mechanism, but the reactions are often worked up *via* hydrolysis to afford an isolated diacid.

14.13.1.3   From thioglycolates and 1,3-dicarbonyl compounds

Thioglycolates react with 1,3-dicarbonyl compounds (or equivalents) to give thiophene-2-carboxylic esters.

In most of the examples of this approach, thioglycolates, as donors of an S–C unit, have been reacted with 1,3-keto-aldehydes, to give intermediates which can be ring closed to give thiophenes as exemplified below.[146]

Alkynylketones react with thioglycolate to generate comparable intermediates by conjugate addition to the triple bond.[147]

14.13.1.4   From α-thiocarbonyl compounds

2-Keto-thiols add to alkenylphosphonium ions, affording ylides which then ring close by Wittig reaction and give 2,5-dihydrothiophenes, which can be dehydrogenated.[148]

14.13.1.5   From thio-diketones

A route[149] in which the 3,4-bond is made by an intramolecular pinacol reaction is nicely illustrated[150] below by the formation of a tricyclic thiophene with two cyclobutane fused rings. The starting materials for this route are easily obtained from sodium sulfide and two mol equivalents of a 2-bromoketone.

14.13.1.5   Using carbon disulfide

The addition of a carbanion to carbon disulfide with a subsequent S-alkylation provides a route to 2-alkylthiothiophenes.[151] In the example below, the carbanion is the enolate of a cyclic 1,3-diketone.

A truly delightful exploitation of this idea is a synthesis of thieno[2,3-*b*]thiophene in which a diyne is lithiated to give a lithio-allene which reacts with carbon disulfide.[152]

When the enolate is that derived from malononitrile,[153] 3-amino-4-cyanothiophenes are the result.[154]

14.13.1.6   From thiazoles

The cycloaddition/cycloreversion sequence which ensues when thiazoles (the best in this context is 4-phenylthiazole) are heated strongly with an alkyne, generates 2,5-unsubstituted thiophenes. Though the conditions are vigorous, excellent yields can be obtained.[81]

14.13.1.7   From thio-nitroacetamides

The *S*-alkylation of thio-nitroacetamides with 2-bromoketones produces 2-amino-3-nitrothiophenes. The scheme below shows how the 3,4-bond making involves the intramolecular interaction of the introduced ketone carbonyl with an enamine/thioenol β-carbon.[155]

14.13.2   Examples of notable syntheses of thiophene compounds
14.13.2.1   Thieno[3,4-b]thiophene[156]

Thieno[3,4-*b*]thiophene was prepared from 3,4-dibromothiophene utilising the two halogens in separate steps: palladium-catalysed coupling and lithiation by transmetallation followed by introduction of sulfur and intramolecular addition to the alkyne

14.13.2.2   2,2′:5′,3″-Terthiophene[157]

This sequence, for the regioselective synthesis of 2,2′:5′,3″-terthiophene uses the reaction of a diyne with sulfide to make the central ring.

14.13.2.3   [6.6]Paracyclophane[158]

Here the thiophene rings were produced using the Hinsberg approach; hydro-genolytic removal of sulfur, having served its purpose to allow construction of the large ring, gave the cyclophane.

Exercises for chapter 14

Straightforward revision exercises (consult chapters 12 and 14)

(a) How could one prepare 2-bromo-, 3-bromo- and 3,4-dibromothiophenes?

(b) What would be the products of carying out Vilsmeier reactions with 2-methyl- and 3-methylthiophenes?

(c) How could one convert 2,5-dimethylthiophene into (i) its S-oxide and (ii) its S,S-dioxide?

(d) What routes could one use to convert thiophene into derivatives carrying at the 2-position: (i) CH(OH)t-Bu; (ii) (CH$_2$)$_2$OH; (iii) Ph?

(e) How could one prepare n-decane from thiophene?

(f) Draw the structures of the thiophenes which would be produced from the following reactant combinations: (i) octane-3,6-dione and Lawesson's reagent; (ii) dimethyl thiodiacetate [S(CH$_2$CO$_2$Me)$_2$], cyclopentane-1,2-dione and base; (iii) pentane-2,4-dione, methyl thioglycolate [HSCH$_2$CO$_2$Me] and base.

More advanced exercises

1. Deduce the structure of the compound, $C_4H_3NO_2S$, produced from thiophene by the following sequence: $ClSO_3H$, then f. HNO_3, then H_2O/heat; the product is isomeric with that obtained by reacting thiophene with acetyl nitrate.

2. Suggest structures for the major and minor, isomeric products, $C_5H_5NO_3S$, from 2-methoxythiophene with HNO_3/AcOH at $-20\,°C$.

3. What compounds would be formed by the reaction of (i) thiophene with propionic anhydride/H_3PO_4; (ii) 3-t-butylthiophene with PhN(Me)CHO/POCl$_3$ then aq. NaOH; (iii) thiophene with Tl(O_2CCF_3)$_3$, then aq. KI $\rightarrow$ C_4H_3IS; (iv) thiophene/succinic anhydride/AlCl$_3$ $\rightarrow$ $C_8H_8O_3S$, then N_2H_4/KOH/heat $\rightarrow$ $C_8H_{10}O_2S$, then SOCl$_2$, then AlCl$_3$ $\rightarrow$ C_8H_8OS.

4. Predict the principle site of deprotonation on treatment of 2- and 3-methoxythiophenes with n-BuLi.

5. Deduce structures for the compounds, C_4HBr_3S and $C_4H_2Br_2S$, produced successively by treating 2,3,4,5-tetrabromothiophene with Mg then H_2O and then the product again with Mg then H_2O.

6. Deduce the structure of the compound, $C_9H_6OS_2$, produced by the sequence: thiophene with BuLi, then CO_2 $\rightarrow$ $C_5H_4O_2S$, then this with thiophene in the presence of P_4O_{10}.

7. Deduce the structure of the thiophenes: (i) $C_6H_4N_4S$, produced by reacting (NC)$_2C = C(CN)_2$ with H_2S; (ii) $C_8H_8O_6S$ from diethyl oxalate, (EtO$_2CCH_2$)$_2S$/NaOEt, aq. NaOH, then Me$_2SO_4$; (iii) $C_{11}H_{16}S$ from 3-acetylcyclononanone with P_4S_{10}.

References

1. 'Recent advances in the chemistry of thiophenes', Gronowitz, S., *Adv. Heterocycl. Chem.*, **1963**, *1*, 1.
2. Deans, F. B. and Eaborn, C., *J. Chem. Soc.*, **1959**, 2303; Taylor, R., *J. Chem. Soc. (B)*, **1970**, 1364.
3. Steinkopf, W. and Köhler, W., *Justus Liebigs Ann. Chem.*, **1937**, *532*, 250; Schreiner, H., *Monatsh. Chem.*, **1951**, *82*, 702.
4. Schwetlick, K., Unverferth, K., and Mayer, R., *Z. Chem.*, **1967**, *7*, 58; Olsson, S., *Arkiv. Kemi*, **1970**, *32*, 89; Baker. R., Eaborn, C., and Taylor, R., *J. Chem. Soc., Perkin Trans. 2*, **1972**, 97.
5. Carmody, M. P., Cook, M. J., Dassanayake, N. C., Katritzky, A. R., Linda, P., and Tack, R. D., *Tetrahedron*, **1976**, *32*, 1767.
6. Curtis, R. F., Jones, D. M., and Thomas, W. A., *J. Chem. Soc. (C)*, **1971**, 234.
7. Butler, A. R. and Hendry, J. B., *J. Chem. Soc. (B)*, **1971**, 102.
8. Babasinian, V. S., *Org. Synth., Coll. Vol. II*, **1943**, 466.
9. Olah, G. A., Kuhn, S., and Mlinko, A., *J. Chem. Soc.*, **1956**, 4257.
10. Östman, B., *Acta Chem. Scand.*, **1968**, *22*, 1687.
11. Blatt, A. H., Bach, S., and Kresch, L. W., *J. Org. Chem.*, **1957**, *22*, 1693.
12. Östman, B., *Acta Chem. Scand.*, **1968**, *22*, 2754.
13. Gronowitz, S. and Gjos, N., *Acta Chem. Scand.*, **1967**, *21*, 2823.
14. Gronowitz, S. and Ander, I., *Chem. Scr.*, **1980**, *15*, 20.
15. Steinkopf, W. and Ohse, W., *Justus Liebigs Ann. Chem.*, **1924**, *437*, 14.
16. Terentev, A. P. and Kadatskii, G. M., *J. Gen. Chem., USSR*, **1952**, 189.
17. Maccarone, E., Musumarra, G., and Tomaselli, G. A., *J. Org. Chem.*, **1974**, *39*, 3286.
18. Söderbäck, E., *Acta Chem. Scand.*, **1954**, *8*, 1851.
19. Steinkopf, W., Jacob, H., and Penz, H., *Justus Liebigs Ann. Chem.*, **1934**, *512*, 136.
20. Marino, G., *Tetrahedron*, **1965**, *21*, 843.
21. Muathen, H. A., *Tetrahedron*, **1996**, *52*, 8863.

22. Buu-Hoï, N. P., *Justus Libeigs Ann. Chem.*, **1944**, *556*, 1; Keegstra, M. A. and Brandsma, L., *Synthesis*, **1988**, 890.
23. Minnis, W., *Org. Synth., Coll. Vol. II*, **1943**, 357; Lew, H. Y. and Noller, C. R., *ibid., Coll. Vol. IV*, **1963**, 545.
24. Antolini, L., Goldini, F., Iarossi, D., Mucci, A., and Schenetti, L., *J. Chem. Soc., Perkin Trans. 1*, **1997**, 1957.
25. Brandsma, L. and Verkruijsse, H. D., *Synth. Commun.*, **1988**, *18*, 1763.
26. Gronowitz, S. and Raznikiewicz, T., *Org. Synth., Coll Vol. V*, **1973**, 149; Hallberg, A., Liljefors, S., and Pedaja, P., *Synth. Commun.*, **1981**, *11*, 25.
27. Gronowitz, S., *Acta Chem. Scand.*, **1959**, *13*, 1045.
28. Xie, Y., Ng, S.-C., Hor, T. S. A., and Chan, H. S. O., *J. Chem. Res., (S)*, **1996**, 150.
29. D'Auria, M. and Mauriello, G., *Tetrahedron Lett.*, **1995**, *36*, 4883.
30. Hoffmann, K. J. and Carlsen, P. H. J., *Synth. Commun.*, **1999**, *29*, 1607.
31. Pearson, D. L. and Tour, J. M., *J. Org. Chem.*, **1997**, *62*, 1376.
32. Johnson, J. R. and May, G. E., *Org. Synth., Coll. Vol. II*, **1943**, 8; Minnis, W., *ibid.*, 520.
33. Hartough, H. D. and Kosak, A. I., *J. Am. Chem. Soc.*, **1947**, *69*, 3093.
34. Pennanen, S. I., *Heterocycles*, , **1976**, *4*, 1021.
35. Weston, A. W. and Michaels, R. J., *Org. Synth., Coll. Vol. IV*, **1963**, 915; Downie, I. M., Earle, M. J., Heaney, H., and Shuhaibar, K. F., *Tetrahedron*, **1993**, *49*, 4015.
36. Finch, H., Reece, D. H., and Sharp, J. T., *J. Chem. Soc., Perkin Trans. 1*, **1994**, 1193.
37. Baraznenok, I. L., Nenajdenko, V. G., and Balenkova, E. S., *Synthesis*, **1997**, 465; Blanchard, P., Brisset, H., Illien, B., Rion, A., and Roncali, J., *J. Org. Chem.*, **1997**, *62*, 2401.
38. Kamitori, Y., Hojo, M., Masuda, R., Izumi, T., and Tsukamoto, S., *J. Org. Chem.*, **1984**, *49*, 4161.
39. Wiberg, K. B. and Shane, H. F., *Org. Synth., Coll. Vol. III*, **1955**, 197; Emerson, W. S. and Patrick, T. M., *ibid., Coll. Vol. IV*, **1963**, 980.
40. Janda, M., *Coll. Czech. Chem. Commun.*, **1961**, *26*, 1889.
41. Blicke, F. F. and Burckhalter, J. F., *J. Am. Chem. Soc.*, **1942**, *64*, 477.
42. Griffing, J. M. and Salisbury, L. F., *J. Am. Chem. Soc.*, **1948**, *70*, 3416.
43. Ballantine, J. A. and Fenwick, R. G., *J. Chem. Soc. (C)*, **1970**, 2264.
44. Tormos, G. V., Belmore, K. A., and Cava, M. P., *J. Am. Chem. Soc.*, **1993**, *115*, 11512.
45. Yang, S.-M. and Fang, J.-M., *J. Chem. Soc., Perkin Trans. 1*, **1995**, 2669.
46. Hartough, H. D. and Meisel, S. L., *J. Am. Chem. Soc.*, **1948**, *70*, 4018.
47. Barker, J. M., Huddleston, P. R., and Wood, M. L., *Synth. Commun.*, **1975**, 59.
48. Dowle, M. D., Hayes, R., Judd, D. B., and Williams, C. N., *Synthesis*, **1983**, 73.
49. Harwood, L. M., Currie, G. S., Drew, M. G. B., and Luke, R. W. A., *Chem. Commun.*, **1996**, 1953; Petasis, N. A., Goodman, A., and Zavialov, I. A., *Tetrahedron*, **1997**, *53*, 16463.
50. Briscoe, H. V. A., Peel, J. B., and Young, G. W., *J. Chem. Soc.*, **1929**, 2589.
51. Steinkopf, W. and Köhler, W., *Justus Liebigs Ann. Chem.*, **1937**, *532*, 250.
52. 'The chemistry of thiophenium salts and thiophenium ylids', Porter, A. E. A., *Adv. Heterocycl. Chem.*, **1989**, *45*, 151.
53. Acheson, R. M. and Harrison, D. R., *J. Chem. Soc. (C)*, **1970**, 1764; Heldeweg, R. F. and Hogeveen, H., *Tetrahedron Lett.*, **1974**, 75.
54. Hashmall, J. A., Horak, V., Khoo, L. E., Quicksall, C. O., and Sun, M. K., *J. Am. Chem. Soc.*, **1981**, *103*, 289.
55. Gillespie, R. J., Murray-Rust, J., Murray-Rust, P., and Porter, A. E. A., *J. Chem. Soc., Chem. Commun.*, **1978**, 83.
56. Gillespie, R. J., Porter, A. E. A., and Willmott, W. E., *J. Chem. Soc., Chem. Commun.*, **1978**, 85.
57. Tranmer, G. K. and Capretta, A., *Tetrahedron*, **1998**, *54*, 15499; Monn, J. A. *et al.*, *J. Med. Chem.*, **1999**, *42*, 1027.
58. Schenk, G. O. and Steinmetz, R., *Justus Liebigs Ann. Chem.*, **1963**, *668*, 19.
59. Cuffe, J., Gillespie, R. J., and Porter, A. E. A., *J. Chem. Soc., Chem. Commun.*, **1978**, 641.
60. Pirrung, M. C., Zhang, J., Lackey, K., Sternback, D. D., and Brown, F., *J. Org. Chem.*, **1995**, *60*, 2112.
61. Nakayama, J., Nagasawa, H., Sugihara, Y., and Ishii, A., *J. Am. Chem. Soc.*, **1997**, *119*, 9077.

62. Melles, J. L. and Backer, H. J., *Recl. Trav. Chim. Pays-Bas*, **1953**, *72*, 314; van Tilborg, W. J. M., *Synth. Commun.*, **1976**, *6*, 583; McKillop, A. and Kemp, D., *Tetrahedron*, **1989**, *45*, 3299.
63. Miyahara, Y. and Inazu, T., *Tetrahedron Lett.*, **1990**, *31*, 5955.
64. Rozen, S. and Bareket, Y., *J. Org. Chem.*, **1997**, *62*, 1457.
65. Melles, J. L., *Recl. Trav. Chim. Pays-Bas*, **1952**, *71*, 869.
66. Nakayama, J. and Hirashima, A., *J. Am. Chem. Soc.*, **1990**, *112*, 7648.
67. Consiglio, G., Spinelli, D., Gronowitz, S., Hörnfeldt, A.-B., Maltesson, B., and Noto, R., *J. Chem. Soc., Perkin Trans. 1*, **1982**, 625.
68. Makosza, M. and Kwast, E., *Tetrahedron*, **1995**, *51*, 8339.
69. Lee, S. B. and Hong, J.-I., *Tetrahedron Lett.*, **1995**, *36*, 8439.
70. Keegstra, M. A., Peters, T. H. A., and Brandsma, L., *Synth. Commun.*, **1990**, *20*, 213.
71. Chadwick, D. J. and Willbe, C., *J. Chem. Soc., Perkin Trans. 1*, **1977**, 887; Feringa, B. L., Hulst, R., Rikers, R., and Brandsma, L., *Synthesis*, **1988**, 316; Furukawa, N., Hoshino, H., Shibutani, T., Higaki, M., Iwasaki, F., and Fujihara, H., *Heterocycles*, **1992**, *34*, 1085.
72. Wu, X., Chen, T. A., Zhu, L., and Rieke, R. D., *Tetrahedron Lett.*, **1994**, *35*, 3673.
73. Rieke, R. D., Kim, S.-H., and Wu, X., *J. Org. Chem.*, **1997**, *62*, 6921.
74. Chadwick, D. J., McKnight, M. V., and Ngochindo, R., *J. Chem. Soc., Perkin Trans. 1*, **1982**, 1343; Chadwick, D. J. and Ennis, D. S., *Tetrahedron*, **1991**, *47*, 9901.
75. Carpenter, A. J. and Chadwick, D. J., *Tetrahedron Lett.*, **1985**, *26*, 1777.
76. Knight, D. W. and Nott, A. P., *J. Chem. Soc., Perkin Trans. 1*, **1983**, 791.
77. Bures, E., Spinazzé, P. G., Beese, G., Hunt, I. R., Rogers, C., and Keay, B. A., *J. Org. Chem.*, **1997**, *62*, 8741; DuPriest, M. T., Zincke, P. W., Conrow, R. E., Kuzmich, D., Dantanarayana, A. P., and Sproull, S. J., *ibid.*, 9372.
78. Gronowitz, S., Hallberg, A., and Frejd, T., *Chem. Scr.*, **1980**, *15*, 1.
79. Hallberg, A. and Gronowitz, S., *Chem. Scr.*, **1980**, *16*, 42.
80. Detty, M. R. and Hays, D. S., *Heterocycles,*, **1995**, *40*, 925.
81. Ye, X.-S. and Wong, H. N. C., *J. Org. Chem*, **1997**, *62*, 1940.
82. Goldberg, Yu., Sturkovich, R., and Lukevics, E., *Synth. Commun.*, **1993**, *23*, 1235.
83. Spagnolo, P. and Zanirato, P., *J. Chem. Soc., Perkin Trans. 1*, **1996**, 963.
84. Frisell, C. and Lawesson, S.-O., *Org. Synth., Coll. Vol. V*, **1973**, 642.
85. Camici, L., Ricci, A., and Taddei, M., *Tetrahedron Lett.*, **1986**, *27*, 5155.
86. Hörnfeldt, A.-B., *Ark. Kem.*, **1964**, *22*, 211.
87. Acheson, R. M., MacPhee, K. E., Philpott, R. G., and Barltrop, J. A., *J. Chem. Soc.*, **1956**, 698.
88. Álvarez, M., Bosch, J., Granados, R., and López, F., *J. Heterocycl. Chem.*, **1978**, *15*, 193.
89. Fikentscher, R., Brückmann, R., and Betz, R, *Justus Liebigs Ann. Chem.*, **1990**, 113.
90. Fuller, L. S., Iddon, B., and Smith, K. A., *J. Chem. Soc., Perkin Trans. 1*, **1997**, 3465.
91. Li, X.-C., Sirringhaus, H., Garnier, F., Holmes, A. B., Moratti, S. C., Feeder, N., Clegg, W., Teat, S. J., and Friend, R. H., *J. Am. Chem. Soc.*, **1998**, *120*, 2206.
92. 'The base-catalysed halogen dance, and other reactions of aryl halides', Bunnett., J. F., *Acc. Chem. Res.*, **1972**, *5*, 139.
93. Brandsma, L. and de Jong, R. L. P., *Synth. Commun.*, **1990**, *20*, 1697.
94. Karlsson, J. O., Svensson, A., and Gronowitz, S., *J. Org. Chem.*, **1984**, *49*, 2018.
95. Melamed, D., Nuckols, C., and Fox, M. A., *Tetrahedron Lett.*,. **1994**, *35*, 8329.
96. Gronowitz, S. and Peters, D., *Heterocycles*, **1990**, *30*, 645.
97. Ohta, A., Akita, Y., Ohkuwa, T., Chiba, M., Fukunaga, R., Miyafuji, A., Makata, T., Tani, N., and Aoyagi, Y., *Heterocycles*, **1990**, *31*, 1951.
98. Pearson, D. L. and Tour, J. M., *J. Org. Chem.*, **1997**, *62*, 1376; Ye, X.-S. and Wong, H. N. C., *Chem. Commun.*, **1996**, 339; Negishi, E., Xu, C., Tan, Z., and Kotora, M., *Heterocycles*, **1997**, *46*, 209.
99. Allred, G. D. and Liebskind, L. S., *J. Am. Chem. Soc.*, **1996**, *118*, 2748; Tamao, K., Nakamura, K., Ishii, H., Yamaguchi, S., and Shiro, M., *ibid.*, 12469.
100. Malm, J., Rehn, B., Hörnfeldt, A.-B., and Gronowitz, S., *J. Heterocycl. Chem.*, **1994**, *31*, 11; Björk, P., Hörnfeldt, A.-B., and Gronowitz, S., *ibid.*, 1161.
101. Chen, T.-A., Wu, X., and Rieke, R. D., *J. Am. Chem. Sooc.*, **1995**, *117*, 233; Rieke, R. D., Kim, S.-H., and Wu, X., *J. Org. Chem.*, **1997**, *62*, 6921.

102. Camazzi, C. M., Leardini, R., Tundo, A., and Tiecco, M., *J. Chem. Soc., Perkin Trans. 1*, **1974**, 271; Bartle, M., Gore, S. T., Mackie, R. K., and Tedder, J. M., *J. Chem. Soc., Perkin Trans. 1*, **1976**, 1636.

103. Camaggi, C.-M., Leardini, R., Tiecco, M., and Tundo, A., *J. Chem. Soc., B*, **1970**, 1683; Vernin, G., Metzger, J., and Párkányi, *J. Org. Chem.*, **1975**, *40*, 3183.

104. Ryang, H. S. and Sakurai, H., *J. Chem. Soc., Chem. Commun.*, **1972**, 594; Allen, D. W., Buckland, D. J., Hutley, B. G., Oades, A. C., and Turner, J. B., *J. Chem. Soc., Perkin Trans. 1*, **1977**, 621; D'Auria, M., De Luca, E., Mauriello, G., and Racioppi, R., *Synth. Commun.*, **1999**, *29*, 35.

105. Chuang, C.-P. and Wang, S.-F., *Synth. Commun.*, **1994**, *24*, 1493; *idem, Synlett*, **1995**, 763.

106. Araneo, S., Arrigoni, R., Bjorsvik, H.-R., Fontana, F., Minisci, F., and Recupero, F., *Tetrahedron Lett.*, **1996**, *37*, 7425.

107. Hansen, S., *Acta Chem. Scand.*, **1954**, *8*, 695.

108. Wynberg, H. and Logothetis, A., *J. Am. Chem. Soc.*, **1956**, *78*, 1958.

109. Gol'dfarb, Ya. L., Taits, S. Z., and Belen'kii, L. I., *Tetrahedron*, **1963**, *19*, 1851.

110. Birch, S. F. and McAllan, D. T., *J. Chem. Soc.*, **1951**, 2556.

111. Blenderman, W. G., Joullié, M. M., and Preti, G., *Tetrahedron Lett.*, **1979**, 4985; Kosugi, K., Anisimov, A. V., Yamamoto, H., Yamashiro, R., Shirai, K., and Kumamoto, T., *Chem. Lett.*, **1981**, 1341; Altenbach, H.-J., Brauer, D. J., and Merhof, G. F., *Tetrahedron*, **1997**, *53*, 6019.

112. 'Applications of ionic hydrogenation to organic synthesis', Kursanov, D. N., Parnes, Z. N., and Loim, N. M., *Synthesis*, **1974**, 633.

113. Kursanov, D. N., Parnes, Z. N., Bolestova, G. I., and Belen'kii, L. I., *Tetrahedron*, **1975**, *31*, 311.

114. Lyakhovetsky, Yu., Kalinkin, M., Parnes, Z., Latypova, F., and Kursanov, D., *J. Chem. Soc., Chem. Commun.*, **1980**, 766.

115. 'Cycloaddition, ring-opening, and other novel reactions of thiophenes', Iddon, B., *Heterocycles*, **1983**, *20*, 1127; 'Cycloaddition reactions with vinyl heterocycles', Sepúlveda-Arques, J., Abarca-González, B., and Medio-Simón, *Adv. Heterocycl. Chem.*, **1995**, *63*, 339.

116. Kotsuki, H., Nishizawa, H., Kitagawa, S., Ochi, M., Yamasaki, N., Matsuoka, K., and Tokoroyama, T., *Bull. Chem. Soc. Jpn.*, **1979**, *52*, 544.

117. Helder, R. and Wynberg, H., *Tetrahedron Lett.*, **1972**, 605; Kuhn, H. J. and Gollnick, K., *Chem. Ber.*, **1973**, *106*, 674.

118. Corral, C., Lissavetzky, J., and Manzanares, I., *Synthesis*, **1997**, 29.

119. Li, Y., Thiemann, T., Sawada, T., Mataka,S., and Tashiro, M., *J. Org. Chem.*, **1997**, *62*, 7926.

120. Furukawa, N., Zhang, S-Z., Sato, S., and Higaki, M., *Heterocycles*, **1997**, *44*, 61.

121. Furukawa, N., Zhang, S.-Z., Horn, E., Takahashi, O., and Sato, S., *Heterocycles*, **1998**, *47*, 793.

122. Wynberg, H., *Acc. Chem. Res.*, **1971**, *4*, 65.

123. Marzinzik, A. L. and Rademacher, P., *Synthesis*, **1995**, 1131.

124. Sato, K., Arai, S., and Yamagishi, T., *J. Heterocycl. Chem.*, **1996**, *33*, 57.

125. Campaigne, E. and Tullar, B. F., *Org. Synth., Coll. Vol. IV*, **1963**, 921; Clarke, J. A. and Meth-Cohn, O., *Tetrahedron Lett.*, **1975**, 4705.

126. Nakayama, J., Kawamura, T., Kuroda, K., and Fujita, A., *Tetrahedron Lett.*, **1993**, *34*, 5725.

127. Gaertner, R., *J. Am. Chem. Soc.*, **1951**, *79*, 3934.

128. Merz, A. and Rehm, C., *J. Prakt. Chem.*, **1996**, *338*, 672; Coffey, M., McKellar, B. R., Reinhardt, B. A., Nijakowski, T., and Feld, W. A., *Synth. Commun.*, **1996**, *26*, 2205.

129. Jakobsen, H. J., Larsen, E. H., and Lawesson, S.-O., *Tetrahedron*, **1963**, *19*, 1867.

130. Allen, D. W., Clench, M. R., Hewson, A. T., and Sokmen, M., *J. Chem. Res. (S)*, **1996**, 242.

131. Gronowitz, S. and Hoffman, R. A., *Ark. Kemi*, **1960**, *15*, 499; Hörnfeldt, A.-B., *ibid.*, **1964**, *22*, 211.

132. Thorstad, O., Undheim, K., Cederlund, B., and Hörnfeldt, A.-B., *Acta Chem. Scand.*, **1975**, *B29*, 647.

133. Lantz, R. and Hörnfeldt, A.-B., *Chem. Scr.*, **1976**, *10*, 126; Hurd, C. D. and Kreuz, K. L. *J. Am. Chem. Soc.*, **1950**, *72*, 5543.

134. Rassu, G., Spanu, P., Pinna, L., Zanardi, F., and Casiraghi, G., *Tetrahedron Lett.*, **1995**, *36*, 1941; 'Furan-, pyrrole-, and thiophene-based siloxydienes for synthesis of densely functionalised homochiral compounds', Casiraghi, G. and Rassu, G., *Synthesis*, **1995**, 607.

135. 'Synthesis of amino derivatives of five-membered heterocycles by Thorpe-Ziegler cyclisation', Granik, V. G., Kadushkin, A. V., and Liebscher, *Adv. Heterocycl. Chem.*, **1998**, *72*, 79.

136. Steinkopf, W., *Justus Liebigs Ann. Chem.*, **1914**, *403*, 17; Steinkopf, W. and Höpner, T., *ibid.*, **1933**, *501*, 174.

137. Meth-Cohn, O. and Narine, B., *Synthesis*, **1980**, 133.

138. Campaigne, E. and Monroe, P. A., *J. Am. Chem. Soc.*, **1954**, *76*, 2447.

139. Brunett, E. W., Altwein, D. M., and McCarthy, W. C., *J. Heterocycl. Chem.*, **1973**, *10*, 1067.

140. 'The preparation of thiophens and tetrahydrothiophens', Wolf, D. E. and Folkers, K., *Org. Reactions*, **1951**, *6*, 410.

141. Shridar, D. R., Jogibhukta, M., Shanthon Rao, P., and Handa, V. K., *Synthesis*, **1982**, 1061; Jones, R. A. and Civcir, P. U., *Tetrahedron*, **1997**, *53*, 11529.

142. Freeman, F., Lee, M. Y., Lu, H., Wang, X., and Rodriguez, E., *J. Org. Chem.*, **1994**, *59*, 3695.

143. Feldkamp, R. F. and Tullar, B. F., *Org. Synth., Coll. Vol. IV*, **1963**, 671.

144. Schulte, K. E., Reisch, J., and Hörner, L., *Chem. Ber.*, **1962**, *95*, 1943; Kozhushkov, S., Hanmann, T., Boese, R., Knieriem, B., Scheib, S., Bäuerle, P., and de Meijere, A., *Angew. Chem., Int. Ed. Engl.*, **1995**, *35*, 781; Alzeer, J. and Vasella, A., *Helv. Chim. Acta*, **1995**, *78*, 177.

145. Wynberg, H. and Kooreman, H. J., *J. Am. Chem. Soc.*, **1965**, *87*, 1739.

146. Taylor, E. C. and Dowling, J. E., *J. Org. Chem.*, **1997**, *62*, 1599.

147. Obrecht, D., Gerber, F., Sprenger, D., and Masquelin, T., *Helv. Chim. Acta*, **1997**, *80*, 531.

148. McIntosh. J. M. and Khalil, H., *Can. J. Chem.*, **1975**, *53*, 209.

149. Nakayama, J., Machida, H., Saito, R., and Hoshino, M., *Tetrahedron Lett.*, **1985**, *26*, 1983.

150. Nakayama, J. and Kuroda, K., *J. Am. Chem. Soc.*, **1993**, *115*, 4612.

151. Prim, D. and Kirsch, G., *Synth. Commun.*, **1995**, *25*, 2449.

152. De Jong, R. L. P. and Brandsma, L., *J. Chem. Soc., Chem. Commun.*, **1983**, 1056; Otsubo, T., Kono, Y., Hozo, N., Miyamoto, H., Aso, Y., Ogura, F., Tanaka, T., and Sawada, M., *Bull. Chem. Soc. Jpn.*, **1993**, *66*, 2033.

153. Gewald, K., Rennent, S., Schindler, R., and Schäfer, H., *J. Prakt. Chem.*, **1995**, *337*, 472.

154. Rehwald, M., Gewald, K., and Böttcher, G., *Heterocycles*, **1997**, *45*, 493.

155. Reddy, K. V. and Rajappa, S., *Heterocycles*, **1994**, *37*, 347.

156. Brandsma, L. and Verkruijsse, H. D., *Synth. Commun.*, **1990**, *20*, 2275.

157. Kagan, J., Arora, S. K., Prakesh, I., and Üstünol, A., *Heterocycles*, **1983**, *20*, 1341.

158. Miyahara, Y., Inazu, T., and Yashino, T., *J. Org. Chem.*, **1984**, *49*, 1177.

15   Furans: reactions and synthesis

furan

Furans[1] are volatile, fairly stable compounds with pleasant odours. Furan itself is slightly soluble in water. It is readily available, and its commercial importance is mainly due to its role as the precursor of the very widely used solvent tetrahydrofuran (THF). Furan is produced by the gas-phase decarbonylation of furfural (2-formylfuran, furan-2-carboxaldehyde), which in turn is prepared in very large quantities by the action of acids on vegetable residues mainly from the manufacture of porridge oats and cornflakes. Furfural was first prepared in this way as far back as 1831 and its name is derived from *furfur* which is the latin word for bran; in due course, in 1870, the word furan was coined from the same root.

perillene

ascorbic acid
(vitamin C)

furfuryl thiol

The aromatic furan ring system, though not found in animal metabolism, occurs widely in secondary plant metabolites, especially in terpenoids: perillene is a simple example. Vitamin C, ascorbic acid, is at the oxidation level of a trihydroxyfuran, though it assumes an unsaturated lactone tautomeric form. Though one normally associates thiols with unpleasant odours, furfuryl thiol is present in the aroma of roasted coffee. Some 5-nitrofurfural derivatives are important in medicine, Nitrofurazone, a bactericide, is a simple example. Ranitidine is one of the most commercially successful medicines ever developed; it is used for the treatment of stomach ulcers.

Nitrofurazone

Ranitidine

15.1   Reactions with electrophilic reagents

Of the three five-membered systems with one heteroatom considered in this book, furan is the 'least aromatic' and as such has the greatest tendency to react in such a way as to give addition products – this is true in the context of its interaction with the usual electrophilic substitution reagents, considered in this section, as well as in Diels-Alder type processes (section 15.9).

15.1.1   Protonation

Furan and the simple alkyl furans are relatively stable to aqueous mineral acids, though furan is instantly decomposed by concentrated sulfuric acid or by Lewis acids such as aluminium chloride. Furan reacts only slowly with hydrogen chloride either as the concentrated aqueous acid or in a non-hydroxylic organic solvent. Hot dilute aqueous mineral acids cause hydrolytic ring-opening.

α-protonated cation which leads to α-exchange   β-protonated cation   O-protonated cation present to minor extent

No pK_a value is available for O-protonation of furan but it is probably much less basic at oxygen than an aliphatic ether. Acid-catalysed deuteration occurs at an α-position;[2] 3/4-deuteriofurans are not obtained because, although β-protonation probably occurs, the cation produced is more susceptible to water, leading to hydrolytic ring opening. An estimate of pK_a −10.0 was made for the 2-protonation of 2,5-di-*t*-butylfuran which implies a value of about −13 for furan itself.[3] An α-protonated cation, stable in solution, is produced on treatment of 2,5-di-*t*-butylfuran with concentrated sulfuric acid.[3,4]

15.1.1.1   *Reactions of protonated furans*

The hydrolysis (or alcoholysis) of furans involves nucleophilic addition of water (or alcohol) to an initially formed cation, giving rise to open-chain 1,4-dicarbonyl compounds or derivatives thereof. This is in effect the reverse of one of the general methods for the construction of furan rings (section 15.14.1.1). Succindialdehyde cannot be obtained from furan itself, presumably because this dialdehyde is too reactive under conditions for hydrolysis, but some alkylfurans can be converted into 1,4-dicarbonyl products quite efficiently, and this can be viewed as a good method for their synthesis, of cyclopentenones derived from them.[5] Other routes from furans to 1,4-dicarbonyl compounds are the hydrolysis of 2,5-dialkoxytetrahydrofurans (section 15.1.4) and by various oxidative procedures (section 15.2).

15.1.2   Nitration

Sensitivity precludes the use of concentrated acid nitrating mixtures. Reaction of furan, or substituted furans[6] with acetyl nitrate produces non-aromatic adducts, in which progress to a substitution product has been interrupted by nucleophilic addition of acetate to the cationic intermediate, usually[7] at C-5.[8] Aromatisation, by loss of acetic acid, to give the nitro-substitution product, will take place under solvolytic conditions, but is better effected by treatment with a weak base like pyridine.[9] Further nitration of 2-nitrofuran gives 2,5-dinitrofuran as the main product.[10]

15.1.3   Sulfonation

Furan and its simple alkyl derivatives are decomposed by the usual strong acid reagents, but the pyridine sulfur trioxide complex can be used, disubstitution of furan being observed even at room temperature.[11]

15.1.4   Halogenation

Furan reacts vigorously with chlorine and bromine at room temperature to give polyhalogenated products, but does not react at all with iodine. More controlled conditions can give 2-bromofuran[12] in a process which probably proceeds *via* a 1,4-dibromo-1,4-dihydro-adduct, indeed such species have been observed at low temperature using ^{1}H NMR spectroscopy.[13] Reaction with bromine in dimethylformamide at room temperature smoothly produces 2-bromo- or 2,5-dibromofurans.[14]

If the bromination is conducted in methanol, trapping of intermediate by C-5 addition of the alcohol then methanolysis of C-2-bromide produces 2,5-dialkoxy-2,5-dihydrofurans, as mixtures of *cis* and *trans* isomers;[15] hydrogenation of these species affords 2,5-dialkoxytetrahydrofurans, extremely useful as 1,4-dicarbonyl synthons – the unsubstituted example is equivalent to succindialdehyde[16] – and heating with benzenethiol or phenyl sulfinic acid in acid gives 2-sulfur-substituted furans.[17] 2,5-Dialkoxy-2,5-dihydrofurans can also be obtained by electrochemical oxidation in alcohol solvents [15,18] or conveniently by oxidation with magnesium monoperoxyphthalate in methanol[19] (see also Oxidation, section 15.2).

The intrinsically high reactivity of the furan nucleus can be further exemplified by the reaction of furfural with excess halogen to produce 'mucohalic acids'; incidentally, mucobromic acid reacts with formamide to provide a useful synthesis of 5-bromopyrimidine.[20] On the other hand, with control, methyl furoate can be cleanly converted into its 5-mono-bromo- or 4,5-dibromo-derivatives; hydrolysis and decarboxylation of the latter then affording 2,3-dibromofuran;[21] bromination of 3-furoic acid produces the 5-monobromo-acid.[22]

mucobromic acid

15.1.5   Acylation

Carboxylic acid anhydrides or halides normally require the presence of Lewis acid (often boron trifluoride) for Friedel-Crafts acylation of furans, though trifluoroacetic anhydride will react alone. The rate of aluminium chloride catalysed acetylation of furan shows the α-position to be 7×10^4 times more reactive than the β-position.[23] 3-Alkylfurans substitute mainly at C-2;[24] 2,5-dialkylfurans can be acylated at a β-position, but generally with more difficulty.

Vilsmeier formylation of furans is a good route to formylfurans,[25] though the ready availability of furfural as a starting material, and methods involving lithiated furans (section 15.4.1) are important. Formylation of substituted furans follows the rule that the strong tendency for α-substitution overrides other factors, thus both 2-methylfuran[26] and methyl furan-3-carboxylate[27] give the 5-aldehyde; 3-methylfuran gives mainly the 2-aldehyde.[28]

15.1.6   Alkylation

Traditional Friedel-Crafts alkylation is not generally practicable in the furan series, partly because of catalyst-caused polymerisation and partly because of polyalkylation. Instances of preparatively useful reactions include: production of 2,5-bis-*t*-butylfuran[29] from furan or furoic acid[30] and the isopropylation of methyl furoate with double substitution, at 3- and 4-positions.[28]

15.1.7   Condensation with aldehydes and ketones

This occurs by acid catalysis, but generally the immediate product, a furfuryl alcohol, reacts further; 2-(3,3,3-trichloro-1-hydroxy)ethylfuran can however be isolated.[31] A macrocycle can be obtained by condensation with acetone[32] *via* a sequence exactly comparable to that described for pyrrole (section 13.1.7).

15.1.8   Condensation with imines and iminium ions

Early attempts to effect Mannich reactions with furan itself failed, though mono-alkylfurans undergo the reaction normally,[33] but by reaction with preformed iminium salt normal 2-substitution of furan itself occurs at room temperature.[34]

More recently the use of furan boronic acids has allowed Mannich substitutions at both α and β positions, and also incidentally with primary amine components.[35]

15.1.9   Mercuration

Mercuration takes place very readily with replacement of hydrogen, or carbon dioxide from an acid.[36]

15.2   Reactions with oxidising agents

The electrochemical or bromine/methanol oxidations of furans to give 2,5-dialkoxy-2,5-dihydrofurans (section 15.1.4), and the cycloaddition of singlet oxygen (section 15.9) are discussed elsewhere. Reaction of furan with lead(IV) carboxylates produces 2,5-diacyloxy-2,5-dihydrofurans in useful yields.[37]

In related chemistry, the ring-opened, Δ-2-unsaturated 1,4-diones can be obtained in *E*- or *Z*-form using reagents such as bromine in aqueous acetone, *meta*-chloroperbenzoic acid, or sodium hypochlorite; an example is given below.[30] Even but-2-en-1,4-dial (maleadehyde) itself, can be produced by oxidation with dimethyldioxirane,[38] and urea/hydrogen peroxide adduct, with catalytic methyltrioxorhenium(VII) has been shown to oxidise a range of furans to *cis* enediones.[39]

Certain oxidants, ruthenium tetraoxide for example, complete destroy furan rings; oxidation of a 2-substituted furan leaves just one carbon of the original heterocycle as a carboxylic acid, without disrupting sensitive functionality in the side-chain.[40]

15.3   Reactions with nucleophilic reagents

Simple furans do not react with nucleophiles by addition or by substitution. Nitro substituents activate the displacement of halogen, as in benzene chemistry and VNS methodology (section 2.3.3) has also been applied to nitrofurans.[41]

15.4 Reactions with bases

15.4.1 Deprotonation of C-hydrogen

Metallation with alkyllithiums proceeds selectively at an α-position, indeed metallation of furan is one of the earliest examples[42] of the now familiar practice of aromatic ring-metallation. The preference for α-deprotonation is nicely illustrated by the demonstration that 3-lithiofuran, produced from 3-bromofuran by metal/halogen exchange at $-78\,^{\circ}C$, equilibrates to the more stable 2-lithiofuran if the temperature rises to $> -40\,^{\circ}C$;[43] more forcing conditions can bring about 2,5-dilithiation of furan.[44]

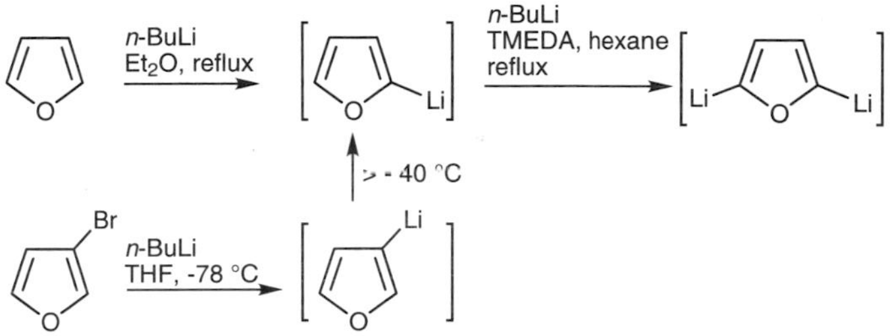

Lithium diisopropylamide can effect C-2-deprotonation of 3-halofurans.[45] With furoic acid and two equivalents of lithium diisopropylamide, selective formation of the lithium carboxylate/5-lithio compound is found,[46] whereas *n*-butyllithium, *via* *ortho*-assistance, produces the lithium carboxylate/3-lithio derivative.[47]

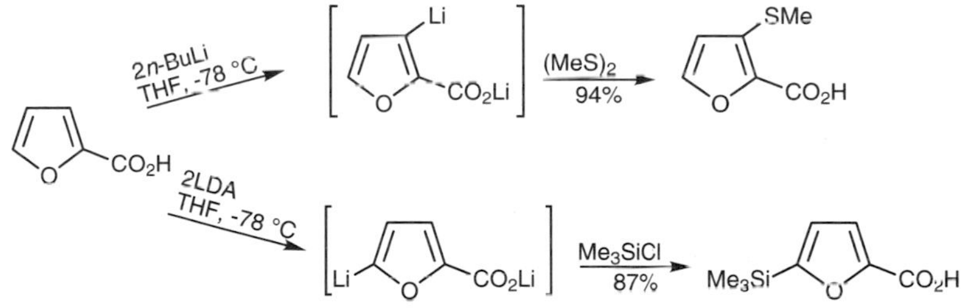

Ortho direction of metallation to C-3 by 2-bis(dimethylamino)phosphate[48] and 2-oxazoline[49] groups, and to C-2 by 3-hydroxymethyl[50] have also been described. 5-Lithiation of furans with non-directing groups at C-2 provides a route to 2,5-disubstituted furans but choice of lithiating conditions can outweigh *ortho* directing effects as illustrated.[51]

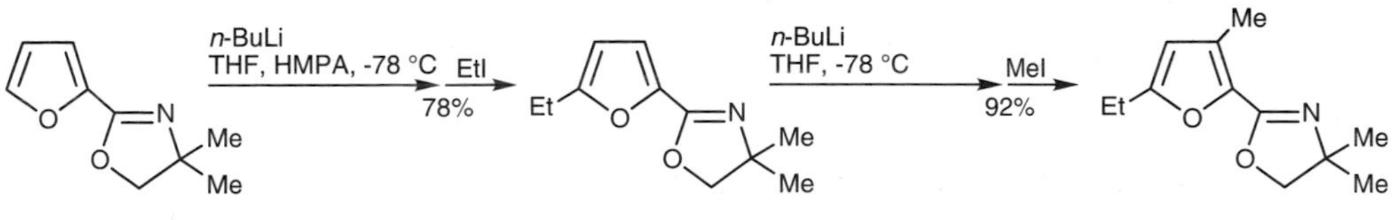

A synthetically useful regioselective 5-lithiation of 3-formylfuran[52] can be achieved by first adding lithium morpholide to the aldehyde and then lithiation at C-5, resulting finally in 2-substituted 4-formylfurans, as is illustrated below.[53]

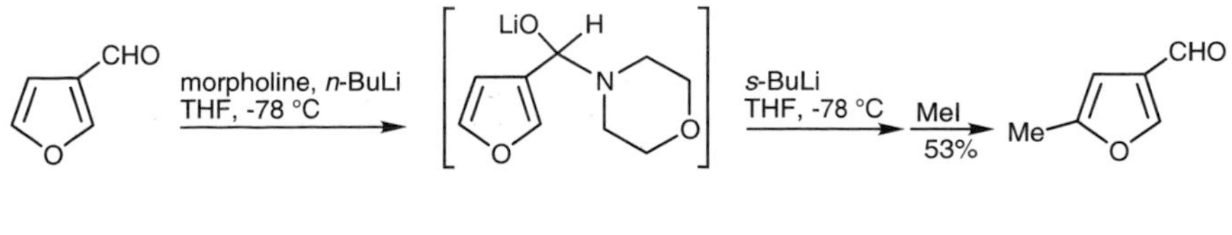

unstable to be isolable, though 2-acylaminofurans have been described and so have more heavily substituted aminofurans.[125] The presence of a 5-ester in conjugation means that methyl 2-aminofuran-5-carboxylate is a relatively stable amino furan; it undergoes Diels-Alder cycloadditions in the usual manner (*cf.* section 15.9).[126]

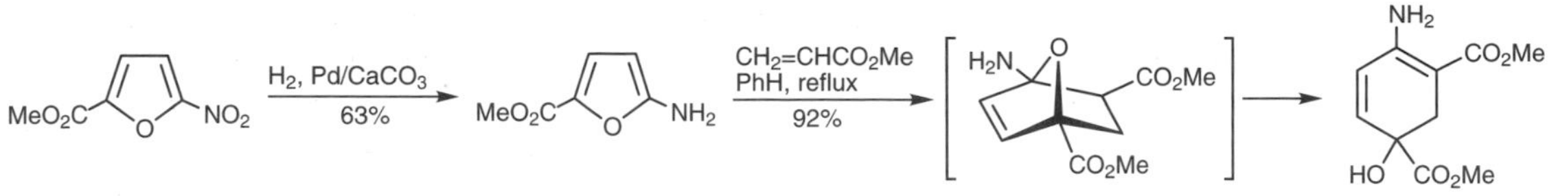

15.14   Synthesis of furans

Furfural and thence furan, by vapour phase decarbonylation, are available in bulk and represent the starting points for many furan syntheses. The aldehyde is manufactured[127] from xylose, obtained in turn from pentosans which are polysaccharides extracted from many plants, e.g. corn cobs and rice husks. Acid catalyses the overall loss of three moles of water in very good yield. The precise order of events in the multistep process is not known for certain, however a reasonable sequence[128] is shown below. Comparable dehydrative ring closure of fructose produces 5-hydroxymethylfurfural.[129]

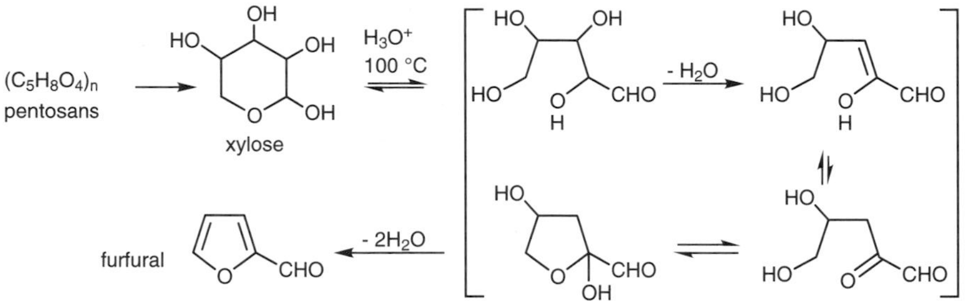

15.14.1   Ring syntheses

Many routes to furans have been described, but the majority are variants on the first general method – the dehydrating ring closure of a 1,4-dicarbonyl substrate.

15.14.1.1   From 1,4-dicarbonyl compounds

1,4-Dicarbonyl compounds can be dehydrated, with acids, to form furans.

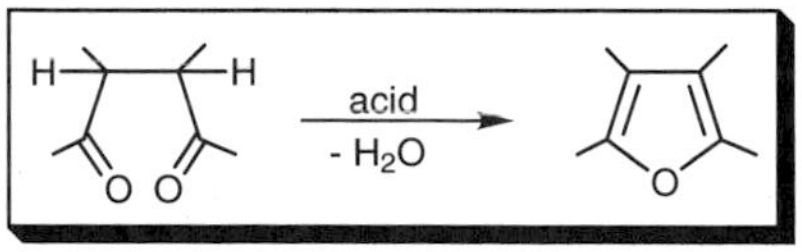

The Paal-Knorr synthesis

The most widely used approach to furans is the cyclising dehydration of 1,4-dicarbonyl compounds, which provide all of the carbon atoms and the oxygen necessary for the nucleus. Usually, non-aqueous acidic conditions[130] are employed to encourage the loss of water. The process involves addition of the enol oxygen of one carbonyl group to the other carbonyl group, then elimination of water.[131]

Access to a 1,4-dicarbonyl substrate has been realised in several ways.[132] Examples include alkylation of imines with 2-alkoxy-allyl halides (equivalents of 2-halo-ketones),[133] addition of β-ketoester anions to nitroalkenes, followed by Nef reaction,[134] and rhodium-catalysed carbonylation of 2-substituted acrolein acetals.[135] The dialdehyde (as a mono-acetal) necessary for a synthesis of diethyl furan-3,4-dicarboxylate was obtained by two successive Claisen condensations between diethyl succinate and ethyl formate, as shown in the sequence below.[136]

15.14.1.2   From γ-hydroxy-α,β-unsaturated carbonyl compounds

γ-Hydroxy-α,β-unsaturated carbonyl compounds can be dehydrated, using mineral or Lewis acids, to form furans.

The simplest example here is the oxidation of *cis*-but-2-ene-1,4-diol, which gives furan *via* the hydroxy-aldehyde.[137]

More elaborate 4-hydroxy-enals and -enones have been generated in a variety of ways, for example *via* alkynes[138] or often *via* epoxides,[139] it being sometimes unecessary to isolate the hydroxy-enone,[140] or via Hörner-Wadsworth-Emmons condensation of β-ketophosphonates with α-acetoxyketones.[141] Acetal[142], thioenol-ether[143] or terminal alkyne[144] can be employed as surrogate for the carbonyl group. Some of these are exemplified below.

acid → $C_5H_5BO_4$; (v) 3-bromofuran/BuLi/–78 °C, then Bu_3SnCl → $C_{16}H_{30}OSn$ and this with $MeCOCl/PdCl_2$ → $C_6H_6O_2$.

7. Write structures for the products of reaction of (i) furfuryl alcohol with $H_2C=C=CHCN$ → $C_9H_9NO_2$ (ii) 2,5-dimethylfuran with $CH_2=CHCOMe/15$ kbar, (iii) furan with 2-chlorocyclopentanone/$Et_3N/LiClO_4$ → $C_9H_{10}O_2$.

8. (i) How could one prepare 2-trimethylsilyloxyfuran? (ii) What product, $C_6H_5NO_2$, would be formed from this with $ICH_2CN/AgOCOCF_3$?

9. What is the product, $C_{11}H_{10}O_3$, formed from the following sequence: 2-t-BuO-furan/n-BuLi, then $PhCH=O$, then TsOH?

10. Decide the structures of the furans produced by the ring syntheses summarised as follows: (i) $CH_2=CHCH_2MgBr/EtCH=O$ then m-CPBA, then CrO_3/pyridine then BF_3; (ii) $CH_2=C(Me)CH_2MgCl/HC(OEt)_3$, then m-CPBA, then aq. H^+; (iii) $(MeO)_2CHCH_2COMe/ClCH_2CO_2Me/NaOMe$ then heat.

11. For the synthesis of tetronic acid summarised as follows, suggest structures for the intermediates: methylamine was added to dimethyl acetylenedicarboxylate (DMAD) → $C_7H_{11}NO_4$, selective reduction with $LiAlH_4$ then giving $C_6H_{11}NO_3$ which with acid cyclised → $C_5H_7NO_2$, aqueous acidic hydrolysis of which produced tetronic acid.

References

1. 'The Furans', Dunlop, A. P. and Peters, F. N., Reinhold, New York, **1953**; 'The development of the chemistry of furans, 1952–1963', Bosshard, P. and Eugster, C. H., *Adv. Heterocycl. Chem.*, **1966**, *7*, 377; 'Recent advances in furan chemistry, Parts 1 and 2', Dean, F. M., *ibid.*, **1982**, *30*, 167 and *31*, 237; 'Regioselective syntheses of substituted furans', Hou, X. L., Cheung, H. Y., Hon, T. Y., Kwan, P. L., Lo, T. H., Tong, S. Y., and Wong, H. N. C., *Tetrahedron*, **1998**, *54*, 1955.

2. Unverferth, K. and Schwetlick, K., *J. Prakt. Chem.*, **1970**, *312*, 882; Salomaa, P., Kankaanperä, A., Nikander, E., Kaipainen, K., and Aaltonen, R., *Acta Chem. Scand.*, **1973**, *27*, 153.

3. Carmody, M. P., Cook, M. J., Dassanayake, N. L., Katritzky, A. R., Linda, P., and Tack, R. D., *Tetrahedron*, **1976**, *32*, 1767.

4. Wiersum, U. E. and Wynberg, H., *Tetrahedron Lett.*, **1967**, 2951.

5. 'Synthesis of 1,4-dicarbonyl compounds and cyclopentenones from furans', Piancatelli, G., D'Auria, M., and D'Onofrio, F., *Synthesis*, **1994**, 867.

6. Michels, J. G. and Hayes, K. J., *J. Am. Chem. Soc.*, **1958**, *80*, 1114.

7. See however Kolb, V. M., Darling, S. D., Koster, D. F., and Meyers, C. Y., *J. Org. Chem.*, **1984**, *49*, 1636.

8. Clauson-Kaas, N. and Faklstorp, J., *Acta Chem. Scand., Ser. B*, **1947**, *1*, 210; Balina, G., Kesler, P., Petre, J., Pham, D., and Vollmar, A., *J. Org. Chem.*, **1986**, *51*, 3811.

9. Rinkes, I. J., *Recl. Trav. Chim. Pays-Bas*, **1930**, *49*, 1169.

10. Doddi, G., Stegel, F., and Tanasi, M. T., *J. Org. Chem.*, **1978**, *43*, 4303.

11. Skully, J. F. and Brown, E. V., *J. Org. Chem.*, **1954**, *19*, 894.

12. Terent'ev, A. P., Belen'kii, L. I., and Yanovskaya, L. A., *Zhur. Obschei Khim.*, **1954**, *24*, 1265 (*Chem. Abs.*, **1955**, *49*, 12327).

13. Baciocchi, E., Clementi, S., and Sebastiani, G. V., *J. Chem. Soc., Chem. Commun.*, **1975**, 875.

14. Keegstra, M. A., Klomp, A. J. A., and Brandsma, L., *Synth. Commun.*, **1990**, *20*, 3371.

15. Ross, S. D., Finkelstein, M., and Uebel, J., *J. Org. Chem.*, **1969**, *34*, 1018.

16. Burness, D. M., *Org. Synth., Coll. Vol. V*, **1973**, 403.

17. Malanga, C., Mannucci, S., and Lardicci, L., *Tetrahedron Lett.*, **1998**, *39*, 5615; Malanga, C., Aronica, L. A., and Lardicci, L., *Synth. Commun.*, **1996**, *26*, 2317.

18. 'Electrochemical oxidation of organic compounds', Weinberg, N. L. and Weinberg, H. R., *Chem. Rev.*, **1968**, *68*, 449.

19. D'Annibale, A. and Scettri, A., *Tetrahedron Lett.*, **1995**, *36*, 4659.

20. Kress, T. J. and Szymanski, E., *J. Heterocycl. Chem.*, **1983**, *20*, 1721.
21. Chadwick, D. J., Chambers, J., Meakins, G. D., and Snowden, R. L., *J. Chem. Soc., Perkin Trans. 1*, **1973**, 1766.
22. Wang, E. S., Choy, Y. M., and Wong, H. N. C., *Tetrahedron*, **1996**, *52*, 12137.
23. Ciranni, G. and Clementi, S., *Tetrahedron Lett.*, **1971**, 3833.
24. Finan, P. A. and Fothergill, G. A., *J. Chem. Soc.*, **1963**, 2723.
25. Traynelis, V. J., Miskel, J. J., and Sowa, J. R., *J. Org. Chem.*, **1957**, *22*, 1269; Downie, I. M., Earle, M. J., Heaney, H., and Shuhaibar, K. F., *Tetrahedron*, **1993**, *49*, 4015.
26. Taylor, D. A. H., *J. Chem. Soc.*, **1959**, 2767.
27. Zwicky, G., Waser, P. G., and Eugster, C. H., *Helv. Chim. Acta*, **1959**, *42*, 1177.
28. Chadwick, D. J., Chambers, J., Hargreaves, H. E., Meakins, G. D., and Snowden, R. C., *J. Chem. Soc., Perkin Trans. 1*, **1973**, 2327.
29. Kamitori, Y., Hojo, M., Masuda, R., Izumi, T., and Tsukamoto, S., *J. Org. Chem.*, **1984**, 49, 4161.
30. Jurczak, J. and Pikul, S., *Tetrahedron Lett.*, **1985**, *26*, 3039; Fitzpatrick, J. E., Milner, D. J., and White, R., *Synth. Commun.*, **1982**, *12*, 489; Williams, P. D. and Le Goff, E., *J. Org. Chem.*, **1981**, *46*, 4143.
31. Willard, J. R. and Hamilton, C. S., *J. Am. Chem. Soc.*, **1951**, *73*, 4805.
32. Chastrette, M. and Chastrette, F., *J. Chem. Soc., Chem. Commun.*, **1973**, 534.
33. Gill, E. W. and Ing, H. R., *J. Chem. Soc.*, **1958**, 4728; Elicl, E. L. and Peckham, P. A., *J. Am. Chem. Soc.*, **1950**, *72*, 1209.
34. Heaney, H., Papageorgiou, G., and Wilkins, R. F., *Tetrahedron Lett.*, **1988**, *29*, 2377.
35. Harwood, L. M., Currie, G. S., Drew, M. G. B., and Luke, R. W. A., *Chem. Commun.*, **1996**, 1953; Petasis, N. A., Goodman, A., and Zavialov, I. A., *Tetrahedron*, **1997**, *53*, 16463.
36. Kutney, J. P., Hanssen, H. W., and Nair, G. V., *Tetrahedron*, **1971**, *27*, 3323; Büchi, G., Kovats, E. Sz., Enggist, P., and Uhde, G., *J. Org. Chem.*, **1968**, *33*, 1227.
37. Elming, N. and Clauson-Kaas, N., *Acta Chem. Scand.*, **1952**, *6*, 535; Trost, B. M. and Shi, Z., *J. Am. Chem. Soc.*, **1996**, *118*, 3037.
38. Adger, B. M., Barrett, C., Brennan, J., McKervey, M. A., and Murray, R. W., *J. Chem. Soc., Chem. Commun.*, **1991**, 1553.
39. Finlay, J., McKervey, M. A., and Gunaratne, H. Q. N., *Tetrahedron Lett.*, **1998**, *39*, 5651.
40. Giovannini, R., and Petrini, M., *Tetrahedron Lett.*, **1997**, *38*, 3781.
41. Makosza, M. and Kwast, E., *Tetrahedron*, **1995**, *51*, 8339.
42. Gilman, H. and Breur, F., *J. Am. Chem. Soc.*, **1934**, *56*, 1123; Ramanathan, V. and Levine, R., *J. Org. Chem.*, **1962**, *27*, 1216.
43. Bock, I., Bornowski, H., Ranft, A., and Theis, H., *Tetrahedron*, **1990**, *46*, 1199.
44. Chadwick, D. J. and Willbe, C., *J. Chem. Soc., Perkin Trans. 1*, **1977**, 887.
45. Ly, N. D. and Schlosser, M., *Helv. Chim. Acta*, **1977**, *60*, 2085.
46. Knight, D. W. and Nott, A. P., *J. Chem. Soc., Perkin Trans. 1*, **1981**, 1125.
47. Carpenter, A. J. and Chadwick, D. J., *Tetrahedron Lett.*, **1985**, *26*, 1777.
48. Näsman, J. H., Kopola, N., and Pensar, G., *Tetrahedron Lett.*, **1986**, *27*, 1391.
49. Chadwick, D. J., McKnight, M. V., and Ngochindo, R., *J. Chem. Soc., Perkin Trans. 1*, **1982**, 1343.
50. Bures, E. J. and Keay, B. A., *Tetrahedron Lett.*, **1988**, *29*, 1247.
51. Lenoir, J.-Y., Ribéreau, and Quéguiner, G., *J. Chem. Soc., Perkin Trans. 1*, **1994**, 2943
52. Hiroya, K. and Ogasawara, K., *Synlett*, **1995**, 175.
53. Lee, G. C. M., Holmes, J. D., Harcourt, D. A., and Garst, M. E., *J. Org. Chem.*, **1992**, *57*, 3126.
54. Sornay, R., Meunier, J.-M., and Fournari, P., *Bull. Soc. Chim. Fr.*, **1971**, 990; Decroix, B., Morel, J., Paulmier, C., and Pastor, P., *ibid.*, **1972**, 1848.
55. Gronowitz, S. and Sörlin, G., *Acta Chem. Scand.*, **1961**, *15*, 1419.
56. Bohlmann, F., Stöhr, F., and Staffeldt, J., *Chem. Ber.*, **1978**, *111*, 3146.
57. Kauffmann, T., Lexy, H., and Kriegesmann, R., *Chem. Ber.*, **1981**, *114*, 3667.
58. Sotoyama, T., Hara, S., and Suzuki, A., *Bull. Chem. Soc. Jpn.*, **1979**, *52*, 1865; Marinelli, E. R. and Levy A. B., *Tetrahedron Lett.*, **1979**, 2313; Akimoto, I. and Suzuki, A., *Synthesis*, **1979**, 146.
59. Fukuyama, Y., Kawashima, Y., Miwa, T., and Tokoroyama, T., *Synthesis*, **1974**, 443.
60. Camici, L., Ricci, A., and Taddei, M., *Tetrahedron Lett.*, **1986**, *27*, 5155.

61. Pelter, A. and Rowlands, M., *Tetrahedron Lett.*, **1987**, *28*, 1203.
62. Florentin, D., Roques, B. P., Fournie-Zaluski, M. C., *Bull. Soc. Chim. Fr.*, **1976**, 1999.
63. Ohta, A., Akita, Y., Ohkuwa, T., Chiba, M., Fukunaga, R., Miyafuji, A., Nakata, T., Tani, N., and Aoyagi, Y., *Heterocycles*, **1990**, *31*, 1951.
64. Yang, Y., *Synth. Commun.*, **1989**, *19*, 1001.
65. Fujiwara, Y., Maruyama, O., Yoshidomi, M., and Taniguchi, H., *J. Org. Chem.*, **1981**, *46*, 851; Tsuji, J. and Nagashima, H., *Tetrahedron*, **1984**, *40*, 2699.
66. Bach, T. and Krüger, L., *Synlett*, **1998**, 1185.
67. Bailey, T. R., *Synthesis*, **1991**, 242; 'Regiospecific synthesis of 3,4-disubstituted furans and thiophenes', Ye, X.-S., Yu, P., and Wong, N. C., *Liebigs Ann./Recl.*, **1997**, 459.
68. Wong, M. K., Leung, C. Y., and Wong, H. N. C., *Tetrahedron*, **1997**, *53*, 3497.
69. Ayres, D. C. and Smith, J. R., *J. Chem. Soc., C*, **1968**, 2737; Camaggi, C. M., Leardini, R., Tiecco, M., and Tundo, A., *J. Chem. Soc., B*, **1969**, 1251; Maggiani, A., Tubul,. A., and Brun, P., *Synthesis*, **1997**, 631.
70. Janda, M., Srogl, J., Stibor, I., Nemec, M., and Vopatrná, P., *Tetrahedron Lett.*, **1973**, 637.
71. Birch, A. J. and Slobbe, J., *Tetrahedron Lett.*, **1975**, 627; Divanford, H. R. and Jouillié, M. M., *Org. Prep. Proc. Int.*, **1978**, *10*, 94; Kinoshita, T., Miyano, K., and Miwa, T., *Bull. Chem. Soc. Jpn.*, **1975**, *48*, 1865; Beddoes, R. L., Lewis, M. L., Gilbert, P., Quayle, P., Thompson, S. P., Wang, S., and Mills, K., *Tetrahedron Lett.*, **1996**, *37*, 9119.
72. Kinoshita, T., Ichinari, D., and Sinya, J., *J. Heterocycl. Chem.*, **1996**, *33*, 1313.
73. Stockmann, H., *J. Org. Chem.*, **1961**, *26*, 2025.
74. Diels, O. and Alder, K., *Chem. Ber.*, **1929**, *62*, 554.
75. Woodward, R. B. and Baer, H., *J. Am. Chem. Soc.*, **1948**, *70*, 1161.
76. Lee, M. W. and Herndon, W. C., *J. Org. Chem.*, **1978**, *43*, 518.
77. Kurtz, P., Gold, H., and Disselnköter, H., *Justus Liebigs Ann. Chem.*, **1959**, *624*, 1; Kozikowski, A. P., Floyd, W. C., and Kuniak, M. P., *J. Chem. Soc., Chem. Commun.*, **1977**, 582.
78. Kienzle, F., *Helv. Chim. Acta*, **1975**, *58*, 1180; Brion, F., *Tetrahedron Lett.*, **1982**, *23*, 5299; Campbell, M. M., Kaye, A. D., Sainsbury, M., and Yavarzadeh, R., *Tetrahedron*, **1984**, *40*, 2461.
79. Dauben, W. G., Gerdes, J. M., and Smith, D. B., *J. Org. Chem.*, **1985**, *50*, 2576; Rimmelin, J., Jenner, G., and Rimmelin, P., *Bull. Soc. Chim. Fr.*, **1978**, II, 461.
80. Kotsuki, H., Mori, Y., Ohtsuka, T., Nishizawa, H., Ochi, M., and Matsuoka, K., *Heterocycles*, **1987**, *26*, 2347.
81. Lazlo, P. and Lucchetti, J., *Tetrahedron Lett.*, **1984**, *25*, 4387.
82. Eberbach, W., Penroud-Argüelles, M., Achenbach, H., Druckrey, E., and Prinzbach, H., *Helv. Chim. Acta*, **1971**, *54*, 2579; Weis, C. D., *J. Org. Chem.*, **1962**, *27*, 3520.
83. Kowarski, C. R. and Sarel, S., *J. Org. Chem.*, **1973**, *38*, 117.
84. Potts, K. T. and Walsh, E. B., *J. Org. Chem.*, **1988**, *53*, 1199.
85. 'Synthetic applications of furan Diels-Alder chemistry', Kappe, C. O., Murphree, S. S., and Padwa, A., *Tetrahedron*, **1997**, *53*, 14179.
86. Lautens, M. and Fillion, E., *J. Org. Chem.*, **1997**, *62*, 4418.
87. Marchionni, C., Vogel, P., and Roversi, P., *Tetrahedron Lett.*, **1996**, *37*, 4149.
88. Herter, R. and Föhlisch, B., *Synthesis*, **1982**, 976.
89. Jin, S.-j., Choi, J.-R., Oh, J., Lee, D., and Cha, J. K., *J. Am. Chem. Soc.*, **1995**, *117*, 10914; Montaña, A. M., Ribes, S., Grima, P. M., García, F., Solans, Y., and Font-Bardia, M., *Tetrahedron*, **1997**, *53*, 11669.
90. Mann, J., Wilde, P. D., and Finch, M. W., *Tetrahedron*, **1987**, *45*, 5431.
91. Gschwend, H. W., Hillman, M. J., Kisis, B., and Rodebaugh, R. K., *J. Org. Chem.*, **1976**, *41*, 104; Parker, K. A. and Adamchuk, M. R., *Tetrahedron Lett.*, **1978**, 1689; Harwood, L. M., Ishikawa, T., Phillips, H., and Watkin, D., *J. Chem. Soc., Chem. Commun.*, **1991**, 527.
92. Metz, P., Meiners, U., Cramer, E., Fröhlich, R., and Wibbeling, B. W., *Chem. Commun.*, **1996**, 431; Metz, P., Seng, D., Fröhlich, R., and Wibbeling, B., *Synlett*, **1996**, 741.
93. Choony, N., Dadabhoy, A., and Sammes, P. G., *Chem. Commun.*, **1997**, 512.
94. 'Photo-oxidation of furans', Fering, B. C., *Recl. Trav. Chim. Pays Bas*, **1987**, *106*, 469; Gorman, A. A., Lovering, G., and Rodgers, M. A. J., *J. Am. Chem. Soc.*, **1979**, *101*, 3050; Iesce, M. R., Cermola, F., Graziano, M. L., and Scarpati, R., *Synthesis*, **1994**, 944.
95. Kernan, M. R. and Faulkner, D. J., *J. Org. Chem.*, **1988**, *53*, 2773.

96. Lee, G. C. M., Syage, E. T., Harcourt, D. A., Holmes, J. M., and Garst, M. E., *J. Org. Chem.*, **1991**, *56*, 7007.
97. Lee, G. C. M., Holmes, J. M., Harcourt, D. A., and Garst, M. E., *J. Org. Chem.*, **1992**, *57*, 3126.
98. Cottier, L., Descotes, G., Eymard, L., and Rapp, K., *Synthesis*, **1995**, 303.
99. White, J. D., Carter, J. P., and Kezar, H. S., *J. Org. Chem.*, **1982**, *47*, 929.
100. 'Cycloaddition reactions with vinyl heterocycles', Sepúlveda-Arques, J., Abarca-González, B., and Medio-Simón, *Adv. Heterocycl. Chem.*, **1995**, *63*, 339.
101. Cornwall, P., Dell, C. P., and Knight, D. W., *J. Chem. Soc., Perkin Trans. 1*, **1993**, 2395.
102. Avalos, L. S., Benítez, A., Muchowski, J. M., Romero, M., and Talamás, F. X., *Heterocycles*, **1997**, *45*, 1795.
103. Nakano, T., Rivas, C., Perez, C., and Tori, K., *J. Chem. Soc., Perkin Trans. 1*, **1973**, 2322; Rivas, C., Bolivar, R. A., and Cucarella, M., *J. Heterocycl. Chem.*, **1982**, 19, 529; Zamojski, A. and Kozluk, T., *J. Org. Chem.*, **1977**, *42*, 1089; Jarosz, S. and Zamojski, A., *J. Org. Chem.*, **1979**, *44*, 3720.
104. Yamamoto, F., Hiroyuki, M., and Oae, S., *Heterocycles*, **1975**, 3, 1; Divald, S., Chun, M. C., and Joullié, M. M., *J. Org. Chem.*, **1976**, *41*, 2835.
105. Kuwajima, I., Atsumi, K., Tanaka, T., and Inoue, T., *Chem. Lett.*, **1979**, 1239.
106. Wilson, W. C., *Org. Synth., Coll. Vol. I*, **1932**, 274; Boyd, M. R., Harris, T. M., and Wilson, B. J., *Synthesis*, **1971**, 545.
107. Ferraz, J. P. and do Amaral, L., *J. Org. Chem.*, **1976**, *41*, 2350.
108. Rinkes, I. J., *Recl. Trav. Chim. Pays-Bas*, **1930**, *49*, 1118.
109. Hill, H. B. and White, G. R., *J. Am. Chem. Soc.*, **1902**, *27*, 193.
110. 'Natural 4-ylidenebutenolides and 4-ylidenetetronic acids', Pattenden, G., *Prog. Chem. Org. Nat. Prod.*, **1978**, *35*, 133.
111. 'The role of heteroatomic substances in the aroma compounds of food stuffs', Ohlaff, G. and Flament, I., *Prog. Chem. Org. Nat. Prod.*, **1979**, *36*, 231.
112. 'Recent advances in the chemistry of unsaturated lactones', Rao, Y. S., *Chem. Rev.*, **1976**, *76*, 625.
113. Grieco, P. A., Pogonowski, C. S., and Burke, S., *J. Org. Chem.*, **1975**, *40*, 542; McMurray, J. E. and Donovan, S. F., *Tetrahedron Lett.*, **1977**, 2869.
114. Yoshii, E., Koizumi, T., Kitatsuji, E., Kawazoe, T., and Kaneko, T., *Heterocycles*, **1976**, *4*, 1663; 'Furan-, pyrrole-, and thiophene-based siloxydienes for synthesis of densely functionalised homochiral compounds', Casiraghi, G. and Rassu, G., *Synthesis*, **1995**, 607.
115. Jefford, C. W., Jaggi, D., and Boukouvalas, J., *Tetrahedron Lett.*, **1987**, *28*, 4037; Jefford, C. W., Sledeski, A. W., and Boukouvalas, J., *J. Chem. Soc., Chem. Commun.*, **1988**, 364; Asaoka, M., Sugimura, N., and Takei, H., *Bull. Soc. Chem. Jpn.*, **1979**, *52*, 1953.
116. Jefford, C. W., Jaggi, D., and Boukouvalas, J., *J. Chem. Soc., Chem. Commun.*, **1988**, 1595.
117. Sornay, R., Neurier, J.-M., and Fournari, P., *Bull. Soc. Chim. Fr.*, **1971**, 990.
118. Kraus, G. A. and Sugimoto, H., *J. Chem. Soc., Chem. Commun.*, **1978**, 30.
119. Tanis, S. P. and Head, D. B., *Tetrahedron Lett.*, **1984**, *40*, 4451.
120. D'Aleilio, G. F., Williams, C. J., and Wilson, C. L., *J. Org. Chem.*, **1960**, *25*, 1028; Sherman, E. and Dunlop, A. P., *ibid.*, 1309.
121. Iten, P. X., Hofmann, A. A., and Eugster, C. H., *Helv. Chim. Acta*, **1978**, *61*, 430.
122. Iten, P. X., Hofmann, A. A., and Eugster, C. H., *Helv. Chim. Acta*, **1979**, *62*, 2202.
123. Brownbridge, P. and Chan, T.-H., *Tetrahedron Lett.*, **1980**, *21*, 3423.
124. Niwa, E., Aoki, H., Tanake, H., Munakata, K., and Namiki, M., *Chem. Ber.*, **1966**, *99*, 3215; Cederlund, B., Lantz, R., Hörnfeldt, A.-B., Thorstad, O. and Undheim, K., *Acta Chem. Scand.*, **1977**, *B31*, 198.
125. 'Synthesis of amino derivatives of five-membered heterocycles by Thorpe-Ziegler cyclisation', Granik, V. G., Kadushkin, A. V., and Liebscher, *Adv. Heterocycl. Chem.*, **1998**, *72*, 79.
126. Cochran, J. E., Wu, T., and Padwa. A., *Tetrahedron Lett.*, **1996**, *37*, 2903.
127. Adams, R. and Voorhees, V., *Org. Synth., Coll. Vol. I*, **1932**, 280.
128. Bonner, W. A. and Roth, M. R., *J. Am. Chem. Soc.*, **1959**, *81*, 5454; Moye, C. J. and Krzeminski, Z. A., *Austr. J. Chem.*, **1963**, *16*, 258; Feather, M. S., Harris, D. W., and Nichols, S. B., *J. Org. Chem.*, **1972**, *37*, 1606.
129. Anet, E. F. C. J., *Chem. Ind.*, **1962**, 262.

130. Nowlin, G., *J. Am. Chem. Soc.*, **1950**, *72*, 5754; Traylelis, V. J., Hergennother, W. L., Hanson, H. T., and Valicenti, J. A., *J. Org. Chem.*, **1964**, *29*, 123; Scott, L. T. and Naples, J. O., *Synthesis*, **1973**, 209.
131. Amarnath, V. and Amarnath, K., *J. Org. Chem.*, **1995**, *60*, 301.
132. Hegedus, L. S. and Perry, R. J., *J. Org. Chem.*, **1985**, *50*, 4955; Mackay, D., Neeland, E. G., and Taylor, N. J., *ibid.*, **1986**, *51*, 2351.
133. Jacobson, R. M., Raths, R. A. and McDonald, J. H., *J. Org. Chem.*, **1977**, *42*, 2545; Jacobson, R. M., Abbaspour, A., and Lahm, G. P., *ibid.* **1978**, *43*, 4650.
134. Boberg, F. and Kieso, A., *Justus Liebigs Ann. Chem.*, **1959**, *626*, 71.
135. Botteghi, C., Lardicci, L., and Menicagli, R., *J. Org. Chem.*, **1973**, *38*, 2361.
136. Kornfeld, E. C. and Jones, R. G., *J. Org. Chem.*, **1954**, *19*, 1671.
137. Clauson-Kaas, N., *Acta Chem. Scand.*, **1961**, *15*, 1177.
138. Seyferth, H. E., *Chem. Ber.*, **1968**, *101*, 619.
139. Cormier, R. A., Grosshans, C. A., and Skibbe, S. L., *Synth. Commun.*, **1988**, *18*, 677.
140. Cormier, R. A. and Francis, M. D., *Synth. Commun.*, **1981**, *11*, 365.
141. Díaz-Cortés, R., Silva, A. L., and Maldonado, L. A., *Tetrahedron Lett.*, **1997**, *38*, 2207.
142. Cornforth, J. W., *J. Chem. Soc.*, **1958**, 1310; Burness, D. M., *Org. Synth., Coll. Vol. IV*, **1963**, 649; Kotake, H., Inomata, K., Kinoshita, H., Aoyama, S., and Sakamoto,Y., *Heterocycles*, **1978**, *10*, 105.
143. Garst, M. E. and Spencer, T. A., *J. Am. Chem. Soc.*, **1973**, *95*, 250.
144. Miller, D., *J. Chem. Soc., C*, **1969**, 12; Schreurs, P. H. M., de Jong, A. J., and Brandsma, L., *Recl. Trav. Chim. Pays-Bas*, **1976**, *95*, 75.
145. Marshall, J. A. and Wang, X., *J. Org. Chem.*, **1991**, *56*, 960.
146. Danheiser, R. L., Stoner, E. J., Koyama, H., and Yamashita, D. S., *J. Am. Chem. Soc.*, **1989**, *111*, 4407.
147. Sheng, H., Lin, S., and Huang, Y., *Tetrahedron Lett.*, **1986**, *27*, 4893.
148. Sheng, H., Lin, S., and Huang, Y., *Synthesis*, **1987**, 1022.
149. Fukuda, Y., Shiragami, H., Utimoto, K., and Nozaki, H., *J. Org. Chem.*, **1991**, *56*, 5816.
150. Marshall, J. E. and DuBay, W. J., *J. Org. Chem.*, **1991**, *56*, 1685.
151. Bisagni, E., Marquet, J.-P., Bourzat, J.-D., Pepin, J.-J., and André-Louisfert, J., *Bull. Soc. Chim. Fr.*, **1971**, 4041.
152. Dann, O., Distler, H., and Merkel, H., *Chem. Ber.*, **1952**, *85*, 457.
153. Howes, P. D. and Stirling, C. J. M., *Org. Synth.*, **1973**, *53*, 1; Batty, J. W., Howes, P. D., and Stirling, C. J. M., *J. Chem. Soc., Perkin Trans. 1*, **1973**, 65; Ojida, A., Tanoue, F., and Kanematsu, K., *J. Org. Chem.*, **1994**, *59*, 5970.
154. Arcadi, A. and Rossi, E., *Tetrahedron*, **1998**, *54*, 15253.
155. MaGee, D. I. and Leach, J. D., *Tetrahedron Lett.*, **1997**, *38*, 8129.
156. Couffignal, R., *Synthesis*, **1978**, 581.
157. Cacchi, S., Fabrizi, G., and Moro, L., *J. Org. Chem.*, **1997**, *62*, 5327.
158. Bew, S. P. and Knight, D. W., *Chem. Commun.*, **1996**, 1007.
159. Garst, M. E. and Spencer, T. A., *J. Org. Chem.*, **1974**, *39*, 584.
160. Jerris, P. J., Wovkulich, P. M., and Smith, A. B., *Tetrahedron Lett.*, **1979**, 4517.
161. Yang, Y. and Wong, H. N. C., *Tetrahedron*, **1994**, *50*, 9583.
162. Song, Z. Z., Ho, M. S., and Wong, H. N. C., *J. Org. Chem.*, **1994**, *59*, 3917.
163. Timko, J. M., Moore, S. S., Walba, D. M., Hiberty, P. C., and Cram, D. J., *J. Am. Chem. Soc.*, **1977**, *99*, 4207.
164. Büchi, G., Demole, E., Thomas, A. F., *J. Org. Chem.*, **1973**, *78*, 123.

16 Typical reactivity of indoles, benzo[*b*]thiophenes, benzo[*b*]furans, isoindoles, benzo[*c*]thiophenes and isobenzofurans

indole   benzothiophene   benzofuran   isoindole   benzo[*c*]thiophene   isobenzofuran
         [benzo[*b*]thiophene] [benzo[*b*]furan]

The fusion of a benzene ring to the 2,3-positions of a pyrrole generates one of the most important heterocyclic ring systems – indole. This chapter develops a description of the chemistry of indole, then discusses modifications necessary to rationalise the chemistry of the benzo[*b*]furan and benzo[*b*]thiophene analogues. Finally, the trio of heterocycles in which the benzene ring is fused at the five-membered ring 3,4-positions are considered.

Typical reactions of indoles

lithiation of *N*-blocked indoles provides means for introduction of substituents into the 2-position

indoles very weakly basic; thermodynamic cation has proton at β-position: a 3*H*-indolium cation

indolyl anion provides means for introduction of substituents onto ring nitrogen

very easy electrophilic substitution at β-position; substitution at 2-position in 3-substituted indoles

The chemistry of indole is dominated by its very easy electrophilic substitution. Of the two rings, the heterocyclic ring is very electron-rich, by comparison with a benzene ring, so attack by electrophiles always takes place in the five-membered ring, except in special circumstances. Of the three positions on the heterocyclic ring, attack at nitrogen would destroy the aromaticity of the five-membered ring, and produce a localised cation; both of the remaining positions can be attacked by electrophiles, leading to *C*-substituted products, but the β-position is preferred by a considerable margin. This contrasts with the regiochemistry shown by pyrrole but again can be well rationalised by a consideration of the Wheland intermediates for the two alternative sites of attack.

intermediate for β attack by X⁺   intermediate for α attack by X⁺

The intermediate for attack at C-2 is stabilised – it is a benzylic cation – but it cannot derive assistance from the nitrogen without disrupting the benzenoid resonance (resonance contributor, which makes a limited contribution, shown in parenthesis). The more stable intermediate from attack at C-3, has charge located adjacent to nitrogen and able to derive the very considerable stabilisation attendant upon interaction with its lone pair of electrons.

The facility with which indoles undergo substitution, and the possiblity for substitution at C-2 can both be illustrated using Mannich reactions – the electrophilic species in such reactions ($C = N^+R_2$) is generally considered to be a 'weak' electrophile, yet substitution occurs easily under mild conditions.

There is a strong preference for attack at C-3, even when that position carries a substituent, and this is nicely shown by examples in which there is the possibility for nucleophilic trapping of the Wheland intermediate: the reaction of indole with sodium hydrogen sulfite is a simple example.

2-Electrophilic substitution of 3-substituted indoles could proceed in three ways: (i) initial attack at a 3-position followed by 1,2-migration to the 2-position; (ii) initial attack at the 3-position followed by reversal (when possible), then (iii); or (iii) direct attack at the 2-position. It has been definitely demonstrated, in the case of some irreversible substitutions, that the migration route operates, but equally it has been demonstrated that direct attack at an α position can occur.

Indoles react with strong bases losing the *N*-hydrogen and forming indolyl anions. When the counterion is an alkali metal these salts have considerable ionic character and react with electrophiles at the nitrogen, affording a practical route for *N*-alkylation (or acylation) of indole nitrogen. Indolyl anions are used, for example, for the synthesis of indoles carrying *N*-blocking substituents. From *N*-blocked indoles, deprotonation (lithiation) can be effected at C-2, often with the additional chelating assistance of the *N*-substituent, though this last is not essential, for even *N*-methylindole lithiates at C-2, where the acidifying effect of the electronegative hetero atom is felt most strongly.

The reactivity of *N*-magnesioindoles, which result from displacement of the active *N*-hydrogen with a Grignard reagent, or the analogous zinc derivatives, are rather different from that of the sodium, potassium and lithium salts. The greater covalent character of the *N*-metal bond means that electrophiles tend to react at C-3, rather than at nitrogen.

As in all heterocyclic chemistry, the advent of palladium(0)-catalysed processes (see section 2.7 for a detailed discussion) has revolutionised the manipulation of indoles, benzothiophens and benzofurans: the example below is typical.

The ready electron availability in the heterocyclic ring means that indoles are rather easily (aut)oxidised in the five-membered ring. Reductions can be made selective for either ring: in acid solution, dissolving metals attack the hetero ring, and the benzenoid ring can be selectively reduced by Birch reduction.

Apart from commenting that substituents on the homocyclic ring of indoles are 'normal', i.e. behave as they would on simpler benzene compounds, the last major aspect of note is the reactivity of indoles which carry leaving groups at benzylic positions, especially C-3, on the heterocyclic ring. Such compounds undergo displacement processes extremely easily, encouraged by stabilisation of positive charge by the nitrogen or, alternatively, in basic conditions, by loss of the indole hydrogen. This last occurs in lithium aluminium hydride reduction of 3-acylindoles which produces 3-alkylindoles. In a sense, 3-ketones behave like vinylogous amides, and reduction intermediates are able to lose oxygen to give species which, on addition of a second hydride, produce the indolyl anion of the 3-alkylindole, converted into the indole during aqueous work up.

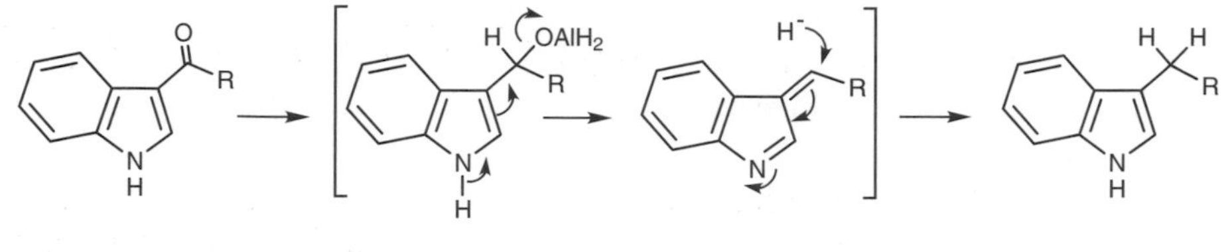

In comparison with indoles, benzo[*b*]furans and benzo[*b*]thiophenes have been studied much less fully, however similarities and some differences have been noted. Each system undergoes electrophilic substitution but the 3-regioselectivity is much lower than for indole, even to the extent that some attack takes place in the benzene ring of benz[*b*]thiophene and that 2-substitution is favoured for benzo[*b*]furan. These changes are consequent upon the much poorer electron donating ability of oxygen and sulfur – the nitrogen of indole is able to make a much bigger contribution to stabilising intermediates, particularly, as was shown above, for β-attack, and consequently to have a larger influence on regioselectivity. In the case of benzo[*b*]furan, it appears that simple benzylic resonance stabilisation in an intermediate from 2-attack outweighs the assistance that oxygen might provide to stabilise an adjacent positive charge.

Oxygen and sulfur systems undergo lithiation at their 2-positions, consistent with the behaviour of furans, thiophenes, and of *N*-blocked pyrroles and indoles.

The chemical behaviour of isoindole, benzo[*c*]thiophene and isobenzofuran is dominated by their lack of a 'complete' benzene ring: these three heterocycles undergo cycloaddition processes across the 1,3-positions with great facility, because the products do now have a regular benzene ring. Often, no attempt is made to isolate these heterocycles but they are simply generated in the presence of the dienophile with which it is desired that they react. As a result of this strong tendency, few of the classical electrophilic and nucleophilic processes have been much studied.

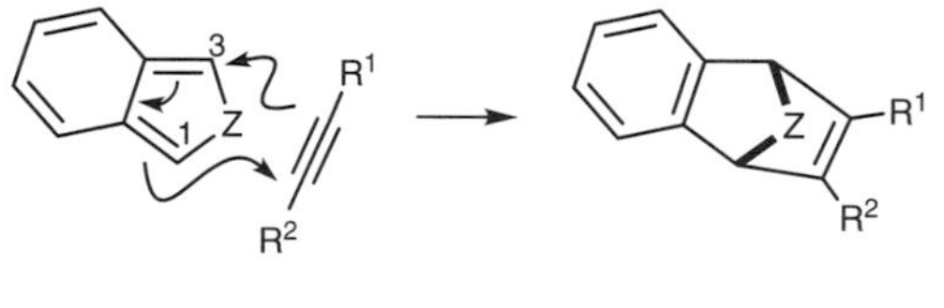

There has probably been more work carried out on the synthesis of indoles than on any other single heterocyclic system and consequently many routes are available; ring syntheses of benzo[*b*]thiophenes, benzo[*b*]furans have been much less studied. It is surprising that the Fischer indole synthesis, now more than a hundred years old, is still widely used – an arylhydrazone is heated with an acid, a multi-step sequence ensues, ammonia is lost and an indole is formed.

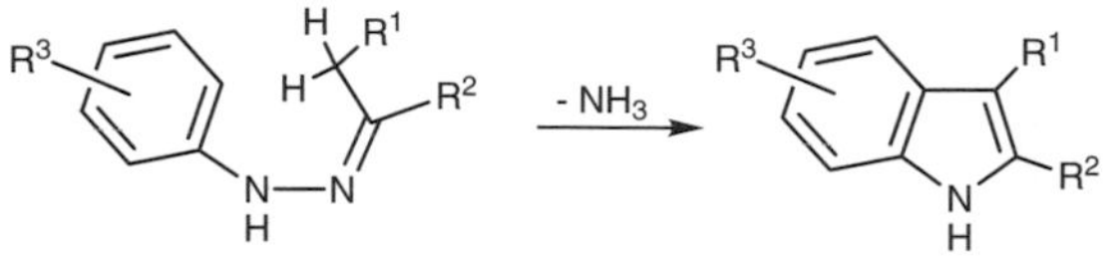

As an illustration of a recently developed and efficient route, 2,3-unsubstituted indoles are obtained from an *ortho*-nitrotoluene by heating with dimethylformamide dimethylacetal generating an enamine which, after reduction of the nitro group, closes with loss of dimethylamine generating the aromatic heterocycle.

Both benzo[*b*]thiophenes and benzo[*b*]furans can be obtained from the thiophenol or phenol respectively, by *S*-/*O*-alkylation with bromoacetaldehyde acetal and then acid-catalysed ring closure involving intramolecular electrophilic attack on the ring.

17 Indoles: reactions and synthesis

indole
[1*H*-indole]

indolenine
[3*H*-indole]

Indole[1] and the simple alkylindoles are colourless crystalline solids with a range of odours from naphthalene-like in the case of indole itself to faecal in the case of skatole (3-methylindole). Many simple indoles are available commercially and all of these are produced by synthesis: indole, for example, is made by the high-temperature vapour-phase cyclising dehydrogenation of 2-ethylaniline. Most indoles are quite stable in air with the exception of those which carry a simple alkyl group at C-2: 2-methylindole autoxidises easily even in a dark brown bottle.

The word indole is derived from the word India: a blue dye imported from India was known as indigo in the sixteenth century. Chemical degradation of the dye gave rise to oxygenated indoles (section 17.14) which were named indoxyl and oxindole; indole itself was first prepared in 1866 by zinc dust distillation of oxindole.

For all practical purposes, indole exists entirely in the 1*H*-form, 3*H*-indole (indolenine) being present to the extent of only *ca.* 1 ppm. 3*H*-indole has been generated in solution and found to tautomerise to 1*H*-indole within about 100 seconds at room temperature.[2]

Indoles are probably the most widely distributed heterocyclic compounds in nature. Tryptophan is an essential amino acid and as such is a constituent of most proteins; it also serves as a biosynthetic precursor for a wide variety of tryptamine-, indole-, and 2,3-dihydroindole-containing secondary metabolites.

tryptophan

tryptamine

serotonin (5-hydroxytryptamine)

melatonin

In animals, serotonin (5-hydroxytryptamine) is a very important neurotransmitter in the central nervous system and also in the cardiovascular and gastrointestinal systems. The structurally similar hormone melatonin is thought to control the diurnal rhythm of physiological functions.

Study and classification of serotonin receptors has resulted in the design and synthesis of highly selective medicines such as Sumatriptan, for the treatment of migraine, Ondansetron for the suppression of the nausea and vomiting caused by cancer chemotherapy and radiotherapy, and Alosetron for treatment of irritable bowel syndrome.

Sumatriptan   Ondansetron   Alosetron

Tryptophan-derived substances in the plant kingdom include indol-3-ylacetic acid, a plant growth-regulating hormone, and a huge number and structural variety of secondary metabolites – the indole alkaloids.[3] In the past, the potent physiological properties of many of these led to their use in medicine, but in most instances these have now be supplanted by synthetic substances, although vincristine, a 'dimeric' indole alkaloid is still extremely important in the treatment of leukemia. Brassinin, isolated from turnips, is a phytoalexin – one of a group of compounds produced by plants as a defense mechanism against attack by microorganisms.

indol-3-ylacetic acid   vincristine   brassinin

The physiological activity of lysergic acid diethylamide (LSD) is notorious. The synthetic indol-3-ylacetic acid derivative Indomethacin is used for the treatment of rheumatoid arthritis.

lysergic acid diethylamide (LSD)   Indomethacin

17.1   Reactions with electrophilic reagents

17.1.1   Protonation

Indoles, like pyrroles, are very weak bases: typical pK_a values are indole, −3.5; 3-methylindole, −4.6; 2-methylindole. −0.3.[4] This means, for example, that in 6M sulfuric acid two molecules of indole are protonated for every one unprotonated, whereas 2-methylindole is almost completely protonated under the same conditions. By NMR and UV examination, only the 3-protonated cation ($3H$-indolium cation) is detectable;[5] it is the thermodynamically stablest cation, retaining full benzene aromaticity (in contrast to the 2-protonated cation) with delocalisation of charge over the nitrogen and α-carbon. The spectroscopically undetectable N-protonated cation must be formed, and formed very rapidly, for acid-catalysed deuterium exchange at nitrogen is 400 times faster than at C-3,[6] indeed the N-hydrogen exchanges rapidly even at pH 7, when no exchange at C-3 occurs: clean conversion of indole into 3-deuterioindole can be achieved by successive deuterio-acid then water treatments.[7] Base-catalysed exchange, *via* the indolyl anion (section 17.4) likewise takes place at C-3.[8]

1*H*-indolium cation
(formed fastest)
2*H*-indolium cation
3*H*-indolium cation
(stablest)
2-methyl stabilises cation

That 2-methylindole is a stronger base than indole can be understood on the basis of stabilisation of the cation by electron release from the methyl group; 3-methylindole is a somewhat weaker base than indole.

Reactions of β-protonated indoles (see also sections 17.1.6 and 17.1.9)

3-*H*-Indolium cations are of course electrophilic species, in direct contrast with neutral indoles, and under favourable conditions will react as such. For example, the 3H-indolium cation itself will add bisulfite at pH 4, under conditions which lead to the crystallisation of the product, the sodium salt of indoline-2-sulfonic acid (indoline is the widely used, trivial name for 2,3-dihydroindole). The salt reverts to indole on dissolution in water, however it can be *N*-acetylated and the resulting acetamide used for halogenation or nitration at C-5, final hydrolysis with loss of bisulfite affording the 5-substituted indole.[9]

When N_b-acyl tryptophans are exposed to strong acid, the 3-protonated indolium cation is trapped by intramolecular cyclisation of the side-chain nitrogen.[10]

17.1.2 Nitration; reactions with other nitrogen electrophiles

Indole itself can be nitrated using benzoyl nitrate as a non-acidic nitrating agent; the usual mixed acid nitrating mixture leads to intractable products, probably because of acid-catalysed polymerisation. This can be avoided by carrying out the nitration using concentrated nitric acid and acetic anhydride at low temperature – under these conditions, *N*-alkylindoles, and indoles carrying electron-withdrawing *N*-substituents, but *not* indole itself, can be satisfactorily nitrated.[11]

2-Methylindole gives a 3-nitro-derivative with benzoyl nitrate,[12] but can also be nitrated successfully with concentrated nitric/sulfuric acids, but with attack at C-5. The absence of attack on the heterocyclic ring is explained by the complete protonation of 2-methylindole under these conditions; the regioselectivity of attack, *para* to the nitrogen, may mean that the actual moiety attacked is a bisulfate adduct of the initial 3-*H*-indolium cation, as shown in the scheme. 5-Nitration of 3-*H*-indolium cations has been independently demonstrated using an authentic 3,3-disubstituted 3-*H*-indolium cation.[13] With an acetyl group at C-3, nitration with nitronium tetrafluoroborate in the presence of tin(IV) chloride takes place at either C-5 or C-6 depending on the temperature of reaction.[14]

Indoles readily undergo electrophilic amination with bis(2,2,2-trichloroethyl)azodicarboxylate, the resulting acylated hydrazine being cleaved by zinc dust to give a 3-acylaminoindole.[15]

17.1.3   Sulfonation; reactions with other sulfur electrophiles

Sulfonation of indole,[16] at C-3, is achieved using the pyridine-sulfur trioxide complex in pyridine as solvent. Gramine is sulfonated in oleum to give 5- and 6-sulfonic acids, attack being on a diprotonated (C-3, side-chain-N) salt.[17] Sulfenylation of indole also occurs readily, at C-3.[18] The reversibility of this process has been demonstrated by desulfenylation in the presence of acid and a trap for the sulfenyl cation,[19] and by the acid-catalysed isomerisation of 3-sulfides to 2-sulfides.[20] Thiocyanation of indole can be achieved in virtually quantitative yield with a combination of ammonium thiocyanate and cerium(IV) ammonium nitrate.[21]

17.1.4 Halogenation

3-Halo, and even more so, 2-haloindoles are unstable and must be utilised as soon as they are prepared. A variety of methods are available for the β-halogenation of indoles: bromine or iodine (the latter with potassium hydroxide) in dimethylformamide[22a] give very high yields; pyridinium tribromide[22b] works efficiently; iodination[22c] and chlorination[22d] tend to be carried out in alkaline solution and, at least in the latter case is believed to involve intial N-chlorination then rearrangement. Reaction of 3-substituted indoles with halogens is more complex; initial 3-halogenation occurs generating a 3-halo-3H-indole,[23] but the actual products obtained then depend upon the reaction conditions, solvent etc. Thus, nucleophiles can add at C-2 in the intermediate 3-halo-3H-indoles when, after loss of hydrogen halide, a 2-substituted indole is obtained as final product, for example in aqueous solvents, water addition produces oxindoles (section 17.14); comparable methanol addition gives 2-methoxindoles.

Direct 2-bromination of 3-substituted indoles can be carried out using N-bromosuccinimide in the absence of radical initiators.[24] 2-Bromo- and 2-iodoindoles can be prepared very efficiently via α-lithiation (section 17.6.1);[25] 2-haloindoles are also available from the reaction of oxindoles with phosphoryl halides.[26] Bromination of methyl indole-3-carboxylate gives a mixture of 5- and 6-bromo derivatives.[27]

The formation of 3-iodoindole shown below, and its advisable immediate stabilisation by formation of a 1-phenylsulfonyl derivative, probably involves initial N-iodination of the indolyl anion and then rearrangement.[28]

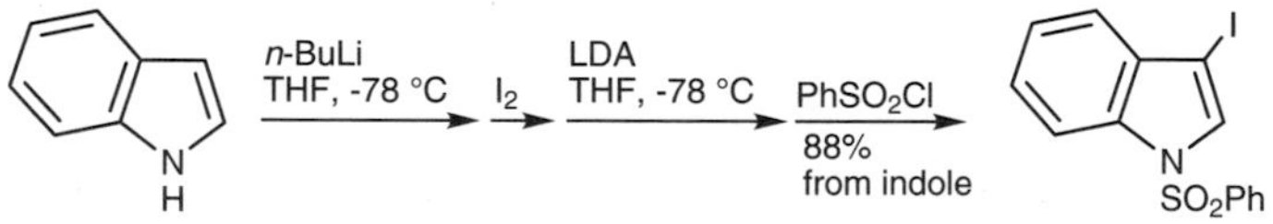

17.1.5 Acylation

Indole only reacts with acetic anhydride at an appreciable rate above 140 °C, giving 1,3-diacetylindole predominantly, together with smaller amounts of N- and 3-acetylindoles; 3-acetylindole is prepared by alkaline hydrolysis of product mixtures.[29] That β-attack occurs first is shown by the resistance of 1-acetylindole to C-acetylation, but the easy conversion of the 3-acetylindole into 1,3-diacetylindole. In contrast, acetylation in the presence of sodium acetate, or 4-dimethylaminopyridine,[30] affords exclusively N-acetylindole, probably via the indolyl anion (section 17.4). Trifluoroacetic anhydride, being much more reactive, acylates at room temperature, in dimethylformamide at C-3, but in dichloromethane at nitrogen.[31]

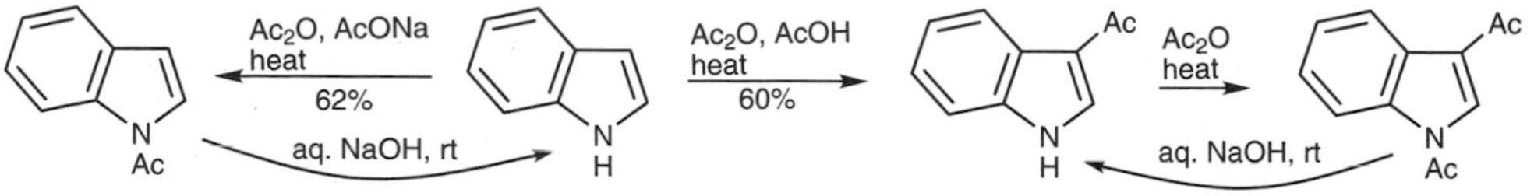

N-Acylindoles are much more readily hydrolysed that ordinary amides, aqueous sodium hydroxide at room temperature being sufficient: this lability is due in part to a much weaker mesomeric interaction of the nitrogen and carbonyl groups, making the latter more electrophilic, and in part to the relative stability of the indolyl anion, which makes it a better leaving group than amide anion.

Simply heating indole with triethyl orthoformate at 160 °C leads to the alkylation of the indole nitrogen introducing a diethoxymethyl (DEM) group which can be used as a reversible *N*-blocking substituent with considerable potential – it allows 2-lithiation (*cf.* section 17.4.2) and can be easily removed with dilute acid at room temperature.[32]

The Vilsmeier reaction is the most efficient route to 3-formylindoles[33] and to other 3-acylindoles using tertiary amides of other acids in place of dimethylformamide.[34] Even indoles carrying an electron-withdrawing group at the 2-position, for example ethyl indole-2-carboxylate, undergo smooth Vilsmeier β-formylation.[35]

A particularly useful and high-yielding reaction is that betwen indole and oxalyl chloride, which gives a ketone-acid chloride convertible into a range of compounds, for example tryptamines; a synthesis of serotonin utilised this reaction.[36]

Indoles, with a side-chain acid located at C-3, undergo cyclising acylation forming cyclic 2-acylindoles.[37] Intramolecular Vilsmeier processes using tryptamine amides lead to the imine, rather than a ketone, as the final product; the cyclic nature of the imine favours its retention rather than hydrolysis to amine plus ketone.[38]

Acylation of 3-substituted indoles is more difficult: 2-acetylation can be effected with the aid of boron trifluoride catalysis.[36] When the 3-substituent is an acetic acid moiety, a subsequent enol-lactonisation produces an indole fused to a 2-pyrone; these can be hydrolysed to the keto-acid, or the diene character of the 2-pyrone (section 8.2.2.4) utilised, as illustrated.[39]

Deactivation of the five-membered ring by electron-withdrawing substituents allows acylation in the six-membered ring. 1-Pivaloylindole gives high yields of 6-substituted ketones on reaction with α-halo acid chlorides and aluminium chloride; simple acid chlorides react only at C-3.[40] Another device which can be used to direct acylation to the benzene ring is to carry out the substitution on an iminium salt intermediate from a Vilsmeier reaction, when attack is at C-5 and C-6 with the former predominating.[41] The sequence below shows how a pivaloyl group can be introduced onto the indole nitrogen of 3-(indol-3-yl)propanoic acid using two mol equivalents of base and then the subsequent Friedel-Crafts cyclisation away from the deactivated heterocyclic ring, to C-4.[42]

17.1.6   Alkylation

Indoles do not react with alkyl halides at room temperature. Indole itself begins to react with iodomethane in dimethylformamide at about 80 °C when the main product is skatole. As the temperature is raised, further methylation occurs until eventually 1,2,3,3-tetramethyl-3H-indolium iodide is formed.

The rearrangement of 3,3-dialkyl-3H-indolium ions by alkyl migration to give 2,3-dialkylindoles, as shown in the sequence above, is related mechanistically to the Wagner-Meerwein rearrangement, and is known as the Plancher rearrangement.[43] It is likely that most instances of 2-alkylation of 3-substituted-indoles by cationic reagents proceed by this route, and this was neatly verified in the formation of 1,2,3,4-tetrahydrocarbazole by boron trifluoride catalysed cyclisation of 4-(indol-3-yl)butan-1-ol. The experiment was conducted with material labelled at the benzylic carbon. The consequence of the rearrangement of the symmetrical spirocyclic intermediate, which results from attack at C-3, was the equal distribution of the label between the C-1 and C-4 carbons of the product.[44] It is important to note that other experiments demonstrated that direct attack at C-2 can and does occur,[45] especially when this position is further activated by a 6-methoxyl group.[46]

In another elegant experiment, the intervention of a 3,3-disubstituted 3H-indolium intermediate in an indole overall α-substitution was proved by cyclisation of the mesylate of an optically active alcohol to give an optically *inactive* product, *via* an achiral, spirocyclic intermediate from initial attack at the β-position.[47]

Indoles react with epoxides and aziridines in the presence of Lewis acids (see section 17.5 for reaction of indolyl anions with such reactants) with opening of the three-membered ring and consequent 3-(2-hydroxyalkylation) and 3-(2-amino-alkylation) of the heterocycle. Both ytterbium triflate and phenylboronic acid are good catalysts for reaction with epoxides under high pressure;[48] silica gel is also an effective catalyst, but slow at normal pressure and temperature.[49] Lewis acid-mediated reaction with aziridines can be catalysed by zinc triflate or boron trifluoride.[50] More reactive alkylating electrophiles react at lower temperatures, at room temperature with dimethylallyl bromide for example.[51]

17.1.7   Reactions with aldehydes and ketones

Indoles react with aldehydes and ketones under acid catalysis – with simple carbonyl compounds, the initial products, indol-3-ylcarbinols are never isolated, for in the acidic conditions they dehydrate to 3-alkylidene-3H-indolium cations; those from aromatic aldehydes have been isolated in some cases;[52] reaction of 2-methylindole with acetone under anhydrous conditions gives the simplest isolable salt of this class.[53] Only where dehydration is not possible have hydroxyalkylindoles been isolated, for example from reaction with diethyl mesoxalate.[54] Reaction with 4-dimethylaminobenzaldehyde (the Ehrlich reaction, see section 13.1.7) gives a mesomeric and highly-coloured cation.

3-Alkylidene-3H-indolium cations are themselves electrophiles and can react with more of the indole, as illustrated for reaction with formaldehyde.[55] Cyclic ketones react with 1,2-dimethylindole producing 3-cycloalkenylindoles.[56]

3-Alkylation of 2-alkyl- and 2-arylindoles can be achieved by trifluoroacetic acid-catalysed condensation with either aromatic aldehydes or aliphatic ketones in the presence of the triethylsilane to reduce the immediate 3-(hydroxyalkyl) products.[57]

17.1.8   Reactions with α,β-unsaturated ketones, -nitriles and -nitro-compounds

Such reactions are usually effected using acid, for catalysis, and can be looked on as an extension of the reactions discussed in 17.1.7 above. In the simplest example indole reacts with methyl vinyl ketone in a conjugate fashion.[58]

The use of Montmorillonite clay, a very efficient 'acidic' catalyst, allows α-alkylation of β-substituted indoles;[59] ytterbium triflate can also be used to catalyse such alkylations.[60] This efficient catalysis contrasts with the different, but very instructive, reaction pathway followed when mesityl oxide and 1,3-dimethylindole are combined in the presence of sulfuric acid – electrophilic attack at the already substituted β-position is followed by intramolecular nucleophilic addition of the enol of the side-chain ketone to C-2.[61]

An extension of this methodology allows the synthesis of tryptophans by aluminium chloride catalysed alkylation with an iminoacrylate, as illustrated[62] (for the same transformation but biocatalysed see section 2.9).

Nitroethene is sufficiently electrophilic to substitute indole without the need for acid catalysis;[63] the employment of 2-dimethylamino-1-nitroethene in trifluoroacetic acid leads to 2-(indol-3-yl)nitroethene – the reactive species is the protonated enamine and the process is similar to a Mannich condensation (section 17.1.9).[64] The use of 3-trimethylsilylindoles, with *ipso* substitution of the silane,[65] is an alternative means for effecting alkylation avoiding the need for acid catalysis.

17.1.9   Reactions with iminium ions: Mannich reactions[66]

Under neutral conditions and at low temperature indole reacts with a mixture of formaldehyde and dimethylamine by substitution at the indole nitrogen;[67] it seems likely that this reaction involves a low equilibrium concentration of the indolyl anion. In neutral solution at higher temperature or in acetic acid, conversion into the thermodynamically more stable, 3-substituted product, gramine, takes place. Gramine is formed directly, smoothly and in high yield, by reaction in acetic acid.[68] The Mannich reaction is very useful in synthesis because not only can the electrophilic iminium ion be varied widely, but the product gramines are themselves intermediates for further manipulation (section 17.12).

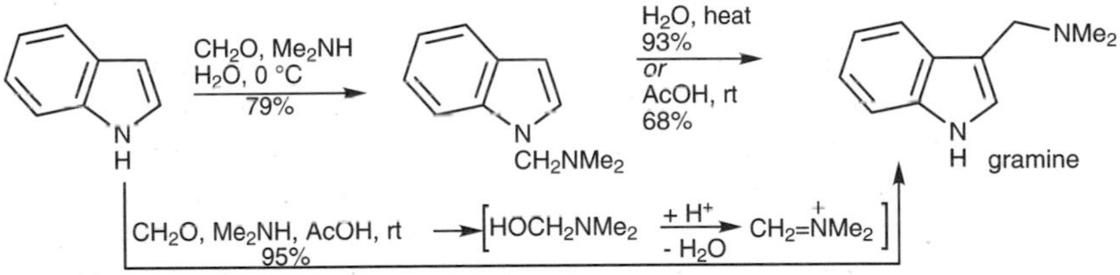

The iminium ion electrophile can be synthesised separately, as a crystalline solid known as 'Eschenmoser's salt' ($Me_2N^+=CH_2$ I^-)[69] and with this reactive electrophile the reaction is normally carried out in a non-polar solvent. Examples which illustrate the variation in iminium ion structure which can be tolerated, include the reaction of indole with pyrimidine,[70] with benzylidene derivatives of arylamines catalysed by lanthanide triflates,[71] and the mineral acid-catalysed dimerisation of indole.[72] In the first example protonated pyrimidine is the electrophile, in the last indole is attacked by *protonated* indole! as shown below.

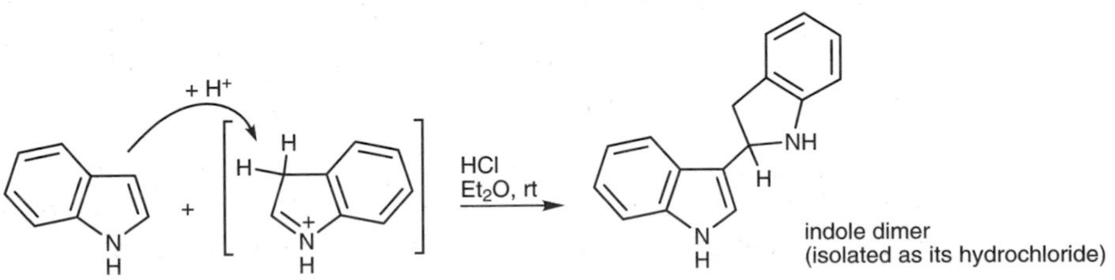

Skatole is converted into an α,α'-linked dimer in acid; 2-methylindole, in contrast, is not susceptible to acid-catalysed dimerisation, reflecting the lower electrophilic character of the 3-protonated 2-substituted $3H$-indolium cation, much as ketones are less reactive than aldehydes.

When protonated 3-bromoindole is employed as electrophile, a final elimination of hydrogen bromide gives rise to rearomatised 2-substituted indoles; pyrrole (illustrated) or indoles will take part in this type of process.[73]

3-bromoindole
CF$_3$CO$_2$H, rt
83%

- HBr

Conducted in an intramolecular sense, both Mannich and Vilsmeier reactions have been much used for the construction of tetrahydro-β-carbolines[74] (dihydro-β-carbolines), such as are found in many indole alkaloids (β-carboline is the widely-used, trivial name for the pyrido[3,4-b]indole).

PhCH$_2$CH=O pH 6
90%

There is still controversy as to whether such cyclisations proceed by direct electrophilic attack at the α-position, or whether by way of β-attack then rearrangement. It may be significant that Mannich processes, as opposed to the alkylations discussed in section 17.1.6, are reversible, which would allow a slower, direct α-substitution to provide the principal route to the α-substituted structure.

It has been shown that tryptamines carrying a 2-carboxylic acid group, which can be conveniently prepared (section 17.17.6.3) but are not easily decarboxylated, undergo cyclising Mannich condensation with aldehydes and ketones, with loss of the carbon dioxide in a final step.[75]

PhCH=O
CF$_3$CO$_2$H
PhH, heat
75%

- CO$_2$

The cyclisation of nitrones derived from tryptamines is a similar process and can be carried out enantioselectively using a chiral Lewis acid.[76]

B(OPh)$_3$, R-BINOL
CH$_2$Cl$_2$, rt
81%

73% ee

17.1.10 Diazo-coupling and nitrosation

The high reactivity of indole is shown up well by the ease with which it undergoes substitution with weakly electrophilic reagents such as benzenediazonium chloride and nitrosating agents. Indoles react rapidly with nitrous acid; indole itself reacts in a complex manner, but 2-methylindole gives a 3-nitroso-product cleanly. This can also be obtained by a base-catalysed process using amyl nitrite as a source of the nitroso group; these basic conditions also allow clean 3-nitrosation of indole itself. 3-Nitrosoindoles exist predominantly in the oximino tautomeric form.[77] Skatole and other 3-substituted indoles give relatively stable *N*-nitroso products with nitrous acid,[78] consistent with kinetic studies on 2-methylindole which show that *N*-substitution precedes *C*-substitution. *N*-Nitrosoindoles may be produced from indoles, following ingestion – such compounds may be mutagenic.

17.1.11 Electrophilic metallation

17.1.11.1 Mercuration

Indole reacts readily with mercuric acetate at room temperature to give a 1,3-disubstituted product.[79] Even *N*-acylindoles are substituted under mild conditions; the 3-mercurated compounds thus produced are useful in palladium-catalysed couplings.[80]

17.1.11.2 Thallation

Thallium trifluoroacetate reacts rapidly with simple indoles, but well defined products cannot be isolated. 3-Acylindoles, however, undergo a very selective substitution at C-4, due to chelation and protection of the heterocyclic ring by the electron-withdrawing 3-substituent.[81] The products are good intermediates for the preparation of 4-substituted indoles, for example 4-iodo- and thence 4-alkoxy-,[82b] 4-alkenyl[82] and 4-methoxycarbonyl,[83] *via* palladium-mediated couplings. The regiochemistry is neatly complemented by thallation of *N*-acetylindoline, which goes to C-7, allowing introduction of substituents at this carbon[84] (cf. sections 17.17.1.8 and 17.3).

17.1.11.3 Palladation

Even indoles bearing acyl or phenylsulfonyl substituents on nitrogen, are easily palladated at moderate temperatures, substitution occuring at C-3, or at the 2-position if C-3 is occupied. The metallated products are seldom isolated but allowed to react with acrylates, other alkenes (Heck reaction), or carbon monoxide[85] *in situ*.[86] Although electrophilic palladation normally requires one equivalent of palladium(II), the incorporation of reoxidants selective for Pd(0), such as *t*-butyl perbenzoate or copper(II) compounds, allows catalytic conversions to be carried out.[87]

17.2 Reactions with oxidising agents

Autoxidation occurs readily with alkyl indoles, thus for example, 2,3-diethylindole gives an isolable 3-hydroperoxy-3H-indole. Generally such processes give more complex product mixtures resulting from further breakdown of the hydroperoxide; singlet oxygen also produces hydroperoxides, but by a different mechanism. If the indole carries a side-chain capable of trapping the indolenine by intramolecular nucleophilic addition, then tricyclic hydroperoxides can be isolated, as shown.[88]

The reagent $MoO_5.HMPA$, known as 'MoOPH', in methanol, brings about addition of the elements of methyl hydrogen peroxide to an N-acylindole, and these adducts in turn, can be utilised: one application is to induce loss of methanol, and thus the overall transformation of an indole into an indoxyl.[89]

Oxidative cleavage of the indole 2,3-double bond has been achieved with ozone, sodium periodate,[90] potassium superoxide,[91] with oxygen in the presence of cuprous chloride,[92] and with oxygen photochemically in ethanolic solution;[93] irradiation in an organic acid as solvent leads to oxidation at 2- and 3-alkyl groups.[94]

The efficient conversion of 3-substituted indoles into their corresponding oxindoles can be brought about by reaction with dimethylsulfoxide in acid; the scheme below shows a reasonable mechanism for the process.[95]

17.3   Reactions with nucleophilic reagents (see also section 17.14.4)

As with pyrroles and furans, indoles undergo very few nucleophilic substitution processes. Those that are known involve special situations: benzene-ring-nitro-indoles, in which the *N*-hydrogen has been removed as well, undergo vicarious nucleophilic substitutions (section 2.3.3).[96]

In chromium carbonyl complexes of indole, the metal is associated with the benzene ring, hence nucleophilic additions take place in that ring, usually at C-4; the example shows the relatively unusual attack at C-7; this regioselectivity can be induced to revert to the usual C-4 if an indole with a bulky *N*-protecting group is utilised.[97]

17.4   Reactions with bases

17.4.1   Deprotonation of *N*-hydrogen

As in pyrroles, the *N*-hydrogen in indoles is much more acidic (pKa 16.2) than that of an aromatic amine, say aniline (pKa 30.7). Any very strong base will effect complete conversion of an *N*-unsubstituted indole into the corresponding indolyl anion, amongst the most convenient being sodium hydride, *n*-butyllithium, or an alkyl Grignard reagent.

Electron-withdrawing substituents, particularly at the β-position, increase the acidity markedly, for example 3-formylindole is about 5 pK_a units more acidic than indole and the α-isomer is some 3 units more acidic.[98]

17.4.2   Deprotonation of *C*-hydrogen

Deprotonation of *C*-hydrogen in indoles requires the absence of the much more acidic *N*-hydrogen i.e. the presence of an *N*-substituent like methyl,[99] or if required, a removable group: phenylsulfonyl,[100] lithium carboxylate,[101] and *t*-butoxycarbonyl[102] have been used widely, dialkylaminomethyl,[103] trimethylsilylethoxymethyl,[104] and methoxymethoxy[105] have also been recommended; clearly in the last case the *N*-substituent cannot be introduced into an indole – it requires a preformed 1-hydroxyindole – but it is possible to reduce it off to leave an *N*-hydrogen-indole. *t*-

Butylaminocarbonyl,[106] methoxymethoxy, are said to be the optimal protecting/activating groups. Each of these removable substituents assists lithiation by intramolecular chelation and in some cases by electron withdrawal, reinforcing the intrinsic tendency for metallation to proceed at the α-position.

Magnesiation at C-2 can be carried out at room temperature; as well as serving in the usual way as nucleophiles, magnesioindoles can also be used directly for palladium-catalysed couplings.[107]

The dimethylamino group of gramine directs lithiation to C-4 when the indolic nitrogen is protected by the bulky tri-i-propylsilyl group but metallation occurs normally at C-2 when this nitrogen bears a simple methyl.[108]

17.5   Reactions of *N*-metallated indoles

The indolyl anion has two main mesomeric structures showing the negative charge to reside mainly on nitrogen and the β-carbon. In its reactions, then, this anion behaves as an ambident nucleophile; the ratio of *N*- to β-substitution with electrophiles depends on the associated metal, the polarity of the solvent, and the nature of the electrophile. Generally, the more ionic sodio- and potassio-derivatives tend to react at nitrogen, whereas magnesio-derivatives have a greater tendency to react at C-3.[109] However, reaction of indolyl Grignards in HMPA leads to more attack at nitrogen, whereas non-polar solvents favour attack at carbon.[110] Complimentarily, more reactive electrophiles show a greater tendency to react at nitrogen than less electrophilic species.

N-Alkylation of indoles can utilise indol-1-ylsodiums,[111] generated quantitatively as above, or it can involve a small concentration of an indolyl anion, produced by phase-transfer methods;[112] indole *N*-acylation[113] and -arylsulfonylation[114] can also be achieved efficiently using phase-transfer methodology (see also section 17.1.4).

Indolyl N-Grignards,[115] or even better their zinc equivalents,[116] undergo reaction predominantly at C-3 with a variety of carbon electrophiles such as aldehydes, ketones and acid halides, and for example 2-bromothiazole.[117] Including aluminium chloride in the reaction mixture produces high yields of 3-acylindoles.[118] The copper-catalysed reactions of indolyl N-Grignards with N-t-butoxycarbonylaziridines also proceed well at C-3.[119] 1-Lithioindoles are equally useful; again, the position of attack depends on both solvent and the nature of the electrophile.[120]

It is important to note that when an N-metallated 3-substituted indole alkylates at carbon, necessarily a 3,3-disubstituted-3H-indole (an indolenine) is formed, which cannot rearomatise to form an indole (see section 17.1.6 for rearrangements of 3,3-disubstituted indolenines).

17.6 Reactions of C-metallated indoles

17.6.1 Lithium derivatives

One of the most convenient N-protecting groups to be used in indole α-lithiations is carbon dioxide[102] because the N-protecting group is installed *in situ* and, further, falls off during normal work-up. This technique has been used to prepared 2-haloindoles[25] and to introduce a variety of substituents by reaction with appropriate electrophiles – aldehydes, ketones, chloroformates etc.

Given below is a selection of α-substitutions achieved with various N-blocking/activating groups.[100–107, 121]

n-BuLi
THF, -78 °C
then PhCOMe
64%

aq. NaOH, heat
80%

t-BuLi
THF, -78 °C
then (CO$_2$Me)$_2$
66%

CF$_3$CO$_2$H, rt

2t-BuLi
THF, -78 °C
then PhCHO
87%

LiOH
85%

3-Lithioindoles can be prepared by halogen exchange;[122] the N-t-butyldimethylsilyl derivative is regiostable, even at 0 °C,[123] whereas 3-lithio-1-phenylsulfonylindole isomerises to the 2-isomer at temperatures above −100 °C though at that temperature, hetero-ring opening and production of an alkyne, with the nitrogen anion acting as a leaving group, (cf. section 18.3) is not a problem.[124,125] The corresponding N-phenylsulfonyl 3-magnesium[126] and 3-zinc[127] species are stable even at room temperaure: they can be prepared from the 3-iodoindole by reaction with ethylmagnesium bromide and lithium trimethylzincate respectively. 3-Lithiation with replacement of a 3-hydrogen has also been accomplished with *ortho* assistance from a 2-(2-pyridyl)[126] or a 2-carboxyl group,[128] which block C-2. Direct 3-lithiation even without a substituent at C-2 can be achieved with an N-di(t-butyl)fluorosilyl substituent in place.[129] Other examples of the directed metallation process in indole chemistry include: 2-lithiation of 1-substituted indole-3-carboxylic acids and amides,[130] and of 3-hydroxymethyl-1-phenylsulfonylindole;[131] 4-lithiation of 5-(dimethylcarbamoyloxy)-1-(t-butyldimethylsilyl)indole and the 6-lithiation of 4-substituted-5-(dimethylcarbamoyloxy)-1-(t-butyldimethylsilyl)indoles.[132]

PhI, Pd(PPh$_3$)$_4$
THF, 0 °C → rt
X = I

2t-BuLi
THF, -78 °C Me$_3$SnCl
X = Br   94%

t-BuLi
Et$_2$O, -105 °C

BF$_3$
Et$_2$O, -105 °C
87%

Me$_3$ZnLi, TMEDA
THF, -78 °C

CH$_2$=CHCH$_2$Br
-78 °C → rt
64%

Amazingly, metal–halogen exchange can be achieved with each of the benzene ring-bromoindoles *without* N-protection; the indole is first converted into its N-potassio-salt.[133]

17.6.2   Palladium-catalysed reactions

2-/3-Bromo- and -iodoindoles, and the similarly reactive 2- and 3-triflates,[134] undergo palladium-catalysed couplings as normal aryl halides. Since 2- and 3-haloindoles are unstable it is expedient to employ their *N*-acyl derivatives.[135] Halogen and triflate on the benzene ring of indoles take part unexceptionally in coupling reactions.

When an organometallic derivative of indole is required for a coupling reaction, boronic acids are to be preferred[136] although 2-zinc and 2-stannyl derivatives can be used.[137] The palladium-catalysed coupling of 6-bromo or 6-iodoindoles with allyl and heteroaryltin compounds does not require masking of the indole *N*-hydrogen.[138]

17.7   Reactions with radicals

Radicals such as benzyl and hydroxyl are unselective in their interaction with indoles resulting in mixtures of products, so such reactions are of little synthetic use. On the other hand, benzoyloxylation of indoles having no *N*-hydrogen gives benzoates of indoxyl,[139] i.e. it effectively oxidises the indole heterocyclic ring, via β-attack by the strongly electrophilic benzoyloxy radical. In contrast, the weakly electrophilic radical derived from malonate reacts selectively at C-2, via an atom transfer mechanism.[140]

Some efficient oxidative[141] and reductive[142] intramolecular carbon radical additions can be carried out. *Ipso* replacement of toluenesulfonyl by tributylstannyl radical occurs readily at C-2[143] (but not C-3) as does intramolecular replacement by carbon radicals.[144]

2-Indolyl radicals can be generated under standard conditions by reacting 2-bromoindole with tributyltin hydride.[145]

17.8   Reactions with reducing agents

The indole ring system is not reduced by nucleophilic reducing agents such as lithium aluminium hydride and sodium borohydride; lithium/liquid ammonia does however reduce the benzene ring; 4,7-dihydroindole is the main product.[146]

Reduction with lithium in the presence of trimethylsilyl chloride, followed by rearomatisation, produces 4-trimethylsilylindole, an intermediate useful for the synthesis of 4-substituted indoles *via* electrophilic *ipso* replacement of silicon.[147]

Reduction of the heterocyclic ring is readily achieved under acidic conditions; formerly, metal-acid combinations[148] were used, but now much milder conditions employ relatively acid-stable metal hydrides such as sodium cyanoborohydride. Triethylsilane in trifluoroacetic acid is another convenient combination; 2,3-disubstituted indoles give *cis* indolines by this method.[149] Such reductions proceed by hydride attack on the β-protonated indole – the 3*H*-indolium cation.[150] Catalytic reduction of indole, again in acid solution, produces indoline initially, further slower reduction completing the saturation.[151] Rhodium-catalysed high pressure hydrogenation of indoles with a *t*-butoxycarbonyl group on nitrogen proceeds smoothly to give 2,3-*cis* indolines.[152]

17.9   Reactions with carbenes

No cyclopropane-containing products have been isolated from the interaction of indoles with carbenes (cf. section 13.10). Methoxycarbonyl-substituted carbenes give only a substitution product.[153]

17.10   Electrocyclic and photochemical reactions

The heterocyclic double bond in simple indoles will take part in cycloaddition reactions with dipolar 4π components,[154] and with electron-deficient dienes (i.e. inverse electron demand), in most reported cases, held close using a tether;[155] a comparable effect is seen in the intermolecular cycloaddition of 2,3-cycloalkyl indoles to *ortho*-quinone generating a 1,4-dioxane.[156] The introduction of electron-withdrawing substituents enhances the tendency for cycloaddition to electron-rich dienes: 3-acetyl-1-phenylsulfonylindole, for example, undergoes aluminium chloride-catalysed cycloaddition with isoprene,[157] and 3-nitro-1-phenylsulfonylindole reacts with 1-acylaminobuta-1,3-dienes without the need for a catalyst.[158]

Some other, apparent cycloadditions probably proceed by non-concerted pathways, for example addition of 1,3-cyclohexadiene in the presence of light and 2,4,6-triphenylpyrylium, probably involves radical intermediates,[159] and reactions of 2-phenylsulfonyl-dienes with indolyl Grignard reagents probably proceed in stepwise fashion as shown.[160]

Both 2- and 3-vinylindoles take part quite readily as 4π components in Diels-Alder cycloadditions;[161] often, but not always,[162] these employ *N*-acyl- or *N*-arylsulfonyl-lindoles, in which the interaction between nitrogen lone pair and π-system has been reduced.[163] The example below shows how this process can be utilised in the rapid construction of a complex pentacyclic indole.[164]

Tethered 1,2,4-triazenes are particularly useful 'dienes' as their interaction with the indole 2,3-double bond generates carbolines. The tether can be incorporated into the product molecule,[165] or be designed to be broken *in situ*, as in the example below.[166] 1,2,4,5-Tetrazines react with the indole 2,3-bond in an intermolecular sense; the initial adduct loses nitrogen and then is oxidised to the aromatic level by a second mol equivalent of the tetrazine.[167]

The Fischer synthesis

The Fischer synthesis,[227] first discovered in 1883, involves the acid- or Lewis acid-catalysed rearrangement of a phenylhydrazone with the elimination of ammonia. The preparation of 2-phenylindole illustrates the process in its simplest form.[228]

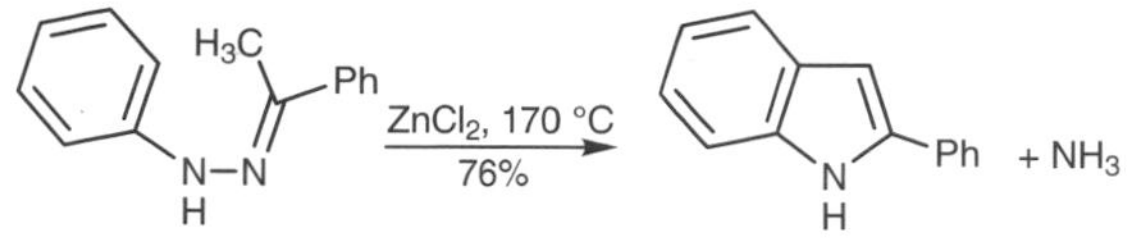

In many instances the reaction can be carried out simply by heating together the aldehyde or ketone and phenylhydrazine in acetic acid;[229] the formation of the phenylhydrazone and its subsequent rearrangement take place without the necessity for isolation of the phenylhydrazone. Toluenesulfonic acid, cation exchange resins, and phosphorus trichloride have each been recommended for efficient cyclisations, sometimes even at or below room temperature.[230] Electron-releasing substituents on the benzene ring increase the rate of Fischer cyclisation whereas electron-withdrawing substituents slow the process down,[231] though even phenylhydrazones carrying nitro-groups can be indolised satisfactorily with appropriate choice of acid and conditions, for example a two-phase mixture of toluene and phosphoric acid,[232] or boron trifluoride in acetic acid.[233] Electron-withdrawing substituents *meta* to the nitrogen give rise to roughly equal amounts of 4- and 6-substituted indoles; electron-releasing groups similarly oriented produce mainly the 6-substituted indole.[165]

The full mechanistic details of the multi-step Fischer sequence are still not completely sure, but there is considerable evidence that the sequence shown below operates, for example labelling studies proved the loss of the β-nitrogen as ammonia, and in some cases intermediates have been detected by ^{13}C and ^{15}N NMR spectroscopy.[234] The most important step – the one in which a carbon–carbon bond is made – is electrocyclic in character and analogous to the Claisen rearrangement of phenyl allyl ethers.

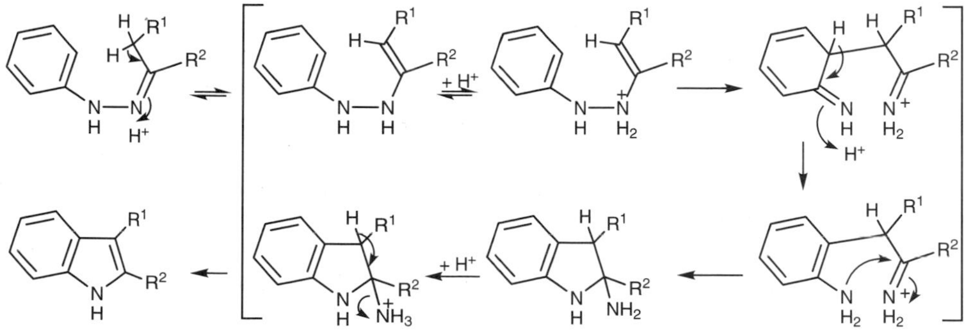

Support for this sequence also comes from the observation that in many cases indolisation can be achieved thermally, at a temperature as low as 110 °C, in the special case of preformed ene-hydrazines, i.e. in which the first step of the normal sequence – acid-catalysed tautomerisation of imine to enamine – has already been accomplished.[235] The reaction does however still occur more rapidly in the presence of acid and this is interpreted as protonation of the β-nitrogen, as shown, facilitating the electrocyclic step.

Fischer cyclisations can be achieved thermally, but generally much higher temperatures are required and proton transfer from solvent (typically a glycol) may be involved. However, using preformed N-trifluoroacetyl ene-hydrazines allows thermal cyclisation at temperatures as low as 65 °C.[236] As the example below shows, in the case of derivatives of cyclopentanones, the 2-aminoindoline intermediates can be isolated at lower temperatures; subsequent elimination of trifluoroacetamide is easy and efficient.

An extreme case of acid catalysis is the indolisation of phenylhydrazones of β-dicarbonyl compounds in concentrated sulfuric acid;[237] in milder acid, only pyrazolones are produced from the interaction of β-keto-esters with hydrazines (section 22.13.1.1).

An aspect of the Fischer reaction which is of considerable practical importance is the ratio of the two possible indoles formed from unsymmetrical ketones; in many instances mixtures result because ene-hydrazine formation occurs in both directions. It appears that strongly acidic conditions favour the least substituted ene-hydrazine.[238]

AcOH	100 : 0
PPA	50 : 50
MeSO$_3$H, P$_4$O$_{10}$	22 : 78

Indolenines (3H-indoles) are formed efficiently on Fischer cyclisation of the phenylhydrazones of branched ketones; note, again, the use of a weaker acid medium to promote formation of the more substituted ene-hydrazine required for indolenine formation.[239] In another example, addition of sodium acetate to the acetic acid reaction medium promoted indolenine formation from the phenylhydrazone of 1-decalone.[240]

An important extension to the Fischer synthesis is the preparation of arylhydrazines by palladium-catalysed coupling of benzophenone hydrazone with aryl halides – this allows the convenient preparation of a much wider range of arylhydrazines than the classical method involving the reduction of diazonium salts. The benzophenone arylhydrazone can be hydrolysed to the hydrazine, but even more conveniently used directly in the Fischer cyclisation where exchange occurs with the ketone. The whole process, from arylhalide to indole, can be carried out in one pot without isolation of any intermediates.[241]

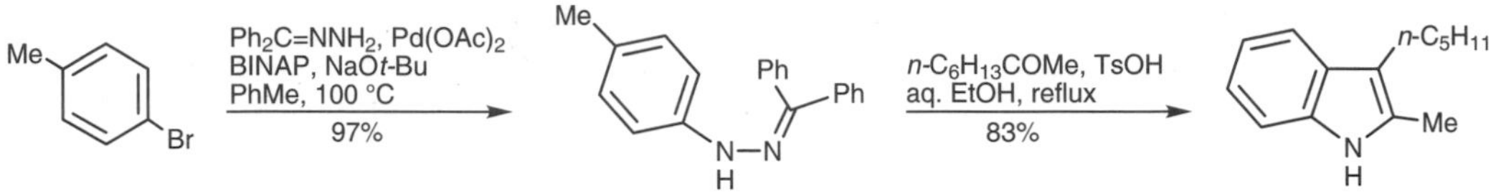

Another way of making arylhydrazines is the electrophilic amination of electron-rich arenes with an azodicarboxylate.[15]

Transformations which are mechanistically analogous to the Fischer, and also produce indoles, use phenylhydroxylamines instead of phenylhydrazines, as shown below.[242]

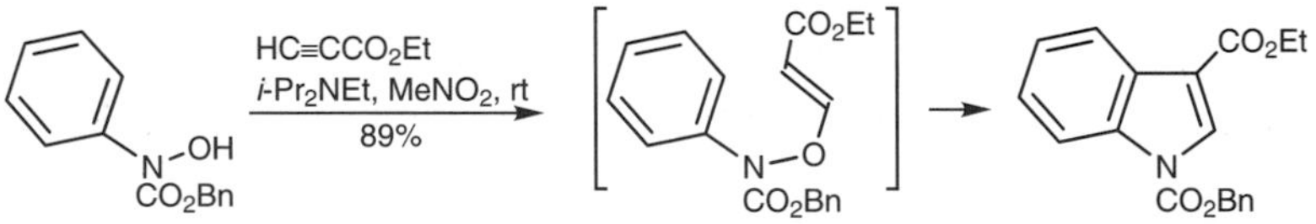

The Grandberg synthesis

An exceptionally useful adaptation is the Grandberg synthesis of tryptamines from 4-halobutanals, or more often in practice their acetals,[243] in which the nitrogen usually lost during the Fischer process is incorporated as the nitrogen of the aminoethyl side-chain.[244]

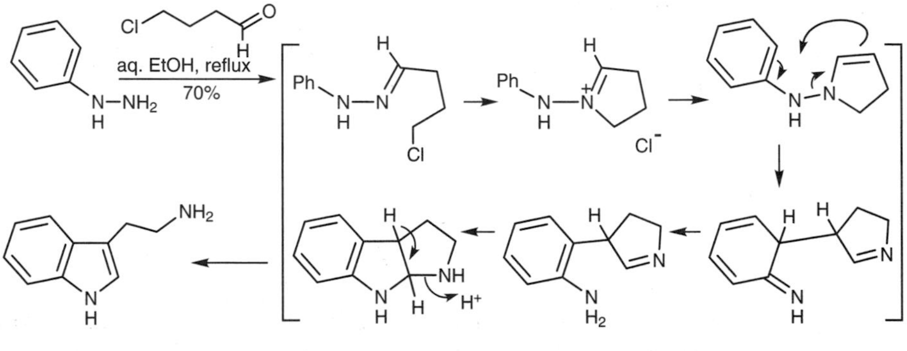

17.17.1.2 From ortho-(2-oxoalkyl)anilines

Cyclisation of *ortho*-(2-oxoalkyl)anilines by simple intramolecular condensation with loss of water, occurs spontaneously. Several new ways of generating the intermediate amino-ketone have been developed; the prototype was the Reissert synthesis.

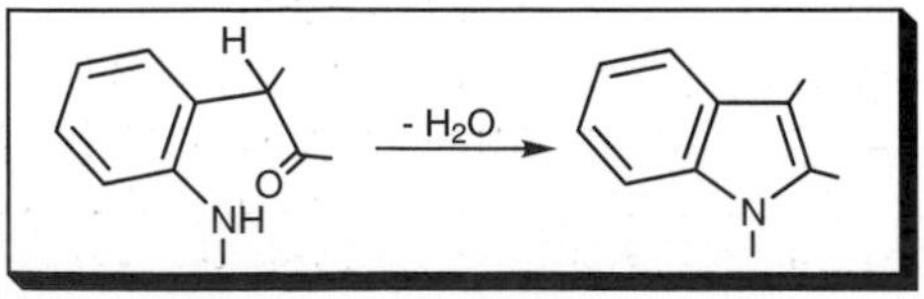

The Reissert synthesis

In the classical Reissert synthesis the acidity of a methyl group *ortho* to nitro on a benzene ring is the means for condensation with oxalate; the nitro group is then reduced to amino.[245]

In a development, the nitrogen is already at the oxidation level of amine, but carries a *t*-butoxycarbonyl group to assist the methyl (alkyl) lithiation, reaction with oxalate as in the classical sequence and removal of the *N*-substituent with acid at the end, again leads to an indole-2-ester.[246] The synthesis of 2-unsubstituted indoles is achieved by reaction of the *N,C*-dilithiated species with dimethylformamide.[247]

In another variant, aromatic nitro compounds can be made to condense[248] with silyl enol ethers using tris(dimethylamino)sulfur (trimethylsilyl)difluoride (TASF); a non-aromatic nitronate intermediate is oxidised up with bromine, without isolation, to provide a 2-(*ortho*-nitroaryl)ketone and thence an indole after nitro group reduction.[249]

Leimgruber-Batcho synthesis

The Leimgruber-Batcho synthesis[250] is one of the most widely used new variations which also depends on the acidity of methyl groups *ortho* to aromatic nitro (or at α or γ positions on a pyridine[251]) to allow introduction of the future indole α-carbon as an enamine. Condensation with hot dimethylformamide dimethyl acetal (DMFDMA) (no added base being necessary) leads to an enamine; subsequent reduction of the nitro group, usually in acid conditions, leads directly to the hetero-ring-unsubstituted indole. Mechanistically this, at first sight extraordinary, process is believed to involve ionisation of the reagent producing methoxide (which deprotonates the aromatic methyl) and an electrophilic component, $MeOCH=N^+Me_2$, which combines with the deprotonated aromatic. Both tris(piperidin-1-yl)methane and bis(dimethylamino)-*t*-butoxymethane are said to function even better than the commercially available DMFDMA.[252] A variety of benzene substituents are tolerated and the approach has been utilised for syntheses of, amongst others, 4- and 7-indole-carboxylic esters.[253]

A Leimbruger-Batcho-type amino-enamine intermediate is likely to be involved on reduction of the base-catalysed condensation product of an *ortho*-nitro araldehydes with nitromethane.[254] Reduction, traditionally with metal/acid combinations, but now with reagents such as palladium/carbon with ammonium formate and formic acid,[255] iron with acetic acid and silica gel,[256] or titanium(III) chloride,[257] gives the indole. The arylacetaldehyde precursors can also be generated by Heck reactions on vinylidene carbonate.[258]

Coupling reactions using *ortho*-haloanilines have been widely used; in these instances no reductive step is required though the carbonyl unit is sometimes incorporated in masked form, such as a 2-ethoxyvinylboronate, requiring deprotection.[259]

17.17.1.3 From ortho-alkynylarylamines

Cyclisation of *ortho*-alkynylarylamines can be achieved in various ways; palladium-catalysed couplings provide the starting alkynylanilines.

Palladium-catalysed coupling methodology now allows easy access to arenes with an alkynyl substituent *ortho* to nitrogen, from *ortho*-iodo- and -bromonitrobenzenes,[260] or *ortho*-iodo- and -bromo-*N*-acyl (or *N*-sulfonyl) arylamines,[261] or even by coupling acetylenes with 2-iodoaniline itself.[262] Conversion of *ortho* alkynyl-nitrobenzenes and -arylamines into indoles has been achieved in various ways. The former react with alkoxides *via* addition to the triple bond and form nitro-acetals, nitro group reduction then acetal hydrolysis bring ring closure. Direct cyclisation of *ortho*-alkynylanilines can be effected simply by treatment with tetrabutylammonium fluoride.[263] Alternatively, palladium or copper salts can be utilised, and in the former cases the organopalladium intermediate can be either protonolysed, or trapped out with consequent insertion of a substituent at the indole β-position.[264]

Disubstituted acetylenes can also be utilised in a palladium-catalysed cyclisation of *ortho* haloanilides; the larger group (or hydroxyl-containing group) finishes at C-2.[265]

17.17.1.4 *From ortho-toluidides*

Base-catalysed cyclo-condensation of an *ortho*-alkylanilide gives an indole.

The Madelung synthesis

In its original form, this route employed very harsh conditions (typically[266] sodium amide or potassium *t*-butoxide at 250–350 °C) to effect base-catalysed intramolecular condensation between an unactivated aromatic methyl and an *ortho* acylamino-substituent, and was consequently limited to situations having no other sensitive groups. With the advent of the widespread use of alkyllithiums as bases, these cyclocondensations can now be brought about under much milder conditions.[267]

Modifications in which the benzylic hydrogens are acidified also allow the use of mild conditions; one example is the generation of a phosphonium ylide and then an intramolecular Wittig-like reaction, involving the amide carbonyl;[268] another variant uses a benzylsilane.[269] The use of an amino-silane permits reaction at both nitrogen and benzylic carbon to take place in one pot.[270]

Finally, in this category there must be included cyclisations of the benzylic anions derived from *ortho*-isocyanotoluenes; the scheme shows the synthesis in its simplest form. However, the synthesis is very flexible, for example the initial benzylic anion can be alkylated with halides or epoxides, before the ring closure thus providing 3-substituted indoles and additionally, the final *N*-lithioindole can be *N*-alkylated by adding a suitable electrophile before work up.[271]

17.17.1.5   From α-arylaminoarbonyl compounds

An α-arylaminoketone is cyclised by electrophilic attack onto the aromatic ring.

The Bischler synthesis

In the original method, the Bischler synthesis, harsh acidic treatment of α-arylaminoketones (produced from the 2-haloketone and an arylamine) was used to bring about electrophilic cyclisation onto the aromatic ring; these conditions often resulted in mixtures of products *via* rearrangements.[272] It is now known that *N*-acylated-α-arylaminoketones can be cyclised under much more controlled conditions, and in contrast to early work, this approach to indoles can even be used to produce hetero-ring-unsubstituted indoles.[273]

17.17.1.6   From pyrroles (see also section 13.9)

Several unrelated strategies have been utilised for the fusion of a benzene ring onto a pyrrole to generate an indole;[274] most follow a route in which a pyrrole, carrying a four-carbon side-chain at the α-carbon, is cyclised *via* an electrophilic attack at the adjacent pyrrole β-position; one of these is shown.[275] Another route involves the electrocyclisation of 2,3-divinylpyrroles.[276]

17.17.1.7 From ortho-substituted nitroarenes
Bartoli synthesis

In the extraordinary, but nonetheless efficient and extremely practically simple process now known as the Bartoli synthesis, *ortho*-substituted nitrobenzenes treated with three mol equivalents of vinylmagnesium bromide give 7-substituted indoles. The process works best when the 7-substitutent is large[277] and it is thought that initial attack by the vinyl Grignard is at the nitro group oxygen with subsequent elimination of magnesium enolate producing the nitroso equivalent of the original – it seems likely that this step is encouraged by non-planarity of the nitro group and the aromatic system forced on the molecule by the large *ortho* substituent. A second mol equivalent of vinyl Grignard then adds, again to oxygen generating an intermediate which undergoes a 3,3-sigmatropic rearrangement, much like that involved in the Fischer sequence, and finally hetero ring closure.[278]

17.17.1.8 From N-arylenamines

It is not clear whether the palladium-mediated cyclisations of anilino-acrylates and related sytems[279] operate via a Heck sequence or *via* an electrophilic palladation of the enamine.

A related palladium-catalysed cyclisation can be used to prepare carbazoles or carbolines (illustrated below) from mono- or dihalo diarylamines.[280] The starting materials for these are also readily prepared by palladium-catalysed reactions.

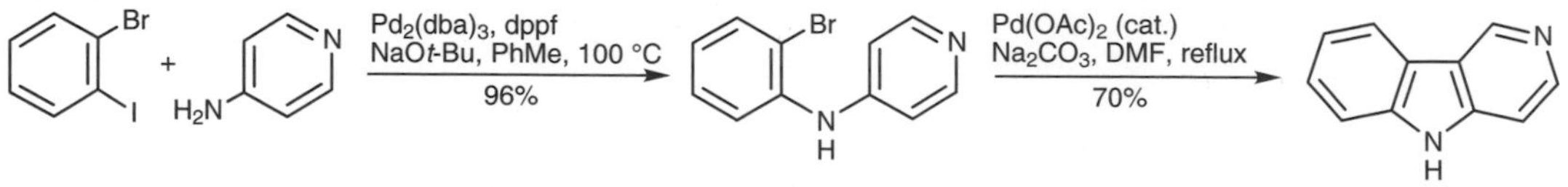

In an important development, simple α-methylene ketones (cyclic ketones work much better than acyclic ketones) and *ortho*-iodoarylamines react under palladium catalysis to give indoles directly. The use of dimethylformamide as solvent and diazabicyclooctane (DABCO) as the base are crucial to the success of the route. Mechanistically, the sequence certainly proceeds through the enamine. As well as being conceptually and practically simple, this method tolerates functional groups which would be sensitive to the acid of the traditional Fischer sequence.[281]

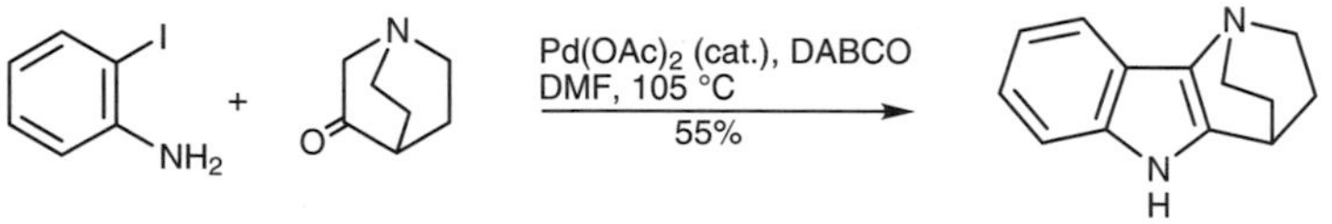

17.17.1.9   From N-allyl ortho-halo arylamines

N-Allyl *ortho*-haloarylamines can be cyclised under a variety of conditions to give either indoles or indolines, the latter being convertible into indoles by dehydrogenation or elimination of hydrogen halide from a suitable intermediate.[282] An intramolecular Heck reaction gives the indole directly via migration of an initially formed exocyclic double bond into the heterocyclic ring; the exo isomer is isolable if silver salts are added to the reaction mixture.[283]

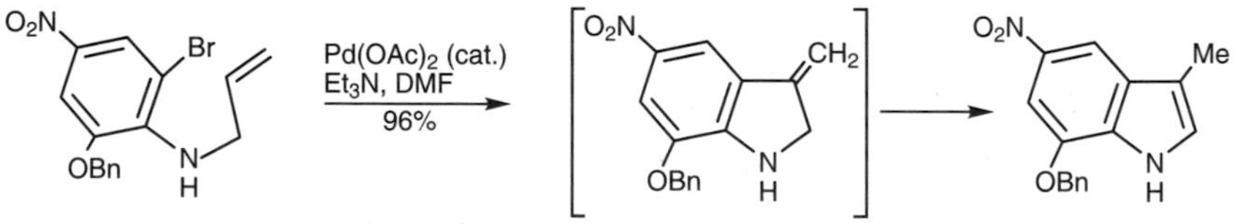

The aryllithiums derived from *N,N*-diallylarylamines with an *ortho* halogen by exchange with an alkyllithium cyclise by addition to an allyl group double bond generating a primary alkyllithium which can be trapped with electrophiles finally producing indolines.[284]

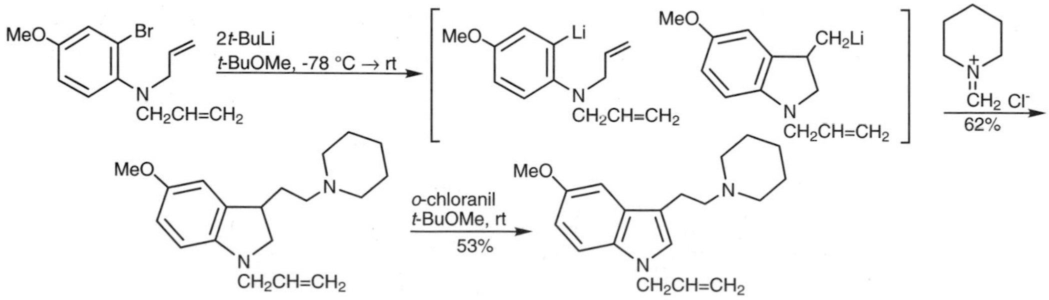

17.17.1.10 From enamines and p-quinones

The Nenitzescu synthesis

The Nenitzescu synthesis[285] is another process about which some of the mechanistic details remain unclear,[286] but which can be used for the efficient synthesis of certain 5-hydroxyindoles.[287]

17.17.1.11 From arylamines

The Gassman synthesis

The Gassman synthesis[288] produces sulfur-substituted indoles, but these can easily be hydrogenolysed if required.

17.17.1.12 From ortho-acyl anilides

The Fürstner synthesis[289]

This flexible synthesis depends on the reductive cyclisation of *ortho*-acylanilides with low valent titanium – the conditions used for the McMurray coupling of ketones. In the example below, the cyclisation precursor was built up via the acylation of trimethylstannylthiazole.[290]

17.17.1.13   *From ortho-isocyano styrenes*
The Fukuyama synthesis[291]

ortho-Isocyano styrenes, which are readily prepared by dehydration of the corresponding formamides, undergo tin-promoted radical cyclisation to give unstable 2-stannylindoles, which can either be hydrolysed to afford the corresponding 2-unsubstituted indole, or used without isolation for coupling with aryl halides using palladium(0)-catalysis, as illustrated below.[292]

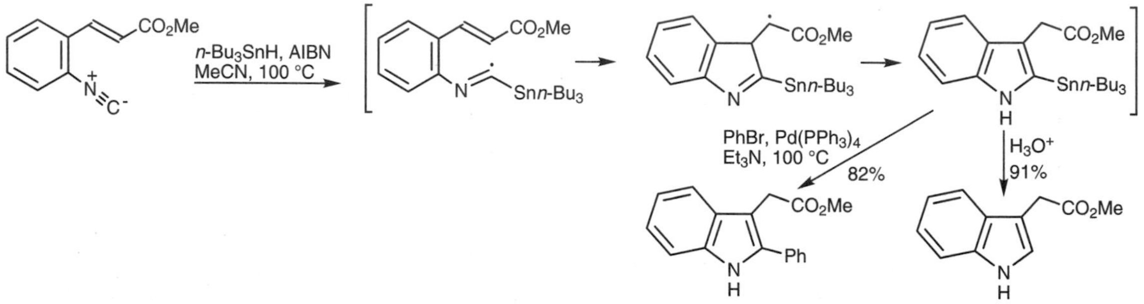

17.17.1.14   *From ortho-chloroacetyl arylamines*
Sugasawa synthesis

Arylamines, without protection of the nitrogen, undergo Friedel-Crafts acylation regioselectively *ortho* to the nitrogen using nitriles and boron trifluoride. Thus, using chloroacetonitrile produces (*ortho*-chloroacetyl)arylamines in which ring closure to give an indole takes place after reduction of the ketone to the alcohol oxidation level.[293]

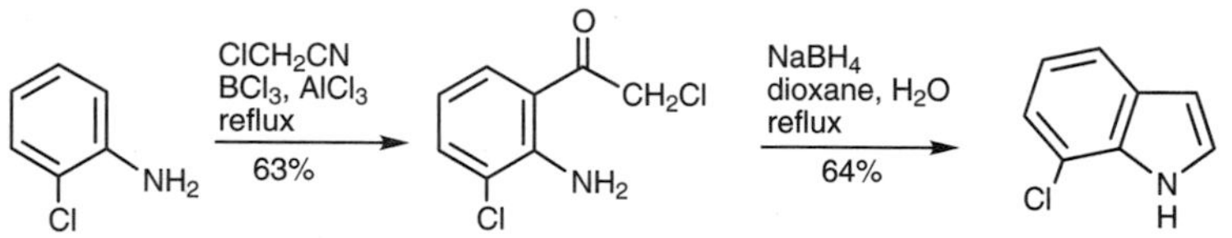

17.17.1.15   *By cyclisation of nitrenes*

Thermolysis of *ortho*-azidostyrenes gives nitrenes which insert into the side chain to form indoles.[294] Similar nitrenes have been generated by reaction of nitro compounds with trialkyl phosphites. The azide thermolysis method can be used to prepare 2-nitroindoles, which are not available by other methods.[295]

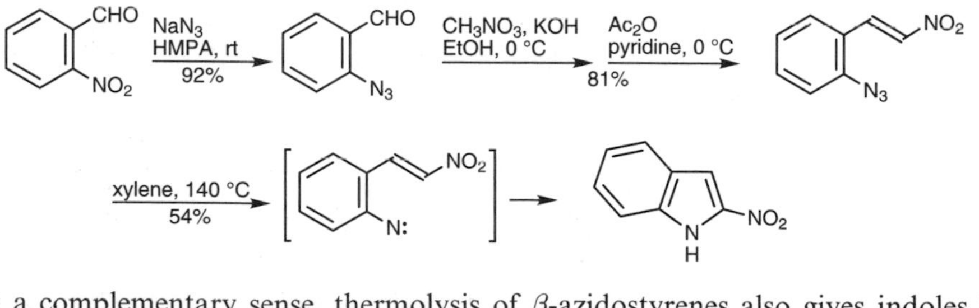

In a complementary sense, thermolysis of β-azidostyrenes also gives indoles but here the intermediate may be an azirine;[296] this method is particularly useful for the fusion of a pyrrole ring onto rings other than a benzene ring, as illustrated below.[297]

17.17.1.16 Cyclisation onto arynes

Indoles can be prepared by intramolecular addition of iminates[298] or vinyllithiums[299] to arynes. In the latter case the intermediate aryllithium can be trapped with electrophiles to give 4-substituted indoles.

17.17.1.17 From ortho-nitro styrenes

ortho-Nitrostyrenes are readily available by a number routes – (1) reaction of an (*ortho*-bromomethyl)nitroarene with a phosphine then Wittig condensation with an aldehyde, (2) Wittig reaction employing an (*ortho*-nitro)araldehyde as the carbonyl component, (3) base-catalysed condensation of a methyl group *ortho* to an aromatic nitro group with an aldehyde, and (4) *ortho* nitration of a styrene. In a palladium-catalysed process, the mechanism of which remains obscure, very efficient ring closure to indoles takes place in one pot, but clearly not by one step.[300]

17.17.1.18 From indolines

Indolines are useful intermediates for the synthesis of indoles with substituents in the carbocyclic ring. In electrophilic substitutions, they behave like anilines; the example shows *N*-acetylindoline undergoing regioselective 7-thallation. Indolines can be obtained easily from indoles by reduction (see section 17.8) and can be cleanly oxidised back to indoles using a variety of methods, including oxygen with cobalt catalysis (salcomine),[301] hypochlorite/dimethylsulfide,[302] Mn(III),[303] and Au(III) compounds.[304]

An attractive variant is to utilise certain products of reversible addition to 3*H*-indolium cations, such as the indole bisulfite adduct (section 17.1.1), or where there has been an intramolecular nucleophilic addition: such compounds, though they are indolines, are still at the oxidation level of indoles, needing only mild acid treatment to regenerate the aromatic system.[305]

Methods involving ruthenium-catalysed condensations of arylamines with alcohols may prove to be useful for the large scale production of indoles. The mechanism involves hydride transfer giving aldehyde intermediates. The process can be carried out intramolecularly as shown[306] or intermolecularly, for example by the reaction of aniline with triethanolamine.[307]

17.17.2   Synthesis of oxindoles

The main synthesis of oxindoles is simple and direct and involves an intramolecular Friedel-Crafts alkylation reaction as the cyclising step.[308] Also straightforward in concept is the displacement of halogen from an *ortho* halonitroarene with malonate, this leading to an oxindole after decarboxylation and reduction of the nitro group with spontaneous lactamisation.[193]

A less orthodox route to oxindoles depends on the intramolecular insertion of a rhodium carbenoid into an adjacent aromatic C–H bond.[309]

Oxindoles can also be prepared by palladium-catalysed enolate cyclisation of *ortho* halo anilides.[310]

17.17.3   Synthesis of indoxyls

Indoxyls are normally prepared from anthranilic acids *via* alkylation with a haloacetic acid followed by a cyclising Perkin condensation.[311] It is also possible to directly chloroacylate an aniline, *ortho* to the nitrogen.[312]

17.17.4   Synthesis of isatins

Isatins are readily prepared *via* the reaction of an aniline with chloral, the resulting product converted into an oxime, and this cyclised in strong acid.[313]

17.17.5   Synthesis of 1-hydroxyindoles

The oxidation of indolines with sodium tungstate/hydrogen peroxide both aromatises and also oxidises the nitrogen, resulting in 1-hydroxyindoles.[106] 1-Hydroxyindoles can also be obtained *via* partial reduction of the nitro group of Leimgruber-Batcho intermediate nitro-enamines (section 17.17.1.2) with zinc then cyclisation.[314]

17.17.6   Examples of notable indole syntheses
17.17.6.1   *Ondansetron*

Ondansetron is a selective, 5-hydroxytryptamine antagonist, used to prevent vomiting during cancer chemotherapy and radiotherapy.

17.17.6.2 Staurosporine aglycone[315]

Staurosporine and related molecules are under active investigation as potential antitumour agents. The synthesis illustrates several aspects of heterocyclic chemistry, including a 2-pyrone acting as a diene in an intramolecular Diels-Alder reaction, and the use of nitrene insertion for the formation of 5-membered nitrogen rings.

17.17.6.3 Serotonin

Serotonin has been synthesised by several routes; the method shown[316] relies on a Fischer indole synthesis, the requisite phenylhydrazone being constructed by a process known as the Japp-Klingemann reaction in which the enol of a 1,3-dicarbonyl compound is reacted with an aryldiazonium salt, with subsequent cleavage of the 1,3-dicarbonyl unit.

17.17.6.4   Chuangxinmycin[317]

This synthesis uses the approach of starting from a pyrrole: the cyclic ketone intermediate is in general a useful intermediate for the synthesis of 4-substituted indoles – in this case a sulfur substituent – it is already at the aromatic oxidation level needing only the loss of the 4-chlorophenylthiol.

17.17.7   Synthesis of azaindoles (see also section 5.5.1)

Most syntheses of azaindoles start from pyridines and parallel the standard indole syntheses discussed above. However, the Fischer reaction using pyridylhydrazones is much less consistent and useful than for phenylhydrazones; the Madelung reaction is also not as useful. The most successful methods involve palladium-catalysed coupling of acetylenes with amino halopyridines either as one-[318] or two-step[319] processes. The starting amino halopyridines are generally available via directed metallations.

Syntheses utilising nitropyridines by Leimgruber-Batcho processes work well[320] but can be limited by the availabilities of the starting nitropyridines – the sequence below shows the assembly of the ring closure precursor using a VNS (section 2.3.3) sequence.[321]

Synthesis from pyrroles is useful in particular cases.[320,322]

Exercises for chapter 17

Straightforward revision exercises (consult chapters 16 and 17)

(a) What is the pK_a of indole as a base and where does it protonate? What is the pK_a of indole as an acid?

(b) At what position is electrophilic substitution of indole fastest? Cite two examples.

(c) What are the structures of the intermediates and final product in the following sequence: indole with $(COCl)_2 \rightarrow C_{10}H_6ClNO_2$ then this with ammonia $\rightarrow C_{10}H_8N_2O_2$ then this with $LiAlH_4 \rightarrow C_{10}H_{12}N_2$? Explain the last transformation in mechanistic terms.

(d) How could one prepare from indole: (i) 3-formylindole, (ii) 3-(2-nitroethyl)indole, (iii) 3-dimethylaminomethylindole, (iv) 1-methylindole.

(e) At what position does strong base deprotonate an *N*-substituted indole? Name two groups which can be used to block the 1-position for such deprotonations and which could be removed later. How would these blocking groups be introduced onto the indole nitrogen?

(f) What is the mechanism of the conversion of 3-dimethylaminomethylindole into 3-cyanomethylindole on reaction with NaCN?

(g) Which phenylhydrazones would be required for the Fischer indole synthesis of (i) 3-methylindole; (ii) 1,2,3,4-tetrahydrocarbazole; (iii) 2-ethyl-3-methylindole; (iv) 3-ethyl-2-phenylindole?

(h) How could one convert 2-bromoaniline into 2-phenylindole (more than one step is required)?

(i) What are the advantages of using an indoline (a 2,3-dihydroindole) as an intermediate for the synthesis of indoles?

More advanced exercises

1. Indole reacts with a mixture of *N*-methyl-2-piperidone and $POCl_3$, followed by NaOH work-up to give $C_{14}H_{18}N_2O$. What is its structure?

2. Suggest a structure for the tetracyclic product, $C_{18}H_{19}NO$, formed when 3-methylindole is treated with 2-hydroxy-3,5-dimethylbenzyl chloride.

3. When indole dimer (section 17.1.9) is subjected to acid treatment in the presence of indole, 'indole trimer', $C_{24}H_{21}N_3$, is produced. Suggest a structure for the 'trimer' (hint: consider which of the two reactants would be most easily protonated, and at which atom).

4. Starting from indole, and using a common intermediate, how could one prepare (i) indol-3-ylacetic acid and (ii) tryptamine?

5. What would be the products from the reactions of 5-bromo-3-iodo-1-phenylsulfonylindole with (i) $PhB(OH)_2/Pd(PPh_3)_4/aq. Na_2CO_3$; (ii) ethyl acrylate/$Pd(OAc)_2/Ph_3P/Et_3N$?

6. Deduce a structure, and write out the mechanism for the conversion of 2-formylindole into a tricyclic compound, $C_{11}H_9N$, on treatment with a combination of NaH and $Ph_3P^+CH=CH_2\ Br^-$.

7. When 3-ethyl-3-methyl-3H-indole is treated with acid, two products, each isomeric with the starting material, are formed – deduce their structures and explain the formation of two products.

8. Suggest a structure for the salt $C_{15}H_{13}N_2^+\ Br^-$ formed by the following sequence: 2-(2-pyridyl)indole reacted first with n-BuLi then $PhSO_2Cl$ ($\rightarrow$ $C_{19}H_{14}N_2O_2S$), then this sequentially with t-BuLi at $-100\,°C$ then ethylene oxide ($\rightarrow C_{21}H_{18}N_2O_3S$), aq. NaOH ($\rightarrow C_{15}H_{14}N_2O$), and this, finally reacted with PBr_3.

9. What are the products formed in the following sequence: indole/n-BuLi, then I_2, then LDA, then $PhSO_2Cl \rightarrow C_{14}H_{10}INO_2S$, then this with LDA, then $I_2 \rightarrow C_{14}H_9I_2NO_2S$?

10. When indol-3-yl-CH_2OH is heated with acid, di(indol-3-yl)methane is formed: suggest a mechanism for this transformation.

11. What product, $C_{10}H_{11}NO$, would be obtained from refluxing a mixture of phenylhydrazine and 2,3-dihydrofuran in acetic acid?

12. Draw structures for the azaindoles resulting from treatment of 2-methyl-3-nitro- and 4-methyl-3-nitropyridines, respectively, with $(EtO_2C)_2/EtONa$, followed by H_2/Pd–C. Both products have the molecular formula $C_{10}H_{10}N_2O_2$.

13. Heating DMFDMA with the following aromatic compounds led to condensation products; subsequent reduction with the reagent shown gave indoles. Draw the structures of the condensation products and the indoles: (i) 2,6-dinitrotoluene then $TiCl_3$ gave $C_8H_8N_2$; (ii) 2-benzyloxy-6-nitrotoluene then H_2/Pt gave $C_{15}H_{13}NO$; (iii) 4-methoxy-2-nitrotoluene then H_2/Pd gave C_9H_9NO; (iv) 2,3-dinitro-1,4-dimethylbenzene then H_2/Pd gave $C_{10}H_8N_2$.

References

1. 'The chemistry of indoles', Sundberg, R. J., Academic Press, New York, **1970**; 'Transition metals in the synthesis and functionalisation of indoles', Hegedus, L. S., *Angew. Chem., Int. Ed. Engl.*, **1988**, *27*, 1113; 'Indoles', Sundberg, R. J., Academic Press, London, **1996**.
2. Gut, I. G. and Wirz, J., *Angew. Chem., Int. Ed. Engl.*, **1994**, *33*, 1153.
3. 'Monoterpenoid indole alkaloids' in 'Indoles' in 'The Chemistry of Heterocyclic Compounds', Series Ed. Taylor, E. C., Vol. 25, Part 4, Ed. Saxton, J. E., Wiley-Interscience, **1983** and supplement, **1994**.
4. Hinman, R. L. and Lang, J., *J. Am. Chem. Soc.*, **1964**, *86*, 3796.
5. Hinman, R. L. and Whipple, E. B., *J. Am. Chem. Soc.*, **1962**, *84*, 2534.
6. Challis, B. C. and Millar, E. M., *J. Chem. Soc., Perkin Trans. 2*, **1972**, 1111; Muir, D. M. and Whiting, M. C., *J. Chem. Soc., Perkin Trans. 2*, **1975**, 1316.
7. Hinman, R. L. and Bauman, C. P., *J. Org. Chem.*, **1964**, *29*, 2437.
8. Challis, B. C. and Millar, E. M., *J. Chem. Soc., Perkin Trans. 2*, **1972**, 1116.
9. Russell, H. F., Harris, B. J., and Hood, D. B., *Org. Prep. Proc. Int.*, **1985**, *17*, 391.
10. Bourne, G. T., Crich, D., Davies, J. W., and Horwell, D. C., *J. Chem. Soc., Perkin Trans. 1*, **1991**, 1693.
11. Pelkey, E. T. and Gribble, G. W., *Synthesis*, **1997**, 1117.
12. Berti, G., Da Settimo, A., and Nannipieri, E., *J. Chem. Soc., C*, **1968**, 2145.
13. Brown, K. and Katritzky, A. R., *Tetrahedron Lett.*, **1964**, 803.
14. Ottoni, O., Cruz, R., and Krammer, N. H., *Tetrahedron Lett.*, **1999**, *40*, 1117.
15. Mitchell, H. and Leblanc, Y., *J. Org. Chem.*, **1994**, *59*, 682.
16. Smith, G. F. and Taylor, D. A., *Tetrahedron*, **1973**, *29*, 669.

17. Fatum, T. M., Anthoni, U. Christophersen, C., and Nielsen, P. H., *Heterocycles*, **1994**, *38*, 1619.
18. Raban, M. and Chern, L.-J., *J. Org. Chem.*, **1980**, *45*, 1688; Gilow, H. M. Brown, C. S., Copeland, J. N., and Kelly, K. E., *J. Heterocycl. Chem.*, **1991**, *28*, 1025.
19. Hamel, P., Zajac, N., Atkinson, J. G., and Girard, Y., *J. Org. Chem.*, **1994**, *59*, 6372.
20. Hamel, P., Girard, Y., and Atkinson, J. G., *J. Org. Chem.*, **1992**, *57*, 2694.
21. Nair, V., George, T. G., Nair, L. G., and Panicker, S. B., *Tetrahedron Lett.*, **1999**, *40*, 1195.
22. (a) Bocchi, V. and Palla, G., *Synthesis*, **1982**, 1096; (b) Piers, K., Meimaroglou, C., Jardine, R. V., and Brown, R. K., *Canad. J. Chem.*, **1963**, *41*, 2399; (c) Arnold, R. D., Nutter, W. M., and Stepp, W. L., *J. Org. Chem.*, **1959**, *24*, 117; (d) De Rosa, M. and Alonso, J. L. T., *J. Org. Chem.*, **1978**, *43*, 2639.
23. Dmitrienko, G. I., Gross, E. A., and Vice, S. F., *Canad. J. Chem.*, **1980**, *58*, 808.
24. Zhang, P., Liu, R., and Cook, J. M., *Tetrahedron Lett.*, **1995**, *36*, 3103.
25. Bergman, J. and Venemalm, L., *J. Org. Chem.*, **1992**, *57*, 2495.
26. Brennan, M. R., Erickson, K. L., Szmalc, F. S., Tansey, M. J., and Thornton, J. M., *Heterocycles*, **1986**, *24*, 2879.
27. Hirano, S., Akai, R., Shinoda, Y., and Nakatsuka, S., *Heterocycles*, **1995**, *41*, 255.
28. Saulnier, M. G. and Gribble, G. W., *J. Org. Chem.*, **1982**, *47*, 757.
29. Saxton, J. E., *J. Chem. Soc.*, **1952**, 3592; Hart, G., Liljegren, D. R., and Potts, K. T., *ibid.*, **1961**, 4267.
30. Nickisch, K., Klose, W., and Bohlmann, F., *Chem. Ber.*, **1980**, *113*, 2036.
31. Cipiciani, A., Clementi, S., Linda, P., Savelli, G., and Sebastiani, G. V., *Tetrahedron*, **1976**, *32*, 2595.
32. Gmeiner, P., Kraxner, J., and Bollinger, B., *Synthesis*, **1996**, 1196.
33. Smith, G. F., *J. Chem. Soc.*, **1954**, 3842; James, P. N. and Snyder, H. R., *Org. Synth., Coll. Vol. IV*, **1963**, 539.
34. Anthony, W. C., *J. Org. Chem.*, **1960**, *25*, 2049.
35. Nogrady, T. and Morris, L., *Canad. J. Chem.*, **1969**, *47*, 1999; Monge, A., Aldana, I., Lezamiz, I., and Fernandez-Alvarez, E., *Synthesis*, **1984**, 160.
36. Speeter, M. E. and Anthony, W. C., *J. Am. Chem. Soc.*, **1954**, *76*, 6208.
37. Ishizumi, K., Shioiri, T., and Yamada, S., *Chem. Pharm. Bull.*, **1967**, *15*, 863.
38. Späth, E. and Lederer, E., *Chem. Ber.*, **1930**, *63*, 2102.
39. 'Indolo-2,3-quinodimethanes and stable analogues for regio- and sterocontrolled syntheses of [*b*]-anelated indoles', Pindur, U. and Erfanian-Abdoust, H., *Chem. Rev.*, **1989**, *89*, 1681.
40. Nakatsuka, S., Teranishi, K., and Goto, T., *Tetrahedron Lett.*, **1994**, *35*, 2699; Teranishi, K., Nakatsuka, S., and Goto, T., *Synthesis*, **1994**, 1018.
41. Demopoulos, V. J. and Nicolaou, I., *Synthesis*, **1998**, 1519.
42. Teranishi, K., Hayashi, S., Nakatsuka, S., and Goto, T., *Tetrahedron Lett.*, **1994**, *35*, 8173.
43. Ciamician, G. and Plancher, G., *Chem. Ber.*, **1896**, *29*, 2475; Jackson, A. H. and Smith, P., *Tetrahedron*, **1968**, *24*, 2227.
44. Jackson, A. H., Naidoo, B., and Smith, P., *Tetrahedron*, **1968**, *24*, 6119.
45. Casnati, G., Dossena, A., and Pochini, A., *Tetrahedron Lett.*, **1972**, 5277.
46. Iyer, R., Jackson, A. H., Shannon, P. V. R., and Naidoo, B., *J. Chem. Soc., Perkin Trans. 2*, **1973**, 872.
47. Ganesan, A. and Heathcock, C. H., *Tetrahedron Lett.*, **1993**, *34*, 439.
48. Kotsuki, H., Teraguchi, M., Shimomoto, N., and Ochi, M. *Tetrahedron Lett.*, **1996**, *37*, 3727.
49. Kotsuki, H., Hayashida, K., Shimanouchi, T., and Nishizawa, H., *J. Org. Chem.*, **1996**, *61*, 984.
50. Sato, K. and Kozikowski, A. P., *Tetrahedron Lett.*, **1989**, *31*, 4073; Dubois, L., Mehta, A., Tourette, E., and Dodd, R. H., *J. Org. Chem.*, **1994**, *59*, 434.
51. Bramely, R. K., Caldwell, J., and Grigg, R., *J. Chem. Soc., Perkin Trans. 1*, **1973**, 1913; Casnati, G., Franciani, M., Guareschi, A., and Pochini, A., *Tetrahedron Lett.*, **1969**, 2485.
52. Burr, G. O. and Gortner, R. A., *J. Am. Chem. Soc.*, **1924**, *46*, 1224.
53. Cook, A. H. and Majer, J. R., *J. Chem. Soc.*, **1944**, 486; Pindur, U. and Flo, C., *Monatsh. Chem.*, **1986**, *117*, 375.

54. Pindur, U. and Kim, M.-H., *Tetrahedron*, **1989**, *45*, 6427.
55. Leete, E. *J. Am. Chem. Soc.*, **1959**, *81*, 6023; Thesing, J., *Chem. Ber.*, **1954**, *87*, 692.
56. Freter, K., *J. Org. Chem.*, **1975**, *40*, 2525.
57. Appleton, J. E., Dack, K. N., Green, A. D., and Stelle, J., *Tetrahedron Lett.*, **1993**, *34*, 1529.
58. Szmuszkovicz, J., *J. Am. Chem. Soc.*, **1957**, *79*, 2819.
59. Iqbal, Z., Jackson, A. H., and Nagaraja Rao, K. R., *Tetrahedron Lett.*, **1988**, *29*, 2577.
60. Harrington, P. E. and Kerr, M. A., *Synlett*, **1996**, 1047.
61. Robinson, B. and Smith, G. F., *J. Chem. Soc.*, **1960**, 4574; Garnick, R. L., Levery, S. B., and Le Quesne, P. W., *J. Org. Chem.*, **1978**, *43*, 1226.
62. Balsamini, C., Diamantini, G., Duranti, A., Spadoni, G., and Tontini, A., *Synthesis*, **1995**, 370.
63. Ranganathan, D., Rao, C. B., Ranganathan, S., Mehrotra, A. K., and Iyengar, R., *J. Org. Chem.*, **1980**, *45*, 1185.
64. Büchi, G. and Mak, C.-P., *J. Org. Chem.*, **1977**, *42*, 1784.
65. Majchrzak, M. W. and Simchen, G., *Synthesis*, **1986**, 956.
66. 'Carbon-carbon alkylations with amines and ammonium salts', Brewster, J. H. and Eliel, E. L., *Org. Reactions*, **1953**, *7*, 99.
67. Swaminathan, S. and Narisimhan, K., *Chem. Ber.*, **1966**, *99*, 889.
68. Kühn, H. and Stein, O., *Chem. Ber.*, **1937**, *70*, 567.
69. Kozikowski, A. P. and Ishida, H., *Heterocycles*, **1980**, *14*, 55.
70. Girke, W. P. K., *Chem. Ber.*, **1979**, *112*, 1.
71. Xie, W., Bloomfield, K. M., Jin, Y., Dolney, N. Y., and Wang, G. P., *Synlett*, **1999**, 498.
72. 'The acid-catalysed polymerisation of pyrroles and indoles', Smith, G. F., *Adv. Heterocycl. Chem.*, **1963**, *2*, 287.
73. Bocchi, V. and Palla, G., *Tetrahedron*, **1984**, *40*, 3251.
74. Hahn, G. and Ludewig, H., *Chem. Ber.*, **1934**, *67*, 2031.
75. Narayanan, K. and Cook, J. M., *Tetrahedron Lett.*, **1990**, *31*, 3397.
76. Kawate, T., Yamada, H., Matsumizu, M., Nishida, A., and Nakagawa, M., *Synlett*, **1997**, 761.
77. Hodson, H. F. and Smith, G. F., *J. Chem. Soc.*, **1957**, 3546.
78. Bonnett, R. and Holleyhead, R., *J. Chem. Soc., Perkin Trans. 1*, **1974**, 962.
79. Kirby, G. W. and Shah, S. W., *J. Chem. Soc., Chem. Commun.*, **1965**, 381.
80. Harrington, P. J. and Hegedus, L. S., *J. Org. Chem.*, **1984**, *49*, 2657.
81. (a) Holins. R. A., Colnaga, L. A., Salim, V. M., and Seidl, M. C., *J. Heterocycl. Chem.*, **1979**, *16*, 993; (b) Somei, M., Yamada, F., Kunimoto, M. and Kaneko, C., *Heterocycles*, **1984**, *22*, 797.
82. Somei, M., Hasegawa, T., and Kaneko, C., *Heterocycles*, **1983**, *20*, 1983; Somei, M. and Yamada, F., *Chem. Pharm. Bull.*, **1984**, *32*, 5064.
83. Yamada, F. and Somei, M., *Heterocycles*, **1987**, *26*, 1173.
84. Somei, M., Yamada, F., Hamada, H., and Kawasaki, T., *Heterocycles*, **1989**, *29*, 643.
85. Itahara, T., *Chem. Lett.*, **1982**, 1151.
86. Itahara, T., Ikeda, M., and Sakakibara, T., *J. Chem. Soc., Perkin Trans. 1*, **1983**, 1361; Itahara, T., Kawasaki, K., and Ouseto, F., *Synthesis*, **1984**, 236.
87. Itahara, I., Ikeda, M., and Sakakibara, T., *J. Chem. Soc., Perkin Trans. 1*, **1983**, 1361.
88. Nakagawa, M., Kato, S., Kataoka, S., and Hino, T., *J. Am. Chem. Soc.*, **1979**, *101*, 3136.
89. Chien, C.-S., Hasegawa, A., Kawasaki, T., and Sakamoto, M., *Chem. Pharm. Bull.*, **1986**, *34*, 1493.
90. Dolby, L. J. and Booth, D. L., *J. Am. Chem. Soc.*, **1966**, *88*, 1049.
91. Balogh-Hergovich, E. and Speier, G., *Tetrahedron Lett.*, **1982**, *23*, 4473.
92. Yukimasa, H., Sawai, H., and Takizawa, T., *Chem. Pharm. Bull.*, **1979**, *27*, 551.
93. Mudry, C. A. and Frasca, A. R., *Tetrahedron*, **1973**, *29*, 603.
94. Itahara, T., Ouya, H., and Kozono, K., *Bull. Chem. Soc. Jpn.*, **1982**, *55*, 3861.
95. Szabó-Pusztay, K. and Szabó, L., *Synthesis*, **1979**, 276.
96. Wojciechowski, K. and Makosza, M., *Synthesis*, **1989**, 106.
97. 'Chemistry and synthetic utility of metal-complexed indoles', Gill, U. S., Moriarty, R. M., Ku, Y. Y., and Butler, I. R., *J. Organomet. Chem.*, **1991**, *417*, 313.
98. Yagil, G., *Tetrahedron*, **1967**, *23*, 2855; Scott, W. J., Bover, W. J., Bratin, K., and Zuman, P., *J. Org. Chem.*, **1976**, *41*, 1952.
99. Shirley, D. A. and Roussel, P. A., *J. Am. Chem. Soc.*, **1953**, *75*, 375.

100. Sundberg, R. J. and Russell, H. F., *J. Org. Chem.*, **1973**, *38*, 3324.
101. Katritzky, A. R. and Akutagawa, K., *Tetrahedron Lett.*, **1985**, *26*, 5935.
102. Hasan, I., Marinelli, E. R., Lin, L.-C. C., Fowler, F. W., and Levy, A. B., *J. Org. Chem.*, **1981**, *46*, 157.
103. Hlasla, D. J. and Bell, M. R., *Heterocycles*, **1989**, *29*, 849; Katritzky, A. R., Lue, P., and Chen, Y.-X., *J. Org. Chem.*, **1990**, *55*, 3688.
104. Edwards, M. P., Doherty, A. M., Ley, S. V., and Organ, H. M., *Tetrahedron*, **1986**, *42*, 3723.
105. Somei, M. and Kobayashi, T., *Heterocycles*, **1992**, *34*, 1295.
106. Gharpure, M., Stoller, A., Bellamy, F., Firnau, G., and Snieckus, V., *Synthesis*, **1991**, 1079.
107. Kondo, Y., Yoshida, A., and Sakamoto, T., *J. Chem. Soc., Perkin Trans. 1*, **1996**, 2331.
108. Iwao, M., *Heterocycles*, **1993**, *36*, 29; Iwao, M. and Motoi, O., *Tetrahedron Lett.*, **1995**, *36*, 5929.
109. Nunomoto, S., Kawakami, Y., Yamashita, Y., Takeuchi, H., and Eguchi, S., *J. Chem. Soc., Perkin Trans. 1*, **1990**, 111.
110. Reinecke, M. G., Sebastian, J. F., Johnson, H. W., and Pyun, C., *J. Org. Chem.*, **1972**, *37*, 3066.
111. Rubottom, G. M. and Chabala, J. C., *Synthesis*, **1972**, 566.
112. Barco, A., Benetti, S., Pollini, G. P., and Baraldi, P. G., *Synthesis*, **1976**, 124.
113. Illi, V. O., *Synthesis*, **1979**, 387; Santaniello, E., Farachi, C., and Ponti, F., *ibid.*, **1979**, 617.
114. Illi, V. O., *Synthesis*, **1979**, 136.
115. 'The indole Grignard reagents', Heacock, R. A. and Kaspárek, S., *Adv. Heterocycl. Chem.*, **1969**, *10*, 43.
116. Bergman, J. and Venemalm, L., *Tetrahedron*, **1990**, *46*, 6061.
117. Ayer, W. A., Craw, P. A., Ma, Y., and Miao, S., *Tetrahedron*, **1992**, *48*, 2919.
118. Yang, C. X., Patel, H. H., Ku, Y.-Y., Shah, R., and Sawick, D., *Synth. Commun.*, **1997**, *27*, 2125.
119. Ezquerra, J. Pedregal, C., Lamas, C., Pastor, A., Alvarez, P., and Vaquero, J. J., *Tetrahedron Lett.*, **1996**, *37*, 683.
120. Fishwick, C. W. G., Jones, A. D., and Mitchell, M. B., *Heterocycles*, **1991**, *32*, 685; Onistschenko, A. and Stamm, H., *Chem. Ber.*, **1989**, *122*, 2397.
121. Katritzky, A. R. and Akutagawa, K., *Synth. Commun.*, **1988**, 1151.
122. Saulnier, M. G. and Gribble, G. W., *J. Org. Chem.*, **1982**, *47*, 757.
123. Amat, M., Hadida, S., Sathyanarayana, S., and Bosch, J., *J. Org. Chem.*, **1994**, *59*, 10.
124. Rubiralta, M., Casamitjana, N., Grierson, D. S., and Husson, H.-P., *Tetrahedron*, **1988**, *44*, 443.
125. Johnson, D. A. and Gribble, G. W., *Heterocycles*, **1986**, *24*, 2127
126. Kondo, Y., Yoshida, A., Sato, S., and Sakamoto, T., *Heterocycles*, **1996**, *42*, 105.
127. Kondo, Y., Takazawa, N., Yoshida, A., and Sakamoto, T., *J. Chem. Soc., Perkin Trans. 1*, **1995**, 1207.
128. Yokoyama, Y., Uchida, M., and Murakami, Y., *Heterocycles*, **1989**, *29*, 1661.
129. Klingebiel, U., Luttke, W., Noltemeyer, M., and Schmidt, H. G., *J. Organometal. Chem.*, **1993**, *456*, 41.
130. Buttery, C. D., Jones, R. G., and Knight, D. W., *J. Chem. Soc., Perkin Trans. 1*, **1993**, 1425; Fisher, L. E., Labadie, S. S., Reuter, D. C., and Clark, R. D., *J. Org. Chem.*, **1995**, *60*, 6224.
131. Saulnier, M. G. and Gribble, G. W., *Tetrahedron Lett.*, **1983**, *24*, 5435.
132. Griffen, E. J., Roe, D. G., and Snieckus, V., *J. Org. Chem.*, **1995**, *60*, 1484.
133. Moyer, M. P., Shiurba, J. F., and Rapoport, H., *J. Org. Chem.*, **1986**, *51*, 5106.
134. Gribble, G. W. and Conway, S. C., *Synth. Commun.*, **1992**, *22*, 2129; Joseph, B., Malapel, B., and Merour, J.-Y., *ibid.*, **1996**, *26*, 3289.
135. Harrington, P. J. and Hegedus, L. S., *J. Org. Chem.*, **1984**, *49*, 2657; Arcadi, A., Burini, A., Cacchi, S., Delmasto, M., and Pietroni, B., *Synlett*, **1990**, 47.
136. Yang, Y. and Martin, A. R., *Heterocycles*, **1992**, *34*, 1395.
137. Danieli, B., Lesma, G., Martinelli, M., Passarella, D., Peretto, I., and Silvani, A., *Tetrahedron*, **1998**, *54*, 14081.
138. Benhida, R., Lecubin, F., Fourrey, J.-L., Casellanos, L. R., and Quintero, L., *Tetrahedron Lett.*, **1999**, *40*, 5701.

139. Kanaoka, T., Aiura, M., and Hariya, S., *J. Org. Chem.*, **1971**, *36*, 458.
140. Byers, J. H., Campbell, J. E., Knapp, F. H., and Thissell. J. G., *Tetrahedron Lett.*, **1999**, *40*, 2677.
141. Chuang, C.-P. and Wang, S.-F., *Synlett*, **1995**, 763.
142. Ziegler, F. E. and Jeroncic, L. O., *J. Org. Chem.*, **1991**, *56*, 3479.
143. Aboutayab, K., Caddick, S., Jenkins, K., Joshi, S., and Khan, S., *Tetrahedron*, **1996**, *52*, 11329.
144. Caddick, S., Aboutayab, K., Jenkins, K., and West, R. I., *J. Chem. Soc., Perkin Trans. 1*, **1996**, 675.
145. Dobbs, A. P., Jones, K., and Veal, K. T., *Tetrahedron Lett.*, **1995**, *36*, 4857.
146. O'Brien, S. and Smith, D. C. C., *J. Chem. Soc.*, **1960**, 4609; Remers, W. A., Gibs, G. J., Pidoocks, C., and Weiss, M. J., *J. Org. Chem.*, **1971**, *36*, 279; Ashmore, J. W. and Helmkamp, G. K., *Org. Prep. Proc. Int.*, **1976**, *8*, 223.
147. Barrett, A. G. M., Dauzonne, D., O'Neil, I. A., and Renaud, A., *J. Org. Chem.*, **1984**, *49*, 4409.
148. Dolby, L. J. and Gribble, G. W., *J. Heterocycl. Chem.*, **1966**, *3*, 124.
149. Lanzilotti, A. E., Littell, R., Fanshawe, W. J., McKenzie, T. C., and Lovell, F. M., *J. Org. Chem.*, **1979**, *44*, 4809.
150. Gribble, G. W. and Hoffman, J. H., *Synthesis*, **1977**, 859.
151. Butula, I. and Kuhn, R., *Angew. Chem., Int. Ed. Engl.*, **1968**, *7*, 208.
152. Coulton, S., Gilchrist, T. L., and Graham, K., *Tetrahedron*, **1997**, *53*, 791.
153. Gillespie, R. J. and Porter, A. E. A., *J. Chem. Soc., Chem. Commun.*, **1979**, 50.
154. Dehaen, W. and Hassner, A., *J. Org. Chem.*, **1991**, *56*, 896.
155. Benson, S. C., Li, J.-H., and Snyder, J. K., *J. Org. Chem.*, **1992**, *57*, 5285.
156. Omote, Y., Harada, K., Tomotake, A., and Kashima, C., *J. Heterocycl. Chem.*, **1984**, *21*, 1841.
157. Wenkert, E., Moeller, P. D. R., and Piettre, S. R., *J. Am. Chem. Soc.*, **1988**, *110*, 7188.
158. Biolatto, B., Kneeteman, M., and Mancini, P., *Tetrahedron Lett.*, **1999**, *40*, 3343.
159. Gieseler, A., Steckhan, E., Wiest, O., and Knoch, F., *J. Org. Chem.*, **1991**, *56*, 1405.
160. Bäckvall, J.-E., Plobeck, N. A., and Juntunen, S. K., *Tetrahedron Lett.*, **1989**, *30*, 2589.
161. 'Cycloaddition reactions with vinyl heterocycles', Sepúlveda-Arques, J., Abarca-González, B., and Medio-Simón, *Adv. Heterocycl. Chem.*, **1995**, *63*, 339.
162. Pindur, U. and Eitel, M., *Helv. Chim. Acta*, **1988**, *71*, 1060; Pindur, U., Eitel, M., and Abdoust-Houshang, E., *Heterocycles*, **1989**, *29*, 11.
163. Saroja, B. and Srinivasan, P. C., *Synthesis*, **1986**, 748; Eberle, M. K., Shapiro, M. J., and Stucki, R., *J. Org. Chem.*, **1987**, *52*, 4661.
164. Simoji, Y., Saito, F., Tomita, K., and Morisawa. Y., *Heterocycles*, **1991**, *32*, 2389.
165. Benson, S. C., Lee, L., and Snyder, J. K., *Tetrahedron Lett.*, **1996**, *37*, 5061
166. Wan, Z.-K., and Snyder, J. K., *Tetrahedron Lett.*, **1998**, *39*, 2487.
167. Benson, S. C., Palabrica, C. A., and Snyder, J. K., *J. Org. Chem.*, **1987**,2 5, 4610; Daly, K., Nomak, R., and Snyder, J. K., *Tetrahedon Lett.*, **1997**, *38*, 8611.
168. Santos, P. F., Lobo, A. M., and Prabhakar, S., *Tetrahedron Lett.*, **1995**, *36*, 8099.
169. Kawasaki, T., Ohtsuka, H., Mihira, A., and Sakamoto, M., *Heterocycles*, **1998**, *47*, 367.
170. Raucher, S. and Klein, P., *J. Org. Chem.*, **1986**, *51*, 123.
171. Davis, P. D. and Neckers, D. C., *J. Org. Chem.*, **1980**, *45*, 456.
172. Weedon, A. C. and Zhang, B., *Synthesis*, **1992**, 95.
173. Julian, D. R. and Tringham, G. D., *J. Chem. Soc., Chem. Commun.*, **1973**, 13; Julian, D. R. and Foster, R., *J. Chem. Soc., Chem. Commun.*, **1973**, 311.
174. Thesing, J. and Semler, G., *Justus Liebigs Ann. Chem.*, **1964**, *680*, 52.
175. Bailey, A. S., Haxby, J. B., Hilton, A. N., Peach, J. M., and Vandrevala, M. H., *J. Chem. Soc., Perkin Trans. 1*, **1981**, 382.
176. Katritzky, A. R. and Akutagawa, K., *J. Am. Chem. Soc.*, **1986**, *108*, 6808.
177. Inagaki, S., Nishizawa, Y., Suguira, T., and Ishihara, H., *J. Chem. Soc., Perkin Trans. 1*, **1990**, 179; Naruse, Y., Ito, Y., and Inagaki, S., *J. Org. Chem.*, **1991**, *56*, 2256.
178. Thesing, J. and Schülde, F., *Chem. Ber.*, **1952**, *85*, 324.
179. Howe, E. E., Zambito, A. J., Snyder, H. R., and Tishler, M., *J. Am. Chem. Soc.*, **1945**, *67*, 38.
180. Allbright, J. D. and Snyder, H. R., *J. Am. Chem. Soc.*, **1959**, *81*, 2239; Baciocchi, E. and Schiroli, A., *J. Chem. Soc (B)*, **1968**, 401.

18 Benzo[b]thiophenes and benzo[b]furans: reactions and synthesis

benzothiophene
(benzothiofuran)
(thiaphthene)
[benzo[b]thiophene]

benzofuran
[benzo[b]furan]

Benzo[b]thiophene[1] and benzo[b]furan,[2] frequently (and in the rest of this chapter) referred to simply as benzothiophene and benzofuran, are the sulfur and oxygen analogues of indole, respectively, but have been much less fully studied. The oxygen system occurs in a range of plant- and microbial-derived natural products, ranging in complexity from 5-methoxybenzofuran, through the orange 'aurones', a group of plant pigments isomeric with co-occurring flavones (section 9.2.3.10), usnic acid, a yellow pigment found in many lichens, to griseofulvin, from *Penicillium griseofulvum*, used in medicine as an antifungal agent. Raloxifene shows potential for preventing osteoporosis and reducing the incidence of breast cancer.

aureusin, an aurone

usnic acid

griseofulvin

Raloxifene

18.1 Reactions with electrophilic reagents

18.1.1 Substitution at carbon

The electrophilic substitution of these systems is much less regioselective than that of indole (effectively complete selectivity for attack at C-3), even to the extent that the hetero-ring positions are only a little more reactive than some of the benzene ring positions. For example, nitration of benzothiophene gives a mixture in which, although more than half the product is the 3-nitro-derivative, 2-nitro-, 4-nitro- 6-nitro- and 7-nitrobenzothiophenes are also all produced, each representing about 10% of the product mixture.[3] Measurements of detritiation of 2- and 3-tritiobenzothiophene in trifluoroacetic acid showed rates which were effectively the same for both hetero-ring positions.[4] Friedel-Crafts alkylation[5] of benzothiophene gives mixtures in which the 3-isomer predominates over the 2-isomer, however in other substitutions the 3-isomer is said to be the only product – iodination[6] falls into this category, as does controlled bromination;[7] the 2,3-dibromide can be selectively reduced to the 3-monobromo derivative with zinc in acetic acid which must relate to the greater stability of a 2- versus a 3-anion.

Benzofuran displays a lesser tendency for 3-substitution: formylation of benzofuran reportedly gives only the 2-formyl derivative,[8] and nitric acid nitration[9] produces 2-nitrobenzofuran, though in all studies where the isolation of a major product is described, particularly those conducted before the advent of modern analytical techniques, one must be aware that the presence of other minor isomers may have gone undetected; a later study using dinitrogen tetroxide found 3-nitrobenzofuran as a major product together with a smaller percentage of the 2-isomer.[10] Treatment of benzofuran with halogens results in 2,3-addition products,[11] with the initial electrophilic attack taking place at C-2; from these addition products, by base-promoted hydrogen halide elimination, 3-monohalo benzofurans can be obtained in high yields.[12] Friedel-Crafts substitution is difficult for hetero-ring unsubstituted benzofurans because typical catalysts tend to cause resinification, but 3-acylations[13] have been reported using ferric chloride.

With substituents already present, the pattern of substitution is even more complex: some examples serve to illustrate this. Nitration of 2-bromobenzothiophene results in *ipso* substitution and thus the formation of 2-nitrobenzothiophene whereas 2-chlorobenzothiophene gives the 3-nitro-substitution product;[14] nitration of 3-bromobenzothiophene proceeds in moderate yield to give the 2-nitro derivative.[15] On the other hand 3-carboxy- or 3-acylbenzothiophenes nitrate mainly in the benzene ring.[16] Bromination[17] and Friedel Crafts substitution[18] of 3-methyl- and 2-methylbenzothiophenes takes place cleanly at the vacant hetero-ring position; similarly 2-bromobenzothiophene undergoes formylation at C-3.[19] 3-Methoxybenzothiophene gives the corresponding 2-aldehyde under Vilsmeier conditions at moderate temperatures but at 95 °C 3-chlorobenzothiophene-2-carboxaldehyde is obtained;[20] 6-ethoxybenzothiophene formylates at C-2.[21]

18.1.2 Addition to sulfur in benzothiophenes

Benzothiophenium salts are produced by the reaction of the sulfur heterocycle with more powerful alkylating combinations such as Meerwein salts;[22] benzothiophenium salts can themselves act as powerful alkylating agents with fission of the $C-S^+$ bond.[23]

S-Oxidation produces 1,1-dioxides which readily undergo cycloadditions as dienophiles,[24] or photodimerisation, the head-to-head dimer (shown) being the major product.[25]

18.2   Reactions with nucleophilic reagents

Halogen at a benzothiophene 2-position is subject to displacement with amine nucleophiles,[26] and, surprisingly, rather more easily than halogen at the 3-position, even though an intermediate for 3-attack carries negative charge at C-2, adjacent to the hetero atom. Equally surprising are reactions in which secondary amine anions add to benzothiophene to give 2-dialkylamino-2,3-dihydrobenzothiophenes;[27] with irradiation, addition of primary amines gives 3-alkylamino-2,3-dihydrobenzothio-phenes.[28] *Ipso* displacement[29] of bromine from 3-bromo-2-nitrobenzothiophene can sometimes be accompanied by rearranged products.[30]

18.3   Reactions with bases; reactions of *C*-metallated benzo[*b*]thiophenes and benzo[*b*]furans

In some of the earliest uses of *n*-butyllithium, 2-lithiobenzofuran was obtained by metal–halogen exchange between the 2-bromo-heterocycle and *n*-butyllithium,[31] or by deprotonation of benzofuran.[32] The generation of 3-metallated benzofurans generally results in fragmentation with the production of 2-hydroxyphenylacetylene at room temperature,[28,33] though the 3-lithio-derivative can be utilised at very low temperature.[34]

In a sequence which may involve a carbene as intermediate, 2-lithiobenzofuran reacts with aryl- or alkyllithiums with ring opening, as shown.[35]

Sodium amide causes ring cleavage of benzothiophene to produce 2-ethynylphe-nylthiol.[36] Ring opening in a rather different manner results from exposure of the heterocycle to lithium dimethylamide, followed by trapping with iodomethane, producing an enamine which must result from initial addition at C-2, perhaps by a minor pathway, but one which then leads to ring-opening elimination.[37]

3-Lithiobenzothiophenes can be generated, and reacted with electrophiles, if the temperature is kept low.[38] Direct deprotonation of benzothiophenes follows the usual pattern for five-membered heterocycles and takes place adjacent to the heteroatom,[39] and in concord with this pattern, metal–halogen exchange processes favour a 2- over a 3-halogen; the sequence below shows how this can be utilised to develop substituted benzothiophenes.[40] 2-Lithiated reagents can be used to react with electrophiles: for example reaction with p-toluenesulfonyl cyanide produces the 2-cyano derivatives.[41] 2-Trialkylstannylbenzofurans[42] and benzofuran-2-[43] and ben-zothiophene-2-boronic acids[44] have been used in palladium-catalysed coupling with aromatic halides, in the last case with morphine triflate.

18.4   Reactions with oxidising and reducing agents

Hydrodesulfurisation of benzothiophenes is conveniently achieved using Raney nickel,[45] and before the advent of modern spectroscopic methods was utilised in the determination of structure of substituted benzothiophenes by conversion to a recognisable derivative.

Reduction of the hetero-rings of both benzofuran and benzothiophene, notably with retention of the sulfur in the latter case, can be achieved using triethylsilane in acidic solution giving 2,3-dihydro-derivatives.[46] 2,3-Dihydroxylation of benzofuran and benzothiophene can be achieved using *Pseudomonas putida*;[47] S-oxidation of the sulfur in the latter heterocycle using the same microbiological method has also been reported.[48]

18.5   Electrocyclic reactions

The fusion of a pyridine ring onto benzothiophene can be achieved using either the 2- or 3-azides which after a Staudinger reaction give ylides which undergo aza-Wittig condensations with unsaturated aldehydes, the ensuing electrocyclisation being followed by spontaneous dehydrogenation.[49]

18.6   Oxy-[50] and amino-benzothiophenes and -benzofurans

Benzothiophen-2-ones can be conveniently accessed by oxidation of 2-lithiobenzothiophenes.[51] Benzothiophen-2-one will condense at the 3-position with aromatic aldehydes;[52] benzothiophen-3-one reacts comparably at its 2-position.[53]

Both benzofuran-2-one, known trivially in the older literature as coumaranone, and best viewed as a lactone, and the isomeric benzofuran-3-one, form ambident anions by deprotonation at a methylene group, the former[54] requiring a stronger base than the latter.[55]

Little is known of simple 2- and 3-amino-derivatives; 2-dialkylaminobenzothiophenes can be obtained by reaction of benzothiophene-2-thiol with secondary amines.[51] In many ways 2-aminobenzothiophene behaves like a normal aromatic amine, but diazotisation leads directly to benzothiophen-2-one.[56]

18.7   Synthesis of benzo[b]thiophenes and benzo[b]furans

18.7.1   Ring synthesis

18.7.1.1   From 2-arylthio- or 2-aryloxyaldehydes, -ketones or -acids

Cyclisation of 2-arylthio- or 2-aryloxyaldehydes, -ketones or -acids *via* intramolecular electrophilic attack on the aromatic ring, with loss of water, creates the heterocyclic ring; this route is the commonest method for benzothiophenes.

In order to produce hetero-ring unsubstituted benzothiophenes[57] an arylthioacetaldehyde acetal is generally employed prepared, in turn, from bromoacetaldehyde acetal and the thiophenol. An exactly parallel sequence produces 2,3-unsubstituted benzofurans.[58]

Comparable acid-catalysed ring closures of 2-arylthio-[59] and 2-aryloxy-[60] -ketones, and -2-arylthio-[61] and -2-aryloxyacetyl[62] chlorides lead to 3-substituted heterocycles and 3-oxygenated heterocycles respectively. Attempted formation of 3-*aryl*benzothiophenes by this route is always accompanied by partial or complete isomerisation to the 2-aryl-heterocycle.[63]

The closure of *O*-allenyl *ortho*-iodophenols using palladium(0) catalysis produces species which can be trapped with azide.[64]

18.7.1.2   From 2-(ortho-hydroxy(or thioxy)aryl)-acetaldehydes, -ketones or -acids

Cyclising dehydration of 2-(*ortho*-hydroxyaryl)-acetaldehydes, -ketones or -acids (and in some cases sulfur analogues) give the heterocycles; this route is important for benzofurans.

Claisen rearrangement of allyl phenolic ethers, followed by oxidation of the alkene generates *ortho*-hydroxyarylacetaldehydes which close to give benzofurans under acid catalysis.[65] The formation of 2-substituted benzofurans from 2-(*ortho*-hydroxyaryl)-ketones is also very easy.[66]

The employment of aryl 2-chloroprop-2-enyl sulfides (or ethers) as thio-Claisen rearrangement substrates neatly eliminates the necessity for an oxidative step thus providing a route to 2-methylbenzothiophenes (-benzofurans).[67]

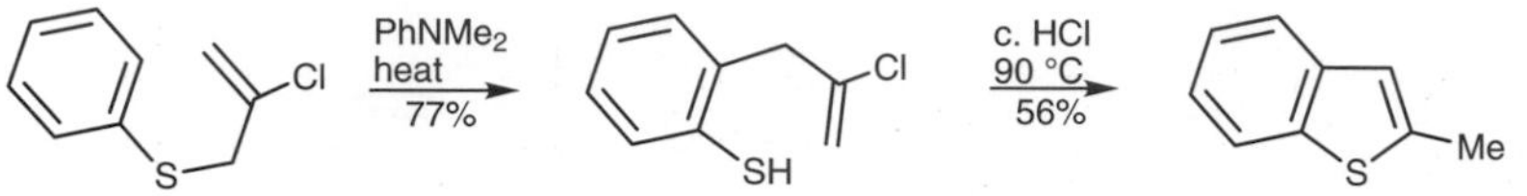

Propargyl aryl ethers undergo a Claisen rearrangement and then ring closure to produce 2-methylbenzofurans.[68]

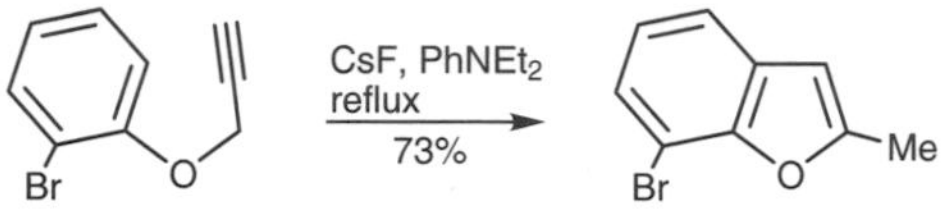

Another route to compounds of the same oxidation level involves palladium-catalysed coupling of enol ethers.[69]

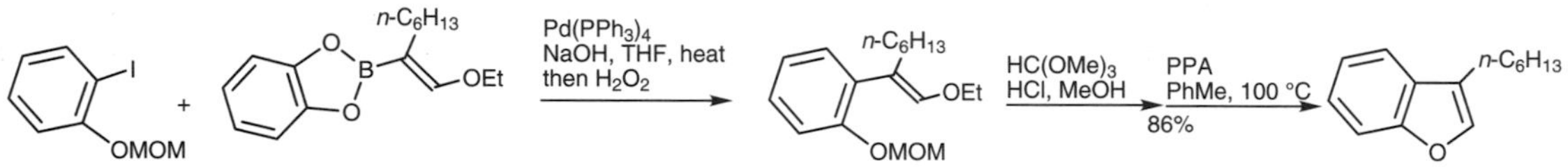

4-Chloromethylcoumarins can be converted into benzofuran-3-acetic acids by exposure to alkali – hydrolysis of the lactone and then reclosure with displacement of chloride by the phenolate leads to the benzofuran.[70]

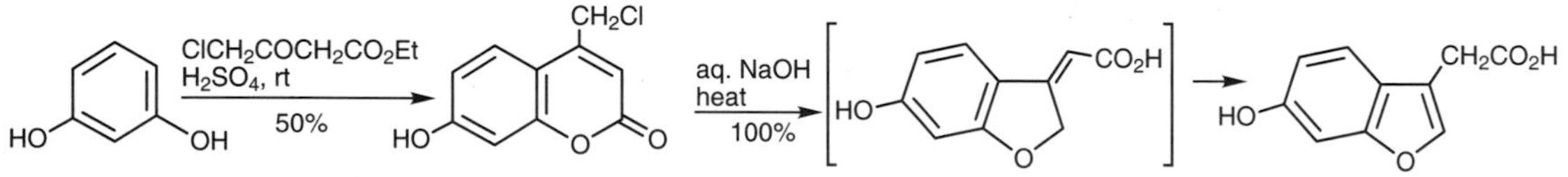

18.7.1.3   From ortho-acylaryloxy- or -arylthioacetic acids (esters) (ketones)

Cyclising condensation of *ortho*-acylaryloxy- or -arylthioacetic acids (esters) or ketones gives the bicyclic heterocycles.

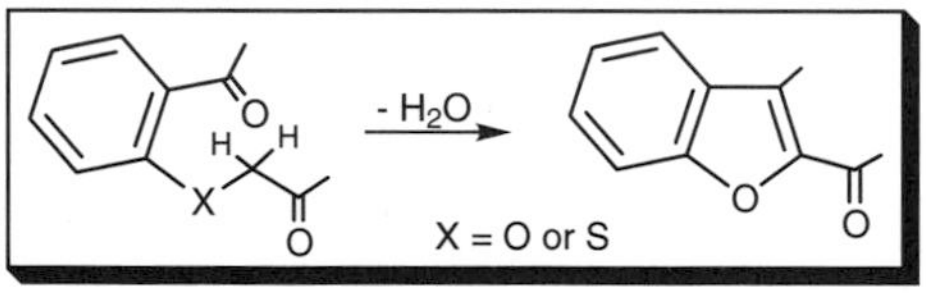

Intramolecular aldol/Perkin type condensation of *ortho*-formylaryloxyacetic acids and arylthioacetic esters produces benzofuran[71] and benzothiophene-2-esters[72] respectively, as illustrated below. *ortho*-Formyl- or *ortho*-acylaryl benzyl ethers, in which the benzyl group carries an electron-withdrawing substituent, can be comparably closed to produce 2-arylbenzofurans, using potassium fluoride or caesium fluoride on alumina.[73]

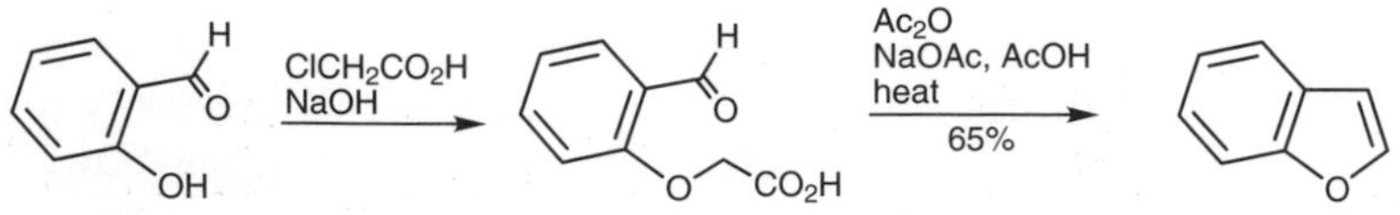

ortho-Hydroxyaryl aldehydes or ketones, by *O*-alkylation with α-haloketones afford substrates which on intramolecular aldol condensation produce 2-acyl benzofurans.[74]

If, instead of an *ortho* carbonyl group, cyclisation is conducted with an *ortho* nitrile, then 3-aminoheterocycles result – the example shows how in appropriately activated situtations, both the introduction of the thioacetate and the cyclisation can take place in one pot.[75]

18.7.1.4   From O-aryl ketoximes

The electrocyclic rearrangement of *O*-aryl ketoximes produces benzofurans.

The acid-catalysed rearrangement of *O*-aryl ketoximes,[76] which produces benzofurans, exactly parallels the rearrangement of phenylhydrazones, which gives indoles – the classical Fischer indole synthesis (section 17.16.1).

18.7.1.5   From ortho-iodophenols

With the advent of palladium(0) catalytic methods, it is now possible to produce the furan ring of a benzofuran by interaction between an *ortho*-iodophenol and an alkyne, the two carbon atoms of the triple bond providing carbons 2 and 3 of the furan ring and the larger substituent of the alkyne ending up at the heterocyclic 2-position.[77] The sequence has been conducted on solid support.[78] Coupling with 2,5-dihydro-2,5-dimethoxyfuran leads to methyl benzofuran-3-acetate.[79]

Coupling between alkynes and *ortho*-methoxy triflates produces precursors which cyclise to benzofurans with hot lithium chloride.[80]

18.7.1.6   Syntheses which involve making the benzene ring

6,7-Dihydrobenzothiophenes react as dienes with alkynes, subsequent retro-Diels-Alder elimination of ethene giving a benzothiophene, as illustrated.[81] In a similar fashion, the silyl enol ether derived from 3-acetylfuran undergoes cycloadditions giving 4-oxygenated benzofurans.[82]

The coupling of furyl and thienyl stannanes to 4-chlorocyclobut-2-enones[83] or the addition of furylcerium reagents to cyclobut-3-ene-1,2-dione monoacetals[84] have been used to synthesise cyclobutenones which on heating ring open to unsaturated ketenes which then undergo an electrocyclic closure producing benzofurans or benzothiophenes with an oxygen substituent at C-4.

18.7.1.7   From partially reduced benzofurans and benzothiophenes

It can be an advantage for the introduction of benzene ring substitutents to operate with hetero-ring-reduced derivatives, the aromatic heterocycle being obtained by a final dehydrogenation. 2,3-Dihydrobenzothiophenes can be oxidised up with sulfuryl chloride or *N*-chlorosuccinimide;[85] 2,3-dichloro-5,6-dicyanobenzoquinone has been employed to dehydrogenate 2,3-dihydrobenzofurans.[86] In the example below a benzene ring substitution is followed by aromatisation via elimination of hydrogen iodide and isomerisation of the double bond into the aromatic position.[87]

Exercises for chapter 18

Straightforward revision exercises (consult Chapters 16 and 18)

(a) In the electrophilic substitution of benzothiophene and benzofuran there is less selectivity than for comparable reactions of indole – why?

(b) What is the principal method for the efficient introduction of substituents to the 2-positions of benzofuran and benzothiophene?

(c) Beginning from a phenol carrying no substituents *ortho* to the hydroxyl, describe two methods for the synthesis of benzofurans.

(d) How can salicaldehydes be used for the synthesis of benzofurans?

More advanced exercises

1. Suggest structures for the compounds formed at each stage in the following sequence: PhSH with $ClCH_2COCH_2CO_2Et$ ($\rightarrow C_{12}H_{14}SO_3$), then PPA/heat ($\rightarrow C_{12}H_{12}SO_2$), then NH_3 ($\rightarrow C_{10}H_9NSO$), then $LiAlH_4$ ($\rightarrow C_{10}H_{11}NS$), then HCO_2H/heat, ($\rightarrow C_{11}H_{11}NSO$) then $POCl_3$/heat giving finally a tricylic substance, $C_{11}H_9NS$.

2. Draw structures for the heterocycles formed from the following combinations: (i) $C_{13}H_{16}O$ from 2,4,5-trimethylphenol with 3-chloro-2-butanone then the product with c. H_2SO_4; (ii) $C_{12}H_8O_4$ from 7-hydroxy-8-methoxycoumarin with $CH_2=CHCH_2Br/K_2CO_3$ then the product heated strongly giving an isomer, then reacted successively with O_3 then H^+; (iii) 4-trifluoromethylfluorobenzene with LDA then DMF ($\rightarrow C_8H_4F_4O$), then with $HSCH_2CO_2Me/NaH$ giving $C_{11}H_7F_3O_2S$; (iv) $C_9H_7NO_3$ from 4-fluoronitrobenzene with $Me_2C=NONa$ then c. HCl/heat.

3. Deduce structures for the bi- and tetracyclic heterocycles formed in the following two steps respectively: 4-chlorophenylthioacetic acid with PCl_3 then $AlCl_3$ ($\rightarrow C_8H_5ClOS$), then this with phenylhydrazine in hot AcOH $\rightarrow C_{14}H_8ClNS$.

References

1. 'Recent advances in the chemistry of benzo[b]thiophenes', Iddon, B. and Scrowston, R. M., *Adv. Heterocycl. Chem.*,**1970**, *11*, 177; 'Recent advances in the chemistry of benzo[b]thiophenes', Scrowston, R. M., *ibid.*, **1981**, *29*, 171.
2. 'Recent advances in the chemistry of benzo[b]furan and its derivatives. Part I: Occurence and synthesis', Cagniant, P. and Cagniant, D., *Adv. Heterocycl. Chem.*, **1975**, *18*, 337.
3. Armstrong, K. J., Martin-Smith, M., Brown, N. M. D., Brophy, G. C., and Sternhell, S., *J. Chem. Soc., C,* **1969**, 1766.
4. Eaborn, C. and Wright, G. J., *J. Chem. Soc., B*, **1971**, 2262.
5. Cooper, J. and Scrowston, R. M., *J. Chem. Soc., Perkin Trans. 1*, **1972**, 414.
6. Van Zyl, G., Bredeweg, C. J., Rynbrandt, R. H., and Neckers, D. C., *Can. J. Chem.*, **1966**, *44*, 2283.
7. Cherry, W. H., Davies, W., Ennis, B. C., and Porter, Q. N., *Austral. J. Chem.*, **1967**, *20*, 313; Forst. Y., Becker, S., and Caubre, P., *Tetrahedron*, **1994**, *50*, 11893.
8. Bisgani, M., Buu-Ho, N. P., and Royer, R., *J. Chem. Soc.*, **1955**, 3688.
9. v. Stoermer, R. and Richter, O., *Chem. Ber.*, **1897**, *30*, 2094; v. Stoermer, R. and Kahlert, B., *ibid.*, **1902**, *35*, 1633.

10. Kaluza, F. and Perold, G., *Chem. Ber.*, **1955**, *88*, 597.
11. Okuyama, T., Kunugiza, K., Fueno, T., *Bull. Chem. Soc. Jpn.*, **1974**, *47*, 1267.
12. Baciocchi, E., Sebastiani, G. V., and Ruzziconi, R., *J. Org. Chem.*, **1979**, *44*, 28.
13. Campaigne, E., Weinberg, E. D., Carlson, G., and Neiss, E. S., *J. Med. Chem.*, **1965**, *8*, 136.
14. Dickinson, R. P., Iddon, B., and Sommerville, R. G., *Int. J. Sulfur Chem.*, **1973**, *8*, 233.
15. Bachelet, J.-P., Royer, R., and Gatral, P., *Eur. J. Med. Chem. Chim. Ther.*, **1985**, *20*, 425.
16. Brophy, G. C., Sternhell, S., Brown, N. M. D., Brown, I., Armstrong, K. J., and Martin-Smith, M., *J. Chem Soc., C*, **1970**, 933; Brown, I., Reid, S. T., Brown, N. M. D., Armstrong, K. J., Martin-Smith, M., Sneader, W. E., Brophy, G. C., and Sternhell, S., *J. Chem. Soc., C*, **1969**, 2755.
17. Dickinson, R. P. and Iddon, B., *J. Chem. Soc., C*, **1971**, 182.
18. Sauter, F. and Golser, L., *Monatsh. Chem.*, **1967**, *98*, 2039; Faller, P. and Cagniant, P., *Bull. Soc. Chim. Fr.*, **1962**, 30.
19. Minh, T. Q., Thibaut, P., Christiaens, L., and Renson, M., *Tetrahedron*, **1972**, *28*, 5393.
20. Ricci, A., Balucani, D., and Buu-Ho, N. P., *J. Chem. Soc., C*, **1967**, 779.
21. Campaigne, E. and Kreighbaum, W. E., *J. Org. Chem.*, , **1961**, *26*, 363
22. Acheson, R. M. and Harrison, D. R., *J. Chem. Soc., C*, **1970**, 1764.
23. Cotruvo, J. A. and Degani, I., *J. Chem. Soc., Chem. Commun.*, **1971**, 436.
24. Davies, W. and Porter, Q. N., *J. Chem. Soc.*, **1957**, 459.
25. Harpp, D. N. and Heitner, C., *J. Org. Chem.*, **1970**, *35*, 3256; *idem*, *J. Am. Chem. Soc.*, **1972**, *94*, 8179.
26. Reinecke, M. G., Mohr, W. B., Adickes, H. W., de Bie, D. A., van de Plas, H. C., and Nijdam, J. *J. Org. Chem.*, **1973**, *38*, 1365; Chippendale, K. E., Iddon, B., Suschitzky, H., and Taylor, D. S., *J. Chem. Soc., Perkin Trans. 1*, **1974**, 1168.
27. Grandclaudon, P. and Lablache-Combier, A., *J. Org. Chem.*, **1978**, *43*, 4379.
28. Grandclaudon, P., Lablache-Combier, A., and Prknyi, C., *Tetrahedron*, **1973**, 29, 651.
29. Guerrera, F. and Salerno, L., *J. Heterocycl. Chem.*, **1995**, *32*, 591.
30. Guerrera, F., Salerno, L., Lamartina, L., and Spinelli, D., *J. Chem. Soc., Perkin Trans. 1*, **1995**, 1243.
31. Gilman, H. and Melstrom, D. S., *J. Am. Chem. Soc.*, **1948**, *70*, 1655.
32. Shirley, D. A. and Cameron, M. D., *J. Am. Chem. Soc.*, **1950**, *72*, 2788.
33. Reichstein, T. and Baud, J., *Helv. Chim. Acta*, **1937**, *20*, 892.
34. Cugnan de Sevricourt, M. and Robba, M., *Bull. Soc. Chim. Fr.*, **1977**, 142.
35. Nguyen, T. and Negishi, E., *Tetrahedron Lett.*, **1991**, *32*, 5903; .
36. Schroth, W., Jordan, H., and Spitzner, R., *Tetrahedron Lett.*, **1995**, *36*, 1421.
37. Beyer, A. E. M. and Kloosterziel, H., *Recl. Trav. Chim. Pays Bas*, **1977**, *96*, 178.
38. Dore, G., Bonhomme, M., and Robba, M., *Tetrahedron*, **1972**, *28*, 2553; Dickinson, R. P. and Iddon, B., *J. Chem. Soc., C*, **1970**, 2592.
39. Kerdesky, F. A. J. and Basha, A., *Tetrahedron Lett.*, **1991**, *32*, 2003.
40. Dickinson, R. P. and Iddon, B., *J. Chem. Soc., C*, **1971**, 2504; Reinecke, M. G., Newsom, J. G. and Almqvist, K. A., *Synthesis*, **1980**, 327; Sura, T. P. and MacDowell, D. W. H., *J. Org. Chem.*, **1993**, *58*, 4360.
41. Nagasaki, I., Suzuki, Y., Iwamoto, K., Higashino, T., and Miyashita, A., *Heterocycles*, **1997**, *46*, 443
42. Clough, J. M., Mann, I. S., and Widdowson, D. A., *Tetrahedron Lett.*, **1987**, *28*, 2645.
43. Blettner, C. G. Knig, W. A., Tenzel, W., and Schotten, T., *Synlett*, **1998**, 295.
44. Hedberg, M. H., Johansson, A. M., Fowler, C. J., Terenius, L., and Hacksell, U., *Biorg. Med. Chem. Lett.*, **1994**, *4*, 2527.
45. Papa, D., Schwenk, E., and Ginsberg, H. F., *J. Org. Chem.*, **1949**, *14*, 723.
46. Kursanov, D. N., Parnes, Z. N., Bolestova, G. I., and Belen'kii, L. I., *Tetrahedron*, **1975**, *31*, 311.
47. Boyd, D. R., Sharma, N. D., Boyle, R., McMurray, B. T., Evans, T. A., Malone, J. F., Dalton, H., Chima, J., and Sheldrake, G. N., *J. Chem. Soc., Chem. Commun.*, **1993**, 49.
48. Boyd, D. R., Sharma, N. D., Haughey, S. A., Malone, J. F., McMurray, B. T., Sheldrake, G. N., Allen, C. C. R., and Dalton, H., *Chem. Commun.*, **1996**, 2363.
49. Degl'Innocenti, A., Funicello, M., Scafato, P., Spagnolo, P., and Zanirato, P., *J. Chem. Soc., Perkin Trans. 1*, **1996**, 2561.
50. 'Chemistry of benzo[*b*]thiophene-2,3-dione', Rajopadhye, M. and Popp, F. D., *Heterocycles*, **1988**, *27*, 1489.

51. Vesterager, N. O., Pedersen, E. B., and Lawesson, S.-O., *Tetrahedron*, **1973**, *29*, 321.

52. Conley, R. A. and Heindel, N. D., *J. Org. Chem.*, **1976**, *41*, 3743.

53. Réamonn, L. S. S. and O'Sullivan, W. I., *J. Chem. Soc., Perkin Trans. 1*, **1977**, 1009.

54. Zaugg, H. E., Dunnigan, D. A., Michaels, R. J., Swett, R. J., Wang, T. S., Sommers, A. H., and DeNet, R. W., *J. Org. Chem.*, **1961**, *26*, 644.

55. v. Auwers, K. and Schtte, H., *Chem. Ber.*, **1919**, *52*, 77.

56. Stacy, G. W., Villaescusa, F. W., and Wollner, T. E., *J. Org. Chem.*, **1965**, *30*, 4074.

57. Tilak, B. D., *Tetrahedron*, **1960**, *9*, 76.

58. Spagnolo, P., Tiecco, M., Tundo, A., and Martelli, G., *J. Chem. Soc., Perkin Trans. 1*, **1972**, 556.

59. Dickinson, R. P. and Iddon, B., *J. Chem. Soc., C*, **1968**, 2733; Chapman, N. B., Clarke, K., and Sawhney, S. N., *J. Chem. Soc., C*, **1968**, 518.

60. Royer, R. and René, L., *Bull. Soc. Chim. Fr.*, **1970**, 1037.

61. Werner, L. H., Schroeder, D. C., and Ricca, S., *J. Am. Chem. Soc.*, **1957**, *79*, 1675.

62. Elvidge, J. A. and Foster, R. G., *J. Chem. Soc.*, **1964**, 981.

63. Banfield, J. E., Davies, W., Gamble, N. W., and Middleton, S., *J. Chem. Soc.*, **1956**, 4791.

64. Gardiner, M., Grigg, R., Sridharan, V., and Vicker, N., *Tetrahedron Lett.*, **1998**, *39*, 435.

65. de Souza, N. J., Nayak, P. V., and Secco, E., *J. Heterocycl. Chem.*, **1966**, *3*, 42.

66. Tinsley, S. W., *J. Org. Chem.*, **1959**, *24*, 1197

67. Anderson, W. K., LaVoie, E. J., and Bottaro, J. C., *J. Chem. Soc., Perkin Trans. 1*, **1976**, 1.

68. Ishii, H., Ohta, S., Nishioka, H., Hayashida, N., and Harayama, T., *Chem. Pharm. Bull.*, **1993**, *41*, 1166; Moghaddam, F. M., Sharifi, A., and Saidi, M. R., *J. Chem. Res. (S)*, **1996**, 338.

69. Satoh, M., Miyaura, N., and Suzuki, A., *Synthesis*, **1987**, 373.

70. Fall, Y., Santana, L., Teijeira, M., and Uriarte, E., *Heterocycles*, **1995**, *41*, 647.

71. Burgstahler, A. W. and Worden, L. R., *Org. Synth., Coll. Vol. V*, **1973**, 251.

72. Bridges, A. J., Lee, A., Maduakor, E. C., and Schwartz, C. E., *Tetrahedron Lett.*, **1992**, *33*, 7499.

73. Hellwinkel, D. and Gke, K., *Synthesis*, **1995**, 1135.

74. Elliott, E. D., *J. Am. Chem. Soc.*, **1951**, *73*, 754.

75. Peinador, C., Veiga, M. C., Vilar, J., and Quintela, J. M., *Heterocycles*, **1994**, *38*, 1299.

76. Sheradsky, T., *Tetrahedron Lett.*, **1966**, 5225; Sheradsky, T. and Elgavi, A., *Isr. J. Chem.*, **1968**, *6*, 895; Kaminsky, D., Shavel, J. and Meltzer, R. I., *Tetrahedron Lett.*, **1967**, 859.

77. Larock, R. C., Yum, E. K., Doty, D. J., and Sham, K. K. C., *J. Org. Chem.*, **1995**, *60*, 3270; Bishop, B. C., Cottrell, I. F., and Hands, D., *Synthesis*, **1997**, 1315; Kundu, N. G., Pal, M., Mahanty, J. S., and De, M., *J. Chem. Soc., Perkin Trans. 1*, **1997**, 2815.

78. Fancelli, D., Fagnola, M. C., Severino, D., and Bedeschi, A., *Tetrahedron Lett.*, **1997**, *38*, 2311.

79. Samizu, K. and Ogasawara, K., *Heterocycles*, **1994**, *38*, 1745.

80. Hiroya, K., Hashimura, K., and Ogasawara, K., *Heterocycles*, **1994**, *38*, 2463.

81. Labadie, S. S., *Synth. Commun.*, **1998**, *28*, 2531.

82. Ben'tez, A., Herrera, F. R., Romera, M., Talams, F. X., and Muchowski, J. P., *J. Org. Chem.*, **1996**, *61*, 1487.

83. Liebskind, L. S. and Wang, J., *J. Org. Chem.*, **1993**, *58*, 3550.

84. Liu, H., Gayo, L. M., Sullivan, R. W., Choi, A. Y. H., and Moore, H. W., *J. Org. Chem.*, **1994**, *59*, 3284.

85. Tohma, H., Egi, M., Ohtsubo, M., Watanabe, H., Takizawa, S., and Kita, Y., *Chem. Commun.*, **1998**, 173.

86. Bchi, G. and Chu, P.-S., *J. Org. Chem.*, **1978**, *43*, 3717; Stanetty, P. and Prstinger, G., *J. Chem. Res.*, **1991**, *(S)* 78; *(M)* 0581.

87. Onito, K., Hatakeyana, T., Takeo, M., Suginome, H., and Tokuda, M., *Synthesis*, **1997**, 23.

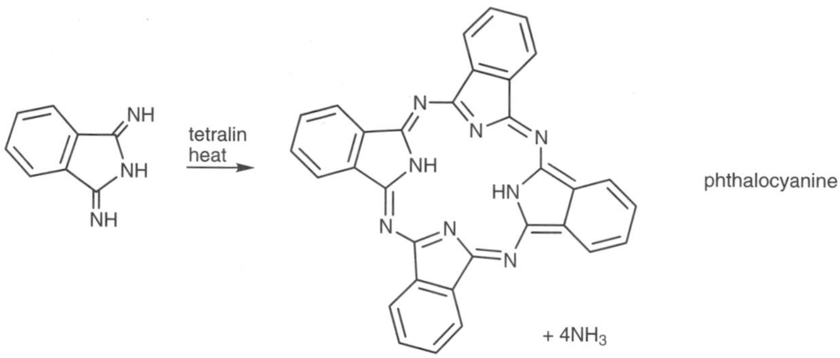

phthalocyanine

+ 4NH$_3$

19.4   Synthesis of isoindoles, benzo[c]thiophenes, and isobenzofurans

19.4.1   Isoindoles

Isoindoles can be produced by eliminations from *N*-substituted isoindolines (1,3-dihydroisoindoles), themselves readily produced by the reaction of a nitrogen nucleophile and a 1,2-bis(bromomethyl)benzene:[22] examples are the pyrolytic elimination of the elements of methyl hydrogen carbonate from the cyclic hydroxylamine carbonate,[4] or, at a much lower temperature, of benzyl alcohol from an *N*-hydroxyisoindoline benzyl ether,[23] or of methanesulfonic acid from a corresponding mesylate.[24]

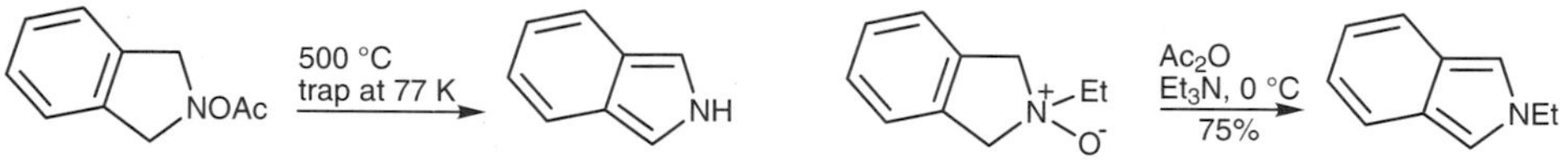

N-substituted isoindoles, too, have generally been made from an isoindoline by elimination processes, thus *N*-oxides can be made to lose water by pyrolysis[25] or better, by treatment with acetic anhydride.[26]

A synthesis of 1-phenylisoindole represents a classical approach to the construction of a heterocycle: a precursor is assembled in which there is an amino group (initially protected in the form of a phthalimide) five atoms away from a carbonyl group with which it must interact and form a cyclic imine.[6]

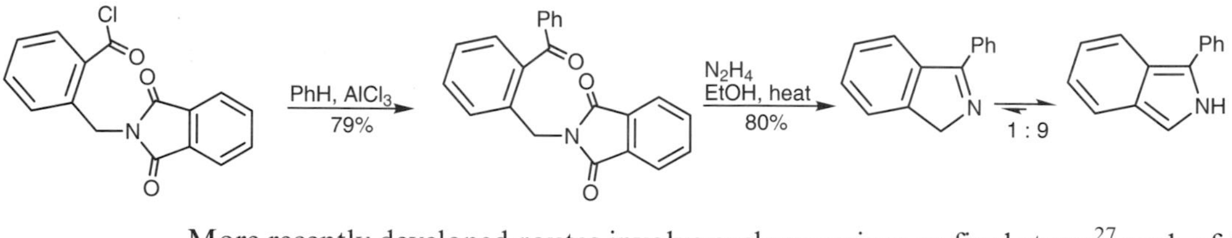

More recently developed routes involve cycloreversions as final steps;[27] each of the starting materials shown below is available from the cycloadduct (cf. section 13.9) of benzyne and 1-methoxycarbonylpyrrole.

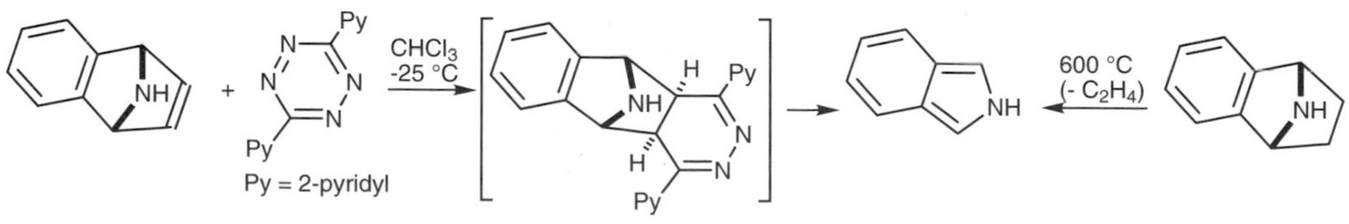

1,3-Diarylisoindoles can be constructed from 1,2-diaroylbenzenes by reaction with an amine and a reducing agent.[28]

19.4.2 Benzo[c]thiophenes

Elimination from dihydrobenzo[c]thiophene S-oxides has been successfully applied, as for isoindoles, for the preparation of benzo[c]thiophenes, including the parent compound.[19,29]

In a neat manipulation of oxidation levels, the reaction of a 1,4-dihydro-1,2-diaroylbenzenes, such as are available from Diels-Alder addition of buta-1,3-dienes with 1,2-diaroylalkynes, with a sulfur source, produces benzo[c]thiophenes;[30] note that no reductant is required as would be necessary if a 1,2-aroylbenzene were utilised. These same Diels-Alder adducts react with primary amines to give 2-substituted isoindoles.

It is comparatively rare for the construction of a benzanellated heterocycle to involve formation of the benzene ring last, however benzo[c]thiophenes can be made by this strategy, utilising a double Friedel-Crafts type alkylation of a 2,5-disubstituted (to prevent attack at α-positions) thiophene with a 1,4-diketone.[31]

19.4.3 Isobenzofurans

Isobenzofuran can be isolated by trapping on a cold finger, following thermolysis of a suitable precursor such as 1,4-epoxy-1,2,3,4-tetrahydronaphthalene,[13,32] but although isolable, for trapping experiments it can be conveniently produced by either acid or base-catalysed elimination of methanol from 1-methoxyphthalan in the presence of the intended dienophile.[33]

1-Methoxyphthalan is obtained by partial oxidation of 1,2-bis(hydroxymethyl)-benzene with hypochlorite in methanol; treatment with lithium diisopropylamide gives isobenzofuran.[34] Conditions have been defined whereby this elimination can be run in such a way as to allow immediate ring lithiation; this species then can be further reacted.[35]

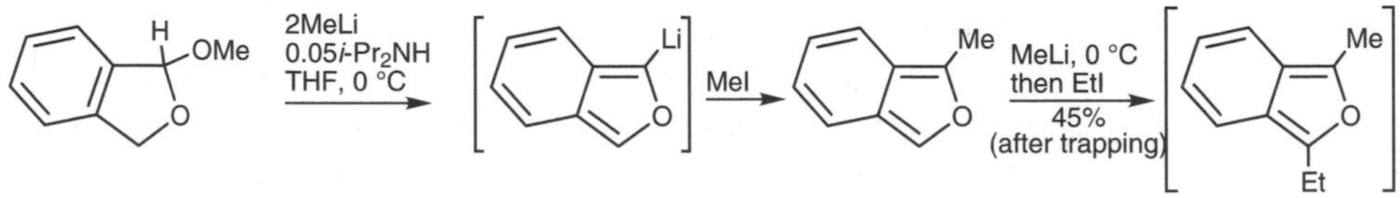

This same oxidation level situation – a disubstituted benzene with aldehyde (ketone) *ortho* to carbinol ready for cyclisation and dehydration to an isobenzofuran – can be achieved in alternative ways: phthalaldehyde can be mono-acetalised, then the remaining aldehyde reduced,[36] or lithiation technology can be utilised, as shown below.[37]

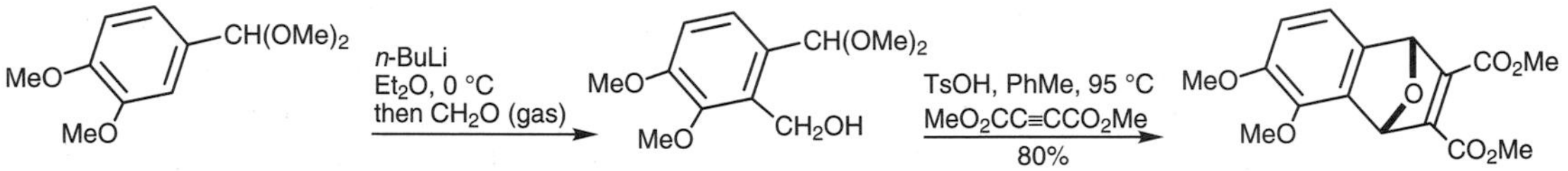

A spectacular demonstration of this approach is provided by the 'stretched' isobenzofuran synthesis shown below.[38] Note that the first cycloaddition is with a dienophile able to provide the *ortho* related carbinol/aldehyde arrangement ready for the formation of another furan ring.

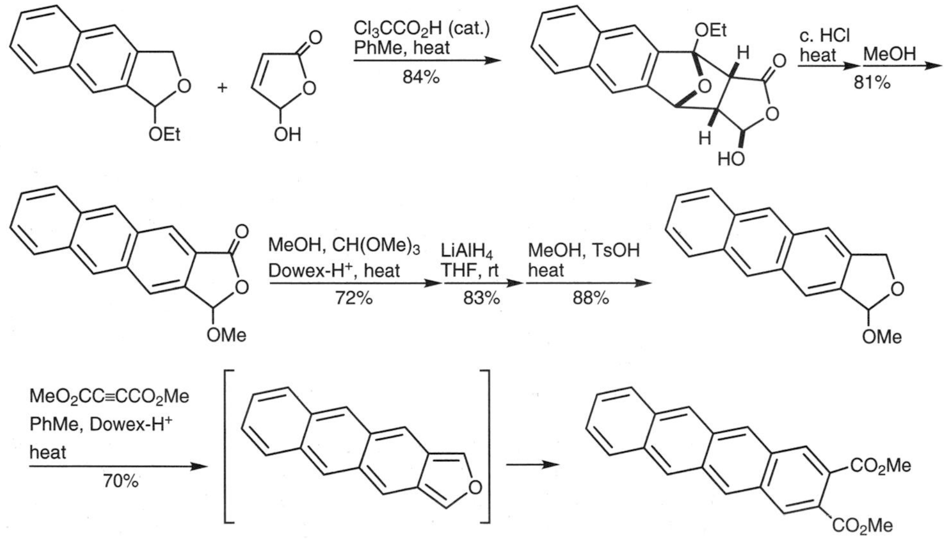

Most of the stable isobenzofurans are 1,3-diaryl substituted, and are deep yellow. Such compounds are available *via* the partial reduction, and dehydrating cyclisation of 1,2-diaroylbenzenes.[39] Both 1-mono- and 1,3-disubstituted isobenzofurans are available from phthalides by Grignard addition then elimination of water.[40]

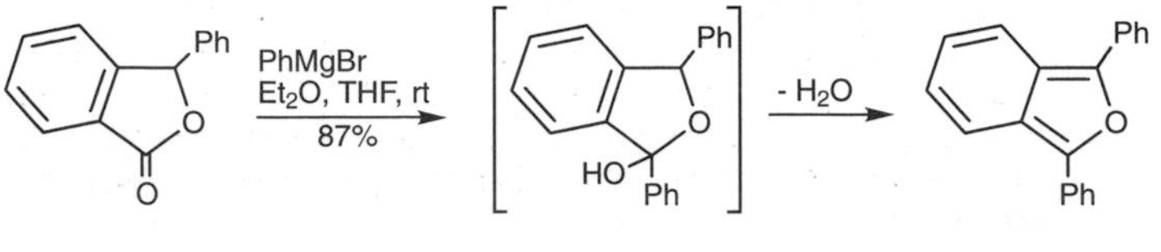

Exercises for chapter 19

Straightforward revision exercises (consult chapters 16 and 19)

(a) The [c]-fused heterocycles considered in this chapter are much less stable than the [b]-fused isomers - why?

(b) What factors favour 1H-isoindoles over 2-H-isoindoles?

(c) What is the most characteristic reactivity of the [c]-fused heterocycles considered in this chapter? Give three examples of this typical reactivity.

(d) Describe one method each for the ring synthesis of isoindoles, benzo[c]thiophenes, and isobenzofurans.

More advanced exercises

1. Deduce structures for the compounds formed at each stage in the following sequences: (i) 1,2-bis(bromomethyl)-4-pivaloylbenzene with $H_2NCH_2C\equiv CH/$ $Et_3N \rightarrow C_{16}H_{19}NO$ which was then heated at 500 °C producing $C_{13}H_{15}NO$, which was trapped with N-phenylmaleimide $\rightarrow C_{23}H_{22}N_2O_3$ (what is the mechanism of the high temperature reaction?); (ii) phthalaldehyde reacted, in sequence, with 2 x $NaHSO_3$, then $MeNH_2$, then 2 x $KCN \rightarrow C_{10}H_8N_2$; (iii) benzoic acid N,N-diethylamide with n-BuLi then $PhCH=O$ then acid $\rightarrow$ $C_{14}H_{10}O_2$ then this with $PhMgBr$ then acid $\rightarrow C_{20}H_{14}O$ and finally this with $O_2/$ methylene blue/$h\nu$/-50 °C $\rightarrow C_{20}H_{14}O_3$; (iv) phthalaldehyde with $HO(CH_2)_2OH/$ $CuSO_4 \rightarrow C_{10}H_{10}O_3$ then this with $NaBH_4$ followed by $TsOH$ with $MeO_2CC\equiv CCO_2Me$ in hot toluene $\rightarrow C_{14}H_{12}O_5$; (v) benzo[c]thiophene with maleic anhydride then hot $NaOH$ then acid $\rightarrow C_{12}H_8O_4$.

References

1. 'Isoindoles', White, J. D. and Mann, M. E., *Adv. Heterocycl. Chem.*, **1969**, *10*, 113; 'The chemistry of isoindoles', Bonnett, R. and North, S. A., *ibid.*, **1982**, *29*, 341.
2. 'Benzo[c]thiophenes', Iddon, B., *Adv. Heterocycl. Chem.*, **1972**, *14*, 331.
3. 'Benzo[c]furans', Friedrichsen, W., *Adv. Heterocycl. Chem.*, **1980**, *26*, 135; 'Isobenzofuran', Haddadin, M. J., *Heterocycles*, **1978**, *9*, 865; 'Progress in the chemistry of isobenzofurans; applications to the synthesis of natural products and polyaromatic hydrocarbons', Rodrigo, R., *Tetrahedron*, **1988**, *44*, 2093; 'Recent advances in the chemistry of benzo[c]furans and related compounds', Friedrichsen, W., *Adv. Heterocycl. Chem.*, **1999**, *73*, 1.
4. Bonnett, R., Brown, R. F. C., and Smith, R. G., *J. Chem. Soc., Perkin Trans. 1*, **1973**, 1432.
5. Bender, C. O. and Bonnett, R., *J. Chem. Soc., Chem. Commun.*, **1966**, 198.
6. Veber, D. F. and Lwowski, W., *J. Am. Chem. Soc.*, **1964**, *86*, 4152.
7. Kreher, R. and Herd, K. J., *Angew. Chem., Int. Ed. Engl.*, **1974**, *13*, 739.
8. Kreher, R., Kohl, N., and Use, G., *Angew. Chem., Int. Ed. Engl.*, **1982**, *21*, 621.
9. Laws, A. P. and Taylor, R., *J. Chem. Soc., Perkin Trans. 2*, **1987**, 591.
10. Kreher, R. P. and Use, G., *Chem. Ber.*, **1989**, *122*, 337.
11. von Dobeneck, H., Reinhard, H., Deubel, H., and Wolkenstein, D., *Chem. Ber.*, **1969**, *102*, 1357.
12. Theilacker, W. and Kalenda, H., *Justus Liebigs Ann. Chem.*, **1953**, *584*, 87.
13. Wiersum, U. E. and Mijs, W. J., *J. Chem. Soc., Chem. Commun.*, **1972**, 347.
14. Berson, J. A., *J. Am. Chem. Soc.*, **1953**, *75*, 1240.
15. Rio, G. and Scholl, M.-J., *J. Chem. Soc., Chem. Commun.*, **1975**, 474.
16. Faragher, R. and Gilchrist, T. L., *J. Chem. Soc., Perkin Trans. 1*, **1976**, 336.
17. Tobia, D. and Rickborn, B., *J. Org. Chem.*, **1987**, *52*, 2611.
18. Mayer, R., Kleinert, H., Richter, S., and Gewald, K., *Angew. Chem., Int. Ed. Engl.*, **1962**, *1*, 115.

19. Cava, M. P., Pollack, N. M., Mamer, O. A., and Mitchell, M. J., *J. Org. Chem.*, **1971**, *36*, 3932.
20. 'Advances in the chemistry of phthalocyanines', Lever, A. B. P., Hempstead, M. R., Leznoff, C. C., Liew, W., Melnik, M., Nevin, W. A., and Seymour, P., *Pure. Appl. Chem.*, **1986**, *58*, 1461.
21. Elvidge, J. A. and Linstead, R. P., *J. Chem. Soc.*, **1955**, 3536.
22. Bornstein, J. and Shields, J. E., *Org. Synth., Coll. Vol. V*, **1973**, 1064.
23. Kreher, R. and Seubert, J., *Z. Naturforsch.*, **1966**, *20b*, 75.
24. Kreher, R. and Herd, K. J., *Heterocycles*, **1978**, *11*, 409.
25. Thesing, J., Schäfer, W., and Melchior, D., *Justus Liebigs Ann. Chem.*, **1964**, *671*, 119.
26. Kreher, R. and Seubert, J., *Angew. Chem., Int. Ed. Engl.*, **1964**, *3*, 639; *ibid.*, **1966**, *5*, 967.
27. Priestley, G. M. and Warrener, R. N., *Tetrahedron Lett.*, **1972**, 4295; Bornstein, J., Remy, D. E., and Shields, J. E., *Chem. Commun.*, **1972**, 1149.
28. Haddadin, M. J. and Chelhot, N. C., *Tetrahedron Lett.*, **1973**, 5185.
29. Holland, J. M. and Jones, D. W., *J. Chem. Soc., C*, **1970**, 536; Kreher, R. P. and Kalischko, J., *Chem.Ber.*, **1991**, 645.
30. Mann, M. E. and White, J. D., *J. Chem. Soc., Chem. Commun.*, **1969**, 420.
31. Dann, O., Kokorudz, M., and Gropper, R., *Chem. Ber.*, **1954**, *87*, 140.
32. Warrener, R. N., *J. Am. Chem. Soc.*, **1971**, *93*, 2346.
33. Mitchell, R. H., Iyer, V. S., Khalifa, N., Mahadevan, R., Venugopalan, S., Weerawarna, S. A., and Zhou, P., *J. Am. Chem. Soc.*, **1995**, *117* 1514.
34. Naito, K. and Rickborn, B., *J. Org. Chem.*, **1980**, *45*, 4061.
35. Crump, S. L. and Rickborn, B., *J. Org. Chem.*, **1984**, *49*, 304.
36. Smith, J. G. and Kruger, G., *J. Org. Chem.*, **1985**, *50*, 5759.
37. Keay, B. A., Plaumann, H. P., Rajapasaka, D., and Rodrigo, R., *Can. J. Chem.*, **1983**, *61*, 1987.
38. Tu, N. P. W., Yip, J. C., and Dibble, P. W., *Synthesis*, **1996**, 77.
39. Zajec, W. W. and Pichler, D. E., *Can. J. Chem.*, **1966**, *44*, 833; Potts, K. T. and Elliott, A. J., *Org. Prep. Proc. Int.*, **1972**, *4*, 269.
40. Newman, M. S., *J. Org. Chem.*, **1961**, *26*, 2630.

20 Typical reactivity of 1,3- and 1,2-azoles

The 1,3- and 1,2-azoles each contain one heteroatom in an environment analogous to that of the nitrogen in pyridine – an imine nitrogen – and one heteroatom in the environment of the nitrogen in pyrrole, the sulfur in thiophene, or the oxygen in furan, respectively. Consequently, their chemical reactions present a fascinating combination and mutual interaction of the types of reactivity which have been described earlier in this book for pyridines on the one hand and for pyrrole, thiophene and furan on the other, with the variation in electronegativity of the five-membered-type heteroatom having a substantial differentiating effect.

Typical reactions of 1,3-azoles

formation of 2-ylids by *C*-deprotonation of azolium salts

electrophilic addition at imine nitrogen produces azolium salts

regioselective 2-lithiation

formation of cycloadducts on reaction of oxazoles with dienophiles

substitution on nitrogen *via* imidazolyl anion

electrophilic substitution preferred at 5-position

Many of the lessons to be learnt apply to both 1,3- and 1,2-azoles, though the direct linking of the two heteroatoms in the latter has a substantial inductive influence, altering properties in degree. The 1,2-azoles tend to be less nucleophilic and less basic at the imine nitrogen than their 1,3-isomers. That such electrophilic additions occur, again illustrates that the imine nitrogen lone-pair is not involved in the aromatic sextet of electrons.

Typical reactions of 1,2-azoles

regioselective 5-lithiation

electrophilic addition at imine nitrogen produces azolium salts

substitution on nitrogen *via* pyrazolyl anion

electrophilic substitution preferred at 4-position

Electrophilic substitution in the azoles is intermediate in facility between pyridine on the one hand and pyrroles, thiophene and furans on the other: the presence of the electron-withdrawing imine unit has an effect on the five-membered aromatic heterocycles just as it does when incorporated into a six-membered aromatic framework, i.e. the comparison is like that between benzene and pyridine (chapter 4). The order of reactivity – pyrrole > furan > thiophene – is echoed in the azoles, though the presence of the basic nitrogen complicates such comparisons. The regiochemistry of electrophilic attack can be seen nicely by comparing the 'character' of the various ring positions – those that are activated in being five-membered in character and those that are deactivated by their similarity to α and γ positions in pyridine.

Relative positional
reactivities of azoles

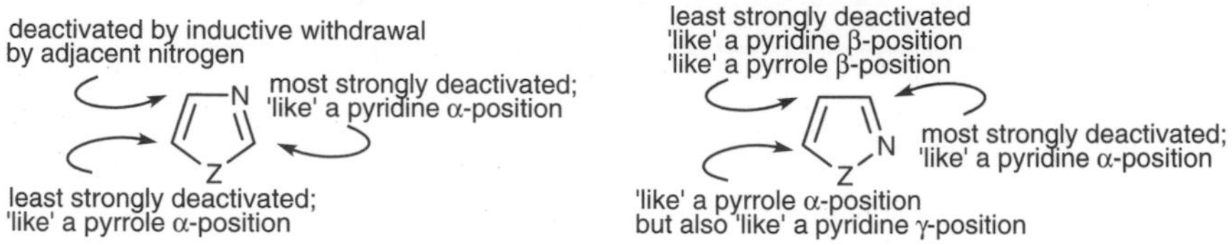

The converse of electrophilic substitution following the five-membered pattern, is that nucleophilic substitution of halogen follows the pyridine pattern i.e. it is much faster at the 2-position of 1,3-azoles and at the 3-position of 1,2-azoles, than at other ring positions. Resonance contributors to the intermediates for such substitutions make the reason for this plain: the imine nitrogen can act as an electron sink for attack only at these positions.

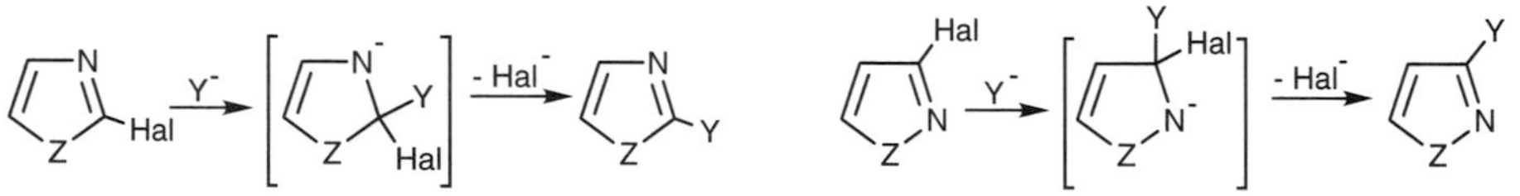

The utility of palladium(0)-catalysed processes (see section 2.7 for a detailed discussion) for the construction of azoles has been extensively developed: one simple example is shown below.

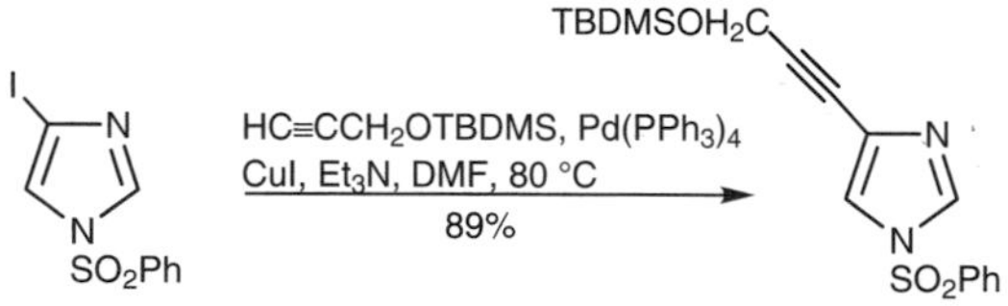

Continuing the analogy with pyridine reactivity, methyl groups at the 2-positions of 1,3-azoles and the 3-positions of 1,2-azoles carry acidified hydrogen atoms and can be deprotonated with strong bases. In further analogy with pyridines, the quaternisation of the imine nitrogen makes such deprotonations even easier; the resulting enamines react with electrophiles at the side-chain carbon.

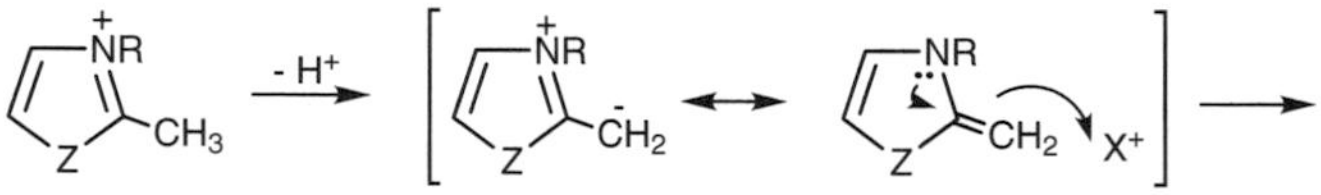

Lithiation is regioselective for the 2-position in the 1,3-azoles and for the 5-position in the 1,2-azoles. The facility with which 1,3-diazolium cations form ylides (carbenes) by 2-deprotonation is at the heart of the biological activity of thiamine pyrophosphate.

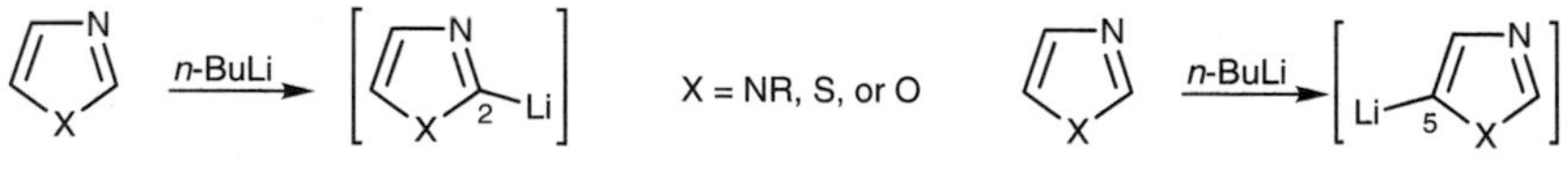

It has long been known that 1,3-azoles can be assembled from a component providing the two heteroatoms – a thioamide or an amidine – and an α-bromoketone. A much more recent route employs the interaction between the anion of an isonitrile and an aldehyde, thioaldehyde or imine.

To produce a 1,2-azole, a 1,3-dicarbonyl compound needs to be condensed with a unit providing the two heteroatoms – a hydrazine or hydroxylamine. The dipolar cycloaddition of alkynes with nitrile oxides or nitrile imines provides a route to isoxazoles and pyrazoles.

21 1,3-Azoles: imidazoles, thiazoles, and oxazoles: reactions and synthesis

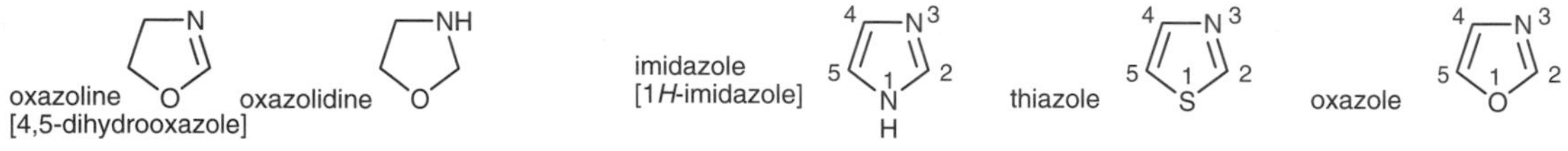

oxazoline O oxazolidine O imidazole [1H-imidazole] thiazole oxazole
[4,5-dihydrooxazole]

The three 1,3-azoles, imidazole,[1] thiazole and oxazole,[2] are all very stable compounds which do not autoxidise. Oxazole and thiazole are water-miscible liquids with pyridine-like odours. Imidazole, which is a solid at room temperature, and 1-methylimidazole are also water-soluble but are odourless. They boil at much higher temperatures (256 °C and 199 °C) than oxazole (69 °C) and thiazole (117 °C); this can be attributed to stronger dipolar association resulting from the very marked permanent charge separation in imidazoles (the dipole moment of imidazole is 5.6D; cf. oxazole, 1.4 D; thiazole, 1.6 D) and for imidazole itself, in addition, extensive intermolecular hydrogen bonding. The dihydro and tetrahydro heterocycles are named imidazoline/imidazolidine, thiazoline/thiazolidine, and oxazoline/oxazolidine.

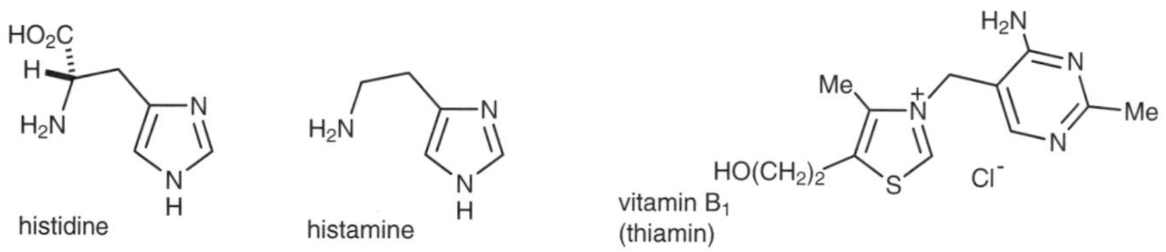

histidine   histamine   vitamin B₁ (thiamin)

Only oxazole, of the trio, does not play any part in normal biochemical processes, though there are secondary metabolites (especially from marine organisms) which incorporate thiazole (and oxazole) units – the antibiotic cystothiazole A, from the myxobacterium *Cyctobacter fuscus* is an example.[3] Imidazole occurs in the essential amino acid histidine; histidines within enzymes are intimately involved in catalysis requiring proton transfers. The structurally related hormone, histamine, is a vasodilator and a major factor in allergic reactions such as hay fever. The thiazolium ring is the chemically active centre in the coenzyme derived from thiamin (vitamin B₁).

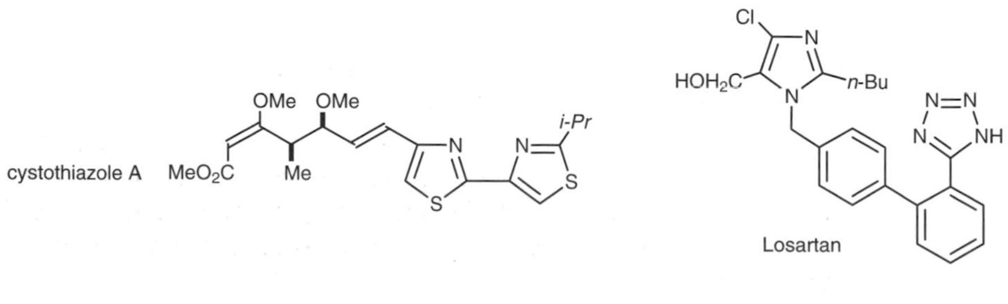

cystothiazole A

Losartan

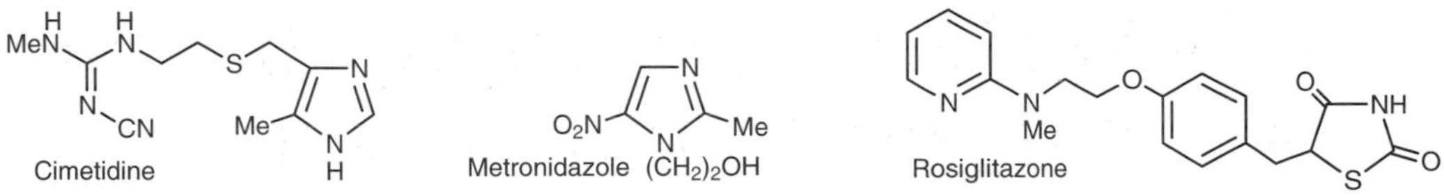

Cimetidine   Metronidazole (CH$_2$)$_2$OH   Rosiglitazone

Amongst synthetic 1,3-azoles in use[4] as therapeutic agents are Cimetidine, for the treatment of peptic ulcers, and Metronidazole, an antibacterial and an antiprotozoal, used for example in the treatment of amoebic dysentry. Rosiglitazone is used in the treatment of type 2 diabetes and Losartan is an angiotensin II antagonist – its use is as an antihypertensive agent.

21.1 Reactions with electrophilic reagents

21.1.1 Addition at nitrogen

21.1.1.1 Protonation

Imidazole, thiazole and alkyloxazoles, though not oxazole itself, form stable crystalline salts with strong acids, by protonation of the imine nitrogen, N-3, known as imidazolium, thiazolium, and oxazolium salts.

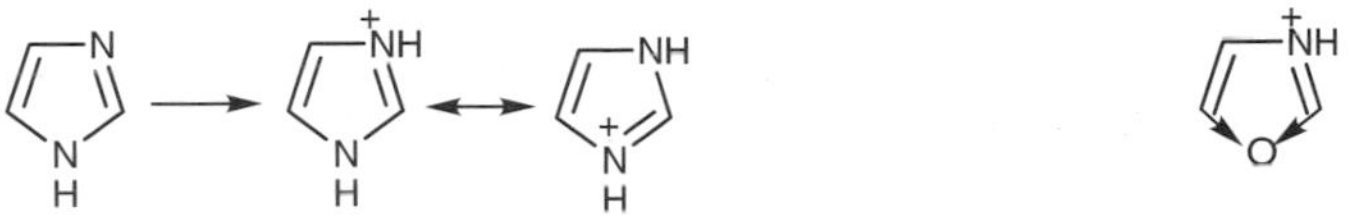

Imidazole, with a pK_a of 7.1 is a very much stronger base than thiazole (pK_a 2.5) or oxazole (pK_a 0.8). That it is also stronger than pyridine (pK_a 5.2) is due to the amidine-like resonance which allows both nitrogens to participate equally in carrying the charge. The particularly low basicity of oxazole can be understood as a combination of inductive withdrawal by the oxygen and weaker mesomeric electron release from it. The 1,3-azoles are stable in hot strong acid.

Hydrogen bonding in imidazoles

Imidazole, like water, is both a good donor and a good acceptor of hydrogen bonds; the imine nitrogen donates an electron pair and the N-hydrogen, being appreciably acidic (section 21.4.1), is an acceptor.

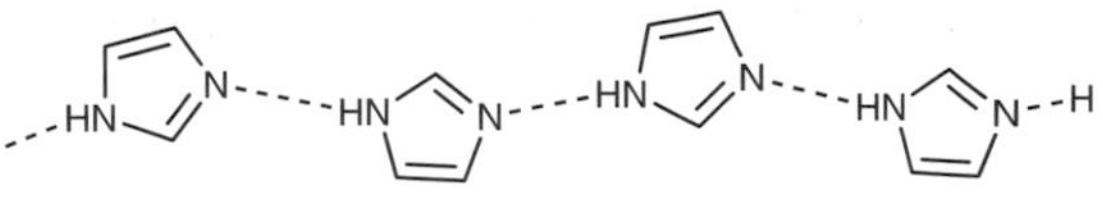

This property is central to the mode of action of several enzymes which utilise the imidazole ring of a histidine. These include the digestive enzyme chymotrypsin, which brings about amide hydrolysis of peptides in the small intestine: the enzyme provides a 'proton' at one site, while it accepts a 'proton' at another, making use of the ambivalent character of the imidazole ring to achieve this. The illustration shows how the heterocycle allows a proton to 'shuttle' from one site to another *via* the heterocycle.

Tautomerism in imidazoles

Imidazoles with a ring *N*-hydrogen are subject to tautomerism which becomes evident in unsymmetrically substituted compounds such as the methylimidazole shown. This special feature of imidazole chemistry means that to write simply '4-methylimidazole' would be misleading, for this molecule is in tautomeric equilibrium with 5-methylimidazole, and quite inseparable from it. All such tautomeric pairs are inseparable and the convention used to cover this phenomenon is to write '4(5)-methylimidazole'. In some pairs, one tautomer predominates, for example 4(5)-nitroimidazole favours the 4-nitro-tautomer by 400:1.

21.1.1.2   *Alkylation at nitrogen*

The 1,3-azoles are quaternised easily at the imine nitrogen with alkyl halides; the relative rates are: 1-methylimidazole:thiazole:oxazole – 900:15:1.[5] Microwave irradiation makes the process particularly rapid.[6] In the case of imidazoles which have an *N*-hydrogen, the immediate product is a protonated *N*-alkylimidazole; this can lose its proton to unreacted imidazole and react a second time, meaning that reactions with alkyl halides give a mixture of imidazolium, 1-alkylimidazolium and 1,3-dialkylimidazolium salts. Furthermore, an unsymmetrically substituted imidazole can give two isomeric 1-alkyl derivatives. The use of a limited amount of the alkylating agent, or reaction in basic solution,[7] when it is the imidazolyl anion (section 21.4.1) which is alkylated, can minimise these complications. Clean formation of doubly alkylated derivatives can be achieved by reacting 1-trimethylsilylimidazole with an alkyl halide.[8] *N*-Arylation of imidazoles, efficient when copper(I)-catalysed, shows the same regioselectivity with 4(5)-substituted imidazoles: generally the 1-aryl-4-substituted imidazole is the major product.[9]

N-Alkylation of oxazoles,[10] or imidazoles carrying, for example, a phenylsulfonyl or acyl[11] group on nitrogen, is more difficult, requiring methyl triflate or a Meerwein salt for smooth reaction. Subsequent simple alcoholysis of the imidazolium-sulfonamide releases the *N*-substituted imidazole;[12] the process can be utilised in another sense for converting alcohols into carbamates.[13] Moreover, since acylation of 4(5)-substituted imidazoles gives the sterically less crowded 1-acyl-4-substituted imidazoles, subsequent alkylation, then hydrolytic removal of the acyl group produces 1,5-disubstituted imidazoles.[14] Complementarily, the 1,4-disubstitution pattern can be achieved by alkylating 1-protected-5-substituted imidazoles (see section 21.6.1) at N-3, then removing the *N*-protection.[15] *N*-Tritylimidazoles can be *N*-alkylated with simple halides, removal of the triphenylmethyl group after alkylation requiring only simple acid treatment.[16] Alkylation with acrylonitrile, via a Michael mechanism, is reversible and can also be made the means for the synthesis of 1,5-disubstituted imidazoles via *N*-alkylation of 1-(2-cyanoethyl)-4-substituted imidazoles then elimination of acrylonitrile.[17]

Another device to control the position of *N*-alkylation is applicable to histidine and histamine: a cyclic urea is first prepared by reaction with carbonyl dimidazole (section 21.1.1.3), forcing the alkylation onto the other nitrogen, ring opening then providing the *N*-1-alkylated, urethane-protected derivative.[18]

Exposure of imidazole to 'normal' Mannich conditions leads to *N*-dimethylaminomethylimidazole, presumably *via* attack at the imine nitrogen, followed by loss of proton from the other nitrogen.[19]

21.1.1.3 Acylation at nitrogen

Acylation of imidazole produces *N*-acylimidazoles *via* loss of proton from the initially-formed *N*-3-acylimidazolium salt.[20] A device which has been employed frequently for the synthesis of 1-acylimidazoles is to use two mol equivalents of the heterocycle for one of the acylating agent, the second mole of imidazole serving to deprotonate the first-formed *N*-acylimidazolium salt.

N-Acylimidazoles are even more easily hydrolysed than *N*-acylpyrroles, moist air is sufficient. The ready susceptibility to nucleophilic attack at carbonyl carbon has been capitalised upon: commercially available 1,1′-carbonyldiimidazole (CDI), prepared from imidazole and phosgene, can be used as a safe, phosgene equivalent, i.e. a synthon for $O = C^{2+}$, and also in the activation of acids for formation of amides and esters *via* the *N*-acylimidazole.[21]

In another application, *N*-acylimidazoles react with lithium aluminium hydride at 0 °C to give aldehydes, providing a route from the acid oxidation level.[22] A related phenomenon is the use of 'imidazylates' as excellent leaving groups in S_N2 reactions.[23] They are also useful precursors for the more reactive fluorosulfonates; such conversions have been carried out on an 800 kg scale.[24]

21.1.2   Substitution at carbon

21.1.2.1   *Protonation*

In acid solution, *via* a proton-addition/proton-loss sequence, hydrogen at the imidazole 5-position exchanges about twice as rapidly as at C-4 and > 100 times faster than at C-2.[25] An altogether faster exchange, which takes place at room temperature in neutral or weakly basic solution, but not in acidic solution, brings about C-2–H exchange;[26] oxazole and thiazole also undergo this regioselective C-2–H exchange, the relative rates being in the order: imidazole > oxazole > thiazole.[27] The mechanism for this special process involves first, formation of a concentration of protonic salt, then C-2–H deprotonation of the salt, producing a transient ylide, to which a carbene form is an important resonance contributor. It follows from this mechanism that quaternary salts of 1-alkylimidazoles and of oxazole and thiazole will also undergo regioselective C-2–H exchange, and this is indeed the case. Most attention[28] has been paid to thiazolium salts (section 21.10) because of the involvement of exactly such an ylide in the mode of action of thiamin in its role as a component of a coenzyme in several biochemical processes.[29] The relative rates of exchange, *via* the ylide mechanism, are in the order: oxazolium > thiazolium > *N*-methylimidazolium, in a ratio of about 10^5:10^3:1.[30]

Ylides at C-5 are thought to intervene in the decarboxylation of 5-acids, where again the order of ease of loss of carbon dioxide is oxazole- > thiazole- > *N*-methylimidazole-5-acids, however comparison with the decarboxylations of the 2-acids, shows the 5-isomers to lose carbon dioxide 10^6 more slowly, implying a much lower stability for the 5-ylides and transition states leading to them.[31]

21.1.2.2   *Nitration*

Imidazole is much more reactive towards nitration than thiazole, substitution taking place *via* the salt,[32] as does nitration of alkylthiazoles.[33] Thiazole itself is untouched by nitric acid/oleum at 160 °C but methylthiazoles are sufficiently activated to undergo substitution, the typical regioselectivity being for formation of more 5-nitro- than 4-nitro derivative;[34] the 2-position is not attacked: 4,5-dimethylimidazole is resistant to nitration. The much less reactive oxazoles do not undergo nitration.

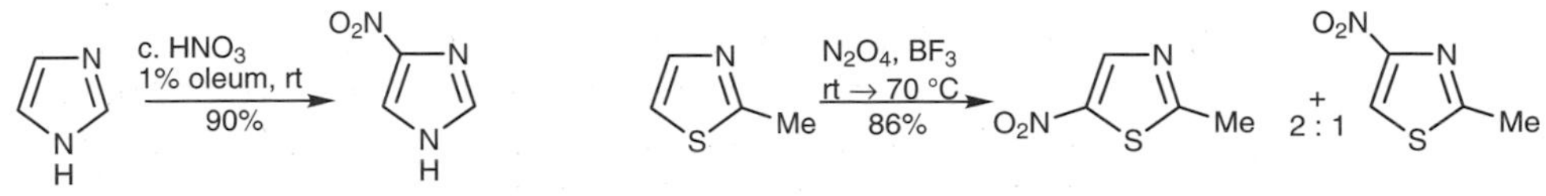

21.1.2.3 Sulfonation

Here again, thiazoles are much less reactive than imidazoles,[35] generally requiring high temperatures and mercury(II) sulfate as catalyst for any reaction to take place;[36] oxazole sulfonations are unknown.

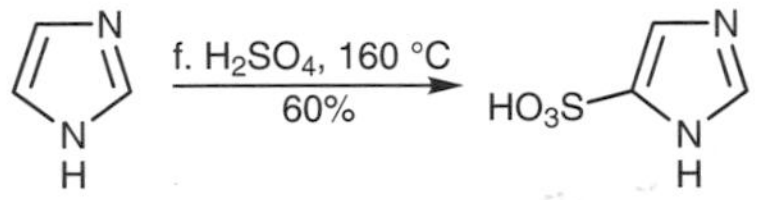

21.1.2.4 Halogenation

Imidazole,[37] and 1-alkyl imidazoles,[38] are brominated with remarkable ease at all free nuclear positions. 4(5)-Bromoimidazole can be obtained by reduction of tribromo-imidazole,[39] *via* regioselective exchange of the 2- and 5-halogens then water quenching,[40] or by bromination with 4,4-dibromocyclohexa-2,5-dienone.[41] Chlor-ination with hypochlorite in alkaline solution effects substitution only at the 4- and 5-positions.[42] Iodination of imidazoles which have a free *N*-hydrogen, in alkaline solution and therefore *via* the imidazolyl anion, can also give fully halogenated products;[43] 4,5-diiodination of imidazole takes place in cold alkaline solution.[44]

It is, at first sight, somewhat surprising that such relatively mild conditions allow bromination of imidazole at C-2, but it must be remembered that the neutral imidazole, not its protonic salt (*cf.* nitration and sulfonation), is available for attack. Electrophilic addition of bromine to nitrogen, then addition of bromide at C-2, and finally elimination of hydrogen bromide may be involved.

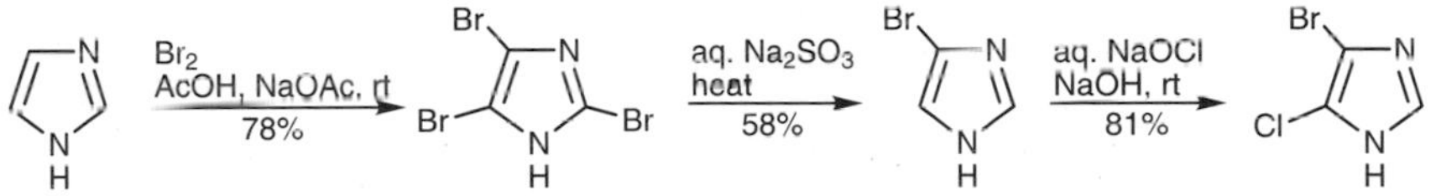

Thiazole does not undergo bromination easily, though 2-methylthiazole bromi-nates at C-5; when the 5-position is not free no substitution occurs, thus 2,5-dimethylthiazole, despite its two activating substituents, is not attacked.[45] Halogenation of simple oxazoles has not been reported.

21.1.2.5 Acylation

Friedel-Crafts acylations are unknown for the azoles, clearly because of interaction between the basic nitrogen and the Lewis acid catalyst. It is, however, possible to 2-aroylate 1-alkylimidazoles[46] or indeed imidazole itself[47] by reaction with the acid chloride in the presence of triethylamine, the substitution proceeding *via* an *N*-acylimidazolium ylide as shown below. It is similarly possible to introduce cyano to the 2-position by reaction with *N*-cyano-4-dimethylaminopyridinium chloride.[48] In the reverse sense, 2-acyl substituents can be cleaved by methanolysis, the mechanism again involving the imidazolium ylide.[49]

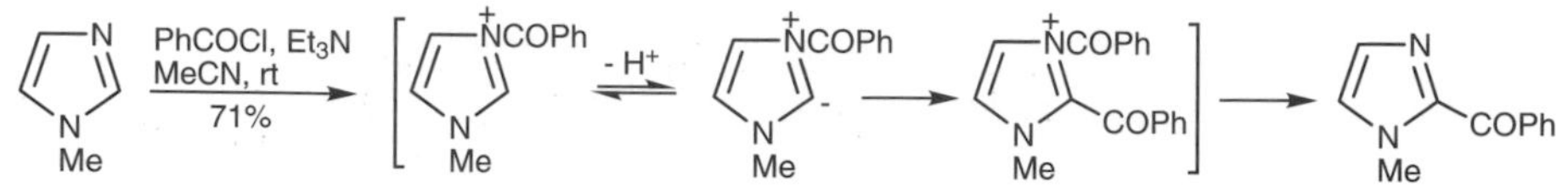

Another fascinating example of the utility of *N*-acylimidazolium ylides provides a means for synthesising 2-formylimidazole efficiently: the electrophile which attacks the ylide is in this case an *N*-benzoylimidazolium cation.[50]

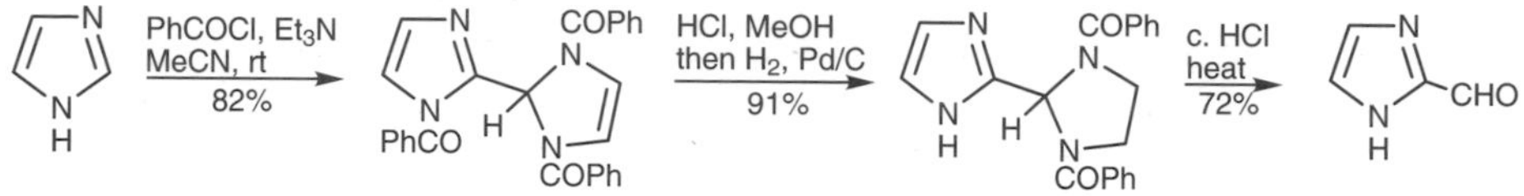

21.1.2.6   Reactions with aldehydes

The discovery of *ipso* displacement of silicon from the thiazole 2-position under mild conditions led to the development of this reaction as an essential component of a route to complex aldehydes. Subsequent quaternisation, saturation of the hetero-cyclic ring using sodium borohydride, and then mercury(II) or copper(II) catalysed treatment leads to the destruction of the thiazolidine and the formation of a new homologous aldehyde; an example is shown below.[51]

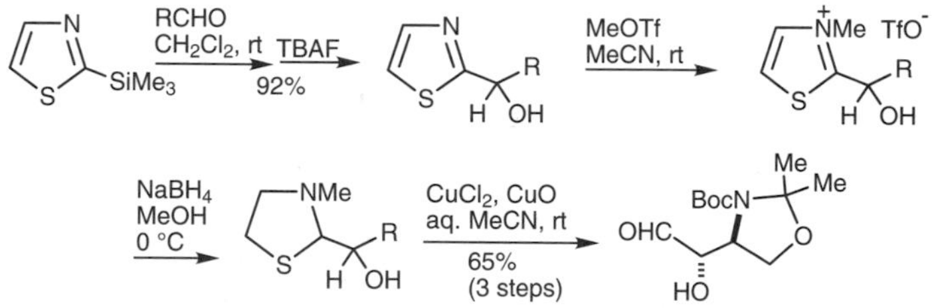

21.1.2.7   Reactions with iminium ions

The standard, acidic Mannich conditions do not allow simple *C*-substitutions of the imidazole. (*cf.* 21.1.1.2), thiazole, or oxazole systems.

21.2   Reactions with oxidising agents

Resistance to oxidative breakdown falls off in the order thiazoles > imidazoles > oxazoles. 2-Substituted thiazoles can be converted into *N*-oxides,[52] however peracids bring about degradation of imidazoles; oxazole *N*-oxides can only be prepared by ring synthesis.

21.3   Reactions with nucleophilic reagents

21.3.1   With ring opening

Generally speaking, the 1,3-azoles do not show the pyridine-type reactions in which hydrogen is displaced, although a Chichibabin substitution on 4-methylthiazole has been reported.[53] There are however reactions in which the heterocyclic ring is opened, for example phenylhydrazine attacks oxazoles leading to osazones.[54] Reaction of an oxazole with hot formamide also leads to a ring opening; a reclosure results in the formation of imidazoles; the example show reasonable intermediates.

21.3.2   With replacement of halogen

There are many examples of halogen at a 2-position undergoing nucleophilic displacement, for example 2-halothiazoles with sulfur nucleophiles[55] (indeed, more rapidly than for 2-halopyridines), 2-halo-1-substituted imidazoles,[56] and 2-chloroox-azoles[57] with nitrogen nucleophiles.

In the special situation where an imidazole nitrogen carries a nitro group which can act as a leaving group (as nitrite) *cine* substitution has been observed.[58]

21.4   Reactions with bases

21.4.1   Deprotonation of N-hydrogen

The pK_a for loss of the N-hydrogen of imidazole is 14.2; it is thus an appreciably stronger acid than pyrrole (pK_a 17.5) because of the enhanced delocalisation of charge onto the second nitrogen in the imidazolyl anion.

21.4.2.   Deprotonation of C-hydrogen

The specific exchange at C-2 in the azoles in neutral solution, *via* an ylide, has already been discussed (section 21.1.2.1). In strongly basic solution, deprotonation takes place by direct abstraction of proton from the neutral heterocycle at the positions adjacent to the oxygen and the sulfur in oxazole and thiazole[59] and, less easily, at C-5 in N-methylimidazole.[60]

21.5   Reactions of N-metallated imidazoles

Salts of imidazoles can be alkylated or acylated on nitrogen. One convenient method is to use the dry sodium/potassium salt obtained by evaporation of an aqueous alkaline solution;[61] sodium hydride in dimethylformamide also serves very well for this purpose. When there is a route for the entering group to be lost again, as in the addition to a carbonyl-conjugated alkene, a 2,4(5)-substituted imidazole will give the less hindered 1,2,4-trisubstituted product rather than the 1,2,5-isomer.[62] The use of

1,3,4,6,7,8-hexahydro-1-methylpyrimido[1,2-*a*]pyridine (MTPD) is particularly effective at promoting the addition of imidazoles to unsaturated esters and nitriles.[63]

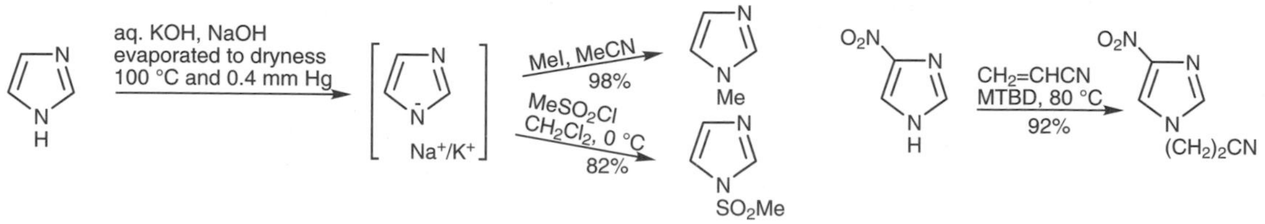

Imidazoles react with Mannich electrophiles at nitrogen, however the overall effect of Mannich *C*-substitution has been found in base-catalysed cyclisation of histamine Schiff bases; closure does not take place in the absence of base and it must be the imidazolyl anion which reacts intramolecularly with the side-chain imine.[64]

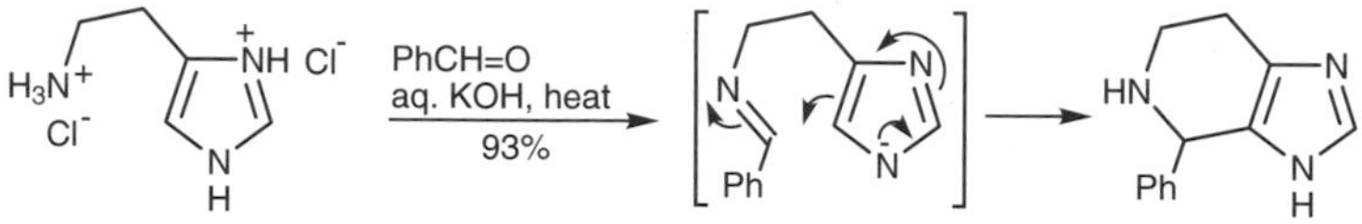

21.6   Reactions of *C*-metallated 1,3-azoles[65]

21.6.1   Lithium derivatives

In line with the exchange processes discussed above, preparative strong base deprotonation of oxazoles,[66] thiazoles,[67] and *N*-methylimidazole[68] takes place preferentially at C-2, or at C-5 if the former position is blocked,[69] and the lithiated derivatives can then be utilised in reactions with electrophiles. A variety of removable *N*-protecting groups have been used to achieve comparable transformations for the eventual synthesis of *N*-unsubstituted imidazoles, including phenylsulfonyl,[70] dimethylaminosulfonyl,[71] dimethylaminomethyl,[19] trimethylsilylethoxymethyl (SEM),[72] diethoxymethyl,[73] 1-ethoxyethyl,[74] and trityl[75] (see also 21.13). The intrinsic tendency to lithiate at C-2, then C-5, taken with metal–halogen exchange processes for the 4-position are a powerful combination for elaborations of the 1,3-azoles. For example, the sequence shown below produces SEM-protected 5-substituted imidazoles,[71,76] with retention of a 2-silyl substituent if required.[77] All three isomeric trimethylsilyl- and all three trimethylstannylthiazoles have been made in similar ways and provide means for subsequent regioselective *ipso* displacement with electrophiles under mild conditions.[78]

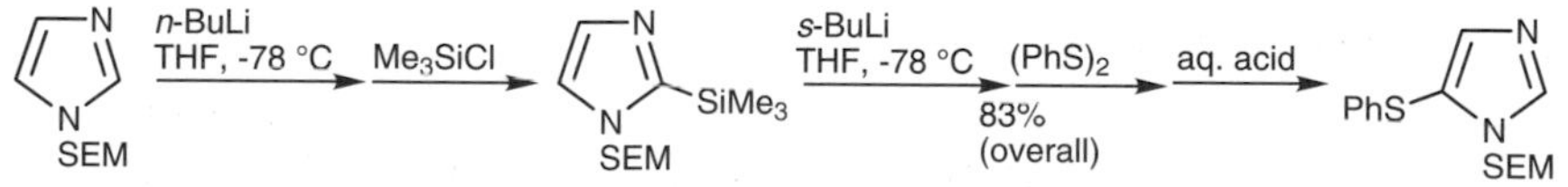

Complementarily, the lithiation of a SEM protected 2-phenylsulfonylimidazole takes place at C-4.[79] Metal–halogen exchange of 4(5)-bromoimidazole is possible without protection.[80]

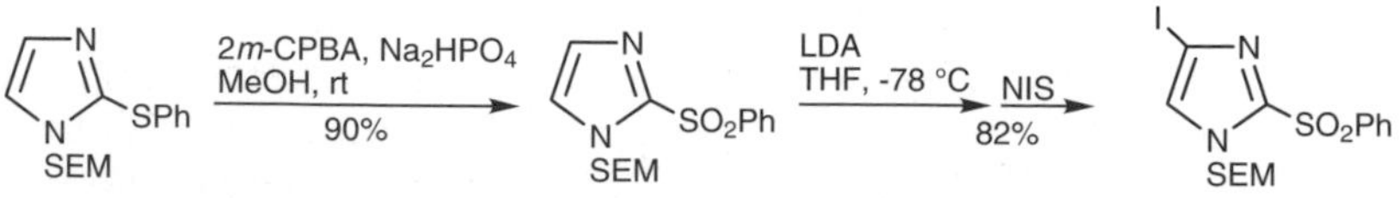

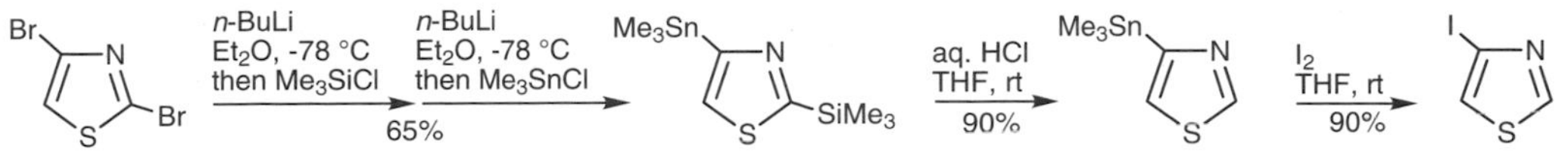

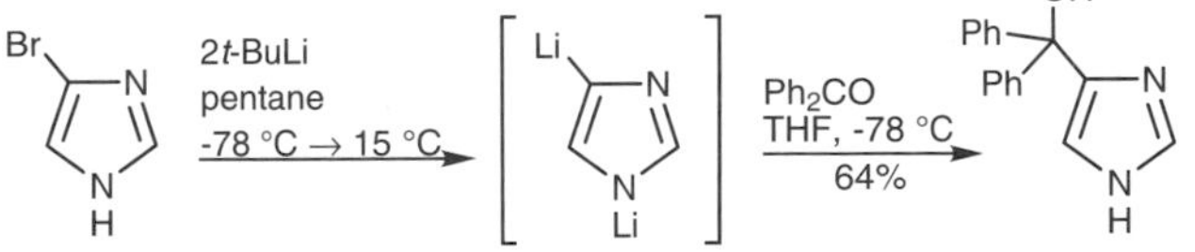

Although oxazoles follow the pattern and lithiate at C-2, 4-substituted products are produced with some electrophiles; this is interpreted by a ring opening of the anion, to produce an enolate, which after *C*-electrophilic attack, recloses. An estimate by NMR spectroscopy showed the ring cleaved tautomer to dominate the equilibrium.[81] The open enolates can be trapped by reaction with chlorotrimethylsilane; the open, enol trimethylsilyl ether will undergo a thermal rearrangement to form a 2-trimethylsilyloxazole.[82]

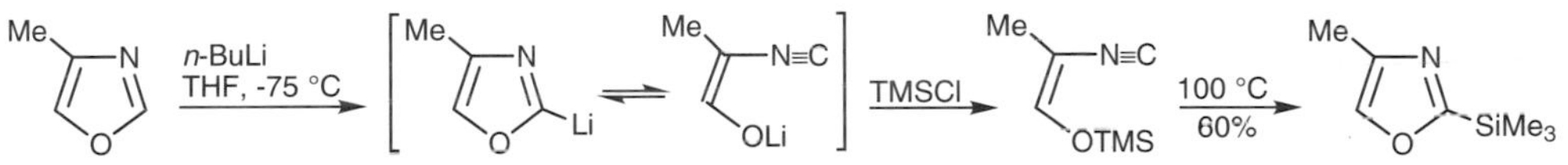

The ring opening of oxazoles can be avoided by transmetallation,[83] or by first forming a borane complex which is then lithiated as shown below.[84]

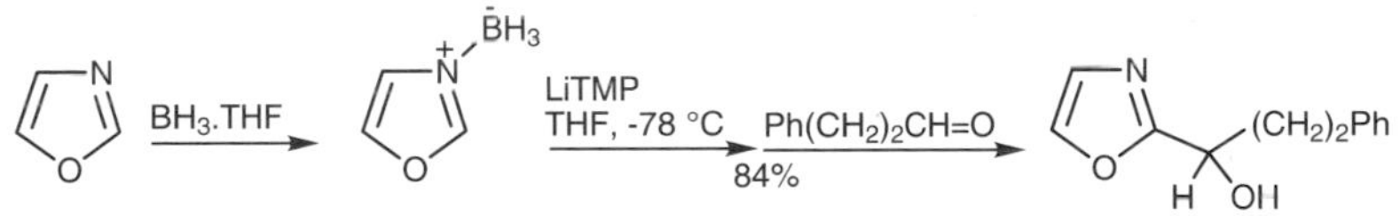

Oxazolylzinc compounds[80,85] and oxazolyl tin compounds[86] take part in coupling processes (see also below) without problems over ring opening.

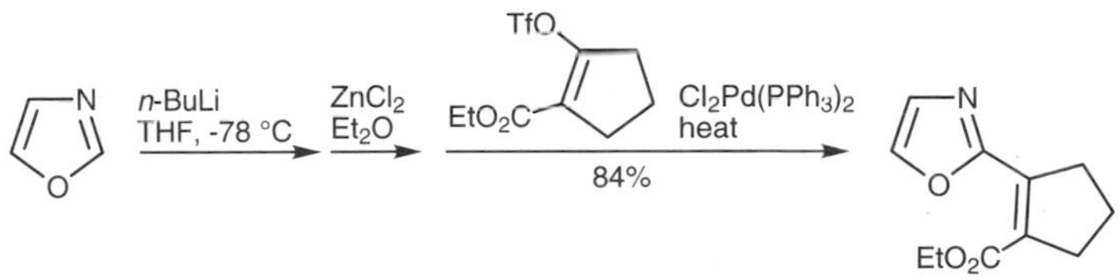

21.6.2 Palladium-catalysed reactions

The palladium(0)-catalysed coupling of *N*-protected imidazoles has been extensively utilised, as illustrated by the examples below.[87] The coupling of 4,5-diiodoimidazole protected with trimethylsilylethoxymethyl on *N*-1, was completely selective for the 5-halogen.[88]

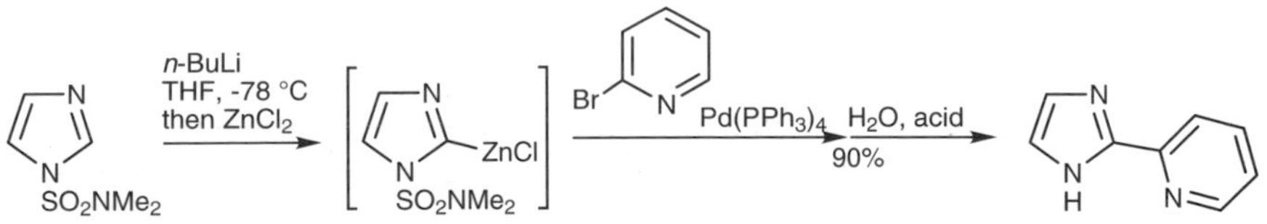

21.10 Alkyl-1,3-azoles

Protons on alkyl groups at the 1,3-azole 2-positions are sufficiently acidic for strong base deprotonation,[105] and are more acidic than methyl groups at other positions; even the assistance of an *ortho*-related carboxylate is usually insufficient to overcome the intrinsic tendency for 2-methyl-lithiation, though an adjacent tertiary amide can do this.[106] The side-chain metallated derivatives can be utilised in reactions with electrophiles. The presence of a 5-nitro group allows much milder, base-catalysed condensations to occur.[33] The condensation at the 2-methyl of thiazoles proceeds in organic acid solution.[107]

N-Acylation also increases the acidity of 2-methyl groups, allowing *C*-acylation *via* a non-isolable enamide.[108]

21.11 Quaternary 1,3-azolium salts

1-Butyl-3-methylimidazolium hexafluorophosphate, which is fluid at room temperature and stable to water, has been recommended as an 'ionic liquid' and shown to serve as the equivalent of a dipolar aprotic solvent in some base-catalysed alkylations; products can be simply extracted with an immiscible organic solvent in the usual way.[109]

N-Alkoxycarbonyl 1,3-azolium salts, generated *in situ* by reaction with chloroformates, will react with allylstannanes,[110] or allylsilanes[111] by addition of the equivalent of an allyl anion. In the same way, silyl enol ethers add the equivalent of an enolate to give 2,3-dihydro-2-substituted imidazoles and thiazoles.[112]

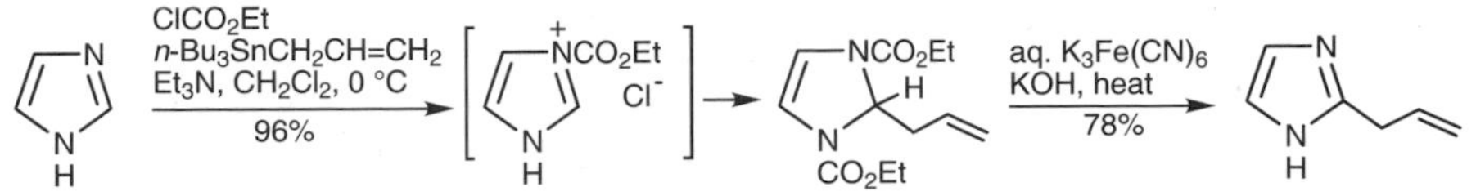

Azolium salts are readily attacked by nucleophiles, for example with hydroxide, addition at C-2 is followed by ring opening.[113]

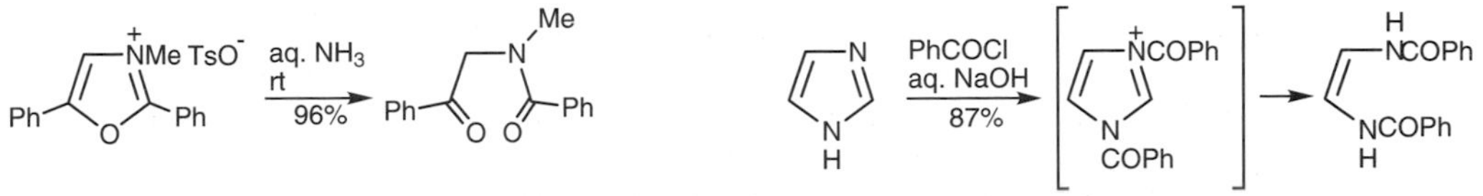

Neat exploitations of this process, include the synthesis of medium-sized heterocycles as products of a three stage sequence involving addition of hydroxide to ω-iodoalkyl thiazolium salts, ring-opening and then reclosure by intramolecular S-alkylation, illustrated below for the formation of an eight-membered ring,[114] and the introduction of sulfur to the imidazole 2-position using phenyl chlorothionoformate.[115]

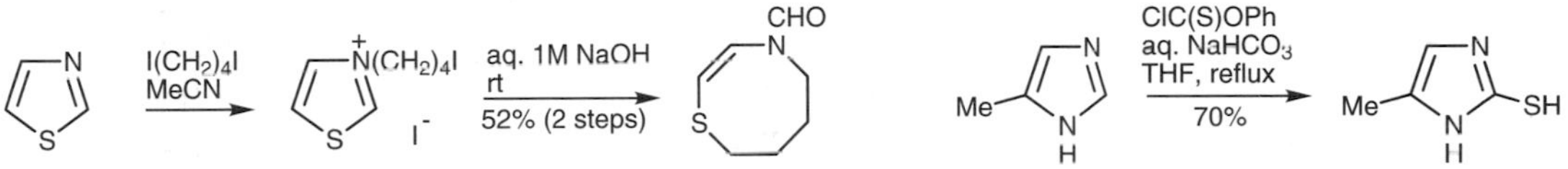

The C-2-exchange of azolium salts *via* an ylide mechanism has already been discussed (section 21.1.2.1). Thiamin pyrophosphate acts as a coenzyme in several biochemical processes and in these, its mode of action also depends on the intermediacy of a 2-deprotonated species. For example, in the later stages of alcoholic fermentation, which converts glucose into ethanol and carbon dioxide, the enzyme pyruvate decarboxylase converts pyruvate into ethanal and carbon dioxide, the former then being converted into ethanol by the enzyme, alcohol dehydrogenase. It is believed, that in the operation of the former enzyme, the coenzyme, thiamin pyrophosphate, adds as its ylide to the ketonic carbonyl group of pyruvate; this is followed by loss of carbon dioxide then the release of ethanal by expulsion of the original ylide.

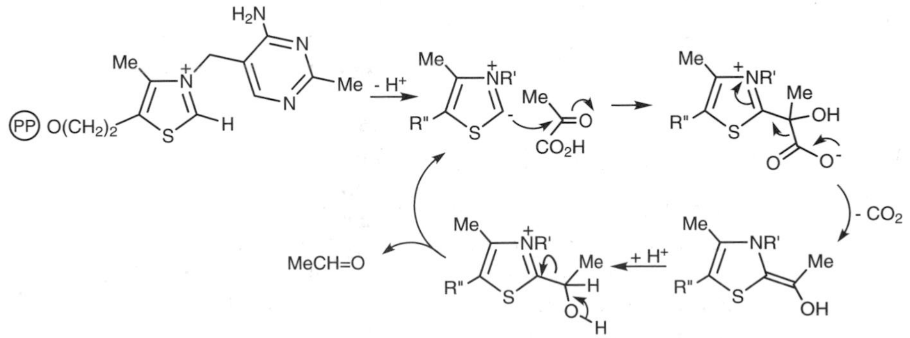

In the laboratory, thiazolium salts (3-benzyl-5-(2-hydroxyethyl)-4-methylthiazolium chloride is commercially available) will act as catalysts for the benzoin condensation, and in contrast to cyanide, the classical catalyst, allow such reactions to proceed with alkanals, as opposed to aryl aldehydes; the key steps in thiazolium ion catalysis for the synthesis of 2-hydroxyketones are shown below. Such catalysis, which also finds other applications, provides acyl anions, in effect.

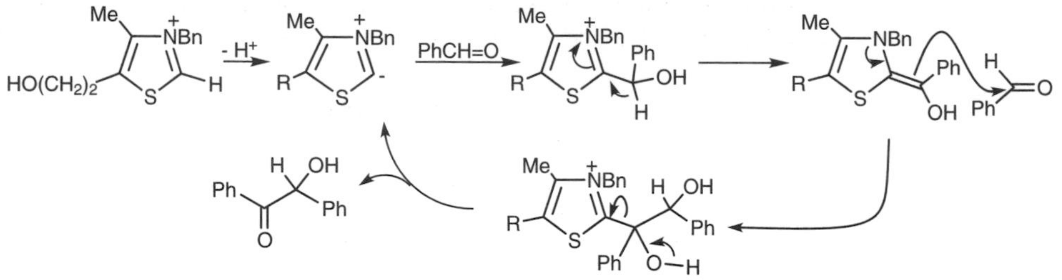

It is very interesting that replacing a thiazolium ring with an oxazolium ring gives a thiamin analogue in which there is no catalytic activity;[116] similarly 3,4-dimethyloxazolium iodide does not catalyse a benzoin condensation.[117] Nature has chosen the heterocyclic system with the correct balance – oxazolium ylides are formed faster, but because of the greater stability that this reflects, do not then add to carbonyl groups as is required for catalytic activity. In keeping with the carbenoid character of thiazolium ylides, they dimerise; the dimers, either in their own right, or by reversion to monomer, are also catalysts for the benzoin condensation.[118]

The 1,3-dimethylimidazolium ylide, generated using sodium hydride, allows the introduction of electrophiles to C-2.[119] Isolable, crystalline carbenes have been derived from 1,3-bis(adamantanyl)imidazolium chloride,[120] and from 1,3,4,5-tetra-phenylimidazolium chloride.[121] A stable thiazol-2-ylidene carried methyl groups at positions 4 and 5 and a 2,6-di-i-propylphenyl substitutent on nitrogen.[122]

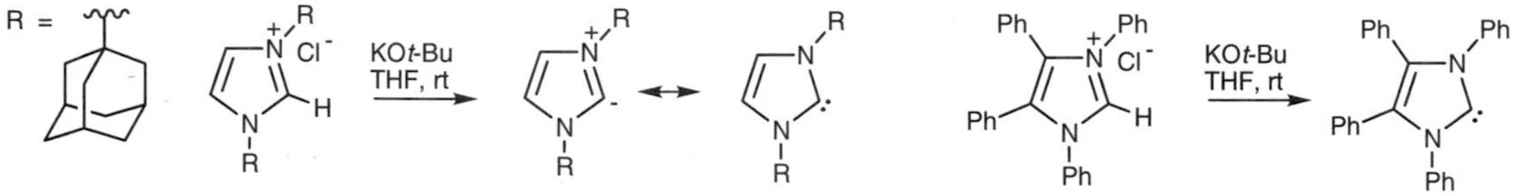

21.12   Oxy-[123,124] and amino-[125] -1,3-azoles

Amino-1,3-azoles exist as the amino tautomers, though 2-arylsulfonylaminothiazoles have been shown to exist as the imino tautomers.[126] 2-Amino-1,3-azoles tend to be more stable than other isomers. All amino-1,3-azoles protonate on the ring nitrogen. 2-Aminothiazole has a pK_a of 5.39 which compares with the value for 2-aminoimidazole of 8.46, reflecting the symmetry of the resonating guanidinium type system in the latter.

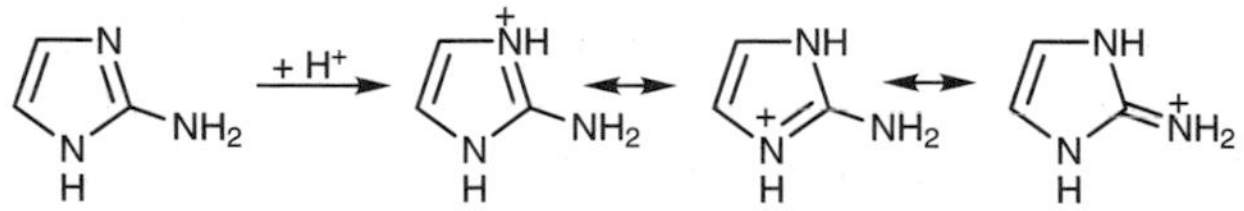

The amino-1,3-azoles behave as normal arylamines, for example undergoing carbonyl condensation reactions, easy electrophilic substitutions,[127] and diazotisation,[128] though 2-aminooxazoles cannot be diazotised,[129] presumably due to the greater electron withdrawal by the oxygen.

The oxygen-substituted 1,3-azoles exist in their carbonyl tautomeric forms. That there is little aromatic character left in such systems is nicely illustrated by the acid-catalysed dimerisation of imidazol-2-one, which acts as an enamide in the process.[130]

The bromination of thiazol-2-one, at C-5, is also a nice demonstration of relative reactivity: here the double bond carries both sulfur and nitrogen, and it is the latter, i.e. the enamide rather than the thioenol ester character, which dictates the site of electrophilic attack.[131]

1,3-Azol-2-ones can be converted into the 2-haloazoles by reaction with phosphorus halides.[56] Thiazolidine-2,4-diones are converted into the dihalothiazoles on exposure to phosphorus halides; when accompanied by dimethylformamide, i.e. under Vilsmeier conditions, ring formylation also occurs and after hydrogenolytic removal of halogen the overall sequence can be seen to be a means for the hydroxyalkylation of a thiazole.[132]

The 5-ones condense in an aldol fashion at C-4.[133] Alkylation of the 1,3-azolones can take place either on the oxygen, giving alkyloxyazoles, or on nitrogen; for example thiazol-2-one reacts with diazomethane giving 2-methoxythiazole, but with methyl iodide/methoxide, to give 3-methylthiazol-2-one.[130]

Erlenmeyer azlactone

4(5)-Oxazolones are simply cyclic anhydrides of N-acyl-α-amino acids, and are constructed in the way that this implies. If the nitrogen also carries an alkyl group, cyclisation[134] can only lead to an overall neutral product by its adopting a zwitterionic structure, for which no neutral canonical form can be written – a mesoionic structure. Mesoionic oxazolones (named 'münchnones' by Huisgen after their discovery at the University of München, Germany) undergo ready dipolar cycloadditions,[135] with loss of carbon dioxide from initial adduct; the examples[136] show the conversion of a münchnone into a mesoionic thiazolone and into an imidazole.

21.13   1,3-Azole N-oxides

The chemistry of azole N-oxides is relatively under developed compared, for example, with that of pyridine N-oxides, largely because of difficulty in their preparation from the azoles themselves. Some ring synthetic methods can be used, for example the reaction of 1,2-dicarbonyl mono-oximes with imines as shown.[137] 1-Substituted imidazole 3-oxides can be converted into nitriles with loss of the oxygen using trimethylsilyl cyanide, careful choice of solvent minimising a tendency for isomeric mixtures to be formed.[138]

1-Hydroxyimidazole N-oxide can be transformed into 1-benzyloxyimidazole and this undergoes useful 2-lithiations; hydrogenolysis produces 2-substituted 1-hydroxyimidazoles and these, in turn, can be converted into the 2-substituted imidazole by reduction with titanium(III) chloride.[139]

21.14   Synthesis of 1,3-azoles[140,141]

21.14.1   Ring synthesis

Considerable parallelism emerges from an examination of the major methods for the construction of oxazole, thiazole and imidazole ring systems.

21.14.1.1   From an α-halocarbonyl component (or an equivalent) and a three-atom unit supplying C-2 and the heteroatoms

Reaction of an α-halocarbonyl component and a three-atom unit supplying C-2 and the heteroatoms gives the five-membered heterocycle; this route is particularly important for thiazoles.

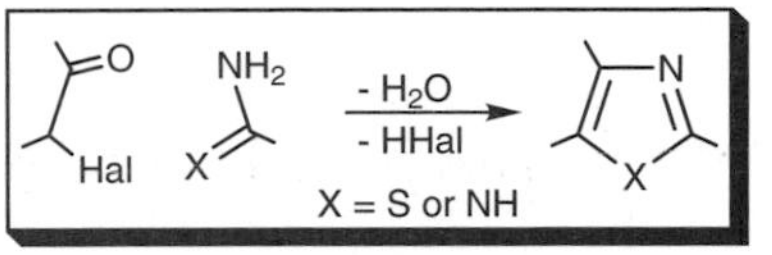

Simple examples of this strategy, which for the synthesis of thiazoles is known as the *Hantzsch synthesis*, are shown below: the syntheses of 2,4-dimethylthiazole where the heteroatoms are provided by thioacetamide,[142] and 2-aminothiazole, in which 1,2-dichloroethyl ethyl ether is utilised as a synthon for chloroethanal and the heteroatoms derive from thiourea.[143] The use of thioureas as the sulfur component with 2-chloroacetamides as the second unit gives rise to 2,4-diaminothiazoles.[144] Conversion of 1,3-diketones into their 2-phenyliodonium derivatives and reaction of these with thioureas produces 2-amino-5-acylthiazoles.[145] The first step in such ring syntheses is S-alkylation.[146] A useful variant is the use of an α-diazo ketone in place of the α-halocarbonyl component.[147]

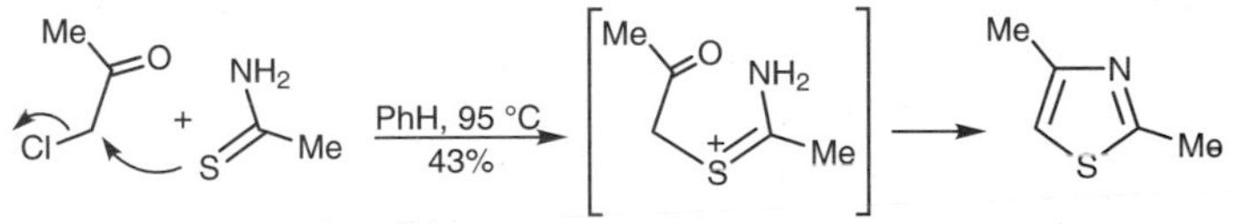

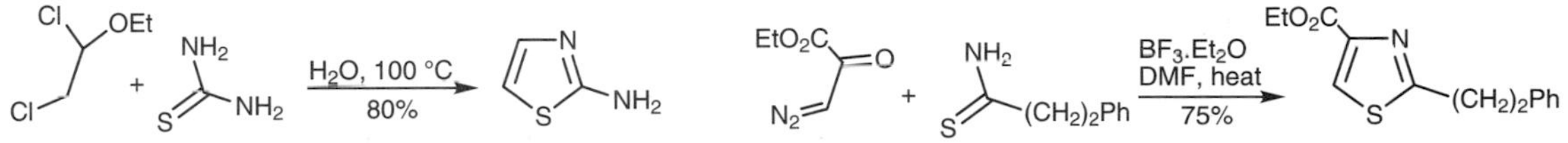

The interaction of ammonia with carbon disulfide produces ammonium dithiocarbamate in solution, which reacts with 2-haloketones to produce thiazol-2-thiones;[148] similarly, methyl dithiocarbamate serves as a component for the construction of 2-methylthiothiazole, reducable to thiazole itself, thus providing a good route to the unsubstituted heterocycle.[149]

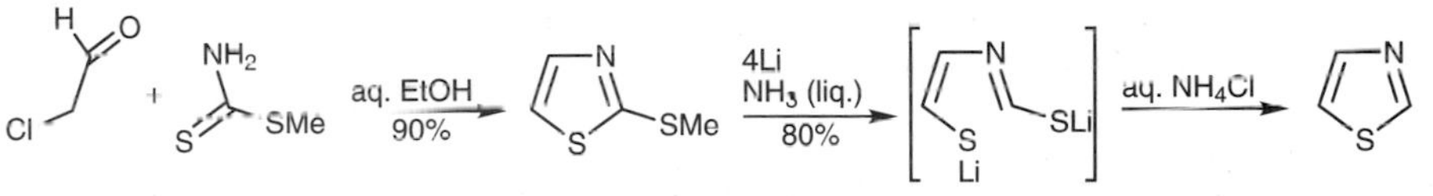

Imidazole itself can be prepared efficiently from bromoethanal ethylene acetal, formamide and ammonia; by analogy it is likely that displacement of halogen by ammonia occurs at an early stage.[150] An enol ether of bromomalonaldehyde reacts with amidines giving 5-formylimidazoles.[151] 2-Acetylaminoimidazoles are formed efficiently from the interaction of 2-bromoketones and N-acetylguanidine.[152]

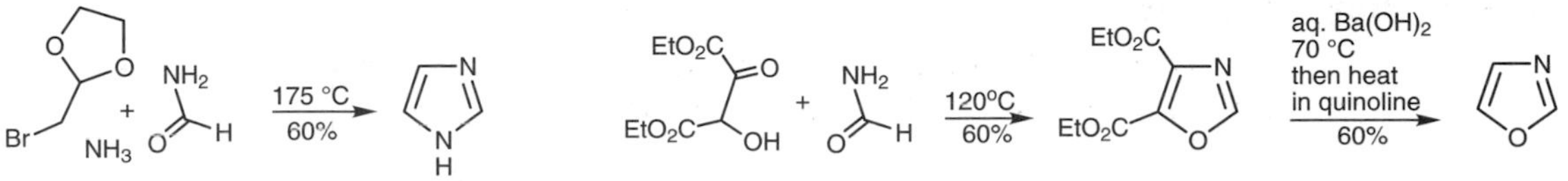

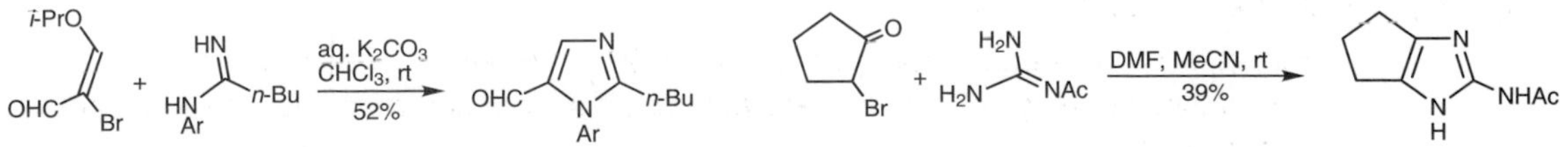

Oxazole itself has been prepared from its 4,5-diester, by hydrolysis then decarboxylation; though this formally falls into the same category of synthesis, it is probable that the ring oxygen derives from the 2-hydroxy-ketone, and not from the formamide;[153] the reaction of acyloins with formamide can be looked on as a general approach to oxazoles.[154] The use of cyanamide gives 2-amino-oxazoles.[155]

21.14.1.2   By cyclising dehydration of α-acylaminocarbonyl compounds

Cyclising dehydration of an α-acylaminocarbonyl compound is particularly important for oxazoles, and can be adapted for thiazole formation.

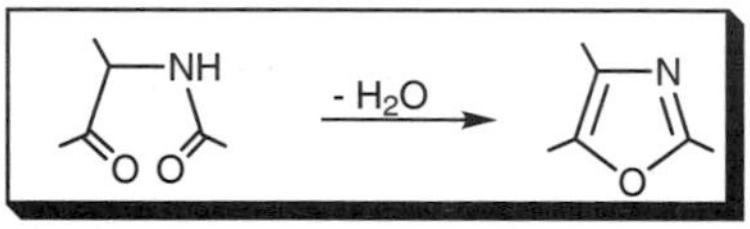

The classical method for making oxazoles, the *Robinson-Gabriel synthesis*, which is formally analogous to the cyclising dehydration of 1,4-dicarbonyl compounds to furans (section 15.13.1.1), is the acid-catalysed closure of α-acylamino carbonyl compounds.[156]

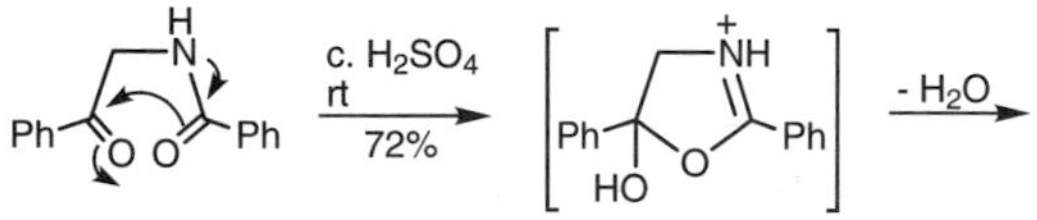

The construction of an amide using aminomalononitrile and a carboxylic acid under typical peptide coupling conditions, is accompanied by ring closure and the production of 5-aminooxazoles *in situ*.[157]

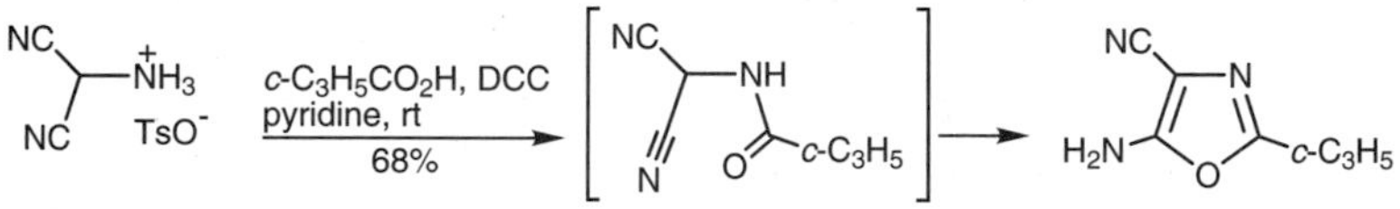

A synthesis shown below of ethyl oxazole-4-carboxylate illustrates a sophisticated use of this strategy.[158]

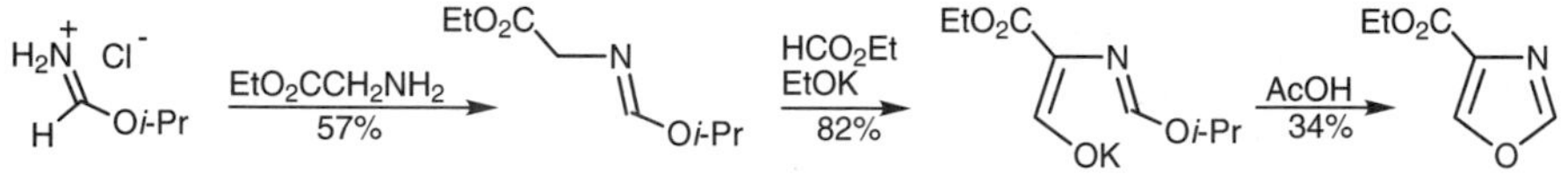

α-Acylthioketones close with ammonia to give thiazoles.[159]

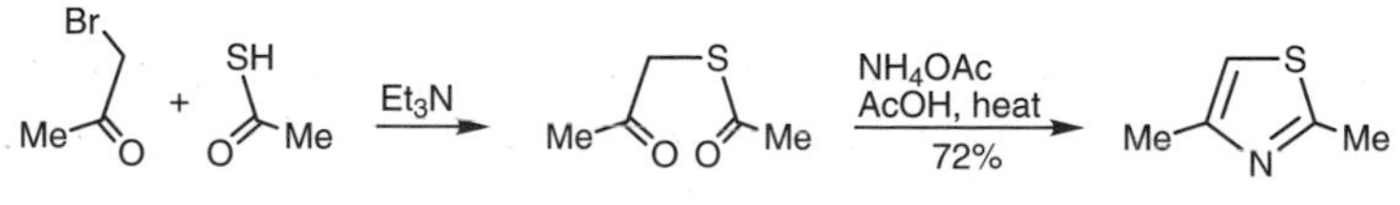

In two modern versions of ring closures in this category of ring synthesis, oxazoles are produced by base-catalysed closure of imino-chloride derivatives of glycine, obtained by acylation of ethyl isocyanoacetate[160] and in the second, by base-catalysed closure of 3-acylamino-2-iodo-1-phenylsulfonylalkenes.[161] In yet another use of an isonitrile, 2-tosylaminoimidazoles can be prepared.[162]

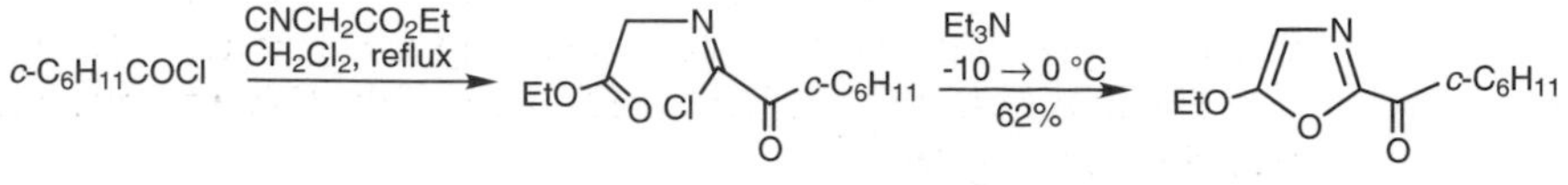

Iminoethers on reaction with aminoacetal give amidines which close in acid to give 2-substituted imidazoles.[163]

21.14.1.3   From isocyanides

Tosylmethylisocyanide (TOSMIC), can be used for the synthesis of all three 1,3-azole types.

Tosylmethylisocyanide has been used in the synthesis of all three 1,3-azole types. It reacts with aldehydes affording adducts, which lose toluenesulfinate on heating giving oxazoles,[164] with carbon disulfide it produces 4-tosyl-5-alkylthiothiazoles (following a subsequent S-alkylation)[165] and, in analogy to its interaction with aldehydes, it adds to N-alkylimines[166] or N-dimethylaminosulfamylimines[167] when, following elimination of toluenesulfinate, imidazoles are formed. The analogous benzotriazolylmethyl isocyanide can serve in the same way and has advantages in some situations.[168]

Anions derived from other isocyanides have been acylated (and thioformylated[169]), the products spontaneously closing to oxazoles[170] (thiazoles).

2-Isocyanoacrylates are proving to be versatile intermediates: they react with amines to give imidazoles, with thiols to give thiazoles, with protected hydrazine to give 2-aminoimidazoles (illustrated) and with O-benzylhydroxylamine to give 1-benzyloxyimidazoles.[171]

21.14.1.4 Oxazoles from α-diazocarbonyl compounds[172]

The carbene or carbenoid derived from an α-diazocarbonyl compound cycloadds to nitriles to produce oxazoles with the nitrile substituent at C-2.

The generation of a carbene (or when using a metal catalyst, a carbenoid) from an α-diazocarbonyl compound, in the presence of a nitrile results in overall cycloaddition and the formation of an oxazole. Both α-diazoketones and α-diazoesters have been used, the examples in the sequence below showing that the result in the latter situation is the formation of a 5-oxygenated oxazole.[173] The exact sequence of events is not certain but may involve a nitrile ylide, the result of electrophilic addition of the carbene to the nitrile nitrogen.

21.14.1.5 By dehydrogenation

The ring synthesis of the tetrahydro-1,3-azoles is simply the formation of N,N-, N,O- or N,S-analogues of aldehyde cyclic acetals; the ring synthesis of the 4,5-dihydro-heterocycles requires an acid oxidation level in place of aldehyde. A good route to the aromatic systems is therefore the dehydrogenation of these reduced and partially reduced systems. Nickel peroxide,[174] manganese(IV) oxide,[175] copper(II) bromide/base,[176] and bromotrichloromethane/diazabicycloundecane[177] have been used. The example shown uses cysteine methyl ester with a chiral aldehyde to form the tetrahydrothiazole.

21.14.2 Examples of notable syntheses involving 1,3-azoles
21.14.2.1 4(5)-Fluorohistamine

4(5)-Fluorohistamine[178] was synthesised via nucleophilic displacement of a side-chain leaving group (cf. pyrroles, section 14.12).

21.14.2.2 Pyridoxine[179]

This synthesis illustrates the use of an oxazole undergoing a Diels-Alder addition, leading on to a pyridine.

21.14.2.3 Thiamin

Thiamin was first synthesised in 1937.[180] It is widely used as a feed/food additive and in pharmaceutical preparations. A modern synthesis[181] of thiamin utilised an α-keto-thiol; the C-2 carbon was neatly delivered as the carbon of an amidine, one of the nitrogens providing the thiazole ring nitrogen and the other being the eventual amino group of the substituent pyrimidine.

21.14.2.4   Thieno[2,3-d]imidazole [182]

The synthesis of thieno[2,3-d]imidazole illustrates again the selectivity in halogen/metal exchange processes in imidazoles. The sequence illustrates the use of vinyl as another N-protecting group, and includes a nucleophilic displacement of bromine from the 4-position, activated by the 5-aldehyde.

21.14.2.5   Grossularine-2 [183]

A synthesis of the tunicate-derived natural product grossularine-2 rested on two key features: the lithiation of a 1-blocked-2-substituted imidazole at C-5, allowing introduction of tin for a subsequent coupling, and the electrocyclic ring closure of an isocyanate (a '2-azatriene' in which one of the double bonds was the imidazole 2,3-double bond).

Exercises for chapter 21

Straightforward revision exercises (consult chapters 20 and 21)

(a) Place the three 1,3-azoles in order of basicity and explain this order.

(b) What problems are associated with the N-alkylation of imidazoles in particular?

(c) What is carbonyl diimidazole (CDI) used for and why does it function well in that role?

(d) What factors must be considered in a discussion of electrophilic substitution of 1,3-azoles? How do they compare in this regard with pyrrole, thiophene, and furan, respectively?

(e) What is the positional order of selectivity for deprotonation of 1,3-azoles? How, then, would you make 5-methylthiazole from thiazole?

(f) What is the most typical electrocyclic process undergone by the 1,3-azoles and which of them show this tendency to the highest degree?

(g) Describe one typical synthesis for (i) an imidazole, (ii) a thiazole, and (iii) an oxazole.

More advanced exercises

1. Suggest structures for the halo compounds formed in the following ways: (i) imidazole with NaOCl $\rightarrow$ $C_3H_2Cl_2N$; (ii) 1-methylimidazole with excess Br_2 in AcOH $\rightarrow$ $C_4H_3Br_3N_2$ then this with EtMgBr followed by water $\rightarrow$ $C_4H_4Br_2N_2$ and this in turn with n-BuLi then $(MeO)_2CO$ gave $C_6H_7BrN_2O_2$.

2. Draw structures for the intermediates and final products which are formed when (i) 4-phenyloxazole is heated with but-1-yn-3-one $\rightarrow$ $C_6H_6O_2$; (ii) 5-ethoxyoxazole is heated with dimethyl acetylenedicarboxylate $\rightarrow$ $C_{10}H_{12}O_6$.

3. When the cyclic acyloin, c-$(CH_2)_{10}COCH(OH)$ was heated with formamide, in the presence of acid, a bicyclic oxazole, $C_{13}H_{21}NO$, was formed; what is its structure? This bicyclic oxazole was converted by exposure to $^1O_2$, then heating, into the acyclic cyano-acid, $HO_2C(CH_2)_{10}CN$; draw a mechanism for the transformation.

4. Deduce structures for the products formed at each stage of the following syntheses: 1,2-dimethyl-5-nitroimidazole heated with $Me_2NCH(Ot\text{-}Bu)_2$ $\rightarrow$ $C_8H_{12}N_4O_2$; this then heated with Ac_2O $\rightarrow$ $C_{10}H_{14}N_4O_3$. This product reacted (i) with guanidine $[H_2NC(NH_2)=NH]$ $\rightarrow$ $C_9H_{10}N_6O_2$, and (ii) with $MeNHNH_2$ $\rightarrow$ $C_9H_{11}N_5O_2$.

5. Deduce structures for the products formed in the following sequences: 1-methylimidazole/n-BuLi/$-30\,^\circ C$ then TMSCl $\rightarrow$ $C_7H_{14}N_2Si$ then n-BuLi/$-30\,^\circ C$ then TMSCl $\rightarrow$ $C_{10}H_{22}N_2Si_2$, then this with MeOH/rt $\rightarrow$ $C_7H_{14}NSi$, which was different to the first product.

6. Explain the following: 4-bromo-1-methylimidazole treated with n-BuLi/$-78\,^\circ C$ then DMF gave $C_5H_6N_2O$. Carrying out the same sequence but allowing the solution to warm to $0\,^\circ C$ before addition of DMF gave an isomeric product.

7. Thiazole-2-thione reacted with $Br(CH_2)_3Br$ to give, mainly, a salt $C_6H_8NS_2^+$ Br^-; suggest a structure and a mechanism for its formation.

8. Deduce structures for the 1,3-azoles which are produced from the following reactant combinations: (i) 1-chlorobutan-2-one and thiourea; (ii) thiobenzamide and chloroethanal; (iii) thioformamide and ethyl bromoacetate.

9. Write structures for the intermediates in the following synthesis of 3,4-bis(acetoxymethyl)furan: phenacyl bromide/NH_4^+ HCO_2^- $\rightarrow$ C_9N_7NO; this then heated with $AcOH_2CC\equiv CCH_2OAc$.

10. What imidazoles would be formed from the following reactant combination: (i) $MeN\equiv C/n$-BuLi and $PhC\equiv N$; (ii) 2-amino-1,2-diphenylethanone and $H_2NC\equiv N$?

References

1. 'Advances in imidazole chemistry', Grimmett, M. R., *Adv. Heterocycl. Chem.*, **1970**, *12*, 103; 'Advances in imidazole chemistry', *ibid.*, **1980**, *27*, 241; 'The azoles', Schofield, K., Grimmett, M. R., and Keene, B. R. T., Cambridge University Press, **1976**; 'Imidazole and benzimidazole synthesis', Grimmett, M. R., Academic Press, London, **1997**.

2. 'Advances in oxazole chemistry', Lakhan, R. and Ternai, B., *Adv. Heterocycl. Chem.*, **1974**, *17*, 99; 'The chemistry of oxazoles', Turchi, I. J. and Dewar, M. J. S., *Chem. Rev.*, **1975**, *75*, 389; 'New chemistry of oxazoles', Hassner, A. and Fischer, B., *Heterocycles*, **1993**, *35*, 1441.

3. 'Synthetic studies with natural oxazoles and thiazoles', Pattenden, G., *J. Heterocycl. Chem.*, **1992**, *29*, 607.

4. Zirngibl, L., 'Azoles. Antifungal active substances - syntheses and uses', Wiley-VCH, **1997**.

5. Deady, L. W., *Aust. J. Chem.*, **1973**, *26*, 1949.

6. Bogdal, D., Pielichowski, J., and Jaskot, K., *Heterocycles*, **1997**, *45*, 715.

7. Baxter, R. A. and Spring, F. S., *J. Chem. Soc.*, **1945**, 232; Roe, A. M., *J. Chem. Soc.*, **1963**, 2195.

8. Harlow, K. J., Hill, A. F., and Welton, T., *Synthesis*, **1996**, 697.

9. Kiyomori, A., Marcoux, J.-F., and Buchwald, S. L., *Tetrahedron Lett.*, **1999**, *40*, 2657.

10. Hayes, F. N., Rogers, B. S., and Ott, D. G., *J. Am. Chem. Soc.*, **1955**, *77*, 1850; Ott, D. G., Hayes, F. N., and Kerr, V. N., *ibid.*, **1956**, *78*, 1941.

11. Ulibarri, G., Choret, N., and Bigg, D. C. H., *Synthesis*, **1996**, 1287.

12. O'Connell, J. F. and Rapoport, H., *J. Org. Chem.*, **1992**, *57*, 4775.

13. Batey, R. A., Yoshina-Ishii, C., Taylor, S. D., and Santhakumar, V., *Tetrahedron Lett.*, **1999**, *40*, 2669.

14. Olofson, R. A. and Kendall, R. V., *J. Org. Chem.*, **1970**, *35*, 2246.

15. Shapiro, G. and Gomez-Lor, B., *Heterocycles*, **1995**, *41*, 215.

16. Daninos-Zeghal, S., Mourabit, A. A., Ahond, A., Poupat, C., and Potier, P., *Tetrahedron*, **1997**, *53*, 7604.

17. Collman, J. P., Bröning, M., Fu, L., Rapta, M., and Schwenninger, R., *J. Org. Chem.*, **1998**, *63*, 8084.

18. Jain, R. and Cohen, L. A., *Tetrahedron*, **1996**, *52*, 5363.

19. Katritzky, A. R., Rewcastle, G. W., and Fan, W.-Q., *J. Org. Chem.*, **1988**, *53*, 5685.

20. Caplow, M. and Jencks, W. P., *Biochemistry*, **1962**, *1*, 883; Reddy, G. S., Mandell, L., and Goldstein, J. H., *J. Chem. Soc.*, **1963**, 1414.

21. Morton, R. C., Mangroo, D., and Gerber, G. E., *Canad. J. Chem.*, **1988**, *66*, 1701.

22. 'Syntheses using heterocyclic amides (azolides)', *Newer Methods of Prep. Org. Chem.*, **1968**, *5*, 61.

23. Vatèle, J.-M. and Hanessian, S., *Tetrahedron*, **1996**, *52*, 10557.

24. Chou, T. S., Becke, L. M., O'Toole, J. C., Carr, M. A., and Parker, B. E., *Tetrahedron Lett.*, **1996**, *37*, 17.

25. Wong, J. L. and Keck, J. H., *J. Org. Chem.*, **1974**, *39*, 2398.

26. Staab, H. A., Irngartinger, H., Mannschreck, A., and Wu, M. T., *Justus Liebigs Ann. Chem.*, **1966**, *695*, 55.

27. Staab, H. A., Wu, M. T., Mannschreck, A., and Schwalbach, G., *Tetrahedron Lett.*, **1964**, *15*, 845; Vaughn, J. D., Mughrabi, Z., and Wu, E. C., *J. Org. Chem.*, **1970**, *35*, 1141; Takeuchi, Y., Yeh, H. J. C., Kirk, K. L., and Cohen, L. A., *J. Org. Chem.*, **1978**, *43*, 3565.

28. Olofson, R. A. and Landesberg, J. M., *J. Am. Chem. Soc.*, **1966**, *88*, 4263; Coburn, R. A., Landesberg, J. M., Kemp, D. S., and Olofson, R. A., *Tetrahedron*, **1970**, *26*, 685.

29. Breslow, R., *J. Am. Chem. Soc.*, **1958**, *80*, 3719; Breslow, R. and McNelis, E., *ibid.*, **1959**, *81*, 3080; Kluger, R., Karimian, K., and Kitamura, K. *ibid.*, **1987**, *109*, 6368

30. Haake, P., Bausher, L. P., and Miller, W. B., *J. Am. Chem. Soc.*, **1969**, *91*, 1113.

31. Haake, P., Bausher, L. P., and McNeal, J. P., *J. Am. Chem. Soc.*, **1971**, *93*, 7045.

32. Austin, M. W., Blackborow, J. R., Ridd, J. H., and Smith, B. V., *J. Chem. Soc.*, **1965**, 1051.

33. Katritzky, A. R., Ögretir, C., Tarhan, H. O., Dou, H. M., and Metzger, J. V., *J. Chem. Soc.*, *Perkin Trans. 2*, **1975**, 1614.

34. Asato, G., *J. Org. Chem.*, **1968**, *33*, 2544.

35. Barnes, G. R. and Pyman, F. L., *J. Chem. Soc.*, **1927**, 2711.

36. Erlenmeyer, H. and Kiefer, H., *Helv. Chim. Acta*, **1945**, *28*, 985.

37. Balaban, I. E. and Pyman, F. L., *J. Chem. Soc.*, **1922**, 947; Stensiö, K.-E., Wahlberg, K., and Wahren, R., *Acta Scand.*, **1973**, *27*, 2179

38. O'Connell, J. F., Parquette, J., Yelle, W. E., Wang, W., and Rapoport, H., *Synthesis*, **1988**, 767.

39. Balaban, I. E. and Pyman, F. L., *J. Chem. Soc.*, **1924**, 1564.
40. Iddon, B. and Khan, N., *J. Chem. Soc., Perkin Trans. 1*, **1987**, 1453.
41. Caló, V., Ciminale, F., Lopez, L., Naso, F., and Todesco, P. E., *J. Chem. Soc., Perkin Trans. 1*, **1972**, 2567.
42. Lutz, A. W. and DeLorenzo, S., *J. Heterocycl. Chem.*, **1967**, *4*, 399.
43. Pauly, H. and Arauner, E., *J. Prakt. Chem.*, **1928**, *118*, 33; Groziak, M. P. and Wei, L., *J. Org. Chem.*, **1991**, *56*, 4296.
44. Naidu, M. S. R. and Bensusan, H. B., *J. Org. Chem.*, **1968**, *33*, 1307.
45. Nagasawa, F., *J. Pharm. Soc. Jpn.*, **1940**, *60*, 433 (*Chem. Abs.*, **1940**, *34*, 5450); Ganapathi, K. and Kulkarni, K. D., *Current Sci.*, **1952**, *21*, 314 (*Chem. Abs.*, **1951**, *45*, 5150).
46. Regel, E. and Büchel, K.-H., *Justus Liebigs Ann. Chem.*, **1977**, 145.
47. Bastiaansen, L. A. M. and Godefroi, E. F., *Synthesis*, **1978**, 675.
48. Whitten, J. P., McCarthy, T. R., and Matthews, D. P., *Synthesis*, **1988**, 470.
49. Antonini, I., Cristalli, G., Franchetti, P., Grifantini, M., Gulini, U., and Martelli, S., *J. Heterocycl. Chem.*, **1978**, *15*, 1201.
50. Bastiaansen L. A. M., Van Lier, P. M., and Godefroi, E. F., *Org. Synth., Coll. Vol. VII*, **1990**, 287.
51. 'The thiazole aldehyde synthesis', Dondoni, A. *Synthesis*, **1998**, 1681.
52. Ochiai, E. and Hayashi, E., *J. Pharm. Soc. Jpn.*, **1947**, *67*, 34 (*Chem. Abs.*, **1951**, *45*, 9533).
53. Ochiai, E. and Nagasawa, H., *Chem. Ber.*, **1939**, *72B*, 1470.
54. Brederick, H., Gompper, R., Reich, F., and Gotsmann, U., *Chem. Ber.*, **1960**, *93*, 2010.
55. Bosco, M., Forlani, L., Liturri, V., Riccio, P., and Todesco, P. E., *J. Chem. Soc. (B)*, **1971**, 1373; Bosco, M., Liturri, V., Troisi, L., Forlani, L., and Todesco, P. E., *J. Chem. Soc., Perkin Trans. 2*, **1974**, 508.
56. de Bie, D. A., van der Plas, H. C., and Guersten, G., *Recl. Trav. Chim. Pays-Bas*, **1971**, *90*, 594.
57. Gompper, R. and Effenberger, F., *Chem. Ber.*, **1959**, *92*, 1928.
58. Suwinski, J. and Swierczek, K., *Tetrahedron Lett.*, **1998**, *39*, 3331.
59. Forlani, L., Magagni, M., and Todesco, P. E., *Chim. Ind. (Milan)*, **1978**, *60*, 348 (*Chem. Abs.*, **1978**, *89*, 179289); Landsberg, J. M., Houk, K. N., and Michelman, J. S., *J. Am. Chem. Soc.*, **1966**, *88*, 4265.
60. Takeuchi, T., Yeh, H. J. C., Kirk, K. L., and Cohen, L. A., *J. Org. Chem.*, **1978**, *43*, 3565.
61. Begtrup, M. and Larsen, P., *Acta Scand.*, **1990**, *44*, 1050.
62. Bhujanga Rao, A. K. S., Rao, C. G., and Singh, B. B., *Synth. Commun.*, **1991**, *21*, 427; *idem*, *J. Org. Chem.*, **1990**, *55*, 3702.
63. Horváth, A., *Tetrahedron Lett.*, **1996**, *37*, 4423.
64. Stocker, F. B., Fordice, M. W., Larson, J. K., and Thorstenson, J. H., *J. Org. Chem.*, **1966**, *31*, 2380.
65. Iddon, B., 'Synthesis and reactions of lithiated monocyclic azoles containing two or more hetero-atoms. Part II: Oxazoles', *Heterocycles*. **1994**, *37*, 1321; 'Part IV: Imidazoles', Iddon, B. and Ngochindo, R. I., *ibid*, **1994**, *38*, 2487; 'Part V: Isothiazoles and thiazoles', Iddon, B., *ibid.*, **1995**, *41*, 533.
66. Hodges, J. C., Patt, W. C., and Conolly, C. J., *J. Org. Chem.*, **1991**, *56*, 449.
67. Dondoni, A., Mestellani, A. R., Medici, A., Negrini, E., and Pedrini, P., *Synthesis*, **1986**, 757; Dondoni, A., Fantin, G., Fagagnolo, M., Medici, A., and Pedrini, P., *J. Org. Chem.*, **1988**, *53*, 1748.
68. 'Metallation and metal-halogen exchange reactions of imidazoles', Iddon, B., *Heterocycles*, **1985**, 23, 417; Ohta, S., Matsukawa, M., Ohashi, N., and Nagayama, K., *Synthesis*, **1990**, 78.
69. Ngochindo, R. I., *J. Chem. Soc., Perkin Trans. 1*, **1990**, 1645.
70. Sundberg, R. J., *J. Heterocycl. Chem.*, **1977**, *14*, 517.
71. Bell, A. S., Roberts, D. A., and Ruddock, K. S., *Tetrahedron Lett.*, **1988**, *29*, 5013.
72. Lipshutz, B. H., Huff, B., and Hagen, W., *Tetrahedron Lett.*, **1988**, *29*, 3411.
73. Curtis, N. J. and Brown, R. S., *J. Org. Chem.*, **1980**, *45*, 4038.
74. Manoharan, T. S. and Brown, R. S., *J. Org. Chem.*, **1988**, *53*, 1107.
75. Kirk, K. L., *J. Org. Chem.*, **1978**, *43*, 4381.
76. Shapiro, G. and Gomez-Lor, B., *Heterocycles*, **1995**, *41*, 215.

77. Vollinga, R. C., Menge, W. M. P. B., and Timmerman, H., *Recl. Trav. Chim. Pays-Bas*, **1993**, *112*, 123.
78. Pinkerton, F. H. and Thames, S. F., *J. Heterocycl. Chem.*, **1972**, *9*, 67; Medici, A., Pedrini, P., and Dondoni, A., *J. Chem. Soc., Chem. Commun.*, **1981**, 655.
79. Phillips, J. G., Fadnis, L., and Williams, D. R., *Tetrahedron Lett.*, **1997**, *38*, 7835.
80. Katritzky, A. R., Slawinski, J. J., Brunner, F., and Gorun, S., *J. Chem. Soc., Perkin Trans. 1*, **1989**, 1139.
81. Crowe, E., Hossner, F., and Hughes, M. J., *Tetrahedron*, **1995**, *52*, 8889.
82. Dondoni, A., Fantin, G., Fogagnolo, M., Medici, A., and Pedrini, P., *J. Org. Chem.*, **1987**, *52*, 3413.
83. Harn, N. K., Gramer, C. J., and Anderson, B. A., *Tetrahedron Lett.*, **1995**, *36*, 9453.
84. Vedejs, E. and Monahan, S. D., *J. Org. Chem.*, **1996**, *61*, 5192.
85. Anderson, B. A. and Harn, N. K., *Synthesis*, **1996**, 583.
86. Vedejs, E. and Monahan, S. D., *J. Org. Chem.*, **1997**, *62*, 4763.
87. Cliff, M. D. and Pyne, S. G., *Tetrahedron*, **1996**, *52*, 13703; *idem*, *Synthesis*, **1994**, 681.
88. Kawasaki, I., Katsuma, H., Nakayama, Y., Yamashita, M., and Ohto, S., *Heterocycles,*, **1998**, *48*, 1887.
89. Jain, R., Cohen, L. A., and King, M. M., *Tetrahedron*, **1997**, *53*, 4539.
90. Aldabbagh, F., Bowman, W. R., and Mann, E., *Tetrahedron Lett.*, **1997**, *38*, 7937.
91. Aldabbagh, F. and Bowman, W. R., *Tetrahedron Lett.*, **1997**, *38*, 3793.
92. Clark, G. M. and Sykes, P., *J. Chem. Soc., (C)*, **1967**, 1269 and 1411.
93. Abbott, P. J., Acheson, R. M., Eisner, U., Watkin, D. J., and Carruthers, J. R., *J. Chem. Soc., Perkin Trans. 1*, **1976**, 1269; Huisgen, R., Giese, B., and Huber, H., *Tetrahedron Lett.*, **1967**, 1883; Acheson, R. M. and Vernon, J. M., *J. Chem. Soc.*, **1962**, 1148.
94. Ohlsen, S. R. and Turner, S., *J. Chem. Soc., (C)*, **1971**, 1632; Hutton, J., Potts, B., and Southern, P. F., *Synth. Commun.*, **1979**, *9*, 789; König, H., Graf, F., and Weberndörfer, *Justus Liebigs Ann. Chem.*, **1981**, 668; Liotta, D., Saindane, M., and Ott, W., *Tetrahedron Lett.*, **1983**, *24*, 2473; Kawada, K., Kitagawa, O., and Kobayashi, Y., *Chem. Pharm. Bull.*, **1985**, *33*, 3670.
95. Whitney, S. E. and Rickborn, B., *J. Org. Chem.*, **1988**, *53*, 5595.
96. Wasserman, H. H., Gambale, R. J., and Pulwer, M. J., *Tetrahedron*, **1981**, *37*, 4059.
97. Wuonola, M. A. and Smallheer, J. M., *Tetrahedron Lett.*, **1992**, *33*, 5697.
98. 'Condensation of oxazoles with dienophiles – a new method for synthesising pyridine bases', Karpeiskii, M. Ya. and Florent'ev, V. L., *Russ. Chem. Rev.*, **1969**, *38*, 540 (*Chem. Abs.*, **1969**, *71*, 91346).
99. Deyrup, J. A. and Gingrich, H. L., *Tetrahedron Lett.*, **1977**, 3115; Ishizuka, T., Osaki, M., Ishihara, H., and Kunieda, T., *Heterocycles*, **1993**, *35*, 901.
100. Dewar, M. J. S. and Turchi, I. J., *J. Am. Chem. Soc.*, **1974**, *96*, 6148.
101. Corrao, S., Macielag, M., and Turchi, I. J., *J. Org. Chem.*, **1990**, *55*, 4484.
102. Nesi, R., Turchi, S., Giomi, D., and Papaleo, S., *J. Chem. Soc., Chem. Commun.*, **1993**, 978.
103. Wan, Z., and Snyder, J. K., *Tetrahedron Lett.*, **1997**, *38*, 7495; Neipp, C. E., Ranslow, P. B., Wan, Z., and Snyder, J. K., *ibid.*, **1997**, *38*, 7499.
104. Deghati, P. Y. F., Wanner, M. J., and Koomen, G.-J., *Tetrahedron Lett.*, **1998**, *39*, 4561.
105. Noyce, D. S., Stowe, G. T., and Wong, W., *J. Org. Chem.*, **1974**, *39*, 2301; Lipshutz, B. H. and Hungate, R. W., *J. Org. Chem.*, **1981**, *46*, 1410; Knaus, G. N. and Meyers, A. I., *J. Org. Chem.*, **1974**, *39*, 1189.
106. Cornwall, P., Dell, C. P., and Knight, D. W., *Tetrahedron Lett.*, **1987**, *28*, 3585.
107. Van Arnum, S. D., Ramig, K., Stepsus, N. A., Dong, Y., and Outten, R. A., *Tetrahedron Lett.*, **1996**, *37*, 8659.
108. Macco, A. A., Godefroi, E. F., and Drouen, J. J. M., *J. Org. Chem.*, **1975**, *40*, 252.
109. Earle, M. J., McCormac, P. B., and Seddon, K. R., *Chem. Commun.*, **1998**, 2245.
110. Itoh, T., Hasegawa, H., Nagata, K., and Ohsawa, A., *J. Org. Chem.*, **1994**, *59*, 1319.
111. Itoh, T., Miyazaki, M., Nagata, K., and Ohsawa, A., *Heterocycles*, **1998**, *49*, 67.
112. Itoh, T., Miyazaki, M., Nagata, K., Matsuya, Y., and Ohsawa, A., *Heterocycles*, **1999**, *50*, 667.
113. Ott, D. G., Hayes, F. N., and Kerr, V. N., *J. Am. Chem. Soc.*, **1956**, *78*, 1941; Ruggli, P., Ratti, R., and Henzi, E., *Helv. Chim. Acta*, **1929**, *12*, 332.
114. Federsel, H.-J., Glasane, G., Högström, C., Wiestå, J., Zinko, B., and Ödman, C., *J. Org. Chem.*, **1995**, *60*, 2597.

115. Yu, J. and Yadan, J.-C., *Synlett*, **1995**, 239.
116. Yount, R. G. and Metzler, D. E., *J. Biol. Chem.*, **1959**, *234*, 738.
117. Hafferl, W., Lundin, R., and Ingraham, L. L., *Biochemistry*, **1963**, *2*, 1298.
118. Castells, J., López-Calahorra, F., Geijo, F., Pérez-Dolz, R., and Bassedas, M., *J. Heterocycl. Chem.*, **1986**, *23*, 715; Teles. J. H., Melder, J.-P., Ebel, K., Schneider, R., Gehrer, E., Harder, W., Brode, S., Enders, D., Breuer, K., and Raabe, G., *Helv. Chim. Acta*, **1996**, *79*, 61.
119. Begtrup, M., *J. Chem. Soc., Chem. Commun.*, **1975**, 334.
120. Arduengo, A. J., Harlow, R. L., and Kline, M., *J. Am. Chem. Soc.*, **1991**, *113*, 361.
121. Arduengo, A. J., Goerlich, J. R., Krafczyk, R., and Marshall, W. J., *Angew. Chem., Int. Ed. Engl.*, **1998**, *37*, 1963.
122. Arduengo, A. J., Goerlich, J. R., Krafczyk, R., and Marshall, W. J., *Liebigs Ann./Recueil*, **1997**, 365.
123. 'Recent advances in oxazolone chemistry', Filler, R., *Adv. Heterocycl. Chem.*, **1965**, *4*, 75.
124. 'The chemistry of 1,3-thiazolinone/hydroxy-1,3-thiazole systems', Barrett, G. C., *Tetrahedron*, **1980**, *36*, 2023.
125. '4-Unsubstituted, 5-amino and 5-unsubstituted, 4-aminoimidazoles', Lythgoe, D. J. and Ramsden, C. A., *Adv. Heterocycl. Chem.*, **1994**, *61*, 1.
126. Forlani, L., *Gazz. Chim. Ital.*, **1981**, *111*, 159.
127. Erlenmeyer, H. and Kiefer, H., *Helv. Chim. Acta*, **1945**, *28*, 985.
128. McLean, J. and Muir, G. D., *J. Chem. Soc.*, **1942**, 383; Kirk, K. L. and Cohen, L. A., *J. Am. Chem. Soc.*, **1973**, *95*, 4619.
129. Gompper, R. and Christmann, O., *Chem. Ber.*, **1959**, *92*, 1944.
130. Zigeuner, G. and Rauter, W., *Monatsh. Chem.*, **1966**, *97*, 33.
131. Klein, G. and Prijs, B., *Helv. Chim. Acta*, **1954**, *37*, 2057.
132. Kerdesky, F. A. and Seif, L. S., *Synth. Commun.*, **1995**, *25*, 2639.
133. Crawford, M. and Little, W. T., *J. Chem. Soc.*, **1959**, 729.
134. Bayer, H. O., Huisgen, R., Knorr, R., and Schafer, F. C., *Chem. Ber.*, **1970**, *103*, 2581.
135. 'Cycloaddition chemistry of anhydro-4-hydroxy-1,3-thiazolium hydroxides' (thioisomunchnones) for the synthesis of heterocycles', Padwa, A., Harring, S. R., Hertzog, D. L., and Nadler, W. R., *Synthesis*, **1994**, 993.
136. Huisgen, R., Funke, E., Gotthardt, H., and Panke, H.-L., *Chem. Ber.*, **1971**, *104*, 1532; Consonni, R., Croce, P. D., Ferraccioli, R., and La Rosa, C., *J. Chem. Res., (S)*, **1991**, 188.
137. Mloston, G., Gendek, T., and Heimgartner, H., *Helv. Chim. Acta*, **1998**, *81*, 1585.
138. Alcázar, J., Begtrup, M., and de la Hoz, A., *J. Org. Chem.*, **1996**, *61*, 6971.
139. Eriksen, B. L., Vedso, P., Morel, S., and Begtrup, M., *J. Org. Chem.*, **1998**, *63*, 12.
140. 'The preparation of thiazoles', Wiley, R. H., England, D. C. and Behr, L. C., *Org. React.*, **1951**, *6*, 367.
141. 'Imidazole and benzimidazole synthesis', Grimmett, M. R., Academic Press, **1997**.
142. Schwarz, G., *Org. Synth., Coll. Vol. III*, **1955**, 332.
143. Vogel's Textbook of Practical Organic Chemistry, 4th Edtn., 929.
144. Flaig, R. and Hartmann, H., *Heterocycles*, **1997**, *45*, 875.
145. Moriarty, R. M., Vaid, B. K., Duncan, M. P., Ley, S. G., Prakash, O., and Goyal, S., *Synthesis*, **1992**, 845; Kamproudi, H., Spyroudis, S., and Tarantili, P., *J. Heterocycl. Chem.*, **1996**, *33*, 575.
146. Babadjamian, A., Gallo, R., Metzger, J., and Chanon, M., *J. Heterocycl. Chem.*, **1976**, *13*, 1205.
147. Kim, H.-S., Kwon, I.-C., and Kim, O.-H., *J. Heterocycl. Chem.*, **1995**, *32*, 937.
148. Buchman, E. R., Reims, A. O., and Sarjent, H., *J. Org. Chem.*, **1941**, *6*, 764.
149. Brandsma, L., de Jong, R. L. P., and VerKruijsse, H. D., *Synthesis*, **1985**, 948.
150. Brederick, H., Gompper, R., Bangert, R., and Herlinger, H., *Angew. Chem.*, **1958**, *70*, 269.
151. Shilcrat, S. C., Makhallati, M. K., Fortunak, J. M. D., and Pridgen, L. N., *J. Org. Chem.*, **1997**, *62*, 8449.
152. Litle, T. L. and Webber, S. E., *J. Org. Chem.*, **1994**, *59*, 7299.
153. Brederick, H. and Bangert, R., *Angew. Chem., Int. Ed. Engl.*, **1962**, *1*, 662.
154. Brederick, H. and Gompper, R., *Chem. Ber.*, **1954**, *87*, 726; Lombardino, J. G., *J. Heterocycl. Chem.*, **1973**, *10*, 697; Wasserman, H. H. and Druckrey, E., *J. Am. Chem. Soc.*, **1968**, *90*, 2440.

155. Cockerill, A. F., Deacon, A., Harrison, R. G., Osborne, D. J., Prime, D. M., Ross, W. J., Todd, A., and Verge, J. P., *Synthesis*, **1976**, 591.

156. Wasserman, H. H. and Vinick, F. J., *J. Org. Chem.*, **1973**, *38*, 2407.

157. Freeman, F., Chen, T., van der Linden, J. B., *Synthesis*, **1997**, 861.

158. Cornforth, J. W. and Cornforth, R., *J. Chem. Soc.*, **1947**, 96.

159. Dubs, P. and Stuessi, R., *Synthesis*, **1976**, 696.

160. Huang, W.-S., Zhang, Y.-Z., and Yuan, C.-Y., *Synth. Commun.*, **1996**, *26*, 1149

161. Short, K. M. and Ziegler, C. B., *Tetrahedron Lett.*, **1993**, *34*, 71.

162. Bissio, R., Marcaccini, S., Pepino, R., and Torroba, T., *J. Org. Chem.*, **1996**, *61*, 2202.

163. Galeazzi, E., Guzmán, A., Nava, J. L., Liu, Y., Maddox, M. C., and Muchowski, J. M., *J. Org. Chem.*, **1995**, *60*, 1090.

164. van Leusen, A. M., Hoogenboon, B. E., and Siderius, H., *Tetrahedron Lett.*, **1972**, 2369; van Leusen, A. M. and Oldenziel, O. H., *ibid.*, 2373; Kulkarni, B. A. and Ganesan, A., *Tetrahedron Lett.*, **1999**, *40*, 5637.

165. van Leusen, A. M. and Wildeman, J., *Synthesis*, **1977**, 501.

166. van Leusen, A. M., Wildeman, J., and Oldenziel, O. H., *J. Org. Chem.*, **1977**, *42*, 1153.

167. ten Have, R., Huisman, M., Meetsma, A., and van Leusen, A. M., *Tetrahedron*, **1997**, *53*, 11355.

168. Katritzky, A. R., Cheng, D., and Musgrave, R. P., *Heterocycles*, **1997**, *44*, 67.

169. Hartman, G. D. and Weinstock, L. M., *Org. Synth., Coll. Vol. VI*, **1988**, 620.

170. Schröder, R., Schöllkopf, U., Blume, E., and Hoppe, I., *Justus Liebigs Ann. Chem.*, **1975**, 533; Hamada, Y., Morita, S., and Shioiri, T., *Heterocycles*, **1982**, *17*, 321; Ohba, M., Kubo, H., Fujii, T., Ishibashi, H., Sargent, M. V., and Arbain, D., *Tetrahedron Lett.*, **1997**, *38*, 6697.

171. Yamada, M., Fukui, T., and Nunami, K., *Synthesis*, **1995**, 1365.

172. Moody, C. J. and Doyle, K. J., *Progr. Heterocycl. Chem.,* **1997**, *9*, 1.

173. Doyle, K. J. and Moody, C. J., *Tetrahedron*, **1994**, *50*, 3761.

174. Evans, D. L., Minster, D. K., Jordis, U., Hecht, M., Mazzu, A. L., and Meyers, A. I., *J. Org. Chem.*, **1979**, *44*, 497.

175. Ninomiya, K., Satoh, H., Sugiyama, T., Shinomiya, M., and Kuroda, R., *Chem. Commun.*, **1996**, 1825; Sowinski, J. A. and Toogood, P. T., *J. Org. Chem.*, **1996**, *61*, 7671.

176. Barrish, J. C., Singh, J., Spergel, S. H., Han, W.-C., Kissick, T. P., Kronenthal, D. R., Mueller, R. H., *J. Org. Chem.*, **1993**, *58*, 4494.

177. Williams, D. R., Lowder, P. D., Gu, Y.-G., and Brooks, D. A., *Tetrahedron Lett.*, **1997**, *38*, 331.

178. Montgomery, J. A., Hewson, K., Struck, R. F., and Sheely, Y. F., *J. Org. Chem.*, **1959**, *24*, 256; Kirk, K. L. and Cohen, L. A., *J. Am. Chem. Soc.*, **1973**, *95*, 4619.

179. Harris, E. E., Firestone, R. A., Pfister, K., Boettcher, R. R., Cross, F. J., Currie, R. B., Monaco, M., Peterson, E. R., and Reuter, W., *J. Org. Chem.*, **1962**, *27*, 2705; Doktorova, N. D., Ionova, L. V., Karpeisky, M. Ya., Padyukova, N. Sh., Turchin, K. F., and Florentiev, V. L., *Tetrahedron*, **1969**, *25*, 3527.

180. Todd, A. R. and Bergel, F., *J. Chem. Soc.*, **1937**, 364.

181. Contant, P., Forzy, L., Heingartner, U., and Moine, G., *Helv. Chim. Acta*, **1990**, *73*, 1300.

182. Hartley, D. J. and Iddon, B., *Tetrahedron Lett.*, **1997**, *38*, 4647.

183. Choshi, T., Yamada, S., Sugino, E., Kuwada, T., and Hibino, S., *J. Org. Chem.*, **1995**, *60*, 5899.

22 1,2-Azoles: pyrazoles, isothiazoles, isoxazoles: reactions and synthesis

pyrazole
[1*H*-pyrazole] isothiazole isoxazole isoxazoline
[4,5-dihydroisoxazole] isoxazolidine

The physical properties of the three 1,2-azoles, pyrazole,[1] isothiazole[2] and isoxazole[3] can be usefully compared and contrasted with those of their 1,3-isomeric counterparts. Echoing the higher boiling point of imidazole, pyrazole, which is the only one of the trio to be solid at room temperature, also has a much higher boiling point (187 °C) than isothiazole or isoxazole (114 °C and 95 °C) again reflecting the intermolecular hydrogen bonding available only to pyrazole. This association probably takes the form of dimers, trimers, and oligomers; dimeric forms are of course not available to imidazole. Each 1,2-azole has a pyridine-like odour but is only partially soluble in water. The dihydro and tetrahydro heterocycles are named pyrazoline/pyrazolidine, isothiazoline/isothiazolidine, and isoxazoline/isoxazolidine.

Rapid tautomerism, involving switching of hydrogen from one nitrogen to the other, as in imidazoles, means that substituted pyrazoles are inevitably mixtures, and a nomenclature analogous to that used for imidazoles, is employed to signify this: 3(5)-methylpyrazole, for example.

3(5)-methylpyrazole

Phenylbutazone has been utilised for some time in the treatment of severe arthritis, which, incidentally, afflicted such notables as Casanova, Goethe, and Luther. Leflunomide is used in the therapy of autoimmune diseases, such as active rheumatoid arthritis. Celecoxib is the first to market of a number of selective cycloxygenase 2 (COX 2) inhibitors which show great promise as anti-inflammatory and analgetic agents, without the undesirable side effects associated with other non-steroidal anti-inflammatories. There are many pyrazole dyestuffs – the food colourant tartrazine is one such substance.

Phenylbutazone Leflunomide Celecoxib Tartrazine

22.1   Reactions with electrophilic reagents

22.1.1   Addition at nitrogen

22.1.1.1   Protonation

Direct linking of two heteroatoms has a very marked base-weakening effect, as in hydrazine and hydroxylamine (pK_as: NH_3, 9.3; H_2NNH_2, 7.9; $HONH_2$, 5.8), and this is mirrored in the 1,2-azoles: pyrazole with a pK_a of 2.5 is some 4.5 pK_a units weaker than imidazole; isothiazole (–0.5) and isoxazole (–3.0) are some 3 pK_a units weaker than their 1,3-isomers. The higher basicity of pyrazole reflects the symmetry of the cation with its two equivalent contributing resonance structures. Clearly, again, oxygen has a larger electron-withdrawing effect than sulfur.

22.1.1.2   Oxidation at nitrogen

The preparation of 1-hydroxypyrazoles can employ peracidic conditions[4] or basic conditions,[5] when it is the pyrazolyl anion which reacts with the oxidising agent, dibenzoyl peroxide.

22.1.1.3   Alkylation at nitrogen

The 1,2-azoles are more difficult to quaternise than their 1,3-analogues: isothiazoles, for example, require reactive reagents such as benzyl halides or Meerwein salts.[6] Additionally, isoxazolium salts are particularly susceptible to ring cleavage (see section 22.11). 3(5)-Substituted pyrazoles which have an *N*-hydrogen, can in principle give rise to two isomeric *N*-alkyl pyrazoles, after loss of proton from nitrogen, and there is the further complication that this initial product can undergo further reaction producing an *N,N'*-disubstituted quaternary salt.[7] However, the quaternisation of an already 1-substituted pyrazole generally requires more vigorous conditions, no doubt because of steric impediment to reaction due to the substituent on the adjacent nitrogen. Microwave irradiation improves the rate of *N*-alkylation.[8]

22.1.1.4   Acylation at nitrogen

The introduction of an acyl[9] or phenylsulfonyl[10] group onto a pyrazole nitrogen is usually achieved in the presence of a weak base such as pyridine; such processes proceed *via* imine nitrogen acylation, then N^+–H-deprotonation. Since acylation, unlike alkylation, is reversible, the more stable product is obtained.

Pyrazole reacts with cyanamide very efficiently to produce an *N*-derivative which can be utilised, by reaction with primary or secondary amines, to synthesise guanidines.[11] Conversion of the pyrazolyl guanidine to a doubly *t*-butoxycarbonyl-protected pyrazolyl guanidine then allows this to be used for the direct synthesis of protected guanidines, as illustrated.[12]

22.1.2 Substitution at carbon

22.1.2.1 Nitration

Pyrazole[13] and isothiazole[14] undergo straightforward nitration, at C-4, but the less reactive isoxazole nitrates in negligible yield; 3-methylisoxazole, however, has sufficient extra reactivity that it can be satisfactorily nitrated, at C-4.[15] With acetyl nitrate or dinitrogen tetraoxide/ozone,[16] 1-nitropyrazole is formed but this can be rearranged to 4-nitropyrazole in acid at low temperature.[17]

22.1.2.2 Sulfonation

Electrophilic sulfonation of isoxazole is of no preparative value; the substitution of only the phenyl substituent of 5-phenylisoxazole with chlorosulfonic acid makes the same point.[18] Both isothiazole[2a,19] and pyrazole[20] can be satisfactorily sulfonated.

22.1.2.3 Halogenation

Halogenation of pyrazole gives 4-monohalopyrazoles, for example 4-iodo-,[21] or 4-bromopyrazole[22] under controlled conditions. Poor yields are obtained on reaction of isothiazole[23] and isoxazole[24] with bromine, again with attack at C-4, but with activating groups present, halogenation proceeds better.[25] 3,4,5-Tribromopyrazole is formed efficiently in alkaline solution, presumably the pyrazolyl anion is the reacting species.[26]

22.1.2.4 Acylation

Only for pyrazole, of the trio, have any useful electrophilic substitutions involving carbon electrophiles been described,[10,27] and even here only N-substituted pyrazoles react well, perhaps because of inhibition of N⁺-salt formation.

22.2   Reactions with oxidising agents

The 1,2-azole ring systems are relatively stable to oxidative conditions, allowing substituent alkyl, or more efficiently, acyl groups to be oxidised up to carboxylic acid.[28] Ozone cleaves the isoxazole ring.[29]

22.3   Reactions with nucleophilic reagents

The 1,2-azoles do not generally react with nucleophiles with replacement of hydrogen; there is a limited range of examples of displacements of leaving groups from the 5-position[30] when it is activated by a 4-keto or similar group, but interestingly, 3-halo groups are less easily displaced; 4-halides behave like halobenzenes.

22.4   Reactions with bases

22.4.1   Deprotonation of pyrazole N-hydrogen

The pK_a for loss of the N-hydrogen of pyrazole is 14.2, compared with 17.5 for imidazole, though there are again two, equally-contributing resonance forms.

22.4.2.   Deprotonation of C-hydrogen[31]

The C-5-deprotonation of pyrazoles requires the absence of the N-hydrogen; removable N-protecting groups which have been used include phenylsulfonyl,[32] trimethylsilylethoxymethyl,[33] hydroxymethyl,[34] methylsulfonyl,[35] and pyrrolidin-1-ylmethyl.[36] The use of 1-benzyloxypyrazole gives 5-substituted-1-hydroxypyrazoles after subsequent hydrogenolytic removal of the benzyl group.[37] Dimethylamino-sulfonyl has been used frequently and the 5-lithiated derivative transformed into the zinc compound and this coupled using palladium(0) catalysis.[38] Isothiazole undergoes rapid exchange at C-5 with sodium deuteroxide in DMSO.[39]

Attempted C-deprotonation of isoxazoles with hydrogen at C-3 leads inevitably to ring opening, with the oxygen as anionic leaving group,[40] indeed this type of cleavage was first recognised as long ago as 1891, when Claisen found that 5-phenylisoxazole was cleaved by sodium ethoxide[41] (see section 22.11 for ring cleavage of isoxazolium salts). Comparable cleavages of isothiazoles can also be a problem.

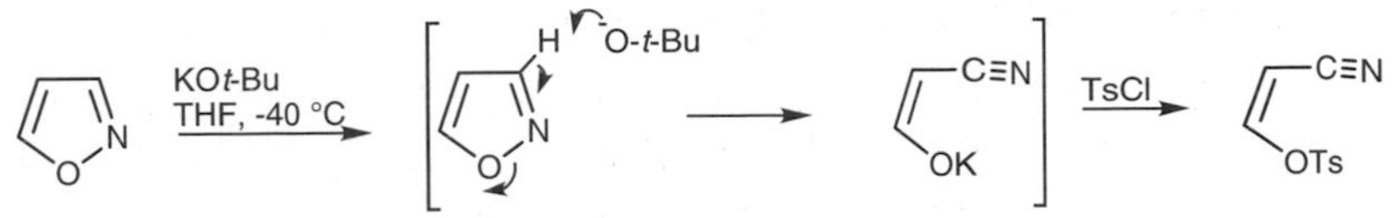

2-Methoxymethoxy-5-phenylisoxazole lithiates at C-4[42] as does 3-amino-5-methylisoxazole, protected as a *t*-butoxycarbonyl urethane, but using two equivalents of *n*-butyllithium.[43]

22.5 Reactions of *N*-metallated pyrazoles

N-Alkylations can be conducted in strongly basic,[44] or phase-transfer conditions[45] or in the presence of 4-dimethylaminopyridine,[46] and it seems likely that under these conditions it is the pyrazolyl anion (section 22.4.1) which is alkylated. The use of sodium hydrogen carbonate, without solvent, but with microwave heating is highly recommended.[47]

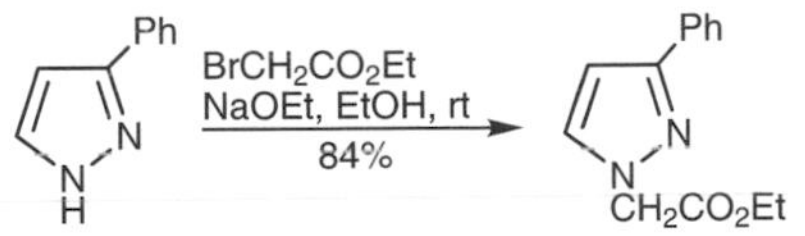

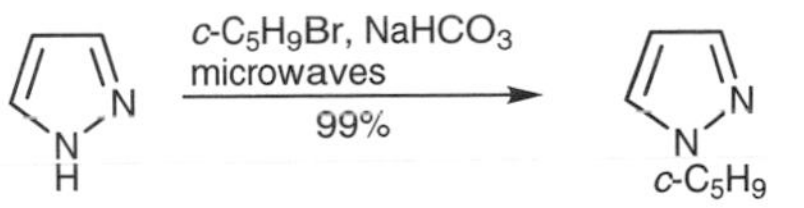

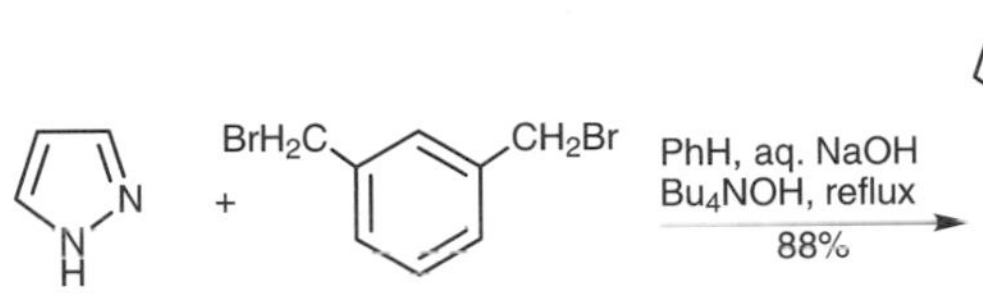

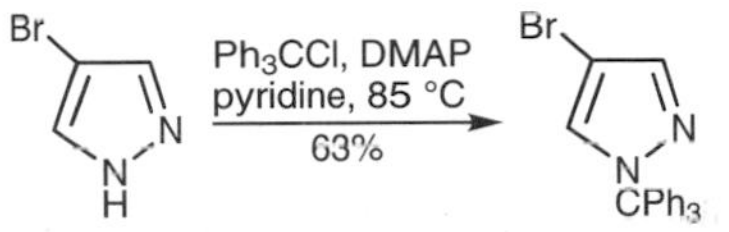

3(5)-Substituted pyrazoles may give a product isomeric with that which is obtained by reaction in neutral solution.[7]

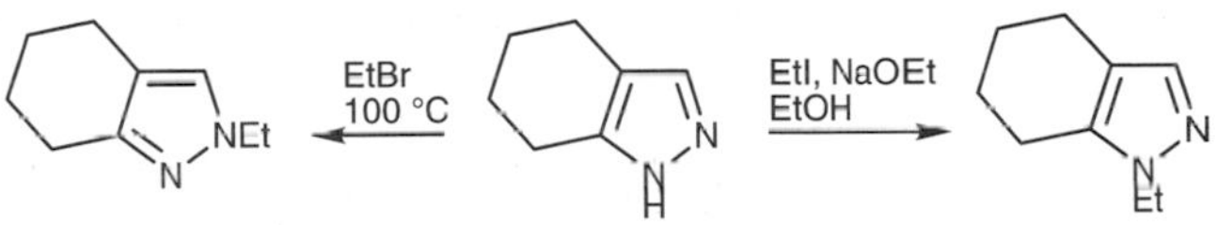

22.6 Reactions of *C*-metallated 1,2-azoles

The reactions of 5-lithiated isothiazoles and of 5-lithiated-1-substituted pyrazoles allow the introduction of substituents at that position by reaction with a range of electrophiles; two examples are shown below.[36,48,49]

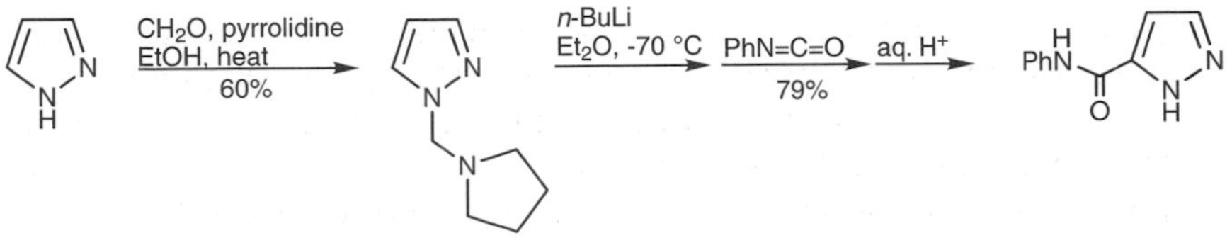

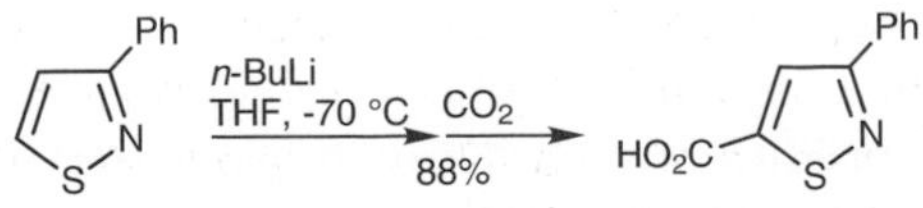

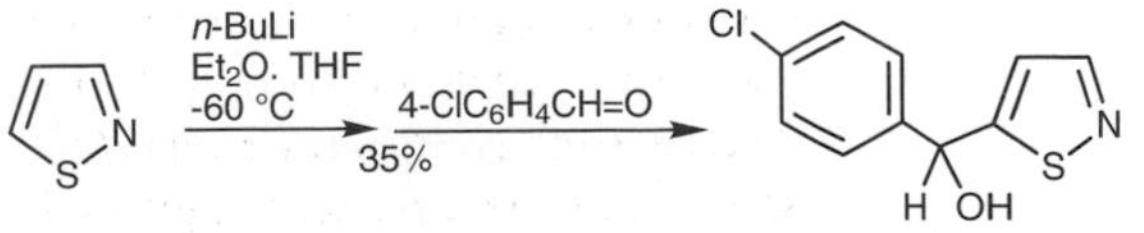

It is significant that treatment of 4-bromo-1-phenylsulfonylpyrazole with *n*-butyllithium results in 5-deprotonation and not metal halogen exchange,[32] however 4-bromo-1-triphenylmethylpyrazole undergoes normal exchange and in this way a tin derivative is obtained which undergoes routine palladium-catalysed couplings.[46]

Metal–halogen exchange has been achieved in the formation of 3-lithio-1-methylpyrazole from the bromopyrazole,[50] and reaction of 4-bromopyrazole with two equivalents of *n*-butyllithium produced a 1,4-dilithiopyrazole which reacts normally with electrophiles at C-4.[51] 4-Iodoisothiazole can be converted into a magnesium compound which shows normal nucleophilic Grignard properties.[52] An intramolecular acylation, involving the lithium salt of an acid, is observed when pyrazoles carrying a suitable length chain on nitrogen are lithiated with two mol equivalents of the strong base.[53]

22.7 Reactions with radicals

The interaction of 1,2-azoles with radical reagents is an area in which little is known so far. Displacement of tosyl from the 5-position of a protected pyrazole shows that there is potential for further development.[54]

22.8 Reactions with reducing agents

Pyrazoles are relatively stable to catalytic and chemical reductive conditions, particularly when there is no substituent on nitrogen, though catalytic reduction can be achieved in acid solution.[55] Isothiazoles are reductively desulfurised using Raney nickel, with loss of the ring.[56] Catalytic hydrogenolysis of the N–O bond in isoxazoles takes place readily over the usual noble metal catalysts,[57] and this process is central to the stratagem in which isoxazoles are employed as masked 1,3-dicarbonyl compounds. The immediate products of N–O hydrogenolysis, β-aminoenones, can often be isolated as such, or further processed. The use of this ring cleavage to provide routes to pyrimidinones,[58] and 3-keto-carboxamides, is illustrated below.[59]

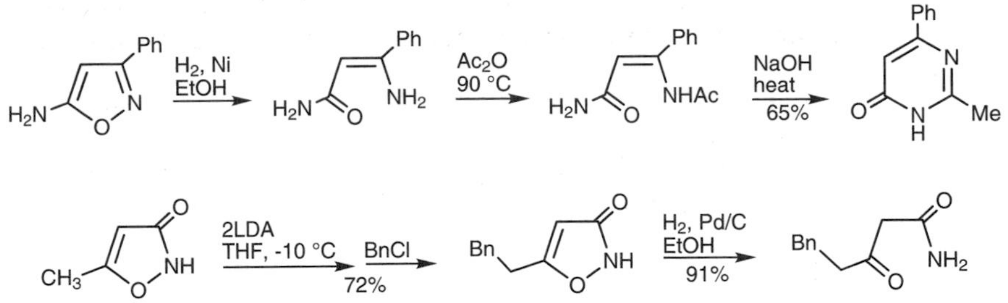

22.9 Electrocyclic reactions

There are examples of 1,2-azoles being converted into their 1,3-isomers by irradiation, though such processes are of limited preparative value. The conversion of cyanopyrazoles into cyanoimidazoles was studied using 3-cyano-5-deuterio-1-methylpyrazole, the resulting mixture of products requiring a duality of mechanism.[60]

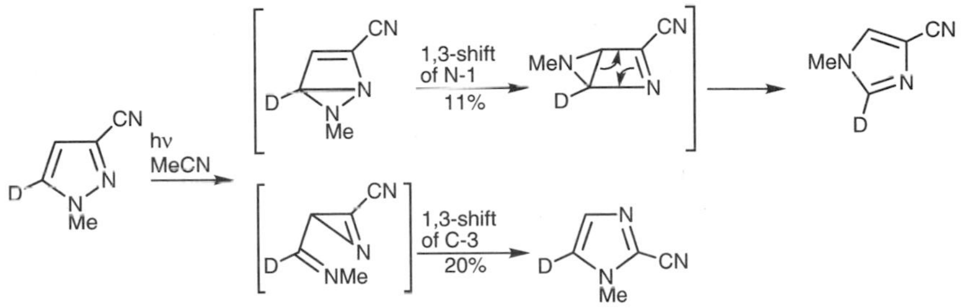

In a similar way, irradiation converts many simpler pyrazoles into imidazoles,[61] phenylisothiazoles[62] and methylisothiazoles[63] partially into the corresponding thiazoles, and 3,5-diarylisoxazoles converted into 2,5-disubstituted oxazoles.[64] 3-Alkoxyisoxazoles undergo an extraordinary ring contraction with iron(II) chloride, producing azirine esters.[65]

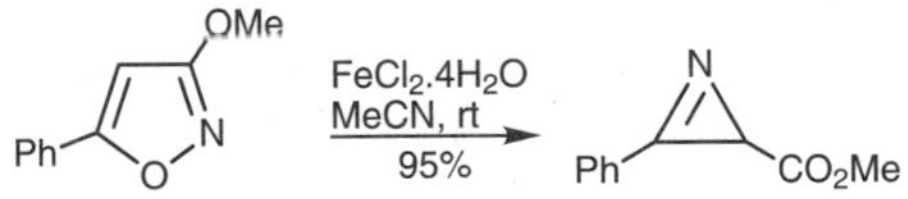

The transformation of 1,2-azoles carrying, at C-3, a side-chain of three atoms terminating in a doubly-bonded heteroatom, into isomeric systems with a new five-membered ring is a general process,[66] though there is no definitive view as to the details of its mechanism.

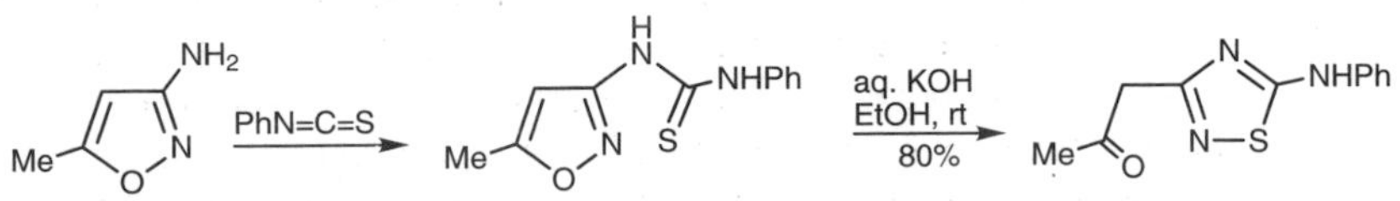

There do not appear to be any examples of 1,2-azoles acting as 1-azadienes in cycloadditions. 4-Nitroisoxazoles react with dienes across the 4,5-bond[67] and in processes useful for the synthesis of purine analogues, 3(5)-aminopyrazoles add to

electron-deficient 1,3,5-triazines, across the pyrazole 4,5-bond, subsequent eliminations giving the final aromatic product.[68]

22.10 Alkyl-1,2-azoles

4-Methylisothiazoles are not especially acidic, but it is rather surprising that 3-
methylisothiazoles are also not reactive whereas 5-methyl substituents will undergo
condensation reactions.[69] This same effect is also found in isoxazoles. In order to
study methyl group acidity in isoxazoles, the 3-position was blocked to prevent ring
degradation (section 22.4.2), thus 3,5-dimethylisoxazole was shown to exchange, with
methoxide in methanol, 280 times faster at the 5- than at the 3-methyl group.
Preparative deprotonations of this same isoxazole proceed exclusively at the 5-methyl
substituent, allowing subsequent reactions with electrophiles at that position. So
strong is this tendency, that reaction of 3,5-dimethylisoxazole with three equivalents
of base and three equivalents of iodomethane produces only 5-t-butyl-3-methylisox-
azole, no alkylation of the 3-methyl being observed, even in competition with the 5-
isopropyl group which is present in a penultimate intermediate.[70] By working at low
temperature, thus avoiding ring degradation, 5-methylisoxazole can be deprotonated
at the methyl, without the 3-deprotonation which would cause ring degradation.[71]
Conversion to N-oxide[72] activates adjacent methyl groups, for example subsequent
reaction with trimethylsilyl iodide permits side-chain iodination.[73]

On subjection of 3-methyl-5-phenylisothiazole or 3-methyl-5-phenylisoxazole to
lithiation conditions, competitive side-chain and C-4 deprotonation is observed
except when lithium i-propyl(cyclohexyl)amide (LICA) is used – this allows exclusive
side-chain lithiation.[74]

22.11 Quaternary 1,2-azolium salts

The base-catalysed degradation of the ring of isoxazolium salts is particularly easy,
requiring only alkali metal carboxylates to achieve it. The mechanism,[75] illustrated
for the acetate-initiated degradation of 2-methyl-5-phenylisoxazolium iodide,
involves initial 3-deprotonation with cleavage of the N–O bond; subsequent
rearrangements lead to an enol acetate which rearranges to a final keto-imide.

22.12 Oxy- and amino-1,2-azoles

Only 4-hydroxy-1,2-azoles can be regarded as being phenol-like.[76] 3- and 5-Hydroxy-1,2-azoles exist mainly in carbonyl tautomeric forms, encouraged by resonance involving donation from a ring heteroatom, and are therefore known as pyrazolones, isothiazolones, and isoxazolones, though for all three systems, and depending on the nature of other substituents, an appreciable percentage of hydroxy tautomer exists in solution.

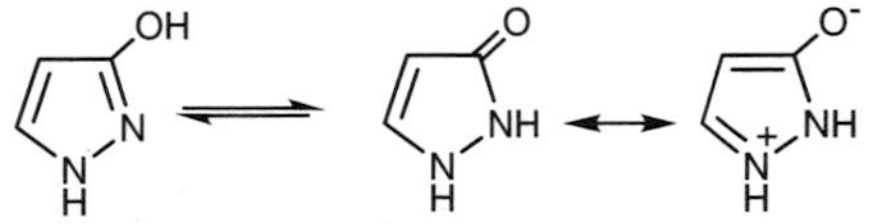

The reactivity of the 3- and 5-azolones centres mainly on their ability to react with electrophiles such as halogens,[77] (giving 4,4-dihalo-derivatives with excess reagent – 4,4-dibromo-3-methylpyrazol-5-one is a *para*-selective brominating agent for phenols and anilines[78]), or to nitrate,[79] or undergo Vilsmeier formylation;[80] the example shown below is the formylation of 'Antipyrine' once used as an analgesic. Many dyestuffs have been synthesised *via* coupling of aryldiazonium cations with 5-pyrazolones at C-4 – tartrazine is such an example.

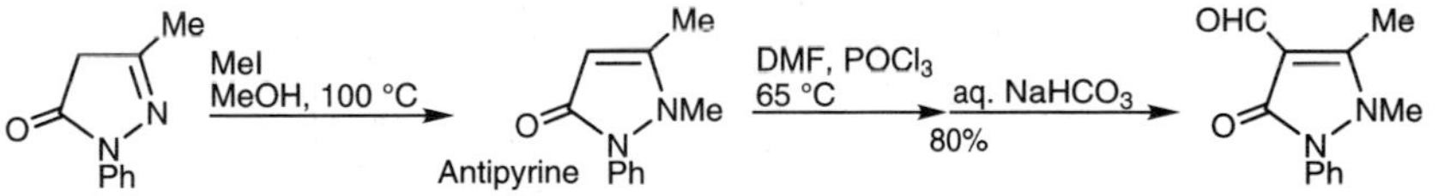

Pyrazolones also condense with aldehydes[81] in aldol-type processes, or react with other electrophiles such as carbon disulfide,[82] in each case reaction presumably proceeding *via* the enol tautomer, or its anion. In basic solution oxazol-3-ones alkylate either on oxygen or nitrogen and the choice of base can influence the ratio.[42]

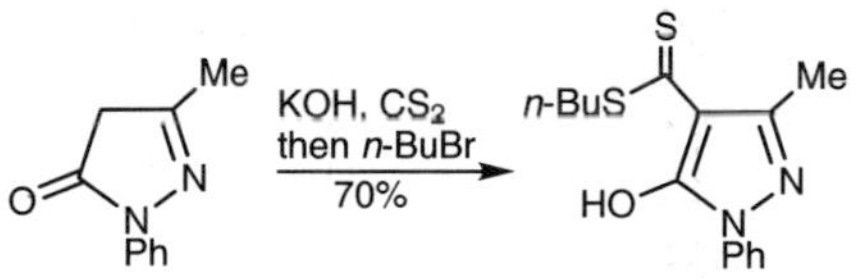

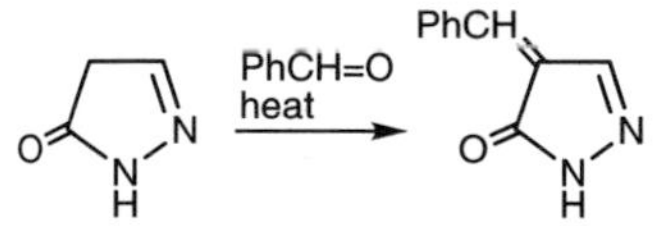

An intriguing and simple synthesis of a useful bromo-allene depends on the lead(IV) acetate oxidation of a bromopyrazolone, as shown.[83]

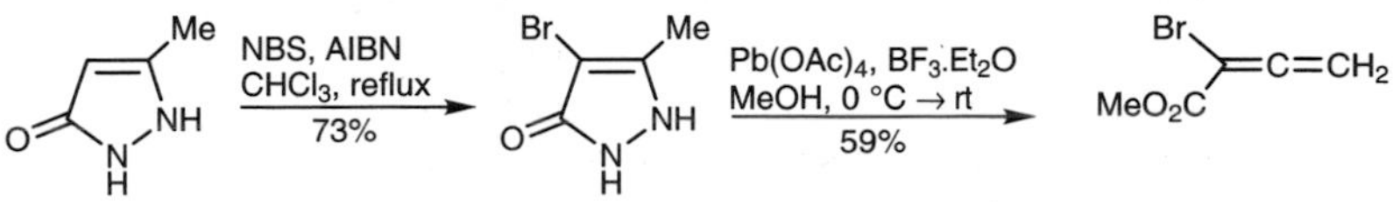

Amino-1,2-azoles exist as the amino tautomers. Aminopyrazoles and amino-isothiazoles are relatively well behaved aromatic amines, for example 3(5)-aminopyrazole undergoes substituent-*N*-acetylation and easy electrophilic bromination at C-4.[84] Diazotisation and a subsequent Sandmeyer reaction provides routes to halo-isothiazoles,[52] and azidopyrazoles.[85]

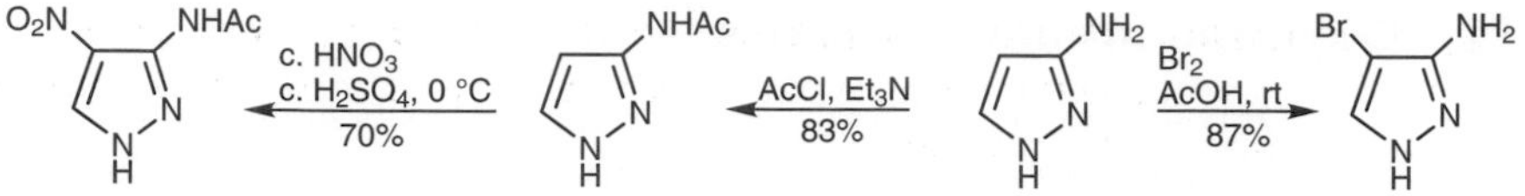

Diazotisation of 4-aminopyrazoles, then deprotonation yields stable diazopyrazoles.[86]

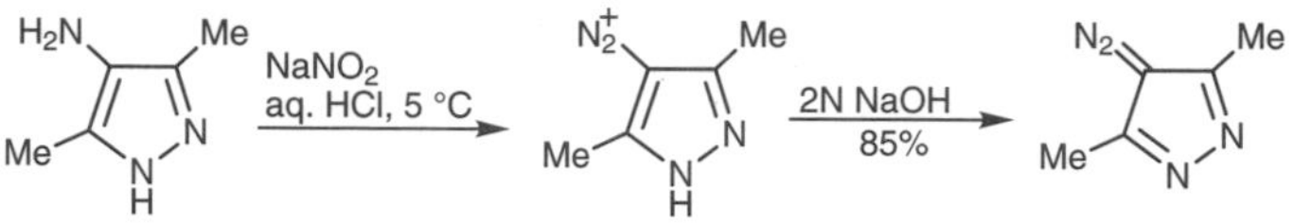

22.13 Synthesis of 1,2-azoles[87]

22.13.1 Ring synthesis

There are parallels, but also methods unique to particular 1,2-azoles, in the principal methods available for the construction of pyrazoles, isothiazoles and isoxazoles: neither the reaction of propene with sulfur dioxide and ammonia at 350 °C which gives isothiazole itself[88] in 65% yield, nor a synthesis[89] from propargyl aldehyde and thiosulfate (shown below) have direct counterparts for the other 1,2-azoles.

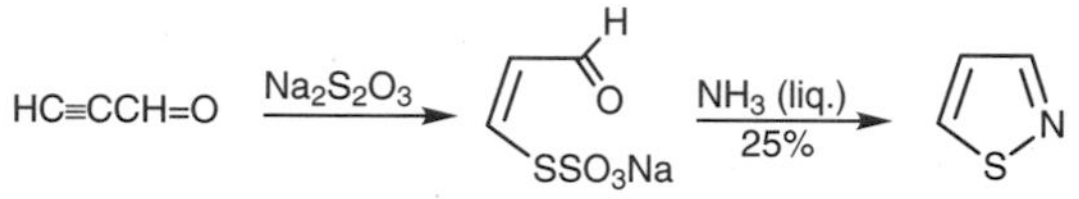

22.13.1.1 From 1,3-dicarbonyl compounds and hydrazines or hydroxylamine

Pyrazoles and isoxazoles can be made from a 1,3-dicarbonyl component and a hydrazine or hydroxylamine respectively.

This, the most widely used route to pyrazoles and isoxazoles rests on the doubly nucleophilic character of hydrazines and hydroxylamines, allowing them to react in turn with each carbonyl group of a 1,3-diketone[90] or 1,3-keto-aldehyde, often with one of the carbonyl groups (especially when aldehyde) masked as enol ether,[91] acetal, imine,[92] or enamine[93] or another synthon for one of these.

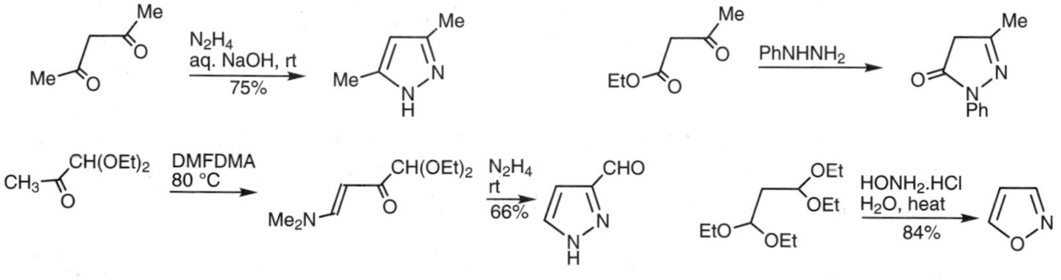

When β-keto-esters are used, the products are pyrazolones[94] or isoxazolones;[95] similarly, β-ketonitriles with hydrazines give 3(5)-aminopyrazoles.[96] 3(5)-Aminopyrazole itself is prepared *via* a dihydro-precursor formed by addition of hydrazine to acrylonitrile then cyclisation;[97] hydrolysis of the first cyclic intermediate in this sequence and dehydrogenation *via* elimination of *p*-toluenesulfinate allows preparation of 3(5)-pyrazolone.[98]

Generally speaking, unsymmetrical 1,3-dicarbonyl components produce mixtures of 1,2-azole products.[76] Sometimes this difficulty can be circumvented by the use of acetylenic-aldehydes or -ketones, for here a hydrazone or oxime can be formed first by reaction at the carbonyl group and this can then be cyclised in a separate, second step.[99] Pyrazole itself can be formed by the reaction of hydrazine with propargyl aldehyde.[8] Using β-chloro-,[100] β-alkoxy-[101] or β-amino-[102] -enones as 1,3-dicarbonyl synthons is another way to influence the regiochemistry of reaction, and in favourable situations this can be effective.[103]

When a β-aminoenethione, which can be produced from an isoxazole *via* hydrogenolysis then reaction of the β-aminoenone with a thionating agent, is treated with a dehydrogenating agent such as chloranil[104] or sulfur,[105] ring closure to an isothiazole results.

The ring closure of β-amino α,β-unsaturated thioamides comparably leads to 5-aminoisothiazoles.[106]

In another oxidative closure, the oximes of chalcones close to isoxazoles using tetrakis(pyridine)cobalt(II) bis(chromate),[107] and in an interesting variant, isoxazoles and pyrazoles are formed from 1,3-diynes.[108]

22.13.2.2 3-Acetylindole[125]

In view of the C-5-selectivity observed in lithiations, the selective 5-destannylation of a 4,5-di(tri-n-butylstannyl)isoxazole is useful. This nice sequence utilised a coupling to a 2-iodonitrobenzene; hydrogenation/hydrogenolysis caused ring cleavage of the isoxazole and produced an intermediate which cyclised with loss of water to give the indole.

Exercises for chapter 22

Straightforward revision exercises (consult chapters 20 and 22)

(a) Compare 1,2- with 1,3-azoles in pairs – which is the more basic? Why?

(b) What is incorrect about the name: '3-methylpyrazole'?

(c) Name some groups which can be used to mask the N-hydrogen in pyrazoles during C-lithiation.

(d) For what functionality are isoxazoles synthons if the N–O bond is cleaved? How could one cleave the N–O bond?

(e) How are 1,3-dicarbonyl compounds used for the synthesis of isoxazoles and pyrazoles?

(f) Describe a method involving an electrocyclic process for the ring synthesis of an isoxazole.

(g) Describe a method for the utilisation of the oxime of a dialkyl ketone, to make an isoxazole.

More advanced exercises

1. Suggest structures for the isomeric products, $C_9H_7N_3O_2$ formed when 1-phenylpyrazole is reacted with (i) c. H_2SO_4/c. HNO_3 or (ii) Ac_2O/HNO_3. Explain the formation of different products under the two conditions.

2. Draw structures for the products obtained by reacting 3,5-dimethylisoxazole with $NaNH_2$ then (i) n-PrBr; (ii) CO_2; or (iii) $PhCO_2Me$.

3. Deduce structures for the products obtained by treating 5-methylisoxazole with $SO_2Cl_2 \rightarrow C_4H_4ClNO$, and this with aqueous sodium hydroxide $\rightarrow C_4H_4ClNO$ (which contains no rings).

4. Draw the structures of the products which would be formed from the reaction of $BnNHNH_2$ with $MeCOCH_2COCO_2Me$.

5. Deduce structures for the products formed in the following sequence: pyrazole/$Me_2NSO_2Cl/Et_3N \rightarrow C_5H_9N_3O_2S$ then this with n-BuLi/$-70\,°C$ then TMSCl $\rightarrow C_8H_{17}N_3O_2SSi$ then this with $PhCH = O/CsF \rightarrow C_{12}H_{15}N_3SO_3$.

6. Draw the structures of the two products which are formed when hydroxylamine reacts with $PhCOCH_2CH=O$; suggest an unambiguous route for the preparation of 5-phenylisoxazole.

7. Deduce the structures of the heterocyclic substances produced: (i) C_7H_9NO, from cyclohexanone oxime with 2 mol equivalents of *n*-BuLi then dimethyl formamide; (ii) $C_{11}H_{15}NOSSi$ from thien-2-ylC($=NOH$)CH_2Br and $Me_3SiC\equiv$-CLi then K_2CO_3/MeOH; (iii) $C_{11}H_{12}N_2$ from MeCOC($=NNHPh$)Me with $(EtO)_2POCH_2SEt$/*n*-BuLi.

8. Suggest a structure for the heterocyclic product, $C_7H_{13}NOSi$, formed by reaction of $Me_3SiC\equiv CC\equiv CSiMe_3$ and hydroxylamine.

References

1. 'Progress in pyrazole chemistry', Kost, A. N. and Grandberg, I. I., *Adv. Heterocycl. Chem.*, **1966**, *6*, 347; 'The Azoles', Schofield, K., Grimmett, M. R., and Keene, B. R. T., Cambridge University Press, **1976**.

2. (a) 'Isothiazoles', Hübenett, F., Flock, F. H., Hansel, W., Heinze, H., and Hofmann, H., *Angew. Chem., Int. Ed. Engl.*, **1963**, *2*, 714; (b) 'Isothiazoles', Slack, R. and Wooldridge, K. R. H., *Adv. Heterocycl. Chem.*, **1965**, *4*, 107; 'Recent advances in the chemistry of mononuclear isothiazoles', Wooldridge, K. R. H., *ibid.*, **1972**, *14*, 1.

3. 'Recent developments in isoxazole chemistry', Kochetkov, N. K. and Sokolov, S. D., *Adv. Heterocycl. Chem.*, **1963**, *2*, 365; 'Isoxazole chemistry since 1963', Wakefield, B. J. and Wright, D. J., *ibid.*, **1979**, *25*, 147; 'Synthetic reactions using isoxazole compounds', Kashima, C., *Heterocycles*, **1979**, *12*, 1343.

4. Begtrup, M. and Vedso, P., *J. Chem. Soc., Perkin Trans. 1*, **1995**, 243.

5. Reuther W., and Baus, V., *Liebig's Ann.*, **1995**, 1563.

6. Chaplen, P., Slack, R., and Wooldridge, K. R. H., *J. Chem. Soc.*, **1965**, 4577.

7. v. Auwers, K., Buschmann, W., and Heidenreich, R, *Justus Liebigs Ann. Chem.*, **1924**, *435*, 277.

8. Pérez, E., Sotelo, E., Loupy, A., Mocelo, R., Suarez, M., Pérez, R., and Autić, M., *Heterocycles*, **1996**, *43*, 539.

9. Hüttel, R. and Kratzer, J., *Chem. Ber.*, **1959**, *92*, 2014; Williams, J. K., *J. Org. Chem.*, **1964**, *29*, 1377.

10. Finar, I. L. and Lord, G. H., *J. Chem. Soc.*, **1957**, 3314.

11. Bernatowicz, M. S., Wu, Y., and Matsueda, G. R., *J. Org. Chem.*, **1992**, *57*, 2497.

12. Drake, B, Patek, M., and Lebl, M., *Synthesis*, **1994**, 579.

13. Hüttel, R., Büchele, F., and Jochum, P., *Chem. Ber.*, **1955**, *88*, 1577.

14. Caton, M. P. L., Jones, D. H., Slack, R., and Woolridge, K. R. H., *J. Chem. Soc.*, **1964**, 446.

15. Quilico, A. and Musante, C., *Gazz. Chim. Ital.*, **1941**, *71*, 327.

16. Suzuki, H. and Nonoyama, N., *J. Chem. Res. (S)*, **1996**, 244.

17. Olah, G. A., Narang, S. C., and Fung, A. P., *J. Org. Chem.*, **1981**, *46*, 2706.

18. Woodward, R. B., Olofson, R., and Mayer, H., *J. Am. Chem. Soc.*, **1961**, *83*, 1010.

19. Pain, D. L. and Parnell, E. W., *J. Chem. Soc.*, **1965**, 7283.

20. Knorr, L., *Justus Liebigs Ann. Chem.*, **1894**, *279*, 188.

21. Hüttel, R., Schäfer, O., and Jochum, P., *Justus Liebigs Ann. Chem.*, **1955**, *593*, 200.

22. Lipp., M., Dallacker, F., and Munnes, S., *Justus Liebigs Ann. Chem.*, **1958**, *618*, 11

23. Finley, J. H. and Volpp, G. P., *J. Heterocycl. Chem.*, **1969**, *6*, 841.

24. Pino, P., Piacenti, F., and Fatti, G., *Gazz. Chim. Ital.*, **1960**, *90*, 356.

25. Blount, J. F., Coffen, D. L., and Katonak, D. A., *J. Org. Chem.*, **1978**, *43*, 3821.

26. Juffermans, J. P. H., and Habraken, C. L., *J. Org. Chem.*, **1986**, *51*, 4656.

27. Tojahn, C. A., *Chem. Ber.*, **1922**, *55*, 291.

28. Benary, E., *Chem. Ber.*, **1926**, *59*, 2198; Holland, A., Slack, R., Warren, T. F., and Buttimore, *J. Chem. Soc.*, **1965**, 7277; Quilico, A., and Stagno d'Alcontres, G., *Gazz. Chim. Ital.*, **1949**, *79*, 654.

29. Kashima, C., Takahashi, K., and Hosomi, A., *Heterocycles*, **1994**, *37*, 1075.

114. Iddon, B., Suschitzky, H., Thompson, A. W., Wakefield, B. J., and Wright, D. J., *J. Chem. Res.* , **1978**, *(S)* 174; *(M)* 2038.
115. Sasaki, T. and Yoshioka, T., *Bull. Chem. Soc. Jpn.*, **1968**, *41*, 2212; Christl, M., Huisgen, R., and Sustmann, R., *Chem. Ber.*, **1973**, *106*, 3275.
116. Hanson, R. N. and Mohamed, F. A., *J. Heterocycl. Chem.*, **1997**, *34*, 345.
117. Sakamaoto, T., Shiga, F., Uchiyama, D., Kondo, Y., and Yamanaka, H., *Heterocycles,* **1992**, *33*, 813.
118. Barber, G. N. and Olofson, R. A., *J. Org. Chem.*, **1978**, *43*, 3015.
119. He, Y. and Liu, N.-H., *Synthesis*, **1994**, 989.
120. Nitz, T. J., Volkots, D. L., Aldous, D. J., and Oglesby, R. C., *J. Org. Chem.,* **1994**, *59*, 5828.
121. Short, K. M. and Ziegler, C. B., *Tetrahedron Lett.*, **1993**, *34*, 75.
122. Church, A. C., Koller, M. U., Hines, M. A., and Beam, C. F., *Synth. Commun.*, **1996**, *26*, 3659.
123. Almirante, N., Cerri, A., Fedrizzi, G., Marazzi, G., and Santagostino, M., *Tetraheron Lett.*, **1998**, *39*, 3287; Almirante, N., Benicchio, A., Cerri, A., Fedrizzi, G., Marazzi, G., and Santagostini, *Synlett*, **1999**, 299.
124. Perez, C., Janin, Y. L., and Grierson, D. S., *Tetrahedron*, **1996**, *52*, 987.
125. Uchiyama, D., Yabe, M., Kameyama, H., Sakamoto, T., Kondo, Y., and Yamanaka, H., *Heterocycles*, **1996**, *43*, 1301.

23 Benzanellated azoles: reactions and synthesis

1H-benzimidazole   benzothiazole   benzoxazole   1H-indazole   1,2-benzisothiazole   1,2-benzisoxazole

2,1-benzisothiazole   (anthranil) 2,1-benzisoxazole

There is only one way in which a benzene ring can be fused to each of the three 1,3-azoles, generating 1H-benzimidazole,[1] benzothiazole, and benzoxazole. Indazole[2] is the only possibility for the analogous fusion to a pyrazole; it exists as a 1-H tautomer – the 2-H-tautomer cannot be detected, though 2-substituted 2H-indazoles are known. Two distinct isomers each are possible for the other two 1,2-azoles: 1,2-benzoisothiazole and 2,1-benzoisothiazole,[3] and 1,2-benzisoxazole and 2,1-benzisoxazole,[4] respectively.

1,2-Benzisothiazolin-3(2H)-one 1,1-dioxide is saccharin the well known sweetening agent. Omeprazole, a gastric proton-pump inhibitor, is an anti-ulcerative, Risperidone is used in the treatment of schizophrenia, and Granisetron, a serotonin receptor antagonist, alleviates the nausea associated with chemotherapy.

saccharin   Granisetron   Omeprazole

Risperidone

23.1 Reactions with electrophilic reagents

23.1.1 Addition at nitrogen

23.1.1.1 Protonation

Benzimidazole is nearly two pK_a units weaker as a base and but somewhat stronger as an acid than imidazole. These trends are echoed in the other benzo-azoles: the bicyclic systems are weaker bases than the corresponding monocyclic heterocycles and indazole is a slightly weaker acid than pyrazole.

1,2-Benzisothiazole and 1,2-benzisoxazole have not been lithiated in the heterocyclic ring, no doubt because attempts to do so would lead to fragmentation of the heterocyclic ring in a way analogous to that observed when indazole-3-acids are heated in quinoline.[23] However, 3-bromoindazole can be converted into an N,C-dithio species – this takes advantage of the fact that following deprotonation of the N-hydrogen, N-1 is no longer a leaving group.[24]

23.4.2 Palladium-catalysed reactions

The comparitively small number of examples so far available suggest that coupling chemistry can be just as important in the benzoazoles as in other areas of heterocyclic chemistry: the two examples below show the use of a zinc benzoxazole[25] and an indazolyl triflate.[26]

23.5 Reactions with reducing agents

The selective hetero ring reduction of the benzo-1,3-azoles or the benzo-1,2-azoles has not been reported.

23.6 Electrocyclic reactions

2,1-Benzisothiazole and 2,1-benzisoxazole seem not to display the tendency to act as aza-dienes which might have been expected on the basis of comparison with the typical reactivity of isoindoles, benzo[c]thiophenes and isobenzofurans (cf. section

19.2). In a different sense, electron-deficient 2-alkenylbenzothiazoles react with electron-rich alkenes as 1-aza-1,3-dienes.[27]

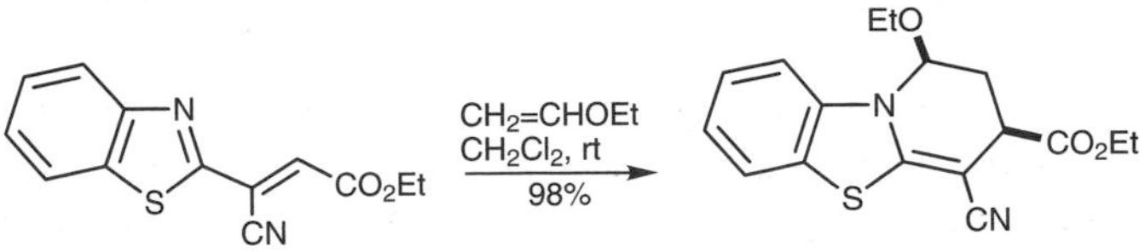

23.7 Quaternary salts

Benzo-1,3-azolium salts are susceptible to nucleophilic addition at C-2, for example, they are converted into the corresponding *ortho* substituted benzene, with loss of C-2, by aqueous base, a process which must involve addition of hydroxide at C-2 as an initiating step.[28] In a more constructive sense, cleavage of 2-(1-hydroxyalkyl)benzothiazolium salts, which can be assembled using 2-lithiobenzothiazole, to liberate the benzothiazolium ylide (*cf.* section 21.11) allows a synthesis of ketones, as illustrated.[29]

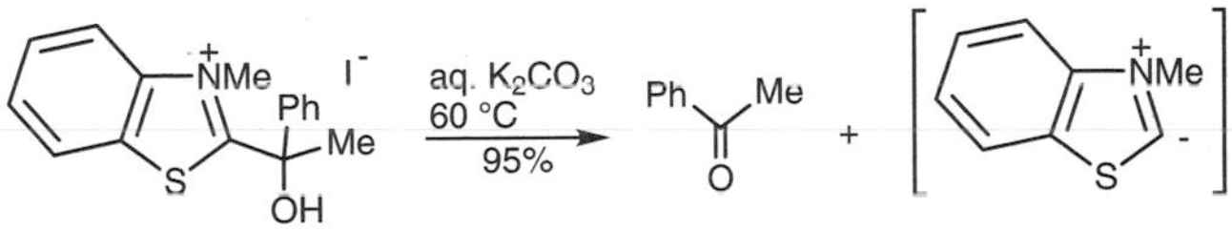

By an alternative sequence: 2-lithiation, reaction with an aldehyde, quaternisation, C-2-*addition* of an alkyllithium, and finally silver-promoted ring cleavage of the resulting dihydrobenzothiazole, the heterocycle can be made the means for the construction of α-hydroxyketones.[30] Lithium enolates also add smoothly to benzothiazolium salts.[31]

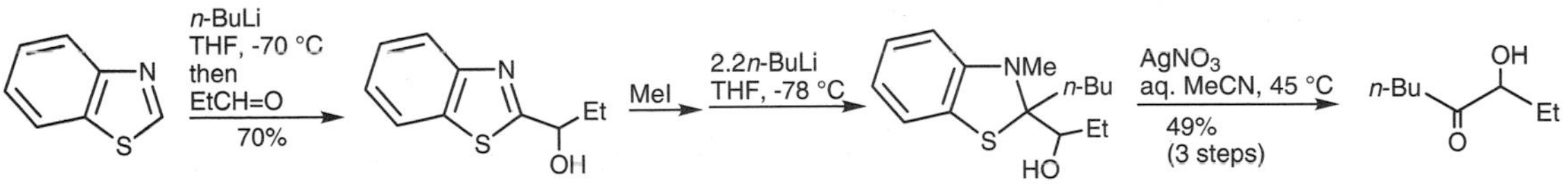

Reissert-type adducts (*cf.* section 6.14) can be obtained from benzothiazole, benzoxazole and indazole, as illustrated below.[32]

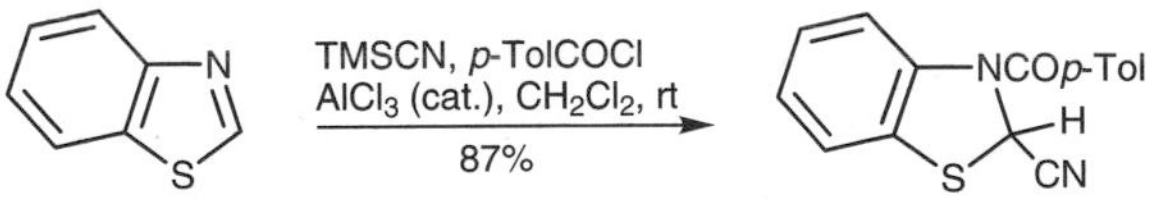

2,1-Benzisothiazolium salts are hydrolysed to *ortho*-aminobenzaldehydes; their use as synthons for such aldehydes is illustrated by the quinolone synthesis below.[33]

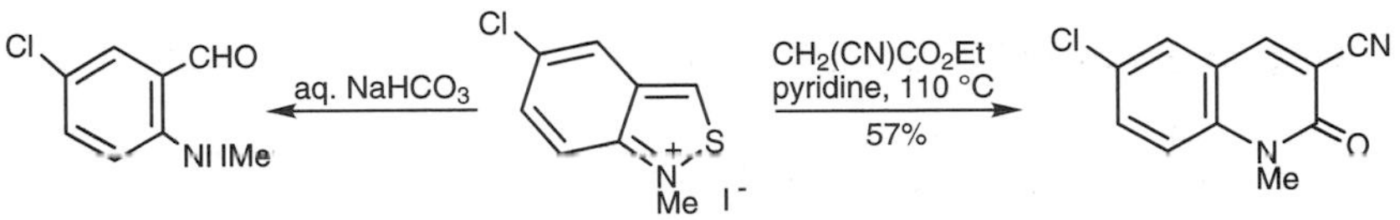

23.8 Oxy- and amino-1,3-azoles

The benzo-1,3-azol-2-ones exist in the carbonyl tautomeric forms. Indazol-3-one however, at least in dimethylsulfoxide solution, is largely in the hydroxy tautomeric form[34] in contrast to 1,2-benzisothiazol-3-one which is wholly in the carbonyl form, at least in the solid.[35]

It is possible to be quite selective in the introduction of alkyl groups onto one of the two nitrogens of 3-hydroxyindazole. Reaction with a halide in neutral solution, no doubt involving the imine tautomer, is selective for N-2.[36] In contrast, in basic solution, the anion reacts at N-1.[26]

Aqueous base brings about hydrolytic cleavage of the heterocyclic ring of benzo-1,3-azol-2-ones giving the corresponding *ortho* substituted benzene.[37] Alkylation in basic medium generally leads to *N*- and *O*-substitution; thiones alkylate on the thione sulfur.[38]

23.9   Synthesis

23.9.1   Ring synthesis of benzo-1,3-azoles

23.9.1.1   *From ortho heteroatom-substituted arenes*

The most important strategy for the synthesis of benzothiazoles, benzimidazoles, and benzoxazoles is the insertion of C-2 into a precursor with *ortho* heteroatoms on a benzene ring. The component which is required for this purpose usually has the future C-2 at the oxidation level of an acid, but many variants on this have been described.

In the standard form of this route, a carboxylic acid is heated with the *ortho* disubstituted benzene. An iminoether will react at much lower temperature.[39] Ortho esters with a KSF clay is a highly recommended variant and can be used for the synthesis of all three unsubstituted benzo-1,3-azoles.[40] An important variant for the synthesis of benzimidazoles, allowing the use of aromatic or aliphatic aldehydes, rather than acids, incorporates nitrobenzene as an oxidant into the reaction mixture.[41]

There are two efficient ways in which to use starting materials which have the carboxylic acid component already installed on *both* heteroatoms: conversion to bis(silyloxy) derivatives,[42] or simply heating with *p*-toluenesulfonic acid as shown.[43] A device which has been frequently used to produce a starting material with just one acyl group in place is to carry out a Beckmann rearrangement on an *ortho*-hydroxyaryl ketone, the Beckmann product cyclising *in situ* when the conditions are acidic; a modern version of this is illustrated below.[44] An excellent route to mono-acylated precursors utilises mixed anhydrides.[37] A very mild method for the dehydrative ring closure of *ortho*-hydroxyarylamino amides, employed in solid-supported benzoxazole syntheses, utilises typical Mitsunobu conditions – triphenyl-phosphine and diethyl azodicarboxylate.[45]

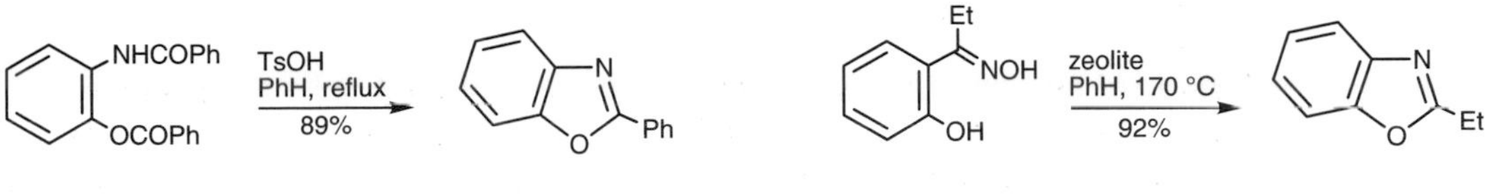

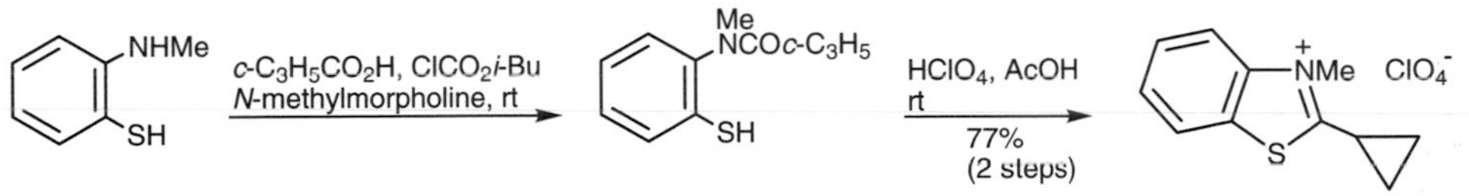

The use of microwaves allows reaction with amides in lieu of acids; when urea or thiourea are used, 2-ones (2-thiones) are obtained, carbon disulfide and potassium hydroxide also leads to 2-thiones, and with isocyanates (or isothiocyanates[46]), 2-acylamino derivatives result.[47] Reaction with cyanogen bromide gives 2-aminoben-zimidazoles.[1b]

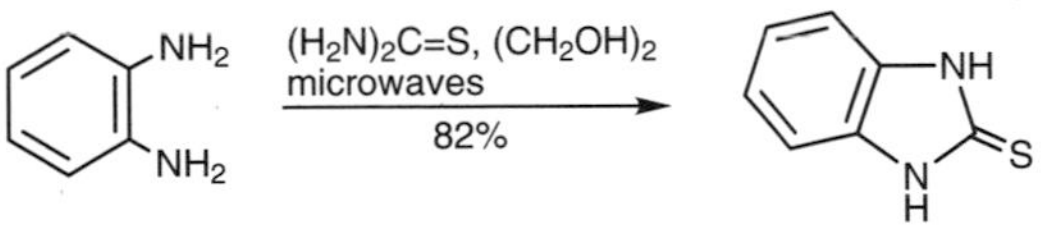

Bis(methylthio)imines (dithiocarbonimidates), readily available from the reaction of an arylamine with carbon disulfide and base and then methyl iodide, react to form heterocyclic rings with a 2-arylamino substitutent.[48]

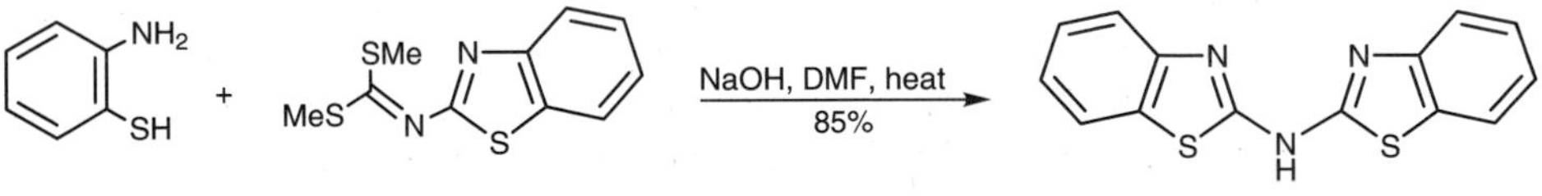

23.9.1.2   *Other methods*

Chief amongst alternative methods, and of importance in that it does not require an *ortho* diheteroatom starting material, is the oxidative ring closure of arylamine thioamides giving benzothiazoles. Typically, potassium ferricyanide or bromine are utilised: examples are shown below.[49]

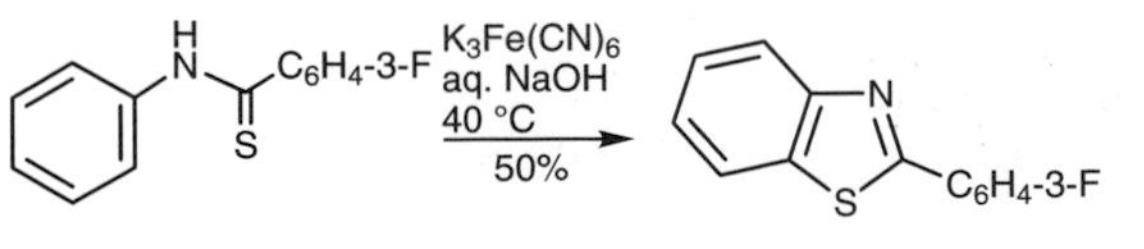

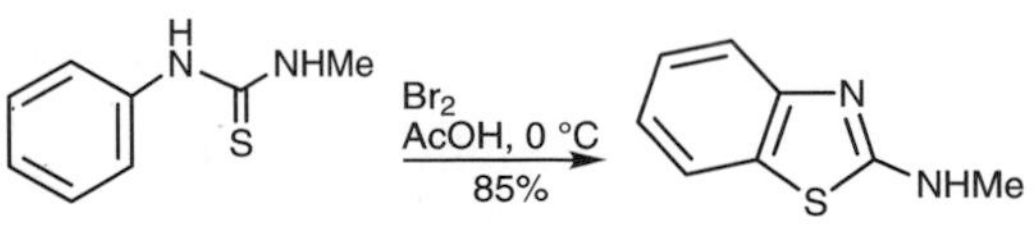

An oxidative ring closure giving benzimidazoles results when *N*-arylamidines are reacted with iodobenzene diacetate; a possible intermediate is shown in the scheme.[50]

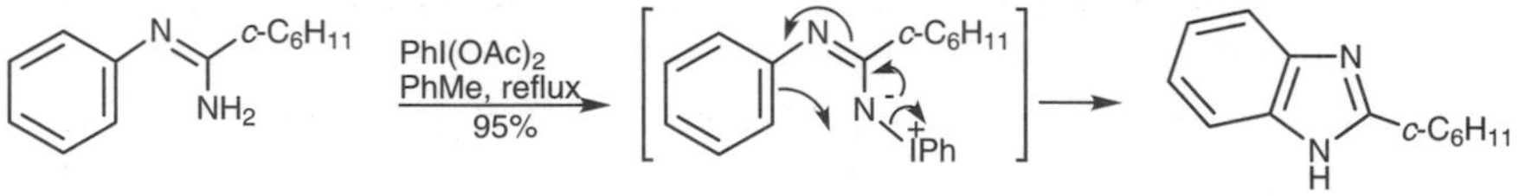

Benzothiazoles can also be produced from thioanilides with a *meta* fluorine by a sequence of *ortho*-assisted lithiation, leading to elimination of fluoride and the formation of an aryne, and then intramolecular addition of the sulfur to generate the heterocycle, and finally trapping with an electrophile placing a substituent at C-7.[51]

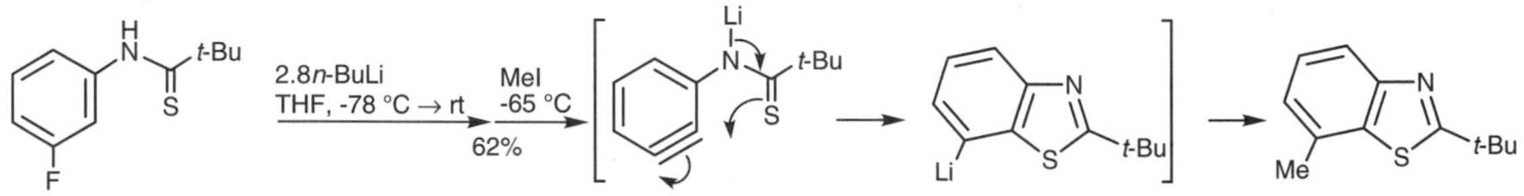

In what is essentially a Friedländer quinoline synthesis (*cf.* section 6.16.1.3), creatinine condenses with an *ortho*-aminoaraldehydes giving polycyclic 2-aminobenzimidazoles, or heteroaryl analogues as illustrated.[52]

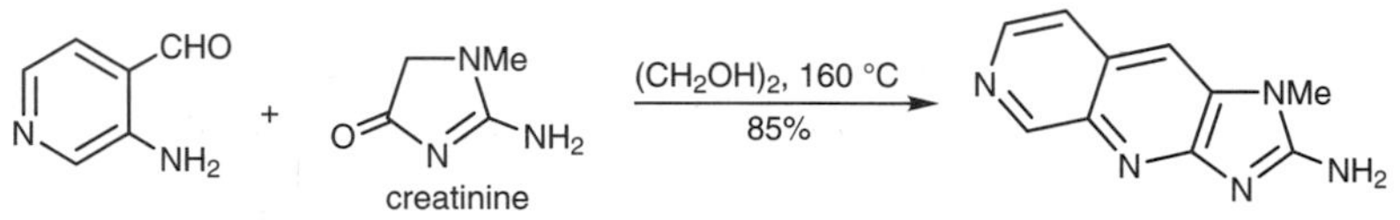

23.9.2   Ring synthesis of benzo-1,2-azoles

23.9.2.1   *Ring synthesis of 1H-indazoles, 1,2-benzisothiazoles, and 1,2-benzisoxazoles*

The earliest syntheses of 1,2-benzisoxazoles depended on the cyclisation of an *ortho*-haloarylketone oxime, typical conditions are shown below.[53] Only one geometrical isomer of the oxime will ring close. Applying this approach to amidoximes is easier, because the two imine geometrical isomers in such compounds are easily interconvertible.[54] Comparable reaction with hydrazones produces indazoles.[55]

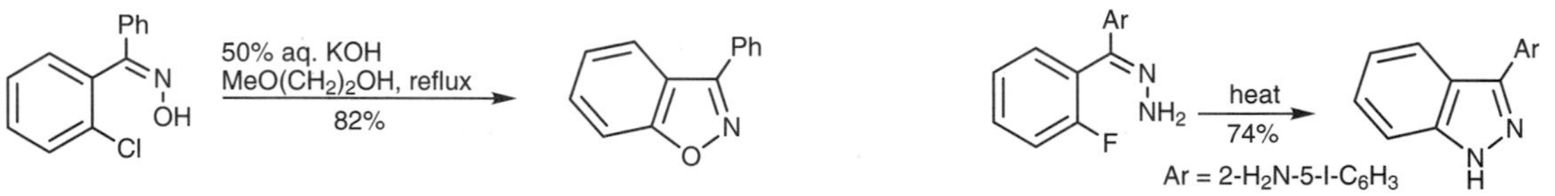

A number of routes involve formation of the bond between the two hetero atoms: typical is the conversion of di(2-cyanophenyl) disulfide into 3-chlorobenzoisothiazole with chlorine[56] and into 3-aminobenzoisothiazoles with magnesium amides, as shown, one 'half' of the starting material is converted directly into the heterocycle, the second 'half' requiring oxidation.[57]

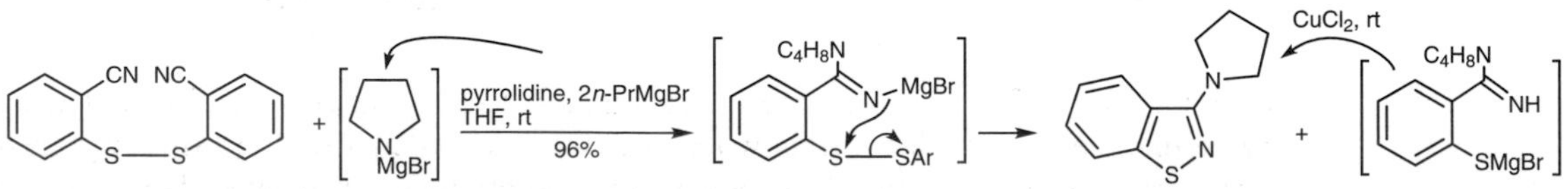

Generally speaking, one of the hetero atoms must carry a leaving group – the hydroxyl of oximes has served this purpose either via protonation[58] or acetylation.[59]

The generation of a chlorosulfide serves the purpose in the opposite sense.[60] Probably the best method for the cyclisation of salicylaldehyde oximes (or *ortho*-hydroxyarylketoximes) is the application of Mitsunobu-type conditions.[61]

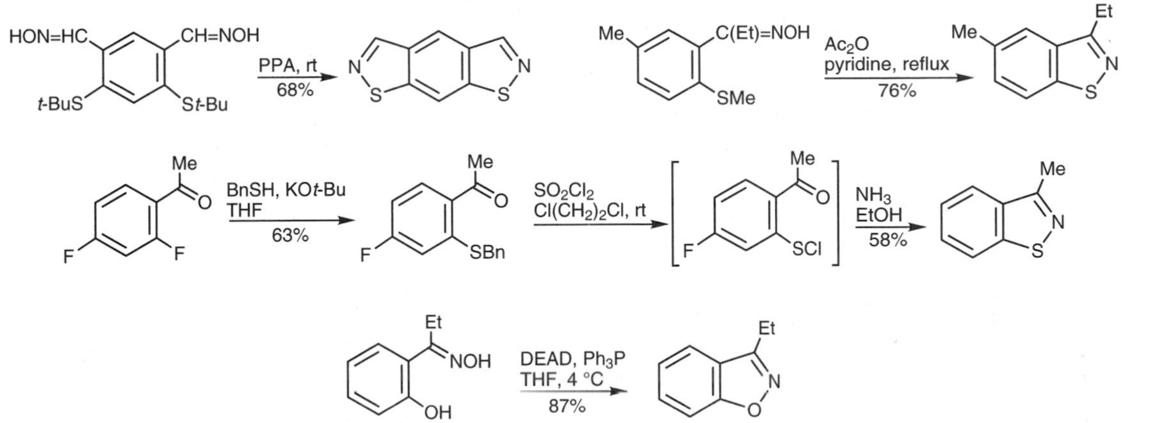

HON=HC / CH=NOH, *t*-BuS / S*t*-Bu → PPA, rt 68% → [benzo-bis-isothiazole]

Me / C(Et)=NOH / SMe → Ac₂O pyridine, reflux 76% → [Me, Et benzisothiazole]

Me / O / F / F → BnSH, KO*t*-Bu THF 63% → Me / O / F / SBn → SO₂Cl₂ Cl(CH₂)₂Cl, rt → [Me / O / F / SCl] → NH₃ EtOH 58% → [Me benzisothiazole]

Et / NOH / OH → DEAD, Ph₃P THF, 4 °C 87% → [Et benzisoxazole]

Electrophilic cyclisation onto the aromatic ring achieves the synthesis of 3-aminoindazoles when arylhydrazines are reacted with *N*-(dichloromethylene)-*N*,*N*-dimethylammonium chloride, as shown.[62] The formation of 1-benzyl-3-hydroxyindazole by heating *N*-benzylphenylhydrazine with urea at 285 °C involves the electrophilic cyclisation of a first formed semicarbazide.[2]

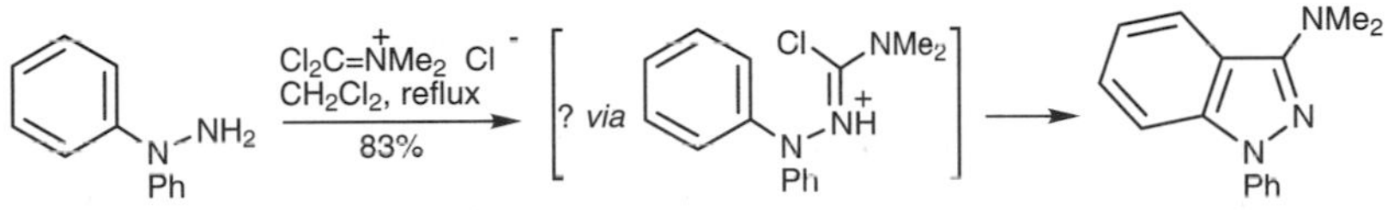

[Ph, N-NH₂] → Cl₂C=NMe₂ Cl⁻ CH₂Cl₂, reflux 83% → ? via [Cl, NMe₂ / N-NH / Ph] → [NMe₂ indazole, N-N, Ph]

A classical,[63] but still used,[5,64] route to indazoles involves *N*-nitrosation of an acetanilide followed by cyclisation onto an *ortho* alkyl group – even unactivated methyl groups enter into reaction, though in the example shown below the *ortho* substituent carries an activating ester. The sequence probably involves a diazoacetate as intermediate, as shown.

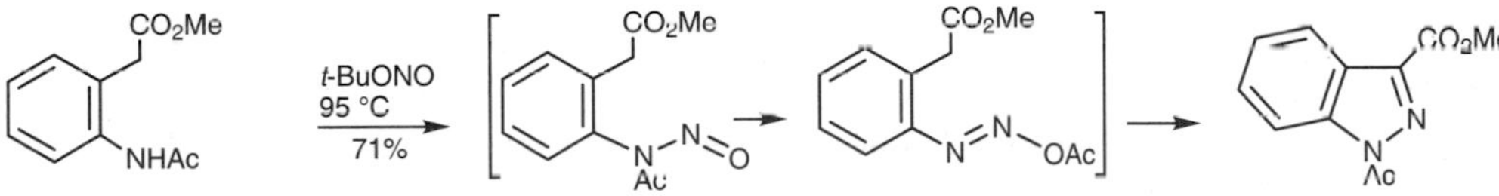

[CO₂Me / NHAc] → *t*-BuONO 95 °C 71% → [CO₂Me / N-N=O / Ac] → [CO₂Me / N=N-OAc] → [CO₂Me indazole / N-N / Ac]

Diazotisation of anthranilic acid and immediate reduction of the diazo group is a very simple route to 3-hydroxyindazole[2] and hydrolysis, diazotisation and reduction starting from isatin produces indazole-3-carboxylic acid.[65]

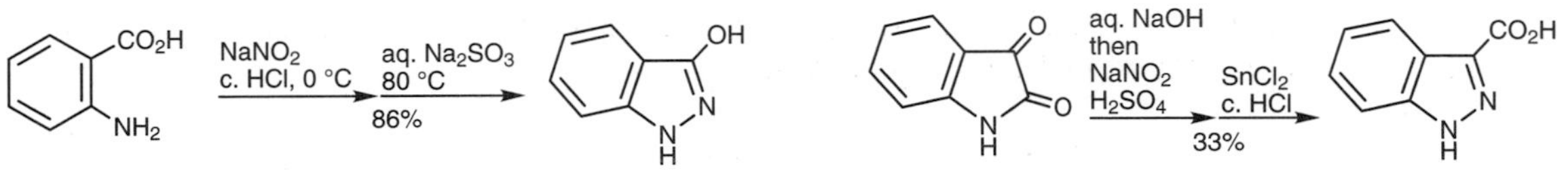

[CO₂H / NH₂] → NaNO₂ c. HCl, 0 °C → aq. Na₂SO₃ 80 °C 86% → [OH indazole, N-N, H]

[isatin O / N, H / O] → aq. NaOH then NaNO₂ H₂SO₄ → SnCl₂ c. HCl 33% → [CO₂H indazole, N-N, H]

A neat use of lithiation methods allows the synthesis of 7-substituted benzisothiazole *S*,*S*-dioxides, thus when two mol equivalents of the lithiating agent are used, two successive lithiations *ortho* to the sulfonamide unit take place, the first leading to ring closure and the second allowing introduction of an added electrophile at C-7.[66]

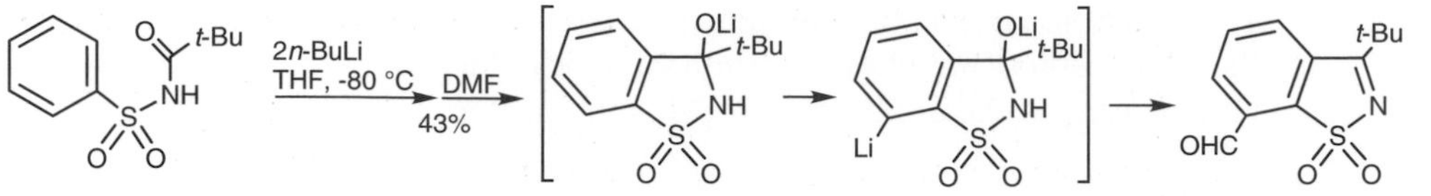

23.9.2.2 Ring synthesis of 2H-indazoles, 2,1-benzisothiazoles, and 2,1-benzisoxazoles

Appropriate oxidations of *ortho*-aminoaryl ketones[67] or esters[68] produce 2,1-benzisoxazoles; these ring closures may involve the intermediate formation of a nitrene, or a conventional sequence like that shown for the formation of the 3-styryl derivative below.

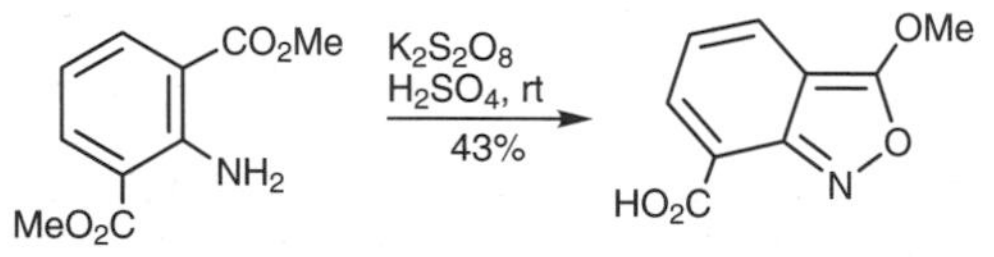

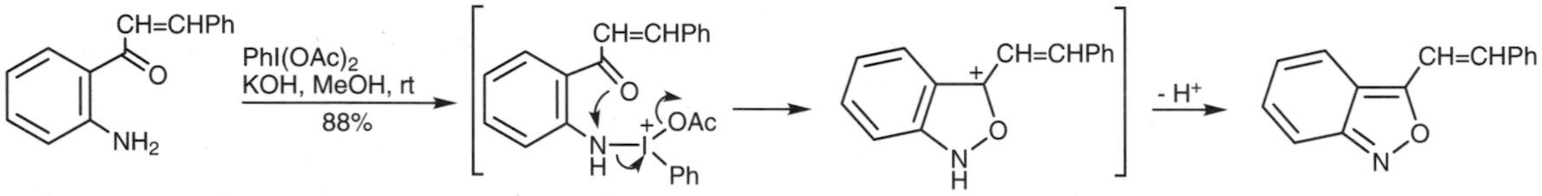

Involvement of a nitrene seems more likely in sequences where an *ortho*-azidoaryl ketone is decomposed thermally producing 3-substituted 2,1-benzisoxazoles,[69] though an intermediate in which the azide has cycloadded in a 1,3-dipolar sense to the carbonyl group has been suggested.[70]

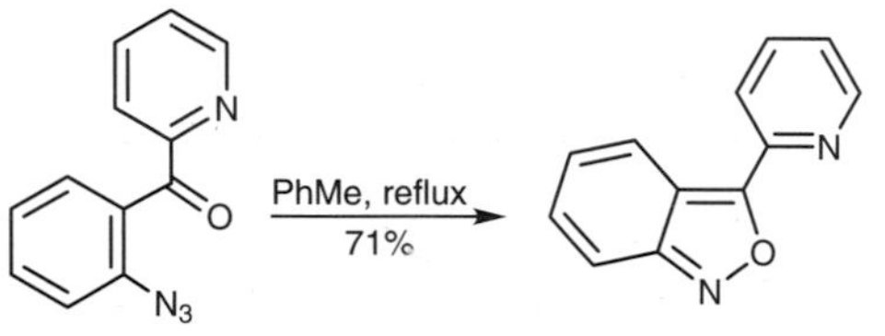

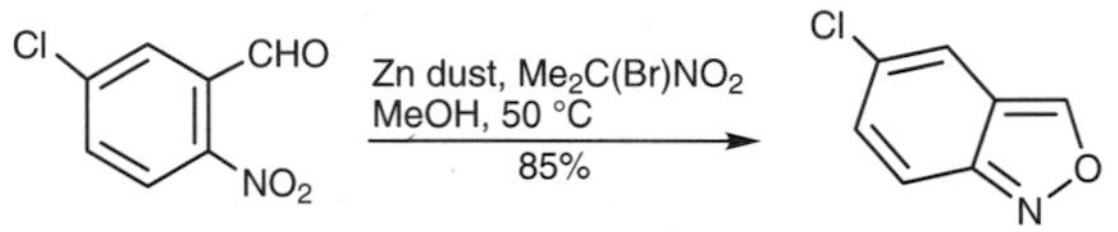

In the opposite sense, reduction of *ortho*-nitroaraldehydes is a very efficient route to 3-unsubstituted 2,1-benzisoxazoles. Both the zinc and the 2-bromo-2-nitropropane are essential components of the reducing mixture, though the mechanistic details of the sequence are as yet not understood. Note the survival of the aromatic halogen in the example shown above.[71]

ortho-Aminotoluenes can be converted into 2,1-benzisothiazoles by reaction with thionyl chloride[72] or with *N*-sulfinylmethanesulfonamide.[73]

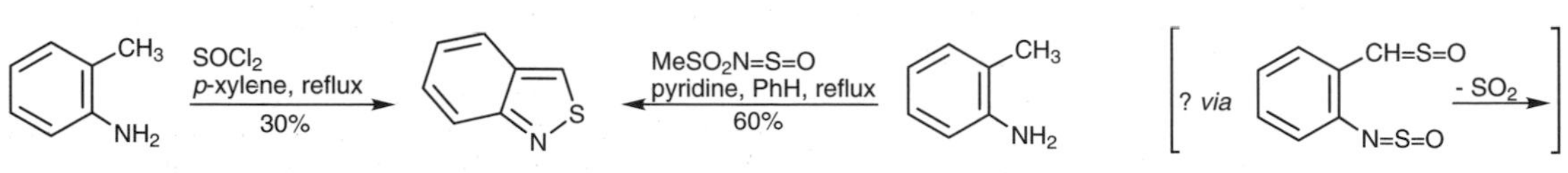

References

1. (a) 'The chemistry of benzimidazoles', Wright, J. B., *Chem. Rev.*, **1951**, *48*, 397; 'Synthesis, reactions and spectroscopic properties of benzimidazoles', Preston, P. N., *ibid.*, **1974**, *74*, 279; (b) '2-Aminobenzimidazoles in organic synthesis', Rastogi, R. and

Sharma, S., *Synthesis*, **1983**, 861; (c) 'Imidazole and benzimidazole synthesis', Grimmett, M. R., Academic Press, London, **1997**.

2. 'Synthesis, properties and reactions of 1*H*-indazol-3-ols and 1,2-dihydro-3*H*-indazol-3-ones', Baiocchi, L., Corsi, G., and Palazzo, G., *Synthesis*, **1978**, 633.

3. 'Benzisothiazoles', Davis, M., *Adv. Heterocycl. Chem.*, **1972**, *14*, 43; 'Advances in the chemistry of benzisothiazoles and other polycyclic isothiazoles', idem, ibid., **1985**, *38*, 105.

4. 'Benzisoxazoles (indoxazenes and anthranils)', Wünsch, K. H. and Boulton, A. J., *Adv. Heterocycl. Chem.*, **1967**, *8*, 277.

5. Sun, J.-H., Teleha, C. A., Yan, J.-S., Rogers, J. D., and Nugiel, D. A., *J. Org. Chem.*, **1997**, *62*, 5627.

6. *e.g.* Harrison, D., Ralph, J. T., and Smith, A. C. B., *J. Chem. Soc.*, **1963**, 2930

7. Mistry, A. G., Smith, K., and Bye, M. R., *Tetrahedron Lett.*, **1986**, *27*, 1051.

8. Jansen, H. E. and Wibaut, J. P., *Recl. Trav. Chim.*, **1937**, *56*, 699.

9. Cohen-Fernandes, P. and Habraken, C. L., *J. Org. Chem.*, **1971**, *36*, 3084.

10. Ellingboe, J. W., Spinelli, W., Winkley, M. W., Nguyen, T. T., Parsons, R. W., Moubanak, I. F., Kitzen, J. M., Engen, D. V., and Bagli, J. F., *J. Med. Chem.*, **1992**, *35*, 705.

11. Jutzi, P. and Gilge, U., *J. Heterocycl. Chem.*, **1983**, *20*, 1011.

12. Harrison, D. and Ralph, J. T., *J. Chem. Soc.*, **1965**, 3132.

13. Hong, Y., Tanoury, G. J., Wilkinson, H. S., Bakale, R. P., Wald, S. A., and Senanayake, C. H., *Tetrahedron Lett.*, **1997**, *38*, 5607.

14. Yamada, M., Sato, Y., Kobayashi, K., Konno, F., Soneda, T., and Watanabe, T., *Chem. Pharm. Bull.*, **1998**, *46*, 445.

15. Kövér, J., Timár, T., and Tompa, J., *Synthesis*, **1994**, 1124.

16. Katritzky, A. R., Ghiviriga, I., and Cundy, D.. J., *Heterocycles,* **1994**, *38*, 1041.

17. Wuts, P. G. M., Gu, R. L., Northuis, J. M., and Thomas, C. L., *Tetrahedron Lett.*, **1998**, *39*, 9155.

18. Adams, D. J. C., Bradbury, S., Horwell, D. C., Keating, M., Rees, C. W., and Storr, R. C., *Chem. Commun.*, **1971**, 828; Kleim, J. T., *et al.*, *J. Med. Chem.*, **1996**, *39*, 570.

19. Harada, H., Morie, T., Hirokawa, Y., Terauchii, H., Fugiwara, I., Yoshida, N., and Kato, S., *Chem. Pharm. Bull.*, **1995**, *43*, 1912.

20. *e.g.* Katritzky, A. R., Drewniak-Deyrup, M., Lan, X., and Brunner, F., *J. Heterocycl. Chem.*, **1989**, *26*, 829.

21. Moore, S. S. and Whitesides, G. M., *J. Org. Chem.*, **1982**, *47*, 1489.

22. Ezquerra, J., Lamas, C., Pastor, A., García-Navío, J. C., and Vaquero, J. J., *Tetrahedron*, **1997**, *53*, 12755.

23. Gale, D. J. and Wilshire, J. F. K., *Aust. J. Chem.*, **1973**, *26*, 2683.

24. Welch, W. M., Hanan, C. E., and Whalen, W. M., *Synthesis*, **1992**, 937.

25. Anderson, B. A. and Harn, N. K., *Synthesis*, **1996**, 583.

26. Gordon, D. W., *Synlett*, **1998**, 1065.

27. Sakamoto, M., Nagano, M., Suzuki, Y., Satoh, K., and Tamura, O., *Tetrahedron*, **1996**, *52*, 733.

28. Quast, H. and Schmitt, E., *Chem. Ber.,* **1969**, *102*, 568.

29. Chikashita, H., Ishihara, M., Takigawa, K., and Itoh, K., *Bull. Chem. Soc. Jpn.*, **1991**, *64*, 3256.

30. Chikashita, H., Ishibaba, M., Ori, K., and Itoh, K., *Bull. Chem. Soc. Jpn.*, **1988**, *61*, 3637.

31. Chikashita, H., Takegami, N., Yanase, Y., and Itoh, K., *Bull. Chem. Soc. Jpn.*, **1989**, *62*, 3389.

32. Uff, B. C., Ho,. Y.-P., Brown, D. S., Fisher, I., Popp, F. D., and Kant, J., *J. Chem. Res.,* **1989** *(S)* 346, *(M)* 2652.

33. Davis, M. and Hudson, M. J., *J. Heterocycl. Chem.*, **1983**, *20*, 1707.

34. Ballesteros, P., Elguero, J., Claramunt, R. M., Faure, R., M. de la Concepción Foces-Foces, Hernández Caro, F., and Rousseau, A., *J. Chem. Soc., Perkin Trans. 2*, **1986**, 1677.

35. Cavalca, L., Gaetani, A., Mangia, A., and Pelizza, G., *Gazz. Chim. Ital.*, **1970**, *100*, 629.

36. Arán, V. J., Diez-Barra, E., de la Hoz, A., and Sánchez-Verdú, P., *Heterocycles*, **1997**, *45*, 129.

37. Quast, H. and Schmitt, E., *Chem. Ber.*, **1969**, *102*, 568.

38. Katritzky, A. R., Aurrecoechea, J. M., Vazquez de Miguel, L. M., *Heterocycles,* , **1987**, *26*, 427.

39. Costanzo, M. J., Maryanoff, B. E., *et al.*, *J. Med. Chem.*, **1996**, *39*, 3039.
40. Villemin, D., Hammadi, M., and Martin, B., *Synth. Commun.*, **1996**, *26*, 2895.
41. Beu-Alloum, A., Bakkas, S., and Soufiaoui, M., *Tetrahedron Lett.*, **1998**, *39*, 4481.
42. Rigo, D., Valligny, D., Taisne, S., and Couturier, D., *Synth. Commun.*, **1988**, *18*, 167.
43. De Luca, M. R. and Kerwin, S. M., *Tetrahedron*, **1997**, *53*, 457.
44. Bhawal, B. M., Mayabhate, S. P., Likhite, A. P., and Deshmukh, A. R. A. S., *Synth. Commun.*, **1995**, *25*, 3315.
45. Wang, T. and Hanske, J. R., *Tetrahedron Lett.*, **1997**, *38*, 6529.
46. Garín, J., Meléndez, E., Merchán, F. L., Merino, P., Orduna, J., and Tejero, T., *J. Heterocycl. Chem.*, **1991**, *28*, 359.
47. Khajavi, M. S., Hajihadi, M., and Nederi, R., *J. Chem. Res. (S)*, **1996**, 92; Khajavi, M. S., Hajihadi, M., and Nikpour, F., *ibid.*, 94.
48. Garín, J., Meléndez, E., Merchán, F. L., Ortiz, D., and Tejero, T., *Synthesis*, **1987**, 368; Garín, J., Meléndez, E., Merchán, F. L., Menino, P., and Tejero, T., *Synth. Commun.*, **1990**, *20*, 2327.
49. Roe, A. and Tucker, W. P., *J. Heterocycl. Chem.*, **1965**, *2*, 148; Dreikorn, B. A. and Unger, P., *J. Heterocycl. Chem.*, **1989**, *26*, 1735; Ambati, N. B., Anand, V., and Hanumanthu, P., *Synth. Commun.*, **1997**, *27*, 1487.
50. Ramsden, C. A. and Rose, H. L., *J. Chem. Soc., Perkin Trans. 1*, **1995**, 615.
51. Stanetty, P. and Krumpak, B., *J. Org. Chem.*, **1996**, *61*, 5130.
52. Grivas, S. and Ronne, E., *J. Chem. Res. (S)*, **1994**, 268; Ronne, E., Olsson, K., and Grivas, S., *Synth. Commun.*, **1994**, *24*, 1363.
53. King, J. F. and Durst, T., *Can. J. Chem.*, **1962**, *40*, 882.
54. Fink, D. M. and Kurys, B. E., *Tetrahedron Lett.*, **1996**, *37*, 995.
55. Walser, A., Flynn, T., and Mason, C., *J. Heterocycl. Chem.*, **1991**, *28*, 1121.
56. Beck, J. R. and Yahner, J. A., *J. Org. Chem.*, **1978**, *43*, 1604.
57. Nakamura, T., Nagata, H., Muto, M., and Saji, I., *Synthesis*, **1997**, 871.
58. Meth-Cohn, O. and Tarnowski, B., *Synthesis,*. **1978**, 58.
59. Saunders, J. C. and Williamson, W. R. N., *J. Med. Chem.*, **1979**, *22*, 1554; McKinnon, D. M. and Lee, K. R., *Can. J. Chem.*, **1988**, *66*, 1405.
60. Fink, D. M. and Strupczewski, J. T., *Tetrahedron Lett.*, **1993**, *34*, 6525.
61. Poissonnet, G., *Synth. Commun.*, **1997**, *27*, 3839.
62. Hervens, F. and Viehe, H. G., *Angew. Chem., Int. Ed. Engl.*, **1973**, *12*, 405.
63. Huisgen, R. and Bast, K., *Org. Synth, Coll. Vol. V*, **1973**, 650; Rüchardt, C. and Hassman, V., *Justus Liebigs Ann. Chem.*, **1980**, 908.
64. Yoshida, T., Matsuura, N., Yamamoto, K., Doi, M., Shimada, K., Morie, T., and Kato, S., *Heterocycles*, **1996**, *43*, 2701.
65. Snyder, H. R., Thompson, C. B., and Hinman, R. L., *J. Am. Chem. Soc.*, **1952**, *74*, 2009; Norman, M. H., Navas, F., Thompson, J. B., and Rigdon, G. C., *J. Med. Chem.*, **1996**, *39*, 4692.
66. Stanetty, P., Krumpak, B., Emerschitz, T., and Mereiter, K., *Tetrahedron*, **1997**, *53*, 3615.
67. Prakarh, O., Saini, R. K., Singh, S. P., and Varma, R. S., *Tetrahedron Lett.*, **1997**, *38*, 3147.
68. Chauhan, M. S. and McKinnon, D. M., *Can. J. Chem.*, **1975**, *53*, 1336.
69. Ning, R. Y., Chen, W. Y., and Sternbach, L. H., *J. Heterocycl. Chem.*, **1974**, *11*, 125; Smalley, R. K., Smith, R. H., and Suschitzky, H., *Tetrahedron Lett.*, **1978**, 2309.
70. Hall, J. H., Behr, R. F., and Reed, R. L., *J. Am. Chem. Soc.*, **1972**, *94*, 4952.
71. Kim, B. H., Jun, Y. M., Kim, T. K., Lee, Y. S., Baik, W., and Lee, B. M., *Heterocycles*, **1997**, *45*, 235.
72. Davis, M. and White, A. W., *J. Org. Chem.*, **1969**, *34*, 2985.
73. Singerman, G. M., *J. Heterocycl. Chem.*, **1975**, *12*, 877.

24 Purines: reactions and synthesis

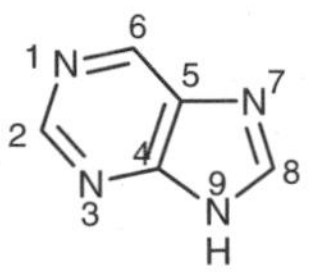

purine
[9*H*-purine]

Purines are of great interest for several reasons, but in particular, together with certain pyrimidine bases, they are constituents of DNA and RNA and consequently of fundamental importance in life processes. Additionally, as nucleosides and nucleotides (see below) they act as hormones and neurotransmitters and are present in some co-enzymes. The interconversion of mono-, di-, and triphosphate esters of nucleosides is at the heart of energy-transfer in many metabolic systems and is also involved in intracellular signalling. This central biological importance, together with medicinal chemists' search for anti-tumour and anti-viral (particularly anti-AIDS) agents have resulted in a rapid expansion of purine chemistry in recent years.

There are significant lessons to be learnt from the chemistry of purines since their reactions exemplify the interplay of its constituent imidazole and pyrimidine rings just as the properties of indole show modified pyrrole and modified benzene chemistry. Thus purines can undergo both electrophilic and nucleophilic attack at carbon in the five-membered ring but only nucleophilic reactions at carbon in the six-membered ring.

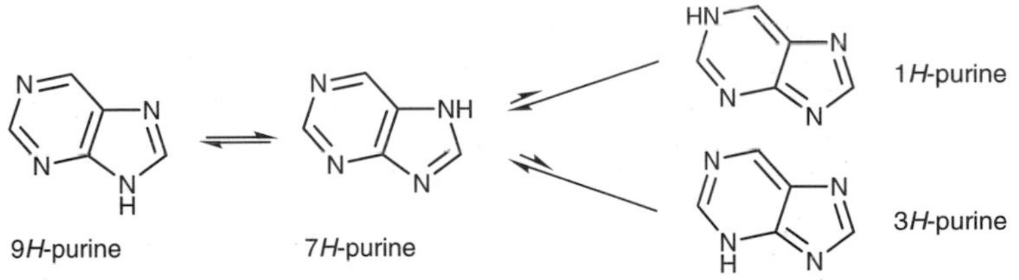

The numbering of the purine ring system is anomolous and reads as if purine were a pyrimidine derivative. There are in principle four possible tautomers of purine containing an *N*-hydrogen; in the crystalline state, purine exists as the 7*H*-tautomer, however in solution both 7*H*- and 9*H*-tautomers are present in approximately equal proportions; the 1*H*- and 3*H*- tautomers are not significant.[1]

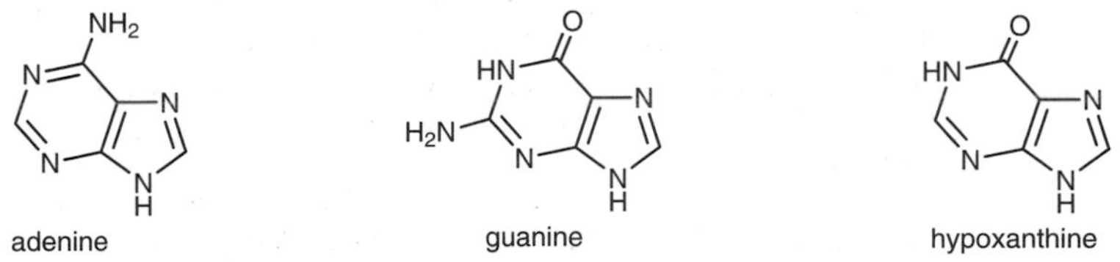

adenine

guanine

hypoxanthine

adenosine   guanosine   inosine   cyclic AMP

Not surprisingly, because the naturally occurring purines are amino and/or oxygenated substances, the majority of reported purine chemistry pertains to such derivatives and, as a consequence, reactions of the simpler examples, such as in other chapters are given as typical, have received limited attention. Since the study of purines stems from interest in the naturally occurring derivatives, a 'trivial' nomenclature has evolved which is in general usage. A nucleoside is a sugar (generally 9-(riboside) or 9-(2′-deoxyriboside)) derivative of a purine base (or pyrimidine base), for example adenosine is the 9-(riboside) of adenine, itself the generally used trivial name for 6-aminopurine. A nucleotide is a 5′-phosphate (or di- or tri-phosphate) of a nucleoside – adenosine 5′-triphosphate (ATP) is an example.

adenosine-5′-monophosphate (AMP)

adenosine-5′-diphosphate (ADP)

adenosine-5′-triphosphate (ATP)

Caffeine (1,3,7-trimethylxanthine) is the well known stimulant present in tea and coffee. In mammals the end product of metabolic breakdown of nucleic acids is urea, but in birds and reptiles it is uric acid; uric acid was one of the first heterocyclic compounds to be isolated as a pure substance, for it was obtained from gallstones by Scheele in 1776.

xanthine   caffeine   uric acid

6-Mercaptopurine is used in the treatment of leukemia and other cancers, Acyclovir is an antiviral agent used in the treatment of *Herpes* infections, and DDI is used in the treatment of AIDS.

6-Mercaptopurine   Acyclovir   Dideoxyinosine (DDI)   allopurinol

Isosteres (i.e. molecules of the same shape but with different atom combinations) of purines are also important as medicines: allopurinol is used to treat gout and Sildenafil achieved international fame, under the trade name Viagra™, for the treatment of impotence. Some natural products can be viewed as purine isosteres: oxanosine and tubercidin, both obtained from *Streptomyces*, have anti-microbial and anti-cancer activity.

Sildenafil   oxanosine   tubercidin

24.1   Nucleic acids, nucleosides, and nucleotides[2]

Nucleic acids are high-molecular-weight, mixed polymers of mononucleotides, in which chains are formed by monophosphate links between the 5′-position of one nucleoside and the 3′-position of the next. The 'backbone' of the chain is thus composed of alternating phosphates and sugars, to which purine and pyrimidine bases are attached at regular intervals. The polymer is known as **ribonucleic acid** (RNA) when the sugar is ribose, and **deoxyribonucleic acid** (DNA) when the sugar is 2-deoxyribose.

The (-phosphate-sugar-phosphate-sugar-) 'backbone' of RNA

B, B', B" represent the purine and pyrimidine bases

cytosine   thymine (DNA only)   uracil

adenine   guanine

DNA contains two purine bases, guanine and adenine, and two pyrimidine bases, cytosine and thymine. In RNA thymine is replaced by uracil and in another form, *t*-RNA, other bases including small amounts of *N*-alkylated derivatives are present.

thymine

The only two effective
hydrogen bonding pairings
are adenine/thymine (AT)
and guanine/cytosine (GC)

adenine

backbone

cytosine

guanine

backbone

Nucleic acids occur in every living cell. DNA carries genetic information and transfers this information, *via* RNA, thus directing protein synthesis. The genetic information embodied in DNA is connected with the close association of two nucleic acid strands, which is based on very specific hydrogen bonding between an adenine (A) residue of one strand and a thymine (T) residue in the precisely opposite section of the other strand, and between a cytosine (C) residue on one strand and a guanine (G) residue on the other. This pairing is absolutely specific – adenine cannot form multiple hydrogen bonds with guanine or cytosine and cytosine cannot form multiple hydrogen bonds with thymine or adenine. It is amazing that all heredity and evolution depend on two sets of hydrogen bonds! The genetic code for the synthesis of a particular amino acid is a sequence of three bases attached to the backbone, read in the $5' \rightarrow 3'$ direction, for example the triplet which codes for the synthesis of tryptophan is UGG, however most amino acids can be coded for by more than one triplet, some having as many as four, the variation coming in the third nucleotide, thus both UAU and UAC code for tyrosine. The genetic information is transmitted when the strands of the DNA separate, replication then being governed by the establishment of the AT and GC sets of hydrogen bonds to a newly developing strand.

24.2   Reactions with electrophilic reagents

24.2.1   Addition at nitrogen

24.2.1.1   *Protonation*

Purine is a weak base, pK_a 2.5. ^{13}C NMR studies suggest that all three protonated forms are present in solution but the predominant cation is formed by N-1-protonation.[3] In strong acid solution a dication is formed by protonation at N-1 and on the five-membered ring.[4]

The presence of oxygen functionality does not seem to affect purine basicity to any great extent, thus hypoxanthine has a pK_a of 2.0. Amino groups increase the basicity, as illustrated by the pK_a of adenine, 4.2, and oxo groups reduce the basicity of amino-purines, thus guanine has a pK_a of 3.3; the position of protonation of the latter in the solid state has been established, by X-ray analysis, as on the five-membered ring – this nicely illustrates the extremely subtle interplay of substituents and ring heteroatoms, for although the 2-amino substituent increases the basicity of the purine to which it is attached, this does not necessarily mean that it is the associated N-3 which is protonated.

Purine itself slowly decomposes in aqueous acid, to the extent of about 10% in 1N sulfuric acid at 100 °C. The stability of oxypurines to aqueous acid varies greatly, for example xanthine is stable to aqueous 1N sulfuric acid at 100 °C whereas 2-oxypurine is completely converted into a pyrimidine in 2 hours under the same conditions.

24.2.1.2 *Alkylation at nitrogen*

As would be expected from systems containing four nitrogen atoms, *N*-alkylation of purines is complex and can take place on the neutral molecule or *via* an *N*-anion. Purine reacts with iodomethane to give a 7,9-dimethylpurinium salt.[5]

Adenine gives mainly 3-alkylated products under neutral conditions but 7/9-substitution when there is a base present. Adenosine derivatives on the other hand usually give 1-alkylated products presumably due to hindrance to *N*-3-attack by the *peri* 9-ribose substituent. That attack can still occur at C-3 is shown by the intramolecular quaternisation of *N*-3 which is an important side reaction when 5′-halides are subjected to displacement conditions.

An effective method for alkylating the 6-amino group of adenosine is to bring about rearrangement of a 1-alkyladenosinium salt; this involves an ANRORC sequence – a Dimroth rearrangement.[6,7]

Another Dimroth rearrangement provides a neat way to isotopically label N-1, starting from adenosine labelled at the amino group.[8]

Alkylation of oxygenated purines in alkaline media, for example hypoxanthine, tends to occur both at amidic nitrogen and also at a five-membered ring nitrogen, making selectivity a problem. Under neutral conditions xanthines give 7,9-dialkylated quaternary salts. The alkylation of 6-chloropurine illustrates the complexity: in basic solution both 7- and 9-substitution occurs,[9] whereas reaction with a carbocation is selective for N-9.[10]

9-*t*-Butyldimethylsilyloxymethyl is a useful protecting group for adenines as it confers good solubility in organic solvents. It is introduced by stepwise conversion into the 9-hydroxymethyl compound by reaction with formaldehyde and base, followed by *O*-silylation.[11]

The ratio of N-9 to N-7 alkylation is also influenced by the size of a 6-substituent, larger groups at C-6 lead to increased percentages of 9- *versus* 7-alkylation.[12] The N-9:N-7 ratio of products varies with time when alkylations employ a Michael acceptor like methyl acrylate, for here the alkylation is reversible and the concentration of thermodynamic product can build up.[13] Regiospecific 7-alkylations can be achieved *via* the quaternisation of a 9-riboside followed by hydrolytic removal of the sugar residue as illustrated.[14] Alkylation on N-7 in nucleic acids is the mechanism of mutagenesis/carcinogenesis by some natural toxins such a aflatoxin.[15]

In suitable cases, where N-7/N-9 selectivity is poor, alkylation can be directed to N-9 by a bulky protecting group installed on a C-6 substituent.[16]

In the ribosylation of purines, in addition to the question of regioselectivity on the purine, there is the possibility of forming epimeric products at the linking C-1′ of the ribose, and this is often the more difficult to control. A great deal of work has been done and many different conditions shown to be effective in specific cases, but conditions which are generally effective have not been defined.[17] These alkylations usually employ acylated or halo ribosides in conjunction with a purine derivative of mercury,[18] silicon,[17] or sodium,[19] and stereoselective displacements of halide can sometimes be achieved.

Other methods of controlling stereochemistry include the use of the size of an isopropylidene protecting group to shield one face of the sugar[20] or, as shown, anchimeric assistance from a 2′-benzoate.[21] Enzymatic catalysis has been used to ribosylate purines and related bases by reaction with a 7-alkylated nucleoside.[22]

24.2.1.3 Acylation at nitrogen

Purines react with acylating agents such as chloroformates or ethyl pyrocarbonate[23] to give non-isolable N^+-acyl salts which can suffer various fates following nucleophilic addition; products of cleavage of either ring have been observed, as have recyclisation products.[24]

24.2.1.4 Oxidation at nitrogen

Peracid *N*-oxidation of purines gives 1- and/or 3-oxides depending on exact conditions.[25] Adenine and adenosine give 1-oxides whereas guanine affords the 3-oxide.[26] The 3-oxide of purine itself has been obtained *via* oxidation of 6-cyanopurine (at N-3) then hydrolysis and decarboxylation,[25] the relatively easy loss of carbon dioxide echoing the analogous process discussed for pyridine-α-acids (section 5.12). The *N*-7-oxide of adenine can be prepared by oxidation of *N*-3-benzyl-3-*H*-adenine, followed by deprotection.[27]

24.2.2 Substitution at carbon

Typical electrophilic aromatic substitution reactions have not been reported for purine or simple alkyl derivatives.

24.2.2.1 Halogenation

Purine itself simply forms an N$^+$-halogen complex but does not undergo *C*-substitution, however adenosine,[28] hypoxanthine and xanthine derivatives[29] undergo fluorination,[30] chlorination and bromination at C-8. There is the possibility that these substitution products arise *via* *N*-halopurinium salts, nucleophilic addition of bromide anion to these at C-8, then elimination of hydrogen halide.

24.2.2.2 Nitration

Xanthines undergo 8-nitration, though under fairly vigorous conditions.[31]

24.2.2.3 Coupling with diazonium salts

Amino and oxypurines couple at their 8-position; a weakly alkaline medium is necessary so it seems likely that the reactive entity is an anion.[32]

24.3 Reactions with radical reagents

Purines react readily with hydroxyl, alkyl, aryl, and acyl radicals, usually at C-6,[33] or at C-8 (or C-2) if the 6-position is blocked. Both reactivity and selectivity for C-8 are increased when the substitution is conducted at lower pH.[34] In nucleosides, 5'-8 radical cyclisation is very efficient, but the 5'-radical can be trapped before cyclisation by using a large excess of acrylonitrile.[35]

24.4 Reactions with oxidising agents

There are few significant oxidations of purines apart from N-oxidations (section 24.2.1.4), but dimethyldioxirane gives good yields of 8-oxo compounds, possibly via the intermediacy of a 9,8- or 7,8-oxaziridine.[36] C-8-Oxidation[37] is an important process in vivo, for example with the oxomolybdoenzyme xanthine oxidase, where oxygen is introduced at C-8 via a mechanism about which there is still debate.

24.5 Reactions with reducing agents

The reduction of substituted purines is very complex and ring-opened products are often obtained. 1,6-Dihydropurine is formed by catalytic or electrochemical[38] reduction of purine, but this is unstable. More stable compounds can be obtained by reduction in the presence of acylating agents.[39] 7/9-Quaternary salts are easily reduced by borohydride in the five-membered ring to dihydro-derivatives.[40]

24.6   Reactions with nucleophilic reagents

The reactions of the 2-, 6-, and 8-halopurines are very important in purine synthesis. Halo-purines can be prepared from oxy-, amino- or thiopurines and the 8-isomers are also available by direct halogenation or *via* lithiated intermediates. Chloropurines have been the most commonly used, but bromo- and iodopurines react similarly, though without any great operational advantage; fluorides, are more reactive.

Relatively easy nucleophilic displacement, via an addition/elimination sequence (section 2.3.1), takes place at all three positions with a wide range of nucleophiles such as alkoxides,[41] sulfides, amines, azide, cyanide, and malonate and related carbanions.[42]

In 9-substituted purines, the relative reactivity is 8 > 6 > 2, but in 9*H*-purines this is modified to 6 > 8 > 2, the demotion of the 8-position being associated with anion formation in the five-membered ring. Conversely, in acidic media the reactivity to nucleophilic displacement at C-8 is enhanced: protonation of the five-membered ring facilitates the nucleophilic addition step.[41] The relative reactivities of 2- and 6-positions is nicely illustrated by the conditions required for the reaction of the respective chlorides with hydrazine, a relatively good nucleophile.[43] It is worth noting the parallelism between the relative positional reactivity here with that in halopyrimidines (4 > 2).

In 2,6-dichloropurine, reactivity at C-6 is enhanced relative to 6-chloropurine by the inductive effect of the second halogen, thus the dihalide will react with simple amines at room temperature where the monochloride would require heating, for example in isopropanol. The presence of electron-releasing substituents, such as amino, somewhat deactivate halogen to displacement, but conversely, oxygenated purines, probably because of their carbonyl tautomeric structures, react easily.[44]

The generation of an *N*-anion by deprotonation in the five-membered ring is given as the reason why 8-chloropurine reacts with sodamide to give adenine (6-aminopurine): inhibition of attack at C-8 allows the alternative addition to C-6 to lead eventually to the observed major product.[45]

Direct conversion of inosines into 6-amino derivatives, without the intermediacy of a halo-purine, can be achieved by heating with a mixture of phosphorus pentoxide and the amine hydrochloride.[46]

Displacement of iodide can be catalysed by copper salts allowing milder reaction conditions, though it is not clear by what mechanism the metal salt brings about its effect.[47]

Other useful leaving groups include sulfoxide,[48] triflate,[49] and aryl- or alkylthio.[50] Sulfones are highly reactive in some nucleophilic substitutions and are also the reactive intermediates in sulfinate-catalysed displacements of halide.[51]

Displacement of halides can be catalysed by amines – trimethylamine, pyridine,[52] and DABCO[53] have been used. Mechanistically, the catalysis involves formation of an intermediate quaternary ammonium salt which is more reactive towards nucleophiles than the starting halide. The intermediate quaternary salts can be isolated, if required. Trimethylamine gives the most reactive quaternary salt but DABCO can be more convenient. The relative reactivities for nucleophilic displacement at C-6 are: trimethylamine : DABCO : chlorine = 100 : 10 : 1.[54] Cyano[55] and fluorine[56] are amongst the groups which have been introduced in this way.

Arylamines can be particularly unreactive as nucleophiles and for these the use of fluorine as a leaving group,[57] or palladium-assisted displacement of bromine[58] may be necessary.

Amino groups can be converted into good leaving groups by incorporation into a 1,3,4-triazole. The isomeric triazoles formed by reaction of the inosine with 1,2,4-triazole in the presence of phosphoryl chloride and triethylamine, are also good leaving groups.[59]

Nucleophilic acylations can be effected using araldehydes and an azolium salt as catalyst[60] (*cf.* section 21.11).

24.7 Reactions with bases

24.7.1 Deprotonation of *N*-hydrogen

Purine, with a pK_a of 8.9, is slightly more acidic than phenol and much more acidic than imidazole or benzimidazole (pK_as 14.2 and 12.3 respectively). This relatively high acidity is probably a consequence of extensive delocalisation of the negative charge over four nitrogens, however alkylation of the anion (section 24.2.1.2) takes place in the five-membered ring since attack at N-1 or N-3 would generate less aromatic products.

Oxypurines are even more acidic, due to more extensive delocalisation involving the carbonyl groups: xanthine has a pK_a of 7.5 and uric acid, 5.75.

24.7.2   Deprotonation of C-hydrogen

The rapid deuteration of purine at C-8[61] in neutral water at 100 °C probably involves 8-deprotonation of a concentration of purinium cation to give a transient ylide (cf. 1,3-azole 2-H-exchange, section 21.1.2.1). 9-Alkylated purines undergo a quite rapid exchange in basic solution involving direct deprotonation of the free heterocycle.

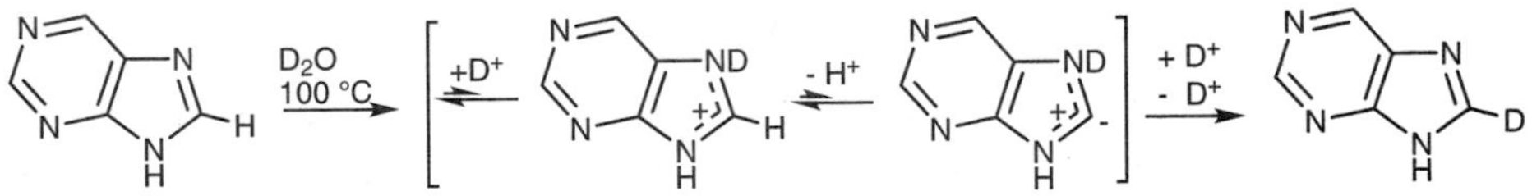

24.8   Reactions of N-metallated purines

These have been dealt with in section 24.2.1.2

24.9   Reactions of C-metallated purines

24.9.1   Lithio derivatives

Preparative lithiation of purines requires the protection of the 7/9-position; lithiation then takes place at C-8.[62] Purines lithiated at C-2 or C-6 can be generated by way of halogen exchange with alkyllithiums, but it is important to maintain a very low temperature in order to avoid subsequent equilibration to the more stable 8-lithiated species.[63]

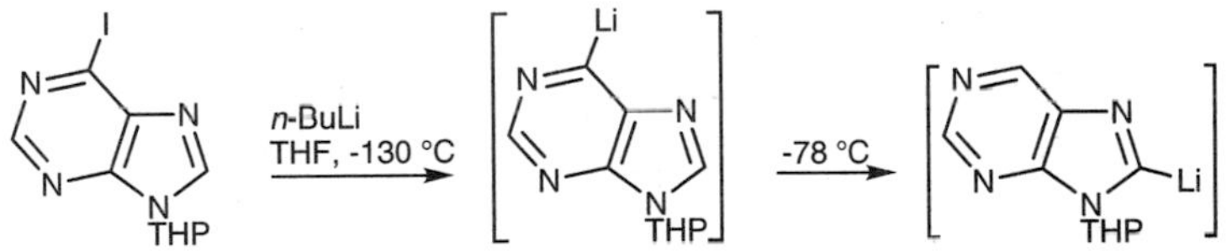

9-Blocked purines can be deprotonated at C-8 with strong bases such as LDA, even in the presence of N-hydrogen in the other ring.[64] Very high yields of 8-halopurines can be obtained by reaction with a variety of halogen donors; 8-lithiation of O-silyl-protected 9-ribofuranosyl purines can be achieved using about three mol equivalents of lithium diisopropylamide.[65]

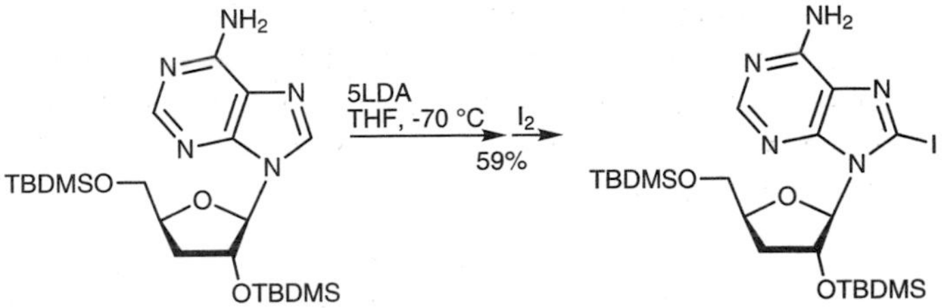

After selective lithiation at C-8 in a 6-chloropurine riboside, quench with a stannyl or silyl chloride leads to the isolation of the 2-substituted compound, via rearrangement of a 2-anion formed by a second lithiation of the initial 8-substituted product, as illustrated below.[66]

24.9.2   Palladium-catalysed reactions

Iodo- and bromopurines undergo the usual palladium-[67] and nickel-catalysed reactions under standard conditions. As with other halo-azines, chloro compounds are usually sufficiently activated to use palladium, though nickel may be the preferred catalyst in certain cases.[68]

Stille couplings with 2,6-dichloropurine occur selectively at C-6, however the selectivity is reversed when chlorine is replaced by bromine or iodine at C-2.[69] A similar pattern is seen for 6,8-dichloropurine, the 6-chlorine again being the more reactive.[70] 8-Bromo-diaminopurines, after prior masking by silylation, undergo normal coupling with heteroaryl stannanes.[71]

24.10   Oxy- and aminopurines

These are tautomeric compounds which exist predominantly as carbonyl and amino structures, thus falling in line with the analogous pyrimidines and imidazoles.

24.10.1   Oxypurines

24.10.1.1   Alkylation

The amide-like *N*-hydrogen in oxypurines is relatively acidic; the acidity is readily understood in terms of the phenolate-like resonance contributor to the anion. Alkylation takes place at nitrogen not oxygen.[72]

24.10.1.2   Acylation

In contrast to alkylation, acylation and sulfonylation frequently occur at oxygen; the resulting *O*-acylated products are relatively unstable but can be utilised, for example, conducting the acylation in pyridine, as solvent, produces a pyridinium salt resulting from displacement of acyloxy by pyridine. Both, *O*-acylated purines, and the corresponding pyridinium salts, can in turn be reacted with a range of nucleophiles[73] to allow the overall replacement of the amide-like oxygen; this is an important alternative to activation of the carbonyl by conversion into halogen (below).

A closely related conversion utilises a silylating agent, in the presence of the desired nucleophile, and presumably involves *O*-silylation then displacement of silyloxy.[74]

24.10.1.3   Replacement by chlorine[75]

This is a very important reaction in purine chemistry and has been widely utilised to allow subsequent introduction of nucleophiles (section 24.5), including replacement with hydrogen by chemical (HI) or catalytic hydrogenolysis. Most commonly, phosphoryl chloride is used, neat, or in solution, especially when there is a ribose present; thionyl chloride is an alternative reagent. 2-Deoxy compounds are more sensitive to acid and with these, milder reagents (carbon tetrachloride with triphenylphosphine) must be used to convert oxo into chloro.[76]

Syntheses of adenine and guanine from uric acid illustrate well the selective transformations to which the halopurines, prepared from a precursor oxypurine,[77] can be put.

24.10.1.4 *Replacement by sulfur*

Replacement by sulfur[78] can be achieved *via* a halopurine, or directly using a phosphorus sulfide.

24.10.2 Aminopurines

24.10.2.1 *Alkylation*

Alkylation under neutral conditions involves attack at a nuclear nitrogen; Dimroth rearrangement (24.2.1.2) of these salts affords side-chain-alkylated purines. Direct introduction of substituents onto a side-chain nitrogen can be achieved by reductive alkylation.[79] A related method involves reduction of an isolated benzotriazolyl intermediate, which allows more control over the reaction.[80]

24.10.2.2 *Acylation*

Aminopurines behave just like anilines with anhydrides and acid chlorides, though the resulting amides are somewhat more easily hydrolysed. Both mono- and diacylation can be utilised as a protecting group strategy.

24.10.2.3 *Diazotisation*

The reaction of 2- and 6-amino groups with nitrous acid is similar to that of 2-aminopyridines, in that diazonium salts are produced, but relative to phenyldiazonium salts, these are unstable. Despite this, they can be utilised for the introduction of groups such as halide[81] or of course oxygen by reaction with water, with loss of nitrogen. 8-Diazonium salts are considerably more stable.[82]

Diazotisation can also be carried out in basic solution and in this way acid-sensitive ribosides can be tolerated.[83] A nucleophilic displacement of amino by hydroxy can be effected enzymatically using adenosine deaminase; this is a useful practical method because it is a very selective transformation under mild conditions.[84] Chemical hydrolysis requires more vigorous conditions.

The related reaction with alkyl nitrites generates purinyl radicals which efficiently abstract halogen from halogenated solvents and this procedure is generally to be preferred for the transformation of aminopurine into halopurine.[85] Comparably, the use of dimethyl disulfide produces methylthiopurines.[86]

24.10.3 Thiopurines

Thiopurines are prepared from halo- or oxypurines or by ring synthesis. In contrast with oxypurines, in alkaline solution they readily alkylate on sulfur, rather than nitrogen.[87]

Thiols are also useful sources of the corresponding bromo compounds, by reaction with bromine and hydrobromic acid.[88] Alkylthio substituents can be displaced by the usual range of nucleophiles, but the corresponding sulfones are more reactive.[52,89]

A useful conversion of a nucleoside 2,6-dithiol into a 6-methylamino adenosine via oxidation with dimethyldioxirane, illustrates several instructive points. The presumed intermediates are sulfinic acids: the 2-sulfinic acid loses sulfur dioxide to leave hydrogen at C-2, and nucleophilic displacement of the 6-sulfinic acid (or possibly the sulfonic acid after further oxidation) introduces the amino group. Similar reactions can be carried out on pyrimidine thiols.[90] The scheme shows intermediates derived from a disulfinic acid – it is not clear in what order oxidations/loss of sulfur dioxide/ displacements take place.

24.11 Alkylpurines

Comparatively little information is available concerning any special reactivity associated with purine alkyl groups, but what is available[91] suggests that their reactivity is comparable to pyridine α-alkyl substituents.

24.12 Purine carboxylic acids

Here again, comparitively little systematic information is available, but a parallel with pyridine α-acids can again be implied in that purine acids undergo decarboxylation on heating.[92]

24.13 Synthesis of purines

Because of the ready availability of nucleosides from natural sources, a frequently used route to substituted purines is *via* the manipulation of one of these.

24.13.1 Ring synthesis

There are two general approaches to the construction of the purine ring system. Additionally, a category which can be defined as 'one pot' methods, are adaptations of the type of process which probably took place in prebiotic times, when simple molecules, such as hydrogen cyanide and ammonia, are believed to have combined to give the first purines.

24.13.1.1 From 4,5-diaminopyrimidines

4,5-Diaminopyrimidines react with carboxylic acids or derivatives to give purines, the 'carboxyl' carbon corresponding to C-8.

Traube synthesis

8-Unsubstituted purines can be prepared simply by heating 4,5-diaminopyrimidines with formic acid,[93] but formamide[94] (or formamidine[95]) are better. The reaction proceeds *via* cyclising dehydration of an intermediate formamide; this usually takes place *in situ* using formamide but generally requires a second, more forcing step when formic acid is employed initially. Purine itself can be prepared by this route.[96]

8-Substituted purines are comparably prepared using acylating agents corresponding to higher acids; in most cases the amide is isolated and separately cyclised.[97] The diaminopyrimidines required are usually prepared by the coupling of a mono-aminopyrimidine with an aryl diazonium ion, then reduction, or by ring synthesis.[98]

Precursors to 9-substituted purines, requiring a substituent on the pyrimidine-4-amino group, are available from the reaction of a 4-chloropyrimidine with the amine.

When milder conditions are required for the cyclisation, perhaps because of the presence of a sugar residue, an ortho ester[99] (often activated[100] with acetic anhydride), or a diacetal-ester[101] (illustrated below), can be used.

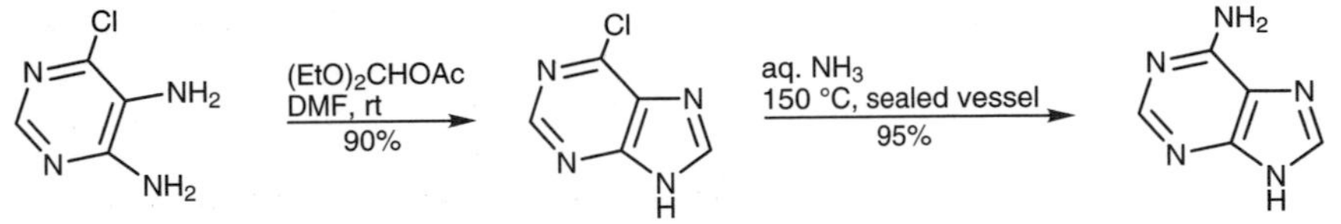

A related reaction is the oxidative cyclisation of anils, originally under vigorous conditions such as heating in nitrobenzene,[102] but now achievable at much lower temperatures using diethyl azodicarboxylate.[103] Amino nitrosopyrimidines can also be converted directly into purines, without the need for reduction to diamine, by reaction with Wittig reagents.[104]

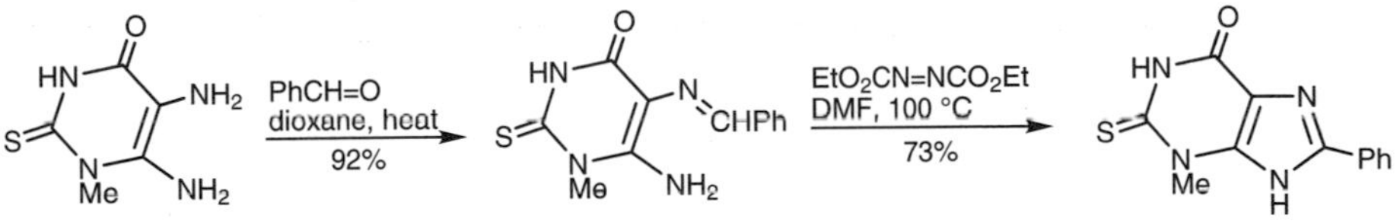

The formation of 8-oxo- or 8-thiopurines requires one-carbon components at a higher oxidation level: urea and thiourea are appropriate. The products of chloroformate five-membered cleavage of purine (section 24.2.1.3) can be recyclised to produce 8-oxopurines.[105]

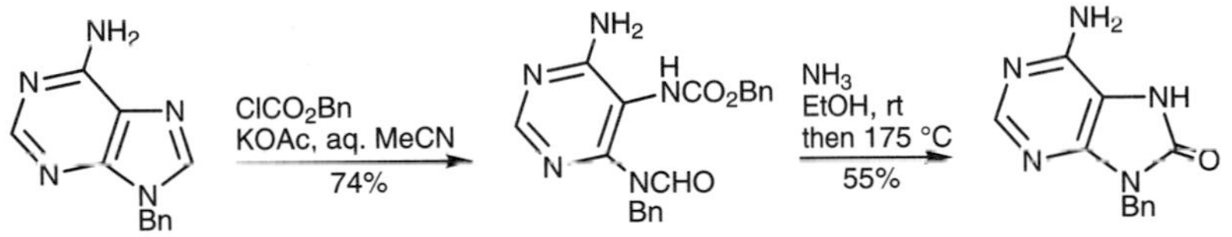

24.13.1.2 *From 5-aminoimidazole-4-carboxamide, or -nitrile*[106]

5-Aminoimidazole-4-carboxamides (or -nitriles) similarly interact with components at the carboxylic acid oxidation level giving purines, the 'carboxyl' carbon becoming C-2.

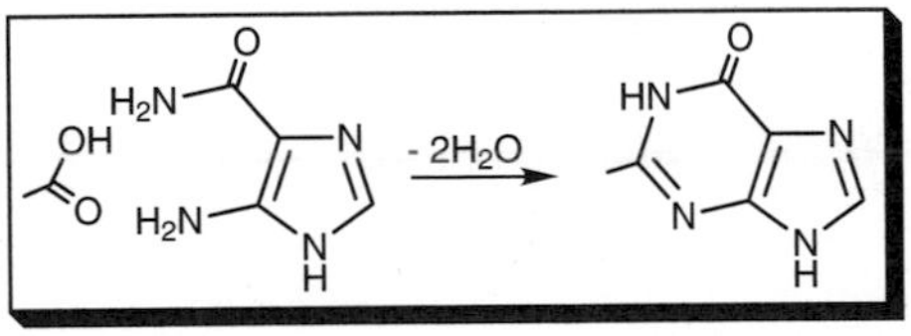

Biosynthetically, purines are built up *via* formation of the imidazole ring first, from glycine and formate, and thence to hypoxanthine and then the other natural purines. In the laboratory, most imidazole-based purine syntheses start with derivatives of 5-aminoimidazole-4-carboxylic acid, particularly its amide (known by the acronym AICA) which together with its riboside are commercially available from biological sources. The use of 5-aminoimidazole-4-carbonitrile in this approach results in the formation of 6-aminopurines, as in a synthesis of adenine itself.[107]

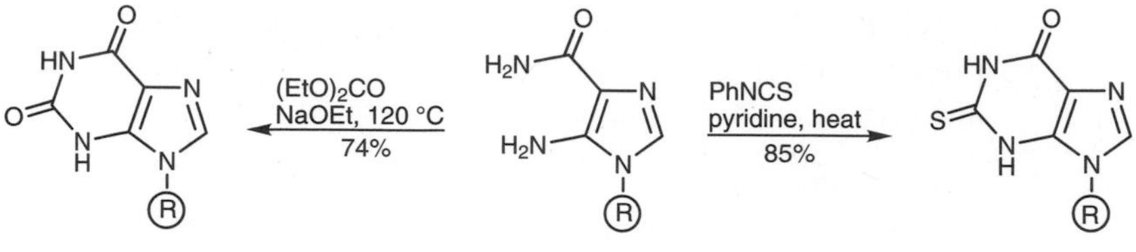

Conversion into 2-alkyl- or -arylpurines requires the insertion of one carbon to create the six-membered ring and this is usually effected by condensation with esters in the presence of base,[108] although amides[109] are occasionally utilised. The use of an isothiocyanate leads to a 2-thiopurine.[110]

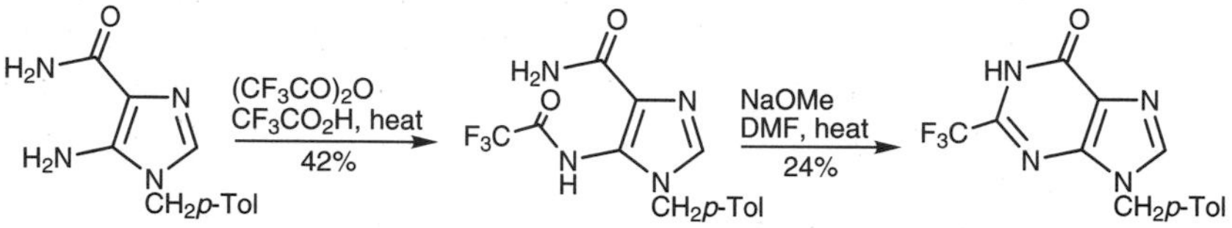

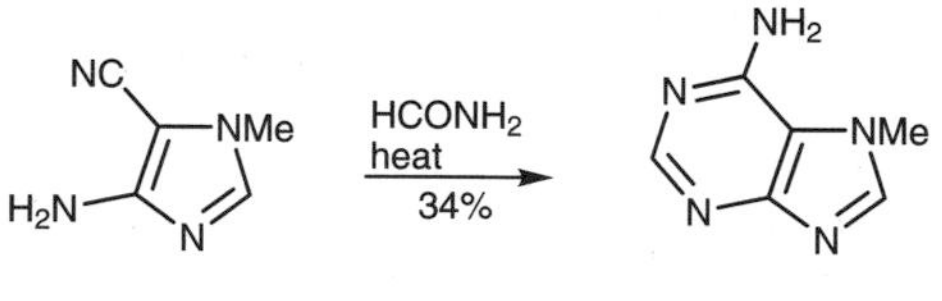

There are a few examples of purine ring syntheses which start from simpler imidazoles, for example a 5-aminoimidazole, generally prepared and utilised *in situ*.[111]

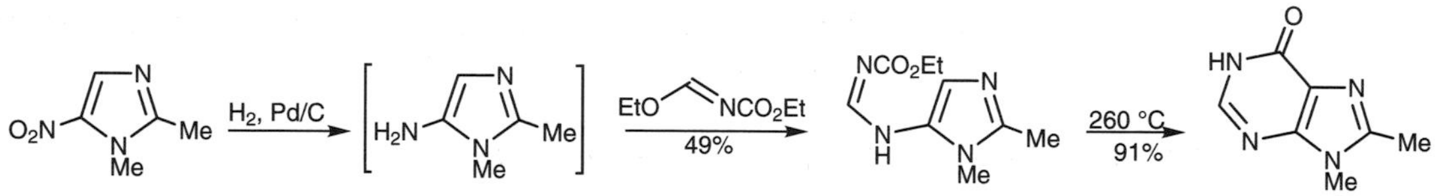

7-Substituted purines can be obtained from 5-aminoimidazole-4-carbaldehyde oximes after conversion into imino ethers and reaction with ammonia as shown below.[112]

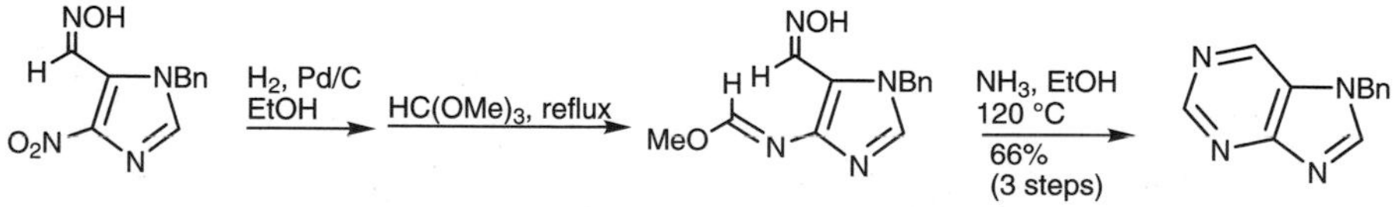

24.13.1.3   *By cycloadditions*

2,4,6-Tris(ethoxycarbonyl)-1,3,5-triazine serves as an azadiene in reaction with 5-aminopyrazoles to produce purine isosteres, pyrazolo[3,4-*d*]pyrimidines.[113] In order to overcome the relative instability of 5-aminoimidazoles, required for analogous synthesis of purines, 5-aminoimidazole-4-carboxylic acids can be used, *in situ* decarboxylation producing the required dienophile.[114]

24.13.1.4 'One-step' syntheses

It is amazing that relatively complex molecules such as purines can be formed by the sequential condensation of very simple molecules such as ammonia and hydrogen cyanide. That the intrinsic reactivity embodied in these simple molecules leads 'naturally' to purines must surely be relevant to the evolution of a natural system which relies on these 'complex' molecules. In other words it seems highly likely that purines existed before the evolution of life and were incorporated into its mechanism because they were there and, of course, because they have appropriate chemical properties.

Adenine, $C_5H_5N_5$, is formally a pentamer of hydrogen cyanide and indeed can be produced in the laboratory by the reaction of ammonia and hydrogen cyanide, although not with great efficiency. A related and more practical method involves the dehydration of formamide.[115] Purine itself can also be obtained from formamide.[116]

Methods derived from this fundamental process involve the condensation of one-, two- and three-carbon units such as amidines, amino-nitriles and carboxamides, which represent intermediate stages of the ammonia/hydrogen cyanide reaction. Pyrimidines or imidazoles are usually intermediates.[117]

24.13.2 Examples of notable syntheses involving purines

24.13.2.1 Aristeromycin

A synthesis of aristeromycin[118] makes use of the displacement of a pyrimidine 4-chloride to allow introduction of the amine and the generation of the 4,5-diaminopyrimidine for subsequent closure of the five-membered ring.

24.13.2.2 Adenosine

Adenosine[119] has also been synthesised using the pyrimidine → purine strategy. In this synthesis the sugar was also introduced at an early stage, but here *via* condensation with a 4-amino group.

24.13.2.3 Sildenafil (Viagra™)

A synthesis of Sildenafil starts with a routine synthesis of a pyrazole (*cf.* section 22.13.1.1) followed by *N*-methylation and ring nitration. Functional group manipulation provides a pyrazole equivalent to AICA (section 24.13.1.2) from which the pyrimidone ring is formed via reaction with an aromatic acid chloride.

Exercises for chapter 24

Straightforward revision exercises (consult chapter 24)

(a) What are the structures of the purine bases involved in DNA and RNA?

(b) How does the Dimroth rearrangement allow the synthesis of 6-alkylaminopurines from 6-aminopurines?

(c) What is the order of reactivity towards nucleophilic displacement of the 2-, 6-, and 8-halopurines? How does the inclusion of a tertiary amine in such nucleophilic displacements facilitate them?

(d) At what position does strong base deprotonation of 9-substituted purines take place?

(e) Name three types of compound which will react with 4,5-diaminopyrimidines to produce purines.

(f) How could one synthesise a 2-thiopurine from 5-aminoimidazole-4-carboxamide?

More advanced exercises

1. What are the structures of the intermediates and final product of the following sequence: guanosine 2′,3′,5′-triacetate reacted with $POCl_3$ → $C_{16}H_{18}ClN_5O_7$ then this with t-BuONO/CH_2I_2 → $C_{16}H_{16}ClIN_4O_7$, this product with NH_3/MeOH → $C_{10}H_{12}IN_5O_4$ and finally this compound with PhB(OH)$_2$/Pd(PPh$_3$)$_4$/Na$_2$CO$_3$ giving $C_{16}H_{17}N_5O_4$. How could this same purine be prepared from AICA-riboside in four steps?

2. Suggest a sequence for the transformation of adenosine into 8-phenyladenosine.

3. Give structures and explain the following: adenosine with Me_2SO_4 → $C_{11}H_{15}N_5O_4$, this with aq. HCl produces $C_6H_7N_5$, and finally aq. NH_3 on this last compound gives an isomer, $C_6H_7N_5$.

4. Write structures for the purines produced by the following reactions: (i) heating 4,5,6-triaminopyrimidine with formamide; (ii) treating 2-methyl-4,5-diaminopyrimidin-6-one with sodium dithioformate, then heating in quinoline.

References

1. Dreyfus, M., Dodin, G., Bensaude, O., and Dubois, J. E., *J. Am. Chem. Soc.*, **1975**, *97*, 2369.
2. 'Nucleic acid chemistry' (2 vols), Townsend, L. B. and Tipson, R. S., Eds., Wiley-Interscience, **1978**; 'Synthetic procedures in nucleic acid chemistry' (2 vols), Zorbach, W. W. and Tipson, R. S., Eds., Wiley-Interscience, **1973**; 'Nucleic acids in chemistry and biology', Blackburn, G. M. and Gait, M. J., Eds., **1996**.
3. Coburn, W. C., Thorpe, M. C., Montgomery, J. A., and Hewson, K., *J. Org. Chem.*, **1965**, *30*, 1110.
4. Schumacher, M. and Gnther, H., *Chem. Ber.*, **1983**, *116*, 2001.
5. Taylor, E. C., Maki, Y., and McKillop, A., *J. Org. Chem.*, **1969**, *34*, 1170.
6. Brookes, P. and Lawley, P. D., *J. Chem. Soc.*, **1960**, 539; Carrea, G., Ottolina, G., Riva, S., Danieli, B., Lesma, G., and Palmisano, G., *Helv. Chim. Acta*, **1988**, *71*, 762.
7. 'The Dimroth rearrangement in the adenine series: a review updated', Fujii, T. and Itaya, T., *Heterocycles*, **1998**, *48*, 359.
8. Pagano, A. R., Zhao, H., Shallop, A., and Jones, R. A., *J. Org. Chem.*, **1998**, *63*, 3213.
9. Montgomery, J. A. and Temple, C., *J. Am. Chem. Soc.*, **1961**, *83*, 630.
10. Lewis, L. R., Schneider, F. H., and Robins, R. K., *J. Org. Chem.*, , **1961**, *26*, 3837.
11. Magnin, G. C., Dauvergne, J., Burger, A., and Biellmann, J.-F., *Tetrahedron Lett.*, **1996**, *37*, 7833.

12. Geen, G. R., Grinter, T. J., Kincey, P. M., and Jarvest, R. L., *Tetrahedron*, **1990**, *46*, 6903.

13. Geen, G. R., Kincey, P. M., and Choudary, B. M., *Tetrahedron Lett.*, **1992**, *33*, 4609.

14. Sessler, J. L., Magda, D., and Furuta, H., *J. Org. Chem.*, **1992**, *57*, 818.

15. Iyer, R. S., Voehler, M. W., and Harris, T. M., *J. Am. Chem. Soc.*, **1994**, *116*, 8863.

16. Breipohl, G., Will, D. W., Peyman, A., and Uhlmann, E., *Tetrahedron*, **1997**, *53*, 14671.

17. 'Nucleoside syntheses, organosilicon methods', Lukevics, E. and Zablocka, A., Ellis Horwood, **1991**, and references therein.

18. Baker, B. R., Hewson, K., Thomas, H. J., and Johnson, J. A., *J. Org. Chem.*, **1957**, *22*, 954; Townsend, L. B., Robins, R. K., Loeppky, F. N., and Leonard, N. J., *J. Am. Chem. Soc.*, **1964**, *86*, 5320.

19. Kazimierczuk, Z., Cottam, H. B., Revankar, G. R., and Robins, R. K., *J. Am. Chem. Soc.*, **1984**, *106*, 6379; Jhingan, A. K. and Meehan, T., *Synth. Commun.*, **1992**, *22*, 3129.

20. Hanna, N. B., Ramasamy, K., Robins, R. K., and Revankar, G. R., *J. Heterocycl. Chem.*, **1988**, *25*, 1899.

21. Imai, K., Nohara, A., and Honjo, M., *Chem. Pharm. Bull.*, **1966**, *14*, 1377.

22. Hennen, W. J. and Wong, C.-H., *J. Org. Chem.*, **1989**, *54*, 4692.

23. Leonard, N. J., McDonald, J. J., Henderson, R. E. L., and Reichman, M. E., *Biochemistry*, **1971**, *10*, 3335.

24. Pratt, R. F. and Kraus, K. K., *Tetrahedron Lett.*, **1981**, *22*, 2431.

25. Giner-Sorolla, A., Gryte, C., Cox, M. L., and Parham, J. C., *J. Org. Chem.*, **1971**, *36*, 1228.

26. Stevens, M. A., Magrath, D. I., Smith, H. W., and Brown, G. B., *J. Am. Chem. Soc.*, **1958**, *80*, 2755.

27. 'The 7-*N*-oxides of purines related to nucleic acids: their chemistry, synthesis, and biological evaluation', Fujii, T., Itaya, T., and Ogawa, K., *Heterocycles*, **1997**, *44*, 573.

28. Ikehara, M. and Kaneko, M., *Tetrahedron*, **1970**, *26*, 4251.

29. Fischer, E., *Justus Liebigs Ann. Chem.*, **1882**, *213*, 316; *ibid.*, *221*, 336; Bruhns, G., *Chem. Ber.*, **1890**, *23*, 225; Beaman, A. G. and Robins, R. K., *J. Org. Chem.*, **1963**, *28*, 2310.

30. Barrio, J. R., Namavari, M., Phelps, M. E., and Satyamurthy, N., *J. Am Chem. Soc.*, **1996**, *118*, 1408.

31. Mosselhi, M. A. and Pfleiderer, W., *J. Heterocycl. Chem.*, **1993**, *30*, 1221.

32. Jones, J. W. and Robins, R. K., *J. Am. Chem. Soc.*, **1960**, *82*, 3773.

33. Desaubry, L. and Bourguignon, J.-J., *Tetrahedron Lett.*, **1995**, *36*, 7875.

34. Zady, M. F. and Wong, J. L., *J. Org. Chem.*, **1979**, *44*, 1450.

35. Maria, E. J., Fourrey, J.-L., Machado, A. S., and Robert-Gero, M., *Synth. Commun.*, **1996**, *26*, 27.

36. Saladino, R., Crestini, C., Bernini, R., Mincione, E., and Ciafrino, R., *Tetrahedron Lett.*, **1995**, *36*, 2665.

37. Madyastha, K. M. and Sridhar, G. R., *J. Chem. Soc., Perkin Trans. 1*, **1999**, 677.

38. Smith, D. L. and Elving, P. J., *J. Am. Chem. Soc.*, **1962**, *84*, 1412.

39. Butula, I., *Justus Liebigs Ann. Chem.*, **1969**, *729*, 73.

40. Hecht, S. M., Adams, B. L., and Kozarich, J. W., *J. Org. Chem.*, **1976**, *41*, 2303.

41. Barlin, G. B., *J. Chem. Soc. (B)*, **1967**, 954.

42. Hamamichi, N. and Miyasaka, T., *J. Heterocycl. Chem.*, **1990**, *27*, 2011; *ibid.*, **1991**, *28*, 397.

43. Montgomery, J. A. and Holum, L. B., *J. Am. Chem. Soc.*, **1957**, *79*, 2185.

44. Focher, F., Hildebrand, C., Freese, S., Ciarrocchi, G., Noonan, T., Sangalli, S., Brown, N., Spadari, S., and Wright, G., *J. Med. Chem.*, **1988**, *31*, 1496.

45. Kos, N. J., van der Plas, H. C., and van Veldhuizen, A., *Recl. Trav. Chim. Pays-Bas*, **1980**, *99*, 267.

46. Motawia, M. S., Meldal, M., Sofan, M., Stein, P., Pedersen, E. B., and Nielsen, C., *Synthesis*, **1995**, 265.

47. Nair, V. and Sells, T. B., *Tetrahedron Lett.*, **1990**, *31*, 807.

48. Xu, Y.-Z., *Tetrahedron*, **1998**, *54*, 187.

49. Edwards, C., Boche, G., Steinbrecher, T., and Scheer, S., *J. Chem. Soc., Perkin Trans. 1*, **1997**, 1887.

50. Flaherty, D., Balse, P., Li, K., Moore, B. M., and Doughty, M. B., *Nucleosides & Nucleotides*, **1995**, *14*, 65.

51. Miyashita, A., Suzuki, Y., Ohta, K., and Higashino, T., *Heterocycles*, **1994**, *39*, 345.

52. De Napoli, L., Montesarchio, D., Piccialli, G., Santacroce, C., and Varra, M., *J. Chem. Soc., Perkin Trans. 1*, **1995**, 15.
53. Linn, J. A., McLean, E. W., and Kelley, J. L., *J. Chem. Soc., Chem. Commun.*, **1994**, 913.
54. Lembicz, N. K., Grant, S., Clegg, W., Grifin, R. J., Heath, S. L., and Golding, B. T., *J. Chem. Soc., Perkin Trans. 1*, **1997**, 185.
55. Herdewin, P., Van Aerschot, A., and Pfleiderer, W., *Synthesis*, **1989**, 961.
56. Kiburis, J. and Lister, J. H., *J. Chem. Soc. (C)*, **1971**, 3942.
57. Lee, H., Luna, E., Hinz, M., Stezowski, J. J., Kiselyov, A. S., and Harvey, R. G., *J. Org. Chem.*, **1995**, *60*, 5604.
58. Lakshman, M. K., Keeler, J. C., Hilmer, J. H., and Martin, J. Q., *J. Am. Chem. Soc.*, **1999**, *121*, 6090.
59. Miles, R. W., Samano, V., and Robins, M. J, *J. Am. Chem. Soc.*, **1995**, *117*, 5951; Clivio, P., Fourrey, J.-L., and Fauvre, A., *J. Chem. Soc., Perkin Trans. 1*, **1993**, 2585.
60. Miyashita, A., Suzuki, Y., Iwamoto, K., and Higashino, T., *Chem. Pharm. Bull.*, **1998**, *46*, 390.
61. Wong, J. L. and Keck, J. H., *J. Chem. Soc., Chem. Commun.*, **1975**, 125.
62. Hayakawa, H., Haraguchi, K., Tanaka, H., and Miyasaka, T., *Chem. Pharm. Bull.*, **1987**, *35*, 72.
63. Leonard, N. J. and Bryant, J. D., *J. Org. Chem.*, **1979**, *44*, 4612.
64. Hayakawa, H., Tanaka, H., Sasaki, K., Haraguchi, K., Saitoh, T., Takai, F., and Miyasaka, T., *J. Heterocycl. Chem.*, **1989**, *26*, 189.
65. Nolsoe, J. M. J., Gundersen, L.-L., and Rise, F., *Synth. Commun.*, **1998**, *28*, 4303.
66. Kato, K., Hayakawa, H., Tanaka, H., Kumamoto, H., Shindoh, S., Shuto, S., and Miyasaka, T., *J. Org. Chem.*, **1997**, *62*, 6833.
67. Matsuda, A., Shinozaki, M., Miyasaka, T., Machida, H., and Abiru, T., *Chem. Pharm. Bull.*, **1985**, *33*, 1766; Nair, V., Turner, G. A., Buenger, G. S., and Chamberlain, S. D., *J. Org. Chem.*, **1988**, *53*, 3051; Jacobson, K. A., Shi, D., Gallo-Rodriguez, C., Manning, M., Mller, C., Daly, J. W., Neumeyer, J. L., Kioriasis, L., and Pfleiderer, W., *J. Med. Chem.*, **1993**, *36*, 2639.
68. Bergstrom, D. E. and Reday, P. A., *Tetrahedron Lett.*, **1982**, *23*, 4191.
69. Langli, G., Gundersen, L.-L., and Rise, F., *Tetrahedron*, **1996**, *52*, 5625.
70. Nolsoe, J. M. J., Gundersen, L.-L., and Rise, F., *Acta Chem. Scand.*, **1999**, *53*, 366.
71. Ozola, V., Persson, T., Gronowitz, S., and Hrnfeldt, A.-B., *J. Heterocycl. Chem.*, **1995**, *32*, 863.
72. Elion, G. B., *J. Org. Chem.*, **1962**, *27*, 2478.
73. Waters, T. R. and Connolly, B. A., *Nucleosides and Nucleotides*, **1992**, *11*, 1561; Fathi, R., Goswani, B., Kung, P.-P., Gaffney, B. L., and Jones, R. A., *Tetrahedron Lett.*, **1990**, *31*, 319.
74. Vorbrggen, H. and Krolikiewicz, K., *Justus Liebigs Ann. Chem.*, **1976**, 745.
75. Gerster, J. F., Jones, J. W., and Robins, R. K., *J. Org. Chem.*, **1963**, *28*, 945; Robins, M. J. and Uznanski, B., *Can. J. Chem.*, **1981**, *59*, 2601.
76. De Napoli, L., Messere, A., Montesarchio, D., Piccialli, G., Santacroce, C., and Varra, M., *J. Chem. Soc., Perkin Trans. 1*, **1994**, 923.
77. Elion, G. B. and Hitchings, G. H., *J. Am. Chem. Soc.*, **1956**, *78*, 3508; Davoll, J. and Lowy, B. A., **1951**, *73*, 2936.
78. Beaman, A. G. and Robins, R. K., *J. Am. Chem. Soc.*, **1961**, *83*, 4038.
79. Kataoka, S., Isono, J., Yamaji, N., Kato, M., Kawada, T., and Imai, S., *Chem. Pharm. Bull.*, **1988**, *36*, 2212.
80. El-Kafrawy, S. A., Zahran, M. A., and Pedersen, E. B., *Acta Chem. Scand.*, **1999**, *53*, 280.
81. Montgomery, J. A. and Hewson, K., *J. Am. Chem. Soc.*, **1960**, *82*, 463.
82. Jones, J. W. and Robins, R. K., *J. Am. Chem. Soc.*, **1960**, *82*, 3773.
83. Moschel, R. C. and Keefer, L. K., *Tetrahedron Lett.*, **1989**, *30*, 1467.
84. Orozco, M., Canela, E. I., and Franco, R., *J. Org. Chem.*, **1990**, *55*, 2630.
85. Nair, V. and Richardson, S. G., *J. Org. Chem.*, **1980**, *45*, 3969; idem, *Synthesis*, **1982**, 670.
86. Nair, V. and Hettrick, B. J., *Tetrahedron*, **1988**, *44*, 7001.
87. Kikugawa, K., Suehiro, H., and Aoki, A., *Chem. Pharm. Bull.*, **1977**, *25*, 2624.
88. Beaman, A. G., Gerster, J. F., and Robins, R. K., *J. Org. Chem.*, **1962**, *27*, 986.

89. Matsuda, A., Nomoto, Y., and Ueda, T., *Chem. Pharm. Bull.*, **1979**, *27*, 183; Wetzel, R. and Eckstein, F., *J. Org. Chem.*, **1975**, *40*, 658; Yamane, A., Matsuda, A., and Ueda, T., *Chem. Pharm. Bull.*, **1980**, *28*, 150.

90. Saladino, R., Mincione, E., Crestini, C., and Mezzetti, M., *Tetrahedron*, **1996**, *52*, 6759.

91. Brown, D. M. and Giner-Sorolla, A., *J. Chem. Soc. (C)*, **1971**, 128.

92. Mackay, L. B. and Hitchings, G. H., *J. Am. Chem. Soc.*, **1956**, *78*, 3511.

93. Traube, *Chem. Ber.*, **1900**, *33*, 1371; *ibid.*, 3035.

94. Robins, R. K., Dille, K. J., Willits, C. H., and Christensen, B. E., *J. Am. Chem. Soc.*, **1953**, *75*, 263.

95. Melguizo, M., Nogueras, M., and Sánchez, A., *Synthesis*, **1992**, 491.

96. Albert, A. and Brown, D. J., *J. Chem. Soc.*, **1954**, 2060.

97. Albert, A. and Brown, D. J., *J. Chem. Soc.*, **1954**, 2066; Montgomery, J. A., *J. Am. Chem. Soc.*, **1956**, *78*, 1928; Young, R. C., Jones, M., Milliner, K. J., Rana, K. K., and Ward, J. G., *J. Med. Chem.*, **1990**, *33*, 2073; Elion, G. B., Burgi, E., and Hitchings, G. H., *J. Am. Chem. Soc.*, **1951**, *73*, 5235.

98. Taylor, E. C., Vogl, O., and Cheng, C. C., *J. Am. Chem. Soc.*, **1959**, *81*, 2442.

99. Párkányi, C. and Yuan, H. L., *J. Heterocycl. Chem.*, **1990**, *27*, 1409.

100. Goldman, L., Marsico, J. W., and Gazzola, A. L., *J. Org. Chem.*, **1956**, *21*, 599.

101. Orji, C. C., Kelly, J., Ashburn, D. A., and Silks, R. A., *J. Chem. Soc., Perkin Trans. 1*, **1996**, 595.

102. Jerchel, D., Kracht, M., and Krucker, K., *Justus Liebigs Ann. Chem.*, **1954**, *590*, 232.

103. Nagamatsu, T., Yamasaki, H., and Yoneda, F., *Heterocycles*, **1992**, *33*, 775.

104. Senga, K., Kanazawa, H., and Nishigaki, S., *J. Chem. Soc., Chem. Commun.*, **1976**, 155.

105. Altman, J. and Ben-Ishai, D., *J. Heterocycl. Chem.*, **1968**, *5*, 679.

106. 'Annelation of a pyrimidine ring to an existing ring', Albert, A., *Adv. Heterocycl. Chem.*, **1982**, *32*, 1.

107. Ferris, J. P. and Orgel, L. E., *J. Am. Chem. Soc.*, **1966**, *88*, 1074 and 3829.

108. Yamazaki, A., Kumashiro, I., and Takenishi, T., *J. Org. Chem.*, **1967**, *32*, 3258.[

109. Kelley, J. L., Linn, J. A., and Selway, J. W. T., *J. Med. Chem.*, **1989**, *32*, 218; Prasad, R. N. and Robins, R. K., *J. Am. Chem. Soc.*, **1957**, *79*, 6401.

110. Imai, K., Marumoto, R., Kobayashi, K., Yoshioka, Y., Toda, J., and Honjo, M., *Chem. Pharm. Bull.*, **1971**, *19*, 576.

111. Al-Shaar, A. H., Gilmour, D. W., Lythgoe, D. J., McClenaghan, I., and Ramsden, C. A., *J. Chem. Soc., Chem. Commun.*, **1989**, 551.

112. Ostrowski, S., *Synlett*, **1995**, 253.

113. Dang, Q., Brown, B. S., and Erion, M. D., *J. Org. Chem.*, **1996**, *61*, 5204.

114. Dang, Q., Liu, Y., and Erion, M. D., *J. Am. Chem. Soc.*, **1999**, *121*, 5833.

115. Ochiai, M., Marumoto, R., Kobayashi, S., Shimazu, H., and Morita, K., *Tetrahedron*, **1968**, *24*, 5731.

116. Yamada, H. and Okamoto, T., *Chem. Pharm. Bull.*, **1972**, *20*, 623

117. Richter, E., Loeffler, J. E., and Taylor, E. C., *J. Am. Chem. Soc.*, **1960**, *82*, 3144.

118. Arita, M., Adachi, K., Ito, Y., Sawai, H., and Ohno, M., *J. Am. Chem. Soc.*, **1983**, *105*, 4049.

119. Kenner, G. W., Taylor, C. W., and Todd, A. R., *J. Chem. Soc.*, **1949**, 1620.

25 Heterocycles containing a ring-junction nitrogen

In addition to the biologically important purines and pteridines and the major benzo-fused heterocycles such as indole, many other aromatic, fused heterocyclic ring systems are known, and of these, the most important are those containing a ring-junction nitrogen – that is, where a *nitrogen is common to two rings*.[1] The vast majority of these systems do not occur naturally, but they have been the subject of many studies from the theoretical viewpoint, for the preparation of potentially biologically active analogues, and for other industrial uses. For reasons of space, only combinations of five- and six-membered rings are considered here, though other combinations are possible and are known.

Indolizine   4*H*-quinolizine   quinolizinium   3*H*-pyrrolizine   pyrrolizine anion

Of the parent systems which have the ring-junction nitrogen as the *only* heteroatom, only indolizine (often 'pyrrocoline' in the older literature) has a neutral, fully conjugated 10-electron π-system, comprising four pairs of electrons from the four double bonds and a pair from nitrogen, much as in indole. 4*H*-Quinolizine is not aromatic – there is a saturated atom interrupting the conjugation – but the cation, quinolizinium, formed formally by loss of hydride from quinolizine, *does* have an aromatic 10-electron system: it is completely isoelectronic with naphthalene, the positive charge resulting from the higher nuclear charge of nitrogen *versus* carbon. Similarly, pyrrolizine, which is already aromatic in being a pyrrole (with an α-vinyl substituent), on conversion into its conjugate anion, attains a 10-electron π-system.

25.1 Indolizines[2]

The aromatic character of indolizine is expressed by three main mesomeric contributors, two of which incorporate a pyridinium moiety; other structures (not shown) incorporating neither a complete pyrrole nor a pyridinium are less important.

swainsonine

Aromatic indolizines are very rare in nature, but the fully reduced (indolizidine) nucleus is widespread, particularly in alkaloids, of which swainsonine is a typical example. Synthetic indolizines have found use in photographic dyes.

25.1.1   Reactions of indolizines

Indolizine is an electron-rich system and its reactions are mainly electrophilic substitutions, which occur about as readily as for indole, and go preferentially at C-3, but may also take place at C-1. Consistent with their similarity to pyrroles, rather than pyridines, indolizines are not attacked by nucleophiles, nor are there examples of nucleophilic displacement of halide.

Indolizine, pK_a 3.9,[3] is much more basic than indole (pK_a -3.5) and the implied relative stability of the cation makes it less reactive and thus indolizines resistant to acid-catalysed polymerisation (*cf.* Section 17.1.9). Indolizine protonates at C-3, but 3-methylindolizine protonates mainly (79%) at C-1; the delicacy of the balance is further illustrated by 1,2,3-trimethyl- and 3,5-dimethylindolizines, each of which protonate exclusively at C-3. Electrophilic substitutions such as acylation,[4] Vilsmeier formylation,[5] and diazo-coupling[6] all take place at C-3.

Nitration of 2-methylindolizine under mild conditions results in substitution at C-3,[7] but under strongly acidic conditions it takes place at C-1,[8] presumably *via* attack on the indolizinium cation.

Indolizine and its simple alkyl derivatives are sensitive to light and to aerial oxidation, which lead to destruction of the ring system. Catalytic reduction in acidic solution – reduction of the indolizinium cation – gives a pyridinium salt;[9] complete saturation, affording indolizidines, results from reductions over platinum.[10]

Despite its 10-electron aromatic π-system, indolizine apparently participates as an 8-electron system in its reaction with diethyl acetylenedicarboxylate, though the process may be stepwise and not concerted. By carrying out the reaction in the presence of a noble metal as catalyst, the initial adduct is converted into an aromatic cyclazine (Section 25.5).[11]

5-Methylindolizine undergoes lithiation at the side-chain methyl;[12] 2-phenylindolizine lithiates at C-5.[13]

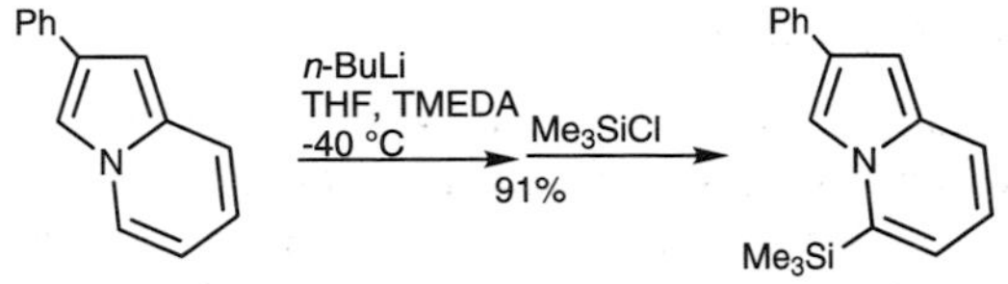

Of its functional derivatives, worth noting is the easy cleavage of carboxyl and acyl groups on heating with aqueous acid, and the instability of amino-derivatives, which cannot be diazotised, but which can be converted into stable acetamides.

25.1.2   Synthesis of indolizines[14]

The most general approach to indolizines is the Chichibabin synthesis[15] which involves quaternisation of a 2-alkylpyridine with an α-haloketone, followed by base-catalysed cyclisation *via* deprotonation of the pyridinium α-methyl[16] which is of course easier if that alkyl groups is further acidified.[17]

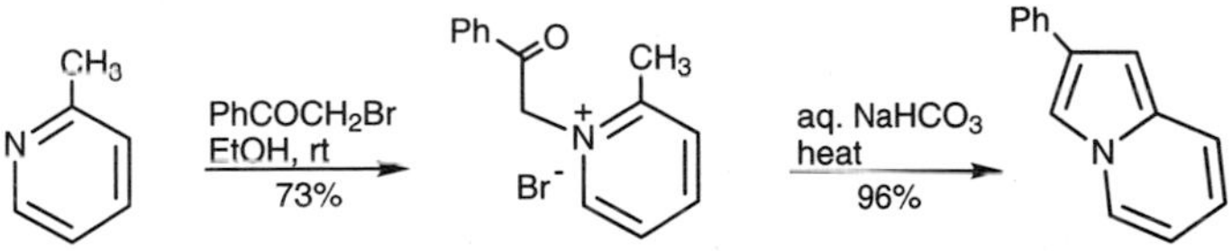

Another useful method involves the intermediacy of a pyridinium ylide as a 1,3-dipole in a cycloaddition.[18]

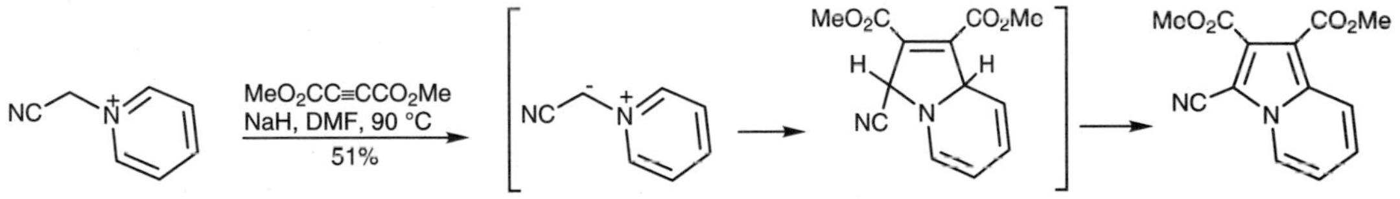

An important feature of this type of reaction (which can also be used in an analogous fashion to prepare aza-indolizines) is that although a dihydroindolizine is the logical product from a mechanistic viewpoint, the fully aromatic compound is usually obtained. The mechanism of aromatisation is not clear: it could be by air oxidation during work up, or *via* hydride transfer to some other component in the reaction mixture. When dihydro-compounds are isolated they can be easily aromatised using the usual reagents such as palladium/charcoal or quinones. An extension of this analysis is the production of aromatic indolizines from reaction of an ylide with an alkene (which would be expected to give the tetrahydro-product) *in the presence* of a suitable oxidant such as the cobalt chromate[19] shown below.[20]

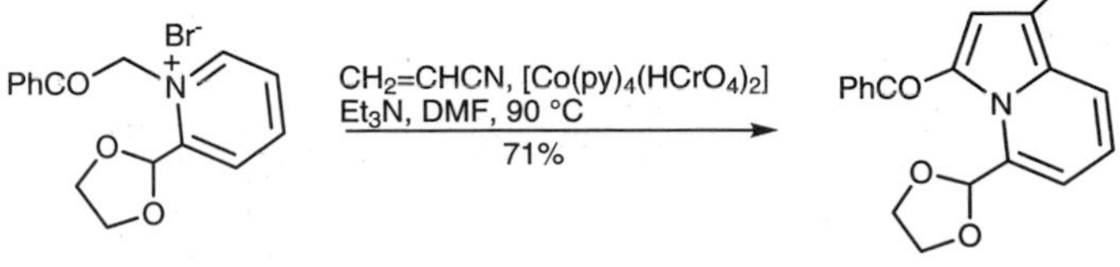

In reaction with alkynes, aromatisation can occur by loss of HX when a leaving group is present in one of the reactants.[21]

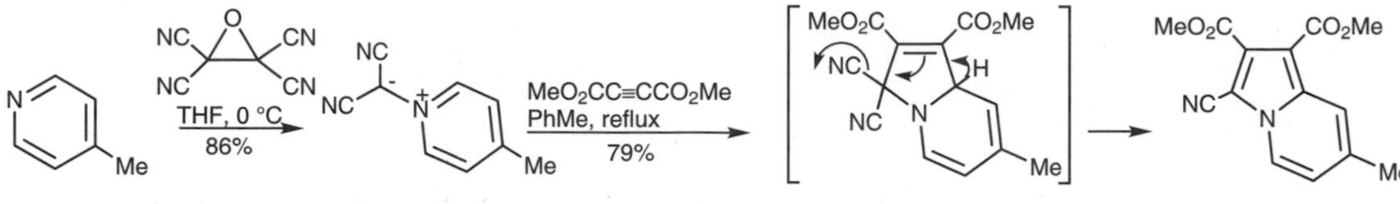

25.2   Aza-indolizines

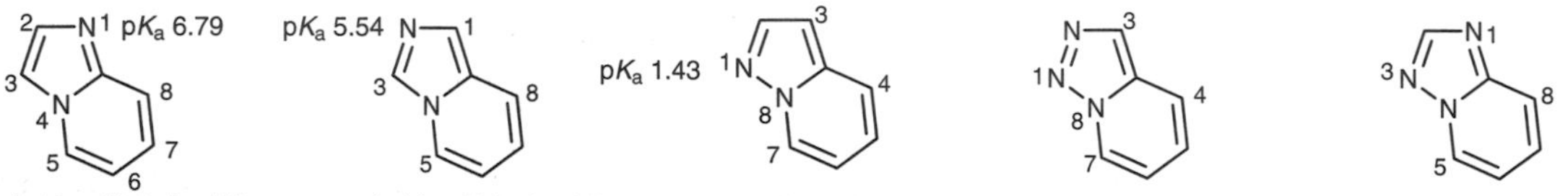

imidazo[1,2-a]pyridine   imidazo[1,5-a]pyridine   pyrazolo[1,5-a]pyridine   1,2,3-triazolo[1,5-a]pyridine 1,2,4-triazolo[1,5-a]pyridine

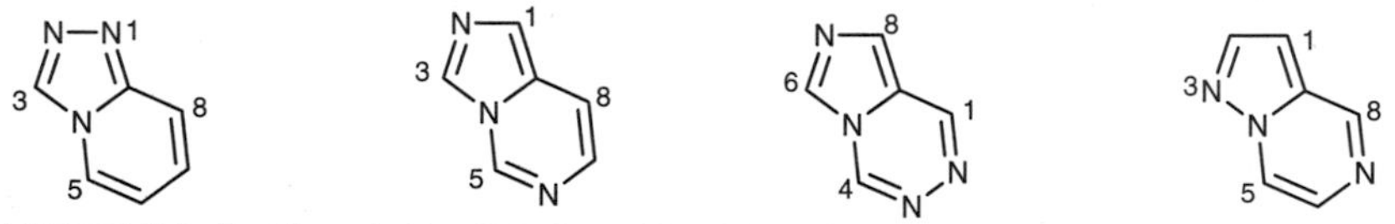

1,2,4-triazolo[4,3-a]pyridine   imidazo[1,5-c]pyrimidine   imidazo[1,5-d][1,2,4]triazine   pyrrolo[1,2-a]pyrazine

Note: as can be seen from the examples above, numbering sequences vary with the number and disposition of the nitrogen atoms.

Seven monoaza- and many more polyaza-indolizines (some are shown above) are possible, indeed compounds with up to six nitrogen atoms have been frequently reported. Despite the great rarity of such systems in nature, there is much interest in aza-indolizines stemming from their structural similarity to both indoles and purines. The imidazopyrazine ring occurs in *Cypridina* luciferin (ch. 11). The antidepressant Trazadone is an example of the large number of aza-indolizines which have been prepared for assessment of their pharmacological activity. Trapadil is a coronary vasodilator, Ibudilast is used in the treatment of asthma and Temozolomide is an anti-cancer agent. Compounds of use in other areas include the plant antifungal agent Pyrazophos.

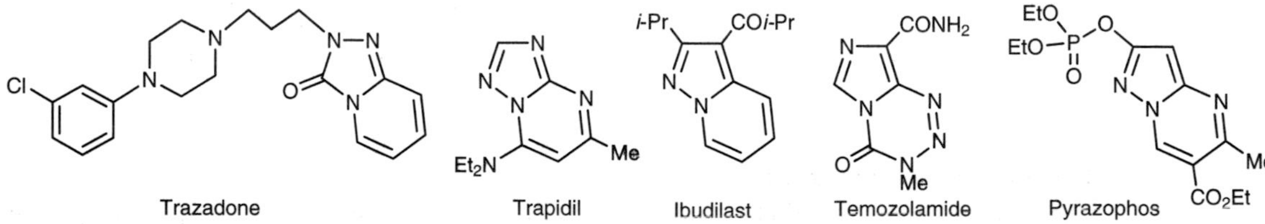

Trazadone   Trapidil   Ibudilast   Temozolamide   Pyrazophos

Apart from pyrrolo[1,2-b]pyridazine, all the monoaza-indolizines protonate on the second (non-ring-junction) nitrogen, rather than on carbon.[3,22] Alkylation similarly goes on nitrogen however other electrophilic reagents attack with regioselectivity similar to indolizine itself – they effect substitution of the five-membered ring at positions 1 and 3 (where these are carbon).

25.2.1   Imidazo[1,2-a]pyridine

Electrophilic substitutions such as halogenation, nitration *etc.* go at C-3, or at C-5 if position 3 is blocked.[23] Acylation does not require a catalyst.[24]

Of all the positional chloro-isomers, nucleophilic displacement reactions are known only for the 5-isomer; 7-chloroindolizine, where one might have anticipated similar activation, is not reactive in this sense.[25]

Base-catalysed deuterium exchange goes at C-3 and C-5;[26] preparative lithiation occurs at C-3, or if C-3 is blocked, at C-5 or C-8 depending on other substituents[27] but the 2,6-dichloro compound reacts selectively at C-5, even in the presence of hydrogen at C-3.[28]

Amino-imidazo[1,2-a]pyridines are even more unstable than aminoindolizines; they exist as amino tautomers, but 2- and 5-oxygenated derivatives are in the keto form. These last react as usual with phosphoryl chloride yielding chloro-compounds.[20]

The ring synthesis of imidazo[1,2-a]pyridines is based on the Chichibabin route to indolizines (Section 25.1.2), but using 2-aminopyridines instead of 2-alkylpyridines. The initial reaction with the halo-ketone is regioselective for the ring nitrogen, so isomerically pure products are obtained.[29] 2-Oxoimidazo[1,2-a]pyridines are the products when an α-bromo-ester is used instead of a ketone.[30]

25.2.2 Imidazo[1,5-a]pyridines

Electrophilic substitution in this system again occurs in the five-membered ring, at C-1, or at C-3 if the former position is occupied.[31,32] Reaction with bromine gives a 1,3-dibromo-product.[33]

Benzoylation provides an instructive example: under normal conditions C-substitution occurs at C-1, however in the presence of triethylamine, 3-benzoylimidazo[1,5-*a*]pyridine is the product.[34] This can be explained by assuming the intermediacy of an ylide formed by deprotonation of an initial N^+-benzoyl salt (*cf.* Section 21.1.2.5):

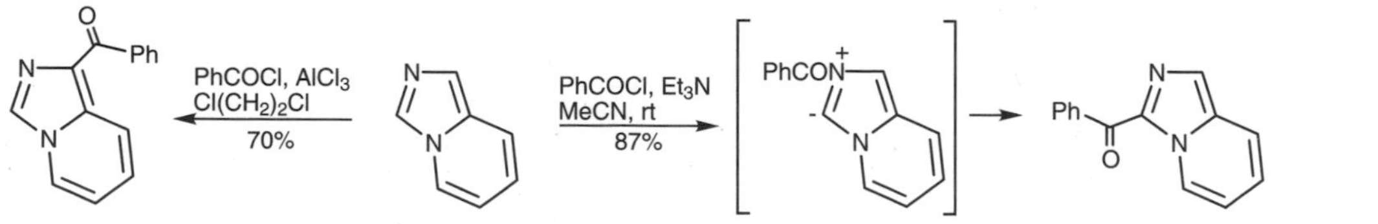

Five-membered ring cleavage occurs relatively easily: hot aqueous acid converts these heterocycles into 2-aminomethylpyridines.

Lithiation, by direct analogy with imidazole, results in loss of the 3-proton,[5] but 5-lithiation occurs on comparable treatment of 3-ethylthio-derivative, the substituent both blocking attack at C-3 *and* assisting lithiation at the *peri* position; the ethylthio group can of course be subsequently easily removed.[35]

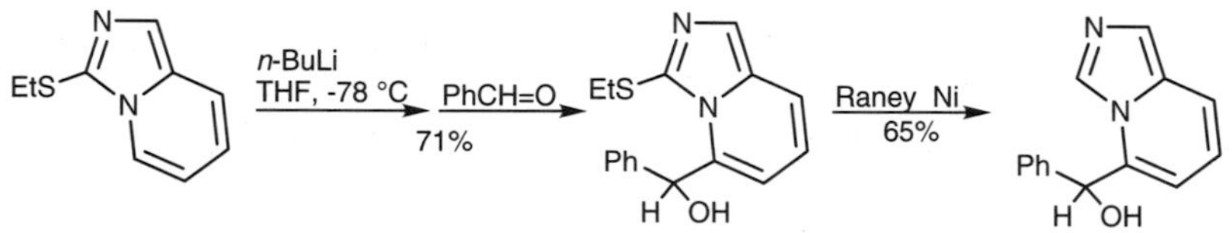

Imidazo[1,5-*a*]pyridines are synthesised by the dehydrative cyclisation of *N*-acyl-2-aminomethylpyridines.[32] 3-Amino-,[36] oxy-[37] and thio-[38] -derivatives are available *via* related cyclisations.

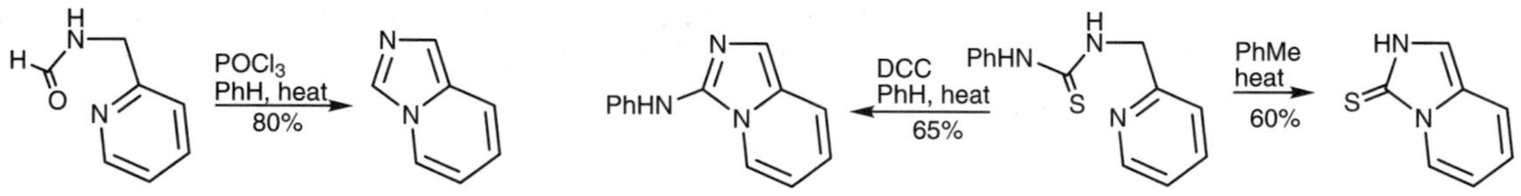

25.2.3 Pyrazolo[1,5-*a*]pyridines

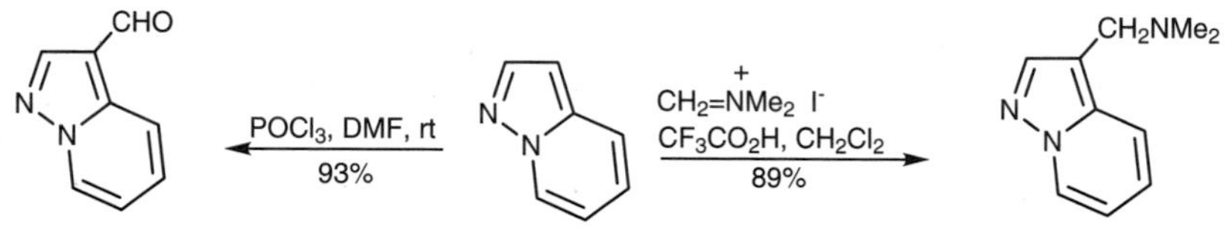

In this system, electrophilic substitution occurs at C-3[39] and lithiation takes place at C-7, though a 3-oxazoline can direct a second metallation to C-2, leading to ring cleavage,[40] as shown below.

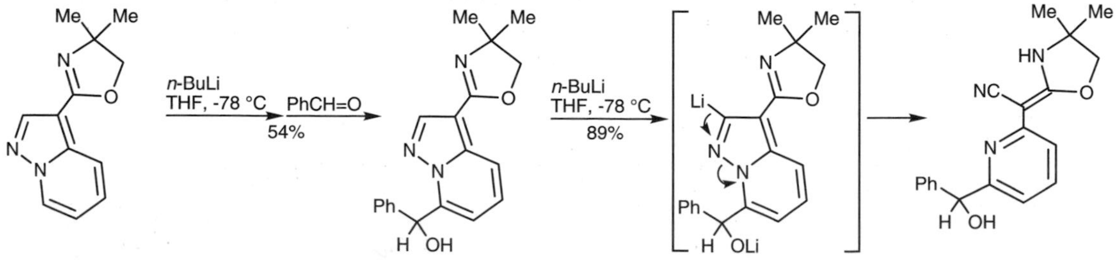

Pyrazolo[1,5-*a*]pyridines can be prepared by cycloaddition of *N*-aminopyridinium ylides with alkynes[41] or *N*-amination of 2-alkynylpyridines.[42] Interestingly, the cycloaddition of *N*-amino-3-benzyloxypyridine goes preferentially to the more hindered C-2.[43]

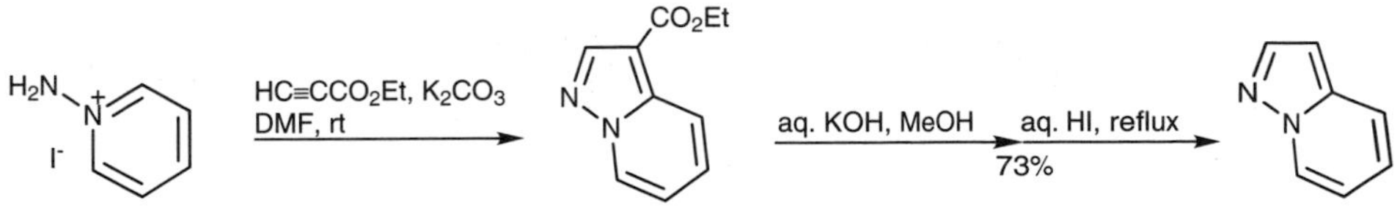

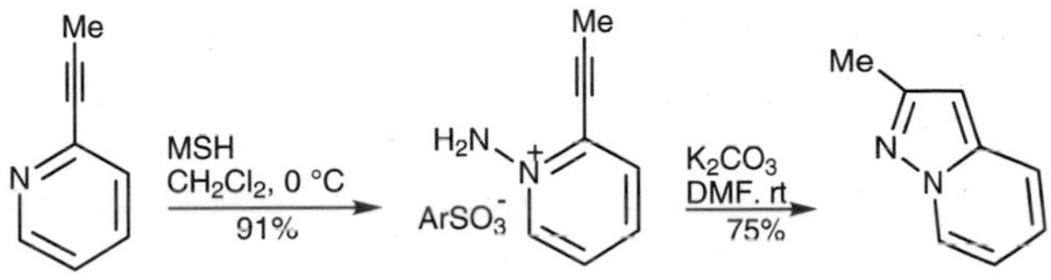

Pyridinium *N*-imide, the ylide produced by removal of a proton from 1-aminopyridinium iodide, serves as a 1,3-dipole and reacts with propiolate (shown above) or fumarate to give bicyclic compounds.[44]

25.2.4 Triazolo-[45] and tetrazolopyridines

1,2,3-Triazolo[1,5-*a*]pyridine can be, in theory, in equilibrium with its ring-opened diazo tautomer;[46] although it actually exists in the closed form, its reactions tend to reflect this potential equilibrium: reaction with electrophiles can take two courses. Acylation and nitration occur normally, at C-1, but reagents such as bromine lead to a very easy ring cleavage.[47] Aqueous acid similarly brings about ring cleavage and the formation in this case of 2-hydroxymethylpyridine.

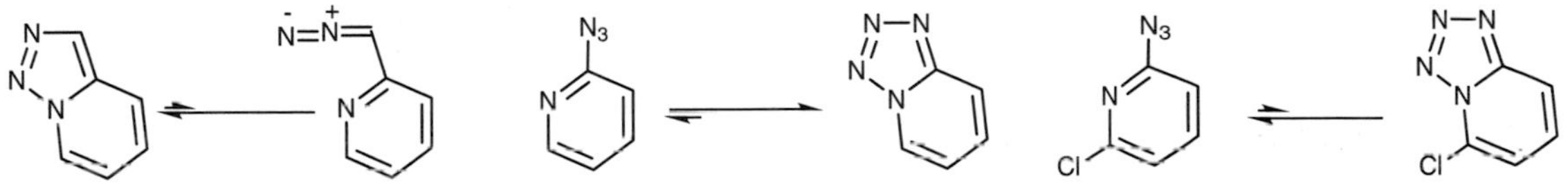

2-Azido-azines are in equilibrium with fused tetrazoles, the position of the equilibrium being very sensitive to substituent influence, for example in the unsubstituted case the equilibrium lies predominantly towards the closed form whereas the analogous 6-chloro-compound is predominantly open.[48]

Direct lithiation of 1,2,3-triazolo[1,5-*a*]pyridines occurs with ease, at C-7, subsequent reaction with electrophiles being unexceptional, for example conversion into the 7-bromo-derivative then allows nucleophiles to be introduced *via* displacement of halide, thus providing, overall, a route to 2,6-disubstituted pyridines.[49] However, lithiation in tetrahydrofuran as solvent leads to formation of a 7,7-linked dimer.[50] The 7-lithio compound is also formed by cleavage of the 7-*N,N*-diethylcarboxamide with *n*-butyllithium.[51]

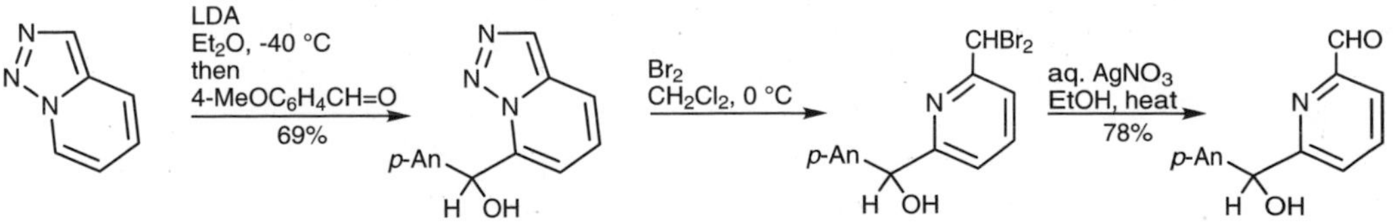

1,2,4-Triazolo[1,5-*a*]pyridine seems to be resistant to electrophilic attack but can be lithiated at C-5; in contrast, 1,2,4-triazolo[4,3-*a*]pyridine readily undergoes electrophilic substitution at C-3.[52]

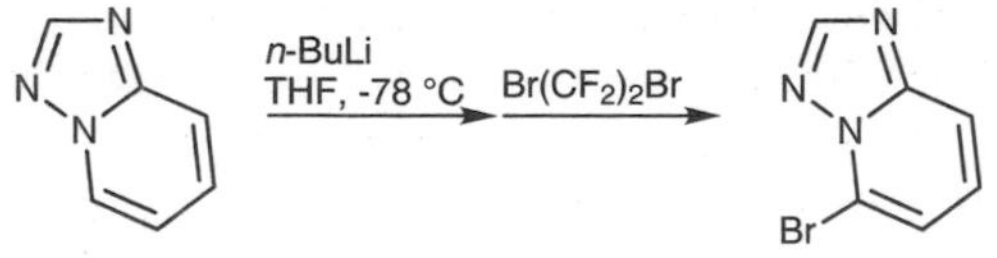

1,2,3-Triazolo[1,5-*a*]pyridines can be synthesised by oxidation of pyridine 2-carboxaldehyde hydrazones, presumably by way of the diazo-species,[53] or by diazo-transfer reactions.[54]

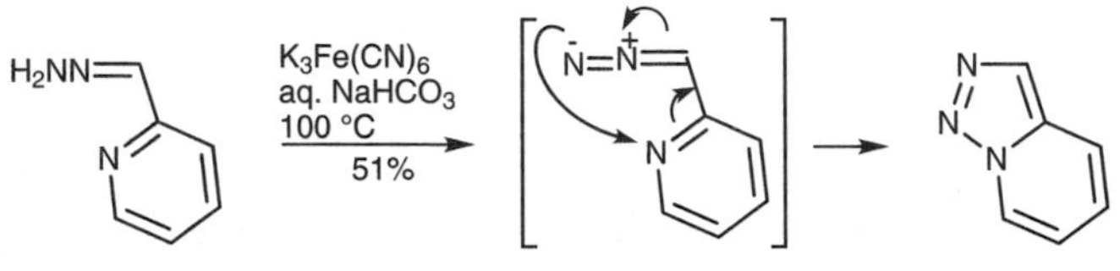

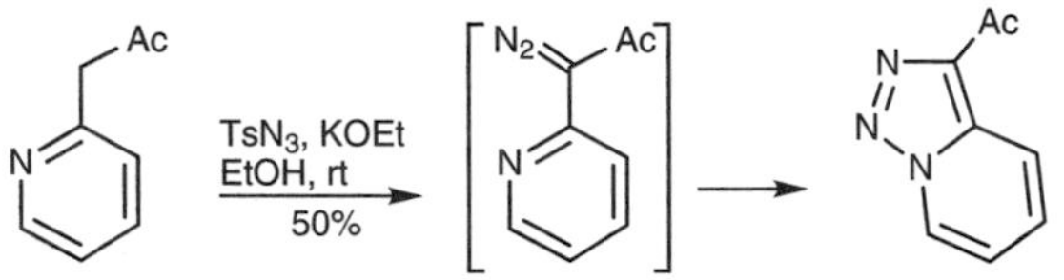

The 1,2,4-triazolo[4,3-*a*]pyridine nucleus can be accessed by cyclocondensation of 2-hydrazinopyridines; the synthesis of Trazadone shown below is an example.[55] Oxidative closure of pyridin-2-ylhydrazones produces 1,2,4-triazolo[4,3-*a*]pyridines.[56]

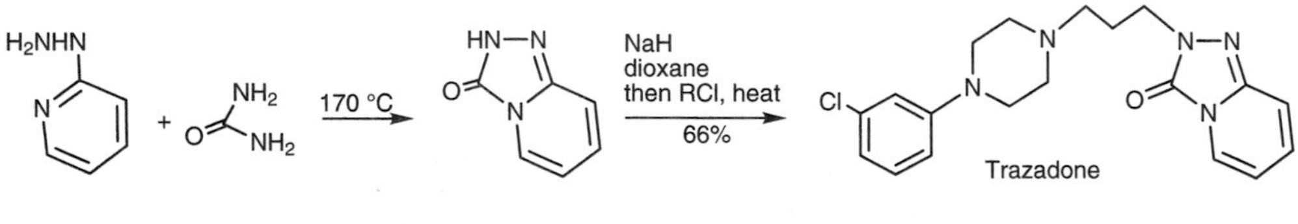

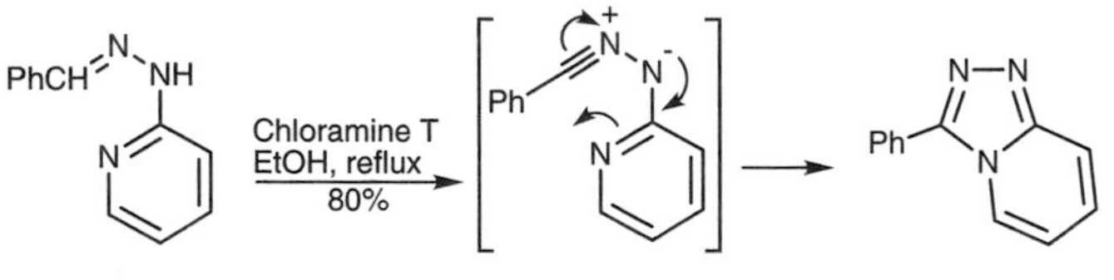

1,2,4-Triazolo[1,5-*a*]pyridines can be prepared *via* oxidative cyclisation of amidines[57] or acid-catalysed cyclisation of amidoximes[58] – each of these is illustrated below.

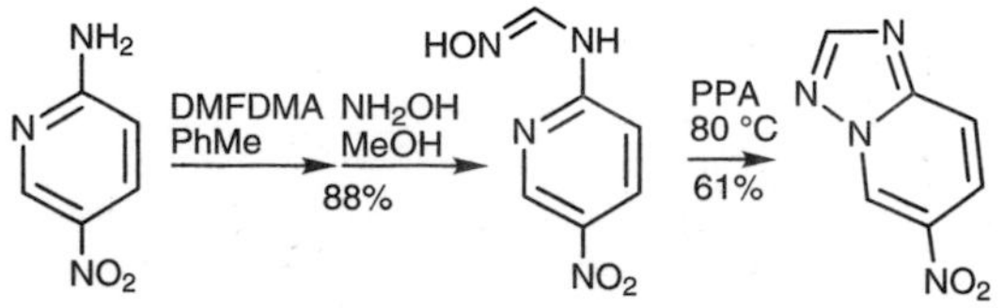

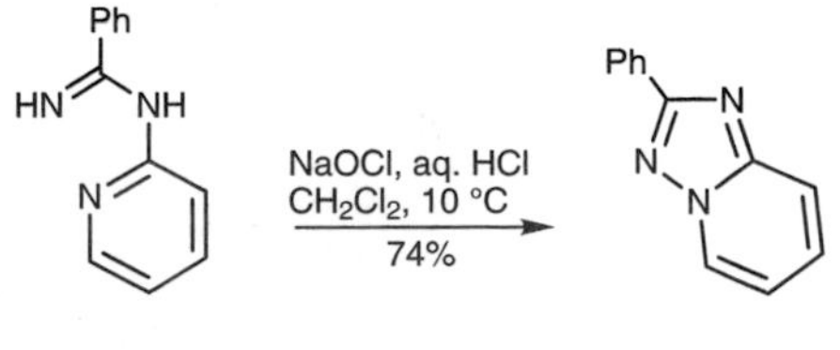

25.2.5   Compounds with an additional nitrogen in the six-membered ring

In addition to the propensity for electrophilic substitution at C-1/C-3 (see above), the main feature of this class of heterocycle is that they undergo relatively easy nucleophilic attack in the six-membered ring,[59] which is now considerably electron-deficient (through the incorporation of imine units) – the analogy with the ease of nucleophilic addition to diazines is obvious – some are so susceptible to nucleophilic addition that they form 'hydrates' even on exposure to moist air.[46] However, preparative lithiations can be carried out using less nucleophilic bases.

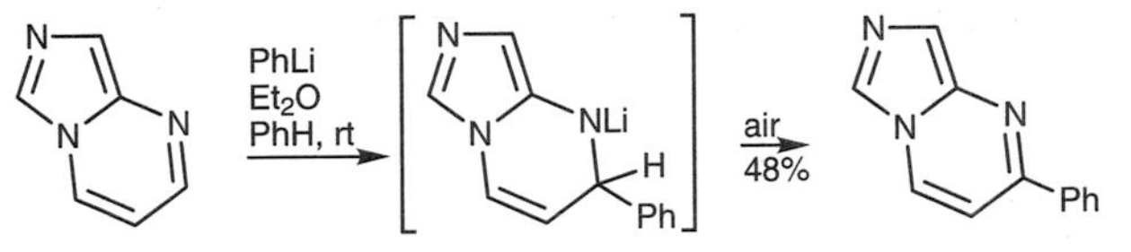

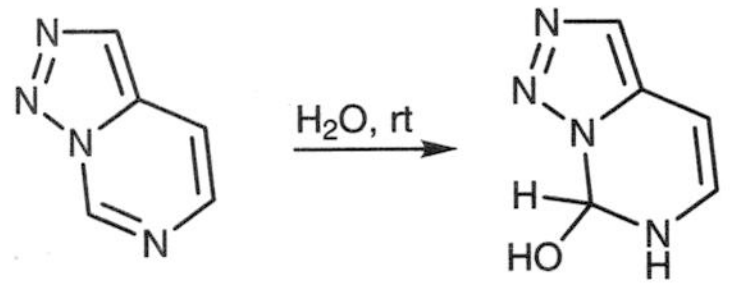

Ring synthesis of such molecules can proceed from diazines[60,61] using methods analogous to those described for the synthesis of azolopyridines from pyridines, or by various methods from the five-membered ring component[62,63] – some representative routes are shown below.

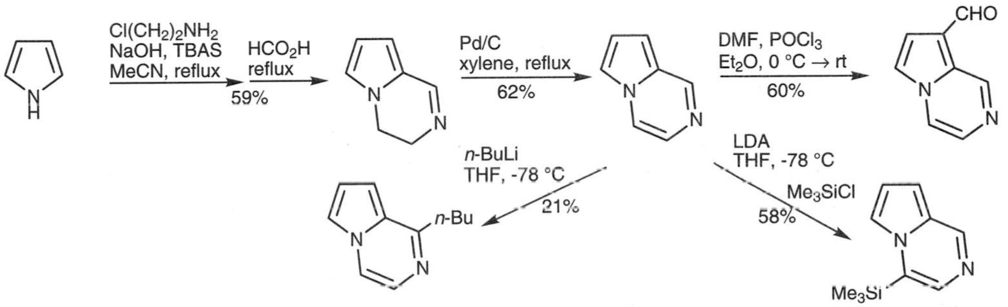

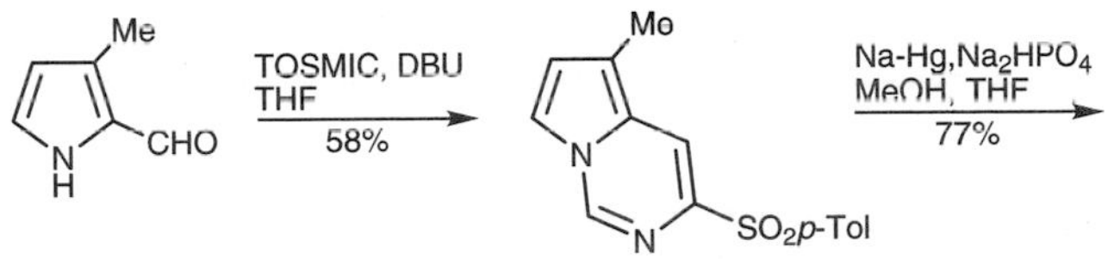

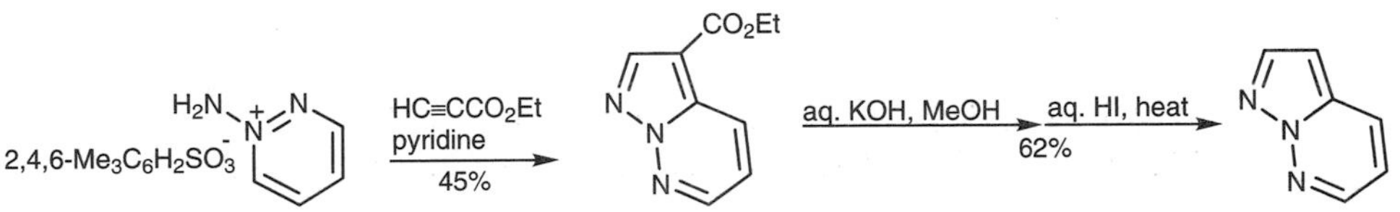

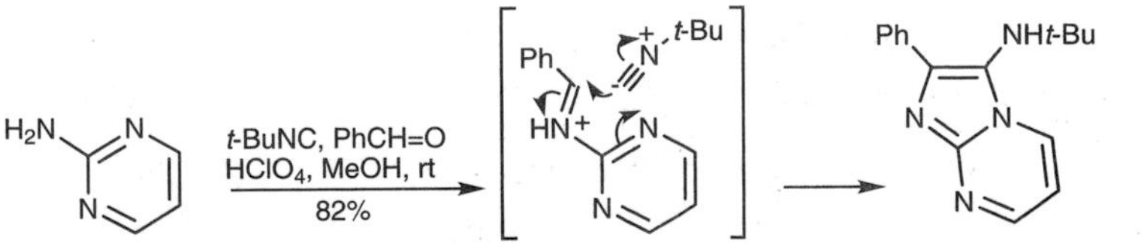

25.3   Quinoliziniums[64] and related systems

The quinolizinium ion occurs naturally only rarely, for example as a fused ylide in the alkaloid sempervirine, however there are hundreds of indole alkaloids which have the same tetracyclic system, but with the quinolizine at an octahydro-level, in addition, many simpler quinolizidine alkaloids, such as lupinine, are known. Amongst synthetic compounds the anti-asthma drug Pemirolast is an aza-analogue.

sempervirine   lupinine   Pemirolast

Practically all the reactions of quinolizinium ion are similar to those of pyridinium salts, thus it is resistant to electrophilic attack but readily undergoes nucleophilic addition, the initial adducts undergoing spontaneous electrocyclic ring opening to afford, finally, 2-substituted pyridines,[65] however the susceptibility of the cation to nucleophiles is not extreme – like simpler pyridinium salts it is stable to boiling water.

Quinazolones *can* be made to undergo electrophilic substitution, at C-1/C-3,[66] there being a clear analogy with the reactivity of pyridones.

Quinolizine derivatives are usually prepared by cyclisations onto the nitrogen in a precursor pyridine.[67]

25.4   Pyrrolizines and related systems

The saturated or partially saturated pyrrolizidine alkaloids are the main naturally occurring pyrrolizines; senecionine is an example.

3H-pyrrolizine   senecionine

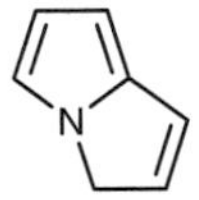

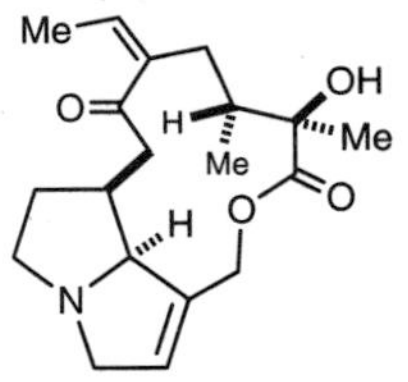

The relatively high pK_a of 29 for deprotonation of 3H-pyrrolizine (cf. indene pK_a 18.5) indicates that formation of the 10-electron pyrrolizine anion adds only minor stabilisation relative to the simple pyrrole originally present. Its reactions are those of a highly reactive carbanion, for example benzophenone condenses to generate a fulvene-like product.[68]

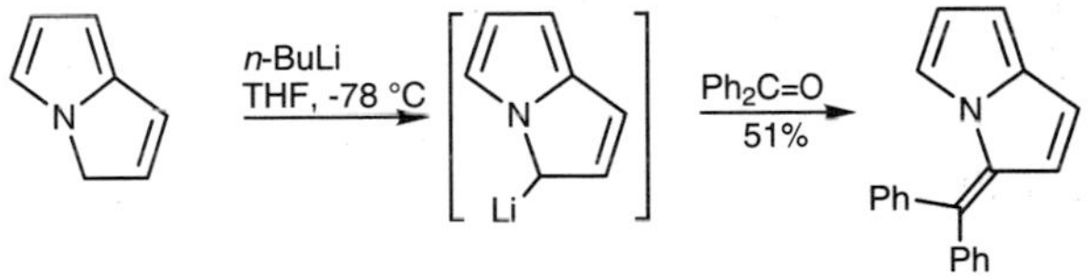

Isoelectronic replacement of a carbanionic carbon by a heteroatom gives much more stable compounds, and such 5,5-bicyclic aromatic systems have received considerable attention. In these compounds, sulfur and oxygen can also be incorporated into fully conjugated systems, unlike the 5,6-compounds where only nitrogen can be used. Because of the variety of such systems, it is difficult to generalise about reactivity but electrophilic substitution, which can take place in either ring, has been most widely reported with occasional examples of nucleophilic displacements and lithiations. Some representative reactions and self-explanatory syntheses are shown below.[69,70]

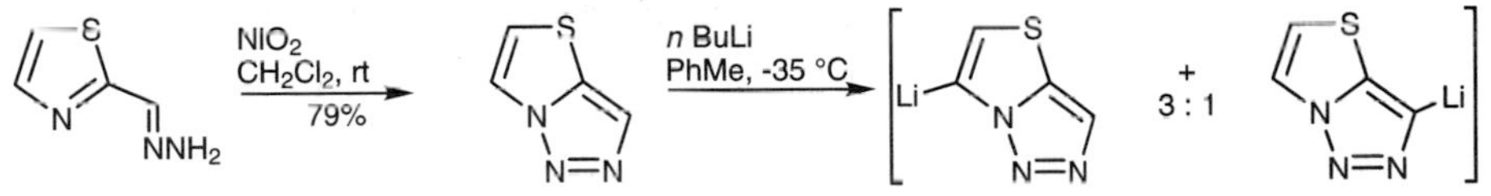

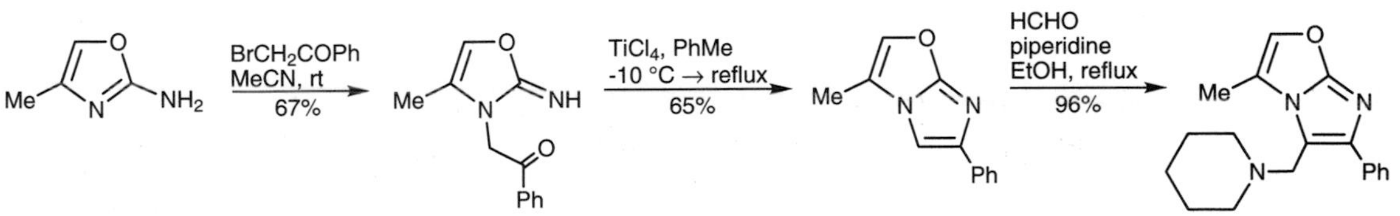

25.5   Cyclazines

The cyclazines (a trivial name) are tricyclic fused molecules containing a central bridgehead nitrogen and a peripheral π-system. The definition of aromaticity in these compounds is not as straightforward as for the simple bicyclic molecules discussed above, and a more detailed analysis of the molecular orbitals may be required.

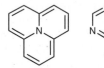

(3.2.2)cyclazine   (3.3.3)cyclazine   hexa-aza-analogue   (2.2.2)azacyclazine
                                                   of (3.3.3)cyclazine

(3.2.2)Cyclazine is a stable aromatic system with a ring current, has a 10-electron annular π-system (excluding nitrogen), and is stable to light and air but, unlike its close analogue indolizine, is non-basic indicating the much weaker interaction between the nitrogen lone pair and the peripheral π-system. It does however react as an electron-rich aromatic, undergoing electrophilic substitution readily.

In contrast, (3.3.3)cyclazine has no aromatic resonance stabilisation and is unstable and highly reactive, displaying some diradical character. However, its hexa-aza-analogue is extremely stable, this stabilisation being attributed to perturbation of the molecular orbitals by the electronegative atoms leading to a much larger separation of the HOMO and LUMO.[71] The double ring-junction nitrogen system, (2.2.2)azacyclazine is isoelectronic with (3.2.2)cyclazine and is similarly a stable system.

Cyclazines can be prepared by cyclisation of bicyclic precursors, for example (3.2.2.)cyclazine is prepared *via* a cycloaddition reaction on indolizine (Section 25.1.1), or by cyclocondensation.[72]

Exercises for Chapter 25

Straightforward revision exercises (consult chapter 25)

(a) At what position(s) does indolizine undergo electrophilic substitution? Why that position(s)?

(b) At what position does indolizine undergo strong base deprotonation?

(c) How could 2-methylpyridine be converted into 3-methylindolizine?

(d) What would be the product of reacting 2-aminopyridine with methyl bromoacetate?

(e) Draw resonance contributors to the quinolizinium cation to rationalise the position at which nucleophiles add to it.

(f) Is pyrrolizine aromatic? If so, how many electrons are there in the aromatic π-system?

More advanced exercises

1. Suggest a structure for the final, monocyclic product of the following sequence: quinolizinium bromide with $LiAlH_4$ and then H_2/Pd giving $C_9H_{13}N$.

2. Write down the structures of the intermediates in the following synthesis of the quinolizinium cation: 2-methylpyridine was reacted with LDA, then $EtO(CH_2)_2$-$CH=O$ to give $C_{11}H_{17}NO_2$ which was heated with HI ($\rightarrow C_9H_{12}NO^+ \ I^-$); this

salt was then heated with Ac_2O ($\rightarrow C_9H_{10}N^+$ I^-) and this, finally heated with Pd-C to afford quinolizinium iodide.

3. Which indolizines would be formed from the following combinations: (i) 2-picoline with (a) $BrCH_2CO.Me/NaHCO_3$, (b) $MeCHBrCHO/NaHCO_3$? (ii) What would be the products if the 2-picoline was replaced by 2-aminopyridine?

4. Deduce the structures of intermediates and final product in the following sequence: 5-methoxy-2-methylpyridine reacted with KNH_2/i-AmONO $\rightarrow$ $C_7H_8N_2O_2$ then this with $Zn/AcOH \rightarrow C_7H_{10}N_2O$, and finally this with HCO_2Me/PPE (polyphosphate ester) $\rightarrow C_8H_8N_2O$.

5. Imidazo[1,5-*a*]pyridine, on reaction with aqueous HNO_2 gave 3-(pyridin-2-yl)-1,2,4-oxadiazole. Suggest a mechanism. What product would be obtained by reaction of indolizine with nitrous acid?

6. Give the structures of the bicyclic compounds formed by the following reactions: (i) 2-hydrazinothiazole with nitrous acid $\rightarrow C_3H_2N_4S$; (ii) 2-aminothiazole with $BrCH_2COPh \rightarrow C_{11}H_8N_2S$.

References

1. 'Heterocyclic systems with bridgehead nitrogen atoms' in 'The Chemistry of Heterocyclic Compounds', Vol. 15, Ed. Mosby, W. L. Series Ed. Weissberger, A., Wiley-Interscience, **1961**; 'Special topics in heterocyclic chemistry' in 'The Chemistry of Heterocyclic Compounds', Vol. 30, Series Eds. Weissberger, A. and Taylor, E. C., Wiley-Interscience, **1977**.
2. 'The chemistry of the pyrrocolines and the octahydropyrrocolines', Borrows, E. T. and Holland, D. O., *Chem. Rev.*, **1948**, *42*, 611; 'Advances in indolizine chemistry', Swinbourne, F. J., Hunt, J. H., and Klinkert, G., *Adv. Heterocycl. Chem.*, **1978**, *23*, 103.
3. Armarego, W. L. F., *J. Chem. Soc.*, **1964**, 4226.
4. Scholtz, M., *Chem. Ber.*, **1912**, *65*, 1718.
5. Rossiter, E. D. and Saxton, J. E., *J. Chem. Soc.*, **1953**, 3654; Fuentes, O. and Paudler, W. W., *J. Heterocycl. Chem.*, **1975**, *12*, 379.
6. Holland, D. O. and Nayler, J. H. C., *J. Chem. Soc.*, **1955**, 1504.
7. Hickman, J. A. and Wibberly, D. G., *J. Chem. Soc., Perkin Trans. 1*, **1972**, 2954.
8. Borrows, E. T., Holland, D. O., and Kenyon, J., *J. Chem. Soc.*, **1946**, 1077; Greci, L. and Ridd, J. H., *J. Chem. Soc., Perkin Trans. 2*, **1979**, 312.
9. Lowe, O. G. and King, L. C., *J. Org. Chem.*, **1959**, *24*, 1200.
10. Walter, L. A. and Margolis, P., *J. Med. Chem.*, **1967**, *10*, 498.
11. Galbraith, A., Small, T., Barnes, R. A., and Boekelheide, V., *J. Am. Chem. Soc.*, **1961**, *83*, 453.
12. Windgassen, R. J., Saunders, W. H., and Boekelheide, V. *J. Am. Chem. Soc.*, **1959**, *81*, 1459.
13. Renard, M. and Gubin, J., *Tetrahedron Lett.*, **1992**, *33*, 4433.
14. 'Methods for the construction of the indolizine nucleus', Uchida, T. and Matsumoto, K., *Synthesis*, **1976**, 209.
15. Chichibabin, A. E., *Chem. Ber.*, **1927**, *60*, 1607.
16. Borrows, E. T., Holland, D. O., and Kenyon, J., *J. Chem. Soc.*, **1946**, 1069.
17. Bragg, D. R. and Wibberly, D. G., *J. Chem. Soc.*, **1963**, 3277.
18. Henrick, C. A., Ritchie, E., and Taylor, W. C., *Aust. J. Chem.*, **1967**, *20*, 2467.
19. Wei, J., Hu, Y., Li, T., and Hu, H., *J. Chem. Soc., Perkin Trans. 1*, **1993**, 2487.
20. Hou, J., Hu, Y., and Hu, H., *Synth. Commun.*, **1998**, *28*, 3397.
21. Matsumoto, K., Ogasawara, A., Kimura, S., Hayashi, N., and Machiguchi, T., *Heterocycles*, **1998**, *48*, 861.
22. Fraser, M. *J. Org. Chem.*, **1971**, *36*, 3087
23. Paolini, J. P. and Robins, R. K., *J. Org. Chem.*, **1965**, *30*, 4085.
24. Chayer, S., Schmitt, M., Collot, V., and Bourgignon, J.-J., *Tetrahedron Lett.*, **1998**, *39*, 9685.
25. Paolini, J. P. and Robins, R. K., *J. Heterocycl. Chem.*, **1965**, *2*, 53.

26. Paudler, W. W. and Helmick, L. H., *J. Org. Chem.*, **1968**, *33*, 1087.
27. Paudler, W. W. and Shin, H. G., *J. Org. Chem.*, **1968**, *33*, 1638; Guildford, A. J., Tometzki, M. A. and Turner, R. W., *Synthesis*, **1983**, 987.
28. Gudmundsson, K. S., Drach, J. C., and Townsend, L. B., *J. Org. Chem.*, **1997**, *62*, 3453.
29. Chichibabin, A. E., *Chem. Ber.*, **1924**, *57*, 1168; Kröhnke, F., Kickhöfen, B., and Thoma, C., *ibid.*, **1955**, *88*, 1117.
30. Chichibabin, A. E., *Chem. Ber.*, **1924**, *57*, 2092.
31. Fuentes, O. and Paudler, W. W., *J. Heterocycl. Chem.*, **1975**, *12*, 379.
32. Bower, J. D. and Ramage, G. R., *J. Chem. Soc.*, **1955**, 2834.
33. Paudler, W. W. and Kuder, J. E., *J. Org. Chem.*, **1967**, *32*, 2430.
34. Hlasta, D. J. and Silbernagel, M. J., *Heterocycles*, **1980**, *48*, 101.
35. Blatcher, P. and Middlemiss, D., *Tetrahedron Lett.*, **1980**, *21*, 2195.
36. Bourdais, J. and Omar, A.-M. M. E., *J. Heterocycl. Chem.*, **1980**, *17*, 555.
37. Iwao, M. and Kuraishi, T., *J. Heterocycl. Chem.*, **1977**, *14*, 993.
38. Bourdais, J., Rajniakova, O., and Povazanec, F., *J. Heterocycl. Chem.*, **1980**, *17*, 1351.
39. Tanji, K., Sasahara, T., Suzuki, J., and Higashino, T., *Heterocycles*, **1993**, *35*, 915; Miki, Y., Yagi, S., Hachiken, H., and Ikeda, M., *ibid*, **1994**, *38*, 1881.
40. Miyashita, A., Sato, Y., Watanabe, S., Tanji, K., and Higashino, T., *Chem. Pharm. Bull.*, **1995**, *43*, 174.
41. Boekelheide, V. and Fedoruk, N. A., *J. Org. Chem.*, **1968**, *33*, 2062.
42. Tsuchiya, T., Sashida, H., and Konoshita, A., *Chem. Pharm. Bull.*, **1983**, *31*, 4568.
43. Miki, Y., Tasaka, J., Uemura, K., Miyazeki, K., and Yamada, J., *Heterocycles,*, **1996**, *43*, 2249.
44. Krischke, R., Grashey, R., and Huisgen, R., *Justus Liebigs Ann. Chem.*, **1977**, 498.
45. 'The chemmistry of the triazolopyridines', Jones, G. and Sliskovic, D. R., *Adv. Heterocycl. Chem.*, **1983**, *34*, 79.
46. Maury, G., Paugam, J.-P., and Paugam, R., *J. Heterocycl. Chem.*, **1978**, *15*, 1041.
47. Jones, G. and Sliskovic, D. R., *J. Chem. Soc.*, *Perkin Trans. 1*, **1982**, 967; Jones, G., Mouat, D. J., and Tonkinson, D. J., *ibid.*, **1985**, 2719.
48. 'Some aspects of azido-tetrazolo isomerisation', Tishler, M., *Synthesis*, **1973**, 123.
49. Abarca, B., Ballesteros, R., Jones, G., and Mojarrad, F., *Tetrahedron Lett.*, **1986**, *27*, 3543.
50. Jones, G., Pitman, M. A., Lunt, E., Lythgoe, D. J., Abarca, B., Ballesteros, R., and Elmasnaouy, M., *Tetrahedron*, **1997**, *53*, 8257.
51. Jones, G., Mouat, D. J., Pitman, M. A., Lunt, E., and Lythgoes, D. J., *Tetrahedron*, **1995**, *51*, 10969.
52. Finkelstein, B. L., *J. Org. Chem.*, **1992**, *57*, 5538.
53. Bower, J. D. and Ramage, G. R., *J. Chem. Soc.*, **1957**, 4506.
54. Regitz, M., *Chem. Ber.*, **1966**, *99*, 2918.
55. Kauffman, T. Vogt, K., Barck, S., and Schulz, J., *Chem. Ber.*, **1966**, *99*, 2593.
56. Bourgeois, P., Cantegril, R., Chene, A., Gelin, J., Mortier, J., and Moyroud, J., *Synth. Commun.*, **1993**, *23*, 3195.
57. Grenda, V. J., Jones, R. E., Gal, G., and Sletzinger, M., *J. Org. Chem.*, **1965**, *30*, 259.
58. Polanc, S., Vercek, B., Sek, B., Stanovnik, B., and Tisler, M., *J. Org. Chem.*, **1974**, *39*, 2143.
59. Paudler, W. W., Chao, C. I. P., and Helmick, L. S., *J. Heterocycl. Chem.*, **1972**, *9*, 1157.
60. Rees, C. W., Stephenson, R. W., and Storr, R. C., *J. Chem. Soc., Chem. Commun.*, **1974**, 941.
61. Bienayme, H. and Bouzid, K., *Angew. Chem., Int. Ed. Engl.*, **1998**, *37*, 2234.
62. Minguez, J. M., Castellote, M. I., Vaquero, J. J., Garcia-Navio, J. L., Alvarez-Builla, J., and Castano, O., *J. Org. Chem.*, **1996**, *61*, 4655.
63. Minguez, J. M., Vaquero, J. J., Garcia-Navio, J. L., and Alvarez-Builla, J., *Tetrahedron Lett.*, **1996**, *37*, 4263.
64. 'Aromatic quinolizines', Thyagarajan, B. S., *Adv. Heterocycl. Chem.*, **1965**, *5*, 291; 'Aromatic quinolizines', Jones, G., *ibid.*, **1982**, *31*, 1.
65. Miyadera, T., Ohki, E., and Iwai, I., *Chem. Pharm. Bull.*, **1964**, *12*, 1344; Kröhnke, F. and Mörler, D., *Tetrahedron Lett.*, **1969**, 3441.
66. Thyagarajan, B. S. and Gopalakrishnan, P. V., *Tetrahedron*, **1964**, *20*, 1051 and **1965**, *21*, 945; Forti, L., Gelmi, M. L., Pocar, D., and Varallo, M., *Heterocycles*, **1986**, *24*, 1401.

67. Boekelheide, V. and Lodge, J. P., *J. Am. Chem. Soc.*, **1951**, *73*, 3681; Glover, E. E. and Jones, G., *J. Chem. Soc.*, **1958**, 3021.
68. Okamura, W. H. and Katz, T. J., *Tetrahedron*, **1967**, *23*, 2941.
69. Jones, G., Ollivierre, H., Fuller, L. S., and Young, J. H., *Tetrahedron*, **1991**, *47*, 2851 and 2861.
70. Mekonnen, B., Crank, G., and Craig, D., *J. Heterocycl. Chem.*, **1997**, *34*, 589; Mekonnen, B. and Crank, G., *Tetrahedron*, **1997**, *53*, 6959.
71. Farquhar, D., Gough, T. T., and Leaver, D., *J. Chem. Soc.*, *Perkin Trans. 1*, **1976**, 341.
72. Windgassen, R. J., Saunders, W. H., and Boekelheide, V., *J. Am. Chem. Soc.*, **1959**, *81*, 1459.

26 Heterocycles containing more than two hetero atoms

In systems which contain more than two hetero atoms in the same ring we find the trends in properties, which this book has described, taken to further extremes. In particular, the additional hetero atoms, in both six- and five-membered systems, lead to a suppression of electrophilic substitution and a slowing of electrophilic addition to nitrogen. On the other hand, further increases in tendencies for nucleophilic substitution and addition, and in the five-membered compounds, further increases in acidities of N-hydrogen are found.

Multihetero atom heterocycles are comparatively rare in nature, dendrodoine, a cytotoxic substance from a marine tunicate, is an example, however in medicinal chemistry they are of considerable significance: Alprazolam is a major drug for the treatment of anxiety, Acetazolamide is an inhibitor of the enzyme carbonic anhydrase and is used principally for the treatment of glaucoma, and Fluconazole, is an antifungal agent.

dendrodoine   Alprazolam   Acetazolamide   Fluconazole

Analogues of the pyrimidine nucleosides have been extensively studied: 5-Azacytidine is antileukemic and Ribavirin, an antiviral agent, is used in the treatment of lung infections in infants. Lamotrigine is used for the treatment of epilepsy. Melamine, which on condensation with formaldehyde produces the melamine resins well known in kitchen utensils, is an important industrial intermediate.

5-Azacytidine   Ribavirin   Lamotrigine   melamine

26.1 Five-membered rings

26.1.1 Azoles[1]

The triazoles are numbered to indicate the relative positions of the nitrogen atoms, tetrazole and pentazole are unambiguous names. 1,2,3-Triazoles are surprisingly stable, when one considers that they contain three directly-linked nitrogen atoms, but on flash vacuum pyrolysis at 500 °C they do lose nitrogen to give 2H-azirines, probably *via* the 1*H*-isomer.[2] Benzotriazole is similarly relatively stable and has been distilled *in vacuo* at 200 °C, though explosions have been reported during this process. Simple tetrazoles are also relatively stable, but the pentazole ring system is only known in a few aryl derivatives which generally decompose (possibly explosively) at or below room temperature.[3]

The additional hetero atoms make these systems less basic but more acidic than comparable 1,2- and 1,3-azoles. Each is subject to the same kind of tautomerism as discussed for the 1,2- and 1,3-azoles (section 21.1.1.1), in which the tautomers are equivalent (not shown) but also, in these systems, to tautomerism which generates different arrangements.

1,2,3-triazole
pK_a (proton added) 1.2
pK_a (proton lost) 9.4

1,2,4-triazole
pK_a (proton added) 2.2
pK_a (proton lost) 10.3

tetrazole
pK_a (proton lost) 4.9

pentazole

26.1.1.1 1,2,3-Triazole[4]

1,2,3-Triazole is fairly resistant to *N*-alkylation under neutral conditions however both acylations and alkylations involving *N*-anions occur readily, but mixtures of 1- and 2-substituted products are often obtained.[1,5] Trimethylsilylation produces 2-trimethylsilyl-1,2,3-triazole and this can be alkylated at N-1 to produce 1-alkylated compounds following loss of silicon.[6] An equilibrium mixture of *N*-acetyl-1,2,3-triazoles contains predominantly the 2-acetyl-isomer,[7] as in the parent: this may reflect unfavourable *ortho* lone-pair/lone-pair interactions in the 1-isomer and is in agreement with calculations which suggest that the 2*H*-isomer is more aromatic.[8] Heating in sulfolane at 150 °C converts the *N*-acyl compounds into oxazoles in a synthetically useful transformation.[9]

1-Methyl-1,2,3-triazole can be brominated at C-4, but the 2-methyl isomer is less reactive, requiring the use of an iron catalyst;[10] the lower reactivity of the latter is probably related to the presence of two imine units. 1,2,3-Triazole itself forms a 4,5-dibromo derivative in high yield with bromine at 50 °C.[11] Nitration of 2-phenyl-1,2,3-triazole proceeds first on the benzene ring, but then does bring about hetero-ring substitution.[12]

The ring system is relatively resistant to both oxidation and reduction as exemplified below.[13]

N-Substituted 1,2,3-triazoles can be lithiated directly at carbon, but low temperatures must be maintained to avoid ring cleavage by cycloreversion.[14,15]

Metal exchange of 5-lithio-1-benzyloxy-1,2,3-triazole with zinc iodide gives a relatively stable zinc derivative which can be used in palladium-catalysed couplings. The corresponding tin compound is less stable and can be used only for palladium-catalysed acylations.[16]

Both 4,5-dibromo-1-methoxymethyl- and 4,5-dibromo-2-methoxymethyl-1,2,3-triazole form 5-lithio componds by exchange with n-butyllithium at −80 °C.[11]

26.1.1.2   1,2,4-Triazole[17]

Alkylations and acylations generally occur at N-1, reflecting the higher nucleophilicity of N–N systems (cf. section 11.1.1.2), however 4-alkyl derivatives can be prepared via quaternisation of 1-acetyl-1,2,4-triazole[18] or the acrylonitrile or crotononitrile adducts[19] (Note that N-1 and N-2 are equivalent until substitution occurs).

Bromination occurs readily in alkaline solution giving 3,5-dibromo-1,2,4-triazole;[20] the 3-monochloro-derivative can be obtained by thermal rearrangement of the N-chloro isomer;[21] an analogous N→C 1,5-sigmatropic shift converts the 1- into the 3-nitro-compound.[22]

C-Lithiations can be easily effected on *N*-1-protected 1,2,4-triazoles, the resulting 5-lithio-derivatives being much more stable than lithiated-1,2,3-triazoles.[23]

3-Amino-1,2,4-triazole can be diazotised normally: the resulting diazonium salt has been used for the production of azo dyes, and also loses nitrogen with easy replacement by nucleophiles. The bromo- and nitrotriazoles which can be thus prepared are themselves substrates for nucleophilic displacement reactions.[18,24] 5-Bromo and 5-nitro[25] groups are good leaving groups in 1-alkyl-1,2,4-triazoles for hetero nucleophiles; for carbon nucleophiles, methanesulfinate is a better leaving group.[26]

The oxidative desulfurisation of 1,2,3-trazole thiones using nitric acid is a type of reaction common to other electron-deficient nitrogen heterocycles. The process involves loss of sulfur dioxide from an intermediate sulfinic acid.[27]

26.1.1.3   *Tetrazole*[28]

Noting the similarity of tetrazole pK_as to those of carboxylic acids, tetrazoles have often been used as bioequivalent replacements for CO_2H in pharmacologically active compounds. Tetrazole and its alkyl and aryl derivatives generally begin significant decomposition at about 180 °C and the chloro and alkylthio derivatives at somewhat lower temperatures; 5-nitrotetrazole explodes unpredicatably on storage.[29]

Tetrazoles alkylate and acylate on N-1 or N-2 depending on substituents at C-5, however selective 1-alkylations by quaternisation of 2-tri-*n*-butylstannyl and 2-*t*-butyl derivatives is possible.[30] Unfortunately, in the latter case the quaternisation fails with alkyl halides and requires alkyl sulfates or more powerful alkylating agents.[31]

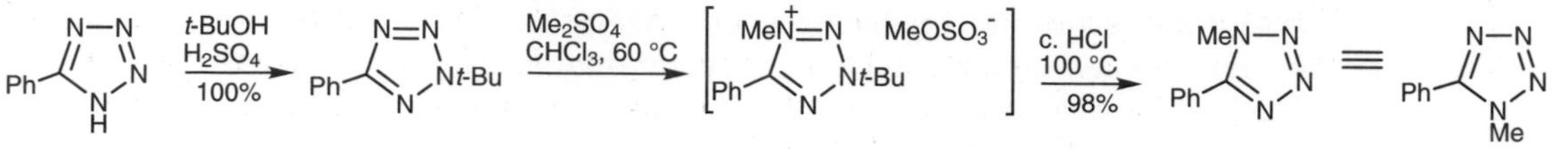

Remarkably, some *C*-electrophilic substitutions such as bromination,[32] mercuration[33] and even Mannich reactions[34] (but not nitration) can be achieved, though the mechanisms for these substitutions may not be of the conventional type.

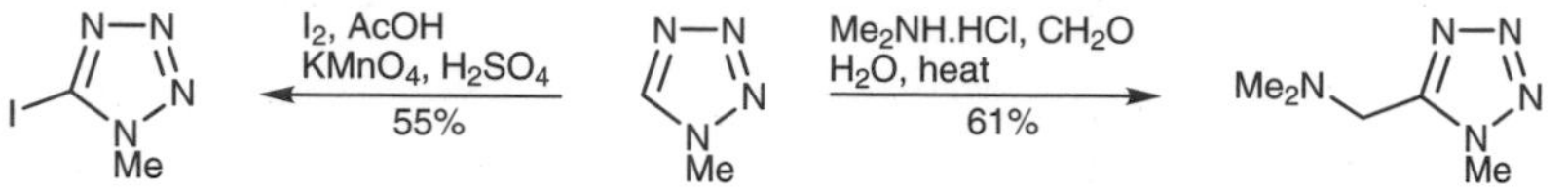

As would be expected from inductive effects, 5-bromo-1-methyltetrazole is more reactive in nucleophilic substitution than are the corresponding halo-1,2,4- and -1,2,3-triazoles, which in turn are more reactive than the corresponding haloimidazoles. 5-Bromo-2-methyltetrazole is significantly less reactive than its 1-methyl isomer due to less effective delocalisation of the negative charge in the intermediate adduct.[35]

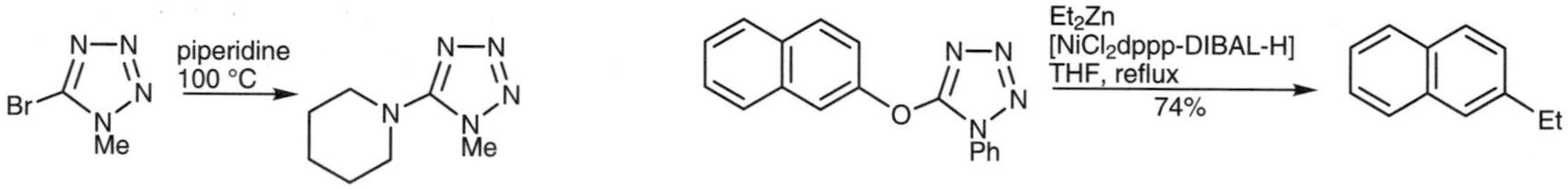

5-Tetrazolyl ethers can be used in two ways: to activate phenols for nickel- or palladium-catalysed coupling reactions, as illustrated above or to allow catalytic hydrogenolysis of the C–O bond of the phenol.[36]

C-Lithiation occurs readily and the resulting lithio-derivatives can be trapped with electrophiles, despite a strong tendency for cycloreversion. Tetrazole can also act as an *ortho*-directing group, as in the lithiation of 5-phenyltetrazole.[37,38] A tetrazolyl ether similarly directs *ortho*-metallation but here, the tetrazole migrates from the oxygen to the lithiated carbon.[39] *para*-Methoxybenzyl is a useful nitrogen protecting group for lithiations and it can be removed finally by hydrogenation, or oxidation as shown.[40]

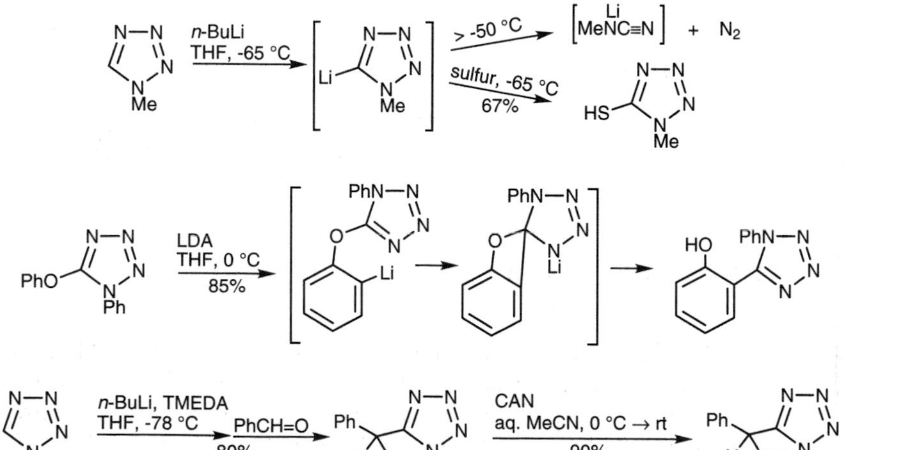

5-Alkyl groups in 1-substituted tetrazoles can be lithiated but in a 5-alkyl-2-methyltetrazole it is the *N*-methyl which is metallated[41] but 5-methyl-2-trityltetrazole lithiates normally at the side chain methyl.[42] The methylene group in 5-(benzotriazolylmethyl)tetrazole is sufficiently activated that *N*-protection is unnecessary – the benzotriazole can then act as a leaving group for displacement by Grignard reagents as shown below.[43]

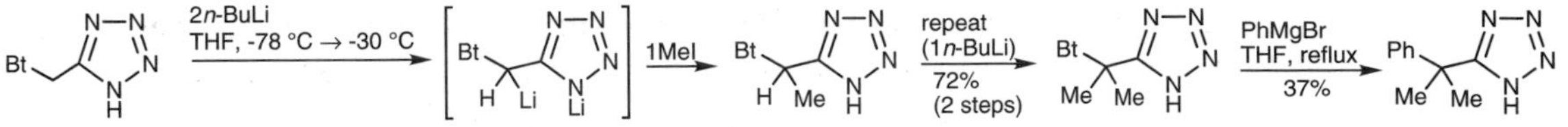

5-Aminotetrazole gives a diazonium salt (with the formula $CN_6.HCl$!) (**CAUTION: EXPLOSIVE**) which has been used to generate atomic carbon![44] 1-Substituted-5-aminotetrazoles seem to give relatively stable *N*-nitroso derivatives.

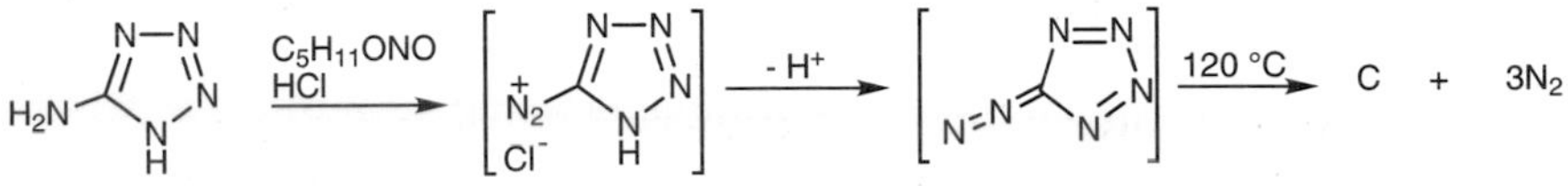

Flash vacuum pyrolysis of 5-aryltetrazoles generates aryl carbenes[45] but heating the pyrimidinyl phenyltetrazole shown below in refluxing decalin (180 °C) results in the formation of an intermediate nitrene which then cyclises onto the pyrimidine nitrogen.[46]

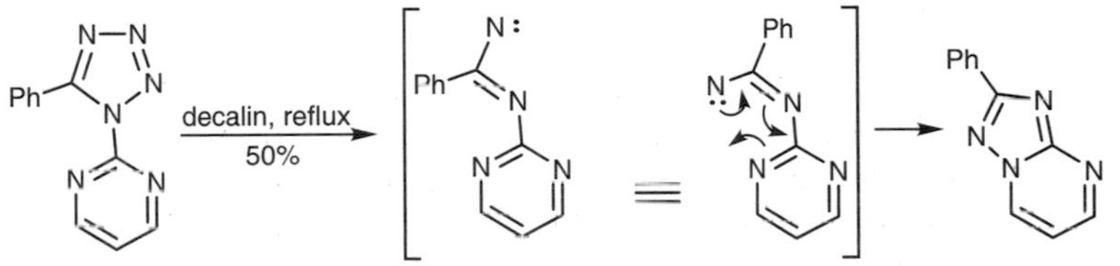

Hypobromite oxidation of 5-benzylaminotetrazoles provides a useful synthesis of benzyl isonitriles, as illustrated below.[47]

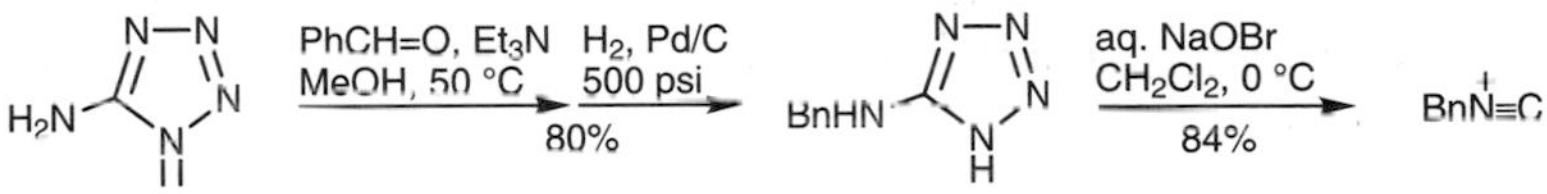

1,3,4-Oxadiazoles are formed on heating tetrazoles with acylating agents, via rearrangement of a first-formed 2-acyl derivatives.[48]

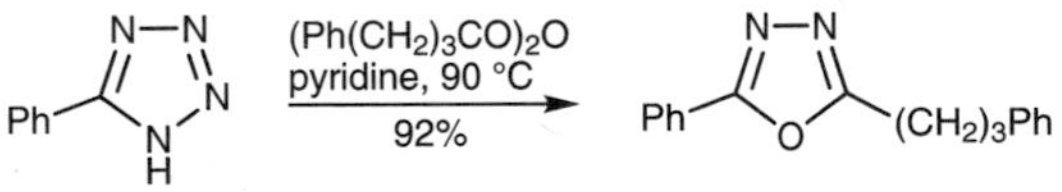

26.1.1.4 Ring synthesis of azoles

1,2,3-Triazoles

1,2,3-Triazoles are generally prepared by the cycloaddition of an alkyne with an azide, but the hazardous nature of some alkyl azides limits the method in these cases. A convenient synthesis which leads to *N*-hydrogen 1,2,3-triazoles utilises the stable (and relatively safe) trimethylsilyl azide.[49] For *C*-unsubstituted 1,2,3-triazoles, ethyne itself would be required but it is much more convenient to use, as starting material, vinyl acetate instead of the gaseous ethyne, or in general, an enamine or an enol ether as alkyne equivalents.[50,51]

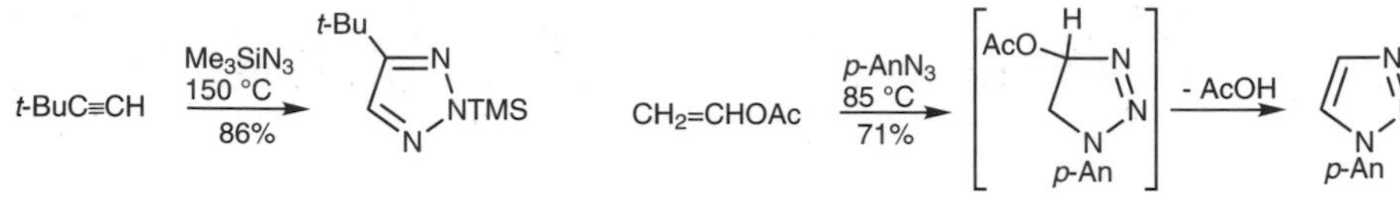

The condensation of azides with acyl-Wittig reagents offers a regiospecific synthesis of 1,5-disubstituted 1,2,3-triazoles.[52]

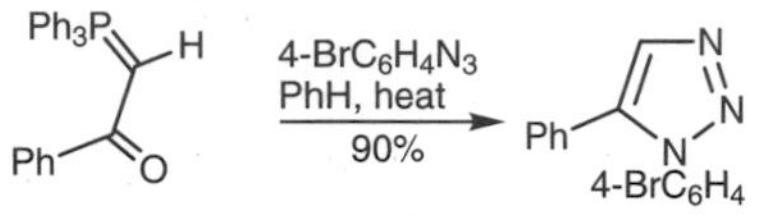

Other useful syntheses of 1,2,3-triazoles include diazo-transfer to enamino-ketones from either sulfonyl azides[53] (or 3-diazo-oxindole),[54] and reaction of dichloroacetaldehyde tosylhydrazone with amines and each of these is illustrated below.[55]

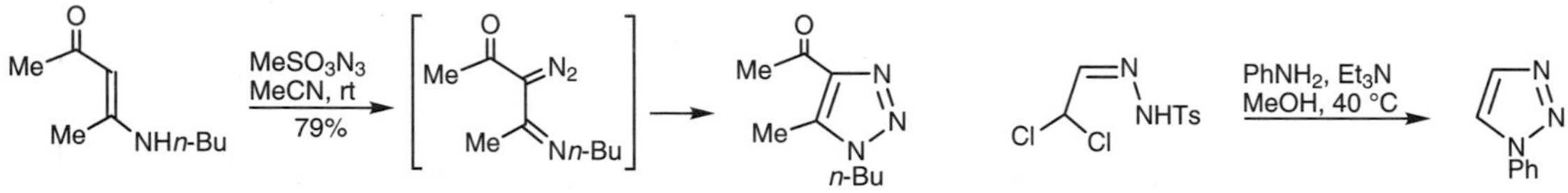

1,2,4-Triazoles

1,2,4-Triazoles are available *via* cyclodehydration reactions of N,N'-diacylhydrazine with amines, although the conditions are often quite vigorous.[56] An interesting variant utilises *sym*-triazine (1,3,5-triazine) as an equivalent of $HN(CHO)_2$.[57] Condensations of aminoguanidine with esters give the versatile 3-amino compounds.[58]

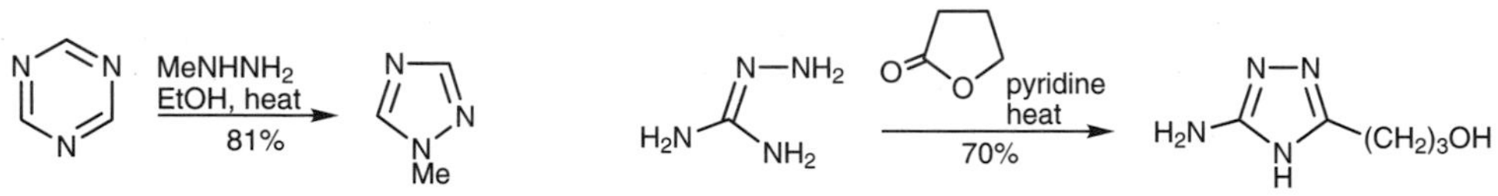

Tetrazoles

Tetrazoles are usually prepared by the reaction of an azide with a nitrile, or an activated amide; tri-*n*-butyltin azide and trimethylsilyl azide are more convenient and safer reagents than azide anion is some cases. The second example shown illustrates the use of a cyanoethyl group as a removable protecting group for amide nitrogen.[59] Other variations on this method from nitriles include the use of triethylammonium chloride (instead of ammonium chloride) to avoid the possible sublimation of potentially explosive azides,[60] and the use of micelles as reaction media.[61] Amides can be activated with trifluoromethanesulfonic anhydride,[62] or via formation of the thioamide,[63] or by the use of triphenylphosphine with diethyl azodicarboxylate; the equivalent imidochloride will react under phase transfer conditions.[64]

Related methods can be used to prepare 5-hetero-substituted compounds: isonitriles with *N*-halosuccinimides and azide give halo derivatives,[65] aryl isothicyanates with azide give the arylthio compounds[66] and isothiocyanates give thiols, as illustrated. The thiols can be converted into 5-unsubstituted tetrazoles by oxidation with hydrogen peroxide[67] or chromium trioxide.[68]

Nitrosation of amidrazones is a method which avoids the use of azide and also offers a regiospecific synthesis of 1- or 2-substituted compounds.[69]

26.1.2 Oxadiazoles and thiadiazoles

Only one divalent hetero atom can be incorporated into a simple five-membered, aromatic heterocycle. These systems are named with the non-nitrogen atom numbered as 1, and the positions of the nitrogen atoms shown with reference to the divalent atom.

1,2,4-oxadiazole   1,3,4-oxadiazole   1,2,5-oxadiazole (furazan)   furoxan   1,2,3-oxadiazole   sydnones

1,2,3-thiadiazole   1,2,4-thiadiazole   1,3,4-thiadiazole   1,2,5-thiadiazole

1,2,4-Oxadiazoles,[70] 1,3,4-oxadiazoles,[71] and 1,2,5-oxadiazoles are well known, but the 1,2,3-oxadiazole system, which calculations indicate to be unstable relative to its ring-open diazoketone tautomer,[72] is known only as a benzo-fused derivative (in solution) and in mesoionic substances, known as 'sydnones',[73] which have been well investigated. 'Furoxans',[74] which are formed by the dimerisation of nitrile oxides, have also been extensively studied. 1,2,3-Thiadiazoles, 1,2,4-thiadiazoles,[75] 1,3,4-thiadiazoles,[76] and 1,2,5-thiadiazoles[77] are all represented by well characterised compounds. Estimates of aromaticity, based on bonds lengths and NMR data produced the following relative order.[78]

As with the azoles, oxa- and thiadiazoles are very weak bases due to the inductive effects of the extra hetero atoms, although *N*-quaternisation reactions can be carried out. For similar reasons, electrophilic substitutions on carbon are practically unknown, apart from a few halogenations and mercurations[79] – it is an intriguing paradox that mercurations, with what is generally thought of as a weak electrophile, are often successful in electron-poor heterocycles. Another important difference from the azoles is of course the absence of *N*-hydrogen, so that *N*-anion-mediated reactions are not possible.

All these systems are susceptible to nucleophilic attack, particularly the oxadiazoles, which often undergo ring cleavage with aqueous acid or base unless both (carbon) positions are substituted. Similarly, leaving groups are generally displaced easily; there is substantial differential positional reactivity: in both 1,2,4-oxa- and -thiadiazoles a 5-chlorine is displaced much more easily that a 3-chlorine, no doubt due to the more effective stabilisation of the intermediate anionic adduct in the former situation. There is a far from complete set of comparisons of relative reactivities, but some data are available.[80]

Relative rates of reaction with piperidine in ethanol

Base-catalysed proton exchange occurs readily, but decomposition *via* cycloreversion or *β*-elimination in the anion often competes.[81] Direct lithiations at carbon are generally easy,[82] but the resulting lithio derivatives vary greatly in stability, some being of no use synthetically.[83] Hydrogens on side-chain alkyl groups are 'acidified' by delocalisation of the charge in the deprotonated species onto ring nitrogens. There is an interesting difference between 1,2,5-oxa- and 1,2,5-thiadiazoles in this context: in the former, smooth metallation of a 3-methyl occurs with *n*-butyllithium, but for the latter, lithium diisopropylamide must be used to avoid competing nucleophilic addition to the sulfur, leading then to ring decomposition.[84]

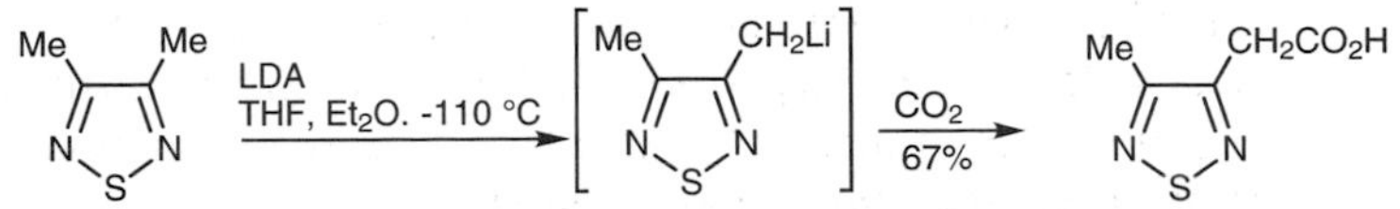

4-Substituted 5-chloro-1,2,3-thiadiazoles react with simple hetero nucleophiles by displacement of the chlorine, but reaction with aryl- and alkyllithiums gives alkynyl thioethers via attack at sulfur and then ring cleavage with loss of nitrogen. A similar ring cleavage occurs, but by a different mechanism, when the 5-unsubstituted analogue is treated with base.[85]

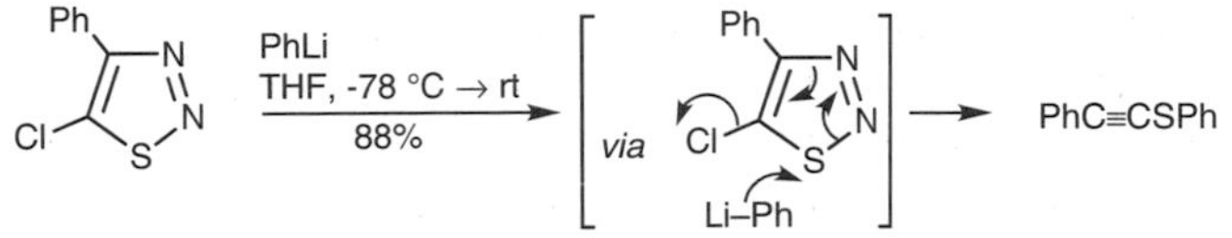

Generally, amines can be diazotised and converted, for example, into halides, but in some cases the intermediate, *N*-nitroso compound is stable, and only then subsequently converted into a diazonium salt by treatment with strong acid – this may reflect the lower stability of a positively charged group attached to an electron-deficient ring.[86]

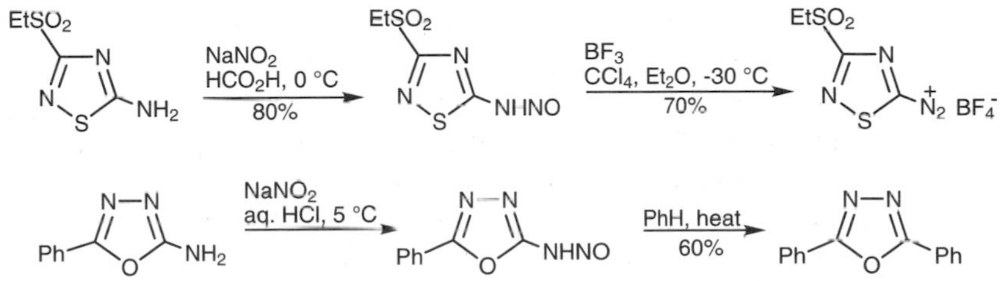

An interesting and fairly general type of reaction in ring systems such as these is ring interconversion via intramolecular attack on nitrogen.[87]

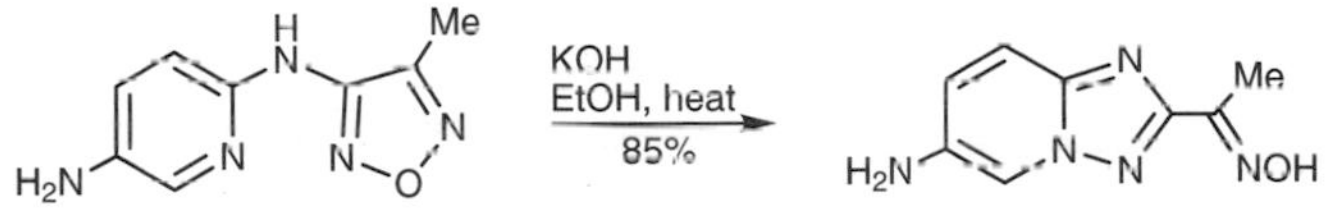

1,2,4-Oxadiazoles are useful as masked amidines where such strongly basic groups would be incompatible with reaction conditions – the amidine is easily liberated by hydrogenation, as illustrated below.[88]

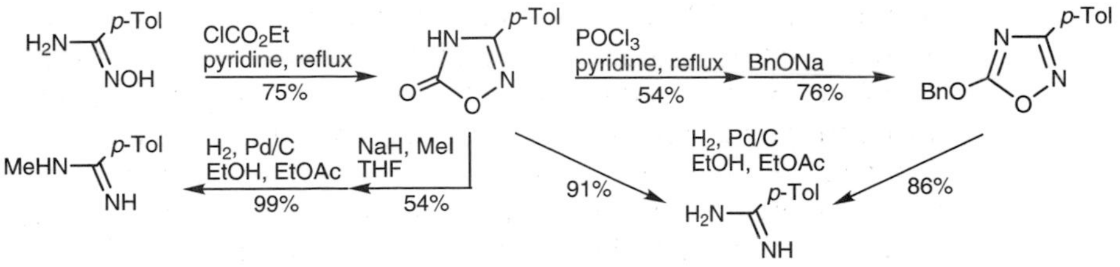

26.1.3 Other systems

Of the higher aza-compounds, only derivatives of 1,2,3,4-thiatriazole[89] are well defined, but even here alkyl derivatives decompose at or below 0 °C, though 5-aryl- and amino derivatives are generally fairly stable. Many other derivatives are, however, dangerously explosive, for example the 5-chloro and thiolate derivatives. The controlled decomposition of 5-alkoxy-1,2,3,4-thiatriazoles (for example the 5-ethoxy-derivative in ether at 20 °C) has been recommended as the best preparation of pure alkyl cyanates; thermal decomposition of 5-aryl compounds gives the corresponding nitrile.[90] An interesting isomerisation cycle interconverts aminothiatriazoles and tetrazole thiols.[91]

Ar = 2-O₂N-C₆H₄

26.1.4   Ring synthesis

26.1.4.1   1,2,4-Oxadiazoles

1,2,4-Oxadiazoles can be prepared by acylation of amidoximes.[92] A variation of this method gives a one-pot synthesis from the amidoxime, an organic acid and a peptide coupling agent; the method is sufficiently mild that there is no racemisation when mandelic acid is used.[93] 1,2,4-Oxadiazoles can also be prepared from amides via acylamidines,[94] or via the cycloaddition of nitrile oxides to nitriles, as illustrated, or to O-methyl oximes.[95]

26.1.4.2   1,3,4-Oxadiazoles

1,3,4-Oxadiazoles are available by cyclodehydration of N,N'-diacylhydrazines or their equivalents.[96] They are also available from tetrazoles (section 26.1.1.3) or by oxidative cyclisation of acyl hydrazones.[97]

26.1.4.3   1,2,5-Oxadiazoles

1,2,5-Oxadiazoles result from the dehydration of 1,2-bisoximes.

a sydnone

26.1.4.4   Sydnones

Sydnones are normally prepared by the dehydration of N-nitroso α-amino acids.[98]

26.1.4.5   1,2,3-Thiadiazoles

1,2,3-Thiadiazoles are prepared by reaction of a hydrazone, containing an acidic methylene group, with thionyl chloride.[99] The 5-thiol can be prepared by reaction of chloral tosylhydrazone with polysulfide, as indicated below.[100]

26.1.4.6   1,2,4-Thiadiazoles

1,2,4-Thiadiazoles carrying identical groups at the 3- and 5-positions are obtained by the oxidation of thioamides;[101] 5-chloro-1,2,4-thiadiazoles result from the reaction of amidines with perchloromethyl mercaptan.[102]

26.1.4.7   1,3,4-Thiadiazoles

1,3,4-Thiadiazoles are available by a number of convenient general routes including cyclisation of N,N'-diacylhydrazines, or 1,3,4-oxadiazoles, with phosphorus sulfides.[103] 3-Amino-1,3,4-thiadiazoles are prepared via acylation of thiosemicarbazides[104] and the parent compound is easily obtained from hydrogen sulfide and dimethylformamide azine.[105]

26.1.4.8 1,2,5-Thiadiazoles

1,2,5-Thiadiazoles can be prepared by the oxidative cyclisation of 1,2-diamines or aminocarboxamides.[106] Condensation of sulfamide ($SO_2(NH_2)_2$) with 1,2-diketones gives 1,2,5-thiadiazole 1,1-dioxides.[107] A good general method is the reaction of trithiazyl trichloride with activated alkenes and alkynes; this method is also useful for the fusion of a 1,2,5-thiadiazole onto other heterocycles such as pyrroles. The main drawback is that the reagent is not commercially available. The reaction possibly proceeds via cycloaddition to an N–S–N unit in the trithiazine ring.[108]

26.2 Six-membered rings

26.2.1 Azines

Neutral six-membered aromatic heterocycles cannot contain a divalent heteroatom. The azines are numbered to indicate the relative positions of the nitrogen atoms. 1,2,3,4-Tetrazine, pentazine and hexazine are unknown. Of the other systems, very little information is available on 1,2,3,5-tetrazine but on the other hand, derivatives of 1,3,5-triazine are very well known and available in large quantities, indeed they are amongst the oldest known heterocycles: the trioxy-compound ('cyanuric acid') was first prepared in 1776 by Scheele by the pyrolysis of uric acid.

The thermal stabilities of the parent systems vary from 1,2,3-triazine, which decomposes at about 200 °C, to 1,3,5-triazine, which is stable to over 600 °C – at this temperature it decomposes to give hydrogen cyanide, of which it is formally a trimer.

In comparison with the diazines, the inductive effects of the 'extra' nitrogen(s) lead to an even greater susceptibility to nucleophilic attack and as a result, all the parent systems and many derivatives react with water, in acidic or basic solution. Similarly, simple electrophilic substitutions do not occur; some apparent electrophilic substitutions, such as the bromination of 1,3,5-triazine probably take place via bromide nucleophilic addition to an N^+–Br salt.[109] Attempted direct N-oxidation of simple tetrazines with the usual reagents generally results in ring cleavage however it can be achieved satisfactorily with methyl(trifluoromethyl)dioxirane.[110]

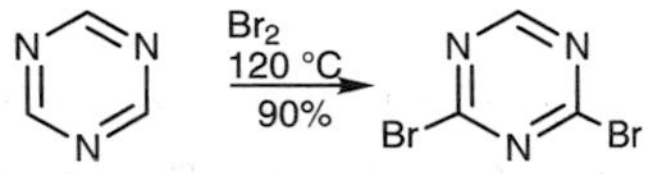

The examples shown below are illustrative of the many easy nucleophilic additions to the polyaza-azines: The reaction of 1,2,4,5-tetrazine with simple amines[111] can be contrasted with the requirement for sodamide (Chichibabin reaction) for the diazines and pyridine.

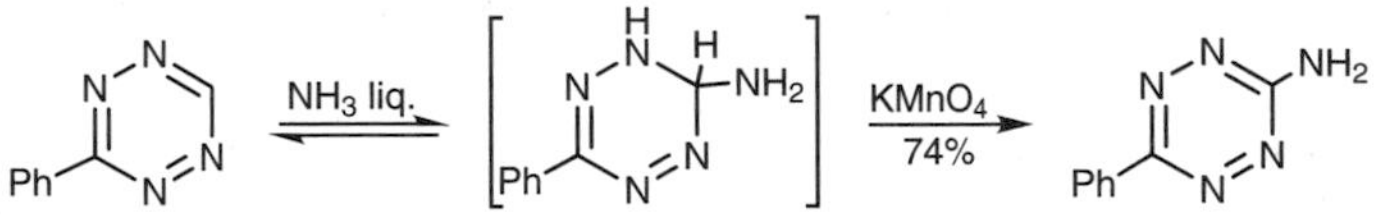

The easy addition at C-5 of 1,2,4-triazines[112] is shown by the VNS (section 2.3.3) reaction of the 3-methylthio-derivatives in the absence of activating groups; a closely related addition of nitroalkanes represents a very useful nucleophilic acylation.[113] The ready displacement of methylthio from the same compound is also indicative.[114] Nucleophilic displacement of methylthio in 1,2,4-triazines and 1,2,4,5-tetrazines by alkoxide and amines is very easy. Mono-displacement can be carried out on 3,6-bis(methylthio)-1,2,4,5-tetrazine but the reaction using methoxide requires careful control of reaction conditions to avoid formation of the dimethoxy derivative.[115] However, reaction of the bis(methylthio) compound with methyllithium resulted[116] in nucleophilic attack at nitrogen!

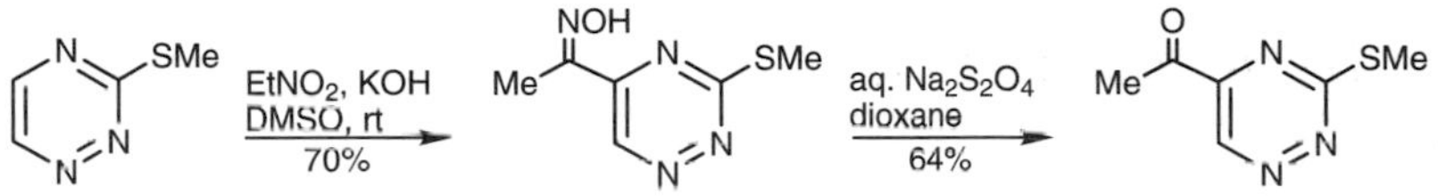

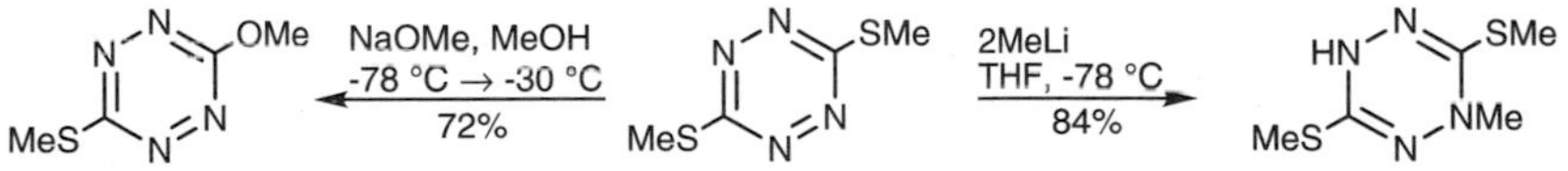

In triazine chemistry, sulfone is a better leaving group than halide for displacement with carbanions.[117]

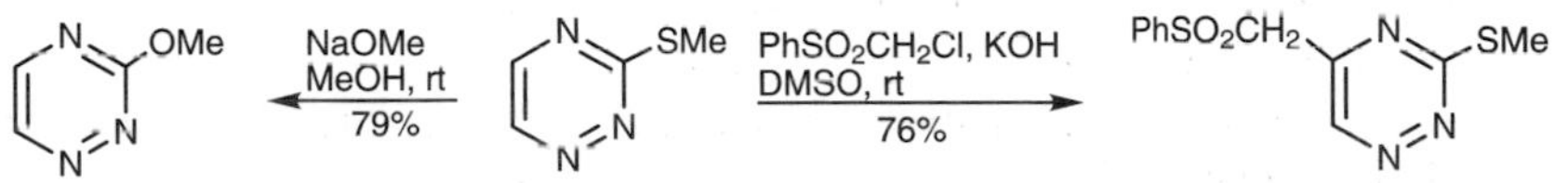

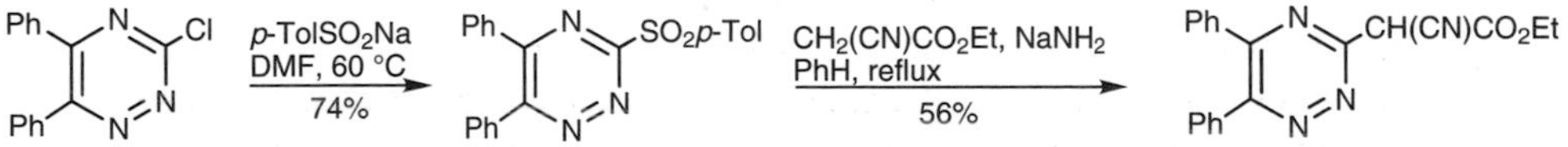

The susceptibility of 1,3,5-triazine to nucleophilic attack with ring-opening makes it a synthetically useful equivalent of formate, or formamide, particularly for the synthesis of other heterocycles such as imidazoles and triazoles[118] (section 26.1.1.4; 1,2,4-triazoles).

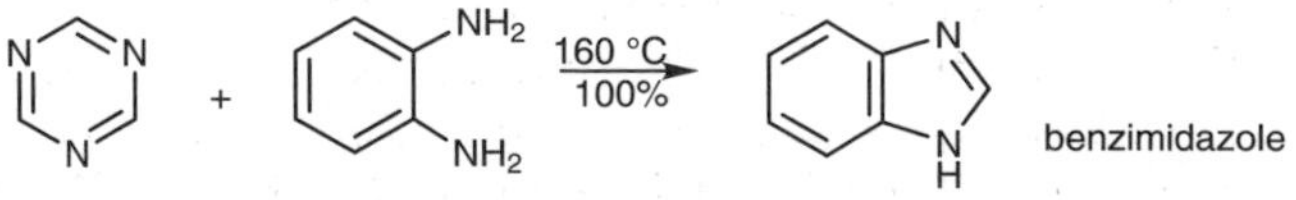

benzimidazole

Despite the high susceptibility of 1,2,4-triazines to nucleophilic addition, 3-substituted-5-methoxy-1,2,4-triazines can be successfully lithiated.[119]

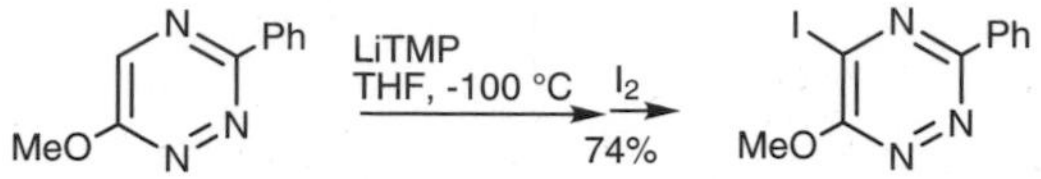

Reaction of 1,2,3-triazine with nucleophiles usually leads to ring opening via attack at C-4. However, silyl enol ethers react with chloroformate/1,2,3-triazine complexes to give 5-substituted 2,5-dihydro-1,2,3-triazines which can be rearomatised using cerium(IV) ammonium nitrate. In this case, initial addition of the electrophile takes place at N-2, leading to the specific activation of C-5.[120]

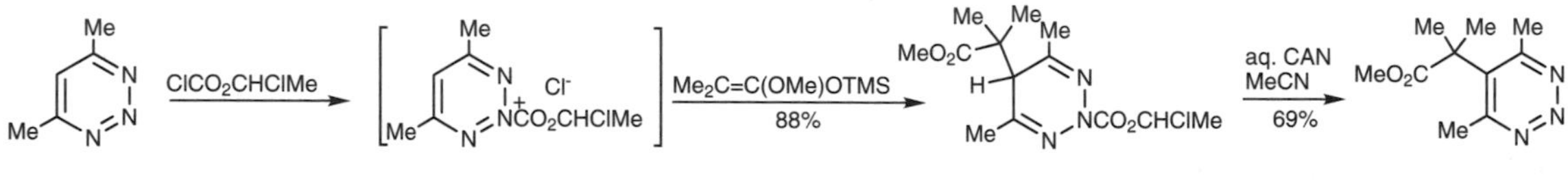

An interesting variant of the Minisci reaction has been reported for 1,2,3-triazine, which is unstable to the usual acidic conditions: here, activation of the heterocycle to attack by the nucleophilic radical is brought about by the agency of a dicyanomethine ylide.[121]

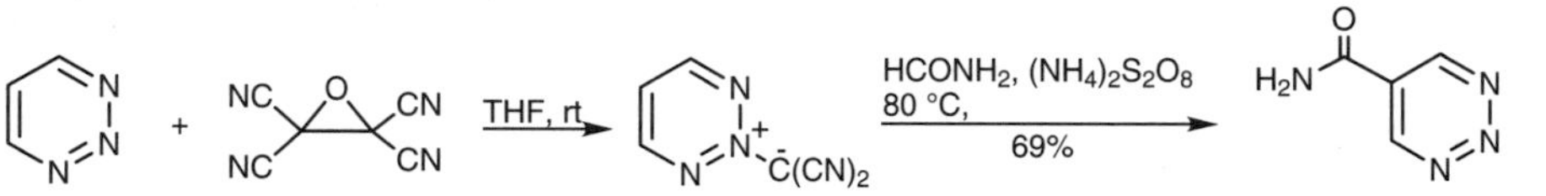

The reductive removal of hydrazine substituents under oxidising conditions is conceptually related to oxidative removal of thiols in other systems (e.g. section 26.1.1.2). In this case, the intermediacy of a diimide seems likely, as illustrated below.[122]

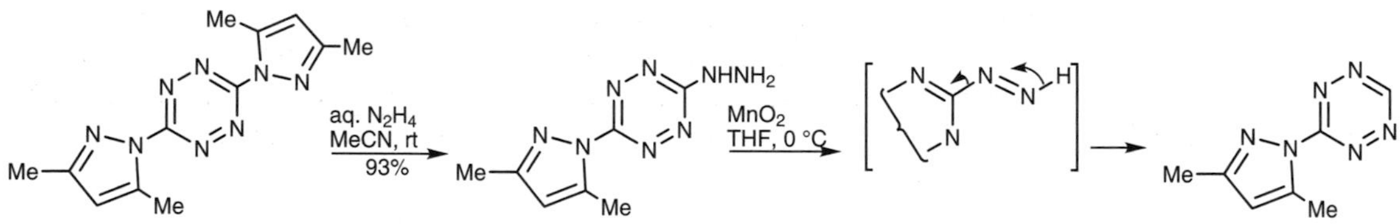

Probably the most useful and general reaction of all these systems is the inverse-electron demand Diels-Alder reaction with acetylenes (or equivalents) to produce either pyridines or diazines *via* elimination of hydrogen cyanide or nitrogen.[123] Abnormal reactions occasionally occur through non-concerted mechanisms, as the last example shows.[124]

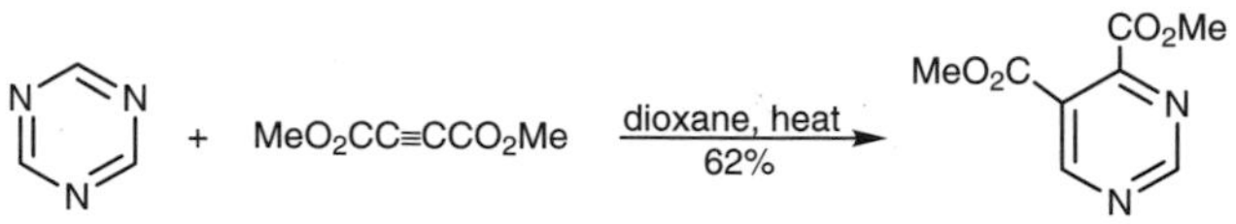

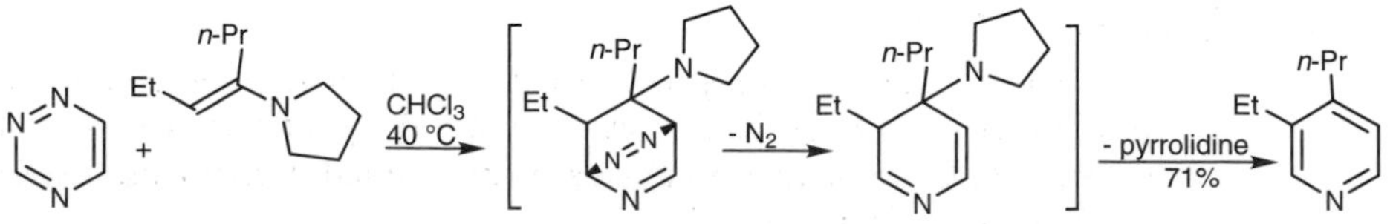

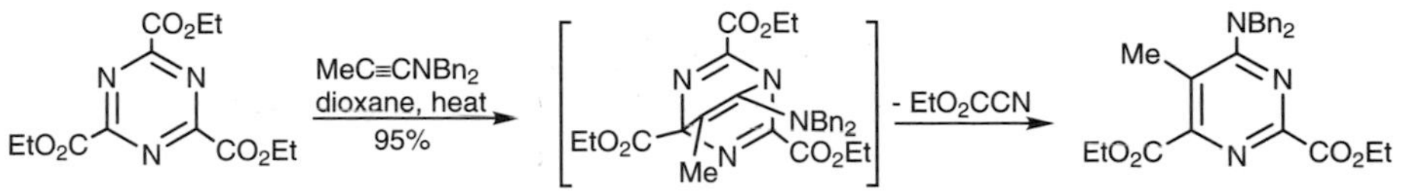

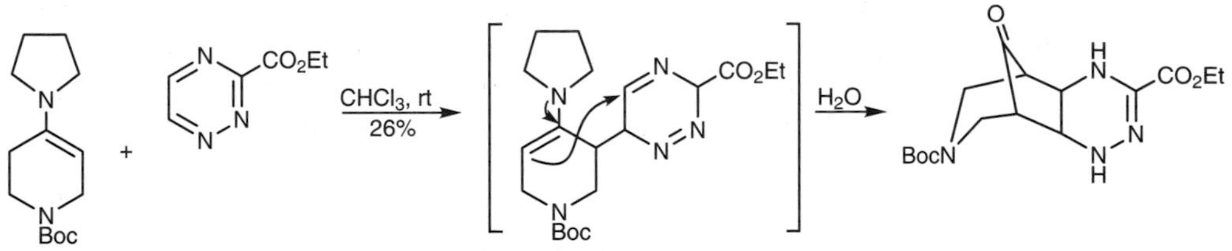

26.2.2 Ring syntheses

26.2.2.1 1,2,3-Triazine

1,2,3-Triazine has been prepared by the oxidation of 1-aminopyrazole.[125]

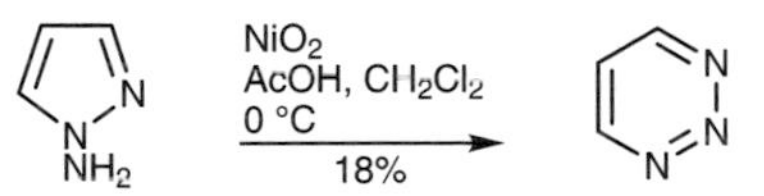

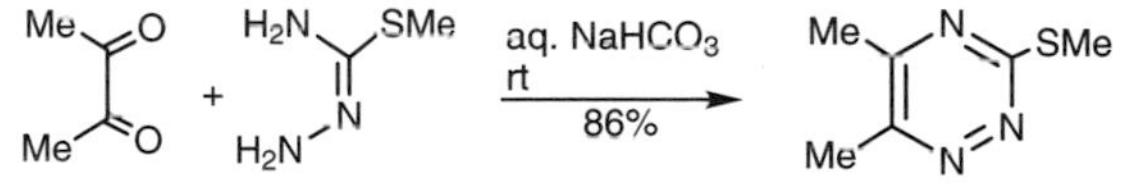

26.2.2.2 1,2,4-Triazines

1,2,4-Triazines have been prepared by the condensation of amidrazones with diketones or halo-ketones, as shown above.[126]

26.2.2.3 1,3,5-Triazines

1,3,5-Triazines are usually most easily obtained by substitution reactions on 2,4,6-trichloro-1,3,5-triazine, but the ring system can also be synthesised by cyclocondensation reactions. Trimerisation of nitriles (a common industrial method) or imidates[127] gives symmetrically substituted compounds; mono-substituted-1,3,5-triazines can be obtained *via* reaction of imidates with 1,3,5-triazine itself.[128]

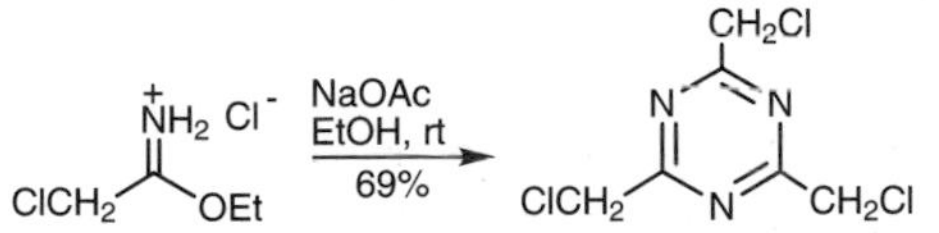

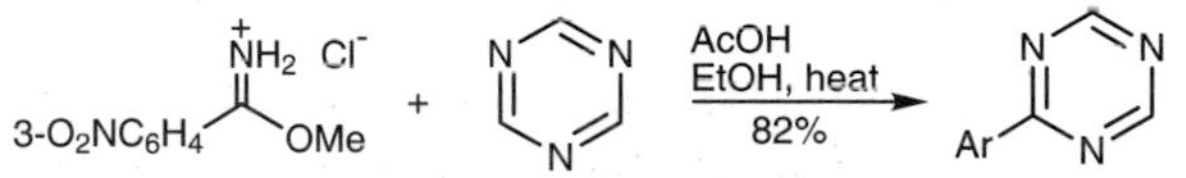

A route which allows the synthesis of 1,3,5-triazines with different substituents at each carbon is exemplified below – an N'-acyl-N,N-dimethylamidine reacts with an amidine (shown) or guanidine to form a 1,3,5-trazine.[129]

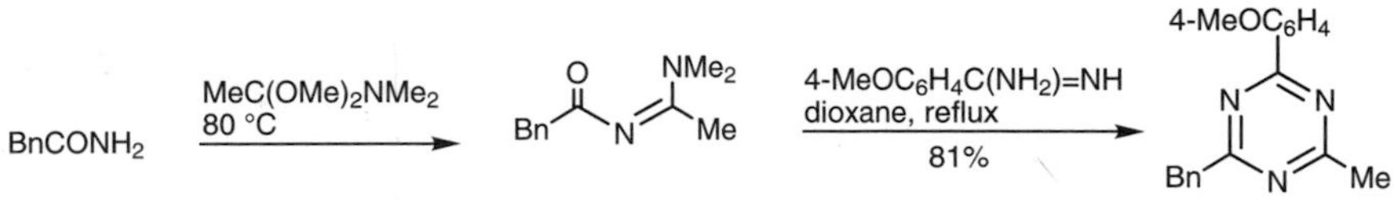

26.2.2.4 1,2,4,5-Tetrazines

1,2,4,5-Tetrazines can be produced by condensation of hydrazine with carbonyl compounds at acid oxidation level, followed by oxidation of the dihydroproducts: this generally produces 3,6-identically-substituted derivatives, crossed condensation reactions being inefficient.[105,130]

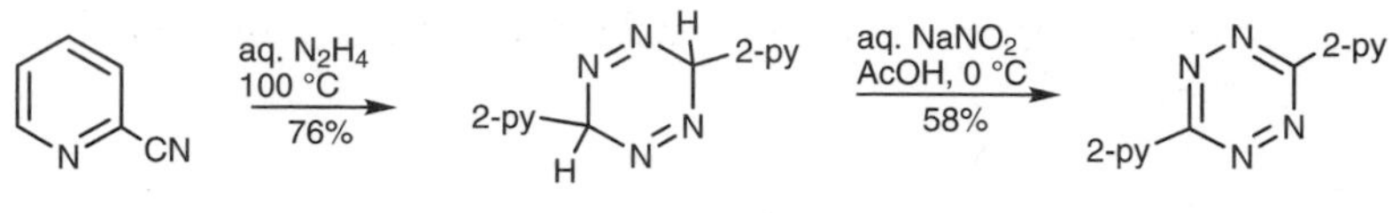

26.3 Benzotriazoles

The chemistry of benzotriazole has been developed to the point where it has now found extensive use for heterocyclic[131] and general synthesis.[132] A useful set of properties give it this role: (i) α-carbanions are stabilised to the same extent as at the benzylic position of a benzene compound; (ii) α-carbocations are also stabilised; (iii) the benzotriazolyl anion is also a good leaving group with a combination of good reactivity and stability/ease of handling. Sequential combinations of these reactivities have been applied to the synthesis of a wide variety of molecules – some illustrative examples are shown below.

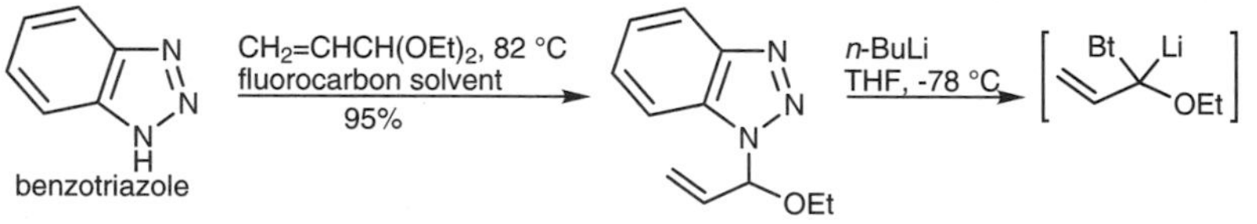

The starting benzotriazole derivatives are usually prepared from the parent heterocycle by N-alkylation with a halide, or via reaction with aldehydes or acetals, which can lead to mixtures of 1- and 2-substituted benzotriazoles, however the reactivities of the two isomers are similar. For clarity, only reactions of 1-substituted compounds are shown.

Alkylation of lithiated 1-(1-ethoxyprop-2-enyl)benzotriazole leads to enones after hydrolytic removal of the heterocycle; addition of the lithiated species to cyclohexenone then hydrolytic cleavage of the heterocycle produces an unsaturated 1,4-diketone.[133] Addition of the same anion to methyl but-2-enoate generates an anion in which the benzotriazole is displaced intramolecularly and a cyclopropane results.[134]

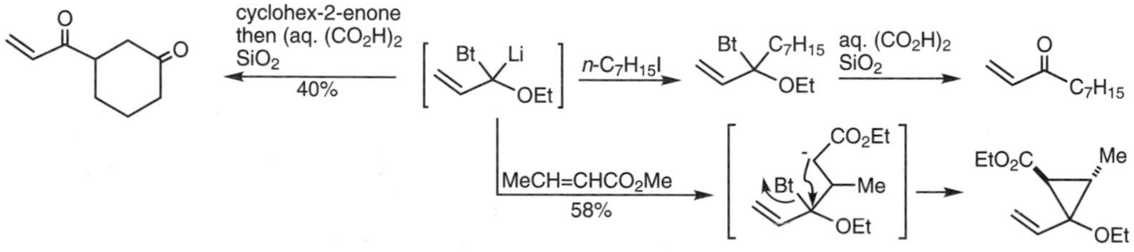

In the next example the benzotriazole unit facilitates benzylic lithiation and in the final step acts as a leaving group.[135]

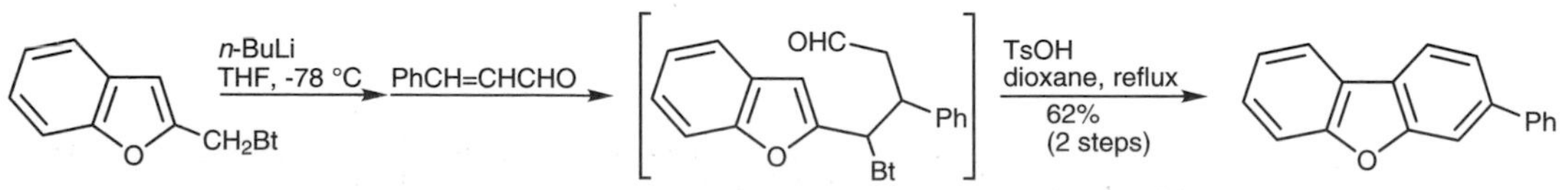

Next, the ability of benzotriazole to stabilise a cation allows 1-hydroxymethyl-benzotriazole to alkylate indole; the product is then lithiated to allow substitution by an electrophile and then finally the benzotriazole unit is displaced by a Grignard reagent.[136]

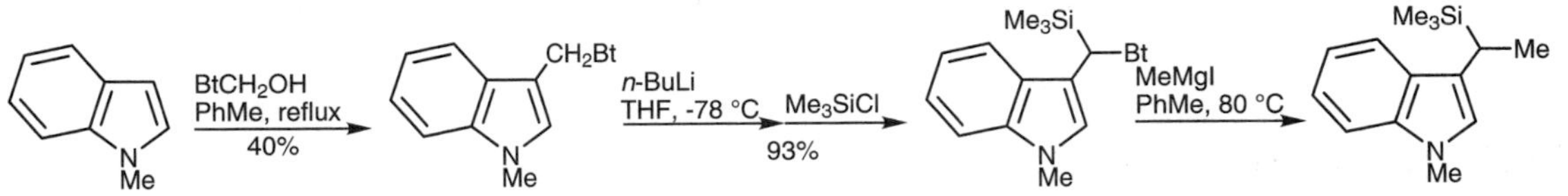

Tris(benzotriazol-1-yl)methane anion will add to nitrobenzene in a VNS process in which the benzotriazole unit is both the anion-stabilising unit and also the leaving group;[137] another use for this compound is illustrated by alkylation of its anion then hydrolysis forming an acid.[138]

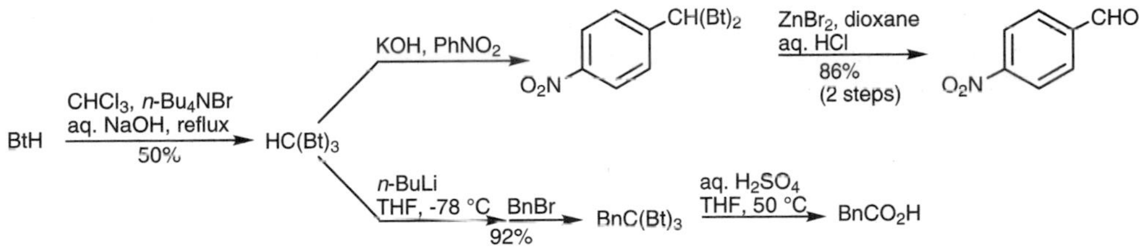

Benzotriazoles also have interesting reactivity in their own right and can be used to form other heterocyclic compounds via various ring cleavage reactions leading to reactive intermediates such as benzynes,[139] aryllithiums,[140] and diradicals, as illustrated below.[141]

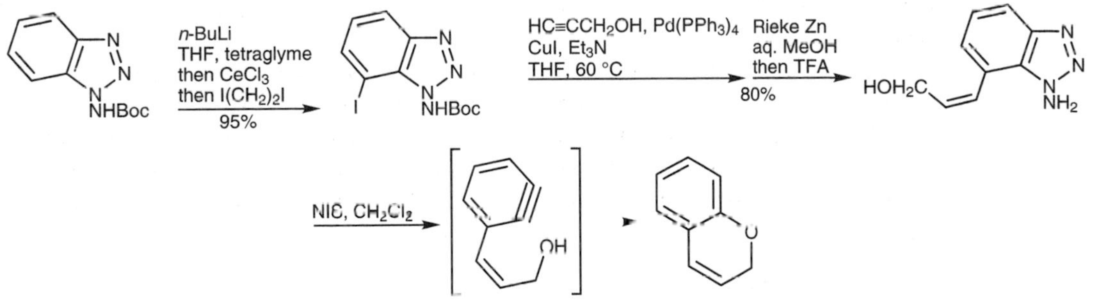

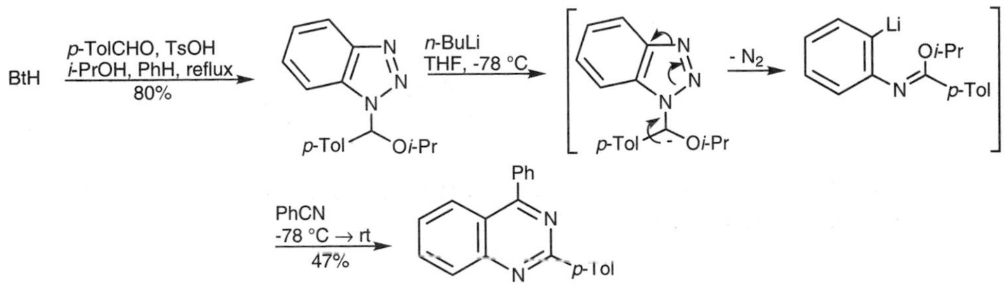

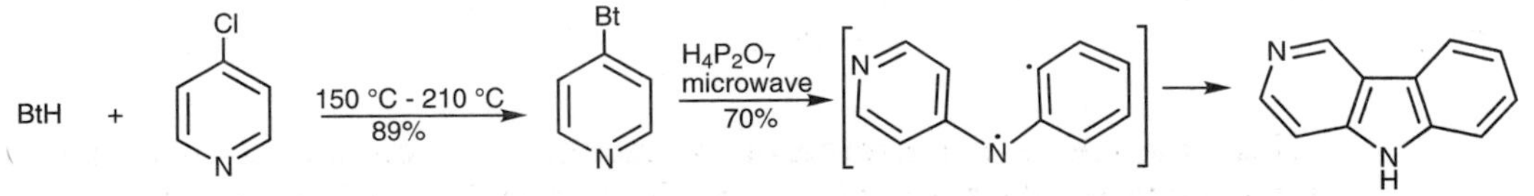

Exercises for chapter 26

1. What are the products of the following (Diels-Alder) reactions: (i) 1-pyrrolidinylcyclopentene with (a) 1,3,5-triazine, (b) 1,2,4-triazine; (ii) 3-phenyl-1,2,4,5-tetrazine with 1,1-diethoxyethene?

2. Thiophosgene ($S = CCl_2$) reacts at low temperature with sodium azide to give a product which contains no azide group; on subsequent reaction with methylamine this compound is converted into $C_2H_4N_4S$ – suggest structures.

3. What are the products of the reaction of $PhCONH_2$ with DMFDMA then (a) N_2H_4 and (b) H_2NOH?

4. 1,3,5-Triazine reacts with (i) aminoguanidine to give 4-amino-1,3,4-triazole and with (ii) diethyl malonate to give ethyl 4-hydroxypyrimidine 5-carboxylate. Write mechanisms for these transformations.

References

1. Grimmett, M. R. and Iddon, B., 'Synthesis and reactions of lithiated monocyclic azoles containing two or more hetero-atoms. Part VI: Triazoles, tetrazoles, oxadiazoles, and thiadiazoles', *Heterocycles*. **1995**, *41*, 1525.
2. Gilchrist, T. L. , Gymer, G. E., and Rees, C. W., *J. Chem. Soc., Perkin Trans. 1*, **1973**, 555; *ibid.*, **1975**, 1.
3. 'Pentazoles', Ugi, I., *Adv. Heterocycl. Chem.*, **1964**, *3*, 373.
4. '1,2,3-Triazoles', Gilchrist, T. L. and Gymer, G. E., *Adv. Heterocycl. Chem.*, **1974**, *16*, 33.
5. Pederdsen, C., *Acta Chem. Scand.*, **1959**, *13*, 888.
6. Ohta, S., Kawasaki, I., Uemura, T., Yamashita, M., Yoshioka, T., and Yamaguchi, S., *Chem. Pharm. Bull,.* **1997**, *45*, 1140.
7. Birkofer, L. and Wegner, P., *Chem. Ber.*, **1967**, *100*, 3485.
8. Bird, C. W., *Tetrahedron*, **1985**, *41*, 1409.
9. Williams, E. L., *Tetrahedron Lett.*, **1992**, *33*, 1033.
10. Hüttel, R. and Welzel, G., *Justus Liebigs Ann. Chem.*, **1955**, *593*, 207.
11. Iddon, B. and Nicholas, M., *J. Chem. Soc., Perkin Trans. 1*, **1996**, 1341.
12. Lynch, B. M. and Chan, T.-L., *Can. J. Chem.*, **1963**, *41*, 274.
13. Wiley, R. H., Hussung, K. F., and Moffat, J., *J. Org. Chem.*, **1956**, *21*, 190; El-Khadem, H., El-Shafei, Z. M., and Meshreki, M. H., *J. Chem. Soc.*, **1961**, 2957.
14. Raap, R., *Can. J. Chem.*, **1971**, *49*, 1792.
15. Ghose, S. and Gilchrist, T. L., *J. Chem. Soc., Perkin Trans. 1*, **1991**, 775.
16. Felding, J., Uhlmann, P., Kristensen, J., Vedso, P., and Begtrup, M., *Synthesis*, **1998**, 1181.
17. 'Approaches to the synthesis of 1-substituted 1,2,4-triazoles', Balasubramanian, M., Keay, J. G., and Scriven, E. F., *Heterocycles*, **1994**, *37*, 1951.
18. Olofson, R. A. and Kendall, R. V., *J. Org. Chem.*, **1970**, *35*, 2246.
19. Horvath, A., *Synthesis*, **1995**, 1183.
20. Kröger, C.-F. and Miethchen, R., *Chem. Ber.*, **1967**, *100*, 2250.
21. Grinsteins, V. and Strazdina, A., *Khim. Geterotsikl. Soedin.*, **1969**, *5*, 1114 (*Chem. Abs.*, **1970**, *72*, 121456).
22. Habraken, C. L. and Cohen-Fernandes, P., *J. Chem. Soc., Chem. Commun.*, **1972**, 37.
23. Anderson, D. K., Sikorski, J. A., Reitz, D. B. and Pilla, L. T., *J. Heterocycl. Chem.*, **1986**, *23*, 1257; Jutzi, P. and Gilge, K., *J. Organomet. Chem.*, **1983**, *246*, 163.
24. Browne, E. J., *Aust. J. Chem.*, **1969**, *22*, 2251; Bagal, L. I., Pevzner, M. S., Frolov, A. N., and Sheludyakova, N. I., *Khim. Geterotsikl. Soedin.*, **1970**, 259 (*Chem. Abs.*, **1970**, *72*, 11383); Bagal, L. I., Pevzner, M. S., Samarenko, V. Ya., and Egorov, A. P., *ibid.*, 1701 (*Chem. Abs.*, **1971**, *74*, 99948).
25. Sano, S., Tanba, M., and Nagao, Y., *Heterocycles*, **1994**, *38*, 481.
26. Zumbrunn, A., *Synthesis*, **1998**, 1357.
27. Kane, J. M., Dalton, C. R., Staeger, M. A., and Huber, E. W., *J. Heterocycl. Chem.*, **1995**, *32*, 183; 'Facile desulfurization of cyclic thioureas by hydrogen peroxide', Grivas, S. and Ronne, E., *Acta Chem. Scand.*, **1995**, *49*, 225.

28. 'Recent advances in tetrazole chemistry', Butler, R. N., *Adv. Heterocycl. Chem.*, **1977**, *21*, 323.
29. 'The high nitrogen compounds', Benson, F. R., Wiley, New York, **1984**.
30. Isida, T., Akiyama, T., Nabika, K., Sisido, K., and Kozima, S., *Bull. Chem. Soc. Jap.*, **1973**, *46*, 2176; Takach, N. E., Holt, E. M., Alcock, N. W., Henry, R. A., and Nelson, J. H., *J. Am. Chem. Soc.*, **1980**, *102*, 2968.
31. Koren, A. O., Gaponik, P. N., Ivashkevich, O. A., and Kovalyova, T. B., *Mendeleev Commun.*, **1995**, 10.
32. Gaponik, P. N., Grigor'ev, Yu. V., and Koren, A. O., *Khim. Geterotsikl. Khim.*, **1988**, 1699 (*Chem. Abs.*, **1989**, *111*, 194674).
33. Gaponik, P. N., Grigor'ev, Y. V., and Karavai, V. P., *Metalloorg. Khim.*, **1988**, *1*, 846 (*Chem. Abs.*, **1989**, *111*, 194937).
34. Karavai, V. P. and Gaponik, P. N., *Khim. Geterotsikl. Khim.*, **1985**, 564 (*Chem. Abs.*, **1985**, *103*, 37426).
35. Barlin, G. B., *J. Chem. Soc., B*, **1967**, 641.
36. Johnstone, R. A. W. and Brigas, A. F., *J. Chem. Soc., Chem. Commun.*, **1994**, 1923; Johnstone, R. A. W. and McLean, W. N., *Tetrahedron Lett.*, **1988**, *29*, 5553; Henderson, P., Kumar, S., Rego, J. A., Ringsdorf, H., and Schuhmacher, P., *J. Chem. Soc., Chem. Commun.*, **1995**, 1059.
37. Raap, R., *Can. J. Chem.*, **1971**, *49*, 2139.
38. Flippin, L. A., *Tetrahedron Lett.*, **1991**, *32*, 6857.
39. Dankwardt, J. W., *J. Org. Chem.*, **1998**, *63*, 3753.
40. Satoh, Y. and Marcopulos, N., *Tetrahedron Lett.*, **1995**, *36*, 1759.
41. Cheney, B. V., *J. Org. Chem.*, **1994**, *59*, 773.
42. Huff, B. E., Le Tourneau, M. E., Staszak, M. A., and Ward, J. A., *Tetrahedron Lett.*, **1996**, *37*, 3653.
43. Katritzky, A. R., Aslan, D., Shcherbakova, I. A., Chen, J., and Belyakov, S. A., *J. Heterocycl. Chem.*, **1996**, *33*, 1107.
44. Kammula, S. and Shevlin, P. B., *J. Am. Chem. Soc.*, **1974**, *96*, 7830.
45. Golden, A. H. and Jones, M., *J. Org. Chem.*, **1996**, *61*, 4460; Kumar, A., Narayanan, R., and Shechter, H., *ibid.*, 4462.
46. Kamala, K., Rao, P. J., and Reddy, K. K., *Bull. Chem. Soc. Jpn.*, **1988**, *61*, 3791.
47. Hoffle, G. and Lange, B., *Org. Synth., Coll. Vol. VII*, **1992**, 27.
48. Jursic, B. S., and Zdravkovski, Z., *Synth. Commun.*, **1994**, *24*, 1575.
49. Birkofer, L. and Wegner, P., *Chem. Ber.*, **1966**, *99*, 2512.
50. Huisgen, R., Möbius, L., and Szeimies, G., *Chem. Ber.*, **1965**, *98*, 1138.
51. Jones, H., Fordice, M. W., Greenwald, R. B., Hannah, J., Jacobs, A., Ruyle, W. V., Walford, G. L., and Shen, T. Y., *J. Med. Chem.*, **1978**, *21*, 1100.
52. Ykman, P., L'Abbé, G., and Smets, G., *Tetrahedron*, **1971**, *27*, 845.
53. Romeiro, G. A., Pereira, L. O. R., deSouza, M. C. B. V., Ferreira, V. F., and Cunha, A. C., *Tetrahedron Lett.*, **1997**, *38*, 5103.
54. Augusti, R. and Kascheres, C., *J. Org. Chem.*, **1993**, *58*, 7079.
55. Harada, K., Oda, M., Matsushita, A., and Shirai, M., *Heterocycles*, **1998**, *48*, 695.
56. Bartlett, R. K. and Humphrey, I. R., *J. Chem. Soc., C*, **1967**, 1664.
57. Grundmann, C. and Rätz, R., *J. Org. Chem.*, **1956**, *21*, 1037.
58. Ried, W. and Valentin, J., *Chem. Ber.*, **1968**, *101*, 2117.
59. Finnegan, W. G., Henry, R. A., and Lofquist, R., *J. Am. Chem. Soc.*, **1958**, *80*, 3908; Duncia, J. V., Pierce, M. E., and Santella, J. B., *J. Org. Chem.*, **1991**, *56*, 2395.
60. Koguro, K., Oga, T., Mitsui, S., and Orita, R., *Synthesis*, **1998**, 910.
61. Jursic, B. S. and LeBlanc, B. W., *J. Heterocycl. Chem.*, **1998**, *35*, 405.
62. Thomas, E. W., *Synthesis*, **1993**, 767
63. Lehnhoff, S. and Ugi, I., *Heterocycles*, **1995**, *40*, 801.
64. Artamonova, T. V., Zhivich, A.B., Dubinskii, M. Yu., and Koldobskii, G. I., *Synthesis*, **1996**, 1428.
65. Collibee, W. L., Nakajima, M., and Anselme, J.-P., *J. Org. Chem.*, **1995**, *60*, 468.
66. LeBlanc, B. W. and Jursic, B. S., *Synth. Commun.*, **1998**, *28*, 3591.
67. Markgraf, J. H. and Sadighi, J. P., *Heterocycles*, **1995**, *40*, 583.
68. Kauer, J. C. and Sheppard, W. A., *J. Org. Chem.*, **1967**, *32*, 3580.
69. Boivin, J., Husinec, S., and Zard, S. Z., *Tetrahedron*, **1995**, *51*, 11737.
70. '1,2,4-Oxadiazoles', Clapp, L. B., *Adv. Heterocycl. Chem.*, **1976**, *20*, 65.

71. 'Recent advances in 1,3,4-oxadiazole chemistry', Hetzhein, A. and Möckel, K., *Adv. Heterocycl. Chem.*, **1966**, *7*, 183.
72. Nguyen, M. T., Hegarty, A. F., and Elguero, J., *Angew. Chem., Int. Ed. Engl.*, **1985**, *24*, 713.
73. 'The chemistry of sydnones', Stewart, F. H. C., *Chem. Rev.*, **1964**, *64*, 129; Huisgen, R., Grashey, R., Gotthardt, H., and Schmidt, R., *Angew. Chem., Int. Ed. Engl.*, **1962**, *1*, 48.
74. 'Furoxans and benofuroxans', Gasco, A. and Boulton, A. J., *Adv. Heterocycl. Chem.*, **1981**, *29*, 251; 'The chemistry of furoxans', Sliwa, W. and Thomas, A., *Heterocycles*, **1985**, *23*, 399.
75. '1,2,4-Thiadiazoles', Kurzer, F., *Adv. Heterocycl. Chem.*, **1982**, *32*, 285.
76. 'Recent advances in the chemistry of 1,3,4-thiadiazoles', Sandström, J., *Adv. Heterocycl. Chem.*, **1968**, *9*, 165.
77. 'The 1,2,5-thiadiazoles', Weinstock, L. M. and Pollak, P. I., *Adv. Heterocycl. Chem.*, **1968**, *9*, 107.
78. Bak, B., Nygaard, L., Pedersen, E. J., and Rastrup-Anderson, J., *J. Mol. Spectrosc.*, **1966**, *19*, 283.
79. Moussebois, G. and Eloy, F., *Helv. Chim., Acta*, **1964**, *47*, 838.
80. Alemagna, A., Bacchetti, T., and Beltrame, P., *Tetrahedron*, **1968**, *24*, 3209.
81. Olofson, R. A. and Michelman, J. S., *J. Org. Chem.*, **1965**, *30*, 1854.
82. Thomas, E. W. and Zimmerman, D. C., *Synthesis*, **1985**, 945.
83. 'Ring-opening of five-membered heteroaromatic anions', Gilchrist, T. I., *Adv. Heterocycl. Chem.*, **1987**, *41*, 41.
84. Boger, D. L., and Brotherton, C. E., *J. Heterocycl. Chem.*, **1981**, *18*, 1247.
85. Voets, M., Smet, M., and Dehaen, W., *J. Chem. Soc., Perkin Trans. 1*, **1999**, 1473.
86. Goerdeler, J. and Deselaers, K., *Chem. Ber.*, **1958**, *91*, 1025; Goerdeler, J. and Rachwalsky, H., *ibid.*, **1960**, *93*, 2190; Butler, R. N., Lambe, T. M., Tobin, J. C., and Scott, F. L., *J. Chem. Soc., Perkin Trans. 1*, **1973**, 1357.
87. 'Ring transformations of five-membered heterocycles', Vivona, N., Buscemi, S., Frenna, V., and Cusmano, G., *Adv. Heterocycl. Chem.*, **1993**, *56*, 49; Cusmano, G., Macaluso, G., and Gruttadauria, M., *Heterocycles*, **1993**, *36*, 1577.
88. Bolton, R.. E., Coote, S. J., Finch, H., Lowden, A., Pegg, N., and Vinader, M. V., *Tetrahedron Lett.*, **1995**, *36*, 4471.
89. '1,2,3,4-Thiatriazoles', Holm, A., *Adv. Heterocycl. Chem.*, **1976**, *20*, 145.
90. Jensen, K. A. and Holm, A., *Acta Chem. Scand.*, **1964**, *18*, 826.
91. Kauer, J. C. and Sheppard, W. A., *J. Org. Chem.*, **1967**, *32*, 3580.
92. Santilli, A. A. and Morris, R. L., *J. Heterocycl. Chem.*, **1979**, *16*, 1197.
93. Liang, G.-B. and Feng, D. D., *Tetrahedron Lett.*, **1996**, *37*, 6627.
94. Lin, Y., Lang, S. A., Lovell, M. F., and Perkinson, N. A., *J. Org. Chem.*, **1979**, *44*, 4160.
95. Neidlein, R. and Li, S., *J. Heterocycl. Chem.*, **1996**, *33*, 1943; *idem, Synth. Commun.*, **1995**, *25*, 2379; Nicolaides, D. N., Fylaktakidou, K. C., Litinas, K. E., Papageorgiou, G. K., and Hadjipavlou-Litina, D. J., *J. Heterocycl. Chem.*, **1998**, *35*, 619.
96. Ainsworth, C., *J. Am. Chem. Soc.*, **1955**, *77*, 1148.
97. Jedlovska, E. and Lesko, J., *Synth. Commun.*, **1994**, *24*, 1879.
98. Thoman, C. J. and Voaden D. J., *Org. Synth., Coll. Vol. V*, **1973**, 962.
99. Hurd, C. D. and Mori, R. I., *J. Am. Chem. Soc.*, **1955**, *77*, 5359.
100. Harada, K., Inoue, T., and Yoshida, H., *Heterocycles*, **1997**, *44*, 459.
101. Cronyn, M. W. and Nakagawa, T. W., *J. Am. Chem. Soc.*, **1952**, *74*, 3693.
102. Goerdeler, J., Groschopp, H., and Sommerlad, U., *Chem. Ber.*, **1957**, *90*, 182.
103. Ainsworth, C., *J. Am. Chem. Soc.*, **1958**, *80*, 5201.
104. Whitehead, C. W. and Traverso, J. J., *J. Am. Chem. Soc.*, **1955**, *77*, 5872; Kress, T. J. and Costantino, S. M., *J. Heterocycl. Chem.*, **1980**, *17*, 607.
105. Föhlisch, B., Braun, R., and Schultze, K. W., *Angew. Chem., Int. Ed. Engl.*, **1967**, *6*, 361.
106. Weinstock, L. M., Davis, P., Handelsman, B., and Tull., R., *J. Org. Chem.*, **1967**, *32*, 2823.
107. Wright, J. B., *J. Org. Chem.*, **1964**, *29*, 1905.
108. Duan, X.-G., Duan, X.-L., and Rees, C. W., *J. Chem. Soc., Perkin Trans. 1*, **1997**, 2831; Duan, X.-G., Duan, X.-L., Rees, C. W., and Yue, T.-Y., *ibid.*, 2597; Duan, X.-G. and Rees, C. W., *ibid.*, 2695.
109. Grundmann, C. and Kreutzberger, A., *J. Am. Chem. Soc.*, **1955**, *77*, 44.
110. Adam, W., van Barneveld, C., and Golsch, D., *Tetrahedron*, **1996**, *52*, 2377.

111. Counotte-Potman, A. and van der Plas, H. C., *J. Heterocycl. Chem.*, **1981**, *18*, 123.
112. 'Behaviour of monocyclic 1,2,4-triazines in reactions with C-, N-, O-, and S-nucleophiles', Charushin, V. N., Alexeev, S. G., Chupakhin, O. N., and van der Plas, H. C., *Adv. Heterocycl. Chem.*, **1989**, *46*, 76.
113. Rykowski, A. and Lipinska, T., *Synth. Commun.*, **1996**, *26*, 4409; Rykowski, A., Branowska, D., Makosza, M., and van Ly, P., *J. Heterocycl. Chem.*, **1996**, *33*, 1567.
114. Makosza, M., Golinski, J., and Rykowski, A., *Tetrahedron Lett.*, **1983**, *24*, 3277; Paudler, W. W. and Chen., T.-K., *J. Heterocycl. Chem.*, **1970**, *7*, 767.
115. Sakya, S. M., Groskopf, K. K., and Boger, D. L., *Tetrahedron Lett.*, **1997**, *38*, 3805.
116. Wilkes, M. C., *J. Heterocycl. Chem.*, **1991**, *28*, 1163.
117. Konno, S., Yokoyama, M., Kaite, A., Yamasuta, I., Ogawa, S., Mizugaki, M., and Yamanaka, H., *Chem. Pharm. Biull.*, **1982**, *30*, 152.
118. 'Synthesis with *s*-triazine', Grundmann, C., *Angew. Chem., Int. Ed. Engl.*, **1963**, *2*, 309.
119. Plé, N., Turck, A., Quéguiner, G., Glassi, B., and Neunhoeffer, H., *Liebigs Ann. Chem.*, **1993**, 583.
120. Itoh, T., Matsuya, Y., Hasegawa, H., Nagata, K., Okada, M., and Ohsawa, A., *J. Chem. Soc., Perkin Trans. 1*, **1996**, 2511.
121. Nagata, K., Itoh, T., Okada, M., Takahashi, H., and Ohsawa, A., *Heterocycles*, **1991**, *32*, 855.
122. Chavez, D. E. and Hiskey, M. A., *J. Heterocycl. Chem.*, **1998**, *35*, 1329.
123. 'Diels-Alder reactions of heterocyclic azadienes: scope and applications', Boger, D. L., *Chem. Rev.*, **1986**, *86*, 781; 'Hetero Diels-Alder methodology in organic synthesis', Ch. 10, Boger, D. L. and Weinreb, S. M., Academic Press, **1987**.
124. Macor, J. E., Kuipers, W., and Lachicotte, R. J., *Chem. Commun.*, **1998**, 983.
125. Ohsawa, A., Arai, H., Ohnishi, H., and Igeta, H., *J. Chem. Soc., Chem. Commun.* **1981**, 1174.
126. Paudler, W. W. and Barton, J. M., *J. Org. Chem.*, **1966**, *31*, 1720.
127. Schaefer, F. W. and Peters, G. A., *J. Org. Chem.*, **1961**, *26*, 2778.
128. Schaefer, F. W. and Peters, G. A., *J. Org. Chem.*, **1961**, *26*, 2784.
129. Chen, C., Dagnino, R., and McCarthy, J. R., *J. Org. Chem.*, **1995**, *60*, 8428.
130. Geldard, J. F. and Lions, F., *J. Org. Chem.*, **1965**, *30*, 318.
131. 'Heterocyclic synthesis with benzotriazole', Katritzky, A. R., Henderson, S. A., and Yang, B., *J. Heterocycl. Chem.*, **1998**, *35*, 1123.
132. 'Properties and synthetic utility of *N*-substituted benzotriazoles', Katritzky, A. R., Lan, X., Yang, J. Z., and Denisko, O. V., *Chem. Rev.*, **1998**, *98*, 409.
133. Katritzky, A. R., Zhang, G., and Jiang, J., *J. Org. Chem.*, **1995**, *60*, 7589.
134. Katritzky, A. R., and Jiang, J., *J. Org. Chem.*, **1995**, *60*, 7597.
135. Katritzky, A. R., Fali, C. N., and Li, J., *J. Org. Chem.*, **1997**, *62*, 8205.
136. Katritzky, A. R., Xie, L., and Cundy, D., *Synth. Commun.*, **1995**, *25*, 539.
137. Katritzky, A. R. and Xie, L., *Tetrahedron Lett.*, **1996**, *37*, 347.
138. Katritzky, A. R., Yang, Z., and Lam, J. N., *Synthesis*, **1990**, 666.
139. Knight, D. W. and Little, P. B., *Tetrahedron Lett.*, **1998**, *39*, 5105.
140. Katritzky, A. R., Zhang, G., Jiang, J., and Steel, P. J., *J. Org. Chem.*, **1995**, *60*, 7625.
141. Molina, A., Vaquero, J. J., Garcia-Navio, J. L., Alvarez-Builla, J., de Pascual-Teresa, B., Gago, F., Rodrigo, M. M., and Ballesteros, M., *J. Org. Chem.*, **1996**, *61*, 5587.

27 Saturated and partially unsaturated heterocyclic compounds: reactions and synthesis

This book is principally concerned with the chemistry of aromatic heterocycles, however mention must be made of the large body of remaining heterocycles, including those with small rings[1] (3- and 4-membered). Most of the reactions of saturated and partially unsaturated heterocyclic compounds are so closely similar to those of acyclic or non-heterocyclic analogues that a full discussion is not appropriate in this book, however in this chapter we discuss briefly those aspects in which they do differ – perhaps the most obvious aspect in which they differ from aromatic heterocycles is in having sp^3 hybridised atoms, i.e. in the exhibition of stereochemistry.[2]

Saturated and partially unsaturated heterocycles are widely distributed as natural products. Some are used as solvents for organic reactions, notably tetrahydrofuran (THF) and dioxane, where diethyl ether is unsuitable. N-Methylpyrrolidone and sulfolane are useful dipolar aprotic solvents, with characteristics like those of dimethylformamide (DMF) and dimethyl sulfoxide (DMSO).

tetrahydrofuran THF — dioxane [1,4-dioxane] — N-methylpyrrolidone [1-methylpyrrolidinone] — sulfolane [tetrahydrothiophene 1,1-dioxide]

The four-membered β-lactam ring is the essential component of the penicillin and cephalosporin antibiotics.

penicillin-G — cephalosporin-C

Epoxides (three-membered saturated oxygen-containing rings), are components of epoxy resins and occur in some natural products, such as the alkaloid scopine. Epoxides, because of their alkylating properties, can be carcinogenic – the biologically active metabolites of carcinogenic hydrocarbons are examples – however they are also found in some anti-tumour agents.

scopine — mitomycin C — thiirane carboxylic acid

Aziridines (three-membered saturated nitrogen-containing rings) are also found in anti-tumour agents, such as the mitomycins. Thiiranes also occur naturally, as plant products such as thiirane-2-carboxylic acid, isolated from asparagus.

27.1   Five- and six-membered rings

27.1.1   Pyrrolidines and piperidines

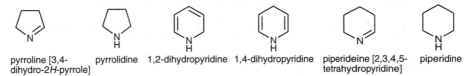

pyrroline [3,4-   pyrrolidine   1,2-dihydropyridine   1,4-dihydropyridine   piperideine [2,3,4,5-   piperidine
dihydro-2*H*-pyrrole]   tetrahydropyridine]

The main chemical aspect in which compounds with a nitrogen in a five- or six-membered ring differ from their acyclic counterparts is in the possibility open to them to be dehydrogenated to the corresponding aromatic system. Dihydroaromatic systems naturally show the greatest tendency to aromatise, indeed one of the important reducing coenzymes, NADPH, makes use of this tendency, as indicated below.

Dihydro-compounds are often useful synthetic intermediates showing different reactivity patterns to the parent, aromatic heterocycle. For example, indolines (2,3-dihydroindoles) can be used to prepare indoles[3] with substituents in the carbocyclic ring, *via* electrophilic substitution then rearomatisation (section 17.16.1.8), and similarly, electrophilic substitutions of dihydropyridines, impossible in pyridines themselves, followed by rearomatisation can give substituted pyridines. Dehydrogenation of tetra- and hexahydro-derivatives requires much more vigorous conditions.

Generally speaking, piperideines and pyrrolines exist predominantly in the imine form and not in the tautomeric enamine form; *N*-alkyl analogues have no alternative but to exist as enamines. These cyclic imines are resistant to hydrolytic fission of the C=N bond, in strong contrast with acyclic imines, but nonetheless they are very susceptible to nucleophilic addition at the azomethine carbon. An example of this is that both piperideine and pyrroline exist as trimers formed by the nucleophilic addition of nitrogen of one molecule to the azomethine carbon of a second molecule, etc.

The presence of some enamine, at equilibrium, is demonstrated by the conversion of piperideine into a dimer, indeed, the ability of these two systems to serve as both imines and enamines in such aldol-like condensations is at the basis of their roles in alkaloid biosynthesis. Formed in nature by the oxidative deamination and decarboxylation of ornithine and lysine, they become incorporated into alkaloid structures by condensation with other precursor units.[4] Hygrine is a simple example in which the 1-pyrroline, as an imine, has condensed with acetoacetate, or its equivalent.

Controlled oxidation of *N*-acylpiperidines and -pyrrolidines can be used to prepare 2-alkoxy-derivatives or the equivalent enamides, which are useful general synthetic intermediates.[5] The former are susceptible to nucleophilic substitution under Lewis acid catalysis, *via* Mannich-type intermediates, and the latter can undergo electrophilic substitution at C-3 or addition to the double bond.

Pyrrolidine and piperidine are better nucleophiles than diethylamine, principally because the lone pair is less hindered – in the heterocycles the two alkyl 'substituents', i.e. the ring carbons, are constrained back and away from the nitrogen lone pair, and approach by an electrophile is thus rendered easier than in diethylamine where rotations of the C–N and C–C bonds hinder approach. The pK_a values of pyrrolidine (11.27) and piperidine (11.29) are typical of amine bases; they are slightly stronger bases than diethylamine (10.98).

Piperidines, like cyclohexanes, adopt a preferred chair conformation. Much controversy centred over the years on the question as to whether in piperidines it is the *N*-substituent (or *N*-hydrogen) or the *N*-lone pair which adopts an equatorial or axial orientation; some confusion arose because of the results from *N*-alkylation reactions, the products from which do not necessarily reflect ground state conformational populations. Both an *N*-hydrogen and an *N*-alkyl substituent adopt an equatorial orientation, though in the former case the equatorial isomer is favoured by only a small margin.[6]

In early days, structure determination of natural products involved degradative methods. Many alkaloids incorporate saturated nitrogen rings, so degradations were used which gave information about the environment of the basic nitrogen atom. The classical method for doing this was the 'Hofmann exhaustive methylation' procedure. This is illustrated as it would be applied to piperidine. What the method does is to cleave N–C bonds and eventually remove the nitrogen; one repetition of the cycle, as in the example, removes the nitrogen – it was originally part of a ring; a third cycle would be necessary if the nitrogen had been originally a component of two rings. At the end of the process a nitrogen-free fragment is left for study to determine the original carbon skeleton.

27.1.2 Pyrans and reduced furans

3,4-Dihydro-2H-pyran and 2,3-dihydrofuran behave as enol ethers, the former being widely used to protect alcohols[7] with which it reacts readily under acidic catalysis, producing acetals which are stable to even strongly basic conditions but easily hydrolysed back to the alcohol under mildly acidic aqueous conditions.

4H-Pyran appears to be somewhat less stable than dihydropyran but reacts similarly, for example it lithiates at C-2 and undergoes Diels-Alder reactions as an enol ether.[8]

A great deal is known about hydroxylated tetrahydrofurans and tetrahydropyrans because such ring systems occur in sugars and sugar-containing compounds – sucrose and RNA (section 24.1) are examples.[9]

Tetrahydropyran, like piperidine, adopts a chair conformation. One of the interesting aspects to emerge from studies of alkoxy-substituted tetrahydropyrans is that when located at C-2, alkoxyl groups prefer an axial orientation (the 'anomeric effect'[10]). The reason for this is that in an equatorial orientation there are unfavourable dipole–dipole interactions between lone pairs on the two oxygen atoms, and the energy gain, when these are relieved in a conformation with the C-2-substituent axial, more than offsets the unfavourable 1,3-diaxial interactions which are introduced at the same time.

Glucose, of which many of the chemical reactions actually involve the small concentration of acyclic polyhydroxyaldehyde in equilibrium with the cyclic forms, hemiacetals containing a tetrahydropyran: this illustrates the inherent stability of chair conformers of saturated six-membered systems. The propensity for cyclisation is a general one: 5-hydroxyaldehydes, -ketones and -acids all easily form six-membered oxygen-containing rings – lactols and lactones respectively.

in water solution
α:β = 37:63

α-(D)-glucopyranose   β-(D)-glucopyranose

Five-membered rings, too, are relatively easy to form: depending on conditions, glucose derivatives can easily be formed in the furanose form i.e. based on tetrahydrofuran.

glucose $\xrightarrow{\text{Me}_2\text{CO, HCl}}$

Saturated cyclic ethers are inert like acyclic ethers, requiring strong conditions for C–O bond cleavage;[11] this contrasts with heterocycles having smaller ring sizes (sections 27.2 and 27.3).

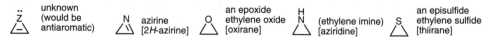

$\xrightarrow[\text{97\%}]{\substack{\text{Ac}_2\text{O, MgBr}_2 \\ \text{MeCN, rt}}}$

27.2   Three-membered rings

27.2.1   Three-membered rings with one hetero atom

Δ-2-Unsaturated three-membered systems are unknown as stable molecules because they would have a 4-electron π-system, and thus be antiaromatic.[12] 1*H*-Azirines occur as reactive intermediates and there is evidence for the existence of 2-thiirene in a low temperature matrix.[13] Azirines,[14] by contrast, are well-known stable compounds. Thiirene *S,S*-dioxides are also stable molecules, probably best likened to cyclopropenones.[15] The chemistry of saturated three-membered heterocycles is, however, very extensive, in particular, epoxides (oxiranes) are vital intermediates in general synthesis.

unknown (would be antiaromatic)   azirine [2*H*-azirine]   an epoxide ethylene oxide [oxirane]   (ethylene imine) [aziridine]   an episulfide ethylene sulfide [thiirane]

A major advance has been the development of an efficient synthesis of epoxides of high optical purity from allylic alcohols and related systems (the *Sharpless epoxidation*) (below); such epoxides have been used extensively for the synthesis of complex natural products in homochiral form.

The pK_a of aziridine (7.98) shows it to be an appreciably weaker base than azetidine (11.29), the four-membered analogue, which is 'normal' for acyclic amines and for five- and six-membered saturated amines. The low basicity is mirrored in the oxygen series, as measured by the ability of oxiranes to form hydrogen bonds. The explanation is probably associated with the strain in the three-membered compounds, meaning that the lone pair is in an orbital with less p-character than a 'normal' sp^3 nitrogen or oxygen orbital, and is therefore held more tightly. The rate of pyramidal inversion of the 'saturated' nitrogen in azirines is very slow compared with simpler amines. This is because there is a further increase in angle strain when the nitrogen rehybridises ($\rightarrow$ sp^2) in the transition state for inversion.

The chemical reactions of three-membered heterocycles are a direct consequence of the strain inherent in such small rings, which, combined with the ability of the heteroatom to act as a leaving group, means that most of the chemical properties involve ring-opening reactions. Most epoxide ring-openings occur by S_N2 nucleophilic displacements at carbon and a very wide range of carbanion and heteroatom nucleophiles have been shown to react in this way, including amines,[16] alcohols, thiols, hydride (LiAlH$_4$), malonate anions,[17] etc. Assistance by protic solvents or O-coordinating metal cations (Lewis acids) which help to further weaken the C–O bond can dramatically increase the rate of reaction. Additives such as alumina,[18] titanium alkoxides,[19] and lithium perchlorate,[20] and reagents such as tributyltin azide,[21] which is itself a Lewis acid (coordination to 'Bu$_3$Sn$^+$'), but also contains a nucleophilic function (N$_3^-$), are useful in this respect.

'Harder' organometallic nucleophiles such as alkyllithiums often give rise to side reactions but their combination (at −78 °C) with boron trifluoride gives very clean and efficient reactions.[22]

The regiochemistry of ring opening is determined mainly by steric and to a lesser extent by inductive and electronic effects. Where strong Lewis acids are used or where a highly stabilised (incipient) carbonium ion can be formed, such as when an α-aryl substituent is present, reaction can occur mainly at the most substituted position, an extreme case being the solvolysis of 2-furyloxirane in neutral methanol;[23] however, selective substitution at the most highly substituted position of even simple, alkyl epoxides has been achieved with an allyltitanium reagent.[24]

The Payne rearrangement of epoxy-alcohols is a special case of an intramolecular nucleophilic opening of epoxides and is of synthetic significance due to its application to Sharpless epoxides.[25]

Ring-opening of epoxides by β-elimination, on reaction with strong bases such as lithium amides, or combinations of trimethylsilyl triflate with diazabicycloundecane,[26] is a useful synthetic method for allylic alcohols, particularly as it can be carried out enantioselectively.[27]

The relative stereochemistry of epoxides can be inverted by equilibration with cyanate anion, as the sequence below shows.[28]

Acid-catalysed opening of aziridines is usually quite rapid, but simple nucleophilic reactions, without acid catalysis, are very slow due to the much poorer leaving ability of negatively charged nitrogen, however N-acyl or N-sulfonyl aziridines have reactivity similar to epoxides.[29] In the nucleophilic ring opening of aziridines with N-nosyl (4-nitrophenylsulfonyl) groups, the excellent leaving group ability of the nitrogen and its substituent can lead to loss of regioselectivity.[30]

Thiiranes similarly undergo ring opening reactions with nucleophiles such as amines,[31] but attack at sulfur can also occur with lithium reagents.[32]

The heteroatom in a three-membered heterocycle can be eliminated via various cycloreversion reactions, for example by nitrosation of aziridines,[33] or by the reaction of thiiranes with trivalent phosphorus compounds.

A related elimination of sulfur dioxide occurs during the Ramberg-Bäcklund synthesis[34] of alkenes, which generates an episulfone as a transient intermediate, although episulfones are isolable under controlled conditions.[35]

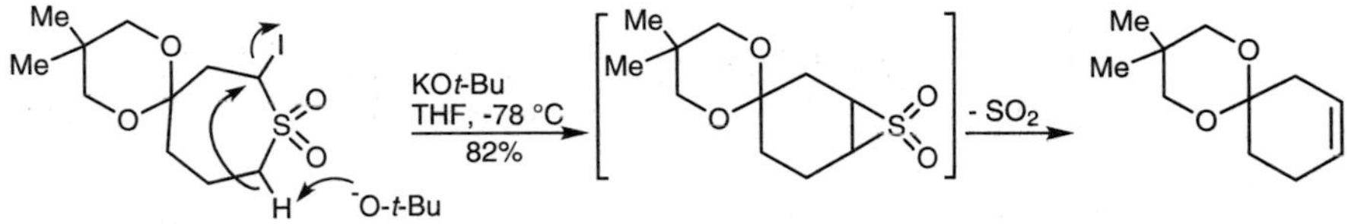

Substituted derivatives of all three systems are able to undergo a highly stereospecific concerted thermal ring opening, generating ylides which can be utilised (trapped) in 3 + 2 cycloaddition reactions. providing a route to pyrrolidines.[36]

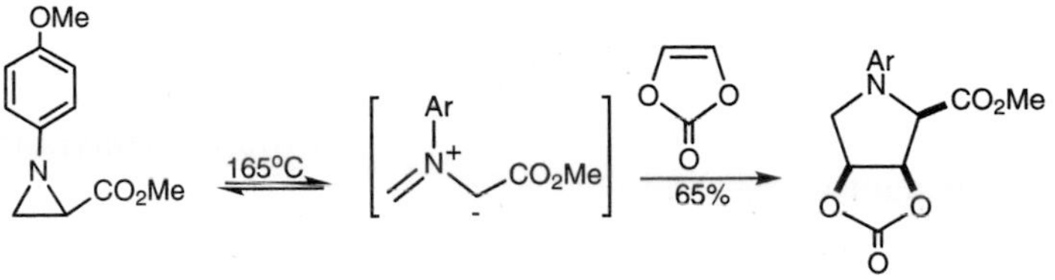

Azirines with an ester group on the imine carbon, will take part in cycloadditions, with the imine unit as the dienophile, as illustrated below.[37]

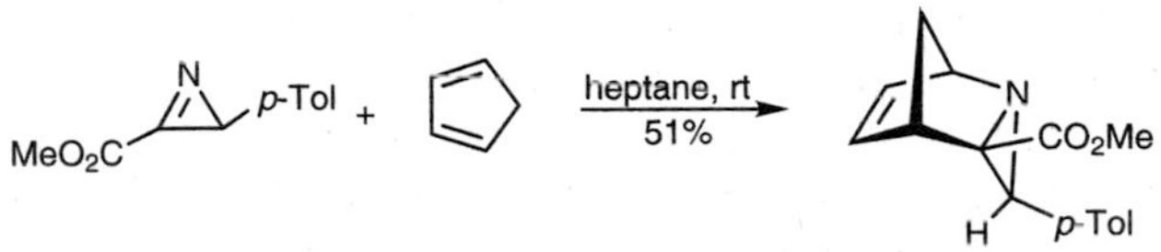

27.2.2 Three-membered rings with two hetero atoms

Diaziridines, diazirines and dioxiranes are all relatively stable isolable systems, although some dioxiranes are explosive.

Three-membered rings with two heteroatoms are usually encountered as reagents. Diazirines are useful carbene precursors[38] – they are generally more stable that the equivalent isomeric diazo compounds, though they are sometimes explosive in the pure state. They can be prepared by oxidation of diaziridines which in turn are available via the condensation of a ketone or aldehyde with ammonia and chloramine.[39] Chlorodiazirines, from the reaction of amidines with hypochlorite, will undergo S_N2 or S_N2' displacement reactions.[40]

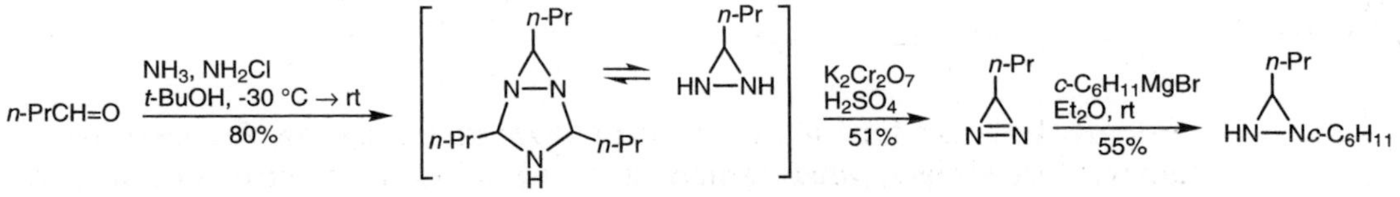

Dimethyldioxirane is a relatively strong oxidant but can show good selectivity: its reactivity is similar to that of a peracid but it has the advantage of producing a neutral byproduct (acetone). Methyl(trifluoromethyl)dioxirane is a more powerful oxidant which can insert oxygen into C–H bonds with retention of configuration, as shown below.[41] Dioxiranes are obtained by reaction of ketones with *OXONE®*.[42] NOTE: Dioxiranes are explosive and are usually handled in dilute solution.

Oxaziridines are selective oxygen-transfer reagents.[43] In particular, the camphor-derived reagent shown below is widely used for enantioselective oxygenation of enolates[44] and other nucleophiles. Oxaziridines are prepared by oxidation of imines.[45]

27.3   **Four-membered rings**

azete   3,4-dihydroazete   azetidine   2*H*-oxete   oxetane   2*H*-thiete   thietane

Derivatives of azete are only known as unstable reaction intermediates. Oxetane and azetidine are considerably less reactive than their three-membered counterparts (oxetane reacts with hydroxide anion 10^3 times more slowly than does oxirane), but nonetheless do undergo similar ring opening reactions, for example oxetane reacts with organolithium reagents,[22] in the presence of boron trifluoride, or with cuprates,[46] and azetidine is opened on heating with concentrated hydrochloric acid.

The most important four-membered system is undoubtedly the β-lactam ring[47] which is present in, and essential for the biological activity of, the penicillin and cephalosporin antibiotics. β-Lactams are very susceptible to ring-opening *via* attack at the carbonyl carbon – in stark contrast to the five-membered analogues (pyrrolidones) or acyclic amides, which are relatively resistant to nucleophilic attack at carbonyl carbon. In addition, β-lactams are hydrolysed by a specific enzyme, β-lactamase, the production of which is a mechanism by which bacteria become resistant to such antibiotics. Although the β-lactam ring is easily cleaved by nucleophiles, both *N*- and *C*-alkylation (α to carbonyl) can be achieved using bases to deprotonate; it is even possible to carry out Wittig reactions at the 'amide' carbonyl without ring-opening.[48] Substitution of the acetoxy group in a 4-acetoxyazetidinone by nucleophiles is an important synthetic method; the reaction proceeds *via* an imine or an iminium intermediate rather than by direct displacement.[49]

β-Lactones (propiolactones)[50] too are readily attacked at the carbonyl carbon, for example they are particularly easily hydrolysed, but a second mode of nucleophilic attack – S_N2 displacement of carboxylate *via* attack at C-4 – occurs with many nucleophiles.[51] The example shows the use of a homochiral lactone, available from serine.

27.4 Metallation

Saturated and mono-unsaturated 5-[52] and 6-membered rings can be metallated in the same way as their acyclic analogues. In the case of tetrahydrofuran however, warming with *n*-butyllithium produces a lithio-derivative which undergoes a cycloreversion generating ethene and the lithium enolate of ethanal.[53] This process represents the most convenient preparation of this enolate but can also be a significant, unwanted side-reaction during lithiation reactions using tetrahydrofuran as solvent.

Three-membered rings have not been metallated directly in the absence of anion-stabilising substituents but simple lithio-derivatives of aziridines have been prepared by exchange from the corresponding stannane.[54]

The conformationally restrained, cyclic nature of dihydrofuran (and dihydropyran) leads to an abnormal sequence during a Heck reaction. The addition of the arylpalladium halide occurs normally but rotation cannot occur so instead of syn β-hydride elimination towards the aryl substituent, elimination takes place towards C-4.[55] In some cases, particularly at higher temperatures, further migration of the double bond occurs.

27.5 Ring synthesis

Five- and six-membered saturated rings can be prepared by reduction of the corresponding aromatic compound, but the most general method for making all ring sizes is by cyclisation of an ω-substituted amine, alcohol, or thiol *via* an intramolecular nucleophilic displacement. As an illustration, the rate of cyclisation of ω-halo-amines goes through a minimum at the four-membered ring size; the five and six-membered rings are by far the easiest to make (relative rates: 72(3-membered ring):1(4):6000(5):1000 (6)).[56] A factor which influences the rate of 3-*exo-tet* cyclisations is the degree of substitution at the carbon carrying the heteroatom: increasing substitution increases the rate of cyclisation, because in the small ring product there is some relief of steric crowding for the substituents compared with acyclic starting material.[57]

Related cyclisations involving hetero atom attachment to an alkene *via* π-complexes with cations such as Br$^+$, I$^+$, Hg$^+$, and Pd$^+$, are useful methods because they give products with functionalised side-chains for further transformations.

27.5.1   Saturated nitrogen heterocycles

There are three main routes to aziridines: they can be prepared by alkali-catalysed cyclisation of 2-haloamines or of a 2-hydroxyamine sulfonate ester, as illustrated,[58] or by additions to alkenes or imines.

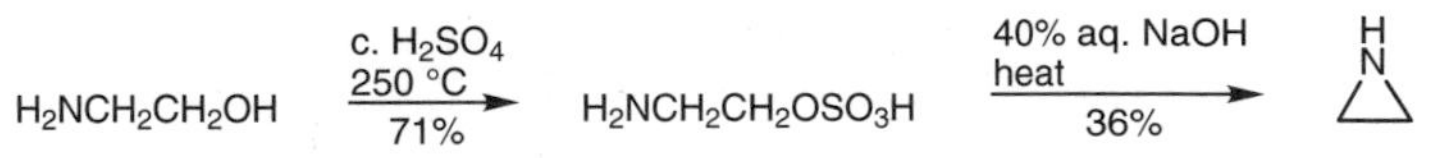

Various homochiral aziridines can be easily obtained from serine;[59] such substances can be transformed into a range of polyfunctional homochiral intermediates and products.

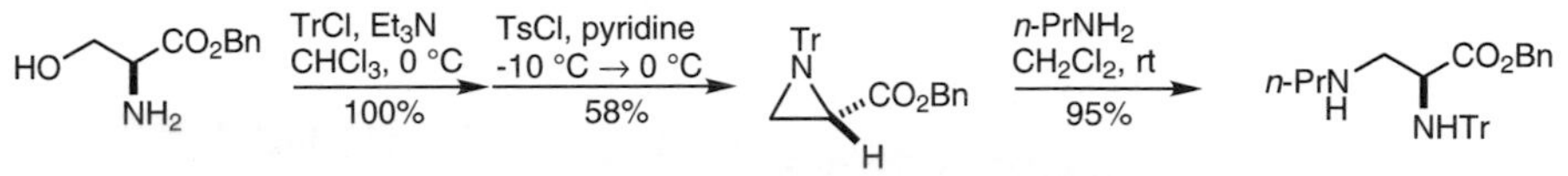

Aziridines can be obtained from alkenes using iodine isocyanate[60] or iodine azide.[61] The product from the latter reaction can be converted into the aziridine *via* reduction, or into an azirine *via* elimination of hydrogen iodide and pyrolysis.[62]

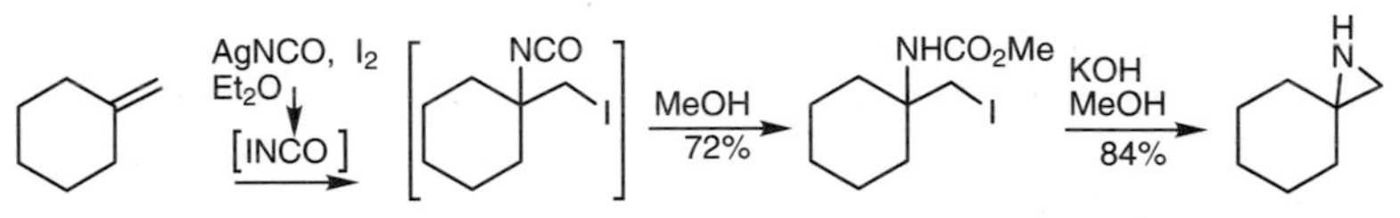

N-Tosylaziridines can be obtained directly from alkenes by reaction with Chloramine T, as shown below.[63]

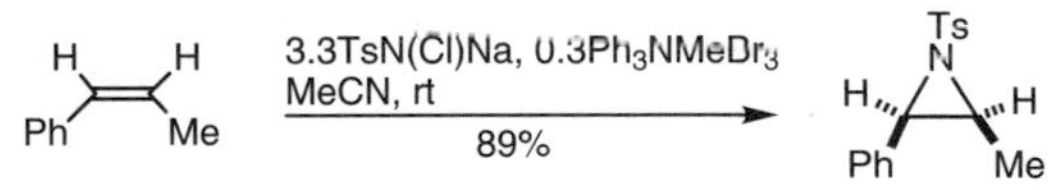

Addition to imines is the third obvious way in which to construct an aziridine, as illustrated below.[64]

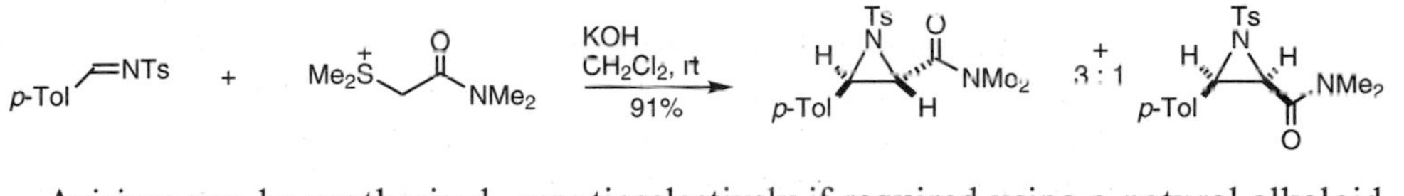

Azirines can be synthesised, enantioselectively if required using a natural alkaloid as base, from the *O*-tosyl derivatives of the oximes of 1,3-keto esters; in this synthesis the carbon is the nucleophilic centre and it is the nitrogen which is attacked with departure of tosylate.[65]

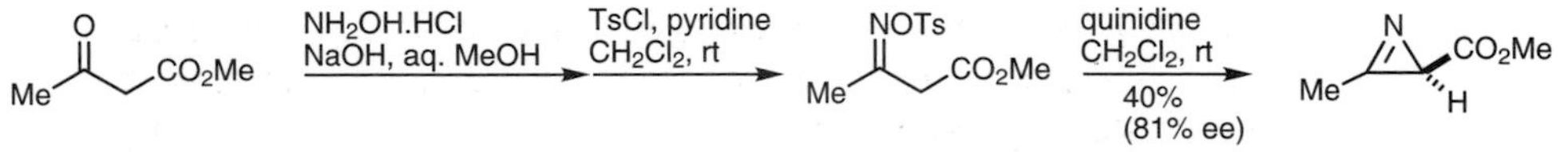

Azetidines can be obtained by cyclisations of 3-halo-amines, but yields are generally not as good as those for the formation of aziridines. The generation of the bifunctional precursors for cyclisation to azetidines has been achieved in a number of ways.[66]

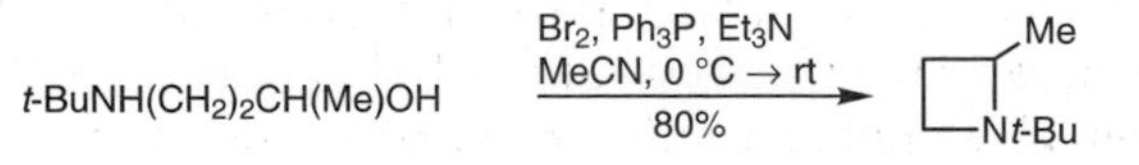

Syntheses of 1-azabicyclo[1.1.0]butane which contains both a four-membered and two three-membered nitrogen-containing rings (!) follow the general route described above.[67] As one would antipate, ring opening reactions, one of which is illustrated, lead to products with an azetidine unit, rather than an aziridine unit.

Aziridines can also be prepared by addition of nitrenes to alkenes,[68] or by the use of nitrogen-transfer agents analogous to epoxidising agents.[69]

Many methods have been developed for β-lactam synthesis,[47,70] including cyclisation of the corresponding amino acids. The most widely used methods are two-component couplings[38,71] which occur *via* concerted cycloaddition or two-step mechanisms.

A very neat method for the synthesis of pyrrolidines does not require a difunctionalised starting material, but relies on the Hofmann-Löffler-Freytag reaction[72] – which is a radical process – to introduce the second functional group. The six-membered size of the cyclic transition state leads selectively to a 1,4-halo-amine, and thence to pyrrolidines.

The cycloaddition of azomethine ylides to alkenes is another elegant entry to pyrrolidines. The required 1,3-dipoles can be produced in a number of ways; the example below is one of the most simple wherein a trimethylsilylmethylamine, an aldehyde and the alkene are simply heated in tetrahydrofuran.[73]

A particularly useful general method for the synthesis of 5- to 7-membered partially unsaturated heterocycles is the Grubbs olefin metathesis applied to acyclic dialkenyl amines, as illustrated by the tetrahydropyridine synthesis shown below.[74]

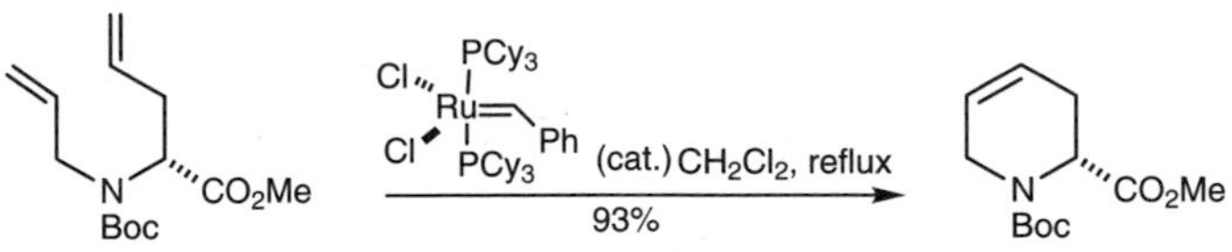

27.5.2   Saturated oxygen heterocycles

The most widely used method for the preparation of epoxides involves oxidation of an alkene by a peracid,[75] *via* a direct one-step transfer of an oxygen atom. More highly (alkyl) substituted alkenes react fastest showing that electronic effects are more important than steric effects in this reaction. Steric effects do, however, control the facial selectivity of epoxidation; conversely hydrogen-bonding groups such as OH and NH can direct the reaction to the *syn* face.

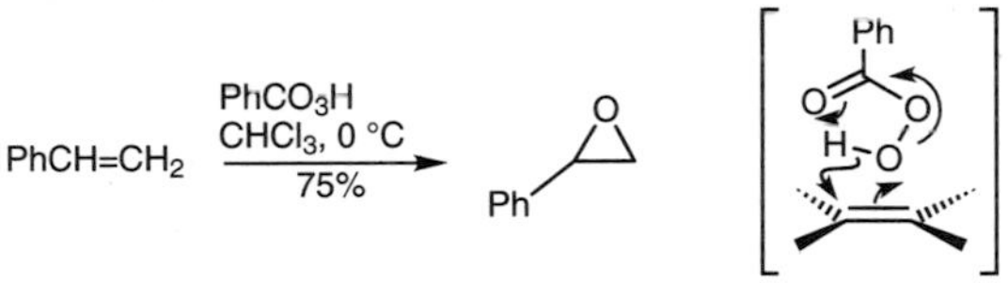

Several other direct oxygen-transfer reagents have been developed of which by far the most important is Sharpless' reagent – a mixture of a hydroperoxide with titanium isopropoxide and an alkyl tartrate.[76] The structure of the reagent is complex but it reacts readily with alkenes containing polar groups, for example allylic alcohols, which can coordinate the metal. The most important feature of this process is that when homochiral tartrate esters are used, a highly ordered asymmetric reactive site results, leading in turn to high optical induction in the product.

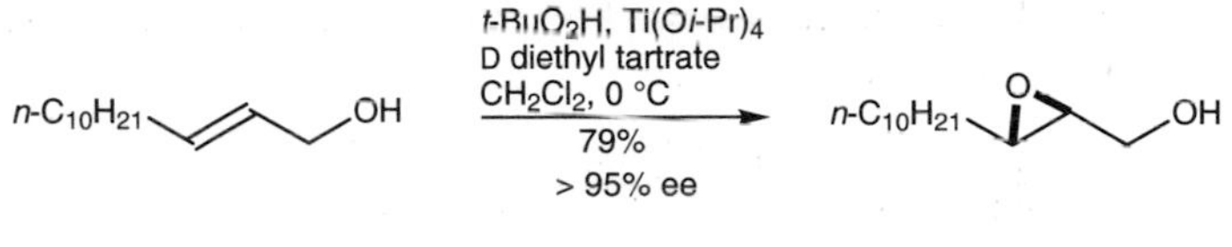

Epoxides and oxetanes can also be prepared by cyclisation of 1,2- (halohydrins) and 1,3-halo-alcohols.[77]

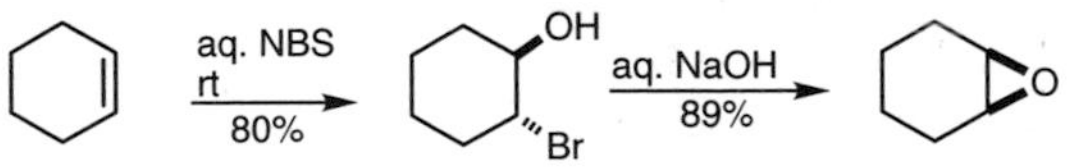

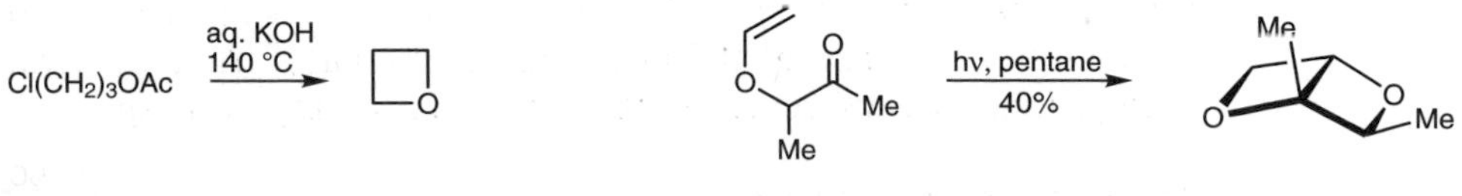

Oxetanes have often been prepared by the *Paterno-Büchi reaction*[78] in which a compound containing a carbon–carbon double bond cycloadds to an aldehyde or ketone under the influence of light.[79]

27.5.3 Saturated sulfur heterocycles

Thiiranes can be prepared by cyclisation of 2-halo-thiols but the most common method is *via* reaction of an epoxide with thiocyanate[80] (cf. section 27.2), thiourea,[81] a phosphine sulfide, or with dimethylthioformamide.[82]

Thietanes, tetrahydrothiophenes and tetrahydrothiapyrans can all be prepared by the reaction of the appropriate 1,ω-dihalide with sulfide anion.

References

1. 'Ethylenimine and other aziridines', Derner, O. C. and Ham, G. E., Academic Press, **1969**; 'Thiiranes', Sander, M., *Chem. Rev.*, **1966**, *66*, 297.
2. 'Stereochemistry of heterocyclic compounds', Parts 1 and 2, Armarego, W. L. F., Wiley-Interscience, New York, **1977**; 'The conformational analysis of heterocyclic compounds', Riddell, F. G., Academic Press, **1980**.
3. Inada, A., Nakamaura, Y., and Morita, Y., *Chem. Lett.*, **1980**, 1287.
4. 'Introduction to the alkaloids. A biogenetic approach', Cordell, G. A., Wiley-Interscience, **1981**.
5. Matsumura, Y., Terauchi, J., Yamamoto, T., Konno, T., and Shono, T., *Tetrahedron*, **1993**, *49*, 8503.
6. 'Conformation of piperidine and of derivatives with additional ring hetero atoms', Blackburne, I. D., Katritzky, A. R., and Takeuchi, Y., *Acc. Chem. Res.*, **1975**, *8*, 300; '*N*-Methyl inversion barriers in six-membered rings', Katritzky, A. R., Patel, R. C., and Riddell, F. G., *Angew. Chem., Int. Ed. Engl.*, **1981**, *20*, 521.
7. 'Protective groups in organic synthesis', Greene, T. W. and Wuts, P. G. M., John Wiley, **1999**.
8. Schlosser, M. and Schneider, P., *Angew. Chem., Int. Ed. Engl.*, **1979**, *18*, 489.
9. 'Stereochemistry of carbohydrates', Stoddart, J. F., Wiley-Interscience, **1971**.
10. 'Stereoelectronic effects in organic chemistry', Deslongchamps, P., Pergamon Press, **1983**.
11. Goldsmith, D. J., Kennedy, E., and Campbell, R. G., *J. Org. Chem.*, **1975**, *40*, 3571.
12. Dewar, M. J. S. and Ramsden, C. A., *J. Chem. Soc., Chem. Commun.*, **1973**, 688.
13. Torres, M., Clement, A., Bertie, J. E., Gunnig, H. E., and Strausz, O. P., *J. Org. Chem.*, **1978**, *43*, 2490.
14. 'Synthesis of heterocycles *via* cycloadditions to 1-azirines', Anderson, D. J. and Hassner, A., *Synthesis*, **1975**, 483; '1-Azirine ring chemistry', Nair, V. and Hyup Kim, K., *Heterocycles*, **1977**, *7*, 353.
15. Carpino, L., McAdams, L. V., Rynbrandt, R. H., and Spiewak., J. W., *J. Am. Chem. Soc.*, **1971**, *93*, 476.
16. Taguchi, T., *J. Pharm. Soc. Jpn.*, **1952**, *72*, 921.
17. Newman, M. S. and VanderWerf, C. A., *J. Am. Chem. Soc.*, **1945**, *67*, 233.
18. Posner, G. H. and Rogers, D. Z., *J. Am. Chem. Soc.*, **1977**, *99*, 8208 and 8214.
19. Chong, J. M. and Sharpless, K. B., *J. Org. Chem.*, **1985**, *50*, 1557.
20. Chini, M., Crotti, P., and Macchia, F., *Tetrahedron Lett.*, **1990**, *31*, 4661.
21. Saito, S., Yamashita, S., Nishikawa, T., Yokoyama, Y., Inaba, M., and Moriwake, T., *Tetrahedron Lett.*, **1989**, *30*, 4153.
22. Eis, M. J., Wrobel, J. E., and Ganem, B., *J. Am. Chem. Soc.*, **1984**, *106*, 3693.
23. Alcaide, B., Biurran, C., and Plumet, J., *Tetrahedron*, **1992**, *48*, 9719.
24. Tanaka, T., Inoue, T., Kamei, K., Murakami, K., and Iwata, C., *J. Chem. Soc., Chem. Commun.*, **1990**, 906.

25. Behrens, C. H., Ko, S. Y., Sharpless, K. B., and Walker, F. J., *J. Org. Chem.*, **1985**, *50*, 5687; Bulman Page, P. C., Rayner, C. M., and Sutherland, I. O., *J. Chem. Soc., Chem. Commun.*, **1988**, 356.
26. Murata, S., Suzuki, M., and Noyori, R., *J. Am. Chem. Soc.*, **1979**, *101*, 2738.
27. Asami, M., *Chem. Lett.*, **1984**, 829.
28. Jankowski, K. and Daigle, J.-Y., *Can. J. Chem.*, **1971**, *49*, 2594.
29. Kozikowski, A. P., Ishida, H., and Isobe, K., *J. Org. Chem.*, **1979**, *44*, 2788; Stamm, H. and Weiss, R., *Synthesis*, **1986**, 392 and 395; Lehmann, J. and Wamhoff, H., *Synthesis*, **1973**, 546.
30. Malignes, P. E., See, M. M., Askin, D., and Reider, P. J., *Tetrahedron Lett.*, **1997**, *38*, 5253.
31. Snyder, H. R., Stewart, J. M., and Ziegler, J. B., *J. Am. Chem. Soc.*, **1947**, *69*, 2672.
32. Trost, B. M. and Ziman, S. D., *J. Org. Chem.*, **1973**, *38*, 932.
33. Clark, R. D. and Helmkamp, G. K., *J. Org. Chem.*, **1964**, *29*, 1316; Lee, K. and Kim, Y. H., *Synth. Commun.*, **1999**, *29*, 1241.
34. 'Recent developments in Ramberg-Bäcklund and episulfone chemistry', Taylor, R. J. K., *Chem. Commun.*, **1999**, 217.
35. Ewin, R. A., Loughlin, W. A., Pyke, S. M., Morales, J. C., and Taylor, R. J. K., *Synlett*, **1993**, 660.
36. De Shong, P., Kell, D. A., and Sidler, D. R., *J. Org. Chem.*, **1985**, *50*, 2309; De Shong, P., Sidler, D. R., Kell, D. A., and Aronson, N. N., *Tetrahedron Lett.*, **1985**, *26*, 3747.
37. Bhuller, P., Gilchrist, T. L., and Maddocks, P., *Synthesis*, **1997**, 271.
38. Bonneau, R.,Liu, M. T. H., and Lapouyade, R., *J. Chem. Soc., Perkin Trans. 1*, **1989**, 1547.
39. Schmitz, E. and Ohme, R., *Chem. Ber.*, **1962**, *95*, 795.
40. Padwa, A. and Eastman, D., *J. Org. Chem.*, **1969**, *34*, 2728.
41. Adam, W., Asensio, G., Curci, R., González-Núñez, M. E., and Mello, R., *J. Org. Chem.*, **1992**, *57*, 953.
42. Murray, R. W. and Jeyaraman, R., *J. Org. Chem.*, **1985**, *50*, 2847.
43. 'Applications of oxaziridines in organic synthesis', Davis, F. A. and Sheppard, A. C., *Tetrahedron*, **1989**, *45*, 5703; 'Asymmetric hydroxylation of enolates with *N*-sulfonyloxaziridines', Davis, F. A. and Chen, B.-C., *Chem. Rev.*, **1992**, *9*, 919.
44. Davis, F. A., Sheppard, A. C., Chen, B.-C., and Haque, M. S., *J. Am. Chem. Soc.*, **1990**, *112*, 6679.
45. Towson, J. C., Weismiller, M. C., Lal, G. S., Sheppard, A. C., and Davis, F. A., *Org. Synth.*, **1990**, *69*, 158.
46. Huynh, C., Derguini-Boumechal, F., and Linstrumelle, G., *Tetrahedron Lett.*, **1979**, *20*, 1503.
47. 'The organic chemistry of β-lactams', Georg, G. I., Ed., VCH, New York, **1993**.
48. Baldwin, J. E., Edwards, A. J., Farthing, C. N., and Russell, A. T., *Synlett*, **1993**, 49; Baldwin, J. E., Adlington, R. M., Godfrey, C. R. A., Gollins, D. W., Smith, M. L., and Russel, A. T., *ibid.*, 51.
49. Clauss, K., Grimm, D., and Prossel, G., *Justus Liebigs Ann. Chem.*, **1974**, 539.
50. 'Recent advances in β-lactone chemistry', Pommur, A. and Pons, J.-M., *Synthesis*, **1993**, 441.
51. Arnold, L. D., Kalantar, T. H., and Vederas, J. C., *J. Am. Chem. Soc.*, **1985**, *107*, 7105.
52. Boeckman, R. K. and Bruza, K. J., *Tetrahedron Lett.*, **1977**, *48*, 4187.
53. Jung, M. E. and Blum, R. B., *Tetrahedron Lett.*, **1977**, 3791.
54. Vedejs, E. and Moss, W. O., *J. Am. Chem. Soc.*, **1993**, *115*, 1607.
55. Lee, T. D. and Daves, G. D., *J. Org. Chem.*, **1983**, *48*, 399.
56. Galli, C., Illuminati, G., Mandolini, L., and Tamborra, P., *J. Am. Chem. Soc.*, **1977**, *99*, 2591.
57. 'Effective molarities for intramolecular reactions', Kirby, A. J., *Adv. Phys. Org. Chem.*, **1980**, *17*, 183.
58. Allen, C. F. H., Spangler, F. W., and Webster, E. R., *Org. Synth., Coll. Vol. IV*, **1963**, 433.
59. 'Serine derivatives in organic synthesis', Kulkarni, Y. S., *Aldrichimica Acta*, **1999**, *32*, 18.
60. Hassner, A., Lorber, M. E., and Heathcock, C., *J. Org. Chem.*, **1967**, *32*, 540.
61. Fowler, F. W., Hassner, A., and Levy, L. A., *J. Am. Chem. Soc.*, **1967**, *89*, 2077; Hassner, A. and Fowler, F. W., *J. Org. Chem.*, **1968**, *33*, 2686.

62. Smolinsky, G., *J. Org. Chem.*, **1962**, *27*, 3557.
63. Jeong, J. U., Tao, B., Sagasser, I., Henninges, H., and Sharpless, K. B., *J. Am. Chem. Soc.* **1998**, *120*, 6844.
64. Li, A.-H., Dai, L.-X., and Hou, X.-L., *J. Chem. Soc., Perkin Trans. 1*, **1996**, 2725; Zhou, Y.-G., Li, A.-H., Hou, X.-L., and Dai, L.-X., *Tetrahedron Lett.*, **1997**, *38*, 7225.
65. Verstappen, M. M. H., Ariaans, G. J. A., and Zwannenburg, B., *J. Am. Chem. Soc.*, **1996**, *118*, 8491.
66. Wadsworth, D. H., *Org. Synth., Coll. Vol. VI*, **1988**, 75; Freeman, J. P. and Mondron, P. J., *Synthesis*, **1974**, 894; Szmuszkovicz, J., Kane, M. P., Laurian, L. G., Chidester, C. G., and Scahill, T. A., *J. Org. Chem.*, **1981**, *46*, 3562.
67. Dave, P. R., *J. Org. Chem.*, **1996**, *61*, 5453; Hayashi, K., Sato, C., Kumagai, T., Tamai, S., Abe, T., and Nagao, Y., *Tetrahedron Lett.*, **1999**, *40*, 3761.
68. 'Nitrenes', Ed. Lwowski, W., Interscience, **1970**.
69. Atkinson, R. S., Coogan, M. P., and Cornell, C. L., *J. Chem. Soc., Chem. Commun.*, **1993**, 1215.
70. 'Synthesis of β-lactams', Mukerjee, A. K. and Srivastava, R. C., *Synthesis*, **1973**, 327.
71. Gluchowski, C., Cooper, L., Bergbreiter, D. E., and Newcomb, M., *J. Org. Chem.*, **1980**, *45*, 3413.
72. 'Cyclisation of *N*-halogenated amines. (The Hofmann-Loffler reaction)', Wolff, M. E., *Chem. Rev.*, **1963**, *63*, 55.
73. Torii, S., Okumoto, H., and Genba, A., *Chem. Lett.*, **1996**, 747.
74. Rutjes, F. P. J. T. and Schoemaker, H. E., *Tetrahedron Lett.*, **1997**, *38*, 677.
75. 'Epoxidation and hydroxylation of ethylenic compounds with organic peracids', Swern, D., *Org. Reactions*, **1953**, *7*, 378; Rebek, J., Marshall, L., McManis, J., and Wolak, R., *J. Org. Chem.*, **1986**, *51*, 1649.
76. Katsuki, T. and Sharpless, K. B., *J. Am. Chem. Soc.*, **1980**, *102*, 5974; 'Asymmetric epoxidation of allylic alcohols: the Sharpless reaction', Pfenninger, A., *Synthesis*, **1986**, 89; 'Mechanism of asymmetric epoxidation; 1. Kinetics', Woodward, S. S., Finn, M. G., and Sharpless, K. B., *J. Am. Chem. Soc.*, **1991**, *113*, 106; '2. Catalyst structure', Finn, M. G. and Sharpless, K. B., *ibid.*, 113.
77. Searles, S. and Gortatowski, M. J., *J. Am. Chem. Soc.*, **1953**, *75*, 3030.
78. Paterno, E. and Chieffi, G., *Gazz. Chim. Ital.*, **1909**, *39*, 341; Büchi, G., Inman, C. G., and Lipinski, E. S., *J. Am. Chem. Soc.*, **1954**, *76*, 4327.
79. Dalton, J. C. and Tremont, S. J., *Tetrahedron Lett.*, **1973**, 4025.
80. Bouda, H., Borredon, M. E., Delmas, M., and Gaset, A., *Synth. Commun.*, **1987**, *17*, 943; Iranpoor, N. and Kazemi, *Synthesis*, **1996**, 821.
81. Bouda, H., Borredon, M. E., Delmas, M., and Gaset, A., *Synth. Commun.*, **1989**, *19*, 491.
82. Ettlinger, M. G., *J. Am. Chem. Soc.*, **1950**, *72*, 4792; Chan, T. H. and Finkenbine, J. R., *J. Am. Chem. Soc.*, **1972**, *94*, 2880; Takido, T., Kobayashi, Y., and Itabashi, K., *Synthesis*, **1986**, 779.

28   Heterocycles at work

The occurrence of heterocyclic compounds in nature is widespread, and the use of natural and synthetic heterocyclic compounds in many commercially important spheres, is enormous. Indeed even a book devoted entirely to 'Heterocycles in Life and Society'[1], can only cover some of the important uses. In the earlier chapters of this book, small selections of some of the significant heterocyclic substances in use as medicines have been cited. In this final chapter we choose to highlight three aspects of heterocyclic chemistry of commercial importance – the increasing use of solid phase chemistry for the discovery of new drugs, the problems associated with the scale up from discovery chemistry to production quantities of new drugs, and the relevance of heterocyclic chemistry to materials of actual or potential value in electroactive and related areas of materials science.

28.1   Solid phase reactions[2]

Reactions on solid phase – that is, where a substrate molecule is attached to an insoluble polymeric support – were originally developed for peptide synthesis but are now widely used in general synthetic and heterocyclic chemistry, particularly for application to combinatorial chemistry.

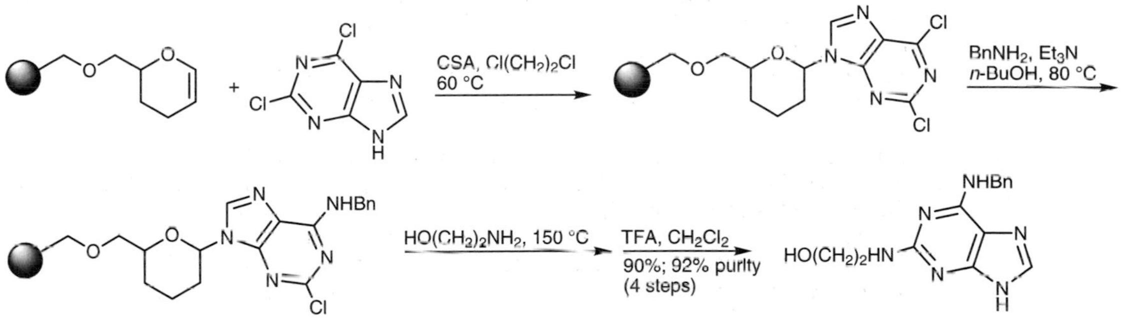

Scheme 1

The particular advantage of solid-supported chemistry is in purification: sequential reactions can be carried out and by-products washed away without the need for extractions, chromatography, or isolation of intermediates. The product is finally cleaved from the support with minimal impurities. Although some individual reactions may be slower on solid phase, overall processes are generally faster because only one isolation step is needed. These features also make reactions suitable for automation.

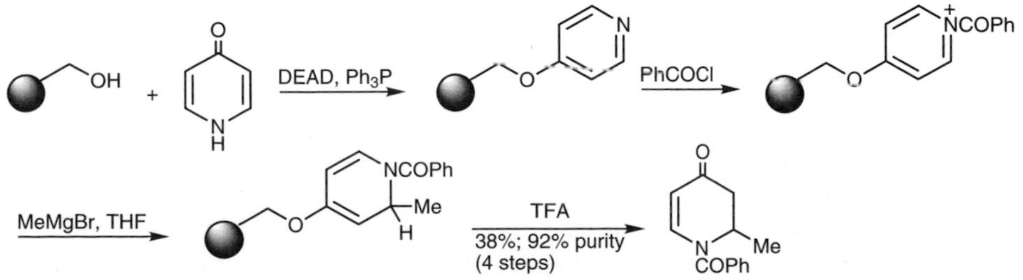

Scheme 2

Practically all types of reaction can be carried out (with choice of the appropriate support) including halogenations, lithiations, transition metal-catalysed reactions and reactions with such aggressive reagents as phosphorus pentachloride. Many standard heterocyclic ring synthesis methods can also be used successfully.

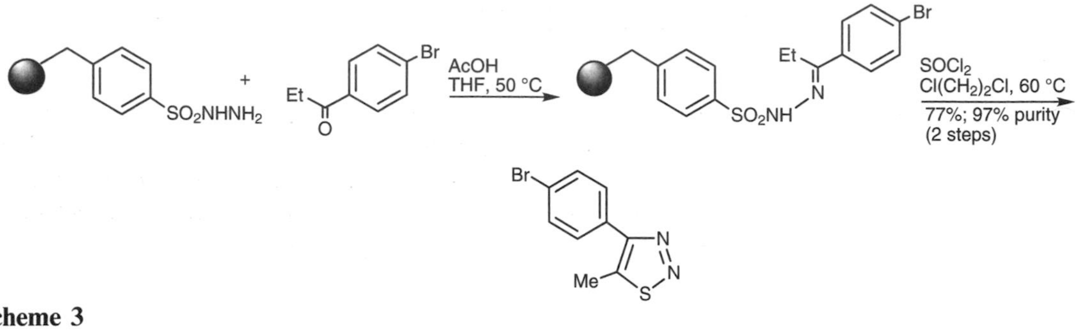

Scheme 3

The solid supports used are often derived from chloromethylated polystyrene resins, which can be used directly (Merrifield) for alkylation or for coupling via a linker, which may be cleavable selectively, by acid or base (*e.g.* Wang).

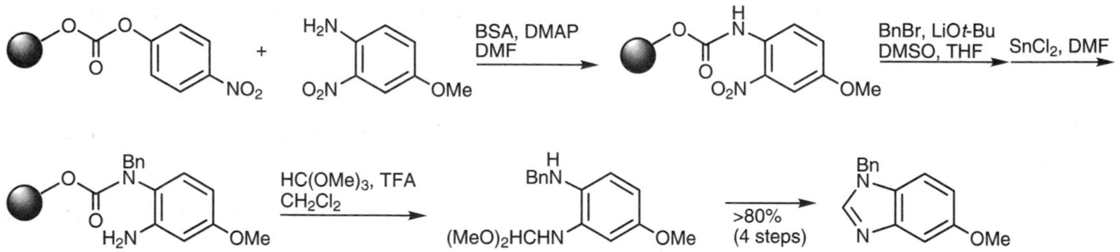

Scheme 4

The main question is how to attach the substrate to the polymer: in carboaromatic and aliphatic chemistry this is usually done through a functional group (Schemes 2 and 4) such as a carboxylic acid or an amine, which can restrict choice of substrate; an alternative method is through a 'traceless link' such as a silane, which can be removed, for example by protonolysis to leave a hydrogen at the point of attachment, but this may not be particularly convenient. Here, heterocycles have the advantage! Attachment to the support can be by methods[3] similar to those described above, but also via the ring heteroatom, particularly nitrogen in azoles[4] (Scheme 1) or by a heteroatom when the heterocyclic ring formation is the final step[5] – it is often easy to incorporate a final cyclisation (heterocycle formation) step in such a way that it results in cleavage of the product from the support (Scheme 3). Sulfur is a useful link for heterocycles because its use as a leaving group (or better, after conversion to sulfoxide[6] or sulfone[7]) can lead to cleavage from the support (Scheme 5). For full discussion of the heterocyclic reactivity involved in the examples shown, the reader should consult earlier chapters.

Scheme 5

Another application of solid phase chemistry is the use of polymer-bound reagents which offer similar advantages in requiring minimal purification: the thiazole synthesis shown below, which involved the use of a polymer-bound brominating agent and secondly a polymer-bound base, gave the intermediate and product in greater than 95% purity without the need for any chromatography.[8]

While these methods have been used very successfully on a small scale, they also have significant potential for rapid synthesis on medium scale (up to several kg) and possibly even larger. High loading resins can bind an equal weight of substrate and so are efficient in terms of volume.[9]

28.2 Heterocycles in the pharmaceutical industry: large scale heterocyclic synthesis

Heterocycles are the basis of the majority of medicines, with combined sales in tens of billions of pounds – about £60 billion for the top 100 drugs in 1998 – so for a chemist in the pharmaceutical and related industries, a thorough knowledge of heterocyclic chemistry is necessary at the discovery, development, and production stages. Most of the total cost is in discovery, development, testing and sales, the cost of production of the actual drug substance being a relatively minor part of the final cost. Production costs are tightly controlled so the synthetic route must be as efficient and cheap as posible – typical production prices are of the order of £1000 to £5000 per kilo for a new drug.

The requirements of scale-up and production are very different from those in medicinal chemistry and drug discovery and may lead to a very different approach to chemistry. The number of steps in the synthesis is a major determinant of cost, as are, of course, the costs of starting materials and reagents. Standard preliminary targets are for 90% hplc yields in each stage, with purification by crystallisation or possibly distillation in some cases; chromatographic purification on a production scale, which may be in ton quantities, is very rare.

A crucial aspect is the quality of the final product, with very rigorous criteria for reproducible levels of purity and of percentages of impurities, particularly metals. All impurities down to ca. 0.1% must be identified and possibly reference samples of the impurities separately synthesised, which provides another interesting challenge to the development chemist. Some metal derivatives such as organotin compounds and mercury must be at almost undetectable levels; others, such as palladium may be acceptable at very low levels depending on dose.

Although some drugs have been produced using organotin reagents, as a general rule such reagents are avoided, particularly in the later steps. These requirements can greatly restrict the synthetic chemistry available for the production route – many elegant and ingenious reactions found, for example, in this book, though suitable for preliminary phases of drug discovery, may be of no use for production due to residual traces of such reagents in the final product.

Minimisation of environmetal impact must also be considered – the composition of gaseous emissions and the content of toxic or prohibited substances in aqueous waste streams. These factors must therefore be taken into account when considering the production synthesis of a drug.

28.3   Electronic applications

An important area where technology and heterocyclic chemistry combine is that of electroactive organic materials.[10] The applications of these materials, which extend beyond simple replacements for metals, include use as conductors, superconductors, semiconductors, batteries, transistors, sensors, light emitting diodes (LEDs), and related electrochromic applications. This area is of great commercial importance.

The use of organic conductors means than an essentially infinite supply of materials is available with minimum ecological impact and without strategic supply problems, which could arise with metals and rare elements. The resistance to corrosion, air and water stability, ease of preparation as very thin films, and suitability for incorporation into materials such as fabrics, are also significant advantages over metallic equivalents.

The mechanism of conduction involves electronic conduction through long chains of conjugated molecules and/or π-stacked structures. A detailed discussion of the theory of conduction is beyond the scope of this book; this section is restricted to demonstating the range of applications of heterocycles in this area.

The types of compound used in organic conductors cover a wide range of unsaturated molecules such as poly(acetylene) and poly(aniline), but of particular significance from the heterocyclic and commercial viewpoints are poly(pyrrole), poly(thiophene), and related polymers and π-stacked structures derived from tetrathiafulvalenes. An advantage of using heterocycles is that a wide range of electron-rich, electron-poor and mixed systems can be easily prepared, allowing for tuning of the electronic and electrical properties of such materials.[11]

28.3.1   Poly(pyrrole) and poly(thiophene)

Pyrrole, thiophene, and their derivatives can be oxidatively polymerised either electrochemically or chemically, for example using iron(III) chloride, to give mainly 2,5-coupled polymers. The initial neutral polymers are non-conducting but on further oxidation are converted partially into cation radicals or dications, with incorporation of counterions from the reaction medium – a process known as 'doping' – giving conducting materials. Reductive doping is also possible in other systems.

The conducting and physical properties can be modified by the use of 3- and/or 4-substitutents, or *N*-substituents in the case of pyrrole. The counterions can be incorporated into a side-chain (self-doping) as in the polymer of 3-(thien-3-yl)propanesulfonic acid. Variation in the size of side-chains allows control of solubility. Mixed polymers with, for example, thiophenes and pyridines, are capable of both oxidative and reductive doping.

Oligo(thiophenes) are also useful in these applications and have been specifically synthesised up to 27 units long by palladium(0)-catalysed couplings or via the diacetylene synthesis (section 14.13.2.2).[12]

28.3.2   Tetrathiafulvalenes

These unusual heterocycles and their analogues have been intensively studied – a search for the parent compound gives more than 1000 literature references – since the discovery[13] that single crystals of TTF.TCNQ (tetrathiafulvalene.tetracyanoquinodimethane) show electrical conductivity. TTF.TCNQ crystallises with segregated stacks, *i.e.* with the electron donor (TTF) and the electron acceptor (TCNQ) molecules aligned separately, in stacks of the planar molecules, rather than as alternating donors and acceptors as is found in conventional, non-conducting charge-transfer stacks. The partial transfer of charge between the segregated stacks is the key feature which allows the flow of current in the stack directions. 'Electrons in crystalline 'organic metals' flow along supermolecular orbitals constructed from molecular orbitals of molecules arranged in columns',[14] though intermolecular communication, not necessarily involving stacks, may be entirely sufficient.[15,16] Thus: 'The structure of any successful organic conductor is probably dictated by two requirements. First, its molecular building blocks must fit closely together, so that the conduction electrons can move easily from one molecule to another. Second, the energy cost of partially filling or opening a valence energy band must be small',[17] for example 'stable open-shell (i.e. free radical) species are needed ... of planar molecules with ... delocalised π-molecular orbitals so that effective overlap ... can occur'.[18] Many variations on both the TTF and the TCNQ units have been prepared and studied, including selenium and tellurium analogues of TTF; and using electron-deficient heterocycles such as tetrazine instead of TCNQ. The bis(ethylenedithio) analogue of TTF, usually known as BEDT-TTF, has been particularly useful. Incidentally, the electron-donating ability of TTF allows its use as a radical initiator for diazonium salts.[19]

tetrathiafulvalene (TTF)   BEDT-TTF   tetracyanoquinodimethane (TCNQ)

Tetrathiafulvalenes can be prepared in a number of ways,[20] for example from 1,3-dithiole-2-thione-4,5-dithiolate – the simple salts of this dianion are not very stable but it can be stored for later use, as a zinc complex, or as a dibenzoate as shown below.

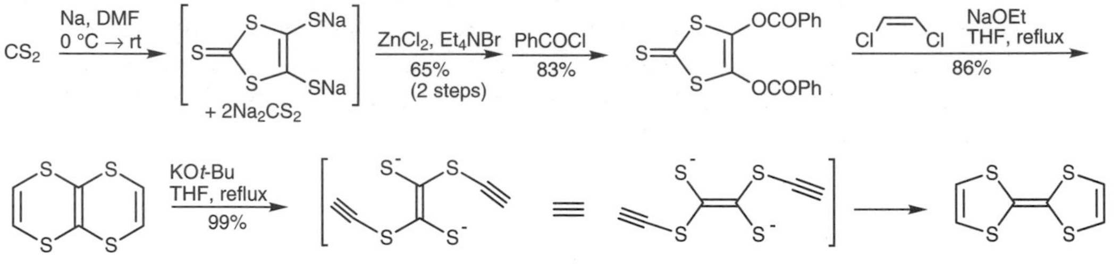

Most chemical transformations of TTF are based on lithiations – even a tetralithio derivative is easily formed. Palladium(0)-catalysed couplings utilising trialkyltin derivatives[21] can also be carried out without difficulty.[22]

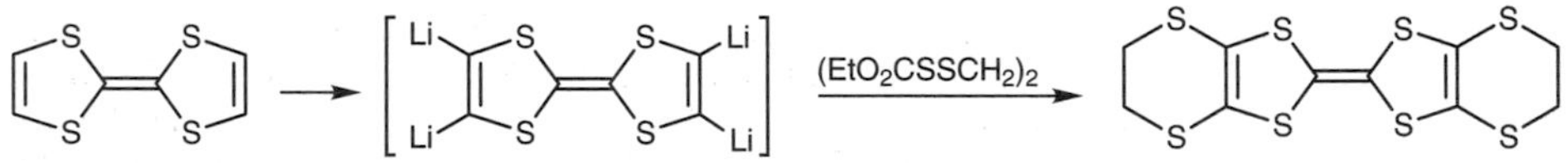

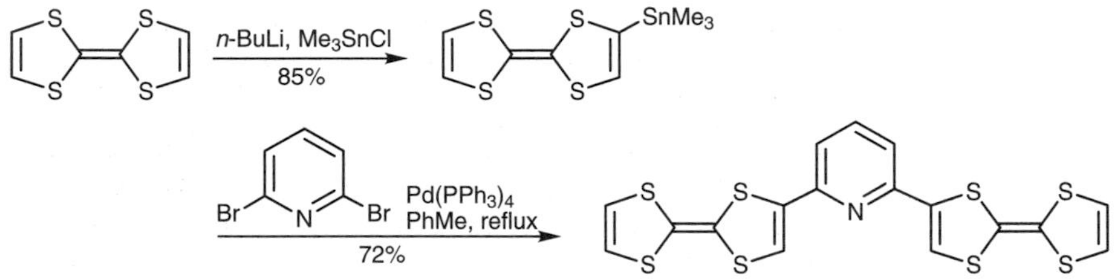

28.3.3 Applications

In addition to the straightforward application of conductors and semi-conductors, there are some interesting examples of other uses. Polymers derived from 3,4-ethylenedioxythiophene are produced commercially as anti-static agents and substituted compounds have found particular application in electrochromic devices – substances which change colour on application of an electric current. Electrochromic devices can be used, for example, for switchable shading or colouring of glass and have been suggested as advanced forms of 'stealth' camouflage for fighter aircraft, where a computer-controlled array of electrochromic coatings would match the colour of the aircraft to that of its immediate surroundings; some of the original stealth coatings contained radar-absorbing poly(pyrroles).[23]

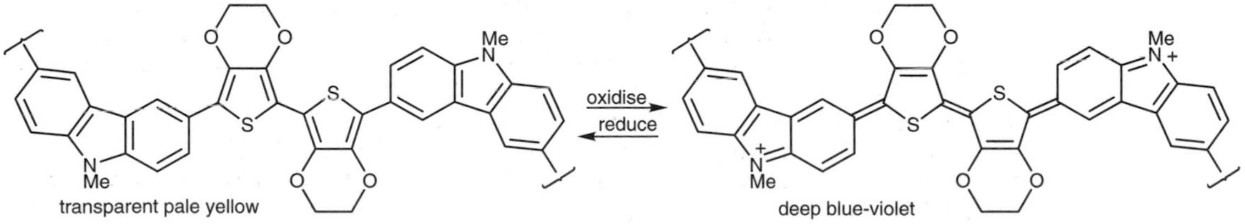

The conductivity of the polymer may change when organic vapours are absorbed and this has been put to use in the delightfully-named 'electronic nose'.[24] In one commercial form, this consists of an array of twelve different polymers, such as poly(pyrrole), with differential responses to absorption of different organic molecules; electronic processing of these responses generates a fingerprint for each compound, allowing identification. One application of the electronic nose is in the food industry, where it can be used to detect spoilage or faulty fermentations; it has a major advantage over current methods such as gas chromatography in being faster and cheaper and much simpler to operate. In a related way, biosensors, for example for the estimation of glucose levels, can be produced by attachment of enzymes to conducting polymer surfaces.

References

1. 'Heterocycles in life and society', Pozharskii, A. F., Soldatenkov, A. T., and Katritzky, A. R., Wiley, **1997**.
2. 'Combinatorial synthesis – the design of compound libraries and their application to drug discovery', Terrett, N. K., Gardner, M., Gordon, D. W., Kobylecki, R. J., and Steele, J., *Tetrahedron*, **1995**, *51*, 8135; 'The current status of heterocyclic combinatorial libraries', Nefzi, A., Ostresh, J. M., and Houghten, R. A., *Chem. Rev.*, **1997**, *97*, 449; 'Solid phase organic reactions, III (for Nov. 1996 - Dec 1997)' [and previous articles in the series], Booth, S., Hermkens, P. H. H., Ottenheijm, H. C. J., and Rees, D. C., *Tetrahedron*, **1998**, *54*, 15385; 'Recent progress in solid phase heterocycle synthesis', Corbett, J. W., *Org. Prep. Proc. Int.*, **1998**, *30*, 489.
3. Chen, C. and Munoz, B., *Tetrahedron Lett.*, **1998**, *39*, 6781.
4. Nugiel, D. A., Cornelius, A. M., and Corbett, J. W., *J. Org. Chem.*, **1997**, *62*, 201.
5. Hu, Y., Baudart, S., and Porco, J. A., *J. Org. Chem.*, **1999**, *64*, 1049; Huang, W. and Scarborough, R. M., *Tetrahedron Lett.*, **1999**, *40*, 2665.
6. Masquelin, T., Meunier, N., Gerber, F., and Rosse, G., *Heterocycles*, **1998**, *48*, 2489.
7. Gayo, L. M. and Suto, M. J., *Tetrahedron Lett.*, **1997**, *38*, 211.
8. Habermann, J., Ley, S. V., Scicinski, J. J., Scott, J. S., Smits, R., and Thomas, A. W., *J. Chem. Soc., Perkin Trans. 1*, **1999**, 2425.
9. Raillard, S. P., Ji, G., Mann, A. D., and Baer, T. A., *Org. Proc. Res. Dev.*, **1999**, *3*, 177.
10. 'Handbook of organic conductive molecules and polymers', Singh Nalwa, H., Wiley, **1997**; 'Polarons, bipolarons, and solitrons in conducting polymers', Brédas, J. L. and Street, G. B., *Acc. Chem. Res.*, **1985**, *18*, 309; 'Recent progress on conducting organic charge-transfer salts', Bryce, M. R., *Chem. Soc. Rev.*, **1991**, *20*, 355.
11. 'Heterocycle-based electroconductive polymers', Berlin, A., *Plastic Eng. (N.Y.) (Electrical and Optical Polymer Systems)*, **1998**, *45*, 47; 'Heterocycle-based electric conductors', Pagani, G. A., *Heterocycles*, **1994**, *37*, 2069; 'Conjugated poly(thiophenes): synthesis, functionalisation and applications', Roncali, J., *Chem. Rev.*, **1992**, *92*, 711.
12. Nakanishi, H., Sumi, N., Aso, Y., and Otsubo, T., *J. Org. Chem.*, **1998**, *63*, 8632.
13. Ferraris, J., Cowan, D. O., Walatka, V. V. and Perlstein, J. H., *J. Am. Chem. Soc.*, **1973**, *95*, 948; Coleman, L. B., Cohen, M. J., Sandman, D. J., Yamagishi, F. G., Garito, A. F., and Heeger, A. J., *Solid State Commun.*, **1973**, *12*, 1125.
14. Wudl, F., *Acc. Chem. Res.*, **1984**, *17*, 227.
15. Parkin, S. S. P., Engler, E. M., Schumaker, R. R., Lagier, R., Lee, V. Y., Scott, J. C., and Greene, R. L., *Phys. Rev. Lett.*, **1983**, *50*, 270; Williams, J. M., Beno, M. A., Wang, H. H., Reed, P. E., Azevedo, L. J., and Schirber, J. E., *Inorg. Chem.*, **1984**, *23*, 1790.
16. Obertelli, S. D., Friend, R. H., Talham, D. R., Kurmoo, M., and Day, P., *J. Phys.: Condens. Matter*, 1989, **1**, 5671; Mallah, T., Hollis, C., Bott, S., Day, P., and Kurmoo, M., *Synth. Met.*, **1988**, *27*, A381.
17. Bechgaard, K. and Jerome, D., *Sci. Amer.*, **1982**, *247*, July, 50.
18. Cowan D. O. and Wiygul, F. M., *Chem. Eng. News*, **1986**, July 21, 28.
19. Fletcher, R. J., Lampard, C., Murphy, J. A., and Lewis, N., *J. Chem. Soc., Perkin Trans. 1*, **1995**, 623.
20. Krief, A., *Tetrahedron*, **1986**, *42*, 1209; Bryce, M. R., *Aldrichimica Acta*, **1985**, *18*, 73; Narita A. and Pittman, C. U., *Synthesis*, **1976**, 489; Wolf, I., Naarmann, H., and Müller, K., *Angew. Chem., Int. Ed. Engl.*, **1988**, *27*, 288' 'The organic chemistry of 1,3-dithiole-2-

thione-4,5-dithiolate (DMIT)', Svenstrup, N. and Becher, J., *Synthesis*, **1995**, 215; Hansen, T. K., Becher, J., and Jorgensen, T., *Org. Synth.*, **1996**, *73*, 270; Meline, E. L. and Elsenbaumer, R. L., *J. Chem. Soc., Perkin Trans. 1*, **1998**, 2467.

21. Iyoda, M., Kuwatani, Y., Ueno, N., and Oda, M., *J. Chem. Soc., Chem. Commun.*, **1992**, 158.

22. Iyoda, M., Fukuda, M., Yoshida, M., and Sasaki, S., *Chem. Lett.*, **1994**, 2369; Lovell, J. M. and Joule, J. A., *J. Chem. Soc., Perkin Trans. 1*, **1996**, 2391.

23. Berlin, A. and Kingston, R., *Chem. Brit.*, **1999**, October, 24.

24. 'Towards an integrated electronic nose using conducting polymer sensors', Hatfield, J. V., Neaves, P., Hicks, P. J., Persaud, K., and Travers, P., *Sensors and Actuators B*, **1994**, *18–19*, 221; Electronic nose detects the nast niffs, Coghlan, A., *New Scientist*, **1994**, *141 (1911)*, 20; 'Electronic noses: principles and applications', Gardner, J. W. and Bartlett, P. N., OUP, Oxford, **1999**.

Appendix: answers to exercises

Chapter 5

1(i) [structure: pyridine with OEt and NO₂] (ii) [structure: CO₂H, Br pyridine] 2 [structure: N NH(CH₂)₂NMe₂] 3(i) (a) [N NHNH₂] (b) [structure] (ii) [structure] 4 [OMe pyridine → OMe pyridinium Me I⁻ → O pyridone Me]

5 [NH₂ pyridine] and [NH₂ pyridine] *via* 3,4-pyridyne 6(i) [OMe pyridine] [structure with Ph Ph lactone] (ii) [I, Cl pyridine] (iii) [Me₂COH, F pyridine] (iv) [N SnMe₃] 7 [O₂N pyridinium Me Br⁻ → O₂N indolizine Me]

8 [pyridinium I⁻ → quinolizidinium I⁻] 9 [H OH Me CH₂OH azetidinone structure] 10 [Me NH₂ pyridine → Me pyridone → Me OMe pyridine] 11 [CH₂CO.CO₂Et, OMe; PhNH₃⁺ has no activating substituent; pyridinium NH₂ does have activating substituent]

12. [OH pyridinium Br⁻ → O⁻ pyridinium → N bicyclic ketone] 13. [N CH₂ structure with H─O] 14. [CH(NHAc)(CO₂Et)₂ pyridine] 15.(i) [Me, CH₂SPh pyridine] (ii) [CH₂SPh, Me pyridine] 16. [O CN pyridone Me]

17. [CN pyridone Me] 17. D [Bu D piperidine] 18. → [N MeO CO pyridine] → [N MeOCO bicyclic] 19. [NO₂ pyridine N⁻O⁻ → NO₂ pyridinium → NH₂ pyridine] 20. Me [MeO₂C, H CH₂Ph, CO₂Me dihydropyridine Me] then oxidation

21. [bicyclic furan] → [(CH₂)₂OH pyridine] 22.(a)(i) [Et, CN HO pyridone] (ii) Me [CN pyridone] (iii) HO [CN pyridone] (b)(i) Me [Me, CO₂Et pyridine]

(ii) Me [CO₂Et pyridine Me] (iii) Me [EtO₂C CO₂Et pyridine Me] 23. [Me O CHOH + H₂N Ac] → Me [Ac pyridine]

Chapter 6

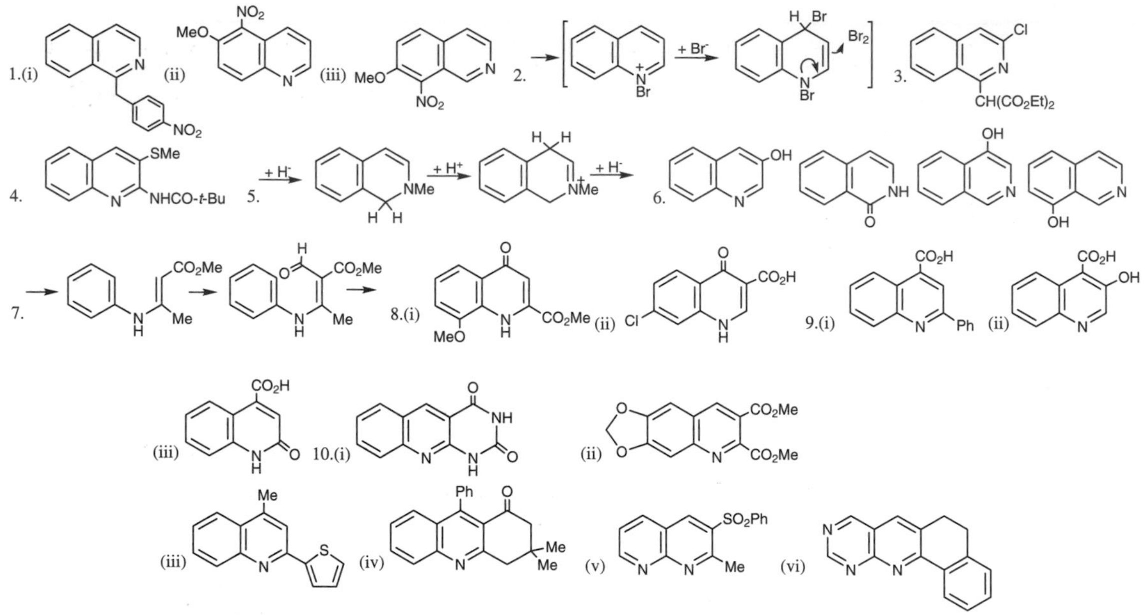

1.(i) (ii) (iii) 2. 3.

4. 5. 6.

7. 8.(i) (ii) 9.(i) (ii)

(iii) 10.(i) (ii)

(iii) (iv) (v) (vi)

Chapter 8

1. 2. 3.

4. 5.(i)

(ii) (iii) 6. 7. 8.(i) (ii)

(iii) (iv) (v)

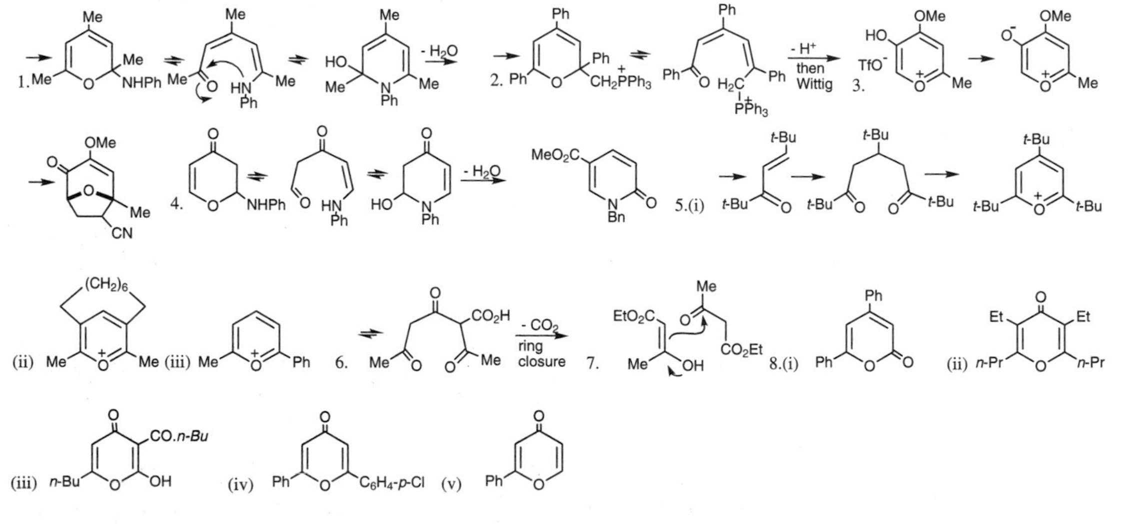

Chapter 9

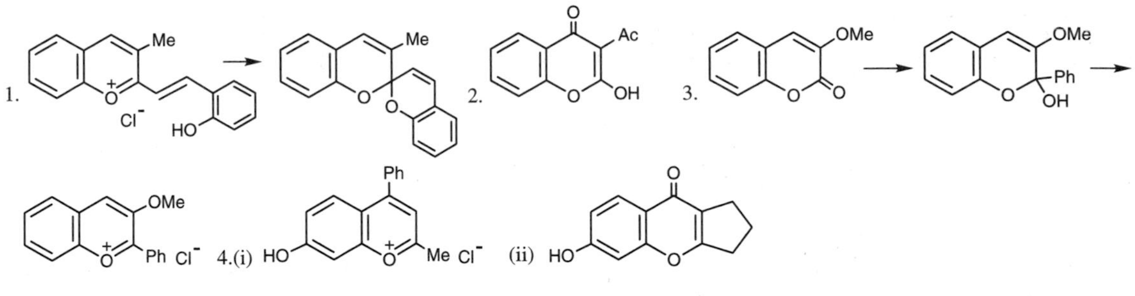

1.

2.

3.

4.(i)

(ii)

Chapter 11

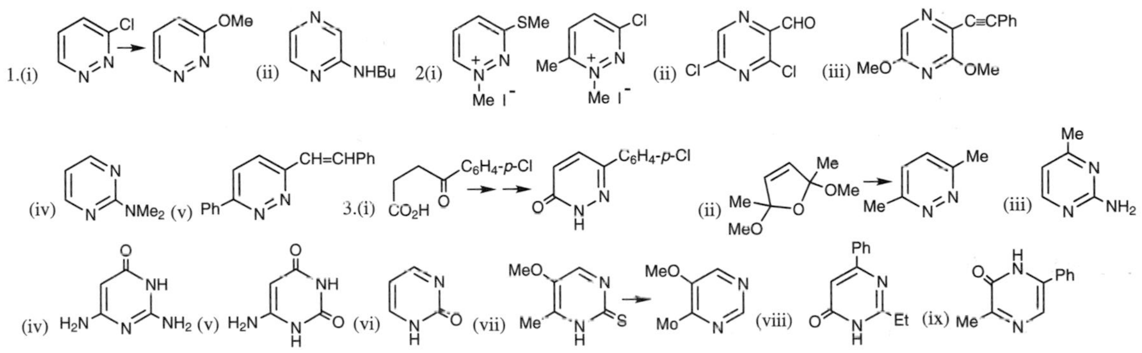

1.(i) → (ii) NHBu

2(i) (ii) (iii)

(iv) NMe$_2$ (v)

3.(i) CO_2H → (ii) (iii)

(iv) H_2N NH_2 (v) H_2N (vi) (vii) → (viii) (ix)

Chapter 13

1. O_2N Me 6:1

2.(i) (ii) (iii)

3.(i) (ii)

(ii)

4. Me Me disubstituted imine much less reactive

5. Me Me

6.(i) then - HC≡CH (ii)

7. →

8. with n-PrNH$_2$, 2-aminothiophen, or PhSO$_2$NH$_2$

9.

hydrolyse, then - CO$_2$, then N$_2$H$_4$/EtONa

10.(i) (ii) + Na$_2$S$_2$O$_4$

(iii) (iv) then NaOEt

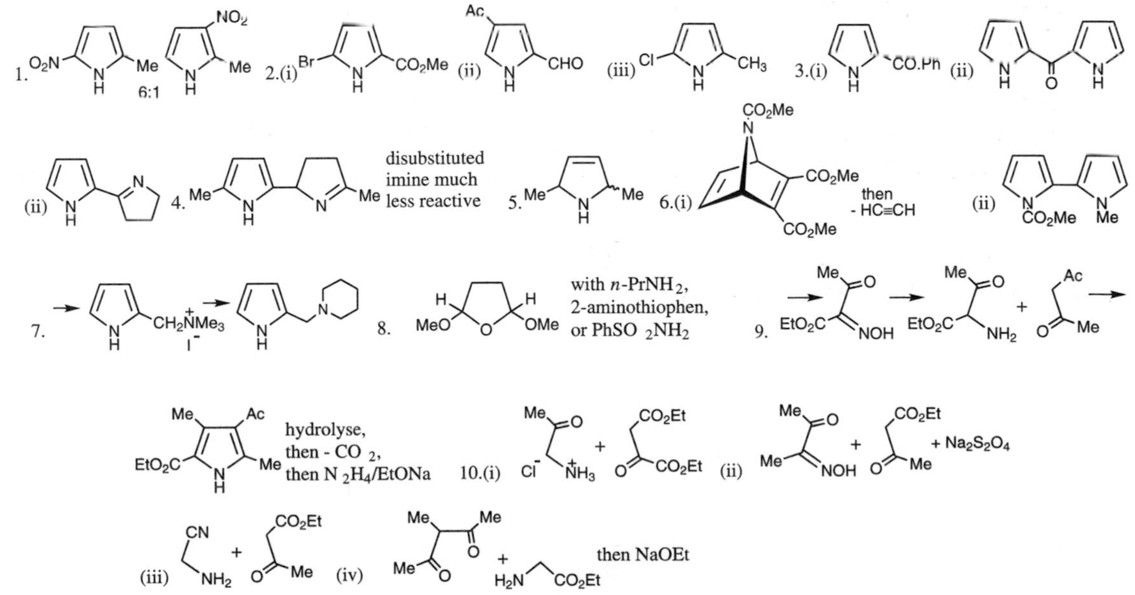

Chapter 14

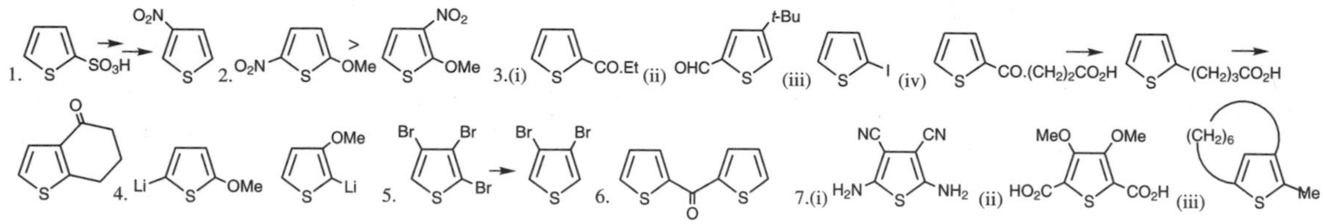

1. → 2. → 3.(i) (ii) (iii) (iv) →

4. 5. → 6. 7.(i) (ii) (iii)

Chapter 15

1. → → → → 2.(i) (ii)

(iii) (iv) 3. →

→ 5.(i) (ii) 6.(i) (ii) (iii) (iv)

(v) 7.(i) (ii) (iii) 8.(i) with TMSCl Et₃N ZnCl₂

(ii) 9. 10.(i) (ii) (iii)

11. →

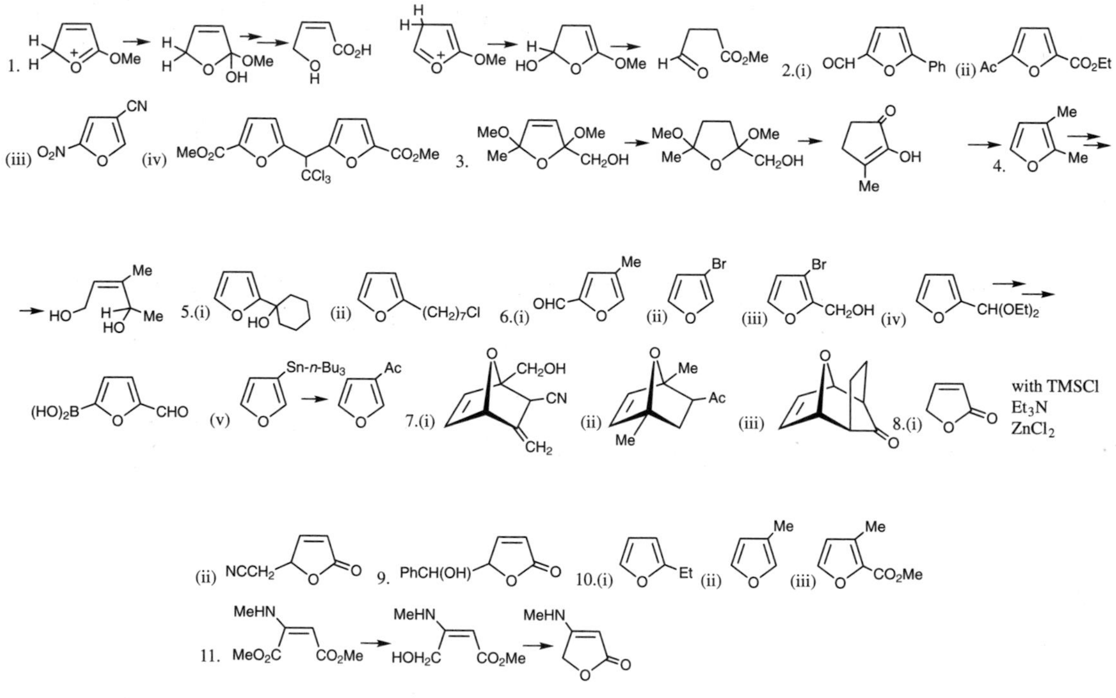

Chapter 17

1. 2. 3. 4.

5.(i) (ii) 6. intramolecular Wittig → 7. and

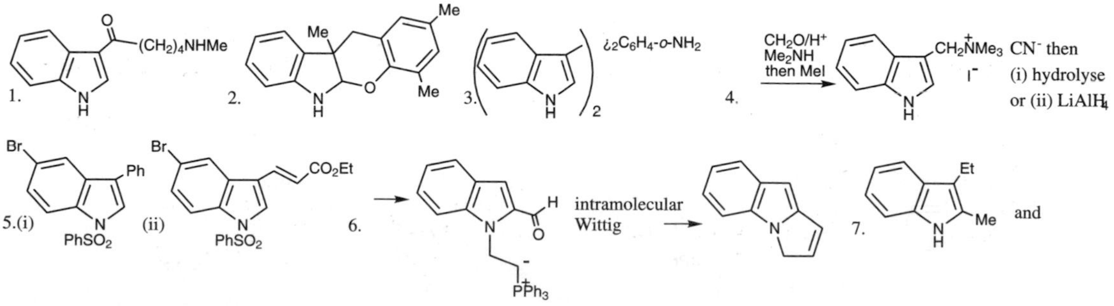

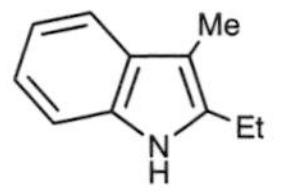

8.

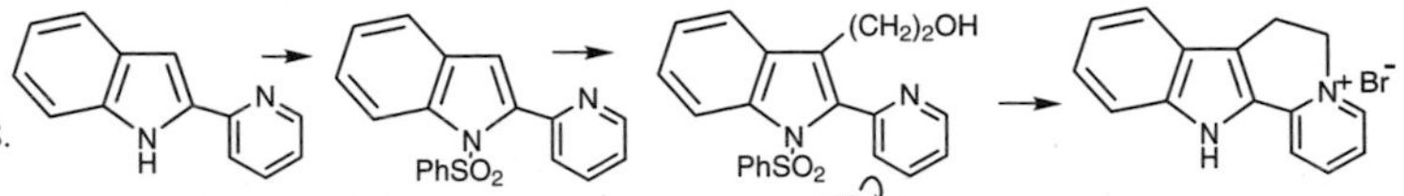

9.

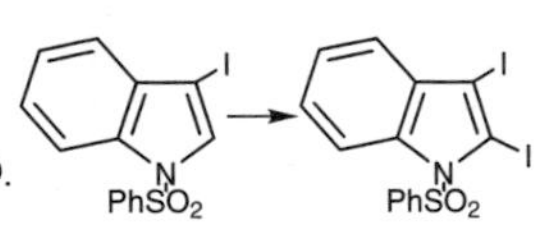

10.

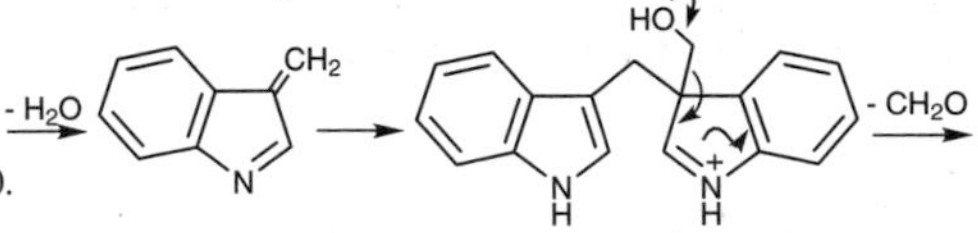

11.(i) PhNHN=CHEt   (ii) 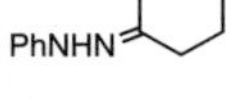 PhNHN=⬡

(iii) PhNHN=C(Et)$_2$ (iv) PhNHN=C(n-Pr)Ph   12.

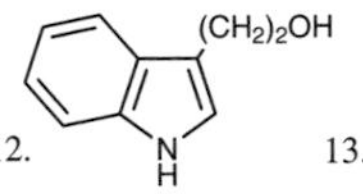

13. 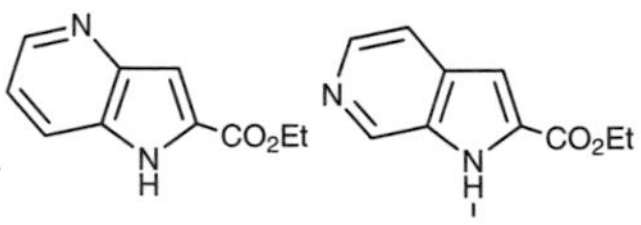   14.(i)

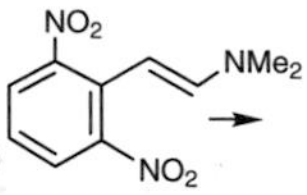

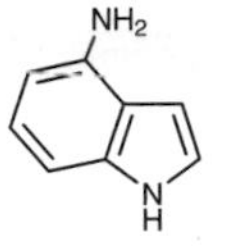

(ii)

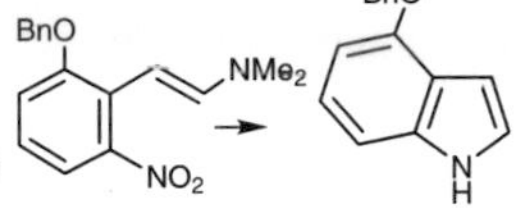

(iii)

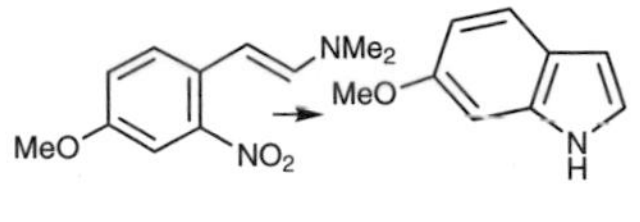

(iv)

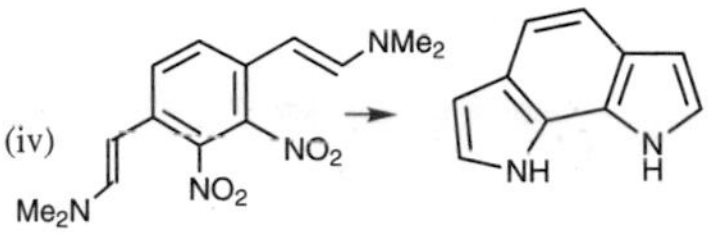

Chapter 18

1.

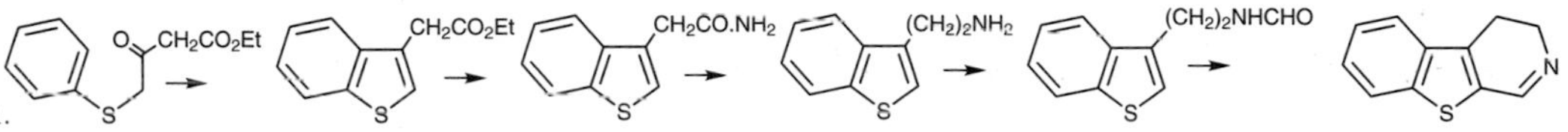

2.(i) 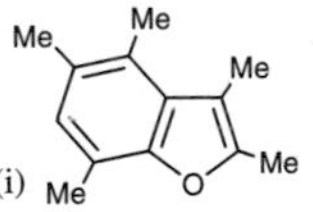   (ii) 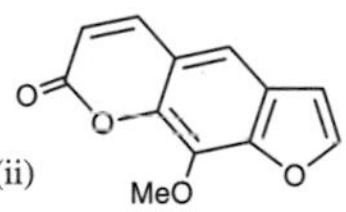   (iii) 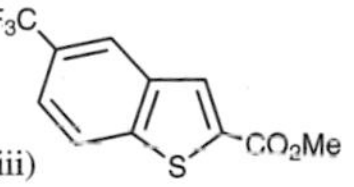   (iv) 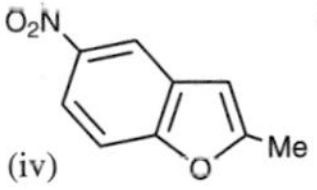   3.

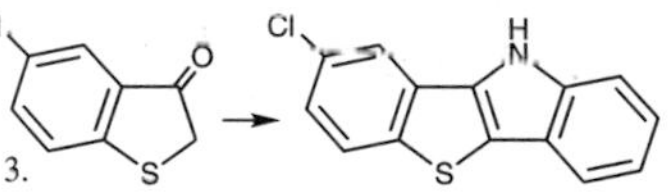

Chapter 19

1.(i)

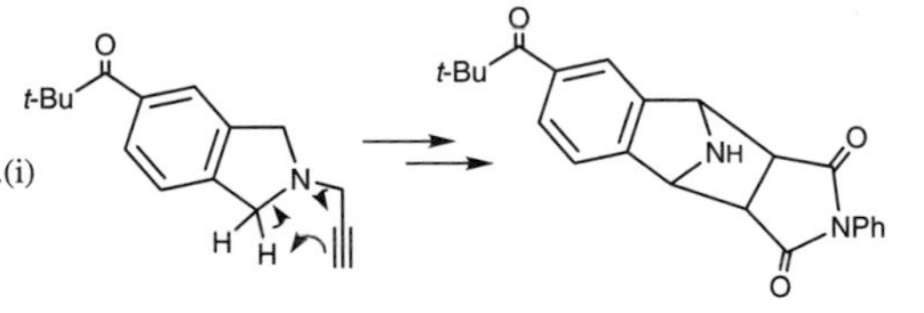

(ii)

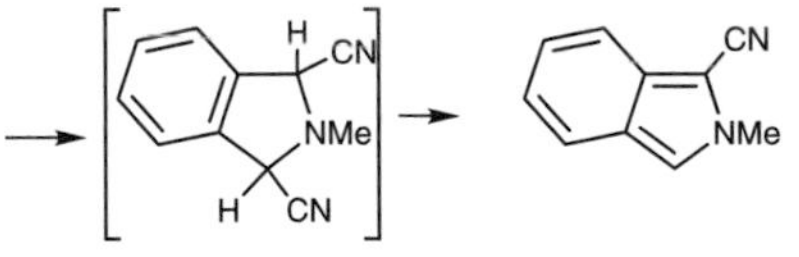

(iii)

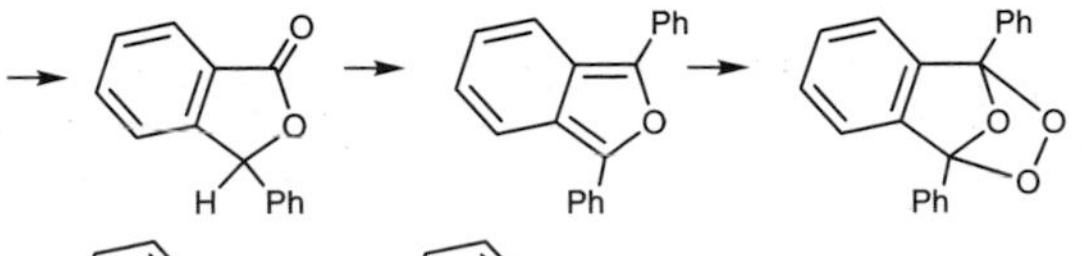

(iv)

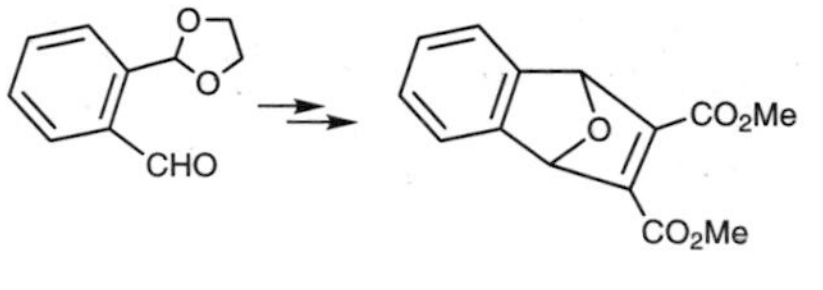

(v)

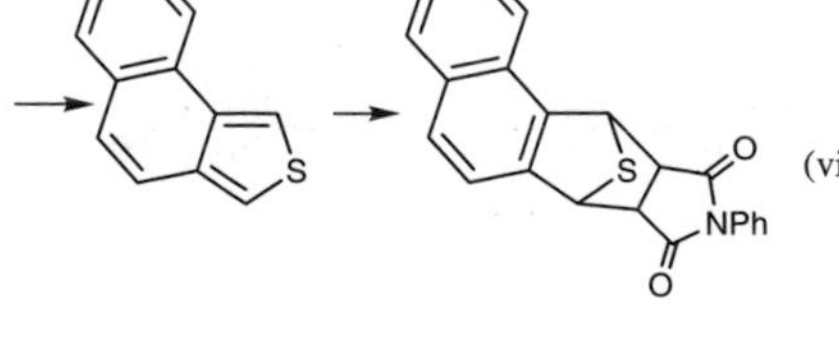

(vi)

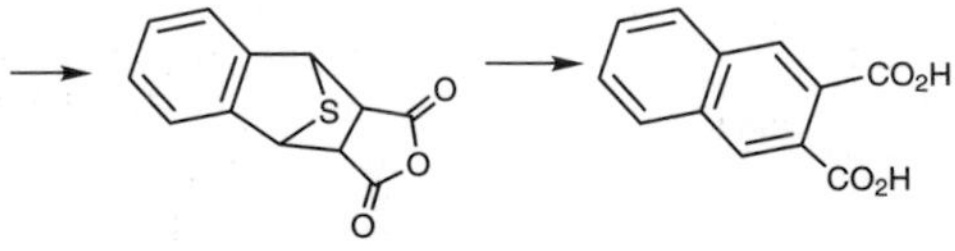

Chapter 21

Chapter 22

Chapter 24

Chapter   25

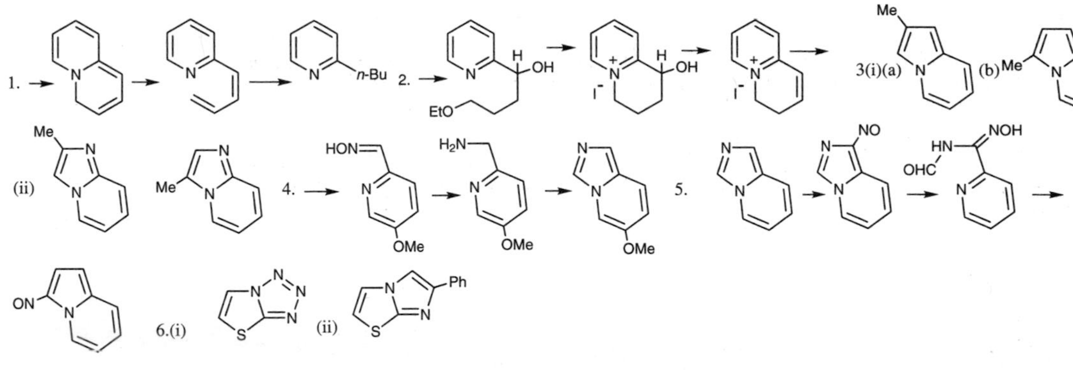

1. →

2.

3(i)(a)   (b)

(ii)

4. →

5.

6.(i)   (ii)

Chapter   26

1.(i)(a)   (b)   (ii)

2.

3.(a)   (b) Ph

4.(i) →   (ii) →

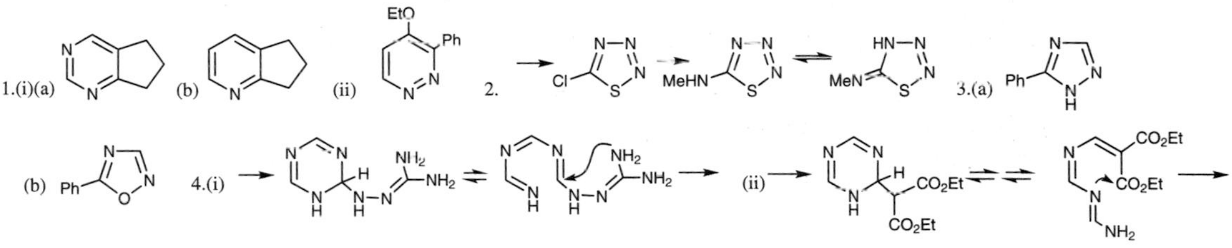

Index

Individual compounds are listed by name when the text gives a ring synthesis, total synthesis, formation from another compound, or where a particular aspect of reactivity is also noted. Metallated species are listed because of their importance in synthesis; products derived from these intermediates are not listed. Some liberties have been taken with chemical names, with the aim of helping the reader seeking chemical information, for example the Index gives: 'Indole, 2-lithio-1-phenylsulfonyl-' but also (incorrectly): 'indole, 2-lithio-1-*t*-butoxycarbonyl'. Also, names emphasise the heterocyclic aspect: 'furan, 2-ethoxycarbonyl' is used rather than 'ethyl furan-2-carboxylate'. The (inorganic) counteranions to heterocyclic cations are omitted: 'Pyrylium' not 'Pyrylium perchlorate'; pyridinium salts are listed as 'Pyridiniums' *etc*.